PROGRESS IN BRAIN RESEARCH

VOLUME 148

CREATING COORDINATION IN THE CEREBELLUM

Other volumes in PROGRESS IN BRAIN RESEARCH

Volume 120: Nucleotides and their Receptors in the Nervous System, by P. Illes and H. Zimmermann (Eds.) – 1999, ISBN 0-444-50082-0.
Volume 121: Disorders of Brain, Behavior and Cognition: The Neurocomputational Perspective, by J.A. Reggia, E. Ruppin and D. Glanzman (Eds.) – 1999, ISBN 0-444-50175-4.
Volume 122: The Biological Basis for Mind Body Interactions, by E.A. Mayer and C.B. Saper (Eds.) – 1999, ISBN 0-444-50049-9.
Volume 123: Peripheral and Spinal Mechanisms in the Neural Control of Movement, by M.D. Binder (Ed.) – 1999, ISBN 0-444-50288-2.
Volume 124: Cerebellar Modules: Molecules, Morphology and Function, by N.M. Gerrits, T.J.H. Ruigrok and C.E. De Zeeuw (Eds.) – 2000, ISBN 0-444-50108-8.
Volume 125: Volume Transmission Revisited, by L.F. Agnati, K. Fuxe, C. Nicholson and E. Syková (Eds.) – 2000, ISBN 0-444-50314-5.
Volume 126: Cognition, Emotion and Autonomic Responses: the Integrative Role of the Prefrontal Cortex and Limbic Structures, by H.B.M. Uylings, C.G. Van Eden, J.P.C. De Bruin, M.G.P. Feenstra and C.M.A. Pennartz (Eds.) – 2000, ISBN 0-444-50332-3.
Volume 127: Neural Transplantation II. Novel Cell Therapies for CNS Disorders, by S.B. Dunnett and A. Björklund (Eds.) – 2000, ISBN 0-444-50109-6.
Volume 128: Neural Plasticity and Regeneration, by F.J. Seil (Ed.) – 2000, ISBN 0-444-50209-2.
Volume 129: Nervous System Plasticity and Chronic Pain, by J. Sandkühler, B. Bromm and G.F. Gebhart (Eds.) – 2000, ISBN 0-444-50509-1.
Volume 130: Advances in Neural Population Coding, by M.A.L. Nicolelis (Ed.) – 2001, ISBN 0-444-50110-X.
Volume 131: Concepts and Challenges in Retinal Biology, by H. Kolb, H. Ripps, and S. Wu (Eds.), – 2001, ISBN 0-444-506772.
Volume 132: Glial Cell Function, by B. Castellano López and M. Nieto-Sampedro (Eds.) – 2001. ISBN 0-444-50508-3.
Volume 133: The Maternal Brain. Neurobiological and neuroendocrine adaptation and disorders in pregnancy and post partum, by J.A. Russell, A.J. Douglas, R.J. Windle and C.D. Ingram (Eds.) – 2001, ISBN 0-444-50548-2.
Volume 134: Vision: From Neurons to Cognition, by C. Casanova and M. Ptito (Eds.) – 2001, ISBN 0-444-50586-5.
Volume 135: Do Seizures Damage the Brain, by A. Pitkänen and T. Sutula (Eds.) – 2002, ISBN 0-444-50814-7.
Volume 136: Changing Views of Cajal's Neuron, by E.C. Azmitia, J. DeFelipe, E.G. Jones, P. Rakic and C.E. Ribak (Eds.) – 2002, ISBN 0-444-50815-5.
Volume 137: Spinal Cord Trauma: Regeneration, Neural Repair and Functional Recovery, by L. McKerracher, G. Doucet and S. Rossignol (Eds.) – 2002, ISBN 0-444-50817-1.
Volume 138: Plasticity in the Adult Brain: From Genes to Neurotherapy, by M.A. Hofman, G.J. Boer, A.J.G.D. Holtmaat, E.J.W. Van Someren, J. Verhaagen and D.F. Swaab (Eds.) – 2002, ISBN 0-444-50981-X.
Volume 139: Vasopressin and Oxytocin: From Genes to Clinical Applications, by D. Poulain, S. Oliet and D. Theodosis (Eds.) – 2002, ISBN 0-444-50982-8.
Volume 140: The Brain's Eye, by J. Hyönä, D.P. Munoz, W. Heide and R. Radach (Eds.) – 2002, ISBN 0-444-51097-4.
Volume 141: Gonadotropin-Releasing Hormone: Molecules and Receptors, by I.S. Parhar (Ed.) – 2002, ISBN 0-444-50979-8.
Volume 142: Neural Control of Space Coding and Action Production, by C. Prablanc, D. Pélisson and Y. Rossetti (Eds.) – 2003, ISBN 0-444-509771.
Volume 143: Brain Mechanisms for the Integration of Posture and Movement, by S. Mori, D.G. Stuart and M. Wiesendanger (Eds.) – 2004, ISBN 0-444-513992.
Volume 144: The Roots of Visual Awareness, by C.A. Heywood, A.D. Milner and C. Blakemore (Eds.) – 2004, ISBN 0-444-50978-X.
Volume 145: Acetylcholine in the Cerebral Cortex, by L. Descarries, K. Krnjević and M. Steriade (Eds.) – 2004, ISBN 0-444-511253.
Volume 146: NGF and Related Molecules in Health and Disease, by L. Aloe and L. Calzà (Eds.) – 2004, ISBN 0-444-51472-4.
Volume 147: Development, Dynamics and Pathology of Neuronal Networks: From Molecules to Functional Circuits, by J. Van Pelt, M. Kamermans, C.N. Levelt, A. Van Ooyen, G.J.A. Ramakers and P.R. Roelfsema (Eds.) – 2005, ISBN 0-444-51663-8.

PROGRESS IN BRAIN RESEARCH

VOLUME 148

CREATING COORDINATION IN THE CEREBELLUM

EDITED BY

C.I. DE ZEEUW

Department of Neuroscience, Erasmus Medical Center, Dr. Molenwerfplein 50, Rotterdam, The Netherlands

F. CICIRATA

Dipartimento di Scienze Fisiologiche, Universitá Catania, Viale A. Doria 6, Catania, Italy

ELSEVIER

AMSTERDAM – BOSTON – HEIDELBERG – LONDON – NEW YORK – OXFORD
PARIS – SAN DIEGO – SAN FRANCISCO – SINGAPORE – SYDNEY – TOKYO

2005

ELSEVIER B.V.
Sara Burgerhartstraat 25
P.O. Box 211,
1000 AE Amsterdam
The Netherlands

ELSEVIER Inc.
525 B Street,
Suite 1900
San Diego,
CA 92101-4495
USA

ELSEVIER Ltd.
The Boulevard,
Langford Lane
Kidlington,
Oxford OX5 1GB
UK

ELSEVIER Ltd.
84 Theobalds Road
London WC1X 8RR
UK

© 2005 Elsevier B.V. All rights reserved.

This work is protected under copyright by Elsevier B.V., and the following terms and conditions apply to its use:

Photocopying
Single photocopies of single chapters may be made for personal use as allowed by national copyright laws. Permission of the Publisher and payment of a fee is required for all other photocopying, including multiple or systematic copying, copying for advertising or promotional purposes, resale, and all forms of document delivery. Special rates are available for educational institutions that wish to make photocopies for non-profit educational classroom use.

Permissions may be sought directly from Elsevier's Rights Department in Oxford, UK: phone (+44) 1865 843830, fax (+44) 1865 853333, e-mail: permissions@elsevier.com. Requests may also be completed on-line via the Elsevier homepage (http://www.elsevier.com/locate/permissions).

In the USA, users may clear permissions and make payments through the Copyright Clearance Center, Inc., 222 Rosewood Drive, Danvers, MA 01923, USA; phone: (+1) (978) 7508400, fax: (+1) (978) 7504744, and in the UK through the Copyright Licensing Agency Rapid Clearance Service (CLARCS), 90 Tottenham Court Road, London W1P 0LP, UK; phone: (+44) 20 7631 5555; fax: (+44) 20 7631 5500. Other countries may have a local reprographic rights agency for payments.

Derivative Works
Tables of contents may be reproduced for internal circulation, but permission of the Publisher is required for external resale or distribution of such material. Permission of the Publisher is required for all other derivative works, including compilations and translations.

Electronic Storage or Usage
Permission of the Publisher is required to store or use electronically any material contained in this work, including any chapter or part of a chapter.

Except as outlined above, no part of this work may be reproduced, stored in a retrieval system or transmitted in any form or by any means, electronic, mechanical, photocopying, recording or otherwise, without prior written permission of the Publisher.
Address permissions requests to: Elsevier's Rights Department, at the fax and e-mail addresses noted above.

Notice
No responsibility is assumed by the Publisher for any injury and/or damage to persons or property as a matter of products liability, negligence or otherwise, or from any use or operation of any methods, products, instructions or ideas contained in the material herein. Because of rapid advances in the medical sciences, in particular, independent verification of diagnoses and drug dosages should be made.

First edition 2005

Library of Congress Cataloging in Publication Data
A catalog record is available from the Library of Congress.

British Library Cataloguing in Publication Data
Creating coordination in the cerebellum. – (Progress in brain research; v. 148)
1. Cerebellum – Physiology – Congresses 2. Cerebellum – Differentiation – Congresses
I. Zeeuw, C. I. de II. Cicirata, F.
612.8′27

ISBN: 0-444-51754-5 (this volume)
ISBN: 0-444-80104-9 (series)
ISSN: 0079-6123

♾ The paper used in this publication meets the requirements of ANSI/NISO Z39.48-1992 (Permanence of Paper).

List of Contributors

J. Affanni, Instituto de Neurosciencia, Facultad de Medicina, Universidad de Morón, Buenos Aires CP 1708, Argentina

P. Alexandre, Régionalisation Nerveuse Niveau 8, CNRS/ENS, UMR 8542, 46 Rue d'Ulm, 75230 Paris Cedex 05, France

P.R. Andjus, Department of Physiology and Biochemistry, School of Biology, Studentski trg 12-16, P.O. Box 52, 11001 Belgrade, Serbia & Montenegro

R. Apps, Department of Physiology, School of Medical Sciences, University of Bristol, University Walk, Bristol BS8 1TD, UK

A. Arata, Laboratory for Memory and Learning, Brain Science Institute, RIKEN, Hirosawa 2-1, Wako, Saitama 351-0198, Japan

T. Arp, Department of Neurology, University of Schleswig-Holstein, Campus Lübeck, Ratzeburger Allee 160, 23538 Lübeck, Germany

T. Baldwinson, Eli Lilly and Company Ltd, Lilly Research Centre, Erl Wood Manor, Sunninghill Road, Windlesham, Surrey GU20 6PH, UK

M.G. Baxter, Department of Experimental Psychology, Oxford University, Oxford, UK

N. Bilovocky, Department of Neurosciences, Case Western Reserve School of Medicine, 10900 Euclid Avenue, Cleveland, OH 44106, USA

R. Bronsing, Department of Neuroscience, Erasmus Medical Center Rotterdam, P.O. Box 1738, 3000 DR Rotterdam, The Netherlands

E. Burguière, Laboratoire de Physiologie de la Perception et de l'Action, UMR CNRS 7124, 11 place Marcellin Berthelot, Collège de France, 75005 Paris, France

B. Carletti, Department of Neuroscience, Rita Levi Montalcini Centre for Brain Repair, University of Turin, Corso Raffaello 30, I-10125 Turin, Italy

P. Cavallari, Dipartimento di Medicina, Chirurgia e Odontoiatria, Università degli Studi, Ospedale S. Paolo, Via di Rudinì 8, I-20142 Milan, Italy

N.L. Cerminara, Department of Physiology, School of Medical Sciences, University of Bristol, University Walk, Bristol BS8 1TD, UK

G. Cerri, Istituto di Fisiologia Umana II, Università degli Studi, Via Mangiagalli 32, 20133 Milan, Italy

R. Cesa, Rita Levi Montalcini Center for Brain Repair, Department of Neuroscience, Corso Raffaello 30, 10125 Turin, Italy

A. Chédotal, CNRS URM7102, Université Pierre et Marie Curie, 75005 Paris, France

G. Chen, Department of Neuroscience, University of Minnesota, Lions Research Building, Room 421, 2001 Sixth St. S.E., Minneapolis, MN 55455, USA

G. Cheron, Laboratory of Neurophysiology, Université Mons-Hainaut, 24 Ave du Champ du Mars, B-7000 Mons, Belgium

D. Cicero, Dipartimento di Scienze Fisiologiche, Università Catania, Viale A. Doria 6, 95125 Catania, Italy

F. Cicirata, Dipartimento di Scienze Fisiologiche, Università Catania, Viale A. Doria 6, 95125 Catania, Italy

B. Dan, Laboratory of Movement Biomechanics, CP 168, Université Libre de Bruxelles, 1070 Brussels, Belgium

E. D'Angelo, Department of Cellular and Molecular Physiology and Pharmacology, University of Pavia and INFM, Via Forlanini 6, I-27100 Pavia, Italy

G. De Filippi, Eli Lilly and Company Ltd, Lilly Research Centre, Erl Wood Manor, Sunninghill Road, Windlesham, Surrey GU20 6PH, UK

E. De Schutter, Laboratory of Theoretical Neurobiology, Born-Bunge Foundation, University of Antwerp, Universiteitsplein 1, B-2610, Belgium

C.I. De Zeeuw, Department of Neuroscience, Erasmus Medical Center, Dr. Molenwaterplein 50, P.O. Box 1738, 3000 DR Rotterdam, The Netherlands

I. Dusart, UMR 7102 NPA, Université de Paris VI, Case 12, Bat B, 6éme étage, 75005 Paris, France

T.J. Ebner, Department of Neuroscience, University of Minnesota, Lions Research Building, Room 421, 2001 Sixth St. S.E., Minneapolis, MN 55455, USA

A.L. Edge, Department of Physiology, School of Medical Sciences, University of Bristol, University Walk, Bristol BS8 1TD, UK

C. Erdmann, Department of Radiology, University of Schleswig-Holstein, Campus Lübeck, Ratzeburger Allee 160, 23538 Lübeck, Germany

R. Esposti, Istituto di Fisiologia Umana II, Università degli Studi, Via Mangiagalli 32, 20133 Milan, Italy

P.B.C. Fenwick, Laboratory for Human Brain Dynamics, Brain Science Institute (BSI), RIKEN, 2-1 Hirosawa, Wako-shi, Saitama 351-0198, Japan

D. Gall, Laboratoire de Neurophysiologie (CP601), Faculté de Medecine, Université de Bruxelles, Route de Lennik 808, B-1070 Brussels, Belgium

W. Gao, Department of Neuroscience, University of Minnesota, Lions Research Building, Room 421, 2001 Sixth St. S.E., Minneapolis, MN 55455, USA

A.M. Ghoumari, INSERM U488, 80 rue du Général Leclerc, 94276 Bicêtre, France

P. Grimaldi, Department of Neuroscience, Rita Levi Montalcini Centre for Brain Repair, University of Turin, Corso Raffaello 30, I-10125 Turin, Italy

F. Gulden, Department of Genetics, Case Western Reserve School of Medicine, 10900 Euclid Avenue, Cleveland, OH 44106, USA

C. Hansel, Department of Neuroscience, Erasmus Medical Center, P.O. Box 1738, 3000 DR Rotterdam, The Netherlands

R. Hawkes, Department of Cell Biology and Anatomy, Faculty of Medicine, University of Calgary, 3330 Hospital Drive N.W., Calgary, AB T2N 4N1, Canada

W. Heide, Department of Neurology, Municipal Hospital Celle, Celle, Germany

K. Herrup, Department of Neurosciences, Case Western Reserve School of Medicine, 10900 Euclid Avenue, Cleveland, OH 44106, USA

H.C. Hulscher, Department of Physiology and Neuroscience, NYU School of Medicine, 550 First Avenue, New York, NY 10016, USA

A.A. Ioannides, Laboratory for Human Brain Dynamics, Brain Science Institute (BSI), RIKEN, 2-1 Hirosawa, Wako-shi, Saitama 351-0198, Japan

M. Ito, RIKEN Brain Science Institute, Wako, Saitama 351-0198, Japan

W.M. Kistler, Department of Neuroscience, Erasmus Medical Center, P.O. Box 1738, 3000 DR Rotterdam, The Netherlands

B. Kuemerle, Department of Neurosciences, Case Western Reserve School of Medicine, 10900 Euclid Avenue, Cleveland, OH 44106, USA

B. Ladd, Department of Psychiatry, Brigham and Women's Hospital, 221 Longwood Avenue, Boston, MA 02115, USA

M. Larouche, Department of Cell Biology and Anatomy, Faculty of Medicine, University of Calgary, Calgary, AB T2N 4N1, Canada

R. Llinás, Department of Physiology and Neuroscience, 550 First Avenue, New York, NY 10016, USA

M. Locatelli, Dipartimento di Scienze Neurologiche, Università degli Studi, Ospedale Policlinic, Via F. Sforza 35, I-20100 Milan, Italy

A. Louvi, Department of Neurobiology, Pharmacology and Physiology, University of Chicago, Chicago, IL 60637, USA

R. Maex, Laboratory of Theoretical Neurobiology, Born-Bunge Foundation, University of Antwerp, Universiteitsplein, B-2610 Antwerp, Belgium

A. Maffei, Department of Neuroscience, Brandeis University, 415 South Street, Waltham, MA 02454-9110, USA

L. Magrassi, Neurosurgery, Department of Surgery, IRCCS Policlinico S. Matteo, University of Pavia, Pavia, Italy

D.E. Marple-Horvat, Institute for Biophysical and Clinical Research into Human Movement, Manchester Metropolitan University, Alsager, UK

H. Marzban, Department of Cell Biology and Anatomy, Faculty of Medicine, University of Calgary, Calgary, AB T2N 4N1, Canada

C. Murcia, Department of Neurosciences, Case Western Reserve School of Medicine, 10900 Euclid Avenue, Cleveland, OH 44106, USA

N. Narboux-Nême, Régionalisation Nerveuse CNRS/ENS, UMR 8542, Ecole Normale Supérieure, 46 rue d'Ulm, 75005 Paris, France

K. Nguon, Department of Psychiatry, Brigham and Women's Hospital, 221 Longwood Avenue, Boston, MA 02115, USA

A. Nicotra, Dipartimento di Scienze Fisiologiche, Università Catania, Viale A. Doria 6, 95125 Catania, Italy

T. Nieus, Department of Cellular and Molecular Physiology and Pharmacology, University of Pavia and INFM, Via Forlanini 6, I-27100 Pavia, Italy

M.F. Nitschke, Department of Neurology, University of Schleswig-Holstein, Campus Lübeck, Ratzeburger Allee 160, 23538 Lübeck, Germany

M.R. Pantò, Dipartimento di Scienze Fisiologiche, Università Catania, Viale A. Doria 6, 95125 Catania, Italy

R. Parenti, Dipartimento di Scienze Fisiologiche, Università Catania, Viale A. Doria 6, 95125 Catania, Italy

F. Prestori, Department of Cellular and Molecular Physiology and Pharmacology, University of Pavia and INFM, Via Forlanini 6, I-27100 Pavia, Italy

K. Reinert, Department of Neuroscience, University of Minnesota, Lions Research Building, Room 421, 2001 Sixth St. S.E., Minneapolis, MN 55455, USA

L. Rondi-Reig, Laboratoire de Physiologie de la Perception et de l'Action, UMR CNRS 7124, 11 place Marcellin Berthelot, Collège de France, 75005 Paris, France

F. Rossi, Department of Neuroscience, Rita Levi Montalcini Centre for Brain Repair, University of Turin, Corso Raffaello 30, I-10125 Turin, Italy

P. Rossi, Department of Cellular and Molecular Physiology and Pharmacology, University of Pavia and INFM, Via Forlanini 6, I-27100 Pavia, Italy

C. Roussel, Laboratory of Neurophysiology, CP 601, Faculté de Medecine, Université de Bruxelles, Route de Lennik 808, B-1070 Brussels, Belgium

T.J.H. Ruigrok, Department of Neuroscience, Erasmus Medical Center Rotterdam, P.O. Box 1738, 3000 DR Rotterdam, The Netherlands

E. Sabel-Goedknegt, Department of Neuroscience, Erasmus Medical Center Rotterdam, 3000 DR Rotterdam, The Netherlands

E.M. Sajdel-Sulkowska, Department of Psychiatry Harvard Medical School, Boston, MA, USA; *Present address*: Brigham and Women's Hospital, 221 Longwood Avenue, Boston, MA 02115, USA

S.N. Schiffmann, Laboratoire de Neurophysiologie (CP601), Faculté de Médecine, Université Libre de Bruxelles, Route de Lennik 808, B-1070 Brusssels, Belgium

M.T. Schmolesky, Department of Neuroscience, Erasmus Medical Center, P.O. Box 1738, 3000 DR Rotterdam, The Netherlands

M.F. Serapide, Dipartimento di Scienze Fisiologiche, Università Catania, Viale A. Doria 6, 95125 Catania, Italy

L. Servais, Laboratory of Neurophysiology CP601, CP 168, Université Libre de Bruxelles, 1070 Brussels, Belgium

E. Sher, Eli Lilly and Company Ltd, Lilly Research Centre, Erl Wood Manor, Sunninghill Road, Windlesham, Surrey GU20 6PH, UK

R.V.. Sillitoe, Department of Cell Biology and Anatomy, Faculty of Medicine, University of Calgary, Calgary, AB T2N 4N1, Canada

J.I. Simpson, Department of Physiology and Neuroscience, NYU School of Medicine, 550 First Avenue, New York, NY 10016, USA

J.E. Slemmer, Department of Neuroscience, Erasmus Medical Center, Dr. Molenwaterplein 50, P.O. Box 1738, 3000 DR Rotterdam, The Netherlands

E. Sola, Department of Biophysics, SISSA, Via Beirut 4, I-34014 Trieste, Italy

C. Sotelo, Cátedra de Neurobiología del Desarrollo 'Remedios Caro Almela' at the Instituto de Neurociencias de la Universidad Miguel Hernández and CSIC San Juan, 03550 Alicante, Spain *and* CNRS URM 7102, Hôpital de la Salpêtrière, Université Pierre et Marie Curie, 46 Boulevard de l'Hôpital, 75651 Paris Cedex 13, France

G. Stavrou, Department of Neurology, University of Schleswig-Holstein, Campus Lübeck, Ratzeburger Allee 160, 23538 Lübeck, Germany

P. Strata, Rita Levi Montalcini Center for Brain Repair, Department of Neuroscience, Corso Raffaello 30, 10125 Turin, Italy

J. Van der Burg, Department of Neuroscience, Erasmus Medical Center Rotterdam, P.O. Box 1738, 3000 DR Rotterdam, The Netherlands

J. Voogd, Erasmus Medical Center, Department of Neuroscience, Box 1738, 3000 DR Rotterdam, The Netherlands

M. Wassef, Régionalisation Nerveuse CNRS/ENS, UMR 8542, Ecole Normale Supérieure, 46 Rue d'Ulm, 75230 Paris Cedex 05, France

S.G. Waxman, Department of Neurology LCI 707, Yale School of Medicine, P.O. Box 208018, 333 Cedar Street, New Haven, CT 06520-8018, USA

J.T. Weber, Department of Neuroscience, Erasmus Medical Center, Dr. Molenwaterplein 50, P.O. Box 1738, 3000 DR Rotterdam, The Netherlands

R. Wehrlé, UMR7102 NPA, CNRS-Université Paris VI, Case 12, Bar B, 6ème étage, 9 Quai Saint Bernard, 75005 Paris, France

S. Zahedi, Department of Cell Biology and Anatomy, Faculty of Medicine, University of Calgary, Calgary, AB T2N 4N1, Canada

A. Zappalà, Dipartimento di Scienze Fisiologiche, Università Catania, Viale A. Doria 6, 95125 Catania, Italy

L. Zhu, Department of Neuroscience, University of Turin, Corso Raffaello 30, 10125 Turin, Italy

Preface

Constantino Sotelo's lifetime work

Surrounded by his former students, friends and colleagues, most of them fitting more than one of these categories, Constantino Sotelo celebrated his more than 40 years of renowned research, last year, at the occasion of his retirement to other positions in France and Spain. The venue at Baia Verde in Catania, in his beloved Sicily, where science mixed with choice bits of Federico Cicirata's hospitality, was wonderful. Recognition for the organization of this successful symposium should go to Annalisa Nicotra and her staff.

I first met Contantino Sotelo in 1965 in Amsterdam, at an early international meeting on the cerebellum. In this case, the prediction of Clem Fox and Ray Snider, the editors of the proceedings of this meeting, that 'the unwritten future contributions which will result from participants who became intellectually motivated by the sessions, and by the genuine enthusiasm of the workers in the field, are unmeasured but unrecognized rewards of the symposium,' certainly came true.

The Amsterdam meeting was an almost purely morphological affair, with Brodal, Fox and Palay as its chief proponents, and a minor role for the physiological reports. The Catania meeting, on the contrary, was dominated by contributions on developmental biology, the main focus of Sotelo's research, and on the physiology of the cerebellum, reflecting the far-reaching influence of his work.

Sotelo's career started with a number of histochemical studies, followed by a series of papers on the synaptology of the cerebellum and related structures, where his meticulous electron microscopical technique revealed the influence of his long-term association with the late Sanford L. Palay. The observations of Sotelo and Llinás on the morphology and the electrophysiology of the gap junctions between the dendrites of the cells of the inferior olive, gave rise to the present research programs of Eric Lang and Chris De Zeeuw in the Neuroscience departments of the New York University School of Medicine and the Erasmus Medical Center in Rotterdam. Sotelo's interest in development probably was raised by his analysis of the cerebellar circuitry in mutant mice. The application of transplantation techniques to study the control of cerebellar development, dates from his collaboration with Dr. Alvarado-Mallart, whose reminiscenses colored the last session of the Catania meeting. Among my favorites are his papers on pattern formation in the cerebellar cortex and the development of the olivocerebellar projection, with Catherine Wassel, Bourrat, and Chédotal, which also started to appear in the late 1980s. The molecular biology of axon guidance and neuronal migration kept his interest till today.

The questions Sotelo asked generally were prompted by careful observation. His creative application of different techniques often provided some of the answers. Not always, though, the precise role of Purkinje cells in cerebellar pattern formation has not yet been resolved. Purkinje cell axons still fail to regenerate and the contribution of the gap junctions in the inferior olive to the function of the cerebellum is still disputed. Fortunately, there is still a lot to do for Constantino Sotelo and all the others who were intellectually motivated by his research. Clearly, his life's work is not finished yet. He will be around for the years to come, looking at us, benevolently but critically over his Cuban cigar.

Jan Voogd
Department of Neuroscience
Erasmus Medical Center, Rotterdam
The Netherlands
Tel.: 31104087308
e-mail: j.voogd@erasmusmc.nl

Acknowledgments

This volume contains the proceedings of the meeting entitled 'Creating Coordination in the Cerebellum' held October 2–4, 2003 in Catania, Italy. The meeting was held to mark the retirement of Constantino Sotelo as Director of the U-106 of the INSERM, in Paris.

The Editors acknowledge the financial support of the Assessorato della Sanità of the Regione di Sicilia, the University of Catania, the Faculty of Pharmacy of Catania, the City of Catania, the Provincia Regionale of Catania, the Banca Agricola Popolare of Ragusa and the Department of Physiological Science of Catania.

The organization of the Symposium was made possible through the excellent and generous collaboration of the technicians and staff of the Department of Physiological Sciences of Catania. The Editors express special thanks to the scholars of the Faculty of Pharmacy who work in the Department of Physiological Sciences, and in particular to Dr. Annalisa Nicotra and Dr. Linda Cicero for their enthusiasm and advice. Without their collaboration some of the activities carried out in the meeting would not have been possible. In addition, we would like to stress our special compliments to Loes Nijs-de Langen of the Department of Neuroscience of Erasmus MC in Rotterdam who did a fantastic job doing all the administrative work for the edition of the current volume of *Progress in Brain Research*.

Federico Cicirata and Chris I. De Zeeuw

Contents

List of Contributors v

Preface by Jan Voogd (The Netherlands) xi

Acknowledgements xiii

Section I.   Development of the Cerebellum

1. Development of the olivocerebellar system: migration and formation
    of cerebellar maps
    C. Sotelo and A. Chédotal (Alicante, Spain and Paris, France) 3

2. The genetics of early cerebellar development: networks not pathways
    K. Herrup, C. Murcia, F. Gulden, B. Kuemerle and
    N. Bilovocky (Cleveland, OH, USA) 21

3. Regionalization of the isthmic and cerebellar primordia
    N. Narboux-Nême, A. Louvi, P. Alexandre and M. Wassef
    (Paris, France and Chicago, IL, USA) 29

4. Bcl-2 protection of axotomized Purkinje cells in organotypic culture is age
    dependent and not associated with an enhancement of axonal regeneration
    A.M. Ghoumari, R. Wehrlé, C. Sotelo and I. Dusart
    (Paris and Bicêtre, France) 37

Section II.   Structural Cerebellar Plasticity

5. Axonal and synaptic remodeling in the mature cerebellar cortex
    R. Cesa and P. Strata (Turin, Italy) 47

6. Fate restriction and developmental potential of cerebellar progenitors.
    Transplantation studies in the developing CNS
    P. Grimaldi, B. Carletti, L. Magrassi and F. Rossi
    (Turin and Pavia, Italy) 57

Section III. Cell Physiological Cerebellar Plasticity

7. Long-term potentiation of synaptic transmission at the mossy fiber–granule
    cell relay of cerebellum
       E. D'Angelo, P. Rossi, D. Gall, F. Prestori, T. Nieus, A. Maffei
       and E. Sola (Pavia and Trieste, Italy and Brussels, Belgium) 71

8. Climbing fiber synaptic plasticity and modifications in Purkinje
    cell excitability
       M.T. Schmolesky, C.I. De Zeeuw and C. Hansel
       (Rotterdam, The Netherlands) 81

9. Bases and implications of learning in the cerebellum — adaptive control
    and internal model mechanism
       M. Ito (Saitama, Japan) 95

Section IV. Imaging of Cerebellar Activity

10. Synaptic transmission and long-term depression in Purkinje cells in an
     in vitro block preparation of the cerebellum isolated from neonatal rats
        A. Arata and M. Ito (Saitama, Japan) 113

11. Optical imaging of cerebellar functional architectures: parallel fiber beams,
     parasagittal bands and spreading acidification
        T.J. Ebner, G. Chen, W. Gao and K. Reinert
        (Minneapolis, MN, USA) 125

12. Imaging cerebellum activity in real time with magnetoencephalographic data
        A.A. Ioannides and P.B.C. Fenwick (Saitama, Japan) 139

13. The cerebellum in the cerebro-cerebellar network for the control
     of eye and hand movements — an fMRI study
        M.F. Nitschke, T. Arp, G. Stavrou, C. Erdmann and
        W. Heide (Lübeck, Germany) 151

**Section V. Oscillations and Synchrony in Cerebellar Cortex and
             Inferior Olive**

14. Fast oscillation in the cerebellar cortex of calcium binding protein-deficient
     mice: a new sensorimotor arrest rhythm
        G. Cheron, L. Servais, B. Dan, D. Gall, C. Roussel and
        S.N. Schiffmann (Mons and Brussels, Belgium) 167

15. Oscillations in the cerebellar cortex: a prediction of their frequency bands
        R. Maex and E. De Schutter (Antwerp, Belgium) 181

16. Gap junctions synchronize synaptic input rather than spike output of
   olivary neurons
   W.M. Kistler and C.I. De Zeeuw (Rotterdam, The Netherlands) 189

Section VI.   Cerebellar Motor Control

17. Is the cerebellum ready for navigation?
   L. Rondi-Reig and E. Burguière (Paris, France) 201

18. The lateral cerebellum and visuomotor control
   N.L. Cerminara, A.L. Edge, D.E. Marple-Horvat and
   R. Apps (Bristol and Alsager, UK) 213

19. Coupling of hand and foot voluntary oscillations in patients suffering
   from cerebellar ataxia: different effect of lateral or medial lesions
   on coordination
   G. Cerri, R. Esposti, M. Locatelli and P. Cavallari (Milan, Italy) 227

20. Modulation of cutaneous reflexes in hindlimb muscles during locomotion
   in the freely walking rat: a model for studying cerebellar involvement
   in the adaptive control of reflexes during rhythmic movements
   R. Bronsing, J. van der Burg and T.J.H. Ruigrok
   (Rotterdam, The Netherlands) 243

Section VII.   Cerebellar Neuro-Anatomical Organization

21. The basilar pontine nuclei and the nucleus reticularis tegmenti pontis
   subserve distinct cerebrocerebellar pathways
   F. Cicirata, M.F. Serapide, R. Parenti, M.R. Pantò, A. Zappalà,
   A. Nicotra and D. Cicero (Catania, Italy) 261

22. Conservation of the architecture of the anterior lobe vermis of the
   cerebellum across mammalian species
   R.V. Sillitoe, H. Marzban, M. Larouche, S. Zahedi,
   J. Affanni and R. Hawkes (Calgary, AB, Canada and
   Buenos Aires, Argentina) 283

23. Pharmacology of the metabotropic glutamate receptor mediated current
   at the climbing fiber to Purkinje cell synapse
   (L. Zhu, P. Strata and P.R. Andjus (Turin, Italy and
   Belgrade, Serbia and Montenegro) 299

Section VIII.   Excitability in Cerebellar Cortex

24. Nicotinic receptor modulation of neurotransmitter release in the cerebellum
   G. De Filippi, T. Baldwinson, E. Sher (Windlesham, UK) 309

25. Role of calcium binding proteins in the control of cerebellar granule
    cell neuronal excitability: experimental and modeling studies
    D. Gall, C. Roussel, T. Nieus, G. Cheron, L. Servais,
    E. D'Angelo and S.N. Schiffmann (Brussels and Mons,
    Belgium and Pavia, Italy) 321

26. Between in and out: linking morphology and physiology of
    cerebellar cortical interneurons
    J.I. Simpson, H.C. Hulscher, E. Sabel-Goedknegt and
    T.J.H. Ruigrok (New York, NY, USA and Rotterdam,
    The Netherlands) 329

Section IX. Cerebellar Pathology

27. Sexual dimorphism in cerebellar structure, function, and response
    to environmental perturbations
    K. Nguon, B. Ladd, M.G. Baxter and E.M. Sajdel-Sulkowska
    (Boston, MA, USA and Oxford, UK) 343

28. Cerebellar dysfunction in multiple sclerosis: evidence for an
    acquired channelopathy
    S.G. Waxman (West Haven, CT, USA) 353

29. Don't get too excited: mechanisms of glutamate-mediated Purkinje cell death
    J.E. Slemmer, C.I. De Zeeuw and J.T. Weber
    (Rotterdam, The Netherlands) 367

Section X. Epilogue

30. Epilogue
    R. Llinás (New York, NY, USA) 393

Subject Index 395

Development of the Cerebellum

Development of the olivocerebellar system: migration and formation of cerebellar maps

Constantino Sotelo[1,2,*] and Alain Chédotal[2]

[1]Cátedra de Neurobiología del Desarrollo 'Remedios Caro Almela' at the Instituto de Neurociencias de la Universidad Miguel Hernández and CSIC San Juan, 03550 Alicante, Spain
[2]CNRS URM7102, Université Pierre et Marie Curie, 75005 Paris, France

Introduction and personal considerations

I warmly thank Federico Cicirata and Chris De Zeeuw for organizing this symposium, celebrating my 40 years of research on the Cerebellum. For me this is a very special occasion, and I want to take this opportunity to express once more my gratitude to my mentors, to whom I am deeply attached.

My interest in the structure of the nervous system started in 1955. During my years of medical study at the University of Madrid, I had the wonderful opportunity to learn from Fernando de Castro, the youngest and the most beloved pupil of Santiago Ramon y Cajal. During those years, I spent my spare time learning the silver methods, and due to Don Fernando's patience, I became a rather accomplished specialist with these important tools. I graduated in 1960 and began my doctoral work under de Castro's direction. Unfortunately, a year later, I had to leave Spain and began working at the Medical School in Paris. Soon, I was attracted to the simplicity and the beauty of the cerebellum, and decided that the subject of my thesis would be the histochemical study of the carbohydrate metabolism in the cerebellum. And so, my earlier papers were devoted to the enzymo-histochemistry of cerebellar glomeruli and Purkinje cells, with an aim to correlate structure and function with energy production. These papers were published in 1962, in a local French journal.

When I finished my thesis, I had a chance meeting with René Couteaux, the unquestionable leader of French Neurocytology. His work on the ultrastructure of the neuromuscular junction is a classic and one of the most important pillars of the 'Neuron Doctrine.' In 1964, with the recommendation of Monsieur Couteaux, I obtained my first job as a research associate in the French national research agency (CNRS). At the Faculty of Sciences of Paris, I began to learn the ultrastructure of the nervous system with Couteaux's closest pupil, Jacques Taxi, on his favorite material, the frog sympathetic system and spinal cord. It was then that I discovered my first electrical synapses between dendrites of motoneurons and that I became interested in comparative neuromorphology. Monsieur Couteaux, knowing my interest for the cerebellum, organized my postdoctoral training at the laboratory of Sanford L. Palay, where I arrived in 1965. There in Boston, I found my final master and a close friend, from whom I learned most of what I know in Neuroscience. My warmest thanks to these people who kindly helped me to develop interest in Neuroscience and who always

*Corresponding author. Tel.: +34965919349; Fax: +34965919555; E-mail: sotelo@umh.es; sotelo@chups.jussieu.fr

supported my work and inspired me during my career.

Of all the work on the cerebellar development done in my laboratory during these last years, I have decided to summarize here, in collaboration with Alain Chédotal, the one concerning the development of the olivocerebellar system. This choice will allow me to acknowledge the important contributions of some of my closest collaborators: Marion Wassef, Franck Bourrat, Antoine Triller, Leonor Arsenio-Nunes, Pierre Angaut, and Evelyne Bloch-Gallego.

Constantino Sotelo

The olivocerebellar system

The inferior olivary nucleus is one of the two major afferent systems of the cerebellum. The importance of this precerebellar nucleus lies in the fact that it is the only source of cerebellar climbing fibers (Desclin, 1974; Sotelo et al., 1975; Brodal and Kawamura, 1980). The olivocerebellar system is organized in a strict topographic manner. Precise clusters of olivary neurons project into the cortical molecular layer in order to contact the dendrites of the Purkinje cells grouped in one or more parasagittal zones that extend throughout several lobules (Groenewegen et al., 1979). In addition, the axons of the olivary neurons give off collateral branches that enter the cerebellar nuclei and establish reciprocal synapses on the neurons receiving their main input from those Purkinje cells in the sagittal zones contacted by the same olivary neurons (Ruigrok and Voogd, 1990). Some of the contacted nuclear neurons, particularly the GABAergic ones, project in turn into the clusters of neurons in the contralateral inferior olive (Fredette and Mugnaini, 1991), constituting a topographically organized olivo-cerebellar-cerebello-olivary feedback loop.

One specific feature of inferior olivary neurons is that their thick proximal dendrites bear large appendages forming the central core of complex synaptic arrangements, called olivary glomeruli. Their peripheral zone is covered by numerous axon terminals, which are in synaptic contact with the central appendages. These appendages are electrotonically coupled through gap junctions (Llinas et al., 1974; Sotelo et al., 1974). Ultrastructural studies with the anti-GAD antibodies revealed that many of the axon terminals synapsing on the dendritic appendages are GABAergic (Fig. 1), although they also receive excitatory synaptic contacts (Sotelo et al., 1986). Moreover, using a double labeling approach (anterograde axon tracing and GABA immunocytochemistry), it was shown that a portion of these GABAergic terminals — many of them straddling the two appendages linked though gap junctions — originated from the cerebellar nuclear neurons (Fig. 1) (Angaut and Sotelo, 1987, 1989). The strategic location of these GABAergic synapses provides the best anatomical situation for a synaptic control of the electrotonic coupling (Llinas et al., 1974). Therefore, an anatomical basis exists for a dynamic decoupling of clusters of olivary neurons that could modulate the functional organization of the inferior olive (Llinas, 1974).

Despite the high level of knowledge in the structural and functional organization of the inferior olive, the function of this system is still a matter of debate. Two main hypotheses, the learning (Ito, 1982) and the timing (Llinas and Sasaki, 1989), have been debated over the last few years. Both hypotheses are supported by the anatomical organization of the inferior olivary circuit as much as by the electrophysiological properties of its neurons (see in De Zeeuw et al., 1998). In addition, since the learning and the timing hypotheses do not exclude each other, it is more likely that the olivocerebellar system is involved in both. In any case, the symposium offered to the supporters of each of these two hypotheses, an opportunity to reach a consensual view of the olivocerebellar function.

Although far from the battlefield of the function of the olivocerebellar system, the results presented here may be of some use to better understanding as to how this system is built during development.

Tangential migration of inferior olivary neurons

The newly generated neurons sometimes move for long distances along specific pathways to reach their final destination. This movement (neuronal migration) is a distinct cellular process essential for the establishment of the normal organization of the CNS.

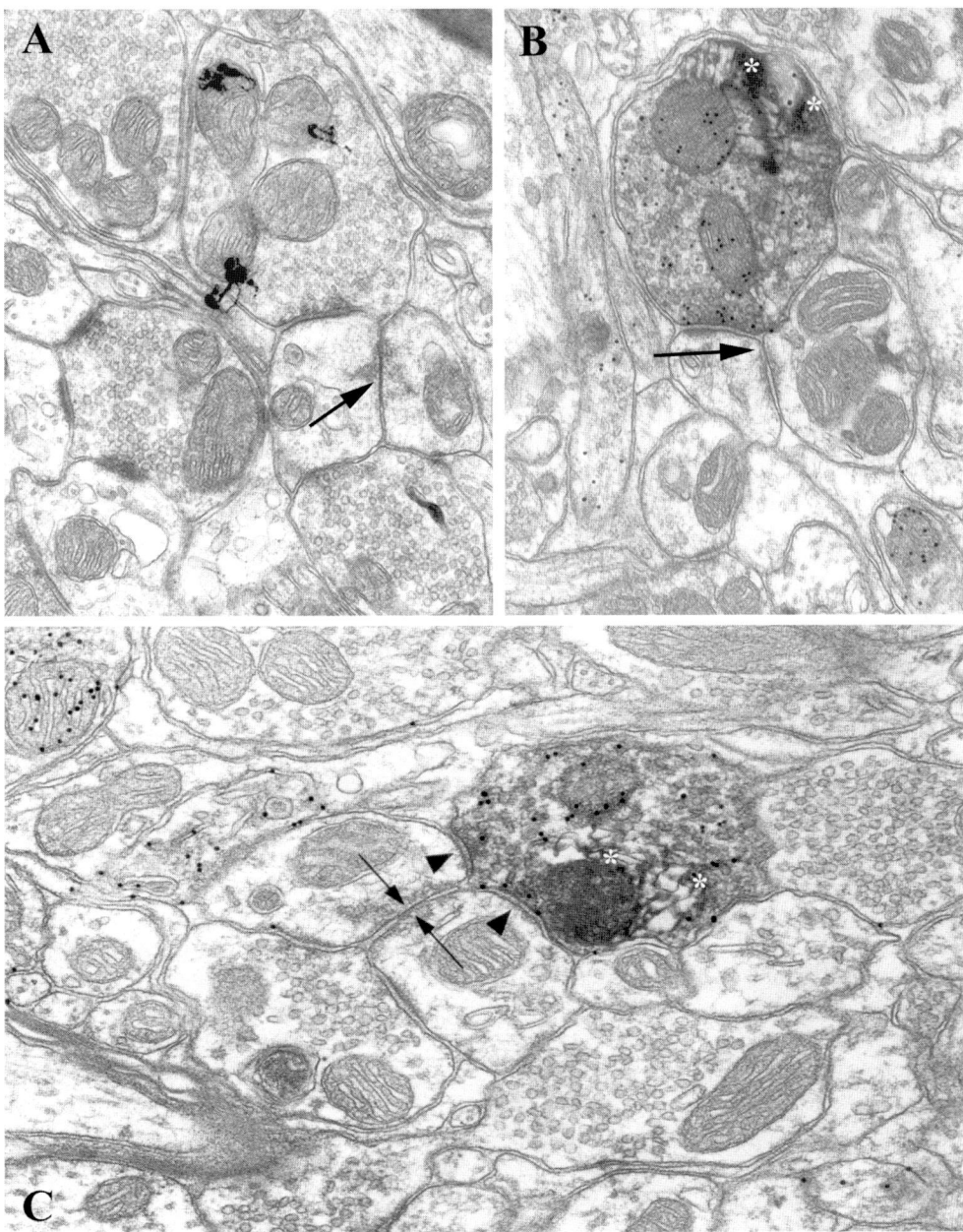

Fig. 1. The cerebello-olivary projection is GABAergic and mainly terminates on dendrites linked by gap junctions within glomeruli. Electron micrographs of olivary glomeruli illustrating dento-olivary axon terminals synapsing upon dendritic appendages linked through gap junctions (arrows). (A) Autoradiographically labeled terminal following the injection of a mixture of tritiated leucine and proline into the cerebellar lateral nucleus. The terminal is filled with pleomorphic vesicles and establishes an asymmetrical synapse with one of the dendritic profiles. (B) Double-labeled axon terminal by WGA-HRP reaction product (asterisks), following an injection of the anterograde axon tracer into the lateral nucleus, and by GABA-immunogold. The ultrastructural features of this axon terminal and its synaptic relationships are similar to those exhibited by the labeled terminal in A. (C) Same material than in B. The double-labeled axon terminal establishes two synaptic junctions (arrowheads), each one on each of the electrotonically coupled dendrites. (Modified from Angaut and Sotelo, 1987, 1989.)

The migration commonly starts when the neuronal precursors have finished dividing in the germinative neuroepithelia or in the subventricular zones. Postmitotic neurons migrate along pre-determined pathways following a precise spatiotemporal order. Since the early work of Rakic in the monkey cerebellum (Rakic, 1971), it is known that most neurons follow a radial direction during their migration, and that they move along radial glial axes (the gliophilic type of migration). However, morphological studies have provided evidence that many other neurons move in a tangential direction, perpendicular to the radial axis, allowing a mixing of cells, which originate from different ventricular regions. This type of migration where the substrate is not longer radial glial axis but the axonal or neuronal surface has been named neurophilic migration (Rakic, 1990). This is particularly the case for neurons generated in the rhombic lip. The rhombic lip emerges from the dorsal border of the alar plate of the rhombencephalon (from rhombomere 1 to upper spinal cord) at the boundary between the neural tube and the roof plate (Wingate, 2001). This neuroepithelium generates cerebellar granule cell precursors and all the brainstem precerebellar neurons (those for inferior olive, lateral reticular, external cuneatus, and basilar pontine nuclei).

With the exception of inferior olivary neurons which originate from the most caudal part of the rhombic lip (r8 to pseudo-r11), the other precerebellar neurons originate from the same neuroepithelium (r2 to r6; Marin and Puelles, 1995), and their birthdates overlap. In the rat, the inferior olivary neurons proliferate on E12 and E13, whereas, the lateral reticular neurons and the external cuneatus neurons proliferate on E13 and E14 (Altman and Bayer, 1987, Bourrat and Sotelo, 1988, 1991). In contrast to this short proliferation period, pontine neurons are generated during a protracted period, extending in the mouse embryo, from E12 to E16 with a late peak at E14–15 (Taber-Pierce, 1966). The more recent use of the FLP recombinase-based fate mapping approach, which allows the permanent labeling of distinct cell lineages, has revealed an unexpected subdivision of the rhombic lip neuroepithelium (Rodriguez and Dymecki, 2000). There is a differential regulation of *Wnt1* expression that divides the neuroepithelium into two genetically different domains, depending on whether or not they have a *Wnt1* expression. The progeny of *Wnt1*-positive progenitors selectively generate precerebellar neurons whose axons will end as mossy fibers. In contrast, the inferior olivary precursors do not express *Wnt1*. Thus, the fate of these precerebellar neurons is already specified in the rhombic lip progenitor cells.

Using thymidine labeling (Altman and Bayer, 1987) or the HRP in vitro axonal tracing method (Bourrat and Sotelo, 1988, 1990a), it was shown that the inferior olivary neurons follow a specific migratory route to reach their destinations called the submarginal stream (Fig. 2A, B). In contrast, the lateral reticular and the external cuneate neurons migrate along the marginal stream and the pontine neurons follow the pontobulbar or pontomedullary stream (Harkmark, 1954; Ono and Kawamura, 1990; Bourrat and Sotelo, 1990b). The migrating olivary neurons adopt a bipolar, elongated shape, with two opposite processes of irregular diameters: a short process or trailing process and a longer one, the leading process (Fig. 2B). The latter is relatively thick at its origin from the cell body, but thereafter, it becomes thin and very long corresponding to the axon (Bourrat and Sotelo, 1988, 1990a). For the precerebellar neurons, axonogenesis is an extremely precocious event, and the oriented growth of the axons is very rapid. We have determined that the entry of the inferior olivary axons, through the inferior peduncle, into the contralateral hemicerebellum coincides with the arrival of the inferior olivary perikarya to the floor plate level (Wassef et al., 1992b). The displacement of the cell bodies occurs by a progressive translocation of the nucleus and the cytoplasmic organelles within the leading processes (nucleokinesis). An important point is that the perikaryon of the olivary neurons, contrarily to their axon, does not cross the midline and settle ipsilaterally to their proliferation side. In contrast, the neurons of the lateral reticular and the external cuneate nuclei and their axons cross the midline (Fig. 2C) (Bourrat and Sotelo, 1990b). More than 14 years ago, we suggested (Bourrat and Sotelo, 1990a) that the floor plate controlled this migration, probably through the secretion of chemoattractive factors. In addition, we had also proposed that the floor plate must produce some kind of stop signals which could instruct the neurons whether or not to cross the midline.

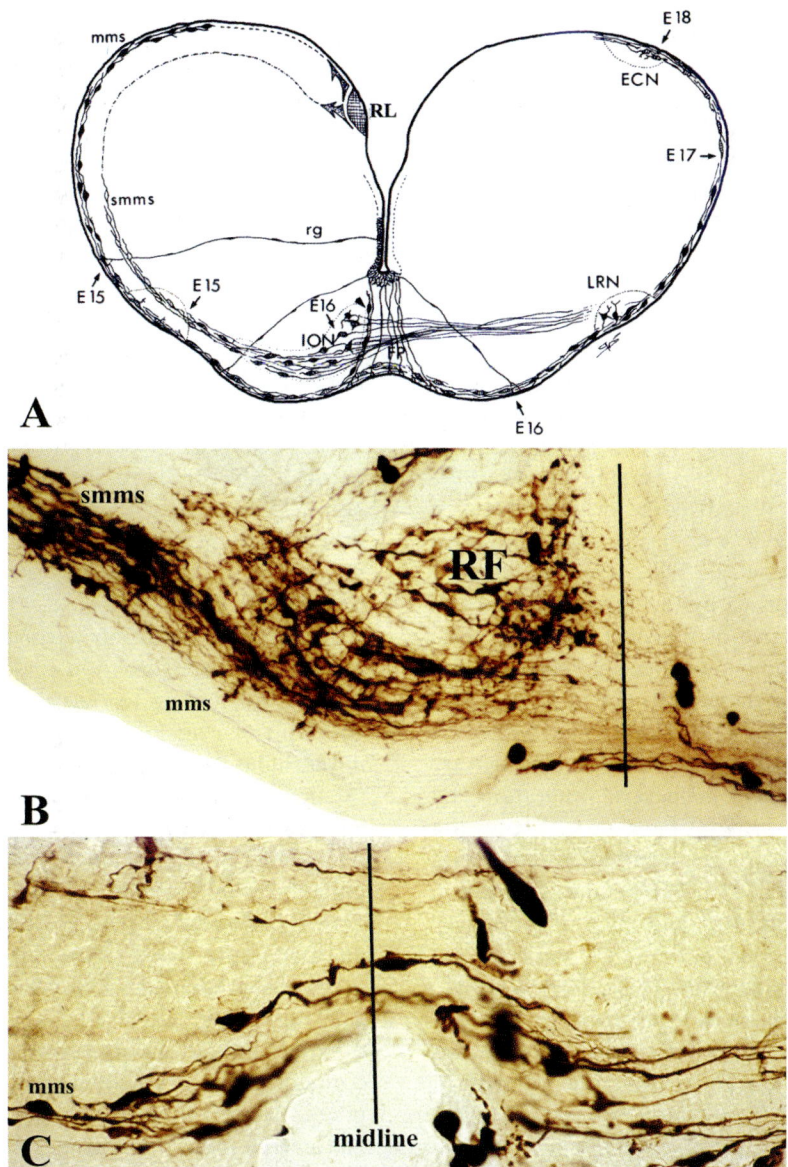

Fig. 2. Migratory pathways of the precerebellar nuclear neurons in the rat embryo identified by retrograde labeling after in vitro HRP injections. (A) Schematic representation illustrating the main events on this migration. Neurons belonging to the inferior olive (ION), lateral reticular (LRN) and external cuneate nuclei (ECN) are generated in the lower rhombic lip (RL). ION neurons migrate along the submarginal stream (smms), while LRN and ECN migrate via the marginal stream (mms). The positions of the fronts of the migration waves are indicated between E15 and E18. ION neurons do not cross the floor plate (FP), whereas those destined to the LRN and the ECN cross the FP. Radial glia (rg). (B) At E16, the HRP labeled olivary neurons start to reach their ultimate domain, contralaterally to the injection site and ipsilaterally to their generation site (the solid line marks the midline). Some neurons belonging to the reticular formation (RF) have been also labeled by the injection. Note ION neurons still migrating in the smms, and some neurons located in the mms crossing the midline. (C) E16 rat embryo. Higher magnification of some of the HRP labeled neurons belonging to the LRN, moving along the mms and crossing the midline. (Modified from Bourrat and Sotelo, 1988, 1990a,b.) A × 300, B × 400.

Chemotactic cues in the migration of inferior olivary neurons

Although the neuronal migration is unidirectional, the migrating neurons can move in both directions, but the potential for bidirectionality is largely restricted by tissue-dependent cues, which could act from a distance. Therefore, chemotactic molecules of both the chemoattractive and the chemorepellant actions may regulate the direction of the migrating neurons.

Some chemotactic molecules governing the directionality of migrating neurons, the slits and the netrins, were previously shown to guide growing axons (Kennedy et al., 1994; Serafini et al., 1994; Bloch-Gallego et al., 1999; Brose et al., 1999; Nguyen Ba-Charvet et al., 1999; Yee et al., 1999; Alcantara et al., 2000; Causeret et al., 2002; De Diego et al., 2002). They are expressed by the floor plate and can provide either attractive or repellent guidance cues, depending on the expression of specific receptors and on the level of second messengers (Ming et al., 1997; Bashaw and Goodman, 1999; Hong et al., 1999).

Accordingly, the inferior olivary neurons are influenced during their tangential migration by netrin-1. The spatiotemporal patterns of expression of netrin-1 and its receptors (Deleted in Colorectal Cancer-DCC-Unc5H2, Unc5H3, and neogenin, expressed by premigratory and migratory olivary neurons), support a role for this molecule in attracting the olivary neurons (Bloch-Gallego et al., 1999). Moreover, the analysis of these neurons in mice deficient for netrin-1 (which die at birth) has shown that in the newborn mice, the number of inferior olivary neurons is greatly reduced. Many of them remain dispersed throughout the submarginal stream grouped into small ectopic clusters and only less than 15% of these neurons reach their ventromedial position (Bloch-Gallego et al., 1999). In addition, the surviving olivary neurons develop ipsilateral projection, instead of the normal contralateral one. Furthermore, in the DCC null mutant mouse, the inferior olivary complex exhibits a similar, although less severe phenotype than in the netrin-1 knock-out mice (Bloch-Gallego, unpublished).

Another evidence of the role of netrin-1 in the tangential migration of olivary neurons was obtained by co-culturing, in collagen gel, lower rhombic lip explants (r8 to pseudo-r11) of E11 mouse embryos with cells secreting netrin-1 (Causeret et al., 2002; De Diego et al., 2002). Under these conditions, the leading processes and the translocating nuclei of the migrating neurons are attracted by netrin-1 (Fig. 3). In addition, the co-expression of *slit-1/2/3* by cells of the floor plate and the expression of *robo-2*, a Slit receptor, by migrating inferior olivary neurons, in particular when they approach the ventral midline, prompted us to look for a possible involvement of Slit in this migration. We have shown (Causeret et al., 2002) that Slit produces a weak repulsive effect on the outgrowth of the leading processes of migrating olivary neurons. However, when combined with netrin-1, Slit is able to antagonize part if not all the netrin-1 effects on axon growth and nucleokinesis in a dose-dependent manner. The molecular mechanisms of such an antagonistic effect remain to be deciphered. It is, however, known that Slit can silence the attraction of netrin-1 for *Xenopus* spinal axons. This silencing action could result from the fact that, upon binding to Slit, robo is able to interact with the netrin-1 receptor DCC through their intracellular domains, blocking netrin function (Stein and Tessier-Lavigne, 2001). These results emphasize that Slit could participate in the ventral positioning of inferior olivary neurons near the floor plate, possibly providing a stop signal.

An important in vitro study aimed mainly at analyzing the early steps in the migration of precerebellar neurons has been published recently. For this study, De Diego et al. (2002) performed organotypic cultures of the portion of the neural tube containing the cerebellum and attached hindbrain taken from E11.5–12.5 mouse embryos. In these cultures, a migration along the submarginal and marginal streams does occur even after ablation of the floor plate, thus, suggesting that the latter is not required for the initiation of the migration, and that dorsoventral cues are distributed along the neural tube. However, through the ectopic grafting of pieces of floor plate, it was shown that this ventral structure exerts — as expected — strong chemoattractive effects on the migration and aggregation of the inferior olivary neurons. In addition, De Diego et al. (2002) have studied the spatiotemporal patterns of expression of some other molecules, acting as short-range cues, that could also be involved in the

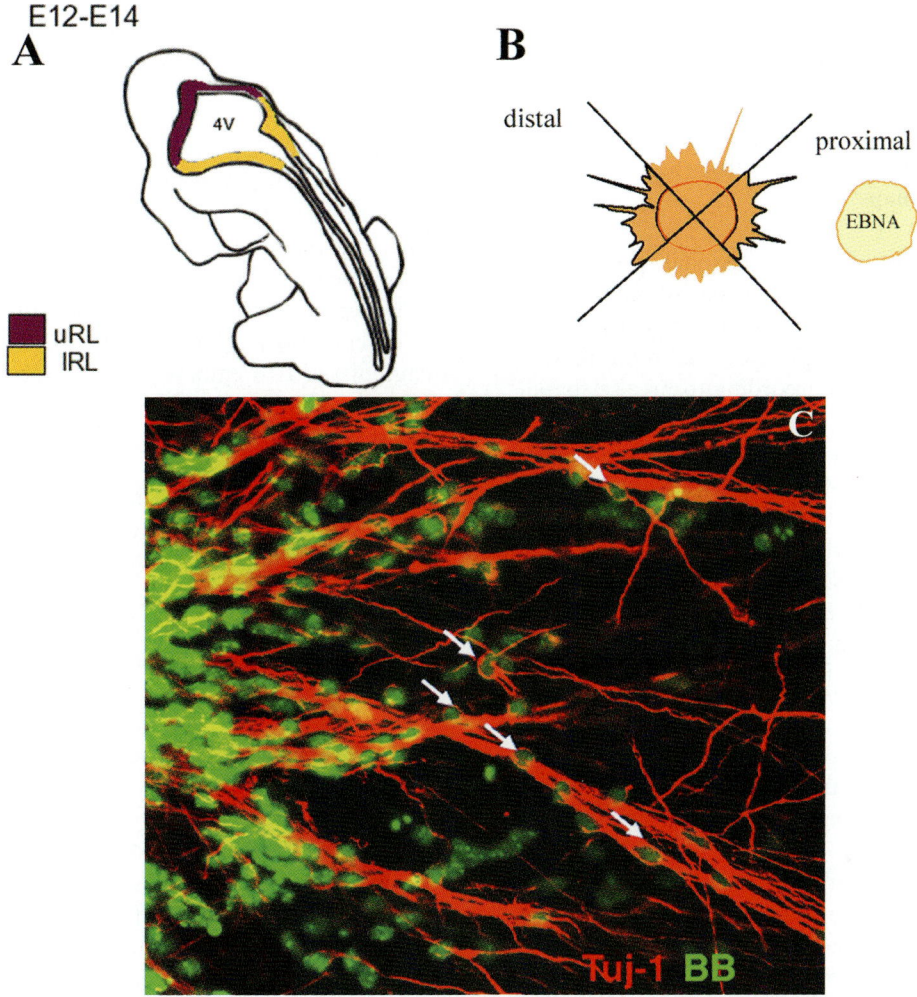

Fig. 3. Attractive action of netrin-1 on the migration of inferior olivary neurons. (A) Schematic drawing showing the anatomical location of inferior rhombic lip (yellow) in a mouse embryo. Pieces of the most caudal portion of the rhombic lip of E11.5 mouse embryos were used as explants for the confrontation assays with aggregates of EBNA-293 cells stably transfected with a construct encoding *netrin-1-cmyc*. (B) The drawing illustrates the method used to quantitatively analyze the amount of neuronal migration. For each explant, the area occupied by the neuritic processes and their translocating cell bodies was measured in the quadrants proximal and distal to the EBNA cell aggregates. (C) Confocal microscopy of a single 0.8 μm thick optic section of the proximal quadrant illustrating neuritic processes of olivary neurons immunostained with class III β-tubulin antibodies Tuj-1. The nuclei of the migrating cells were stained with bis-benzimide. The overlay shows the nuclei (white arrows) translocating within leading axonal processes, attracted by the source of the netrin-1. (A, B: modified from Alcantara et al., 2000; C: from Causeret et al., 2002.) C × 400.

migration. For instance, EphA4 is expressed by olivary neurons approaching the ventral midline (E13.5), while the ligands for this receptor (ephrins-B3, -B1) are expressed by the floor plate. Taking into account the occurrence of other local cues, particularly adhesion molecules (Backer et al., 2002; Kyriakopoulou et al., 2002) that might also be involved in this tangential migration, the allocation of precerebellar neurons should be considered to be a high complex and dynamic developmental process. These new insights on the tangential migration of inferior olivary neurons allow us to underline the similarity of the molecular mechanisms involved in the guidance of growing axons and those involved in the neuronal migration.

Formation of olivocerebellar maps: Coarse-grained maps

The establishment of orderly axonal projections, which precedes synaptogenesis, is one of the essential steps in the formation of neuronal networks. In the cerebellum, as in other regions of the brain, afferent projections are arranged according to precise spatial order, resulting from axonal segregation. The segregation of incoming fibers into distinct compartments is one of the most common ways for building up neural maps. This is the case for the cerebral cortex and its modular columnar organization (Rakic, 1977), the dorsal lateral geniculate nucleus and its laminar arrangement (Sretavan and Shatz, 1984), or the striatum with its two intermingled compartments, named the patches and the matrix (Gerfen, 1984).

Organization of projection topography and heterogeneity of Purkinje cells in adult cerebellum

The striking uniformity of the neural elements constituting the cerebellar cortex is in apparent contradiction with the well-known functional parcellation of this cortex into narrow sagittal zones (Oscarsson, 1979). This longitudinal-zonal organization is based upon differences in the arrangement of afferent and efferent connections (see in Ito, 1984).

The work of Hawkes and his collaborators has unequivocally shown that the Purkinje cells are a heterogeneous neuronal population (Gravel et al., 1987; Hawkes and Gravel, 1991). This evidence has been mainly gathered through the use of monoclonal antibodies that recognize antigens expressed according to a parasagittal band pattern, with subsets of Purkinje cells, either possessing, or lacking such antigens (Fig. 4A). To determine whether the compartmentation of the Purkinje cells was correlated with the olivocerebellar map, we used tritiated amino acids to trace olivary fibers and immunostaining for zebrin I (Wassef et al., 1992a). Under these conditions, we corroborated and extended previous results (Gravel et al., 1987), by showing that the boundaries of labeled climbing fibers and those of immunoreactive Purkinje cell subsets were strongly correlated. Hence, both the maps are congruent. In addition, to discard the possibility that the distribution of the zebrin-positive and zebrin-negative compartments could emerge as a result of the topographic organization of the olivocerebellar projection, instead of being intrinsic to the cerebellar cortex, we performed grafting experiments (Wassef et al., 1990). Pieces of the anlage of the E14 embryonic rat cerebellum, before the arrival of olivary axons, were transplanted to the neocortex of adult rats. In such transplants, devoid of precerebellar afferent fibers, clusters of negative and positive Purkinje cells still formed (Fig. 4B, C). In conclusion, there is an almost exact overlap between the olivocerebellar projection map and the parasagittal compartmentation of the Purkinje cell heterogeneity, parcellation that is independent of the occurrence of inferior olivary projections.

The development of the olivocerebellar map: the 'matching hypothesis'

Unfortunately, the markers of the adult Purkinje cell heterogeneity are expressed too late in development to allow us to investigate the early phases of the olivocerebellar map formation. Indeed, using the anterograde axon tracing technique in postnatal rat cerebellum (Fig. 5A), we showed that the olivocerebellar projection was already organized in its characteristic banding pattern several days before the expression of adult Purkinje cell heterogeneity markers (Sotelo et al., 1984). To investigate the early phases of the formation of the projection, we used selective Purkinje cell markers that, although not directly involved in the matching between olivary axons and their cellular targets, were useful tools to study the development of the olivocerebellar projection. In the cerebellum, we have shown that antibodies which selectively stain all Purkinje cells in the adult (calbindin, GMP-cyclic dependent protein kinase, Purkinje cell specific glycoprotein and PEP-19) start being expressed at different ages and, not simultaneously by all Purkinje cells. The combination of these markers delimits the

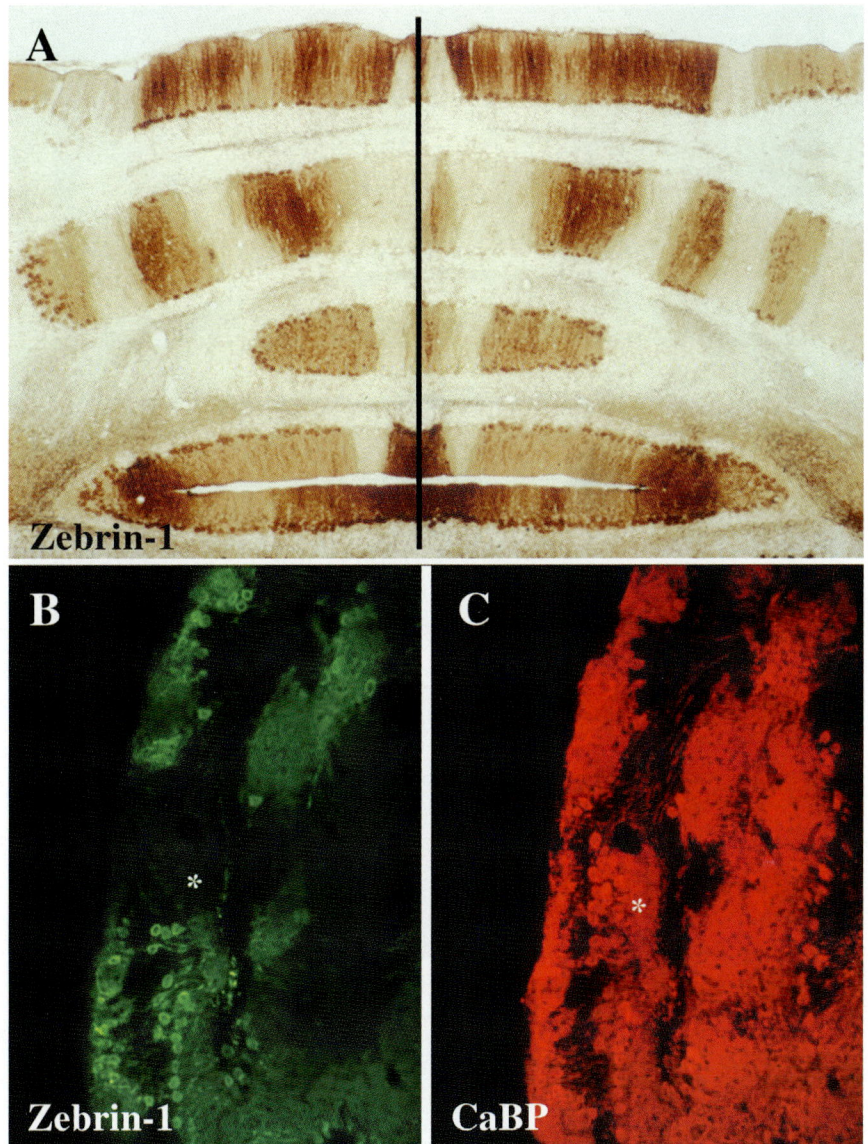

Fig. 4. The embryonic heterogeneity of Purkinje cells is independent from their extracerebellar afferent fibers. (A) Frontal section through the posterior vermal cortex of an adult rat cerebellum immunostained for zebrin-I, illustrating the alternated expression pattern in parasagittal bands of Purkinje cells. (B and C) Double labeled section of an E13 cerebellar transplant two months after grafting into a cerebral cortex cavity. Despite the absence of extracerebellar afferent fibers, the Purkinje cells are organized into zebrin-1 positive and negative (asterisks) clusters, mimicking the parasagittal banding organization of control cerebellum. (Micrographs provided by Marion Wassef.) A ×40, B, C ×120.

biochemically defined zones in the cerebellar cortex containing Purkinje cells of similar identity, which constitute the basic cortical compartment (Wassef et al., 1985). Similarly, in the inferior olive, the antibodies against calbindin, parvalbumin, and calcitonin gene-related peptide (CGRP) are also asynchronically expressed by subsets of inferior olivary neurons. Parvalbumin and CGRP were only transiently expressed. In addition, a subset of CGRP labeled inferior olivary axons can transiently

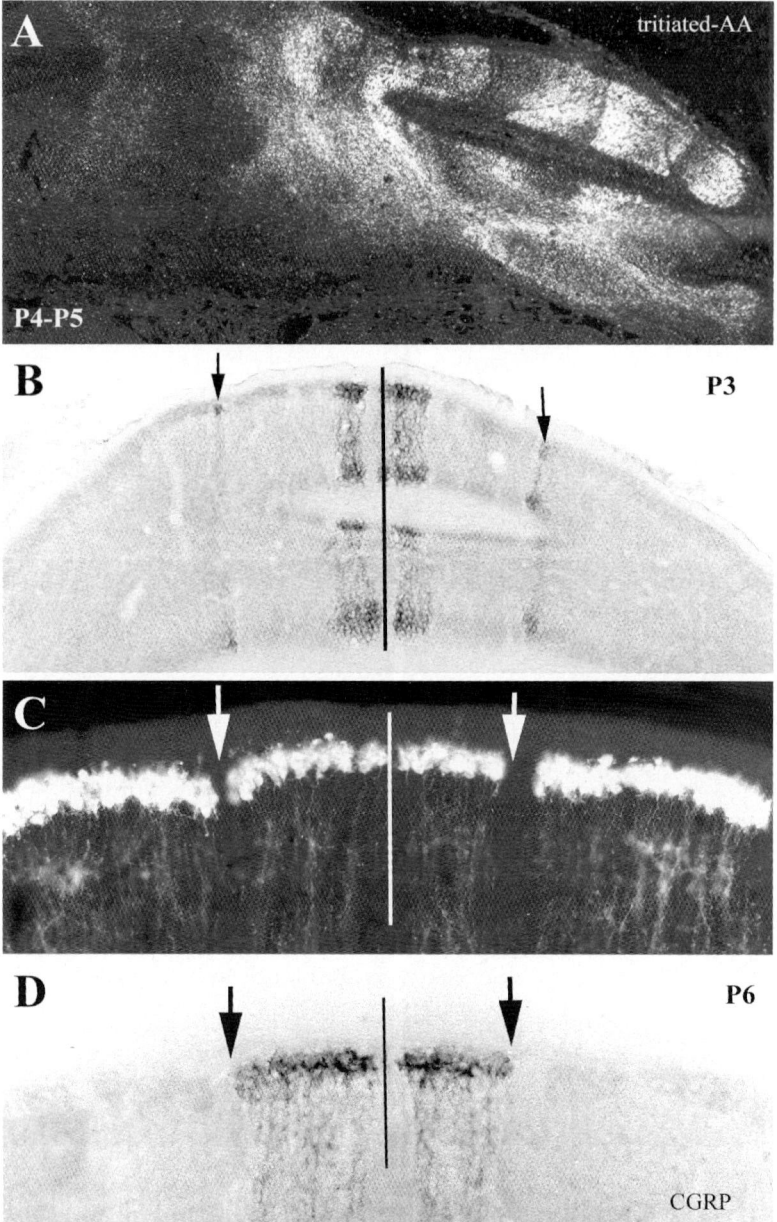

Fig. 5. Development of the topographic organization of the olivocerebellar projection. (A) Darkfield micrograph of an autoradiogram of the hemispheric cortex of a P4 rat that received an injection of tritiated aminoacids in the inferior olive and was fixed 20 h later. This micrograph illustrates the banding pattern of the terminal distribution of labeled climbing fibers. (B) Topographic distribution of CGRP-IR climbing fibers in the rat P3 cerebellum. Note the presence of two large bands, each one on each side of the midline (solid line) which is free of labeling. The arrows point to two thin intermediate bands symmetrically located at both edges of the vermal cortex. (C and D) Double immunostaining (CaBP immunofluorescent and CGRP immunoperoxidase). In the P6 rat cerebellum, the Purkinje cells are already arranged in a monolayer. The solid line marks the midline and the arrows point to two raphes in the Purkinje cell layer. CGRP positive climbing fibers are concentrated at the basal pole of the Purkinje cell bodies. Note that the midline region is devoid of climbing fibers despite the presence of a Purkinje cell, and that the CGRP-climbing fibers stop at the raphes. A × 35, B × 50, C–D × 150.

(from E16 to P21 in rats) be visualized in the cerebellar cortex. With this approach, we were able to show that from their, arrival throughout the inferior peduncle, the CGRP+ olivocerebellar axons reach highly selective entry points at the surface of the rostral cerebellar plate, forming early and late appearing stripes confined to precise cerebellar territories (Fig. 5B–D) (Chedotal and Sotelo, 1992).

We have, therefore, hypothesized that the acquisition of subpopulation-type identities in inferior olivary neurons and their targets, the Purkinje cells, would allow the formation of the projection map by matching their shared positional cues (the 'matching hypothesis,' Sotelo and Wassef, 1991; Sotelo and Chedotal, 1997). This hypothesis is supported by three observations, two of them: (1) the Purkinje cells and the inferior olivary neurons are biochemically heterogeneous and (2) both biochemical heterogeneities start before inferior olivary axons reach the cerebellum (Wassef et al., 1985, 1992b; Chedotal and Sotelo, 1992), have been discussed above. The third one, (3) the developmental compartments related to the topography of the olivocerebellar projection, is based upon the fact that the neurons in the dorsal cap of the inferior olive, which transiently express the calcium binding protein parvalbumin, have their axons within their proper vermal compartments (in lobule IX and X), confined to sagittal clusters of Purkinje cells PEP 19 negative. This demonstrates that the inferior olivary and Purkinje cell compartments are congruent, i.e., they match each other through their projection map (Wassef et al., 1992c).

Intrinsic nature of cerebellar compartmentation: the mediolateral and anteroposterior compartments

As already discussed, the discovery of the biochemical heterogeneity of the embryonic Purkinje cells (Wassef and Sotelo, 1984) provided the first evidence for an early parcellation of the developing cerebellar cortex along its mediolateral axis. The intrinsic nature of this parcellation was demonstrated almost 10 years later by Jim Morgan and his collaborators (Oberdick et al., 1993; Morgan and Smeyne, 1997) in their studies of the transgenic mice carrying the *L7/Pcp-2/lacZ* hybrid gene, a Purkinje cell specific transgene. They have shown that the transgene is expressed in parasagittally oriented stripes, reminiscent of those observed with the Purkinje cell markers. Moreover, this pattern can be perturbed by promoter mutation, and the resulting parcellations disclosed that, in addition to the parasagittal alignment of the compartments, the Purkinje cell compartments distributed in the anteroposterior axis also occurred. These results are of importance because they show that the allocation of parasagittal and/or transverse compartments is intrinsic to the Purkinje cells and, therefore, that this neuronal population contains, early in development, encoded positional information.

More recently, the work of Joyner and her collaborators (Millen et al., 1995) has provided some insight into the genetic regulation of the Purkinje cell compartmentation. They have shown that from E17.5, the mouse embryonic cerebellum is parcellated into 11 parasagittal domains defined by differential expression of specific genes (*En-2*, *En-1*, *Wnt-7B*, and *Pax-2*). An important confirmation of the major role of *En-2* in controlling the compartmentation of Purkinje cells has been obtained in mice with ectopic expression of *En-2* under the *L7/Pcp2* promoter. In these transgenic mice, the parasagittal banding pattern of zebrin-positive Purkinje cells is not complete, and the boundaries between compartments are blurred (Baader et al., 1999). Moreover, the cerebellar parcellation in the transverse axis has also been reported, particularly for *En-2*, *En-1*, *Gli*, *HNK-1*, and *Wnt-7B* genes. This anteroposterior compartmentation can be substantiated by the existence of some spontaneous mutations in the mouse, such as *tottering/leaner* (Hess and Wilson, 1991), *meander tail* (Ross et al., 1990) and *swaying* (Thomas et al., 1991), that only induce defects in the anterior lobe. The occurrence of such mutants suggests that, there must be differences in the developmental process between the anterior and the posterior cerebellum (see in Herrup and Kuemerle, 1997). As summarized by Joyner and collaborators, all these studies have shed light on the genetic regulation of spatial cues in the developing cerebellum, that divide this nervous structure 'into a grid of positional information required for patterning foliation and afferent fibers.'

The existence of positional cues required for the establishment of the olivocerebellar projections: evidence obtained with in vitro studies

We have developed an in vitro model to demonstrate the existence of Purkinje cell guidance cues for inferior olivary axons (Chedotal et al., 1997). This model was easily accessible to experimental manipulations and allowed us to perform somewhat similar experiments than those used much earlier in the study of the *Xenopus* retinotectal topography (see in Willshaw et al., 1983). Large hindbrain explants of E7.5–E8 chick embryos, containing the cerebellum and the brainstem from the spino-medullary junction to the optic tectum, were cultured on inserts (Fig. 6A) (Chedotal et al., 1997). The explanted cerebella and the brainstems were manipulated after half an hour in vitro and in most experiments, the axon tracers were injected after 7 days in vitro and fixed 24 h later. In the chick embryo, olivary neurons are generated between E3 and E5 (Armstrong and Clarke, 1979). During E5–E7, they migrate from the dorsal position of their precursors in the rhombic lip to the ventral aspect of the medulla (Harkmark, 1954). The axons of the olivary neurons enter the cerebellar plate by E9 and contact the Purkinje cells by E10 (Chedotal et al., 1996). Therefore, the olivocerebellar axons were still growing in the peripheral aspect of the medulla of our explants when the cerebellar manipulations were done.

The axonal tracing methods in these cultures have shown that the normal anteroposterior topography of the olivocerebellar projection was achieved even when the cerebellar lamella was sectioned, detached from the brainstem, and placed back into its original position. In vitro, neurons are located in the caudolateral inferior olive project to the anterior cerebellum, whereas rostromedian olivary neurons project to the posterior cerebellum (Fig. 6B), as in the adult chicken (Furber, 1983). Two kinds of surgical operations were performed in these experiments.

The first one involved either a simple rotation of the cerebellar lamella (type-1 rotation of 180° along the anteroposterior axis) or a similar rotation followed by a mediolateral inversion, type-2 rotation. In the later case, the external surface of the cerebellar lamellae abutted the brainstem. In both cases, the rostromedian olivary neurons project to the posterior vermis and the caudolateral neurons to the anterior vermis, but with now inverted locations (Fig. 6C). Thus, the Purkinje cells receive axons of the same inferior olivary neurons, regardless of their final position. In addition, using short term survival experiments, we were able to show that the olivary axons project to their correct terminal domains a few hours after they have entered the cerebellum, in intact and in rotated experiments. This demonstrates that the olivocerebellar projection does not emerge from an initially more diffuse projection, contrary to what has been reported to occur with the rodent retinocollicular projection (Simon et al., 1994).

The second form of the surgical operation consisted of ablations of the posterior or anterior halves of each cerebellar plate, giving rise to a single posterior or a single anterior cerebellum. A single posterior cerebellum receives projections from the entire contralateral inferior olive. These projections are not randomly organized. A 'compressed map' is established with the caudolateral olivary neurons projecting to the anterior-most portion of the posterior cerebellum and rostromedian olivary neurons projecting to the posterior-most portion (Fig. 6D). In contrast, only neurons from the caudolateral olive project to a single anterior cerebella (Fig. 6D). Thus, either the growth of axons from the rostromedian olivary neurons is promoted by molecules restricted to the posterior cerebellum, or there are repellent molecules present in the anterior cerebellum which prevent the entry of axons, emerging from the rostromedian olive. This last explanation seems to be the correct one (see next section below), although a recent study using our in vitro model suggests that the cerebellum also secretes an attractant for inferior olivary axons (Zhu et al., 2003).

These observations strongly indicated that olivocerebellar fibers recognize, within their target region, positional cues which organize their anteroposterior topography. In addition, the type-2 rotation experiments indicate that these positional cues are independent of the normal pathway of the projection. Even when the olivary axons enter the cerebellar plate through its external surface, they are allocated to their corresponding anterior or posterior vermal lobe. The presence of a superficial Purkinje cell plate just below the external granular layer at this stage of

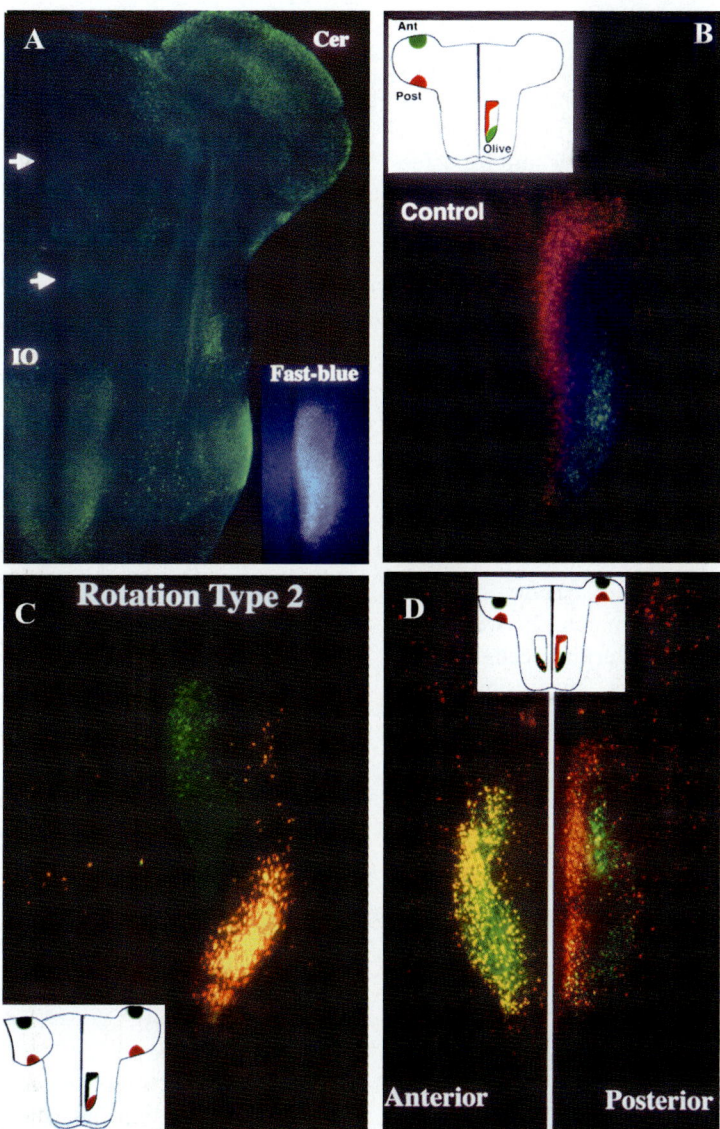

Fig. 6. In vitro demonstration of the presence of topographic cues in the chick embryonic cerebellum that guide growing inferior olivary axons. (A) Calbindin immunostaining of the hindbrain and attached cerebellum of an E7.5 chick embryo, opened through its dorsal midline and flat explanted on an insert for 15 DIV. The arrows point to the midline. Note the occurrence of CaBP positive cells in the cerebellum (Cer) and in the inferior olive (IO). Inset: Retrograde labeling of the IO 24 h after injection of Fast-Blue in the contralateral hemicerebellum. (B) Olivocerebellar topography in a control explant, retrograde axon tracing. A DiA crystal (green in the inset) is inserted close to the anterior cerebellar margin, while a DiI crystal (red in the inset) is placed close to the posterior margin. Fast-Blue is injected in the same hemicerebellum. The whole contralateral inferior olive is labeled by the Fast-Blue marker, while DiI labels rostromedian territory and DiA caudolateral territory. (C) This micrograph illustrates the results obtained after a rotation type 2 experiment. The staining of the two IO subdivisions, caudolateral and rostromedian, are inverted in comparison with controls. (D) Illustrations of the results of a single anterior and a single posterior cerebellum experiments. The single posterior received projections from both IO territories. A compressed map is established, with the rostromedian IO neurons projecting to the most posterior region of the posterior cerebellum, and the caudolateral IO projecting to the most anterior region of the posterior cerebellar half. In single anterior cerebellum, intermixed DiI and DiA-labeled neurons were found exclusively in the caudolateral IO, whereas no labeled neurons was observed in the rostromedian IO territory. (Modified from Chedotal et al., 1997.) A × 20, B–D × 80.

development, suggests that the Purkinje cells themselves could be the carriers of the positional cues, which agrees with our matching hypothesis. Nevertheless, the olivocerebellar map is more complex and fragmented than a simple anteroposterior arrangement. As indicated above, olivary axons must not only recognize their anterior–posterior position, but also the sagittal stripe in which they project. Therefore, positional information is also required for the mediolateral position of growing olivary axons. The possible molecular nature of the positional cues will be addressed in the next section.

Molecular candidates for providing positional information during the formation of cerebellar maps

Our 'matching hypothesis' is only a new variant of the Sperry's (1963) chemoaffinity theory, but the molecules involved in the match between the precerebellar neurons and the Purkinje cells are still unknown. However, the fact that similar or complementary molecules must be expressed in precise patterns (clusters of inferior olivary neurons, parasagittal bands of Purkinje cells) provides a way to identify potential candidates. The first were adhesion molecules (BEN, cadherins), and more recently, Eph receptor tyrosine kinases and their ligands, the ephrins. The adhesion molecules might act as binding devices between the axonal growth cones and their postsynaptic partners. In contrast, Eph expressing growth cones would prevent neurons bearing the membrane bound corresponding ephrin, acting as repellent devices, as hypothesized in our in vitro experiments with single anterior cerebella (see preceding section).

Cell adhesion molecules

From the immunoglobulin-like cell adhesion molecules, we found that in the chick embryos BEN/SC1/DM-GRASP was expressed by the two neuronal partners, olivary neurons and Purkinje cells, before and during the formation of the projection. Only subsets of inferior olivary neurons and Purkinje cells express BEN, and the latter are arranged in regularly occurring parasagittal bands. The expression was transient, started by E7.5 and by hatching completely disappeared (Chedotal et al., 1996). Moreover, the pattern of BEN expression in the olivocerebellar system appears to be preserved in rodents (Chédotal and Sotelo, unpublished observations). These observations based on spatiotemporal correlations, are in themselves insufficient to prove the functional implication of BEN in the formation of the olivocerebellar map. The genetic manipulations (knock-out and overexpressing mice) should be done to obtain more solid evidence.

Cadherins form a large family of cell surface glycoproteins with cell adhesion properties and are regulators of the multiple developmental processes (Takeichi et al., 2000). These calcium-dependent adhesion molecules establish homophilic binding, although heterophilic binding may also occur, particularly, between cells expressing a large repertoire of cadherins (Nakagawa and Takeichi, 1995). Redies and collaborators (see in Redies, 2000) analyzed the cerebellar expression patterns of mRNA and proteins for several members of the cadherin family. They found that many cadherins are expressed according to a parasagittal band pattern. In addition, inferior olivary neurons also differentially express numerous member of the cadherin family. Thus, with the same uncertainties than for BEN, cadherins might also be positional information molecules involved in the organization of the cerebellar functional modularity.

Eph and ephrins

The Eph are receptor tyrosine kinases and have ligands named ephrins. There are at least 14 known receptors and eight ligands (see in Flanagan and Vanderhaeghen, 1998). The expression patterns of many Eph and ephrin have been analyzed in the developing cerebellum, particularly in chick embryos (Rogers et al., 1999; Karam et al., 2000; Blanco et al., 2002). Thus, EphA3, EphA4, ephrin-A2, and ephrin-A5 are expressed by subsets of Purkinje cells organized in a mediolateral alternating parasagittal binding pattern. Precerebellar nuclei, including the basilar pons and the inferior olive, also express EphA4. Therefore, the Eph–ephrin system might contribute to the development of the topographic organization of extracerebellar projections and participate in the modular organization of the cerebellar cortex.

The first results obtained with this knock-out technology were rather disappointing. With the exception of a small reduction in size, the phenotype of the cerebellum of mice with inactivation of the *EphA4* gene appeared normal. Even Purkinje cells compartmentation analyzed with Zebrin-II antibodies was unaltered, as was the striped expression of other Eph receptor family members in Purkinje cells (Karam et al., 2002). However, the organization of the olivocerebellar projection in these null-mutants await careful analysis. More interesting results were obtained in the chick embryos by Nishida et al. (2002) using our in vitro model (Chedotal et al., 1997). Overexpression of *ephrin-A2* in the cerebellum with retroviral vectors and transfection technology disrupted the olivocerebellar projection. Inferior olivary axons with high receptor activity were prevented from entering into areas with ectopic ephrin-A2 expression. These data represent the first evidence that the Eph–ephrin system may provide specific positional information and participate in the organization of cerebellar maps providing repulsive signals.

From a coarse-grained to a fine-grained olivocerebellar map

The results reviewed here can be summarized as follows: Inferior olivary axons enter the cerebellar parenchyma exactly when their cell bodies reach their final destinations on both sides of the floor plate. This entry is not chaotic. From the beginning, the olivary axons project to their correct cerebellar domains a few hours after their arrival to the cerebellum. The shared positional information between the inferior olivary neurons and the Purkinje cells strongly suggests that target invasion and crude map formation are not activity-dependent processes. In contrast, the acquisition of a fine-grained map, in which the neuron-to-neuron specific connectivity is attained, occurs during a last step of the climbing fiber/Purkinje cell synapse formation. This step is characterized by a process of 'synapse elimination and refinement of synaptic connections,' which is out of the scope of this review (see in Rabacchi et al., 1992; Sugihara et al., 2000; Scelfo et al., 2003).

The required synaptic remodeling will most probably occur within the framework of a given cerebellar compartment, defined by the heterogeneity of the Purkinje cells and seems to be activity-dependent.

References

Alcantara, S., Ruiz, M., De Castro, F., Soriano, E. and Sotelo, C. (2000) Netrin 1 acts as an attractive or as a repulsive cue for distinct migrating neurons during the development of the cerebellar system. Development, 127: 1359–1372.

Altman, J. and Bayer, S.A. (1987) Development of the precerebellar nuclei in the rat: II. The intramural olivary migratory stream and the neurogenetic organization of the inferior olive. J. Comp. Neurol., 257: 490–512.

Angaut, P. and Sotelo, C. (1987) The dentato-olivary projection in the rat as a presumptive GABAergic link in the olivo-cerebello-olivary loop. An ultrastructural study. Neurosci. Lett., 83: 227–231.

Angaut, P. and Sotelo, C. (1989) Synaptology of the cerebello-olivary pathway. Double labelling with anterograde axonal tracing and GABA immunocytochemistry in the rat. Brain Res., 479: 361–365.

Armstrong, R.C. and Clarke, P.G. (1979) Neuronal death and the development of the pontine nuclei and inferior olive in the chick. Neuroscience, 4: 1635–1647.

Baader, S.L., Vogel, M.W., Sanlioglu, S., Zhang, X. and Oberdick, J. (1999) Selective disruption of 'late onset' sagittal banding patterns by ectopic expression of engrailed-2 in cerebellar Purkinje cells. J. Neurosci., 19: 5370–5379.

Backer, S., Sakurai, T., Grumet, M., Sotelo, C. and Bloch-Gallego, E. (2002) Nr-CAM and TAG-1 are expressed in distinct populations of developing precerebellar and cerebellar neurons. Neuroscience, 113: 743–748.

Bashaw, G.J. and Goodman, C.S. (1999) Chimeric axon guidance receptors: the cytoplasmic domains of slit and netrin receptors specify attraction versus repulsion. Cell, 97: 917–926.

Blanco, M., Pena-Melian, A. and Nieto, M.A. (2002) Expression of EphA receptors and ligands during chick cerebellar development. Mech. Dev., 114: 225–229.

Bloch-Gallego, E., Ezan, F., Tessier-Lavigne, M. and Sotelo, C. (1999) Floor plate and netrin-1 are involved in the migration and survival of inferior olivary neurons. J. Neurosci., 19: 4407–4420.

Bourrat, F. and Sotelo, C. (1988) Migratory pathways and neuritic differentiation of inferior olivary neurons in the rat embryo. Axonal tracing study using the *in vitro* slab technique. Brain Res., 467: 19–37.

Bourrat, F. and Sotelo, C. (1990a) Early development of the rat precerebellar system: migratory routes, selective aggregation and neuritic differentiation of the inferior olive and lateral reticular nucleus neurons. An overview. Arch. Ital. Biol., 128: 151–170.

Bourrat, F. and Sotelo, C. (1990b) Migratory pathways and selective aggregation of the lateral reticular neurons in the rat embryo: a horseradish peroxidase *in vitro* study, with special reference to migration patterns of the precerebellar nuclei. J. Comp. Neurol., 294: 1–13.

Bourrat, F. and Sotelo, C. (1991) Relationships between neuronal birthdates and cytoarchitecture in the rat inferior olivary complex. J. Comp. Neurol., 313: 509–521.

Brodal, A. and Kawamura, K. (1980) Olivocerebellar projection: a review. Adv. Anat. Embryol. Cell Biol., 64: 1–140.

Brose, K., Bland, K.S., Wang, K.H., Arnott, D., Henzel, W., Goodman, C.S., Tessier-Lavigne, M. and Kidd, T. (1999) Slit proteins bind Robo receptors and have an evolutionarily conserved role in repulsive axon guidance. Cell, 96: 795–806.

Causeret, F, Danne, F., Ezan, F., Sotelo, C. and Bloch-Gallego, E. (2002) Slit antagonizes netrin-1 attractive effects during the migration of inferior olivary neurons. Dev. Biol., 246: 429–440.

Chedotal, A and Sotelo, C. (1992) Early development of olivocerebellar projections in the fetal rat using CGRP immunocytochemistry. Eur. J. Neurosci., 4: 1159–1179.

Chedotal, A, Pourquie, O., Ezan, F., San Clemente, H. and Sotelo, C. (1996) BEN as a presumptive target recognition molecule during the development of the olivocerebellar system. J. Neurosci., 16: 3296–3310.

Chedotal, A, Bloch-Gallego, E. and Sotelo, C. (1997) The embryonic cerebellum contains topographic cues that guide developing inferior olivary axons. Development, 124: 861–780.

De Diego, I., Kyriakopoulou, K., Karagogeos, D. and Wassef, M. (2002) Multiple influences on the migration of precerebellar neurons in the caudal medulla. Development, 129: 297–306.

Desclin, J.C. (1974) Histological evidence supporting the inferior olive as the major source of cerebellar climbing fibers in the rat. Brain Res., 77: 365–384.

De Zeeuw, C.I., Simpson, J.I., Hoogenraad, C.C., Galjart, N., Koekkoek, S.K.E. and Ruigrok, T.J.H. (1998) Microcircuitry and function of the inferior olive. Trends Neurosci., 21: 391–400.

Flanagan, J.G. and Vanderhaeghen, P. (1998) The ephrins and Eph receptors in neural development. Annu. Rev. Neurosci., 21: 309–345.

Fredette, B.J. and Mugnaini, E. (1991) The GABAergic cerebello-olivary projection in the rat. Anat. Embryol., 184: 225–243.

Furber, S.E. (1983) The organization of the olivocerebellar projection in the chicken. Brain Behav. Evol., 22: 198–211.

Gerfen, C.R. (1984) The neostriatal mosaic: compartmentalization of corticostriatal input and striatonigral output systems. Nature, 311: 461–464.

Gravel, C., Eisenman, L.M., Sasseville, R. and Hawkes, R. (1987) Parasagittal organization of the rat cerebellar cortex: direct correlation between antigenic Purkinje cell bands revealed by mabQ113 and the organization of the olivocerebellar projection. J. Comp. Neurol., 265: 294–310.

Groenewegen, H.J., Voogd, J. and Freedman, S.L. (1979) The parasagittal zonation within the olivocerebellar projection. II. Climbing fiber distribution in the intermediate and hemispheric parts of cat cerebellum. J. Comp. Neurol., 183: 551–601.

Harkmark, W. (1954) Cell migrations from the rhombic lip to the inferior olive, the nucleus raphe and the pons. A morphological experimental investigation in chick embryos. J. Comp. Neurol., 100: 115–209.

Hawkes, R. and Gravel, C. (1991) The modular cerebellum. Prog. Neurobiol., 36: 309–327.

Herrup, K. and Kuemerle, B. (1997) The compartmentation of the cerebellum. Annu. Rev. Neurosci., 20: 61–90.

Hess, E.J. and Wilson, M.C. (1991) Tottering and leaner mutations perturb transient developmental expression of tyrosine hydroxylase in embryologically distinct Purkinje cells. Neuron, 6: 123–132.

Hong, K., Hinck, L., Nishiyama, M., Poo, M.M., Tessier-Lavigne, M. and Stein, E. (1999) A ligand-gated association between cytoplasmic domains of UNC5 and DCC family receptors converts netrin-induced growth cone attraction to repulsion. Cell, 97: 927–941.

Ito, M. (1982) Cerebellar control of the vestibulo-ocular reflex — around the flocculus hypothesis. Ann. Rev. Neurosci., 5: 275–296.

Ito, M. (1984) The Cerebellum and Neural Control. Raven Press, New York.

Karam, S.D., Burrows, R.C., Logan, C., Koblar, S., Pasquale, E.B. and Bothwell, M. (2000) Eph receptors and ephrins in the developing chick cerebellum: relationship to sagittal patterning and granule cell migration. J. Neurosci., 20: 6488–6500.

Karam, S.D., Dottori, M., Ogawa, K., Henderson, J.T., Boyd, A.W., Pasquale, E.B. and Bothwell, M. (2002) EphA4 is not required for Purkinje cell compartmentation. Dev. Brain Res., 135: 29–38.

Kyriakopoulou, K., de Diego, I., Wassef, M. and Karagogeos, D. (2002) A combination of chain and neurophilic migration involving the adhesion molecule TAG-1 in the caudal medulla. Development, 129: 287–296.

Llinas, R. (1974) Eighteenth Bowditch lecture. Motor aspects of cerebellar control. Physiologist, 17: 19–46.

Llinas, R. and Sasaki, K. (1989) The Functional Organization of the Olivo-Cerebellar System as Examined by Multiple Purkinje Cell Recordings. Eur. J. Neurosci., 1: 587–602.

Llinas, R., Baker, R. and Sotelo, C. (1974) Electrotonic coupling between neurons in cat inferior olive. J. Neurophysiol., 37: 560–571.

Marin, F. and Puelles, L. (1995) Morphological fate of rhombomeres in quail/chick chimeras: a segmental analysis of hindbrain nuclei. Eur. J. Neurosci., 7: 1714–1738.

Millen, K.J., Hui, C.C. and Joyner, A.L. (1995) A role for En-2 and other murine homologues of Drosophila segment polarity genes in regulating positional information in the developing cerebellum. Development, 121: 3935–3945.

Ming, G.L., Song, H.J., Berninger, B., Holt, C.E., Tessier-Lavigne, M. and Poo, M.M. (1997) cAMP-dependent growth cone guidance by netrin-1. Neuron, 19: 1225–1235.

Morgan, J.I. and Smeyne, R.J. (1997) Transgenic approaches to cerebellar development. Perspect. Dev. Neurobiol., 5: 33–41.

Nakagawa, S. and Takeichi, M. (1995) Neural crest cell-cell adhesion controlled by sequential and subpopulation-specific expression of novel cadherins. Development, 121: 1321–1332.

Nguyen Ba-Charvet, K.T., Brose, K., Marillat, V., Kidd, T., Goodman, C.S., Tessier-Lavigne, M., Sotelo, C. and Chedotal, A. (1999) Slit2-Mediated chemorepulsion and collapse of developing forebrain axons. Neuron, 22: 463–473.

Nishida, K., Flanagan, J.G. and Nakamoto, M. (2002) Domain-specific olivocerebellar projection regulated by the EphA-ephrin-A interaction. Development, 129: 5647–5658.

Oberdick, J., Schilling, K., Smeyne, R.J., Corbin, J.G., Bocchiaro, C. and Morgan, J.I. (1993) Control of segment-like patterns of gene expression in the mouse cerebellum. Neuron, 10: 1007–1018.

Oscarsson, O. (1979) Functional units of the cerebellum: sagittal zones and microzones. Trends Neurosci., 2: 143–145.

Rabacchi, S.A., Bailly, Y., Delhaye-Bouchaud, N., Herrup, K. and Mariani, J. (1992) Role of the target in synapse elimination: studies in cerebellum of developing lurcher mutants and adult chimeric mice. J. Neurosci., 12: 4712–4720.

Rakic, P. (1971) Neuron-glia relationship during granule cell migration in developing cerebellar cortex. A Golgi and electronmicroscopic study in Macacus Rhesus. J. Comp. Neurol., 141: 283–312.

Rakic, P. (1977) Prenatal development of the visual system in rhesus monkey. Philos. Trans. R. Soc. Lond. B Biol. Sci., 278: 245–260.

Rakic, P. (1990) Principles of neural cell migration. Experientia, 46: 882–891.

Redies, C. (2000) Cadherins in the central nervous system. Prog. Neurobiol., 6: 611–648.

Rodriguez, C.I. and Dymecki, S.M. (2000) Origin of the precerebellar system. Neuron, 27: 475–486.

Rogers, J.H., Ciossek, T., Menzel, P. and Pasquale, E.B. (1999) Eph receptors and ephrins demarcate cerebellar lobules before and during their formation. Mech. Dev., 87: 119–128.

Ross, M.E., Fletcher, C., Mason, C.A., Hatten, M.E. and Heintz, N. (1990) Meander tail reveals a discrete developmental unit in the mouse cerebellum. Proc. Natl. Acad. Sci. USA, 87: 4189–4192.

Ruigrok, T.J.H. and Voogd, J. (1990) Organization of projections from the inferior olive to the cerebellar nuclei in the rat. J. Comp. Neurol., 298: 315–333.

Scelfo, B., Strata, P. and Knopfel, T. (2003) Sodium imaging of climbing fiber innervation fields in developing mouse Purkinje cells. J. Neurophysiol., 89: 2555–2563.

Serafini, T., Kennedy, T.E., Galko, M.J., Mirzayan, C., Jessell, T.M. and Tessier-Lavigne, M. (1994) The netrins define a family of axon outgrowth-promoting proteins homologous to C. elegans UNC-6. Cell, 78: 409–424.

Simon, D.K., Roskies, A.L. and O'Leary, D.D. (1994) Plasticity in the development of topographic order in the mammalian retinocollicular projection. Dev. Biol., 162: 384–393.

Sotelo, C. and Chedotal, A. (1997) Development of the olivocerebellar projection. Perspect. Dev. Neurobiol., 5: 57–67.

Sotelo, C. and Wassef, M. (1991) Cerebellar development: afferent organization and Purkinje cell heterogeneity. Philos. Trans. R. Soc. Lond. B Biol. Sci., 331: 307–313.

Sotelo, C., Llinas, R. and Baker, R. (1974) Structural study of inferior olivary nucleus of the cat: Morphological correlates of electrotonic coupling. J. Neurophysiol., 37: 541–559.

Sotelo, C., Hillman, D.E., Zamora, A.J. and Llinas, R. (1975) Climbing fiber deafferentation: its action on Purkinje cell dendritic spines. Brain Res., 98: 574–581.

Sotelo, C., Bourrat, F. and Triller, A. (1984) Postnatal development of the inferior olivary complex in the rat. II. Topographic organization of the immature olivocerebellar projection. J. Comp. Neurol., 222: 177–199.

Sotelo, C., Gotow, T. and Wassef, M. (1986) Localization of glutamic-acid-decarboxylase-immunoreactive axon terminals in the inferior olive of the rat, with special emphasis on anatomical relations between GABAergic synapses and dendrodendritic gap junctions. J. Comp. Neurol., 252: 32–50.

Sperry, R.W. (1963) Chemoaffinity in the orderly growth of nerve fiber patterns and connections. Proc. Natl. Acad. Sci. USA, 50: 703–710.

Sretavan, D. and Shatz, C.J. (1984) Prenatal development of individual retinogeniculate axons during the period of segregation. Nature, 308: 845–848.

Stein, E. and Tessier-Lavigne, M. (2001) Hierarchical organization of guidance receptors: silencing of netrin attraction by slit through a Robo/DCC receptor complex. Science, 291: 1928–1938.

Sugihara, I., Bailly, Y. and Mariani, J. (2000) Olivocerebellar climbing fibers in the granuloprival cerebellum: morphological study of individual axonal projections in the X-irradiated rat. J. Neurosci., 20: 3745–3760.

Taber-Pierce, E. (1966) Histogenesis of the nuclei griseum pontis, corporis pontobulbaris and reticularis tegmenti pontis (Bechterew) in the mouse. An autoradiographic study. J. Comp. Neurol., 126: 219–254.

Takeichi, M., Nakagawa, S., Aono, S., Usui, T. and Uemura, T. (2000) Patterning of cell assemblies regulated by adhesion receptors of the cadherin superfamily. Philos. Trans. R. Soc. Lond. B Biol. Sci., 355: 885–890.

Thomas, K.R., Musci, T.S., Neumann, P.E. and Capecchi, M.R. (1991) Swaying is a mutant allele of the proto-oncogene Wnt-1. Cell, 67: 969–976.

Wassef, M. and Sotelo, C. (1984) Asynchrony in the expression of guanosine $3':5'$-phosphate-dependent protein kinase by clusters of Purkinje cells during the perinatal development of rat cerebellum. Neuroscience, 13: 1217–1241.

Wassef, M., Sotelo, C., Thomasset, M., Granholm, A.C., Leclerc, N., Rafrafi, J. and Hawkes, R. (1990) Expression of compartmentation antigen zebrin I in cerebellar transplants. J. Comp. Neurol., 294: 223–234.

Wassef, M., Angaut, P., Arsenio-Nunes, L., Bourrat, F. and Sotelo, C. (1992a) Purkinje cell heterogeneity: its role in organizing the topography of the cerebellar cortex connections. In: Llinas R. and Sotelo C. (Eds.), The Cerebellum Revisited. Springer-Verlag, New York, pp. 5–21.

Wassef, M., Chedotal, A., Cholley, B., Thomasset, M., Heizmann, C.W. and Sotelo, C. (1992b) Development of the olivocerebellar projection in the rat: I. Transient biochemical compartmentation of the inferior olive. J. Comp. Neurol., 323: 519–536.

Wassef, M., Cholley, B., Heizmann, C.W. and Sotelo, C. (1992c) Development of the olivocerebellar projection in the rat: II. Matching of the developmental compartmentations of the cerebellum and inferior olive through the projection map. J. Comp. Neurol., 323: 537–550.

Willshaw, D.J., Fawcett, J.W. and Gaze, R.M. (1983) The visuotectal projections made by Xenopus 'pie slice' compound eyes. J. Embryol. Exp. Morphol., 74: 29–45.

Wingate, R.J. (2001) The rhombic lip and early cerebellar development. Curr. Opin. Neurobiol. 11: 82–88.

Yee, K.T., Simon, H.H., Tessier-Lavigne, M. and O'Leary, D.M. (1999) Extension of long leading processes and neuronal migration in the mammalian brain directed by the chemoattractant netrin-1. Neuron, 24: 607–622.

Zhu, Y., Khan, K. and Guthrie, S. (2003) Signals from the cerebellum guide the pathfinding of inferior olivary axons. Dev. Biol., 257: 233–248.

CHAPTER 2

The genetics of early cerebellar development: networks not pathways

Karl Herrup[1,*], Crystal Murcia[1], Forrest Gulden[2], Barbara Kuemerle[1] and Natalie Bilovocky[1]

[1]Department of Neurosciences, Case Western Reserve School of Medicine, 10900 Euclid Avenue, Cleveland, OH 44106, USA
[2]Department of Genetics, Case Western Reserve School of Medicine, 10900 Euclid Avenue, Cleveland, OH 44106, USA

Constantino Sotelo begins his retirement in an era when the nucleotide sequence of a different organism's genome is published with astonishing regularity. It is tempting in times such as these to seek the elegance and simplicity of this one-dimensional view of life (the nucleotide sequence of the DNA double helix) in the biological problems facing the field of neuroscience. This urge is encouraged by the concepts of biochemical 'pathways' and 'cycles' that are taught to help students understand complex processes ranging from glucose metabolism to cell division. Pathways and cycles are linear concepts, well described by simple schemes such as A → B → C → D. Even descriptions of developmental pathways tend to focus on linear models. For example: Wnt → frizzled → GSK3β → β-catenin → TCF/Lef. The more we learn about the impact of developmental genes on the developing CNS, however, the less satisfying simple linear schemes become. An excellent example of this situation is the developmental program of the early cerebellar anlage.

The cerebellum arises from an area just posterior to the midbrain/hindbrain boundary (MHB; for a recent review see Liu and Joyner, 2001). The position of the MHB in the early embryo is defined by the interface between the anterior expression domain of *Otx2* and the posterior expression domain of *Gbx2*. These two DNA binding proteins mutually repress each other's expression resulting in a sharp boundary. At this boundary, the expression of *Fgf8* and *Pax2* appear on the rostral and caudal sides respectively. Shortly after this a transverse band of *Wnt1* appears (in addition to a thin strip of expression along most of the dorsal midline of the embryo). *Engrailed* gene expression begins a short time later. There are two *Engrailed* genes in the mouse, homologues of the fruit fly *engrailed*. *Engrailed-1* appears first during development in symmetric domains of the mid/hindbrain region of the one-somite embryo (Joyner et al., 1985; Joyner and Martin, 1987). *Engrailed-2* expression is seen shortly thereafter in a slightly more anterior position at the 5-somite stage. After this staggered start, the domains of both genes remain largely similar during early development. Expression is high in the metencephalon with a decreasing gradient of expression in the mesencephalic vesicle. After ventricular neurogenesis is complete, the CNS expression patterns diverge with *En1* uniquely appearing in cells of the spinal cord and ventral midbrain while *En2* expression appears in cerebellar granule cells. The function of the persistent expression in the adult is

*Corresponding author. Tel.: 1-216-368-3435; Fax: 1-216-368-3079; E-mail: kxh26@cwru.edu

unknown, but engineered *Engrailed* mutations make it plain that normal cerebellar development requires both.

The *Engrailed-2* mutant

Joyner et al. (1991) created a null allele of the *Engrailed-2* gene by engineering a deletion of the homeobox DNA binding region. Surprisingly, the homozygous mutants ($En2^{-/-}$) are viable and fertile with a normal life expectancy. They have a mild behavioral abnormality (Gerlai et al., 1996) and subtle changes in the size and structure of the cerebellum. Cerebellar volume is reduced and the pattern of folding in the posterior regions is altered from the wild-type. The midline vermis is illustrated in Fig. 1; similar changes are found in the hemispheres. Folding patterns seen in sagittal sections are representative of anterior–posterior changes in the cerebellar structure. The medio-lateral pattern is best seen using immunostaining of a variety of different substances that show alternating bands of high and low staining (e.g., the antigens known as Zebrin I and Zebrin II) or the pattern of cerebellar cortical afferents (mossy fibers to the granule cells; climbing fibers to the Purkinje cells). Both are disrupted in the *Engrailed-2* mutant. Bands of Zebrin II are reduced or fractured (Kuemerle et al., 1997) and the pattern of mossy fiber projections is unusually diffuse (Vogel et al., 1996). This loss of patterning is made all the more intriguing by the observation that the cell surface glycolipid, recognized by the ppath antigen, while usually expressed only in Zebrin-negative zones of the cerebellum, is found to co-exist with Zebrin II inside single Purkinje cells in the $En2^{-/-}$ mutant (Kuemerle et al., 1997).

One final observation attesting the significant loss of pattern regulation in the absence of Engrailed-2 protein is the projection patterns of the cortico-nuclear projections of the Purkinje cells to the deep nuclear neurons. Normally, the Purkinje cells in the anterior, Crus I, folium project to the center of nucleus lateralis (the most lateral of the three rodent nuclei), while cells in the more posterior Crus II and paramedian (pml) lobules project to a lateral region of nucleus medialis. A preliminary analysis of these projection patterns in the $En2^{-/-}$ mutant, however, reveals a curious transformation (Kuemerle et al., unpublished). The Crus I projections are no different from those found in the wild-type. The fused Crus II/pml projections abandon their normal targets, however, and project instead to the medial portion of nucleus lateralis. One way of interpreting these results is to say that the more posterior cells are behaving as if they were more anterior (a posterior to anterior transformation — the exact phenotype of the cells in the fly wing in the absence of the Drosophila homologue, *engrailed*).

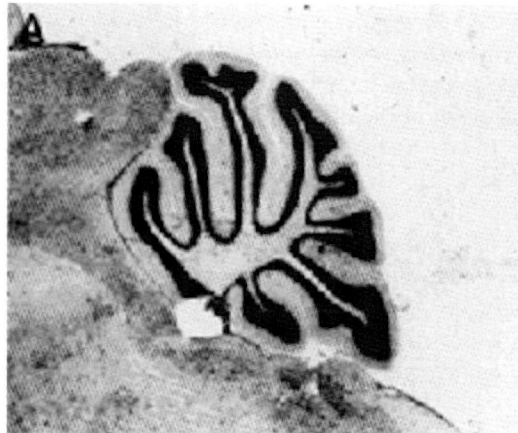 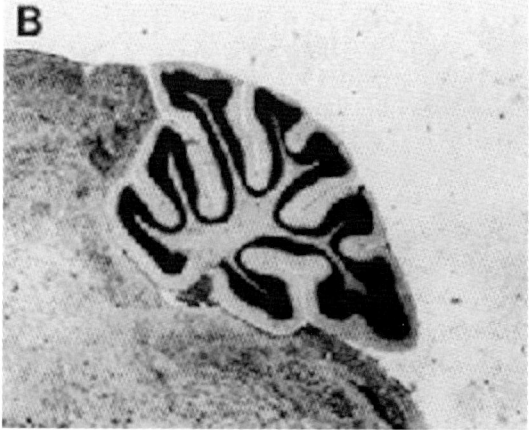

Fig. 1. The phenotype of the *Engrailed-2* mutant mouse. Relative to the wild-type cerebellum (A), the $En2^{-/-}$ cerebellar cortex (B) is reduced in size with a distorted folial pattern on the more posterior folia.

The reduction in the size of the $En2^{-/-}$ cerebellum noted above is reflected in the loss of significant numbers of cells in all of the major populations of the olivocerebellar circuit. The numbers of Purkinje, granule, deep nuclear and olive neurons are reduced by roughly similar amounts (~35%). A more detailed analysis of the pattern of loss in the cerebellar cortex is revealing. As shown in Fig. 2, the Purkinje cell loss appears to be more or less evenly distributed over the entire structure. This is somewhat surprising in light of the regional distortions in pattern — the apparent fusion of lobules VII and VIII in the vermis and Crus II with PML in the hemisphere and the loss of Purkinje cell biochemical (Zebrin/ppath) and anatomical (aberrant projection) identity. The suggestion is that the regulation of cell number is independent of the forces that lead to the structural, biochemical and anatomical morphogenesis of the cerebellum.

Engrailed-1 mutant

In 1994, Wurst et al. using a strategy similar to that for *Engrailed-2*, created an engineered mutation in the *Engrailed-1* gene (Wurst et al., 1994). $En1^{-/-}$ mice are dramatically different from the $En2^{-/-}$ cousins. Beginning as early as mid-gestation there are significant abnormalities in the developmental processes in the brain and skeleton (Wurst et al., 1994). The skeletal deformities have been well described elsewhere. In the brain there is a dramatic failure of midbrain–hindbrain development (Fig. 3B). Viewed in midline parsagittal sections, there is a complete absence of cerebellar structures anterior to the choroid plexus of the fourth ventricle. Further, there is a near total loss of dorsal midbrain leaving only a stub of what we may presume is superior colliculus jutting into a fourth ventricle. This dramatic loss of CNS tissue with relative sparing of structures anterior and posterior to the deficiencies is a satisfying phenotype when viewed against the expression pattern of *Engrailed-1* in the developing embryo. The missing regions correspond almost exactly to those areas marked by *Engrailed-1* expression at embryonic day 10.5.

The effect of genetic background

The problem with this simple explanation is that recent experiments have illustrated how much more complex the regulatory networks of the early embryo must be. While the dramatic loss of cerebellum and midbrain structures is a proper description of the mutation, it is only accurate when the mutation is maintained on a 129/S1 genetic background (formerly known as 129/Sv). When simple breeding techniques (repeated backcrossing) are used to transfer the $En1^-$ mutation to a C57BL/6 background, the phenotype changes dramatically (Bilovocky et al., 2003). Instead of a perinatal lethal condition with loss of substantial portions of the mesencephalic and metencephalic descendents, the entire CNS anlage

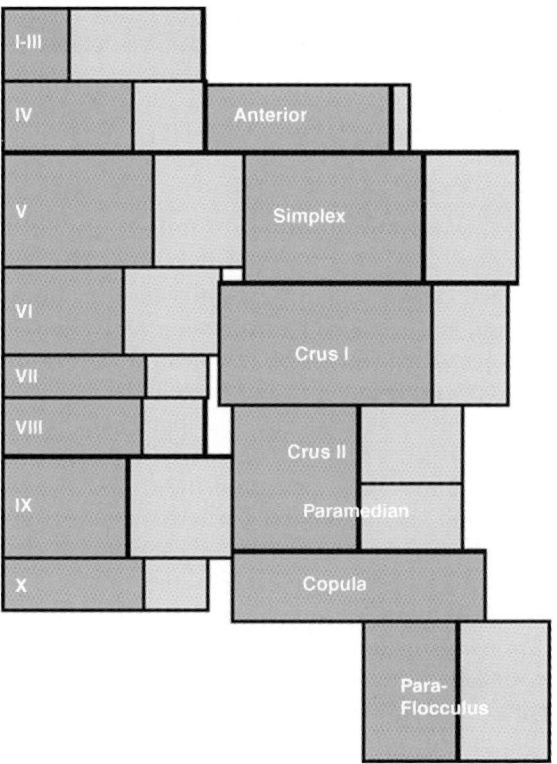

Fig. 2. Graphical representation of the Purkinje cell loss in the various cerebellar lobules. The size of each light grey box is proportional to the number of Purkinje cells in this lobule in the wild-type mouse. The smaller inner grey box is proportional to the number of Purkinje cells in the homologous lobules of the $En2^{-/-}$ mutant. Note that the loss of this cell type is relatively uniform in all locations. (After *Kuemerle et al. (1997)*.)

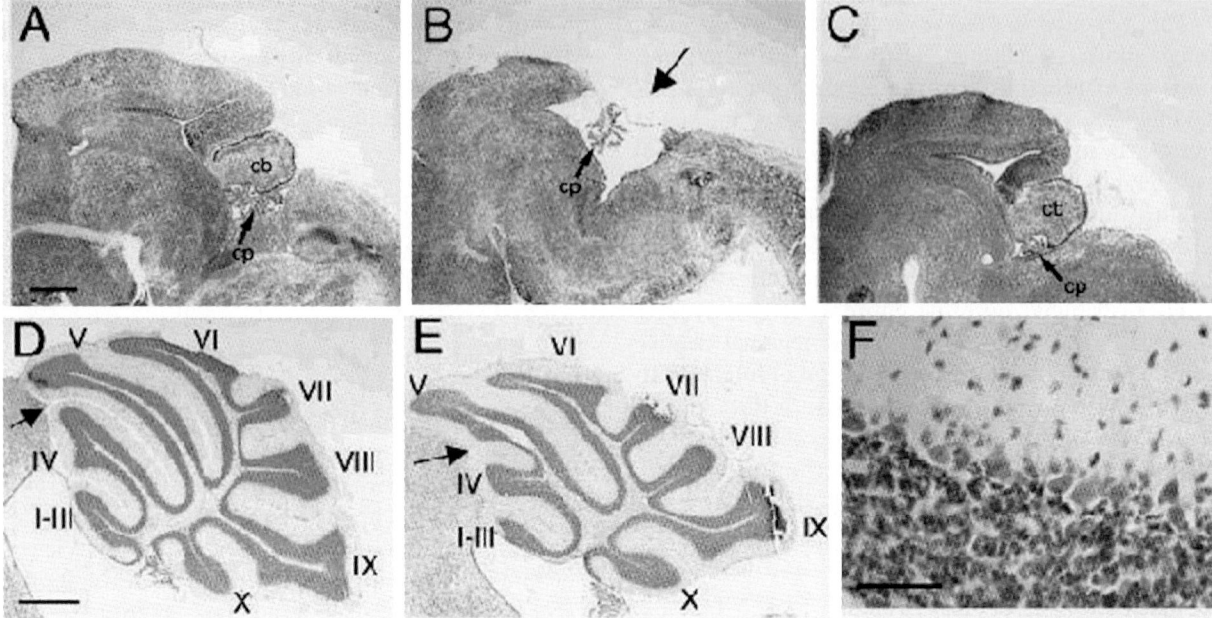

Fig. 3. The loss of *Engrailed-1* function leads to a peri-natal lethal condition. Midline sagittal sections of the wild-type mouse brain (A) show a clear cerebellar anlage (cb) with the attached choroids plexus on the posterior edge. In the $En1^{-/-}$ mutant (B) neither cerebellar nor caudal midbrain structures develop properly, if the mutation is carried on a 129/S1 genetic background. The same $En1^{-/-}$ mutation on a C57BL/6 background (C) is far less disruptive. These animals can live into adulthood with cerebellar structure (E) and cell populations (F) that closely resemble their wild-type littermates (D). (From *Bilovocky et al. (2003)*.)

looks nearly normal at birth (Fig. 3C). This is a stunning example of the importance of the genetic background in modulating the expression of a mutant gene and, by extension, its wild-type counterpart. If the 'rescued' $En1^{-/-}$ pups are allowed to mature they are sub-vital, but they produce a remarkably normal cerebellum. All of the cell types found in a wild-type cerebellar cortex are present (Fig. 3F). The expression of Purkinje cell markers such as calbindin is largely normal and total Purkinje cell counts are also apparently unaffected by the total absence of a functional Engrailed-1 protein. This is further evidence for the tremendous affect of the change in genetic background.

A key question that emerges from these observations is how the dramatic change in phenotype is brought about. One immediate candidate for the change in the $En1^{-/-}$ phenotype is a compensatory response of the *Engrailed-2* gene. We explored the possibility that the switch from 129/S1 to C57BL/6 background simply reversed the relative importance of the roles of the two Engraileds. All of the published phenotypes reported for the Engrailed-2 mutation are based on its expression in a 129/S1 or mixed genetic background e.g., Joyner et al. (1991). It seemed plausible that a small switch in the regulatory environment would tip the developmental balance of the two genes in favor of *Engrailed-2*. Thus, we transferred the $En2^-$ gene onto the C57BL/6 background and intercrossed the heterozygous mutants after 5 backcross generations (3 backcross generations were sufficient to observe the original $En1$ 'rescue'). With the exception of a few modest changes in the precise pattern of folding of the cortex, which are similar to those found between the two wild-type strains, there was no apparent change in the $En2^{-/-}$ phenotype after this manipulation (Bilovocky et al., 2003). From this observation it is apparent that the 'rescue' of the *Engrailed-1* phenotype is not due to its switching places with *Engrailed-2*.

While the tendency might be to dismiss the role of *Engrailed-2* at this point, it would be an error to do so. The effects of genetic background on the *Engrailed-1* phenotype require the presence of

Engrailed-2 and thus must work through changing the spatial/temporal regulation of *Engrailed-2*. This is apparent when one examines the mice that carry mutations in both *Engrailed* genes. Of the four alleles of the two genes, one wild type copy of *Engrailed-1* ($En1^{+/-}$; $En2^{-/-}$) is sufficient to generate a complete cerebellar anlage — on either a 129/S1 or a C57BL/6 genetic background. By contrast, in the absence of *En1*, the loss of even one of the wild-type *En2* alleles ($En1^{-/-}$; $En2^{+/-}$) leads to the failure of cerebellar development — even on a C57BL/6 background. The implication of these findings is that rather than the total amount of Engrailed protein, it is the identity of that protein, Engrailed-1 or Engrailed-2, that is crucial to the correct development of the cerebellum in the permissive, C57BL/6, background.

The nature of the genes in the C57BL/6 background that alter the $En1^{-/-}$ phenotype so dramatically are clearly of considerable interest to us. As a first step in uncovering their identity, we have begun a genome-wide scan to locate the modifier genes involved. We crossed C57BL/6-$En1^{+/-}$ mice by 129/S1 wild-type animals and then either crossed the resulting F1 heterozygotes to each other (intercross) or mated them with a heterozygote from the original C57BL/6-$En1^{+/-}$ parental line (backcross). Animals were scored as late-stage embryos to avoid loss of progeny at birth; only homozygous, C57BL/6-$En1^{-/-}$, animals were scored. In both the intercross and backcross, the number of 'rescued' mutants was far too small to be compatible with a model in which a single gene was responsible for the background effect. Instead, both types of crosses predicted that three recessive C57BL/6 genes are required to be homozygous to change the mutant phenotype from no-cerebellar-anlage to normal cerebellum at birth. Finding three modifier genes is a significant challenge, but using SSLP markers that differ between the C57BL/6 and 129/S1 strains, we have located two chromosomal regions that show significant linkage with the 'rescue.' The identities of the responsible genes are still uncertain, and the third locus is, as of this writing, only statistically 'suggestive.' One intriguing outcome of this exercise to date, however, is that in neither the backcross nor the intercross does the region on mouse chromosome 5 containing the *Engrailed-2* locus emerge as a candidate locus. The implication of this is clear. The background effect is accomplished through *Engrailed-2* (see above), but the genetic change responsible for the effect is not in *Engrailed-2*. The effect must therefore be accomplished either by altering the mix of transcription factors and other trans-acting elements that regulate *Engrailed-2* transcription or by changing the factors with which the Engrailed-2 protein itself interacts thereby changing the downstream read-out of the developmental program.

Background effects in other genes of the developing cerebellum

Were the effects of genetic background on the *Engrailed-1* mutation an isolated example, one might be tempted to dismiss the finding as being of little more than esoteric interest. But Bilovocky et al. (2003) list five other examples for the developing nervous system and their listing is incomplete at best. Often the existence of the effect of genetic background is only apparent from a careful reading of the paper. One gene, however, that has been carefully examined is *Pax2*. Recall that *Pax2* expression is integral to the development of the midbrain/hindbrain region — its expression preceding that of the two *Engrailed* genes. Null alleles of *Pax2* have been examined for their effects on brain development (Favor et al., 1996; Torres et al., 1996). Severe alleles that lead to failed neural tube development manifested as an exencephalic phenotype in 100% of the mutant embryos. Once again, this statement is only correct when the genetic background is 129/S1; when the gene is transferred to a C57BL/6 background, only 30% of embryos are exencephalic and those that survive appear to develop a normal appearing cerebellar rudiment (Schwarz et al., 1997). The details of this transformation are complicated but this finding offers a second example of a gene with a major role in midbrain/hindbrain development, whose mutant phenotype is made less severe by moving the gene from 129/S1 to C57BL/6. Given our study of the *Engrailed-1* gene and its modifiers, the identity of these *Pax2* modifier genes is of considerable interest. If they are the same as those involved in the *Engrailed-1* 'rescue' then we will have identified a small number of genes whose concerted action can reform the early MHB landscape such that the role of the major genes is dramatically altered. If, however, the *Pax2* modifiers are different from those at work in the *Engrailed-1* example, then it would be an additional evidence that development of the MHB will have to be viewed as a complex network of genes whose final readout is sensitive to the genetic context formed by a number of genes involved in the detailed regulation and function of the key players.

Our work on the *Pax2* modifiers has just begun, but the direction of the answer is becoming clear. Crosses similar to those outlined above for the *Engrailed-1* gene lead to the recovery of 'rescued' embryos in numbers that are fully consistent with a single C57BL/6 recessive gene (or three dominants — an unlikely alternative). Further, preliminary genome scan results appear to rule out several of the regions that we identified in the *Engrailed-1* screen. These findings point to the second alternative outlined above, to which, the development of the MHB is the result of a complex network of genes. Successful completion of the developmental program can be accomplished even in the absence of certain key genes as long as the exact nature of the rest of the network allows for a compensatory response.

Concluding remarks

The work to refine the definition of the genes involved in cerebellar development continues unabated from its beginnings over 40 years ago. Constantino Sotelo is among a small number of individuals who can truly be said to have watched the growth of the field nearly from the beginning. In that relatively short period of time we have progressed from a struggle to learn the complete phenotype of a given mutation (e.g., *weaver*) including its cellular site(s) of gene action, to a detailed hunt for the nucleotide changes that underlie a given mutation and the predicted consequences in protein structure and function. In the process, researchers have shown, with greater and greater clarity, that no gene can be understood as functioning alone. Each one works in a context that is created by the properties of other genes in the genome as well as by the organism's environment and physiology. This notion of context makes simple linear schemes such as 'pathways' and 'cycles' less satisfying as ways of describing the full range of gene actions — despite the obvious appeal of the one-dimensional simplicity. The demonstrated complexities of the interactions that can significantly change gene function are better grasped as networks. Direct analogies are difficult to find, but a relational database/spreadsheet seems a much more realistic model than a simple linear pathway. Change a cell in the interior and the changes reverberate throughout the spreadsheet. Sometimes the output 'cells' return an error message; sometimes the interactions among the interior cells adjust and the output 'cell' contains a useful value. The establishment of the MHB in the early vertebrate neural tube is an example of this emerging view and the responses of genes such as *Engrailed-1* and *Pax-2* to their genetic background serve as reminders of how important it is to consider the validity of each phenotype in both its isolated and interactive modes.

References

Bilovocky, N.A., Romito-DiGiacomo, R.R., Murcia, C.L., Maricich, S.M. and Herrup, K. (2003) Factors in the genetic background suppress the engrailed-1 cerebellar phenotype. J. Neurosci., 23: 5105–5112.

Favor, J., Sandulache, R., Neuhauser-Klaus, A. et al. (1996) The mouse Pax2(1Neu) mutation is identical to a human PAX2 mutation in a family with renal-coloboma syndrome and results in developmental defects of the brain, ear, eye, and kidney. Proc. Natl. Acad. Sci. USA, 93: 13870–13875.

Gerlai, R., Millen, K.J., Herrup, K., Fabien, K., Joyner, A.L. and Roder, J. (1996) Impaired motor learning performance in cerebellar En-2 mutant mice. Behav. Neurosci., 110: 126–133.

Joyner, A., Auerbach, B., Davis, C., Herrup, K. and Rossant, J. (1991) Subtle cerebellar phenotype in mice homozygous for a targeted deletion of the En-2 homeobox. Science, 259: 1239–1243.

Joyner, A., Kornberg, T., Coleman, K., Cox, D. and Martin, G. (1985) Expression during embryogenesis of a mouse gene with sequence homology to the Drosophila *engrailed* gene. Cell, 43: 29–37.

Joyner, A. and Martin, G. (1987) *En-1* and *En-2*, two mouse genes with sequence homolgy to the *Drosophila engrailed* gene: expression during embryogenesis. Genes Dev., 1: 29–38.

Kuemerle, B., Zanjani, H., Joyner, A. and Herrup, K. (1997) Pattern deformities and cell loss in *Engrailed-2* mutant mice suggest two separate patterning events during cerebellar development. J. Neurosci., 17: 7881–7889.

Liu, A. and Joyner, A.L. (2001) Early anterior/posterior patterning of the midbrain and cerebellum. Annu. Rev. Neurosci., 24: 869–896.

Schwarz, M., Alvarez-Bolado, G., Urbanek, P., Busslinger, M. and Gruss, P. (1997) Conserved biological function between Pax-2 and Pax-5 in midbrain and cerebellum development: evidence from targeted mutations. Proc. Natl. Acad. Sci. USA, 94: 14518–14523.

Torres, M., Gomez-Pardo, E. and Gruss, P. (1996) Pax2 contributes to inner ear patterning and optic nerve trajectory. Development, 122: 3381–3391.

Vogel, M.W., Ji, Z., Millen, K. and Joyner, A.L. (1996) The *Engrailed-1* homeobox gene and patterning of spinocerebellar mossy fiber afferents. Dev. Brain Res., 96: 210–218.

Wurst, W., Auerbach, A.B. and Joyner, A.L. (1994) Multiple developmental defects in *Engrailed-1* mutant mice: an early mid-hindbrain deletion and patterning defects in forelimbs and sternum. Development, 120: 2065–2075.

CHAPTER 3

Regionalization of the isthmic and cerebellar primordia

Nicolas Narboux-Nême, Angeliki Louvi, Paula Alexandre and Marion Wassef*

Régionalisation Nerveuse CNRS/ENS, UMR 8542, Ecole normale supérieure, 46 rue d'Ulm, 75005 Paris, France

Keywords: cerebellum; development; regionalization; Purkinje cells

Abstract: The complex migrations of neurons born in the dorsal neural tube of the isthmic and rhombomere 1 (r1) domains complicate the delineation of the cerebellar primordium. We show that Purkinje cells (P) are likely generated over a wide territory before gathering in the future cerebellar primordium under the developing external granular layer. Later expansion of the cerebellum over a restricted ependymal domain could rely on mutual interations between P cells and granule cell progenitors (GCP). P are attracted by GCP and in turn stimulate their proliferation, increasing the surface of the developing cortex. At later stages, regionalization of the developing and adult cerebellar cortex can be detected through regional variations in the distribution of several P cell markers. Whether and how the developmental and adult P subtypes are related is still unknown and it is unclear if they delineate the same sets of cerebellar subdivisions. We provide evidence that the early P regionalization is involved in intrinsic patterning of the cerebellar primordium, in particular it relate to the organization of the corticonuclear connection. We propose that the early P regionalization provides a scaffold to the mature P regionalization but that the development of functional afferent connections induces a period of P plasticity during which the early regional identity of P could be remodeled.

The cerebellar cortex has long been used as a model system and is often considered a simple neuronal structure due to its regular and modular architecture and the crystalline arrangement of a small number of cell types, each easily identifiable under the light and electron microscopes. Patterning of this comparatively simple neural structure during development involves unexpected complications. First, the cerebellar primordium is not easily outlined during development due to the complex migrations of cohorts of cells generated in the dorsal part of rhombomere 1 (r1) which are destined either to remain in the cerebellum or to populate more ventral regions of r1. Second, there is a gap in our understanding of cerebellar cortex developmental and adult regionalizations. Both can be detected on the basis of distribution of Purkinje cell (P) subtypes. Whether and how the developmental and adult P subtypes are related is still unknown and it is unclear if they delineate the same sets of cerebellar subdivisions.

Relating the neuroepithelium of the early primordium to the ependymal surface of the mature cerebellum

During the early stages, HH10–12 in chick and E8.5–E9 in mouse, the cerebellar primordium is split into a pair of discrete neuroepithelial stripes that lie on each side of the midline in the dorsal part of rhombomere 1. First extended along the anteroposterior axis, the cerebellar primordia later rotate to adopt a mediolateral orientation (Fig. 1A–C) allowing the fusion of their anterior parts on the midline and bridging the two halves of the vermis together.

*Corresponding author. Tel.: +33 1 44 32 35 32; Fax: +33 1 44 32 39 58; E-mail: wassef@wotan.ens.fr

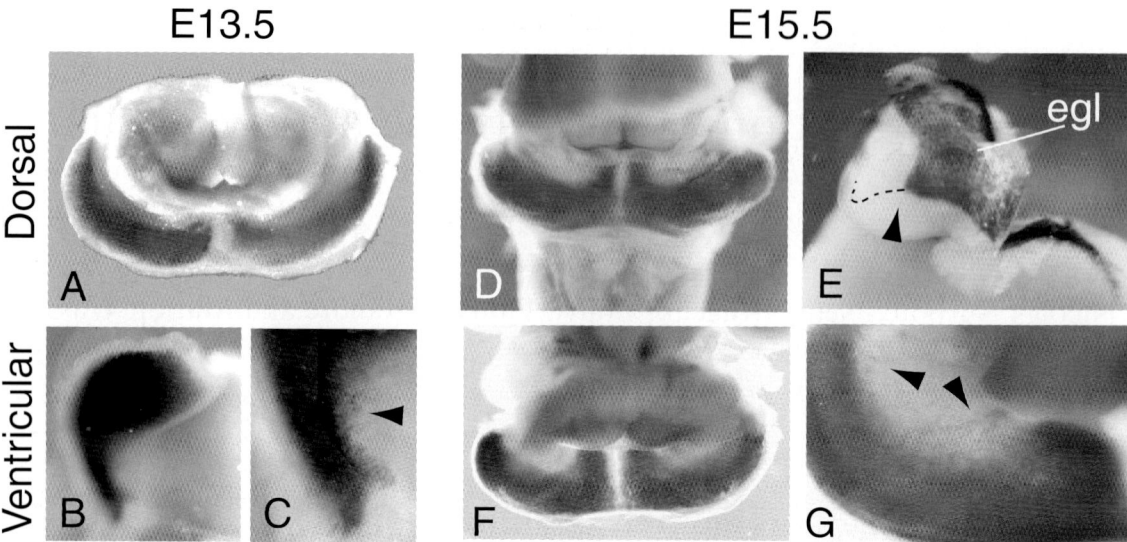

Fig. 1. Comparison of the dorsal and ventral aspects of the developing mouse cerebellum. Dorsal (A,B) ventricular (B, C, F, G) and medial views (the section is delineated by a dotted line of E13.5 (A–C) and E15.5 (D–G) cerebellum treated in toto by in situ hybridization to detect P cells (RORa, all except E) and the external granular layer (Math1, E). Note that the extent of RORa labeling is wider on the ventral than the dorsal aspect of the cerebellar plate and that dispersed P cells can be detected outside the main labeled domain (arrowheads in C and G). At E15.5, the EGL avoids the anterior vermis region (compare D and E) the anterior limit of the EGL is labeled with an arrowhead.

The morphogenetic movements mentioned above and the early regional variations in neuroepithelial growth both complicate the interpretation of fate maps. The representation of graft sizes and shapes in the ependymal layer of mature animals is also affected by an unexplained bias. It is generally assumed, even though not always explicitly stated, that a label corresponding to the marking process used for fate mapping remains in the ependymal layer of the mature cerebellum (quail cells in avian chimeras or the retrovirus tag in infected mouse embryos). This label is generally used to compare the relative sizes and positions of the grafts or the origin of the clone in the neuroepithelium. Small quail to chick grafts, that produce distinct populations of P and other neurons in the cerebellum often lack quail cells in the ependymal layer. In contrast, the surface of the velum medullaris, a sheet of cells linking the cerebellar vermis with the inferior colliculus expands dramatically around birth while the ependymal surface of the vermis decreases in parallel. The vermis, that previously abutted the MHB junction (Fig. 1A,D), retracts caudally and bulges over a narrow ependymal domain (Louvi et al., 2003). We have supposed that this process frees an ependymal surface that is relinquished to the velum medullaris and we proposed that the vermis and velum medullaris share a common ependymal surface. Recent results described below suggest that local remodeling of the ependymal surface may be common during cerebellar development and may blur the topographic relationship between the cerebellar ventricular surface and the cerebellar neurons.

The cerebellar neurons are produced over a protracted period in overlapping temporal patterns. The early-generated neurons, deep cerebellar nuclei (DCN) neurons and P cells arise from ventricular layer progenitors of the cerebellar primordium and become postmitotic over a short time period (E10.5–E12.5 in mouse). The Golgi cells that are produced over a longer period are thought to share a similar origin. The molecular layer interneurons and the granule cells begin to be produced perinatally from two secondary proliferation zones in the subventricular (Zhang and Goldman, 1996) and external granular layers (EGL, Wingate and Hatten, 1999) respectively. Clonal analysis (Ryder and Cepko, 1994; Mathis et al., 1997) has suggested that the EGL derives from a distinct pool of dedicated progenitors that is small or quiescent at early stages. They also showed that unlike spinal cord or rhombencephalic neurons (Edlund and Jessell, 1999), the other cerebellar cell types are not generated from distinct subdivisions of the neuroepithelium.

There is no way to delineate unambiguously the cerebellar primordium, therefore, we tried to outline the domain in the neuroepithelium that produces the DCN and P. The presumptive DCN were identified as the early population of TUJ-1 labeled neurons at E10.5 and P were detected by in situ hybridization for the orphan nuclear receptor RORα. Both labeling methods detected a wide crescent shaped territory on both sides of the midline. Although a subset of P are already labeled at E12.5, it is not before E13.5 (Fig. 1A) that the bulk of P can be detected through their expression of RORα. Because the mature cerebellum is a closed structure, the cerebellar primordium is often represented as a well delimited pool of 'dedicated progenitors.' It is however noteworthy that isolated P are detected with RORα at E13.5 (Fig. 1C, G) suggesting either that the limit of the cerebellar primordium is fuzzy even at this relatively late stage of cerebellar development or that P have migrated outside it. Lineage studies have also detected noncerebellum-committed progenitors of P, even if the stage at which they were picked up is unknown. Analyzing 115 small cerebellar clones Mathis et al. (1997) found that 1/7 contained noncerebellar cells and Lin et al. (2001) identified common progenitors of the cerebellum and pons. In fact, it is common to find isolated P at a long distance from the cerebellum in some mouse strains as well as in various mutants. P in these cases have generally been considered as escaping from their site of origin. An alternative interpretation of all these observations, not considered in previous studies, is that P could be produced over a relatively wide territory and need to home in under the EGL to form the bulging cerebellum. In this scenario, except in the anterior vermis, the EGL would not actively migrate to cover the cerebellar primordium but would maintain a fixed anterior edge while P actively accumulate beneath them, stimulating EGL proliferation and expansion. Consistent with this interpretation is the fact that P are attracted by EGL cells migrating in ectopic locations in unc5H3 mutants (Ackerman et al., 1997; Przyborski et al., 1998; Goldowitz et al., 2000) and in the absence of EGL, in Math1 mutants, they are no more confined to the main cerebellar territory (Jensen et al., 2002). Indeed, at early stages of its development the anterior edge of the EGL is not detected at various distances from the rhombic lip as it should if it were progressing anteriorly, rather dispersed EGL cells rapidly accumulate within fixed limits. If our interpretation is correct, the cerebellar primordium could correspond to a wide neuroepithelial domain. Both the DCN and P neurons it produces in a slime mold type of behavior would migrate posteriorly and converge resulting in the culmination of the cerebellum over a narrow ependymal domain.

Two other modes of cell dispersion could modify the ventricular surface of the cerebellum. First, it could shrink as a result of multiple symmetric divisions producing DCN and P neurons over a short period. Recent findings suggest that P cells could be produced through a symmetric last division. Hashimoto and Mikoshiba (2003) injected a replication defective lacZ-adenovirus into the ventricle of E10.5 or E11.5 mice embryos and found the P progenitors that pass through the S/M phase at the time of injection to be preferentially targeted. Interestingly, the P that become postmitotic one day after injection do not express lacZ indicating that the lacZ DNA is not retained in progenitors. This suggests that during the last mitosis giving birth to P cells the two daughter cells become postmitotic. Alternatively, during an asymmetric division, a bias could occur in the distribution of the viral DNA to the postmitotic daughter cell. Later in development, another cause of ventricular surface modification could be the detachment of radial glia from the ventricular surface first described by Ramon y Cajal (1911). He observed the dorsal migration of radial glia cell bodies and proposed that they accumulate in the vicinity of the P layer where they transform into Bergman glia. Progenitors destined to proliferate in the subventricular and intermediate zones also detach from the ventricular surface. If these processes are more common in specific regions, they could also modify the topography of the ventricular surface. The interplay of the various factors that result in ventricular surface remodeling during cerebellar development is schematically represented in Fig. 2E.

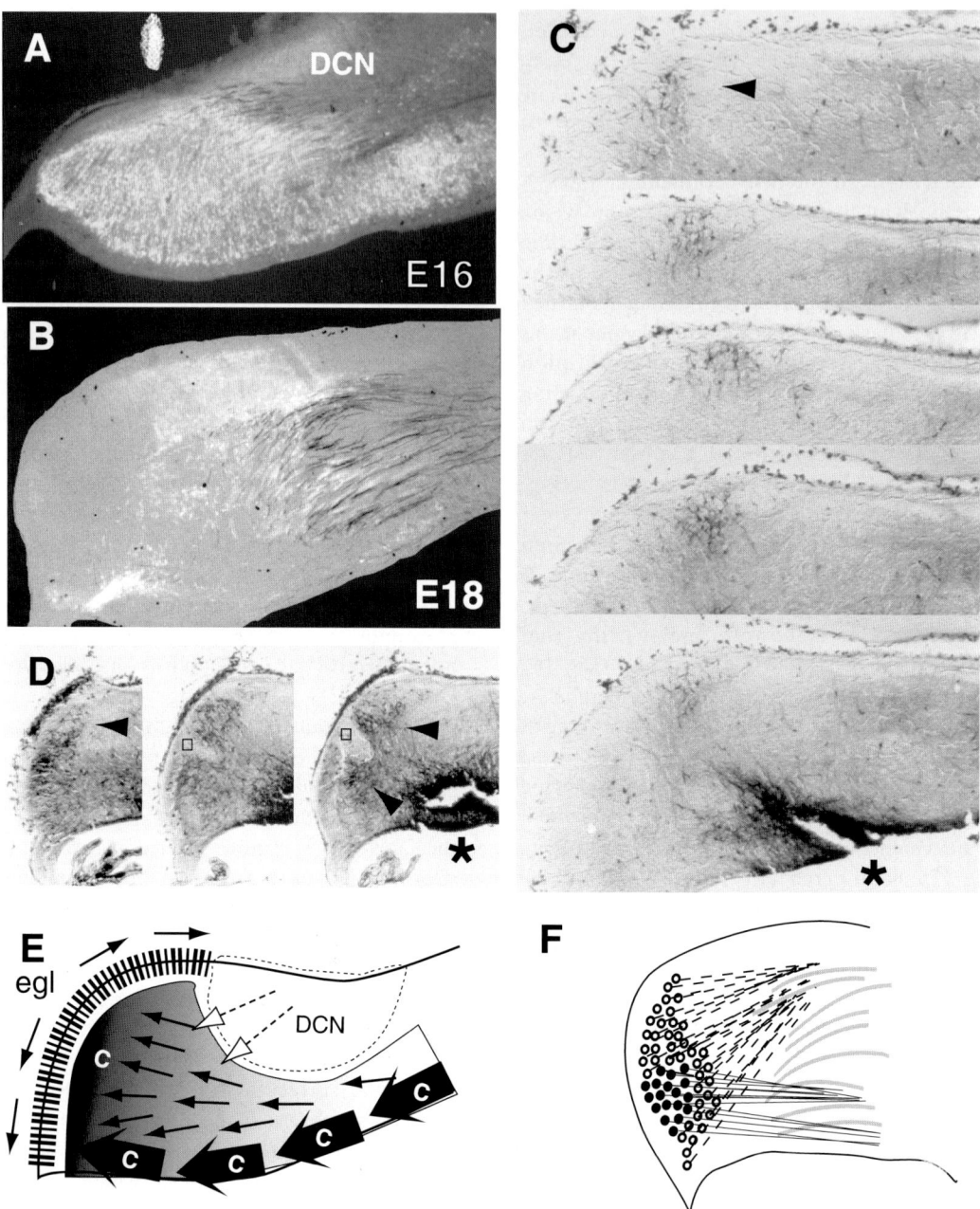

Fig. 2. **Development of Purkinje cell axons.** (A, B) Transverse sections through the cerebellar plate of E16 (A) and E18 (B) rat embryos labeled for the immunodetection of calbindin (white) and parvalbumin (black). The white axon fascicles of P cells run between the black fascicles of the primary vestibular fibers. (C, D) Piles of successive sections through two cerebellums of E19 rat embryos injected with HRP in vitro and allowed to survive for several hours. The asterisks mark the injection sites, the arrowheads point to the retrogradely labeled P cells and the squares point to an unlabeled cluster of P cells. (E) Schematic representation of the cell migrations that provoke cerebellar bulging. (F) Patterning of the cerebellar cortex efferent projection. The continuous and dotted lines represent the efferent projections from the black and white P compartments respectively. The axons from the adjacent clusters follow different paths.

Cerebellum regionalization and P subtype specification

The mature cerebellum is organized into anatomical and functional sagittal domains called zones or microzones (Oscarsson, 1980). In adult rodents a family of markers called 'zebrin' label alternate sagittal stripes of P cells (Hawkes and Leclerc, 1987; Hawkes, 1997). The tight correlation between the pattern of zebrin expression and the anatomical subdivisions in the afferent connections of the climbing and mossy fibers provides a clear link with the functional organization of the cerebellum (Gravel et al., 1987; Wassef et al., 1991; Sotelo and Wassef, 1991; Hallem et al., 1999). The development of positive and negative P cells and their sorting out according to their zebrin expression occurs even in the absence of cerebellar afferent axons (Wassef et al., 1990; Oberdick et al., 1993). In vivo, the pattern of expression of certain zebrins could nevertheless be influenced by the afferent connections (see next section). Thus, regionalization of the mature cerebellum is underlain by the distribution of P cells subtypes. The differences between P subtypes are not anecdotal (at least from the point of view of P cells) as, in mouse mutants, they may control their survival (Wassef et al., 1987) and their migration (Beierbach et al., 2001).

Distinct populations of P cells were detected in the developing cerebellum of several mammals and birds on the basis of a typical clustering (Korneliussen, 1967; Feirabend, 1983). It was later shown that these clusters differ in their expression of early P cells markers (Wassef and Sotelo, 1984; Wassef et al., 1985; Oberdick et al., 1993; Millen et al., 1995; Lin and Cepko, 1998; Karam et al., 2000). Interestingly, a given marker cannot be used to follow a group of P cells from early development to adulthood. All the P subtype markers tested so far present a phase of uniform expression at about P12 in rat (Wassef et al., 1985; Leclerc et al., 1988), P8–P10 in mouse (Rivkin and Herrup, 2003) thus preventing a direct comparison between the two P compartments. To circumvent this problem, the early P compartmentation has been related to the pattern of climbing fiber projections (Wassef et al., 1992a,b; Paradies et al., 1996), a comparison that is necessarily crude or partial at the early stages. It was also noted that the early clusters of P cells differ in their birthdates both in chick and mouse (Wassef, 1992; Hashimoto and Mikoshiba, 2003). In an early study (Wassef, 1992), we detected a variation in birthdates of P subtypes but found that it was blurred at later stages and the distribution of P born on a given day could not be unambiguously related to the adult zebrin pattern. This puzzling observation questioned the significance of our birth-dating experiments and reinforced the conclusion that the early and late patterns were somehow distinct. The recent study by Hashimoto and Mikoshiba (2003) shed a new light on cerebellar compartmentation and beautifully illustrated the fate of the early compartmentation. Because they contain β-galactosidase, the P cells labeled on a given day can be detected by in toto staining of the cerebellum allowing a direct comparison of their distribution patterns at successive stages. A trace of the early P compartmentation is still detectable in the adult cerebellum but it becomes salt and pepper and less accurate than the zebrin pattern. This could occur because birthdating is only partially discriminative of P subtypes and the striking developmental pattern observed by Hashimoto and Mikoshiba (2003) results from not only variations in birthdates but also from the mode of P cell migration. Alternatively, P cells belonging to adjacent developmental clusters, that are partially overlapping (Wassef et al., 1985), could tend to mix with neighbors when they lose their tight clustering to form a monolayer. Finally, late migrating P cells could interperse within early clusters without sharing the same birthdate/label. In any case, a smoothening process would be needed to establish the adult zebrin pattern based on the distribution of developmental P subtypes. Such refinement is indeed observed during the second postnatal week when the exuberant innervation of P cells by climbing fiber is replaced by the one to one adult organization (Crepel et al., 1976) and zebrin I and II adopt their adult expression pattern (Leclerc et al., 1988). How the early P cell regionalization contributes to cerebellar patterning is still unclear. We suggest that one of the first roles of P subtypes differentiation is the patterning of an axonal plexus that prefigures the corticonuclear projection and the main subdivisions of the white matter and cerebellar cortex.

Development and patterning of the cerebellar cortex efferent pathway

The majority of DCN neurons are produced at E10.5–E11.5 in mouse. They migrate superficially towards the isthmic region (Altman and Bayer, 1978; Goffinet, 1983; Bourrat and Sotelo, 1986) where, at E11.5, they have already elongated axons that exit the cerebellar plate through two distinct pathways, the prospective superior peduncle that decussates in the pontine region and the uncinate bundle (Cholley et al., 1989) that crosses the cerebellar midline. At this stage, the DCN and the newly post-mitotic P are separated by a fibrous layer containing DCN axons, primary vestibular fibers (Morris et al., 1988) and possibly other unidentified afferents. In E16 rats, while migrating to the cortex, the calbindin labeled P cells elongate axons (Fig. 2A) that reach the fibrous layer and bundle in small fascicles alternating with the vestibular fiber fascicles detected by parvalbumin immunostaining. By E18.5, when the DCN neurons sink below the intermediate sheet with a characteristic switch in their dendritic polarity (Bourrat and Sotelo, 1986), the P axons remain bundled between the vestibular fascicles (Fig. 2B) that eventually ensheath the deep cerebellar nuclei. Interestingly, the topography of the P axons in this 'white matter' compartment reflects the early P compartmentation. Injections of HRP in the medullar region of the cerebellar plate trace back clusters of P cells often arranged in rows (Fig. 2C). In some cases (Fig. 2D) the labeled cells delineate a distinct cluster of unlabeled P indicating that the axons of the unlabeled cells avoid the site of HRP injection and probably project in a different location from those of the labeled clusters.

Thus, in the early cerebellar plate, a crowded environment of migrating cells, one of the first roles of the P compartmentation could be the establishment of an ordered layer of axons prefiguring the cortico-nuclear projection and the sagittal organization of the white matter where columns of P and afferent axons have been shown to alternate (Voogd, 1967). Perinatally, the developmental P compartmentation could also provide distinct addresses recognized by climbing fibers (Wassef et al., 1992a,b; Paradies et al., 1996). Distinct surface and secreted molecules (Lin and Cepko, 1998; Karam et al., 2000; Luckner et al., 2001) are expressed by the developmental clusters of P cells in mouse and chick. They provide a blueprint for afferent fibers (Chedotal et al., 1997) and funnel granule cells into thin sagittal raphes that course between the P cell stripes through the whole thickness of the cerebellar cortex.

In vitro, the developmental P compartmentation is probably able to provide a strong bias allowing the alternate expression of adult zebrin markers. In vivo, the functional connections are likely to induce a period of P cell plasticity during which the early informations could be partially erased. The uniformization of adult P cell markers (Wassef et al., 1985; Leclerc et al., 1988) during the first and second postnatal weeks could reflect this process.

In conclusion, the developmental P compartmentation is likely to be sufficient to allow for a relatively elaborate patterning of the cerebellar cortex afferent and efferent connections before the P cell monolayer forms. This pattern is likely to require a further step of local refinement during which the P subtype is reassessed.

Acknowledgments

Many of the illustrations in this chapter were obtained while the senior author was in the laboratory of C. Sotelo. Work in MW group is supported by grants from ACI Developpement and ARC. NNN and PA are recipients of MRT (France) and FCT (Portugal) fellowships, respectively.

References

Ackerman, S.L., Kozak, L.P., Przyborski, S.A., Rund, L.A., Boyer, B.B. and Knowles, B.B. (1997) The mouse rostral cerebellar malformation gene encodes an UNC-5-like protein. Nature, 386: 838–842.

Altman, J. and Bayer, S.A. (1978) Prenatal development of the cerebellar system in the rat I. Cytogenesis and histogenesis of the deep nuclei and the cortex of the cerebellum. J. Comp. Neurol., 179: 23–48.

Beierbach, E., Park, C., Ackerman, S.L., Goldowitz, D. and Hawkes, R. (2001) Abnormal dispersion of a Purkinje cell subset in the mouse mutant cerebellar deficient folia (cdf). J. Comp. Neurol., 436: 42–51.

Bourrat, F. and Sotelo, C. (1986) Neuronal migration and dendritic maturation of the medial cerebellar nucleus in rat embryos: an HRP in vitro study using cerebellar slabs. Brain Res., 378: 69–85.

Chedotal, A., Bloch-Gallego, E. and Sotelo, C. (1997) The embryonic cerebellum contains topographic cues that guide developing inferior olivary axons. Development, 124: 861–870.

Crepel, F., Mariani, J. and Delhaye-Bouchaud, N. (1976) Evidence for a multiple innervation of Purkinje cells by climbing fibers in the immature rat cerebellum. J. Neurobiol., 7: 567–578.

Cholley, B., Wassef, M., Arsenio-Nunes, L., Brehier, A. and Sotelo, C. (1989) Proximal trajectory of the brachium conjunctivum in rat fetuses and its early association with the parabrachial nucleus. A study combining in vitro HRP anterograde axonal tracing and immunocytochemistry. Brain Res. Dev. Brain Res., 45: 185–202.

Edlund, T. and Jessell, T.M. (1999) Progression from extrinsic to intrinsic signaling in cell fate specification: a view from the nervous system. Cell, 96: 211–224.

Feirabend, H.K.P. Anatomy and development of longitudinal patterns in the architecture of the cerebellum of the White Leghorn (*Gallus domesticus*). Thesis, Leiden.

Goldowitz, D., Hamre, K.M., Przyborski, S.A. and Ackerman, S.L. (2000) Granule cells and cerebellar boundaries: analysis of Unc5h3 mutant chimeras. J. Neurosci., 20: 4129–4137.

Goffinet, A.M (1983) The embryonic development of the cerebellum in normal and reeler mutant mice. Anat. Embryol. (Berl.), 168(1): 73–86.

Gravel, C., Eisenman, L.M., Sasseville, R. and Hawkes, R. (1987) Parasagittal organization of the rat cerebellar cortex: direct correlation between antigenic Purkinje cell bands revealed by mabQ113 and the organization of the olivocerebellar projection. J. Comp. Neurol., 265: 294–310.

Hallem, J.S., Thompson, J.H., Gundappa-Sulur, G., Hawkes, R., Bjaalie, J.G. and Bower, J.M. (1999) Spatial correspondence between tactile projection patterns and the distribution of the antigenic Purkinje cell markers anti-zebrin I and anti-zebrin II in the cerebellar folium crus IIA of the rat. Neuroscience, 93: 1083–1094.

Hashimoto, M. and Mikoshiba, K. (2003) Mediolateral compartmentalization of the cerebellum is determined on the 'birth date' of Purkinje cells. J. Neurosci., 23: 11342–11351.

Hawkes, R. (1997) An anatomical model of cerebellar modules. Prog. Brain Res., 114: 39–52.

Hawkes, R. and Leclerc, N. (1987) Antigenic map of the rat cerebellar cortex: the distribution of parasagittal bands as revealed by monoclonal anti-Purkinje cell antibody mabQ113. J. Comp. Neurol., 256: 29–41.

Jensen, P., Zoghbi, H.Y. and Goldowitz, D. (2002) Dissection of the cellular and molecular events that position cerebellar Purkinje cells: a study of the math1 null-mutant mouse. J. Neurosci., 22: 8110–8116.

Karam, S.D., Burrows, R.C., Logan, C., Koblar, S., Pasquale, E.B. and Bothwell, M. (2000) Eph receptors and ephrins in the developing chick cerebellum: relationship to sagittal patterning and granule cell migration. J. Neurosci., 20: 6488–6500.

Korneliussen, H.K. (1967) Cerebellar corticogenesis in cetacea with special reference to regional variations. J. Hirnforsch., 9: 151–185.

Leclerc, N., Gravel, C. and Hawkes, R. (1988) Development of parasagittal zonation in the rat cerebellar cortex: MabQ113 antigenic bands are created postnatally by the suppression of antigen expression in a subset of Purkinje cells. J. Comp. Neurol., 273: 399–420.

Lin, J.C. and Cepko, C.L. (1998) Granule cell raphes and parasagittal domains of Purkinje cells: complementary patterns in the developing chick cerebellum. J. Neurosci., 18: 9342–9353.

Lin, J.C., Cai, L. and Cepko, C.L. (2001) The external granule layer of the developing chick cerebellum generates granule cells and cells of the isthmus and rostral hindbrain. J. Neurosci., 21: 159–168.

Louvi, A., Alexandre, P., Métin, C., Wurst, W. and Wassef, M. (2003) The isthmic neuroepithelium is essential for cerebellar midline fusion. Development, 130: 5319–5330.

Luckner, R., Obst-Pernberg, K., Hirano, S., Suzuki, S.T. and Redies, C. (2001) Granule cell raphes in the developing mouse cerebellum. Cell Tissue Res., 303: 159–172.

Mathis, L., Bonnerot, C., Puelles, L. and Nicolas, J.F. (1997) Retrospective clonal analysis of the cerebellum using genetic laacZ/lacZ mouse mosaics. Development, 124: 4089–4104.

Millen, K.J., Hui, C.C. and Joyner, A.L. (1995) A role for En-2 and other murine homologues of Drosophila segment polarity genes in regulating positional information in the developing cerebellum. Development, 121: 3935–3945.

Morris, R.J., Beech, J.N. and Heizmann, C.W. (1988) Two distinct phases and mechanisms of axonal growth shown by primary vestibular fibres in the brain, demonstrated by parvalbumin immunohistochemistry. Neuroscience, 27: 571–596.

Oberdick, J., Schilling, K., Smeyne, R.J., Corbin, J.G., Bocchiaro, C. and Morgan, J.I. (1993) Control of segment-like patterns of gene expression in the mouse cerebellum. Neuron, 10: 1007–1018.

Oscarsson, O. (1980) Sagittal zones and microzones — The functional units od the cerebellum. In: Szentagothai, J., Hamori, J., and Palkovits (Eds.), Adv. Physiol. Sci. Vol 2. Regulatory Functions in the CNS Subsystems. Pergamon Press, Oxford, New York, pp. 21–28.

Paradies, M.A., Grishkat, H., Smeyne, R.J., Oberdick, J., Morgan, J.I. and Eisenman, L.M. (1996) Correspondence between L7-lacZ-expressing Purkinje cells and labeled olivocerebellar fibers during late embryogenesis in the mouse. J. Comp. Neurol., 374: 451–466.

Przyborski, S.A., Knowles, B.B. and Ackerman, S.L. (1998) Embryonic phenotype of Unc5h3 mutant mice suggests chemorepulsion during the formation of the rostral cerebellar boundary. Development, 125: 41–50.

Ramon y Cajal, S. (1911) Histologie du Système Nerveux de l'Homme et des Vertébré. Maloine, Paris.

Rivkin, A. and Herrup, K. (2003) Development of cerebellar modules: extrinsic control of late-phase Zebrin II pattern and the exploration of rat/mouse species differences. Mol. Cell Neurosci., 24: 887–901.

Ryder, E.F. and Cepko, C.L. (1994) Migration patterns of clonally related granule cells and their progenitors in the developing chick cerebellum. Neuron, 12: 1011–1028.

Sotelo, C. and Wassef, M. (1991) Cerebellar development: afferent organization and Purkinje cell heterogeneity. Philos. Trans. R. Soc. Lond. B Biol. Sci., 331: 307–313.

Voogd, J. (1967) Comparative aspects of the structure and fiber connections of the mammalian cerebellum. Prog. Brain Res., 25: 94–135.

Wassef, M (1992) The birthdates of Purkinje cells determine their biochemical subtypes and eventually underly the pattern of connections of the cerebellar cortex. Int. J. Neurosci., 10(suppl 1): 115 abstract.

Wassef, M. and Sotelo, C. (1984) Asynchrony in the expression of guanosine 3′:5′-phosphate-dependent protein kinase by clusters of Purkinje cells during the perinatal development of rat cerebellum. Neuroscience, 13: 1217–1241.

Wassef, M., Zanetta, J.P., Brehier, A. and Sotelo, C. (1985) Transient biochemical compartmentalization of Purkinje cells during early cerebellar development. Dev. Biol., 111: 129–137.

Wassef, M., Sotelo, C., Cholley, B., Brehier, A. and Thomasset, M. (1987) Cerebellar mutations affecting the postnatal survival of Purkinje cells in the mouse disclose a longitudinal pattern of differentially sensitive cells. Dev. Biol., 124: 379–389.

Wassef, M., Sotelo, C., Thomasset, M., Granholm, A.C., Leclerc, N., Rafrafi, J. and Hawkes, R. (1990) Expression of compartmentation antigen zebrin I in cerebellar transplants. J. Comp. Neurol., 294: 223–234.

Wassef, M., Angaut, P., Arsenio-Nunes, L., Bourrat, F. and Sotelo, C. (1991) Purkinje cell heterogeneity: its role in organizing the topography of the cerebellar cortex connections. In: Llinás, R. and Sotelo, C. (Eds.), The Cerebellum Revisited. Springer, New York, pp. 5–21.

Wassef, M., Chedotal, A., Cholley, B., Thomasset, M., Heizmann, C.W. and Sotelo, C. (1992a) Development of the olivocerebellar projection in the rat: I. Transient biochemical compartmentation of the inferior olive. J. Comp. Neurol., 323: 519–536.

Wassef, M., Cholley, B., Heizmann, C.W. and Sotelo, C. (1992b) Development of the olivocerebellar projection in the rat: II. Matching of the developmental compartmentations of the cerebellum and inferior olive through the projection map. J. Comp. Neurol., 323: 537–350.

Wingate, R.J. and Hatten, M.E. (1999) The role of the rhombic lip in avian cerebellum development. Development, 126: 4395–4404.

Zhang, L. and Goldman, J.E. (1996) Generation of cerebellar interneurons from dividing progenitors in white matter. Neuron, 16: 47–54.

CHAPTER 4

Bcl-2 protection of axotomized Purkinje cells in organotypic culture is age dependent and not associated with an enhancement of axonal regeneration

A.M. Ghoumari[1,2], R. Wehrlé[1,3], C. Sotelo[1,3] and I. Dusart[1,3],*

[1]INSERM U106, Hôpital de la Salpêtrière, 47 boulevard de l'Hôpital, 75013 Paris, France
[2]INSERM U488, 80, rue du Général Leclerc, 94276 Bicêtre, France
[3]UMR7102 CNRS-Université Paris VI, 9 Quai Saint Bernard, 75005 Paris, France

Keywords: Bcl-2; Purkinje cell; cerebellum; axotomy; neuronal death; axonal regeneration

Introduction

There are evidences reported to suggest that the intracellular signaling pathways leading to regeneration and cell death may be interrelated (Herdegen et al., 1997; Wehrlé et al., 2001). Neuronal systems with high regenerative capacity after injury are also those that undergo a more severe neuronal degeneration (Richardson et al., 1982; Misantone et al., 1984; Villegas-Pérez et al., 1988; Doster et al., 1991; Tetzlaff et al., 1991; Herdegen et al., 1997; Buffo et al., 1998). For example, most of the axotomized olivary neurons die after axotomy in the absence of a permissive environment (Barmack and Simpson, 1980; Ito et al., 1980; Buffo et al., 1998; Wehrlé et al., 2001), whereas they are able to regenerate their axons in the presence of permissive environments (Rossi et al., 1995; Bravin et al., 1997). Herdegen et al. (1997) have proposed that the transcription factor c-Jun could be one bipotential mediator of regeneration and cell death. GAP-43 (growth associated protein of 43 kD), the protein that has been the most frequently associated with growth during axonal regeneration (Benowitz and Routtenberg, 1997), has been involved in cell death (Herdegen et al., 1997; Wehrlé et al., 2001). Indeed, Purkinje cells with GAP-43 overexpression in vivo (Buffo et al., 1997), and in vitro (Wehrlé et al., 2001) are extremely vulnerable to a lesion, whereas wild type Purkinje cells are not. The death of adult GAP-43 overexpressing motoneurons has also been reported after axotomy of these neurons (Harding et al., 1999). Last, in GAP-43 null mutant mice, the survival of E18 sensory neurons in vitro and of Purkinje cells in P2 cerebellar explants is improved (Gagliardini et al., 2000). These studies suggest that

*Corresponding author. Tel.: +33 1 44 27 21 29; Fax: +33 1 44 27 26 69; E-mail: isabelle.dusart@snv.jussieur.fr

DOI: 10.1016/S0079-6123(04)48004-2

GAP-43 could be involved in the essential choice of the signaling pathway that will provoke either axonal growth or neuronal death (Herdegen et al., 1997, Wehrlé et al., 2001). Thus, cell death and axonal regeneration were though to be regulated in part by common molecular mechanisms.

In mouse and rat organotypic cultures of cerebellum, Purkinje cell survival and axonal regeneration are age-dependent (Dusart et al., 1997; Ghoumari et al., 2000). Until birth (P0), Purkinje cells survive axotomy and regenerate their axons. Between P1 and P7, the vast majority of the Purkinje cells die by apoptosis in cultured slices. Finally, P10 Purkinje cells are very resistant to axotomy but do not regenerate. Thus, organotypic cerebellar cultures offer a good model to determine whether experimentally induced changes in Purkinje cell death vulnerability could be correlated with changes in their regenerative capability. To this aim we have been using several approaches to delay the Purkinje cell death in the P1–P5 explants such as: (1) Bcl-2 overexpression, (2) inhibition of PKC, and (3) administration of mifepristone (RU486) (Ghoumari et al., 2000, 2002, 2003). In PKC inhibition, we have already shown that enhanced Purkinje cell survival does not potentiate the regenerative capability of P3, P5, and P7 Purkinje cells. In addition, in PKC inhibition the switches for survival and axonal regeneration do not occur at the same ages (Ghoumari et al., 2002). In the present study, we show that Bcl-2 protection of Purkinje cell from axotomy is not constant but age dependent, and does not enhance Purkinje cell axonal regeneration.

Slice cultures

Postnatal day 0 (P0), P1, P3, P5, P7, and P10 *Hu-bcl-2* transgenic mice (generated by microinjection of the NSE-*bcl-2* DNA construct into the pronuclei of fertilized C57BL/6J ova, as described by Farlie et al. (1995)) were used in this study. P0 was the day of birth. For each experiment at least 3 animals and 12 slices have been used. After decapitation, the brains were dissected out into cold Gey's balanced salt solution containing 5 mg/ml glucose (GBSS-Glu) and meninges were removed. Cerebellar parasagittal slices (350 μm thick) were cut on a McIlwain tissue chopper and transferred onto membranes of 30 mm Millipore culture inserts with 0.4 μm pore size (Millicell, Millipore, Bedford, MA, USA). The slices were maintained in culture in 6-well plates containing 1 ml or in 10 cm culture dishes containing 3 ml of medium at 35°C in an atmosphere of humidified 5% CO_2. The medium was composed of 50% basal medium with Earle's salts (Invitrogen, Cergy Pontoise, France), 25% Hanks' balanced salts solution (Invitrogen), 25% horse serum (Invitrogen), L-glutamine (1 mM) and 5 mg/ml glucose (Stoppini et al., 1991).

The cultures were transected at their dorsoventral half with two needles under a dissecting microscope. The two parts were gently separated to ensure a complete axotomy before being reapposed. Two thin cotton threads were put to delineate the transection line (Fig. 1E).

Staining procedures

After 5 days in vitro, the cultures were fixed in 4% paraformaldehyde in phosphate buffer (0.1 M, pH 7.4) for 1 h at room temperature. After washing in phosphate buffered saline, the slices were taken off the Millicell and processed for immunocytochemistry to reveal the Purkinje cells. The slices were incubated for 1 h in 0.12 M (pH 7.4) phosphate buffer containing 0.9% NaCl, 0.25% Triton-X, 0.2% gelatin, 0.1% sodium azide (PBSGTA) and lysine (0.1 M), before applying the rabbit polyclonal antibody against CaBP (diluted 1/5000; Swant, Bellinzona, Switzerland) diluted in PBSGTA overnight. After several washes and incubation for 2 h in a buffer containing the sheep anti-rabbit FITC (1/200 dilution, Eurobio, Les Ulis, France), the slices were mounted in mowiol (Calbiochem, La Jolla, USA) and analyzed using a Leica DMR microscope.

Quantification of Purkinje cell survival and axonal regeneration

The determination of the Purkinje cell survival and axonal regeneration was done in the CaBP-immunostained cultures under a fluorescence Leica DMR microscope. Under these conditions, we counted either the total number of surviving Purkinje cells in

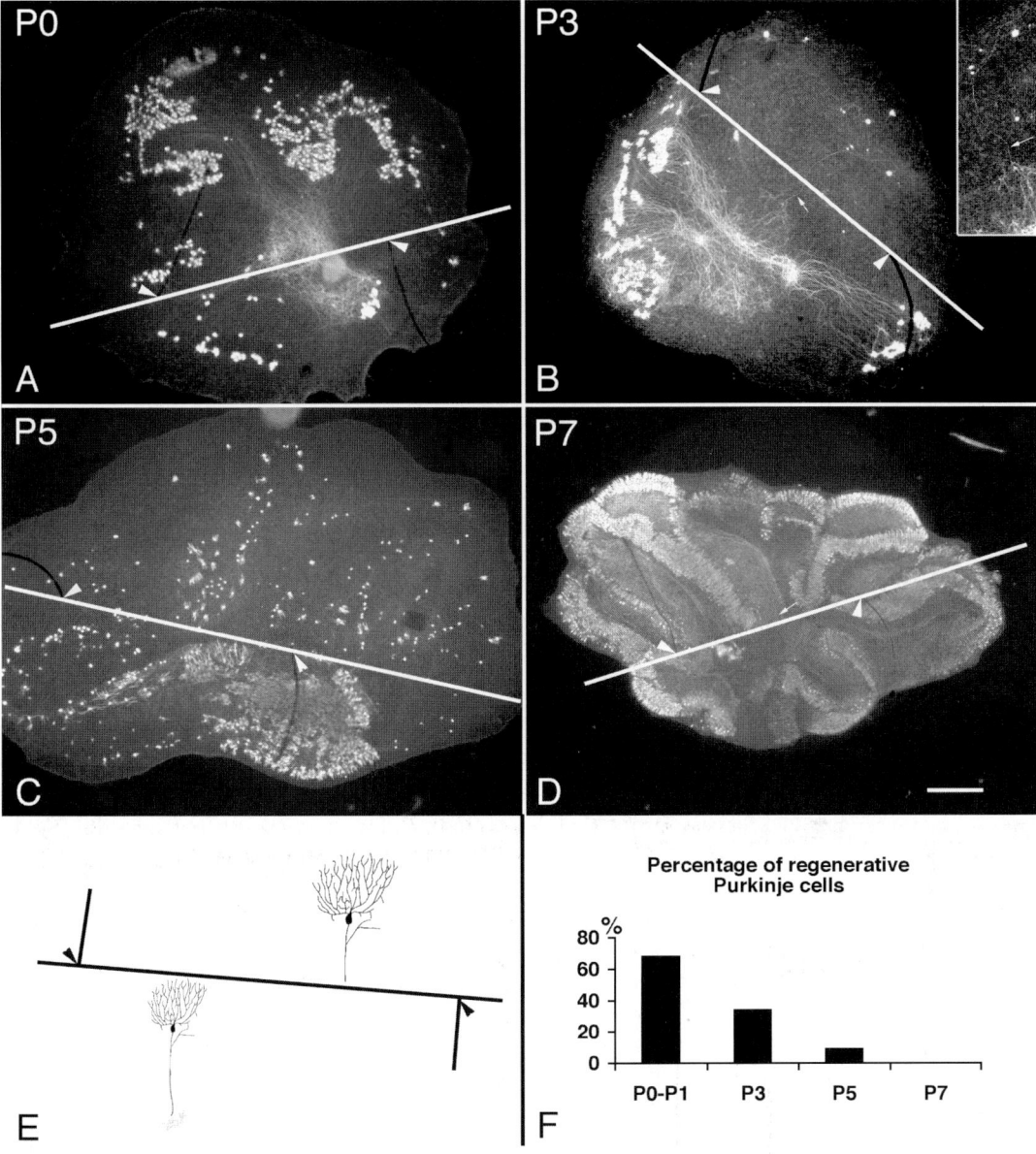

Fig. 1. Effect of Bcl-2 expression on survival and axonal regeneration of axotomized Purkinje cells in organotypic culture. A–D: Photomicrographs of P0 (A), P3 (B), P5 (C), and P7 (D) *hu-bcl-2* transgenic cerebellar slices, axotomized at the time of the cultures. These slices were maintained for 5 days in vitro and immunostained with anti-CaBP antibodies to label Purkinje cells. E: Schematic representation of the lesion that can be observed on the different slices (A–D). The lesion line (white line in A, B, C, D) is determined by the two more internal points, labeled with arrowheads on each figure, of thin cotton threads placed on the slice at the time of the axotomy. Thus, each slice is divided into two parts and axotomized Purkinje cells were present in the dorsal parts, while the ventral ones contain intact Purkinje cells. Note that numerous axotomized Purkinje cells (dorsal part) survive and regenerate their axons at P0 (A). Very few survive and regenerate at P3 (B). An example of a regenerative Purkinje cell is indicated in B by a small arrow and is enlarged in the box. At P5, few Purkinje cells survive and they do not regenerate their axons (C). At P7, numerous axotomized Purkinje cells survive and maintain their axons without regeneration or retraction at the lesion site (small arrow in D). F: Quantitative analysis of the regenerative *hu-bcl-2* overexpressing Purkinje cells between P0 and P7. The percentage of regenerative *hu-bcl-2* overexpressing Purkinje cells decreases progressively with age. Bar in D is 250 μm in A and C, 300 μm in B and 600 μm in D.

the dorsal half of the slices or at least 100 of these cells and among them the number of Purkinje cells whose axons cross the line of axotomy (Fig. 1E). The line of axotomy was the line that passed through the two points defined by the most internal part of the two thin cotton threads (Fig. 1A–E). For each age, we calculated the means and standard error of the means (SEM) of the Purkinje cell number present in the dorsal half and the percentage of those that regenerate.

The Purkinje cell deaths consecutive to either organotypic culture or to axotomy are different

Most Purkinje cells in murine cerebellar organotypic cultures die when taken from 1 to 5-day-old mice (P1–P5), but they survive when taken before and after these ages. In addition, while Purkinje cells in P10 slices are extremely resistant to axotomy, those in E18 to P7 slices are, in contrast, very sensitive and die after axotomy (Ghoumari et al., 2002). Therefore survival in slices does not always imply resistance to axotomy. For instance, immature Purkinje cells in P0 slices, despite their high survival rate in intact slices, are extremely sensitive to axotomy. In the present study, we have investigated whether Bcl-2 that protects Purkinje cells from death in P3 slices, as shown with *hu-bcl-2* transgenic mice (Ghoumari et al., 2000), could also prevent axotomy-induced death.

For this study, we used slices taken from the cerebella of *hu-bcl-2* transgenic mice aged from P0 to P10. Each cerebellar explant was divided into two parts by a complete cut separating the dorsal from the ventral vermal lobules. Thus, axotomized Purkinje cells were only present in the dorsal parts, while the ventral ones contain intact Purkinje cells (Fig. 1B, E). When the axotomy was performed in P0 or P1 slices, the results differed from slice to slice. In half of them (46%, $n = 24$), the survival of Purkinje cells was much better than in the others, and much more than 100 Purkinje cells remained in the dorsal half (Fig. 1A), whereas in 54% of the slices, only a mean of 30 ± 6 (S.E.M.) Purkinje cells per slice survived. In P3 dorsal slices the Purkinje cell survival was much less important than in the younger slices. In none of them ($n = 26$), were we able to count more than 100 Purkinje cells, and the mean was 25 ± 4 Purkinje cells per dorsal slice. At P5 ($n = 16$), the number of surviving Purkinje cells was significantly higher than at P3 (Fig. 1C), with a mean of 69 ± 9 (S.E.M.) Purkinje cells per dorsal slice, and even in three of them we encountered more than 100 (133, 118, and 102) Purkinje cells. As already described for the wild type cerebellum (Ghoumari et al., 2002), the few Purkinje cell survivors retracted their proximal axonic stumps, and exhibited atrophy and/or disruption of their dendritic trees. From P7, Bcl-2 overexpressing Purkinje cells become very resistant to axotomy (Fig. 1D), in contrast to wild type Purkinje cells (Ghoumari et al., 2002). Moreover, in the dorsal halves of the P7 slices, the numerous surviving Purkinje cells did not retract their axons and dendrites (Fig. 1D).

These results confirmed that Bcl-2 protects Purkinje cells from the age-related cell death occurring under organotypic culture conditions (Dusart et al., 1997). Indeed, whatever the age of the mice providing the explants, many Purkinje cells located in the ventral portions of the divided slices (the intact Purkinje cells) were able to survive. Furthermore, the results also revealed that Bcl-2 was not able to constantly protect Purkinje cells from death consecutive to the axotomy, since the protection varied with age: it was partial at P0–P1, absent at P3–P5 and important at P7 (Fig. 1F).

Bcl-2 does not enhance Purkinje cell axonal regeneration

Numerous surviving Purkinje cells in the *hu-bcl-2* cerebellar slices taken from P0 were able to regenerate their axons after 5 DIV (Fig. 1A). Regenerating axons reentered into the ventral part of the slices. In contrast, when P3 or P5 explants were used, from the rare Purkinje cell survivors, very few were able to regrow axons (Fig. 1B). In P7 slices, even if, as reported above, the survival of Purkinje cells to axotomy was highly improved, the severed axons did not regenerate (Fig. 1D). Finally, Purkinje cells in P10 slices were extremely resistant to axotomy, and all of them survived without retraction of their proximal stumps, but did not regenerate (data not shown). We have evaluated for each age the percentage of surviving Purkinje cells that were able to regenerate their axons after axotomy. We found that while at P0 and P1 more than 68% of the surviving axotomized Purkinje cells regenerated their axons, only 34% were able at P3 and 9% at P5 (Fig. 1F). From P7 on, only exceptional ones were able to regenerate their axons, and the percentage of the regenerating Purkinje cells dropped to less than 0.1%. Thus, the percentage of Purkinje cells able to regenerate their axons after axotomy decreased with age as previously reported (Ghoumari et al., 2002). In addition, the fact that Bcl-2 protected some Purkinje cells from axotomy did not enhance their ability to regenerate.

hu-bcl-2 protects mouse Purkinje cell from programmed cell death

It is well established that overexpression of *bcl-2* can protect many classes of neurons from programmed cell death (Dubois-Dauphin et al., 1994; Martinou et al., 1994; Farlie et al., 1995; Cenni et al., 1996; Zanjani et al., 1996; Bernard et al., 1997; Offen et al., 1998; Nakamura et al., 1999). Despite the absence of a precise Purkinje cell quantification during development (due to the lack of Purkinje cell markers for tracing all the elements of this neuronal population from young postmitotic to the adult neurons; Wassef et al., 1985); increasing evidence suggests that Purkinje cells undergo — as most other neurons — programmed cell death. The developmental death of mouse Purkinje cells has been inferred from increases in the number of adult Purkinje cells in transgenic mice either overexpressing the anti-apoptotic Bcl-2 protein or deficient in the pro-apoptotic *bax* gene (Zanjani et al., 1996; Heaton et al., 1999; Fan et al., 2001). We have previously reported that Purkinje cells in organotypic culture taken from P3 mouse cerebellum die massively during the first 36 h in vitro. This death occurs with the expected features of programmed cell death (apoptosis), including DNA fragmentation, caspase-3 activation and strong dependency on expression of the anti-apoptotic Bcl-2 protein (Ghoumari et al., 2000). We have interpreted these results as indirect evidence unmasking the less important programmed Purkinje cell death that should occur in in vivo cerebellum. More direct evidence for the engagement of Purkinje cell in programmed cell death was provided recently by TUNEL and active caspase-3 staining in P3 mouse CaBP-positive Purkinje cells (Marin-Teva et al., 2004). These investigators have shown that apoptosis of Purkinje cells in P3 cerebellar slices was strongly reduced following selective elimination of microglia or neutralization of microglia-derived superoxide ions (Marin-Teva et al., 2004). Furthermore, they provided evidences that in P3 cerebellum in vivo, Purkinje cells die by the same mechanisms as those reported in organotypic culture. Thus, the age-related massive cell death of Purkinje cells in organotypic cultures during a precise time-window (P1–P5) could reflect, as already hypothesized, that these neurons undergo programmed cell death at this period in vivo. In the present study, we confirm that Bcl-2 protects Purkinje cell death in vitro all over this lapse of time.

The protection of *hu-bcl-2* from cell death due to axotomy is age-dependent

Bcl-2 protects adult retinal ganglion cells (RGCs) (Bonfanti et al., 1996; Cenni et al., 1996; Chierzi et al., 1998), Clarke's nucleus (Takahashi et al., 1999), red nucleus (Zhou et al., 1999; Shibata et al., 2000), thalamic (Caleo et al., 2002), and nigral (Winter et al., 2002) neurons. Although fewer experiments have been undertaken in young animals, it has been shown that motoneurons and RGCs are protected by Bcl-2 against axotomy retrograde death respectively at P2 and P5 (Dubois-Dauphin et al., 1994; De Bilbao and Dubois-Dauphin, 1996; Lodovichi et al., 2001). De Bilbao and Dubois-Dauphin (1996) suggested that the apoptosis following axotomy and the naturally occurring cell death may occur via the same mechanisms. Here in contrast, we propose that neuronal death following axotomy and the naturally occurring cell death do not follow similar signaling pathways. More interestingly, we propose that depending on their state of differentiation, Purkinje cells do not respond to the same traumatism, e.g., the axotomy, in the same way. This hypothesis is sustained not only by the fact that the protection by the overexpression of Bcl-2 is age dependent but, in addition, by the observation that different morphologies and time courses of axotomized Purkinje cell death occur at different ages. For instance, at P3 and to a lesser extent at P5, the axotomized Purkinje cells retracted both their axon and dendritic tree, whereas they survived and did not retract their axon and dendrites at P7. However, the expression of *hu-bcl-2* does not seem sufficient to override these two types of traumatisms, since *hu-bcl-2* is able to protect axotomized Purkinje cells from death outside their presumptive period of natural programmed cell death (i.e., before P1 and after P5), suggesting that these two types of death are additive.

hu-bcl-2 does not promote axonal Purkinje cell regeneration

Purkinje cell regenerative capacity and Purkinje cell death do not follow the same temporal fate. It has been clearly established that *bcl-2* overexpression enhances survival of RGCs both in neonatal and adult mice (Chen et al., 1997; Chierzi et al., 1999; Lodovichi et al., 2001; Inoue et al., 2002). Concerning the ability of Bcl-2 to enhance axonal regeneration, the results are more controversial. It has been reported that in neonatal *bcl-2* transgenic mice, axonal regeneration of RGCs was promoted when the axotomy was performed in the tectum (Chen et al., 1997) and not when it was performed in the optic nerve (Lodovichi et al., 2001). However, in the adult, the regeneration was not enhanced even in the presence of a permissive environment (Chierzi et al., 1999; Inoue et al., 2002). Here we report that in another system (the cerebellar Purkinje cells), *hu-bcl-2* was not able to promote axonal regeneration. More interestingly, we showed that by promoting Purkinje cell survival at P7 by two different approaches, namely the overexpression of Bcl-2 (present study) and the inactivation of PKC (Ghoumari et al., 2002), the axonal regeneration was not enhanced. Bcl-2 and inactivation of PKC most probably act via different mechanisms, since they do not provide the same pattern of protection: inhibition of PKC protects axotomized Purkinje cells whatever their ages be, whereas the protection of Bcl-2 is age dependent. It is obvious that more work is required to understand the different responses of Purkinje cells to axotomy. However, it seems likely that survival and axonal regeneration are largely independent. The regulation of the loss of Purkinje cell capacity to regenerate during development is still unknown. We have recently demonstrated that it is independent of myelin (Bouslama-Oueghlani et al., 2003). However, further work is necessary to understand the intrinsic and the extrinsic factors involved in the developmental switch that convert immature regenerating Purkinje cells into mature neurons devoid of regenerative capacity.

Conclusions

The occurrence of a partial temporal overlapping between the sensitivity of neurons to die after axotomy and their ability to regenerate has suggested that cell death and axonal regeneration might be regulated by common signaling pathways. GAP-43 (growth associated protein of 43 kD) has been proposed to be part of the regulation pathway controlling these two processes. Recently, by applying PKC inhibitor in cerebellar organotypic cultures, we were able to protect Purkinje cells but not to promote their axonal regeneration. To determine whether an increase in cell death vulnerability is or not correlated with increased regenerative capability, we used *hu-bcl-2* overexpressing mice. We have previously shown that overexpression of *hu-bcl-2*, an anti-apoptotic molecule, prevents the massive Purkinje cell death in P3 organotypic cultures. In the present study, we show that this *hu-bcl-2* overexpression is not enough to prevent the cell death induced by axotomy at least in cerebellar explants taken from P0 to P5 mouse pups, suggesting that the Purkinje cell death following either organotypic culture or axotomy are different. Finally, we have shown that among the few *hu-bcl-2* overexpressing Purkinje cells that survive axotomy, the percentage able to regenerate their axons decreases with time. These results confirm that Purkinje cell death and axonal regeneration are differentially regulated, and that the intracellular signaling pathways implied in these two cell decisions do not overlap.

Acknowledgments

We thank Dr. O. Bernard for providing *hu-bcl-2* transgenic mice. Grant sponsors: Institut National de la Santé et de la Recherche Médicale (INSERM, grant U106); European Community (EC), Biotechnology Programme (CT98-0293).

References

Barmack, N.H. and Simpson, J.I. (1980) Effects of microlesions of dorsal cap of inferior olive of rabbits on optokinetic and vestibuloocular reflexes. J. Neurophysiol., 43: 182–206.

Benowitz, L.I. and Routtenberg, A. (1997) GAP-43: an intrinsic determinant of neuronal development and plasticity. Trends Neurosci., 20: 84–91.

Bernard, R., Farlie, P. and Bernard, O. (1997) NSE-bcl-2 transgenic mice, a model system for studying neuronal death and survival. Dev. Neurosci., 19: 79–85.

Bonfanti, L., Strettoi, E., Chierzi, S., Cenni, M.C., Liu, X.H., Martinou, J.C., Maffei, L. and Rabacchi, S.A. (1996) Protection of retinal ganglion cells from natural and axotomy induced cell death in neonatal transgenic mice overexpressing bcl-2. J. Neurosci., 16: 4186–4194.

Bouslama-Oueghlani, L., Wehrlé, R., Sotelo, C. and Dusart, I. (2003) The developmental loss of Purkinje Cell ability to regenerate their axons occurs in the absence of myelin: an *in vitro* model to prevent myelination. J. Neurosci., 23: 8318–8329.

Bravin, M., Savio, T., Strata, P. and Rossi, F. (1997) Olivocerebellar axon regeneration and target reinnervation following dissociated Schwann cell grafts in surgically injured cerebella of adult rats. Eur. J. Neuroscience, 9: 2634–2649.

Buffo, A., Holtmaat, A.J., Savio, T., Verbeek, J.S., Oberdick, J., Oestreicher, A.B., Gispen, W.H., Verhaagen, J., Rossi, F. and Strata, P. (1997) Targeted overexpression of the neurite growth-associated protein B-50/GAP-43 in cerebellar Purkinje cells induces sprouting after axotomy but not axon regeneration into growth-permissive transplants. J. Neurosci., 17: 8778–8791.

Buffo, A., Fronte, M., Oestreicher, A. and Rossi, F. (1998) Degenerative phenomena and reactive modifications of the adult rat inferior olivary neurons following axotomy and disconnection from their targets. Neuroscience, 85: 587–604.

Caleo, M., Cenni, M.C., Costa, M., Menna, E., Zentilin, L., Giadrossi, S., Giacca, M. and Maffei, L. (2002) Expression of Bcl-2 via adeno-associated virus vectors rescues thalamic neurons after visual cortex lesion in the adult rat. Eur. J. Neurosci., 15: 1271–1277.

Cenni, M.C., Bonfanti, L., Martinou, J.C., Ratto, G.M., Strettoi, E. and Maffei, L. (1996) Long-term survival of retinal ganglion cells following optic nerve section in adult bcl-2 transgenic mice. Eur. J. Neurosci., 8: 1735–1745.

Chen, D.F., Schneider, G.E., Martinou, J.C. and Tonegawa, S. (1997) Bcl-2 promotes regeneration of severed axons in mammalian CNS. Nature, 385: 434–439.

Chierzi, S., Cenni, M.C., Maffei, L., Pizzorusso, T., Porciatti, V., Ratto, G.M. and Strettoi, E. (1998) Protection of retinal ganglion cells and preservation of function after optic nerve lesion in bcl-2 transgenic mice. Vision Res., 38: 1537–1543.

Chierzi, S., Strettoi, E., Cenni, M.C. and Maffei, L. (1999) Optic nerve crush: axonal responses in wild type and bcl-2 transgenic mice. J. Neurosci., 19: 8367–8376.

De Bilbao, F. and Dubois-Dauphin, M. (1996) Time course of axotomy-induced apoptotic cell death in facial motoneurons of neonatal wild type and bcl-2 transgenic mice. Neuroscience, 71: 1111–1119.

Doster, S.K., Lozano, A.M., Aguayo, A.J. and Willard, M.B. (1991) Expression of the growth-associated protein GAP-43 in adult rat retinal ganglion cells following axon injury. Neuron, 6: 635–647.

Dubois-Dauphin, M., Frankowski, H., Tsujimoto, Y., Huarte, J. and Martinou, J.C. (1994) Neonatal motoneurons overexpressing the bcl-2 protooncogene in transgenic mice are protected from axotomy-induced cell death. Proc. Natl. Acad. Sci. USA, 91: 3309–3313.

Dusart, I., Airaksinen, M.S. and Sotelo, C (1997) Purkinje cell survival and regeneration are age dependent: an in vitro study. J. Neurosci., 17: 3710–3726.

Fan, H., Favero, M. and Vogel, M.W. (2001) Elimination of Bax expression in mice increases cerebellar purkinje cell numbers but not the number of granule cells. J. Comp. Neurol., 436: 82–91.

Farlie, P.G., Dringen, R., Rees, S.M., Kannourakis, G. and Bernard, O. (1995) Bcl-2 transgene expression can protect neurons against developmental and induced cell death. Proc. Natl. Acad. Sci. USA, 92: 4397–4401.

Gagliardini, V., Dusart, I. and Fankhauser, C. (2000) Absence of GAP-43 can protect neurons from death. Mol. Cell Neurosci., 16: 27–33.

Ghoumari, A.M., Wehrlé, R., Bernard, O., Sotelo, C. and Dusart, I. (2000) Implication of Bcl-2 and Caspase-3 in the age-related Purkinje cell death in murine organotypic culture: An *in vitro* model to study apoptosis. Eur. J. Neurosci., 12: 2935–2949.

Ghoumari, A.M., Wehrle, R., De Zeeuw, C.I., Sotelo, C. and Dusart, I. (2002) Inhibition of protein kinase C prevents Purkinje cell death but does not affect axonal regeneration. J. Neurosci., 22: 3531–3542.

Ghoumari, A.M., Dusart, I., Tronche, F., El-Etr, M., Sotelo, C., Schumacher, M. and Baulieu, E.E. (2003) Mifepristone (RU486) protects Purkinje cells from cell death in organotypic slice cultures of postnatal rat and mouse cerebellum. Proc. Natl. Acad. Sci. USA, 86: 848.

Harding, D.I., Greensmith, L., Mason, M., Anderson, P.N. and Vrbova, G. (1999) Overexpression of GAP-43 induces prolonged sprouting and causes death of adult motoneurons. Eur. J. Neurosci., 11: 2237–2242.

Heaton, M.B., Moore, D.B., Paiva, M., Gibbs, T. and Bernard, O. (1999) Bcl-2 overexpression protects the neonatal cerebellum from ethanol neurotoxicity. Brain Res., 817: 13–18.

Herdegen, T., Skene, P. and Baehr, M. (1997) The c-Jun transcription factor–bipotential mediator of neuronal death, survival and regeneration. Trends Neurosci., 20: 227–231.

Inoue, T., Hosokawa, M., Morigiwa, K., Ohashi, Y. and Fukuda, Y. (2002) Bcl-2 overexpression does not enhance in vivo axonal regneration of retinal ganglion cells after peripheral nerve transplantation in adult mice. J. Neurosci., 22: 4486–4477.

Ito, M., Jastreboff, P.J. and Miyashita, Y. (1980) Retrograde influence of surgical and chemical flocculectomy upon dorsal cap neurons of the inferior olive. Neurosci. Lett., 20: 45–48.

Lodovichi, C., Di Cristo, G., Cenni, M.C. and Maffei, L. (2001) Bcl-2 overexpression per se does not promote regeneration of neonatal crushed optic fibers. Eur. J. Neurosci., 13: 833–838.

Marin-Teva, J.L., Dusart, I., Colin, C., Gervais, A., van Rooijen, N. and Mallat, M. (2004) Microglia promote the death of developing Purkinje cells. Neuron, 41: 535–547.

Martinou, J.C., Dubois-Dauphin, M., Staple, J.K., Rodriguez, I., Frankowski, H., Missotten, M., Albertini, P., Talabot, D., Catsicas, S., Pietra, C. and Huarte, J. (1994) Overexpression of BCL-2 in transgenic mice protects neurons from naturally occurring cell death and experimental ischemia. Neuron, 13: 1017–1030.

Misantone, L.J., Gershenbaum, M. and Murray, M. (1984) Viability of retinal ganglion cells after optic nerve crush in adult rats. J. Neurocytol., 13: 449–465.

Nakamura, M., Raghupathi, R., Merry, D.E., Scherbel, U., Saatman, K.E. and McIntosh, T.K. (1999) Overexpression of bcl-2 is neuroprotective after experimental brain injury in transgenic mice. J. Comp. Neurol., 412: 681–692.

Offen, D., Beart, P.M., Cheung, N.S., Pascoe, C.J., Hochman, A., Gorodin, S., Melamed, E., Bernard, R. and Bernard, O. (1998) Transgenic mice expressing human Bcl-2 in their neurons are resistant to 6-hydroxydopamine and 1-methyl-4-phenyl-1,2,3,6-tetrahydropyridine neurotoxicity. Proc. Natl. Acad. Sci. USA, 95: 5789–5794.

Richardson, P.M., McGuinness, U.M. and Aguayo, A.J. (1982) Peripheral nerve autografts to the rat spinal cord: studies with axonal tracing methods. Brain Res., 237: 147–162.

Rossi, F., Jankovski, A. and Sotelo, C. (1995) Differential regenerative response of Purkinje cell and inferior olivary axons confronted with embryonic grafts: environmental cues versus intrinsic neuronal determinants. J. Comp. Neurol., 359: 663–677.

Shibata, M., Murray, M., Tessier, A., Ljubetic, C., Connors, T. and Saavedra, R.A. (2000) Single injections of a DNA plasmid that contains the human Bcl-2 gene prevent loss and atrophy of distinct neuronal populations after spinal cord injury in adult rats. Neurorehabil. Neural Repair., 14: 319–330.

Stoppini, L., Buchs, P.A. and Muller, D. (1991) A simple method for organotypic cultures of nervous tissue. J. Neurosci. Meth., 37: 173–182.

Takahashi, K., Schwarz, E., Ljubetic, C., Murray, M., Tessier, A. and Saavedra, R.A. (1999) DNA plasmid that codes for human Bcl-2 gene preserves axotomized Clarke's nucleus and reduces atrophy after spinal cord hemisection in adult rats. J. Comp. Neurol., 404: 159–171.

Tetzlaff, W., Alexander, S.W., Miller, F.D. and Bisby, M.A. (1991) Response of facial and rubrospinal neurons to axotomy: changes in mRNA expression for cytoskeletal proteins and GAP-43. J. Neurosci., 11: 2528–2544.

Villegas-Pérez, M.P., Vidal-Sanz, M., Bray, G.M. and Aguayo, A.J. (1988) Influences of peripheral nerve grafts on the survival and regrowth of axotomized retinal ganglion cells in adult rats. J. Neurosci., 8: 265–280.

Wassef, M., Zanetta, J.P., Brehier, A. and Sotelo, C. (1985) Transient biochemical compartmentalization of Purkinje cells during early cerebellar development. Dev. Biol., 111: 129–137.

Wehrlé, R., Caroni, P., Sotelo, C. and Dusart, I. (2001) Role of GAP-43 in mediating the responsiveness of cerebellar neurons to axotomy. Eur. J. Neurosci., 13: 857–870.

Winter, C., Weiss, C., Martin-Villalba, A., Zimmermann, M. and Schenkel, J. (2002) JunB and Bcl-2 overexpression results in protection against cell death of nigral neurons following axotomy. Mol. Brain Res., 104: 194–202.

Zanjani, H.S., Vogel, M.W., Delhaye-Bouchaud, N., Martinou, J.C. and Mariani, J. (1996) Increased cerebellar Purkinje cell numbers in mice overexpressing a human bcl-2 transgene. J. Comp. Neurol., 374: 332–341.

Zhou, L., Connors, T., Chen, D.F., Murray, M., Tessier, A., Kambin, P. and Saavedra, R.A. (1999) Red nucleus neurons of Bcl-2 over-expressing mice are protected from cell death induced axotomy. Neuroreport, 10: 3417–3421.

SECTION II

Structural Cerebellar Plasticity

CHAPTER 5

Axonal and synaptic remodeling in the mature cerebellar cortex

Roberta Cesa* and Piergiorgio Strata

Rita Levi Montalcini Center for Brain Repair, Corso Raffaello 30, 10125 Turin, Italy
IRCCS Santa Lucia Foundation, Via Ardeatina 306, 00179, Rome, Italy

Keywords: cerebellum; parallel fiber; climbing fiber; plasticity; activity-dependent competition; glutamate receptor $\delta 2$ subunit

Abstract: By blocking electrical activity in the cerebellar cortex the Purkinje cell dendrites become a uniform territory with a high density of spines all bearing the glutamate receptor $\delta 2$ subunit (GluR$\delta 2$) and being mainly innervated by parallel fibers. Such a subunit, which is constitutively targeted specifically to the parallel fiber synapses, appears in the spines contacted by the climbing fibers before they disconnect from the target. A similar pattern of hyperspiny transformation and innervation occurs a few days after a subtotal lesion of the inferior olive, the source of climbing fibers.

During the climbing fiber reinnervation process which follows the removal of the electrical block or by collateral sprouting of surviving inferior olive neurons, the new active climbing fibers establish synaptic contacts with proximal dendritic spines that bear GluR$\delta 2$s. After, they repress these subunits and displace the parallel fibers to the distal dendritic territory.

These findings suggest the following operational principle in the axonal competition for a common target. The Purkinje cells have an intrinsic phenotypic profile which is compatible with the parallel fiber innervation, this mode being operational in targets innervated by a single neuronal population, like the neuromuscular system. An additional input, the climbing fibers, in order to achieve its own territory on the proximal dendrite needs the ability to displace the competitor. Such an inhibition is activity-dependent and the activity needs to be present in order to allow the climbing fiber to maintain its territory, even when the developmental period is over.

Introduction

The function of the nervous system critically relies on the establishment of precise synaptic connections between a neuron and its specific target cells. The specificity of neural circuitry is reached through three developmental steps of *pathway selection, target selection,* and *address selection* (Goodman and Shatz, 1993).

First, the axonal growth cones traverse long distances where they are confronted with a series of choice points and yet they choose the correct *pathways* in an unerring way. Once they reach the correct region, they recognize and thus contact their proper *target* neural population. Subsequently, these initial patterns of connections are refined by axonal retraction and extension to select a specific subset of target cells or a single neuron within the overall target, a process called *address* selection. Many studies in both vertebrates and invertebrates suggest that the early events of axon outgrowth, pathfinding and target selection are relatively accurate and do not require any electrical activity and synaptic transmission. These processes rely largely on the intrinsic cellular and molecular mechanisms that guide axons and facilitate

*Corresponding author. Tel.: +39-011-6707733;
Fax: +39-011-6707708; E-mail: roberta.cesa@unito.it

a correct synaptic partnership. Cues include cell adhesion molecules that regulate the interactions between axons and the surfaces upon which they grow, diffusible molecules that attract growing axons, and the important family of neurotrophins that promote and maintain stable synapses between the axons and their target. From this point onwards, patterns of neuronal activity represent the main mechanism that gradually increase the precision of synaptic connectivity by the selective addition or the removal of connections throughout the developing brain. This remodeling relies on homologous competition with surrounding inputs and is capable of transforming an overlapping projection into a refined and highly tuned wiring.

A subtle reorganization of terminal arbors and synaptic remodeling are thought to occur throughout life in the intact brain and may underlie some aspects of learning and memory (Buonomano and Merzenich, 1998; Yuste and Bonhoeffer, 2001; Kaas, 2002; Knott et al., 2002).

Another important issue in development is to understand how different afferent fibers distribute to the different postsynaptic domains of the same target cell and to find out whether any electrical activity is involved in this process that may be defined '*postsynaptic selection.*'

The cerebellum provides a suitable model to study how different neuronal populations find their proper postsynaptic territory during development and how they maintain their target region in the mature brain. This is due to the fact that in the cerebellar cortex, two different excitatory inputs impinge upon two separate domains of the Purkinje cell dendrites. The topographical distributions of the two excitatory inputs reflect a precise program of segregation, which is the result of the developmental processes of homologous competition inside each neuronal population and of the heterologous competition between them. The Purkinje cell dendrites are characterized by two compartments, a proximal one, on which a single climbing fiber, the terminal arbor of an olivocerebellar neuron, terminates and a distal one, on which many parallel fibers, the axons of the granule cells, impinge (Fig. 1).

During development, the Purkinje cell is initially innervated by several climbing fibers (Crepel, 1982) but, upon maturation, it becomes singly innervated. An extensive literature shows that the regression of the supernumerary climbing fibers is under the influence of the parallel fibers. A multiple climbing fiber

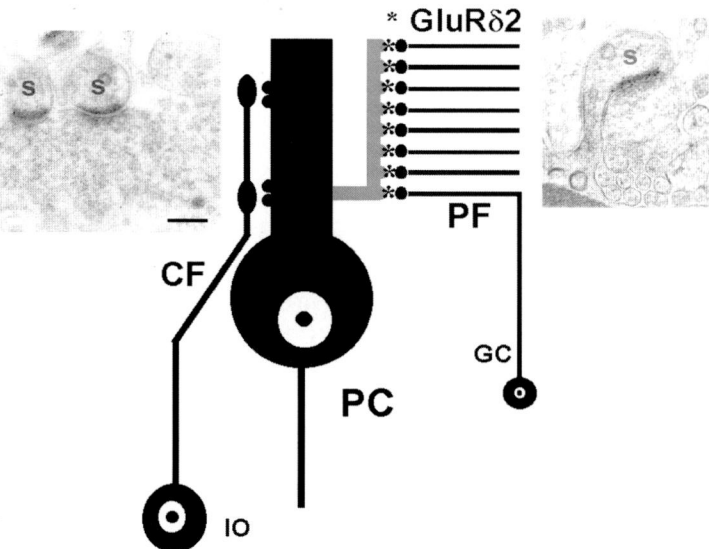

Fig. 1. Excitatory connections under control conditions. Sparse clustered spines of the Purkinje cell (PC) proximal dendrite contacted by the climbing fiber (CF) varicosities and densely packed spiny branchlet spines, with GluRδ2s (*), making contact with parallel fiber (PF) varicosities. On the left: electronmicrograph showing two spines, devoid of labeling for the GluRδ2, contacted by a climbing fiber terminal. On the right: electronmicrograph of a parallel fiber-innervated spine expressing the GluRδ2s (black dots) in its postsynaptic density. IO, inferior olive; GC, granule cells; s, spine. Scale bar, 0.25 μm.

innervation has been reported to persist in animal models that have significant defects in a granule cell survival or in a parallel fiber-Purkinje cell synaptogenesis (Crepel et al., 1976, 1980, 1981; Mariani et al., 1990; Sotelo, 1990; Bravin et al., 1995; Kashiwabuchi et al., 1995). It has been proposed that in the normal animal, the early phase of the regressive process follows a rule like that of the neuromuscular junction where the homologous competition among climbing fibers is intrinsic to the system whereas the late phase is strictly under the control of the parallel fibers (Crepel, 1982).

The impairment of climbing fiber regression by an abnormally weak parallel fiber input suggests that during development, there is a competitive mechanism between the two different excitatory inputs. In fact, in mutant mice with deficient parallel fibers to Purkinje cell synapses, there is even an extension of the climbing fiber territory of innervation to the distal dendritic compartment (Ichikawa et al., 2002).

Structural and synaptic plasticity of the mature cerebellum

Both inferior olivary neurons and granule cells maintain, in the mature state, a constitutive expression of several genes related to plasticity, such as GAP-43 (Kruger et al., 1993), Krox-24 transcription factor (Buffo et al., 1998), and the protein kinase C myristoylated alanine-rich C kinase substrate (MARCKS) (McNamara and Lenox, 1997). It is likely that the constitutive presence of these growth-associated proteins in both inputs is the requirement for a lifelong competition between climbing fibers and parallel fibers in the activity-dependent synaptic remodeling. Usually, the growth-associated proteins are downregulated at the end of development, but some of them persist in adulthood in defined regions of the mature brain that retain the capacity of synaptic remodeling in response to patterns of physiological activity (Kruger et al., 1993; McNamara and Lenox, 1997; Strata and Rossi, 1998).

Indeed, a high degree of structural plasticity in the climbing and parallel fibers has been demonstrated by changes in target size and after blocking the electrical activity.

Target deletion and extension

After the Purkinje cell degeneration induced by intracerebellar injections of kainic acid, the climbing fibers progressively undergo remarkable regressive modifications with the disappearance of most of the terminal branches, sparing the proximal thick ones. One month after the injection, the most peripheral branches are completely absent and the second order ones are largely reduced, while the number of varicosities drops to 12% of normal (Rossi et al., 1993). Similar results have been observed when the Purkinje cells are deleted by propidium iodide (Rossi et al., 1993) and in mutant mice with Purkinje cell degeneration (Rossi et al., 1995). Interestingly, these regressive modifications are reversed when new Purkinje cells are available, for example when embryonic Purkinje cells are transplanted into a kainic acid lesioned (Armengol et al., 1989) or into a mutant cerebellar cortex (Sotelo and Alvarado-Mallart, 1987). These experiments of target deletion show that synaptic connectivity requires intimate interaction between pre- and postsynaptic cells and that the highly plastic climbing fiber terminal arborization and the Purkinje cell show remarkable reciprocal trophic interactions (Rossi and Strata, 1995).

Intact climbing fibers are also able to sprout new collateral branches in the absence of a lesion, following the grafting of embryonic cerebellar tissue onto the surface of an intact cerebellum (Rossi et al., 1993, 1994). After several weeks, the graft organizes itself into a minicerebellum-like structure attached to the host cerebellar surface. The intact host climbing fibers maintain their normal distribution with their original adult target, but develop collateral sprouts through the pia to innervate supernumerary Purkinje cells inside the graft. Here, they give rise to many ramifications with varicosities, which terminate upon the somatodendritic region.

The rearrangement of the climbing fiber terminal arbor, in response to a newly available target, following a partial lesion of its pathway, has also been studied. A subtotal lesion of the inferior olivary neurons induced by intraperitoneal injection of a neurotoxin, 3-acetylpyridine (3-AP) leads to many Purkinje cells losing their climbing fiber input (Desclin and Escubi, 1974; Sotelo et al., 1975; Rossi et al., 1991 a,b). The first evidence that surviving climbing fibers are able to sprout and reinnervate these denervated Purkinje cells was obtained by an electrophysiological recording (Benedetti et al., 1983; Benedetti and Strata, 1989). A few days after the lesion, the number of Purkinje cells showing climbing fiber responses is congruent with the number of surviving neurons in the inferior olive. However, 2–6 months after the lesion, the number of Purkinje cells showing this kind of activity is 5–10 times higher. Therefore, it was concluded that the increased number of innervated Purkinje cells is due to a new innervation coming from collateral sprouting from the surviving olivocerebellar axons. Subsequently, these results have been confirmed by morphological studies. Rossi et al. (1991b) showed that new collaterals emerge from the surviving arbors within three days after a 3-AP injection and, in the following days, elongate for several hundred micrometers in the molecular layer, parallel to the cerebellar surface. Once these axons contact the dendritic tree of a deafferentated Purkinje cell, they give rise to a profuse branching along the proximal dendrite and form new synapses. A quantitative evaluation shows that a single surviving climbing fiber can extend its territory of innervation by more than six times.

Furthermore, the climbing fiber terminal arbor deletion is also accompanied by sprouting of nearby parallel fibers that invade its territory of innervation (Sotelo et al., 1975; Rossi et al., 1991b). The capacity of reactive synaptogenesis of this input had been previously demonstrated by experiments of partial parallel fiber transection (Chen and Hillman, 1982). However, after 3-AP, the two excitatory inputs coexist on the proximal Purkinje cell dendrite only as a transitory phenomenon that disappears when the new climbing fibers reach their full maturity. The regenerating climbing fibers eliminate the competitor parallel fiber afferents and restore the original synaptic organization on the Purkinje cell (Rossi et al., 1991a,b).

The high degree of plasticity of the two competing inputs that converge on two different regions of the Purkinje cells suggests that the wiring of the cerebellar cortex might be modulated by changes in the electrical activity and the resulting alterations might provide important information on the mechanisms of heterologous competition in the mature brain and in setting up the architecture of the cerebellar cortex during development.

Block of electrical activity

After infusing tetrodotoxin (TTX) for 7 days in vivo in the adult cerebellar parenchyma, the climbing fiber terminal arbor retracts from the Purkinje cell proximal dendritic domain and becomes atrophic in a manner similar to that seen after Purkinje cell degeneration (Bravin et al., 1999). The process of the climbing fiber withdrawal is accompanied by the generation of thin branches almost devoid of varicosities which arise from the thick climbing fiber trunks in the molecular layer, and elongated unbranched toward the upper granular layer. The presence of such new ramifications indicates that in the absence of electrical activity there is no impairment of the growing capacities of the climbing fiber input and that the regressive phenomena may be caused by an active repressing action of the competing parallel fibers. This input, normally distributed upon the spines of the distal dendritic domain, under TTX presents a competitive advantage over the climbing fiber and forms synapses on the proximal dendrites (Morando et al., 2001). The changes are reversible 2–4 weeks after removal of the electrical activity block: the climbing fiber arbor reacquires its normal morphology expanding its territory to reinnervate the proximal dendritic domain (Bravin et al., 1999). This is a further support for the view that the adult cerebellum is able to undergo remarkable changes in its architecture showing a strong intrinsic plastic potentiality.

Activity-dependent regulation of dendritic spines in Purkinje cells

The activity-dependent remodeling of the parallel fibers and the climbing fibers, is accompanied by a profound restructuring of the distribution of spines in the two dendritic compartments of the Purkinje cells. In the rat, the proximal compartment has a very low number of spines, about 0.7 spines/µm of dendritic length and 0.07 spines/µm^2 of the dendritic surface, whereas the distal compartment, the spiny branchlets, has a very high spine density, around 17.6 spines/µm of dendritic length and 3.65 spines/µm^2 of dendritic surface (Fig. 1; Strata, 2002).

After a subtotal lesion of the inferior olivary neurons by 3-AP, the Purkinje cells become hyperactive for several weeks (Montarolo et al., 1982) and a large number of new spines emerge from the proximal dendritic domain (Rossi et al., 1991a). This result is in line with the main general view that increases in neuronal activity produce more spines (Popov and Bocharova, 1992; Engert and Bonhoeffer, 1999; Segal, 2001; Knott et al., 2002; Nimchinsky et al., 2002). However, a similar pattern of hyperspiny transformation occurs following block of the electrical activity by infusing TTX for 7 days when the climbing fiber retracts from its domain. The block does not affect the number of branchlet spines (Bravin et al., 1999). Therefore, although a role of spontaneous transmitter release cannot be excluded (Bravin et al., 1999; McKinney et al., 1999), these results demonstrate that, even with blocked electrical activity, neurons are able to form new spines but only when the climbing fiber input does not exert its action on the Purkinje cell.

Thus, the Purkinje cell spine formation is the expression of an intrinsic program along the entire dendritic tree, regardless of its level of firing. However, in the distal compartment, the spines are maintained independently of the activity of the afferent input, whereas, in the proximal dendrite, spines are repressed by the activity of the climbing fibers (Sotelo et al., 1975; Bravin et al., 1999).

The glutamate receptor $\delta 2$ subunit (GluRδ2)

The glutamate receptor $\delta 2$ subunit (GluRδ2) plays a crucial role in the differential distribution and stabilization of the climbing fiber and parallel fiber synapses on the proximal and distal domains of the Purkinje dendritic arbor (Guastavino et al., 1990; Kurihara et al., 1997; Ichikawa et al., 2002). During development, GluRδ2s are expressed in both parallel fiber- and climbing fiber-innervated spines. In contrast, at the end of the developmental period, these subunits remain exclusively targeted to the parallel fiber synapses (Fig. 1; Takayama et al., 1996; Landsend et al., 1997; Zhao et al., 1998; Morando et al., 2001). Although this subunit, on the basis of aminoacid similarity, was classified as an ionotropic glutamate receptor (Araki et al., 1993; Lomeli et al., 1993), it has been referred to as an 'orphan' receptor because it does not bind to glutamate or to any known ligands and it does not form a functional glutamate-gated ion channel when expressed either alone or with other iGluRs in transfected cells (Araki et al., 1993; Lomeli et al., 1993). Nevertheless, GluRδ2 plays a crucial role in cerebellar function: mice that lack the gene encoding GluRδ2 display ataxia and impaired synaptic plasticity (Kashiwabuchi et al., 1995). Owing to the lack of specific pharmacological tools to manipulate GluRδ2, its mechanism of function is not well understood. However, studies of the mutant mice such as *lurcher*, *hotfoot*, and GluRδ2 knock-out mice have provided insights into the GluRδ2 signaling in neurons (Yuzaki, 2003 a,b). The morphological and electrophysiological analysis in the knock-out mice revealed a significant reduction of parallel fiber synapses on the Purkinje cell spines, suggesting that GluRδ2 is involved in stabilization and strengthening of synaptic connectivity between the parallel fibers and the Purkinje cells. At the end of the first postnatal week, no significant difference had been detected in the synaptic contact percentage of Purkinje cell spines: in both wild type and mutant mice about 80% of spines form immature synapses with parallel fibers. However, at the end of the third postnatal week, approximately 40% of the Purkinje cell spines in the knock-out mouse remained unattached to any nerve terminals, whereas all spines in the distal dendrites of wild-type Purkinje cells are innervated by parallel fibers (Kurihara et al., 1997; Lalouette et al., 2001). This reduction in the number of innervated spines is in line with the observation that parallel fiber excitatory post-synaptic current (EPSC)-amplitude in knock-out mouse Purkinje cells is significantly smaller than those in wild-type Purkinje cells (Kurihara et al., 1997). In addition, in the parallel fiber-innervated spines, the length of the postsynaptic density is sometimes longer than the juxtaposed presynaptic active zone in the *hotfoot* and in the knock-out mice. These results suggest that GluRδ2 is a key postsynaptic molecule involved in stabilization and strengthening of synaptic connectivity between parallel fibers and Purkinje cells, leading to the formation of a functionally mature synapse.

The impaired parallel fiber-Purkinje cell synaptogenesis in the *hotfoot* and knock-out mice could explain another abnormality in their cerebellar development, that is the persistence of multiple climbing fiber innervation to adulthood (Kashiwabuchi et al., 1995). In addition, a serial electron microscopy demonstrated that in the knock-out mouse, climbing fibers extend distally to the spiny branchlets, 'invading' the parallel fiber territory, where nearly half of the spines are free of innervation (Ichikawa et al., 2002). Therefore it has been suggested that the presence of GluRδ2 may not only induce or stabilize parallel fiber synapses, but also restrict the climbing fiber innervation to the Purkinje cell proximal dendritic domain.

The role of the GluRδ2 in the heterologous afferent fiber competition

The experimental block of electrical activity revealed the involvement of the GluRδ2 in the competition between climbing and parallel fibers in acquiring and maintaining their postsynaptic domain on the Purkinje cells (Morando et al., 2001). After TTX, all proximal dendritic spines express this subunit but, surprisingly, not only those ectopically contacted by parallel fibers, but also those innervated by GABAergic neurons and those still in contact with the climbing fibers before their withdrawal (Fig. 2).

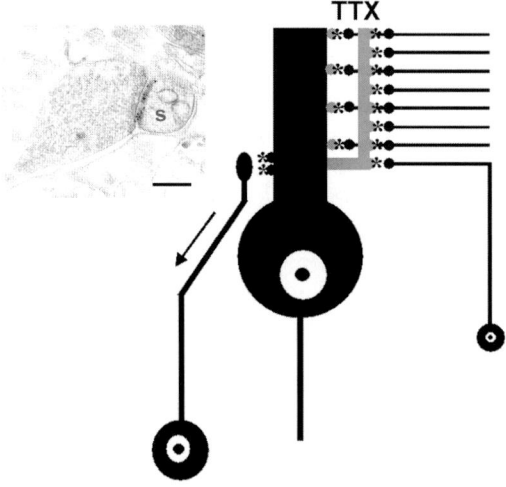

Fig. 2. Changes induced in the Purkinje cell by blocking electrical activity. New spines emerge from the proximal dendrite and are innervated by the parallel fibers. Note the appearance of GluRδs not only in these spines but also in those still contacted by the climbing fiber before its withdrawal. On the left: electronmicrograph of a climbing fiber-innervated spine (s) expressing the GluRδ2s (black dots). Scale bar, 0.25 μm.

Thus, as for the acetylcholine receptor clustering in neuromuscular synapse differentiation (Yang et al., 2001), we suggest that GluRδ2 expression is an intrinsic activity-independent property of all Purkinje cell spines, and independent from the afferent type. Moreover, since the GluRδ2s are involved in stabilizing and strengthening parallel fiber–Purkinje cell synapses (Kurihara et al., 1997), the competitive advantage of the parallel fibers over the climbing fibers in the absence of activity is attributable to the expression of these subunits by the Purkinje cell proximal dendrite that become receptive to the competitive input. Therefore, the expression of the GluRδ2 in the inactive climbing fiber synapses suggests that, normally, climbing fiber activity prevents the subunit targeting to proximal dendrites and that the appearance of this subunit in the postsynaptic density of the spine may cause withdrawal of the silent terminal afferents.

This possibility has been verified by other experiments aimed at observing the distribution of the GluRδ2 during the recovery period after electrical block, and during the sprouting of climbing fibers that follows a subtotal lesion of the inferior olive by 3-AP (Cesa et al., 2003). Forty-five days after removal of the TTX infusion there is a significant increase in the number of climbing fiber-innervated spines expressing GluRδ2s while the GluRδ2 density significantly decreases relative to the condition under block. After a recovery period of 135 days, the whole picture is similar to the control. An interpretation of this finding is that, along with the process of recruiting new spines, the growing climbing fiber exerts a repressive action on the targeting of these receptors to those spines with which it establishes synaptic contacts (Cesa et al., 2003). In the experimental model consisting of subtotal lesion of the inferior olive by 3-AP, it has been demonstrated that up to 90 days, when the extension of surviving climbing fibers reaches a peak (Benedetti et al., 1983; Rossi et al., 1991a,b; Strata and Rossi, 1998), almost all climbing fiber-innervated spines bear GluRδ2 (Fig. 3). The expression of GluRδ2s in these synapses might be due to an abnormal activity of the surviving inferior olive neurons. The 3-AP interferes with the oxidative metabolism and it is possible that surviving neurons undergo a period of decreased or absent activity.

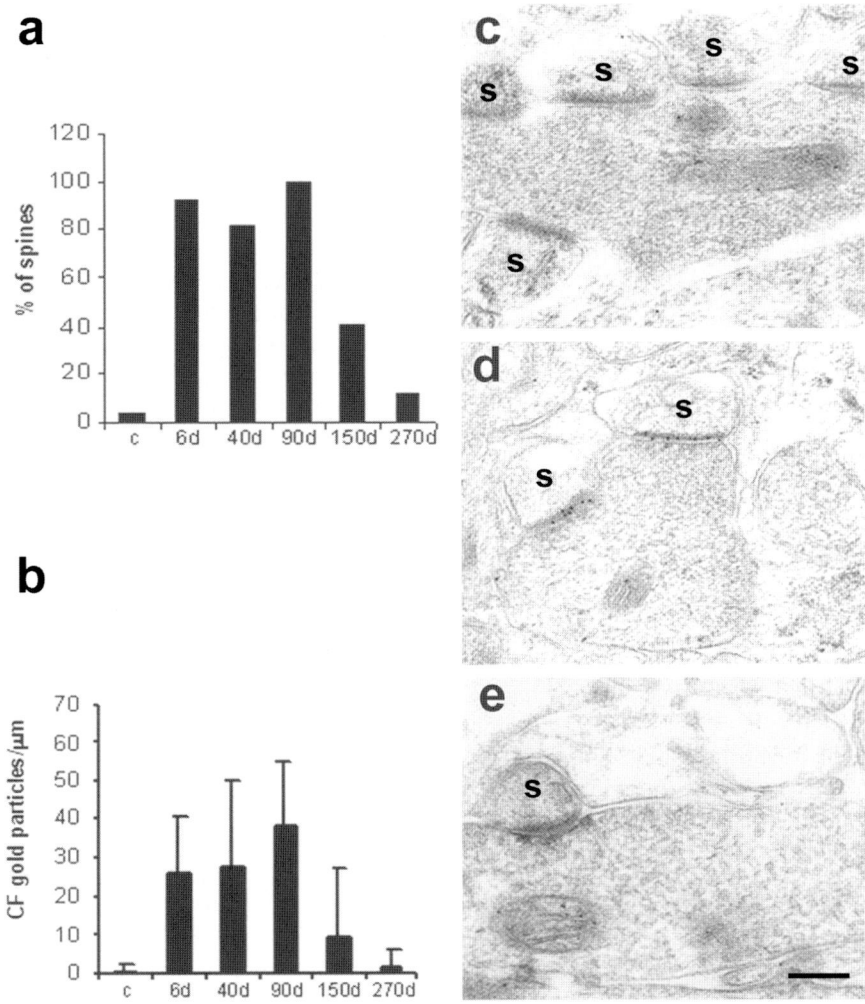

Fig. 3. Quantitative evaluation of proximal dendritic spines innervated by climbing fiber (CF) and their GluRδ2 expression in rats with inferior olive lesion at different periods in days (d). (a) Percentage of climbing fiber-innervated spines bearing GluRδ2. (b) Number of gold particles per micrometers of postsynaptic density length in climbing fiber synapses. The percent value and the gold particle density increase remarkably up to 90 days after the lesion, then recover at 150 days and are similar to control (c) at 270 days. Climbing fiber-innervated spines (s) in control (c), at 6 (d), and 270 (e) days after the lesion. A high density of GluRδ2s is present in (d) and not in (c) and (e). Scale bar, 0.25 μm.

Therefore, a likely reduction of the repressive action of climbing fibers on the GluRδ2 expression may explain the significant increase of the receptor density. This hypothesis is supported by the observation of spines, with a high GluRδ2 density, contacted by many dark degenerating climbing fiber terminals six days after the lesion. Only after 270 days from the lesion the incidence of climbing fiber synapses expressing the GluRδ2 is close to normal and the density is reduced to a value not significantly different from controls (Fig. 3) (Cesa et al., 2003). This fact strongly supports the results of the TTX experiments in which new climbing fiber-innervated spines express the GluRδ2s and, in addition, support the view that these subunits are not responsible for displacing silent climbing fibers. As during development, GluRδ2 is present in the climbing fiber–Purkinje cell synapses during early stages of climbing fiber regeneration or sprouting and it is repressed during the process of synapse maturation.

Conclusions

In the mature cerebellar cortex, a profound alteration of connectivity occurs by blocking electrical activity. In the Purkinje cells, spinogenesis is under a double control. An activity-independent, probably intrinsic, mechanism (Sotelo et al., 1975; Sotelo, 1978) promotes spine growth over the whole dendritic territory while an activity-dependent spine pruning is exerted by only one input, the climbing fiber, specifically at the proximal dendrites (Bravin et al., 1999). Thus, different excitatory afferents operate differently on the same neuron to create a phenotypic profile, which is aimed at optimizing their specific local functions. The low spine density of the Purkinje-cell proximal dendrites is the result of the repressing action of the active climbing fiber. This effect is aimed at inducing large interspine intervals to accommodate a high number of voltage dependent Ca^{2+} channels. This allows the generation of a large all-or-none current in the cell that is typical of the climbing fiber response. In contrast, a high spine density in the branchlets is necessary to discriminate the high number of parallel fibers (Strata et al., 2000). Such an activity-dependent mechanism is different from the synaptic plasticity changes, which occur locally during long-term potentiation and long-term depression (Yuste and Bonhoeffer, 2001), which is related to the mechanisms of learning and memory and of adaptive modifications.

During development and in the mature cerebellum in the absence of activity, all spines bear GluRδ2. Therefore, expression of this subunit appears to be an intrinsic property of the Purkinje cells (Morando et al., 2001) and it is important for the maturation and stabilization of the parallel fiber synapses (Kashiwabuchi et al., 1995; Kurihara et al., 1997). Active climbing fibers are able to innervate GluRδ2-bearing spines, limited to their dendritic domain and, after achieving the mature state, to downregulate this subunit in those spines. In addition, they displace the competitor afferents, the parallel fibers, to the distal dendritic territory (Cesa et al., 2003). By this mechanism the active climbing fiber establishes in its dendritic domain the characteristic synaptic profile that consists of sparse clusters of spine innervated by a single afferent axon.

Indeed, based on the remarkable activity-dependent competition between the two excitatory synaptic inputs to the Purkinje cells, we propose the following principle in building the architecture of the cerebellar cortex. In a simple system, like the neuromuscular junction, a target cell, the muscle fiber, is provided with intrinsic cues to allow the incoming fiber, the motoneuron, to form a specific synaptic organization. The Purkinje cell, with its dual excitatory input, represents a simple central system, where the climbing fiber synaptic organization is reminiscent of a neuromuscular system. However, the addition of a further afferent population requires competition between the two inputs. The intrinsic development of spines with GluRδ2 suggests that the Purkinje cell has the intrinsic property to be innervated by the parallel fibers. However, the additional input, made by the climbing fiber, has the ability to inhibit the competitor in order to achieve its own territory on the proximal dendrite. Such an inhibition is activity-dependent and the activity needs to be present in order to allow the climbing fiber to maintain its territory, even when the developmental period is over.

Abbreviations

3-AP	3-acetylpyridine
TTX	tetrodotoxin
GluRδ2	glutamate receptor δ2 subunit

References

Araki, K., Meguro, H., Kushiya, E., Takayama, C., Inoue, Y. and Mishina, M. (1993) Selective expression of the glutamate receptor channel delta 2 subunit in cerebellar Purkinje cells. Biochem. Biophys. Res. Commun., 197: 1267–1276.

Armengol, J.A., Sotelo, C., Angaut, P. and Alvarado-Mallart, R.M. (1989) Organization of host afferents to cerebellar grafts implanted into kainate lesioned cerebellum in adult rats. Eur. J. Neurosci., 1: 75–93.

Benedetti, F. and Strata, P. (1989) Functional synaptogenesis in the cerebellar cortex after inferior olive lesion. In: Bonavita, V. and Piccoli, F. (Eds.), Biological Aspects of Neuron Activity. Fidia Biomedical Information, Padova, pp. 97–107.

Benedetti, F., Montarolo, P.G., Strata, P. and Tosi, L. (1983) Collateral reinnervation in the olivocerebellar pathway in the rat. Birth Defects Orig. Artic. Ser., 19: 461–464.

Bravin, M., Rossi, F. and Strata, P. (1995) Different climbing fibres innervate separate dendritic regions of the same Purkinje cell in hypogranular cerebellum. J. Comp. Neurol., 357: 395–407.

Bravin, M., Morando, L., Vercelli, A., Rossi, F. and Strata, P. (1999) Control of spine formation by electrical activity in the adult cerebellum. Proc. Natl. Acad. Sci. USA, 96: 1704–1709.

Buffo, A., Fronte, M., Oestreicher, A.B. and Rossi, F. (1998) Degenerative phenomena and reactive modifications of the adult rat inferior olivary neurons following axotomy and disconnection from their targets. Neuroscience, 85: 587–604.

Buonomano, D.V. and Merzenich, M.M. (1998) Cortical plasticity: from synapses to maps. Annu. Rev. Neurosci., 21: 149–186.

Cesa, R., Morando, L. and Strata, P. (2003) Glutamate receptor $\delta 2$ subunit in activity dependent heterologous synaptic competition. J. Neurosci., 23: 2363–2370.

Chen, S. and Hillman, D.E. (1982) Marked reorganization of Purkinje cell dendrites and spines in adult rat following vacating of synapses due to deafferentation. Brain Res., 245: 131–135.

Crepel, F. (1982) Regression of functional synapses in the immature mammalian cerebellum. Trends Neurosci., 266–269.

Crepel, F., Mariani, J. and Delhaye-Bouchaud, N. (1976) Evidence for a multiple innervation of Purkinje cells by climbing fibers in the immature rat cerebellum. J. Neurobiol., 7: 567–578.

Crepel, F., Delhaye-Bouchaud, N., Guastavino, J.M. and Sampaio, I. (1980) Multiple innervation of cerebellar Purkinje cells by climbing fibres in staggerer mutant mouse. Nature, 283: 483–484.

Crepel, F., Delhaye-Bouchaud, N. and Dupont, J.L. (1981) Fate of the multiple innervation of cerebellar Purkinje cells by climbing fibers in immature control, x-irradiated and hypothyroid rats. Dev. Brain Res., 1: 59–71.

Desclin, J.C. and Escubi, J. (1974) Effects of 3-acetylpyridine on the central nervous system of the rat, as demonstrated by silver methods. Brain Res., 77: 349–364.

Engert, F. and Bonhoeffer, T. (1999) Dendritic spine changes associated with hippocampal long-term synaptic plasticity. Nature, 399: 66–70.

Goodman, C.S. and Shatz, C.J. (1993) Developmental mechanisms that generate precise patterns of neuronal connectivity. Cell, 72: 77–98.

Guastavino, J.M., Sotelo, C. and Damez-Kinselle, I. (1990) Hot-foot murine mutation: behavioral effects and neuroanatomical alterations. Brain Res., 523: 199–210.

Ichikawa, R., Miyazaki, T., Kano, M., Hashikawa, T., Tatsumi, H., Sakimura, K., Mishina, M., Inoue, Y. and Watanabe, M. (2002) Distal extension of climbing fiber territory and multiple innervation caused by aberrant wiring to adjacent spiny branchlets in cerebellar Purkinje cells lacking glutamate receptor $\delta 2$. J. Neurosci., 22: 8487–8503.

Kaas, J.H. (2002) Sensory loss and cortical reorganization in mature primates. Prog. Brain Res., 138: 167–176.

Kashiwabuchi, N., Ikeda, K., Araki, K., Hirano, T., Shibuki, K., Takayama, C., Inoue, Y., Kutsuwada, T., Yagi, T., Kang, Y., Aizawa, S. and Mishina, M. (1995) Impairment of motor coordination, Purkinje cell synapse formation, and cerebellar long-term depression in GluR$\delta 2$ mutant mice. Cell, 81: 245–252.

Knott, G.W., Quairiaux, C., Genoud, C. and Welker, E. (2002) Formation of dendritic spines with GABAergic synapses induced by whisker stimulation in adult mice. Neuron, 34: 265–273.

Kruger, L., Bendotti, C., Rivolta, R. and Samanin, R. (1993) Distribution of GAP-43 mRNA in the adult rat brain. J. Comp. Neurol., 333: 417–434.

Kurihara, H., Hashimoto, K., Kano, M., Takayama, C., Sakimura, K., Mishina, M., Inoue, Y. and Watanabe, M. (1997) Impaired parallel fiber → Purkinje cell synapse stabilization during cerebellar development of mutant mice lacking the glutamate receptor $\delta 2$ subunit. J. Neurosci., 17: 9613–9623.

Lalouette, A., Lohof, A., Sotelo, C., Guenet, J. and Mariani, J. (2001) Neurobiological effects of a null mutation depend on genetic context: comparison between two hotfoot alleles of the delta-2 ionotropic glutamate receptor. Neuroscience, 105: 443–455.

Landsend, A.S., Amiry-Moghaddam, M., Matsubara, A., Bergersen, L., Usami, S., Wenthold, R.J. and Ottersen, O.P. (1997) Differential localization of delta glutamate receptors in the rat cerebellum: coexpression with AMPA receptors in parallel fiber-spine synapses and absence from climbing fiber-spine synapses. J. Neurosci., 17: 834–842.

Lomeli, H., Sprengel, R., Laurie, D.J., Kohr, G., Herb, A., Seeburg, P.H. and Wisden, W. (1993) The rat delta-1 and delta-2 subunits extend the excitatory amino acid receptor family. FEBS Lett., 315: 318–322.

Mariani, J., Benoit, P., Hoang, D.M., Thompson, M. and Delhaye-Bouchaud, N. (1990) Extent of multiple innervation of Purkinje cells by climbing fibers in adult X-irradiated rats. Comparison of different schedules of irradiation during the first postnatal week. Dev. Brain Res., 57: 63–70.

McKinney, R.A., Capogna, M., Durr, R., Gähwiler, B.H. and Thompson, S.M. (1999) Miniature synaptic events maintain dendritic spines via AMPA receptor activation. Nat. Neurosci., 2: 44–49.

McNamara, R.K. and Lenox, R.H. (1997) Comparative distribution of myristoylated alanine-rich C kinase substrate (MARCKS) and F1/GAP-43 gene expression in the adult rat brain. J. Comp. Neurol., 379: 48–71.

Montarolo, P.G., Palestini, M. and Strata, P. (1982) The inhibitory effect of the olivocerebellar input on the Purkinje cell in the rat. J. Physiol., 332: 187–202.

Morando, L., Cesa, R., Rasetti, R., Harvey, R. and Strata, P. (2001) Role of glutamate δ-2 receptors in activity-dependent competition between heterologous afferent fibers. Proc. Natl. Acad. Sci. USA, 98: 9954–9959.

Nimchinsky, E.A., Sabatini, B.L. and Svoboda, K. (2002) Structure and function of dendritic spines. Annu. Rev. Physiol., 64: 313–353.

Popov, V.I. and Bocharova, L.S. (1992) Hibernation-induced structural changes in synaptic contacts between mossy fibres and hippocampal pyramidal neurons. Neuroscience, 48: 53–62.

Rossi, F. and Strata, P. (1995) Reciprocal trophic interactions in the adult climbing fibre-Purkinje cell system. Prog. Neurobiol., 47: 341–369.

Rossi, F., van der Want, J.J., Wiklund, L. and Strata, P. (1991a) Reinnervation of cerebellar Purkinje cells by climbing fibres surviving a subtotal lesion of the inferior olive in the adult rat. II. Synaptic organization on reinnervated Purkinje cells. J. Comp. Neurol., 308: 536–554.

Rossi, F., Wiklund, L., van der Want, J.J. and Strata, P. (1991b) Reinnervation of cerebellar Purkinje cells by climbing fibres surviving a subtotal lesion of the inferior olive in the adult rat. I. Development of new collateral branches and terminal plexuses. J. Comp. Neurol., 308: 513–535.

Rossi, F., Borsello, T., Vaudano, E. and Strata, P. (1993) Regressive modifications of climbing fibres following Purkinje cell degeneration in the cerebellar cortex of the adult rat. Neuroscience, 53: 759–778.

Rossi, F., Borsello, T. and Strata, P. (1994) Embryonic Purkinje cells grafted on the surface of the adult uninjured rat cerebellum migrate in the host parenchyma and induce sprouting of intact climbing fibres. Eur. J. Neurosci., 6: 121–136.

Rossi, F., Jankovski, A. and Sotelo, C. (1995) Target neuron controls the integrity of afferent axon phenotype: a study on the Purkinje cell-climbing fiber system in cerebellar mutant mice. J. Neurosci., 15: 2040–2056.

Segal, M. (2001) Rapid plasticity of dendritic spine: hints to possible functions? Prog. Neurobiol., 63: 61–70.

Sotelo, C. (1978) Purkinje cell ontogeny: formation and maintenance of spines. Progr. Brain Res., 48: 149–170.

Sotelo, C. (1990) Cerebellar synaptogenesis: what we can learn from mutant mice. J. Exp. Biol., 153: 225–249.

Sotelo, C. and Alvarado-Mallart, R.M. (1987) Reconstruction of the defective cerebellar circuitry in adult Purkinje cell degeneration mutant mice by Purkinje cell replacement through transplantation of solid embryonic implants. Neuroscience, 20: 1–22.

Sotelo, C., Hillman, D.E., Zamora, A.J. and Llinás, R. (1975) Climbing fiber deafferentation: its action on Purkinje cell dendritic spines. Brain Res., 98: 574–581.

Strata, P. (2002) Dendritic spines in Purkinje cells. Cerebellum, 1: 230–232.

Strata, P. and Rossi, F. (1998) Plasticity of the olivocerebellar pathway. Trends Neurosci., 21: 407–413.

Strata, P., Morando, L., Bravin, M. and Rossi, F. (2000) Dendritic spine density in Purkinje cells. Trends Neurosci., 23: 198.

Takayama, C., Nakagawa, S., Watanabe, M., Mishina, M. and Inoue, Y. (1996) Developmental changes in expression and distribution of the glutamate receptor channel delta2 subunit according to the Purkinje cell maturation. Brain Res. Dev. Brain Res., 92: 147–155.

Yang, X., Arber, S., William, C., Li, L., Tanabe, Y., Jessell, T.M., Birchmeier, C. and Burden, S.J. (2001) Patterning of muscle acetylcholine receptor gene expression in the absence of motor innervation. Neuron, 30: 399–410.

Yuste, R. and Bonhoeffer, T. (2001) Morphological changes in dendritic spines associated with long-term synaptic plasticity. Annu. Rev. Neurosci., 24: 1071–1089.

Yuzaki, M. (2003a) The delta2 glutamate receptor: 10 years later. Neurosci. Res., 46: 11–22.

Yuzaki, M. (2003b) New insights into the structure and function of glutamate receptors: the orphan receptor delta2 reveals its family's secrets. Keio J. Med., 52: 92–99.

Zhao, H.M., Wenthold, R.J. and Petralia, R.S. (1998) Glutamate receptor targeting to synaptic populations on Purkinje cells is developmentally regulated. J. Neurosci., 18: 5517–5528.

CHAPTER 6

Fate restriction and developmental potential of cerebellar progenitors. Transplantation studies in the developing CNS

Piercesare Grimaldi[1], Barbara Carletti[1], Lorenzo Magrassi[2] and Ferdinando Rossi[1],*

[1]Department of Neuroscience and 'Rita Levi Montalcini Centre for Brain Repair', University of Turin, Corso Raffaello 30, I-10125 Turin, Italy
[2]Neurosurgery, Department of Surgery, IRCCS Policlinico S. Matteo, University of Pavia, Italy

Keywords: cell specification; commitment; in utero transplantation; differentiation; neural progenitor; cerebellar development; neural graft

Abstract: The generation of cell diversity from undifferentiated progenitors is regulated by interdependent mechanisms, including cell intrinsic programs and environmental cues. This interaction can be investigated by means of heterochronic/heterotopic transplantation, which allows to examine the behaviour of precursor cells in an ususual environment. The cerebellum provides an ideal model to study cell specification, because its neurons originate according to a well-defined timetable and they can be are readily recognised by morphological features and specific markers. Cerebellar progenitors transplanted to the embryonic cerebellum develop fully mature cerebellar neurons, which often integrate in the host circuitry in a highly specific manner. In extracerebellar locations, cerebellar progenitors preferentially settle in caudal CNS regions where they exclusively acquire cerebellar identities. By contrast, neocortical precursors preferentially settle in rostral regions and fail to develop hindbrain phenotypes. The phenotypic repertoire generated by transplanted cerebellar progenitors is strictly dependent on their age. Embryonic progenitors originate all mature cerebellar cells, whereas postnatal ones exclusively generate later-born types, such as molecular layer interneurons and granule cells. Together, these observations foster the hypothesis that neural progenitors are first specified towards region-specific phenotypes along the rostro-caudal axis of the neural tube. Thereafter, the developmental potential of progenitor cells is progressively restricted towrds later generated types. Such a progressive specification of precursor cells in space and time is stably transmitted to their progeny and it cannot be modified by local cues, when these cells are confronted with heterotopic and/or heterochronic environments.

Introduction

One of the most fascinating topics in developmental biology is the generation of cellular diversity. How is it possible that all the different cell types in an adult organism originate from a single fertilized cell? This problem is particularly challenging in the vertebrate brain. The wide variety of neuronal and glial phenotypes that populate the adult nervous system originate from a seemingly homogeneous population of proliferating progenitor cells. So, what factors are responsible for specification and differentiation of neural cells?

When studying specification, one must consider two different kinds of information, i.e., which act

*Corresponding author. Tel.: +39/011/670 7705; Fax: +39/011/670 7708; E-mail: ferdinando.rossi@unito.it

synergistically to assign cell identity: cell-intrinsic programs, determined by cell lineage, and extrinsic signals, provided by the surrounding milieu. The latter include a variety of molecules such as morphogens, trophic factors and inductors, which are best studied in simple systems, such as drosophila, or at the very early stages of vertebrate neural development (see Ephrussi and St Johnston, 2004). Molecules such as bone morphogenetic proteins or sonic hedgehog and others induce the subdivision of the neural tube into distinct domains that will give rise to the different regions of the mature CNS (Lee and Jessell, 1999; Wilson and Rubenstein, 2000). In this way, extracellular signals act on the immature progenitors inducing or selecting defined phenotypic repertoires. Nevertheless, the acquisition of mature identities is not exclusively determined by extrinsic instructions, but also by the intrinsic competence of the progenitor cells, which, in turn, depends on their past history.

In general, early embryonic progenitors have a broader potential and may generate many different phenotypes. Subsequently, as development proceeds, this potential narrows and cells get specified towards their final mature identity (McConnell, 1995). Alternatively, as shown for retinal development, progenitors are multipotent until the last mitotic division, and they go through waves of competence, during which they can only generate certain phenotypes (Cepko et al., 1996).

One of the most commonly used approaches to dissect the relative contribution of cell-autonomous mechanisms and extrinsic cues is heterochronic and/or heterotopic transplantation of progenitor cells. By this approach, small amounts of dissociated donor cells are exposed to the host environment, allowing one to evaluate their actual competence in responding to chronological or positional foreign cues. In this context, the pioneering experiments by Barbe and Levitt (1991) have shown that E12 precursors displaced from limbic to somato-motor cortex or vice-versa can switch the expression of the limbic system-associated protein in a host-specific manner. This plasticity is no longer possible for E14–17 cells, which maintain the original donor phenotype in the new heterotopic location. Similarly, McConnell and colleagues demonstrated that layer specification of neocortical progenitors occurs at the time of the last mitotic division (McConnell and Kaznowski, 1991).

Early proliferating progenitors, destined to populate inner cortical layers, can change their fate and settle into outer layers when grafted into older hosts. Nevertheless, in the reverse experiment late proliferating cells placed into a younger environment fail to incorporate into the earlier-generated inner layers (Desai and McConnell, 2000), showing that the progenitor capability for integrating into different layers is progressively restricted during cortical ontogenesis.

Together, these observations indicate that the sequence of generation of the different phenotypes that populate a given brain region is regulated by a progressive narrowing of progenitors' potential. However, later experiments, in which different populations of precursor cells were heterotopically/heterochronically transplanted into the embryonic brain in utero, yielded somewhat conflicting results (Brüstle et al., 1995; Campbell et al., 1995; Fishell, 1995; Olsson et al., 1998). Indeed, these studies showed that donor cells can incorporate into wide regions of the host CNS, and often acquire site-specific identities. This suggested that, at least to a certain extent, progenitor cells may retain a wide developmental potential, being capable of responding to heterotopic and/or heterochronic signals.

The cerebellum as a model to investigate the generation of phenotypic diversity in the CNS

The cerebellum represents the most favorable ground to investigate the relative contribution of cell-autonomous mechanisms and extrinsic signaling in the specification and differentiation of neural progenitors. The adult cerebellum comprises a limited number of neuronal phenotypes, characterized by distinctive morphological features and expression of type-specific markers. In addition, the different cell populations have well-established reciprocal interactions and a highly specific distribution within the cerebellar architecture (Palay and Chan-Palay, 1974; Ito, 1984). As a consequence, transplanted cerebellar neurons can be securely identified even in ectopic locations (e.g., De Camilli et al., 1984; Rossi and Borsello, 1993), and it is also possible to determine whether they can be specifically integrated into the recipient network (Sotelo and Alvarado-Mallart, 1991).

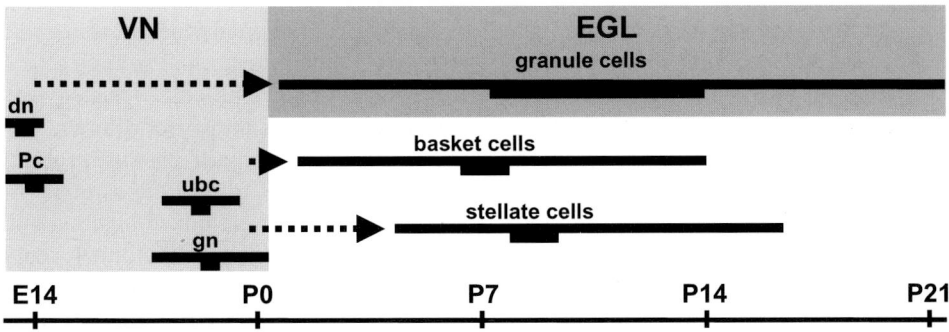

Fig. 1. Sequential generation of neuronal phenotypes during cerebellar development. The different types of neurons that populate the adult cerebellum are generated according to a precise time-sequence by two distinct germinative layers. During embryonic life, the ventricular neuroepithelium (VN, light gray) generates deep nuclei neurons (dn), Purkinje cells (Pc), and granular layer interneurons, such as Golgi (gn) and unipolar brush cells (ubc). Proliferating precursors from this neuroepithelium migrate in the cerebellar parenchyma during early postnatal life and generate basket and stellate neurons. Progenitors from a restricted region of the ventricular neuroepithelium, the rhombic lip, are committed to a granule cell fate during early embryonic development. These cells migrate over the cortical surface, where they form a secondary germinative layer, the external granular layer (EGL, dark gray), which will generate granule cells during the first three postnatal weeks. The bars indicate the birth period of the different phenotypes (double bars represent the peak generation time). Neuron birthdates, which refer to rat development, have been derived from Altman and Bayer (1997).

Most importantly, during development all cerebellar neurons and glia originate from two different neuroepithelia according to a well-defined spatio-temporal schedule (Fig. 1; Ramón y Cajal, 1911; Miale and Sidman, 1961; Altman and Bayer, 1997). The cerebellar ventricular zone that proliferates during embryonic life generates deep nuclear neurons, Purkinje cells and granular layer interneurons (e.g., Golgi neurons or unipolar brush cells) according to a precise time-sequence (Altman and Bayer, 1997). Its proliferative activity ceases after birth, but dividing progenitor cells emigrate into the cerebellar parenchyma, where they remain during the first three postnatal weeks to generate molecular layer interneurons (basket and stellate cells) and glia (Zhang and Goldman, 1996a,b). The other cerebellar neuroepithelium, the external granular layer (EGL), is formed by a well-defined subset of progenitor cells from the rhombic lip, which migrate to cover the entire cortical surface and proliferate up to the third postnatal week (Altman and Bayer, 1997). In rodents, these progenitors appear to be strictly committed to a granule cell fate from early developmental stages (Gao and Hatten, 1994; Alder et al., 1996). This well-defined sequence of phenotype generation, together with the time mismatch between the proliferative periods of the two germinative neuroepithelia, offer a unique opportunity to investigate the mechanisms of progenitor cell specification, by confronting cerebellar precursors with heterotopic and/or heterochronic environments.

The initial experiments aimed at elucidating the developmental potential of cerebellar precursors were primarily focused on EGL cells. In this context, Renfranz et al. (1991) showed that oncogene-immortalized cells, derived from the embryonic hippocampus, acquired host-specific morphological phenotypes following transplantation to the neonatal hippocampus or cerebellum. Similar results were obtained with another cell line derived from EGL progenitors (Snyder et al., 1992): when implanted in the postnatal cerebellum, these cells developed a variety of cell types, including granule cells, molecular layer interneurons, astrocytes and oligodendrocytes. Together, these observations suggested that immortalized precursors are highly plastic and take their fate decisions according to local instructive cues available in their integration site.

A subsequent report suggested that this concept might be extended to primary progenitors, freshly dissociated from the donor brain (Vicario-Abejón et al., 1995). Indeed, EGL cells grafted to the neonatal hippocampus appeared to switch their fate to granule neurons of the dentate gyrus, recognized by calbindin expression or c-fos activation after a seizure-inducing treatment. This result, however, could not be confirmed in another transplantation paradigm (Alder et al., 1996), and direct comparison of primary and immortalized EGL precursors revealed very different developmental potentialities (Gao and Hatten, 1994). Following implantation to the neonatal cerebellum, immortalized cells produced a variety of neuronal and glial types, whereas primary progenitors exclusively generated granule cells, indicating that immortalization procedures induce profound alterations in the commitment of precursor cells and in their sensitivity to extrinsic instructive cues.

These conflicting results left several unanswered questions about the actual developmental potential of cerebellar progenitors. Namely, since most experiments were focused on the fate of EGL progenitors grafted to the neonatal brain, it was not clear whether late proliferating precursors exposed to the heterochronic environment of the embryonic cerebellar primordium can be induced to differentiate into earlier generated phenotypes. Furthermore, the real capability of cerebellar progenitors to integrate in extracerebellar CNS regions and to adopt local phenotypes also remained to be established. Over the last few years, we addressed these issues using different paradigms of heterochronic and/or heterotopic transplantation of genetically tagged cerebellar progenitors. The results of these experiments will be reviewed hereafter. We will first examine the phenotypic choices and integration of cerebellar precursors transplanted to the cerebellar primordium and, then, describe their fate in foreign regions of the embryonic CNS.

Fate of cerebellar progenitors following heterochronic transplantation to the embryonic cerebellum

It is crucial to understand the mechanisms underlying the sequential generation of neural phenotypes. Elucidating whether progenitor cells remain sensitive to heterochronic signaling or whether they progressively lose the ability for responding to the instructive cues active during earlier ontogenetic phases (Barbe and Levitt, 1991; McConnell, 1995; Desai and McConnell, 2000). In our cerebellar model, this issue can be addressed by asking whether postnatal precursors, destined to generate granule cells and molecular layer interneurons, can switch to earlier-generated identities, e.g., Purkinje cells or deep nuclei neurons, when exposed to the environment of the embryonic cerebellar primordium.

In a first approach to this problem, morsels of cerebellar tissue from P4 postnatal mice and E12 mouse embryos were mixed together and cografted to the cerebellum or to the fourth ventricle of adult syngenic hosts (Jankovski et al., 1996). To allow their identification within the grafts, postnatal donor cells were isolated from several lines of transgenic mice in which various populations of cerebellar neurons expressed the *LacZ* reporter gene. Both in the ventricles and in the intraparenchymal locations, the transplants developed according to the time course of the E12 primordium and formed typical trilayered cortical folia and deep nuclei, in all similar to those observed with solid grafts of embryonic cerebellar tissue (see Sotelo and Alvarado-Mallart, 1991; Rossi et al., 1992).

Postnatal cells appeared well integrated in these 'mixture' transplants. During the first few days after implantation, they proliferated abundantly, being frequently incorporated in germinal regions such as the EGL of cortical folia. At later stages, they generated fully mature cell types dispersed throughout the graft texture. On the whole, the transplant developed according to the ontogenetic rhythm of the embryonic tissue and postnatal cells appeared to be incorporated in this framework. Nonetheless, the analysis of the mature phenotypes, carried out by ultrastructural examination or expression of cell-specific markers, showed that these precursors exclusively generated granule cells and molecular layer interneurons. These observations thus indicated that postnatal cerebellar progenitors are strictly committed to late-generated phenotypes and they are not responsive to embryonic neurogenic cues.

Although the results of these experiments were clear-cut, the possibility remained that postnatal precursors, which were about half of the donor cells in the 'mixture' grafts, were not fully exposed to the embryonic environment. Thus, we readdressed this issue by comparing the phenotypic repertoires generated by embryonic (E12) or postnatal (P4) mouse cerebellar progenitors after in utero transplantation (Carletti et al., 2002). By this method, small amounts of dissociated cells can be grafted to the embryonic brain, so that individual donor elements are fully embedded into the recipient milieu and effectively exposed to local signals (Cattaneo et al., 1994; Magrassi et al., 1998). In our experiments, single cell suspensions, obtained from transgenic mice expressing the Enhanced Green Fluorescent Protein (EGFP) under control of the β-actin promoter, were injected into the cerebral ventricles of E15 rat embryos. Live-born recipient animals were examined at different postnatal ages, from P7 to P40.

Quite surprisingly, the cerebellum was not a preferential integration site for EGFP-positive cells, which settled in wide regions of the recipient brain (see below). Cerebellar engraftment only occurred in about 25% of the cases with successful transplantation, and in many instances only a few cells were found, scattered throughout the cerebellar parenchyma. A similar low frequency of cerebellar incorporation has been also observed after in utero transplantation of other cell types (Lim et al., 1997; Yang et al., 2000), and it may be attributed to different concurrent factors related to the applied experimental procedure: (i) donor cells were injected into the forebrain ventricles, at a distance from the cerebellar primordium, (ii) at E15, the time of implantation, the size of the recipient cerebellum is still very small, thus reducing the probability that donor cells navigating through the ventricular system eventually incorporate in this region, (iii) species-specific differences between mouse donors and rat environment may also hinder the integration and/or survival of grafted cells. Indeed, in recent experiments, in which rat progenitors were implanted into rat embryos, EGFP-positive cells were observed in the cerebellum in 80% of the recipient animals (our unpublished observations). Whatever the case, the low rate of cerebellar engraftment actually allowed us to examine the fate of scattered donor cells fully exposed to the host environment.

The mature identities adopted by donor cells were assessed by morphological criteria, expression of population-specific markers and localization in the host cerebellar architecture. The phenotypic repertoires derived from E12 or P4 donor cells were clear-cut and in full accordance with the previous conclusions of Jankovski et al. (1996). Embryonic progenitors generated all major cerebellar neuron phenotypes, including Purkinje cells, deep nuclei neurons, cortical interneurons, and granule cells, but they failed to adopt glial identities. However, the relative amounts of the different cell types did not match those observed during cerebellar development in situ; Purkinje cells were the most represented type, whereas deep nuclei neurons or cortical interneurons were less frequent. In addition, granule cells were present only in half of the cases with successful cerebellar engraftment. On their part, postnatal progenitors exclusively generated granule cells, molecular layer interneurons and astrocytes. In all instances, granule neurons greatly outnumbered the other cell types, representing more that 95% of the whole donor cell population.

Together, these observations further corroborated the conclusion that postnatal progenitors are strictly committed to late-generated types and their fate cannot be changed by exposure to the embryonic milieu. However, because of technical constraints, we could not transplant into embryos younger than E15. Since deep nuclei neurons and Purkinje cells are generated at earlier ages (E12–E13 in the mouse, E14–E15 in the rat, Altman and Bayer, 1997) it may be possible that specific instruction to generate these phenotypes was no longer available in the host embryos. Hence, we designed an additional experimental approach in which embryonic or postnatal cerebellar cells were grafted to organotypic explants of brainstem-cerebellum derived from E12 mouse embryos. These explants, originally set up by Chédotal et al. (1997), can be maintained in vitro for several weeks, where they develop a peripherally located cortex surrounding the deep nuclei. Despite the rather high degree of maturation achieved by these organotypic cultures, they remain partially underdeveloped compared to the cerebellum in situ, with reduced amounts of granule cells and molecular layer interneurons (Carletti et al., 2002). Nevertheless, the cortico-nuclear circuitry appears well organized, indicating that these explants contain efficient neurogenic information to generate Purkinje cells and deep nuclei neurons.

The results of these in vitro transplantation experiments were in line with the previous observations: E12 donor cells generated all major cerebellar cell types, whereas their postnatal counterparts exclusively adopted granule cell or basket/stellate neuron identities. Thus, three alternative approaches consistently showed that embryonic and postnatal cerebellar progenitors have different developmental potentialities: cells derived from E12 mouse embryos are able to generate all major cerebellar phenotypes, whereas those from P4 donors are strictly committed to late-generated identities. In no case, can the fate of P4 cells be changed by exposure to a younger cerebellar environment, indicating that the precise sequence of generation of the different cerebellar phenotypes is regulated by the progressive restriction of the progenitor cell developmental potential.

Integration of donor cells in the host cerebellar architecture

In addition to survival and differentiation, another major issue of neural transplantation refers to the ability of donor cells to incorporate themselves in a specific manner into the recipient architecture and, more precisely, to establish appropriate connections with host neurons. Numerous reports about cell grafts in the adult CNS show that donor elements often achieve specific identities, whereas their effective integration into the recipient circuitries is much more difficult (Rossi and Cattaneo, 2002). In this context, one of the best examples is represented by the pioneering work by Constantino Sotelo and Rosa Magda Alvarado-Mallart, who showed that embryonic Purkinje cells which survive in the adult cerebellum, develop fully mature structural and functional features, but fail to achieve their correct position in the cerebellar cortex and, consequently, are unable to restore the cortico-nuclear projection (Sotelo and Alvarado-Mallart, 1987, 1991). Such a defective distribution of donor cells cannot be attributed to the tissue injury produced by intraparenchymal implantation, since the same outcome is observed when embryonic grafts are placed onto the surface of the intact cerebellum (Rossi et al., 1992, 1994). Rather, these observations suggest that the degree of donor cell engraftment primarily depends on properties of the recipient microenvironment.

Some interesting considerations about this problem can also be drawn from transplantation into the immature cerebellum. In the different experimental conditions described in the previous sections, the vast majority of E12 donors achieved their typical location in the cortical layers or in the deep nuclei and established specific interactions with neighboring elements, at least at the morphological level (Jankovski et al., 1996; Carletti et al., 2002). A good degree of integration may be still achieved in early postnatal hosts (Gao and Hatten, 1994; Gianola and Rossi, unpublished observations), suggesting that incorporation of donor cells into the host network primarily depends on specific environmental conditions that are no longer available in the adult cerebellum.

Nevertheless, transplantation of P4 cerebellar precursors indicates that the interaction between donor cells and recipient milieu might be more subtle. For instance, in the 'mixture' grafts (Jankovski et al., 1996), basket and stellate cells derived from postnatal progenitors were almost exclusively located in the internal granular layer of cortical folia. This was not due to a general defect of graft development, since the molecular layer of the same folia did contain a normal complement of interneurons derived from embryonic progenitors. Rather, this situation suggests that postnatal progenitors destined to generate basket and stellate cells differentiate before the molecular layer is formed. In other words, it is most likely that such late proliferating precursors are not able to adapt to the ontogenetic 'tempo' imposed by the embryonic tissue, but develop according to their inner rhythm.

The situation is slightly different for granule cells. In both mixture and in utero grafts, many such neurons are correctly positioned in the internal granular layer and send properly oriented parallel fibers in the molecular layer (Jankovski et al., 1996; Carletti et al., 2002). Nonetheless, ectopic granule cells can be often found in the deep nuclei region or in the surrounding white matter. The analysis of donor cells at different times posttransplantation suggests that granule cell precursors may be able to properly integrate in the mature cerebellar architecture, provided, they succeed in incorporating into the nascent EGL and then follow the typical inward migratory route. In contrast, the same progenitors that enter the host cerebellum through the ventricular surface are still able to survive and generate mature granule neurons, but fail to attain their natural cortical location. Thus, the correct integration of new elements in the recipient circuitry does not solely depend on the properties of the host environment, but also on the ability of donor cells to integrate into specific neurogenic niches and to adapt their intrinsic programs to the ontogenetic schedules imposed by the surrounding tissue.

Fate of cerebellar progenitors transplanted to heterotopic regions of the embryonic CNS

During CNS embryogenesis distinct domains of the neural tube become specified and proceed to develop different regions of the mature brain in a largely autonomous manner (Altmann and Brivanlou, 2001). Several lines of evidence indicate that cells belonging to such domains become committed to region-specific fates from early ontogenetic phases (Lumsden et al., 1994; McCarthy et al., 2001). Alternatively, however, it is possible that the phenotypic choices of multipotent progenitors are determined by a precise spatio-temporal arrangement of environmental instructive cues. A direct approach to this question consists in moving progenitors from one region of embryonic CNS to another and asking whether they can settle into the foreign site and acquire local identities. Several reports about such heterotopic transplantation experiments with different populations of neural precursors indicate that forebrain cells have a stronger ability to settle into foreign sites than their counterparts from more caudal regions (Olsson et al., 1998; Carletti et al., 2003). In addition, transdifferentiation of donor cells towards host-specific phenotypes frequently occurs when cells are displaced along the dorso-ventral axis of the neural tube, whereas it is more difficult along the rostro-caudal axis (Campbell et al., 1995; Na et al., 1998; Olsson et al., 1998; Carletti et al., 2003). Quite surprisingly, despite the large amount of work devoted to this issue, little attention has been paid to the behavior of cerebellar cells in heterotopic positions.

After implantation into the cerebral ventricles of rat embryos in utero, both embryonic and postnatal cerebellar progenitors showed a similar pattern of engraftment in the host brain. Namely, donor cells were found in many recipient regions, including lower brainstem, mesencephalon, and diencephalon. In addition, some grafted cells settled in the basal ganglia, but consistently failed to incorporate into the other telencephalic regions, such as neocortex, hippocampus, and olfactory bulb (Carletti et al., 2002). This peculiar distribution pattern is not due to species-specific differences between mouse donors and the host rat environment or due to a bias related to the implantation procedure. Indeed, rat cerebellar progenitors, grafted according to the same procedure, showed the same distribution pattern (our unpublished observations), whereas neocortical donors efficiently engrafted in all telencephalic sites (see below; Carletti et al., 2003). Thus, although cerebellar cells can engraft many foreign regions of the host CNS, they fail to penetrate and/or survive in rostral telencephalic domains.

Cerebellar progenitors that settled in heterotopic sites generated mature neurons and glia. However, the repertoire of mature types was fully consistent with a strict regional commitment and a progressive restriction of developmental potentialities. Indeed, in any ectopic location, donor cells invariably generated cerebellar phenotypes, which could be readily recognized for their morphology and expression of distinctive markers (see Carletti et al., 2002). Embryonic progenitors differentiated into all major categories of cerebellar neurons, whereas postnatal cells produced a vast majority of granule cells and a small amount of molecular layer interneurons. Nevertheless, while postnatal cells generated the same phenotypic repertoire in all heterotopic locations, embryonic donors showed a more variable differentiation pattern. Purkinje cells and deep nuclei were present in almost all engraftment sites, whereas granule cells were only found in a few defined sites, such as the dorsal cochlear nucleus.

To further characterize the properties of cerebellar progenitors and to ask whether they can be forced to integrate into the regions where they fail to engraft when implanted into the brain ventricles, we transplanted cerebellar progenitors to organotypic cultures of E15 mouse neocortex in vitro (Carletti et al., 2002). Also in this condition, postnatal donors differentiated into granule cells and basket/stellate neurons. In contrast, embryonic cells generated a peculiar phenotypic repertoire comprising almost exclusively of granule cells and Purkinje neurons.

On the whole, in line with the conclusions drawn for grafts to the cerebellar primordium (see above), these heterotopic transplantation experiments show that both embryonic and postnatal cerebellar cells are strictly committed to regional phenotypes and cannot change their fate after exposure to a foreign milieu. Their developmental potential is age-specific. Postnatal precursors are committed to late-generated identities and appear to differentiate according to intrinsic mechanisms irrespective of their location. In contrast, embryonic cells have a wider range of differentiative capabilities, but the actual assortment of mature cell types that they generate varies according to the integration site. In other words, if embryonic cells appear to have the potential for generating all cerebellar types, they are not already competent to produce all phenotypes in a cell-autonomous manner and require additional instructive information, particularly to adopt late-generated identities.

Fate of neocortical progenitors transplanted to heterotopic regions of the embryonic CNS

To ask whether the peculiar integration pattern and differentiative behavior of cerebellar cells was shared by other populations of neural progenitors, we performed additional in utero transplantation experiments using E12 neocortical cells as donors (Carletti et al., 2003). These telencephalic progenitors settled virtually in all regions of the recipient brain. However, they showed a peculiar pattern of integration and differentiation that changed according to the engraftment site. In different telencephalic regions these cells adopted a variety of local identities, such as pyramidal or stellate neurons in the neocortex, granule cells in the olfactory bulb, pyramidal neurons in the hippocampus, and medium-sized spiny neurons in the corpus striatum. In contrast, in more caudal regions, donor cells predominantly generated glial types, whereas the few neurons did not acquire any overt host-specific feature, nor showed any sign of integration into the host architecture. In addition, in extratelencephalic sites neocortical cells often formed large aggregates that remained sharply separated from the host parenchyma, being frequently attached to the ventricular walls or to the brain surface. Thus, the behavior of telencephalic cells somewhat reciprocates that of their cerebellar counterparts, showing a strong ability for integrating along the dorso-ventral axis of the neural tube, but essentially failing to engraft when displaced along different rostro-caudal domains.

Mechanisms of generation of cerebellar phenotypes

Transplantation experiments revealed a number of properties of cerebellar progenitors and also yielded some clues about the mechanisms that regulate the ordered generation of different cell populations during cerebellar development. Our observations show that mouse cerebellar cells are strictly committed to local fates at least from E12. A previous report indicates that heterotopically grafted E10.5 mid-hindbrain progenitors can switch to telencephalic phenotypes, but they fail to change their fate at E13 (Olsson et al., 1997). However, the differentiation of such donor cells was not followed longer than a few days after grafting and, hence, it remained to be established whether they were actually able to develop into mature forebrain neurons. Whatever the case, cerebellar cells appear to be regionally specified since early stages of cerebellar embryogenesis: they retain the ability for engrafting and surviving into wide regions of the host CNS, but exclusively generate cerebellar identities.

Despite such a strict commitment towards local fates, embryonic and postnatal progenitors show different developmental potentialities (Fig. 2). In both heterotopic and heterochronic conditions embryonic precursors appear to be able to generate all major types of cerebellar neurons, whereas postnatal progenitors only adopt late-generated identities. This indicates that the sequential generation of the different types of cerebellar neurons is regulated through a progressive restriction of progenitor cell developmental potential, similar to what occurs for the development of neocortical layers (Desai and McConnell, 2000). Nonetheless, the results of heterotopic transplants suggest that other mechanisms may also be involved in this process.

In all conditions postnatal precursors consistently produced a vast majority of granule cells and a small fraction of molecular layer interneurons, suggesting that these cells differentiate according to cell-autonomous programs that are minimally

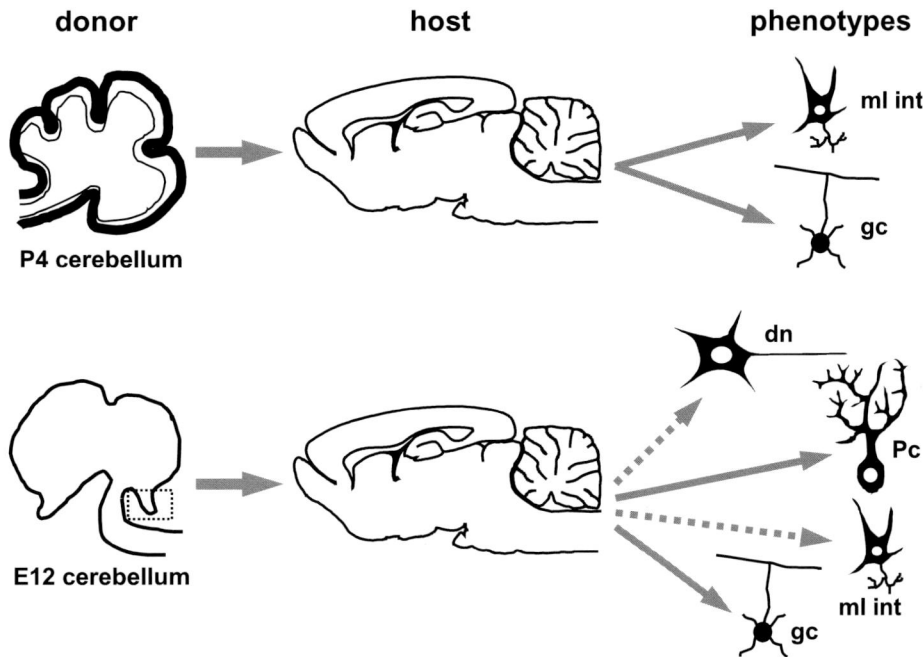

Fig. 2. Developmental potential of postnatal and embryonic cerebellar progenitors following heterotopic/heterochronic transplantation. In all transplantation conditions, postnatal cerebellar progenitors exclusively differentiate granule cells (gc) and molecular layer interneurons (ml int), suggesting that they developed in a cell autonomous manner, irrespective of the recipient milieu. In contrast, their embryonic counterparts have the potential for generating all major cerebellar types, but the actual repertoire of mature phenotypes is conditioned by the integration site in the host brain. Pc, Purkinje cells; dn, deep nuclei neurons.

influenced by local conditions. The same conclusion cannot be drawn for embryonic donors. Although these cells showed wider developmental potentialities, the relative amounts of mature types did not match those occurring during cerebellar development in situ (see Ito, 1984), and the phenotypic repertoire was often different in distinct engraftment sites (Fig. 2). The high frequency of Purkinje cells and deep nuclei neurons, at the expense of granule cells and molecular layer interneurons, may be attributed to the fact that grafted embryonic progenitors do not proliferate long enough to generate appropriate quantities of late types. However, other mechanisms should be envisaged to explain the peculiar phenotypic distributions observed in some recipient regions.

It has been shown that rhombic lip precursors are committed to a granule cell fate from early embryonic development (around E13 in rodents), but they are not able to differentiate cell-autonomously into mature granule neurons (Alder et al., 1996). In line with this view, our observations indicate that E12 progenitors from the whole cerebellar primordium are not competent to generate all phenotypes, and require additional information to adopt late-generated identities. The peculiar phenotypic repertoires generated by these progenitors in some engraftment sites — e.g., neocortical explants, where they mostly produced Purkinje and granule cells — indicate that different brain regions may provide peculiar patterns of environmental signals, which favor or select the development of specific cell types. For instance, the differentiation of granule cells might be conditioned by the local availability of Sonic Hedgehog (Wechsler-Reya and Scott, 1999; Solecki et al., 2000). Hence, although embryonic cerebellar precursors have the potential to produce all cerebellar identities, the actual repertoire of mature phenotypes that will be generated is conditioned by local cues.

The question remains open as to whether such cues select distinct subsets of precursors already oriented

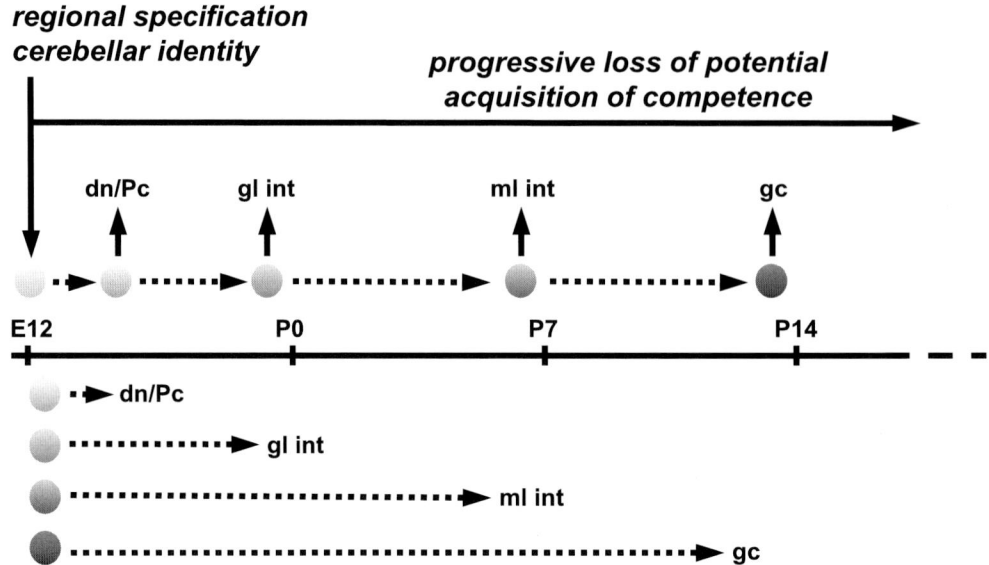

Fig. 3. Possible mechanisms underlying the generation of cerebellar phenotypes. Transplantation experiments show that mouse cerebellar progenitors are regionally specified towards local identities already at E12. Thereafter, precursor cells undergo a progressive restriction of potential and acquisition of competence leading to the sequential generation of the different types of neurons. This outcome can be achieved by initially multipotent progenitors that sequentially generate mature neurons during precise ontogenetic phases (upper part of the figure). Alternatively, distinct lineages may get separated during early embryogenesis and differentiate into different adult types at appropriate times (lower part of the figure). Indeed, there is evidence that granule cell precursors already become specified during early embryonic life (Alder et al., 1996). Pc, Purkinje cells; dn, deep nuclei neurons; gl int, granular layer interneurons; ml int, molecular layer interneurons; gc, granule cells.

towards specific identities or whether they provide instructive information to multipotent progenitors (Fig. 3). Some studies indicate that individual cerebellar lineages become separated during early embryonic development (Alder et al., 1996; Baader et al., 1999). In addition, there is evidence that different cell types may originate from distinct regions of the cerebellar primordium (Maricich and Herrup, 1999; Louvi et al., 2003). On the other hand, the observation that multiple cerebellar phenotypes may be clonally related (Mathis et al., 1997) suggests that individual progenitors may differentiate along different lineages. Thus, it is possible that the cerebellar anlage is actually made up of a mosaic of microdomains or restricted cell populations destined to produce defined phenotypes. Alternatively, the generation of different neuron populations, together with the progressive restriction of progenitors' potential, may be regulated by a set of specific instructive/selective signals precisely distributed in space and time throughout cerebellar ontogenesis.

Abbreviation

EGL   external granular layer

Acknowledgments

The work reported in this chapter was supported by grants from the Italian Ministry of University and Research, MIUR-FIRB 2001; Italian Ministry of Health, Alzheimer Project 2000; Italian Institute for Health, National Program on Stem Cells.

References

Alder, J., Cho, N.K. and Hatten, M.E. (1996) Embryonic precursor cells from the rhombic lip are specified to a cerebellar granule neuron identity. Neuron, 17: 389–399.

Altman, J. and Bayer, S.A. (1997) Development of the Cerebellar System in Relation to its Evolution, Structure and Functions. CRC Press, New York.

Altmann, C.R. and Brivanlou, A.H. (2001) Neural patterning in the vertebrate embryo. Int. Rev. Cytol., 203: 447–482.

Baader, S.L., Bergmann, M., Merz, K., Fox, P.A., Gerdes, J., Oberdick, J. and Schilling, K. (1999) The differentiation of cerebellar interneurons is independent of their mitotic history. Neuroscience, 90: 1243–1254.

Barbe, M.F. and Levitt, P. (1991) The early commitment of fetal neurons to the limbic cortex. J. Neurosci., 2: 519–533.

Brüstle, O., Maskos, U. and McKay, R.D.G. (1995) Host-guided migration allows targeted introduction of neurons into the embryonic brain. Neuron, 15: 1275–1285.

Campbell, K., Olsson, M. and Björklund, A. (1995) Regional incorporation and site-specific differentiation of striatal precursors transplanted to the embryonic forebrain ventricle. Neuron, 15: 1259–1273.

Carletti, B., Grimaldi, P., Magrassi, L. and Rossi, F. (2002) Specification of cerebellar progenitors following heterotopic/heterochronic transplantation to the embryonic CNS *in vivo* and *in vitro*. J. Neurosci., 22: 7132–7146.

Carletti, B., Grimaldi, P., Magrassi, L. and Rossi, F. (2003) Specification of cortical progenitors after transplantation in the embryonic hindbrain. VI IBRO Congress, abstract nr. 1029.

Cattaneo, E., Magrassi, L., Butti, G., Santi, L., Giavazzi, A. and Pezzotta, S. (1994) A short term analysis of the behaviour of conditionally immortalized neuronal progenitors and primary neuroepithelial cells implanted into the fetal rat brain. Brain. Res. Dev. Brain. Res., 83: 197–208.

Cepko, C.L., Austin, C.P., Yang, X., Alexiades, M. and Ezzeddine, D. (1996) Cell fate determination in the vertebrate retina. Proc. Natl. Acad. Sci. (USA), 93: 589–595.

Chédotal, A., Bloch-Gallego, E. and Sotelo, C. (1997) The embryonic cerebellum contains topographic cues that guide developing inferior olivary axons. Development, 124: 861–870.

De Camilli, P., Miller, P.E., Levitt, P., Walter, U. and Greengard, P. (1984) Anatomy of cerebellar Purkinje cells in the rat determined by a specific immunohistochemical marker. Neuroscience, 11: 761–817.

Desai, A.R. and McConnell, S.K. (2000) Progressive restriction in fate potential by neural progenitors during cerebral cortical development. Development, 127: 2863–2872.

Ephrussi, A. and St Johnston, D. (2004) Seeing is believing: the bicoid morphogen gradient matures. Cell, 116: 143–152.

Fishell, G. (1995) Striatal precursors adopt cortical identities in response to local cues. Development, 121: 803–812.

Gao, W.-Q. and Hatten, M.E. (1994) Immortalizing oncogenes subvert the establishment of granule cell identity in developing cerebellum. Development, 120: 1059–1070.

Ito, M. (1984) The Cerebellum and Neural Control. Raven Press, New York.

Jankovski, A., Rossi, F. and Sotelo, C. (1996) Neuronal precursors in the postnatal mouse cerebellum are fully committed cells: evidence from heterochronic transplantation. Eur. J. Neurosci., 8: 2308–2320.

Lee, K.J. and Jessell, T.M. (1999) The specification of dorsal cell fates in the vertebrate central nervous system. Annu. Rev. Neurosci., 22: 261–294.

Lim, D.A., Fishell, G.J. and Alvarez-Buylla, A. (1997) Postnatal mouse subventricular zone neuronal precursors can migrate and differentiate within multiple levels of the developing neuraxis. Proc. Natl. Acad. Sci. (USA), 94: 14832–14836.

Louvi, A., Alexandre, P., Metin, C., Wurst, W. and Wassef, M. (2003) The isthmic neuroepithelium is essential for cerebellar midline fusion. Development, 130: 5319–5330.

Lumsden, A., Clarke, J.D.W., Keynes, R. and Fraser, S. (1994) Early phenotypic choices by neuronal precursors, revealed by clonal analysis of the embryonic chick hindbrain. Development, 120: 1581–1589.

Maricich, S.M. and Herrup, K. (1999) Pax-2 expression defines a subset of GABAergic interneurons and their precursors in the developing murine cerebellum. J. Neurobiol., 41: 281–294.

McCarthy, M., Turnbull, D.H., Walsh, C.A. and Fishell, G. (2001) Telencephalic neural progenitors appear to be restricted to regional and glial fates before the onset of neurogenesis. J. Neurosci., 21: 6772–6781.

McConnell, S.K. (1995) Constructing the cerebral cortex: neurogenesis and fate determination. Neuron, 15: 761–768.

McConnell, S.K. and Kaznowski, C.E. (1991) Cell cycle dependence of laminar determination in developing neocortex. Science, 254: 282–285.

Magrassi, L., Ehrlich, M.E., Butti, G., Pezzotta, S., Govoni, S. and Cattaneo, E. (1998) Basal ganglia precursors found in aggregates following embryonic transplantation adopt a striatal phenotype in heterotopic locations. Development, 125: 2847–2855.

Mathis, L., Bonnerot, C., Puelles, L. and Nicolas, J.F. (1997) Retrospective analysis of the cerebellum using genetic laacZ/lacZ mouse mosaics. Development, 124: 4089–4104.

Miale, I.L. and Sidman, R.L. (1961) An autoradiographic analysis of histogenesis in mouse cerebellum. Exp. Neurol., 4: 277–296.

Na, E., McCarthy, M., Neyt, C., Lai, E. and Fishell, G. (1998) Telencephalic progenitors maintain anteroposterior identities cell autonomously. Curr. Biol., 8: 987–990.

Olsson, M., Campbell, K. and Turnbull, D.H. (1997) Specification of mouse telencephalic and mid-hindbrain progenitors following heterotopic ultrasound-guided embryonic transplantation. Neuron, 19: 761–772.

Olsson, M., Bjerregaard, K., Winkler, C., Gates, M., Björklund, A. and Campbell, K. (1998) Incorporation of mouse neural progenitors transplanted in the rat embryonic forebrain is developmentally regulated and dependent on regional adhesive properties. Eur. J. Neurosci., 10: 71–85.

Palay, S. and Chan-Palay, V. (1974) Cerebellar Cortex. Springer-Verlag, Berlin, Heidelberg, New York.

Ramón y Cajal, S. (1911) Histologie du Système Nerveux de l'Homme et des Vertébrés. Maloine, Paris.

Renfranz, P.J., Cunningham, M.G. and McKay, R.D.G. (1991) Region-specific differentiation of the hippocampal stem cell line HiB5 upon implantation into the developing mammalian brain. Cell, 66: 713–729.

Rossi, F. and Borsello, T. (1993) Ectopic Purkinje cells in the adult rat: olivary innervation and different capabilities of migration and development after grafting. J. Comp. Neurol., 377: 70–82.

Rossi, F. and Cattaneo, E. (2002) Neural stem cell therapy for neurological diseases: dreams and reality. Nat. Rev. Neurosci., 3: 401–409.

Rossi, F., Borsello, T. and Strata, P. (1992) Embryonic Purkinje cells grafted on the surface of the cerebellar cortex integrate in the adult unlesioned cerebellum. Eur. J. Neurosci., 4: 589–593.

Rossi, F., Borsello, T. and Strata, P. (1994) Embryonic Purkinje cells grafted on the surface of the adult uninjured rat cerebellum migrate in the host parenchyma and induce sprouting of intact climbing fibres. Eur. J. Neurosci., 6: 121–136.

Solecki, D.J., Liu, X., Tomoda, T., Fang, Y. and Hatten, M.E. (2000) Activated *Notch2* signaling inhibits differentiation of granule neuron precursors by maintaining proliferation. Neuron, 31: 557–568.

Sotelo, C. and Alvarado-Mallart, R.M. (1987) Embryonic and adult neurons interact to allow Purkinje cell replacement in mutant cerebellum. Nature, 327: 421–423.

Sotelo, C. and Alvarado-Mallart, R.M. (1991) The reconstruction of cerebellar circuits. Trends Neurosci., 8: 350–355.

Snyder, E.Y., Deitcher, D.L., Walsh, C., Arnold-Aldea, S., Hartweig, E.A. and Cepko, C.L. (1992) Multipotent neural cell lines can engraft and participate in development of mouse cerebellum. Cell, 68: 33–51.

Vicario-Abejón, C., Cunningham, M.G. and McKay, R.D.G. (1995) Cerebellar precursors transplanted to the neonate dentate gyrus express features characteristic of hippocampal neurons. J. Neurosci., 15: 6351–6363.

Wechsler-Reya, R.J. and Scott, M.P. (1999) Control of neuronal precursor proliferation in the cerebellum by Sonic Hedgehog. Neuron, 22: 103–114.

Wilson, S.W. and Rubenstein, J.L. (2000) Induction and dorsoventral patterning of the telencephalon. Neuron, 28: 641–651.

Yang, H., Mujtaba, T., Venkatraman, G., Wu, Y.Y., Rao, M.S. and Luskin, M.B. (2000) Region-specific differentiation of neural tube-derived neuronal restricted progenitor cells after heterotopic transplantation. Proc. Natl. Acad. Sci. (USA), 97: 13366–13371.

Zhang, L. and Goldman, J.E. (1996a) Generation of cerebellar interneurons by dividing progenitors in the white matter. Neuron, 16: 47–54.

Zhang, L. and Goldman, J.E. (1996b) Developmental fates and migratory pathways of dividing progenitors in the postnatal rat cerebellum. J. Comp. Neurol., 370: 536–550.

SECTION III

Cell Physiological Cerebellar Plasticity

CHAPTER 7

Long-term potentiation of synaptic transmission at the mossy fiber–granule cell relay of cerebellum

Egidio D'Angelo[1,*], Paola Rossi[1], David Gall[2], Francesca Prestori[1], Thierry Nieus[1], Arianna Maffei[3] and Elisabetta Sola[4]

[1]*Department of Cellular and Molecular Physiology and Pharmacology, University Pavia and INFM, Via Forlanini 6, I-27100 Pavia, Italy*
[2]*Laboratoire de Neurophysiologie (CP601), Faculte de Mediciné, Université de Bruxelles, Route de Lennik 808, B-1070 Brussels, Belgium*
[3]*Department of Neuroscience, Brandeis University, 415 South St., Waltham, MA 02454-9110, USA*
[4]*Department of Biophysics, SISSA, via Beirut 4, I-34014 Trieste, Italy*

Keywords: cerebellum; mossy fiber; granule cell; LTP; NMDA receptors; calcium

Abstract: In the last decade, the physiology of cerebellar neurons and synapses has been extended to a considerable extent. We have found that the mossy fiber–granule cell relay can generate a complex form of long-term potentiation (mf–GrC LTP) following high-frequency mf discharge.

Induction. Mf–GrC LTP depends on NMDA and mGlu receptor activation, intracellular Ca^{2+} increase, PKC activation, and NO production. The preventative action of intracellular agents (BAPTA, PKC-inhibitors) and of membrane hyperpolarization, and the correlated increase in intracellular Ca^{2+} observed using florescent dyes, indicate that induction occurs postsynaptically.

Expression. Expression includes three components: (a) an increase of synaptic currents, (b) an increase of intrinsic excitability in GrC, and (c) an increase of intrinsic excitability in mf terminals. Based on quantal analysis, the EPSC increase is mostly explained by enhanced neurotransmitter release. NO is a candidate retrograde neurotransmitter which could determine both presynaptic current changes and LTP. NO cascade blockers inhibit both presynaptic current changes and LTP. The increase in intrinsic excitability involves a raise in apparent input resistance in the subthreshold region and a spike threshold reduction.

Together with other forms of cerebellar plasticity, mf–GrC LTP opens new hypothesis on how the cerebellum processes incoming information.

The cerebellum and the mossy fiber–granule cell relay

The cerebellum is a brain structure of primary importance for the coordination of movement, and is also probably involved in processing higher brain functions (Ghez and Thach, 2003). The basic architecture of the cerebellar circuitry is well known (Eccles et al., 1967; Ito, 1984). The cerebellum receives two main inputs through mossy fibers (mfs) and climbing fibers (cfs). Cfs originate in the inferior olivary nucleus and innervate Purkinje cells (PCs). Mfs are the largest cerebellar afferent system and originate from various regions in the spinal cord,

*Corresponding author. Tel.: +39 0382 507-606/-794; Fax: +39 0382 507-527; E-mail: dangelo@unipv.it

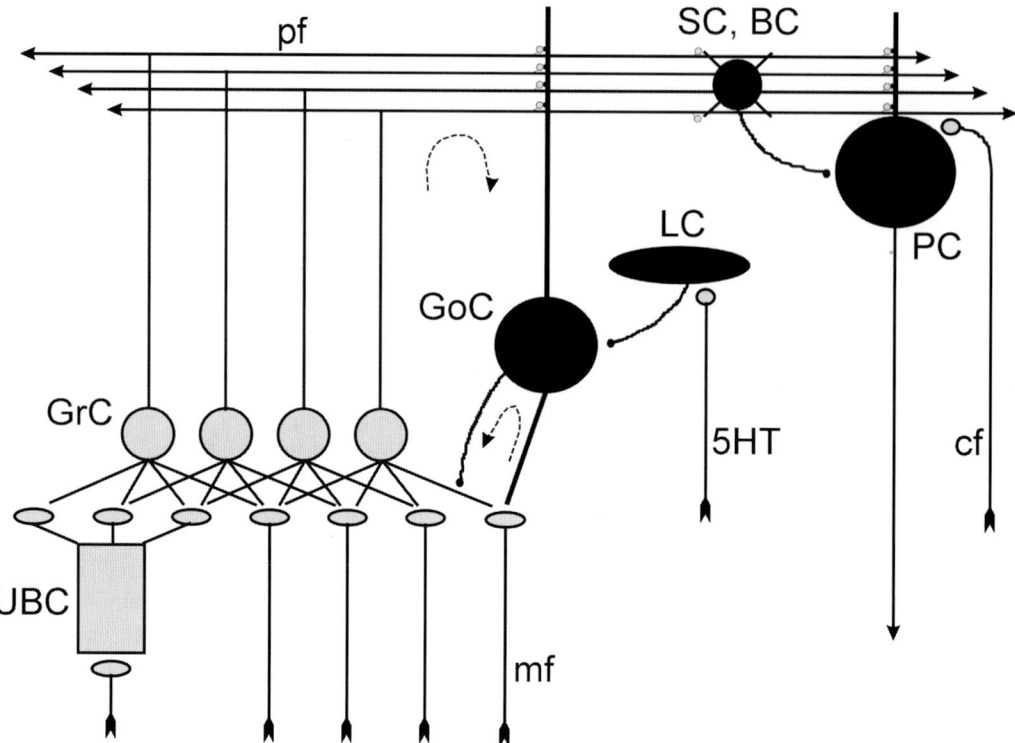

Fig. 1. Schematic representation of the neuronal circuit of the cerebellar cortex: mf, mossy fiber; pf, parallel fiber; cf, climbing fiber; GrC, granule cell; GoC, Golgi cell; LC, Lugaro Cell; UBC, unipolar brush cell; PC, Purkinje cell; SC, stellate cell; BC, basket cell. The granular layer comprises GrC, GoC, LC, and UBC. Note divergence and convergence at the mf–GrC relay, and amplification of mf input determined by UBC. Note also the double feed-back and feed-forward inhibition of GrC through GoC (dashed arrows). LC inhibits GoC when it is activated by serotonin (5HT). SC, BC, PC are outside the granular layer. Gray indicates excitation, black inhibition. Arrows indicate the direction of information flow.

brain-stem, and cerebral cortex. Mfs activate granule cells (GrC) and Golgi cells (GoC), the main inhibitory interneurons of the granular layer. Golgi cells are also excited by parallel fibers (pf), the GrC axons. Thus, GoCs inhibit GrCs through a double feed-forward and feed-back loop. The granular layer contains also Lugaro cells (LC) and unipolar brush cells (UBC), the former inhibiting GoCs (Dieudonne and Damoulin, 2000), the latter emitting mfs and activating groups of GrCs (Nunzi et al., 2001). The output from the granular layer is conveyed through the parallel fibers to PCs, which inhibit cerebellar nucleus neurons providing the final output from the cerebellum (Fig. 1).

The excitatory nature of the mf–GrC synapse and its inhibitory control by Golgi cells were recognized early (Eccles et al., 1967). Mfs diverge onto numerous GrCs (about 28 in the rat), which in turn receive just four different mf inputs on as many independent dendrites. This convergence/divergence ratio suggested that the mf–GrC relay performs a spatial pattern separation of the mf input (Marr, 1969). Details about synaptic receptor organization and function had to wait until the 1990s, when the advent of patch-clamping in cerebellar slices allowed single GrC recordings in situ. Electrophysiological recordings revealed that GrC have a complex functional specialization. Perhaps the most striking aspect is that the mf–GrC relay shows multiple mechanisms of synaptic and nonsynaptic plasticity. The present review will focus on mf–GrC synaptic transmission and its plastic changes.

GrCs dendrites terminate with 3–4 bulbs endowed with postsynaptic densities aligned with releasing sites in the mf terminal (Jakab and Hamori, 1988). Miniature synaptic currents are uniquantal (Chatala et al., 2003) and EPSCs are determined by release of 1 to 3 quanta, as expected from ultrastructural investigations (Sola et al., 2004). The probability of release estimated with binomial models ranges from 0.2 to 0.7. Mf are glutamatergic and activate AMPA, NMDA, and mGlu receptors in GrCs. AMPA receptors are located in clusters facing the releasing sites (Silver et al., 1996; DiGregorio and Silver, 2002; Xu-Friedman and Regher, 2003), while NMDA receptors are in part synaptic and in part extra-synaptic (Petralia et al., 2002; Rossi et al., 2002). Glutamate spillover in the glomerulus can activate both AMPA receptors located at different postsynaptic sites and extrasynaptic NMDA receptors. Analysis of the spillover current indicates that neither AMPA nor NMDA receptors are likely to be saturated by synaptically released glutamate. EPSPs generated by a single mf are usually not sufficient to activate an action potential from rest, and the synchronous activation of 2–3 synapses is needed (D'Angelo et al., 1995). During repetitive stimulation, the ongoing depression in AMPA receptor-mediated responses is compensated by a large increase in the NMDA response enhancing temporal summation. Glutamate also activates class-I mGlu receptors coupled to the PIP_2 pathway. Both NMDA and mGlu receptors play an important role in regulating GrC intracellular Ca^{2+} increase (Gall et al., 2004).

Mossy fiber–granule cell long-term potentiation (LTP)

Marr (1969) explicitly negated the possibility that mf–GrC synaptic weights could be modified by activity. He noted that '*sooner or later all weights would be saturated*' so that plasticity would not be useful. Thus, Marr's model did not include any mf–GrC synaptic plasticity, although the subsequent extension due to Albus (1971) was more permissive. However, the observation that mf discharge consists of high-frequency bursts (Kase et al., 1980) and that GrCs express NMDA receptors, which trigger the induction of LTP/LTD at other central synapses (see Bliss and Collingridge, 1993; Hawkins et al., 1993; Edwards, 1995; Malenka and Nicoll, 1999), led us to revisit the question. In fact, once theta-burst stimulation or prolonged (1 s) high-frequency (100 Hz) stimuli were applied to mfs in cerebellar slices,

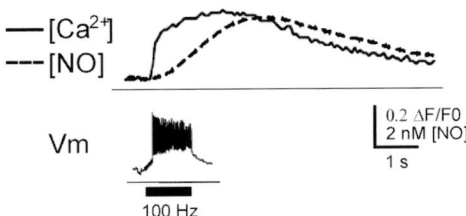

Fig. 2. LTP induction. Intradendritic $[Ca^{2+}]$ and extracellular [NO] increase during LTP induction. Mfs are stimulated with a 100 Hz 1-s impulse train and generate a burst of action potentials (bottom tracing). Adapted with permission from Maffei et al. (2003).

granule cell synaptic excitation was persistently strengthened indicating the occurrence of long-term potentiation (LTP). Mf–GrC LTP has been investigated using extracellular recordings (Maffei et al., 2002, 2003), whole-cell recordings (D'Angelo et al., 1999; Rossi et al., 2002), and perforated-patch recordings (Armano et al., 2000). Thus LTP can be induced without altering the cytoplasm and can be revealed both in neuronal ensembles and in single cells. In all these cases potentiation of synaptic currents is between 30% and 50% and affects both the AMPA and NMDA receptor-mediated component of the response. At present, several aspects of the induction and expression mechanism of mf–GrC LTP have been clarified (Figs. 2 and 3). The current model of mf–GrC LTP is shown in Fig. 4.

LTP induction

Repetitive high-frequency stimulation activates a cascade of events leading to mf-GrC LTP (Fig. 2). AMPA receptor activation causes membrane depolarization and NMDA receptor unblock, which, in addition to reinforcing depolarization, determines a remarkable cytoplasmic Ca^{2+} increase in the dendrites (Gall, Prestori, Sola, Rossi and D'Angelo, unpublished observation). The Ca^{2+} increase is reinforced by mGlu receptor activation, probably through a PLC-mediated pathway initiated by mGlu receptor stimulation and leading to DAG and IP_3 production (Monti et al., 2002). These events are critical for LTP, since either preventing membrane depolarization, blocking NMDA or mGlu receptors, or buffering intracellular Ca^{2+} prevented LTP (Rossi et al., 1996; D'Angelo et al., 1999; Armano

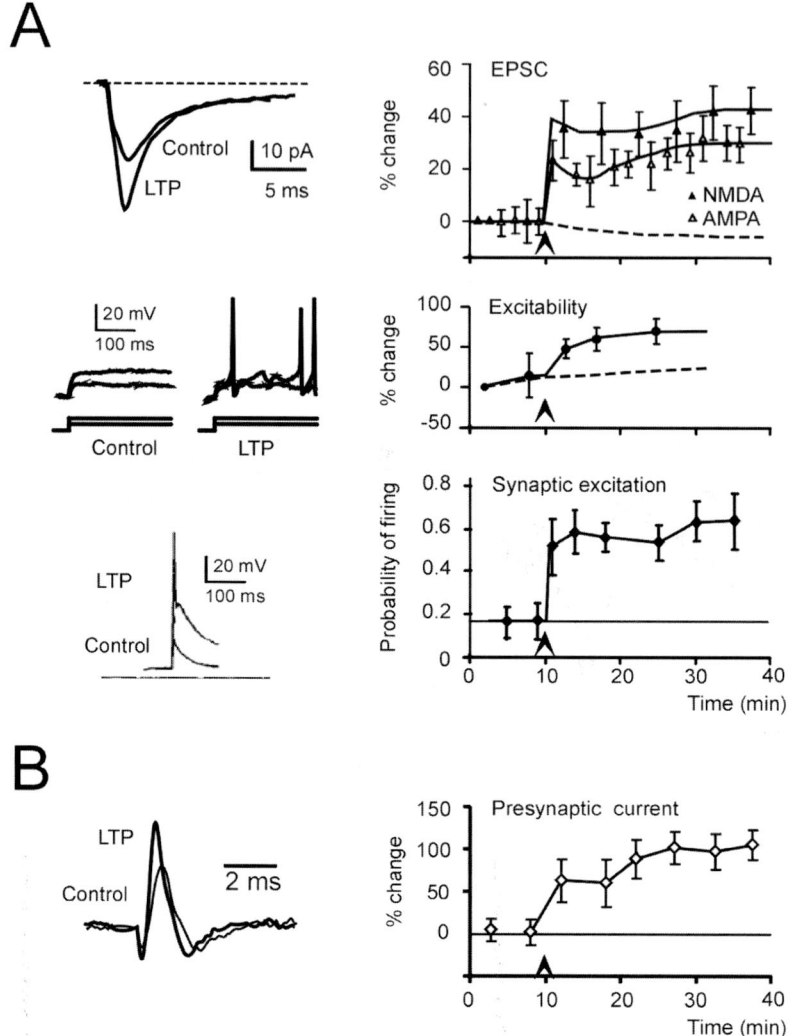

Fig. 3. LTP expression. The main electrophysiological changes and their time course are illustrated. LTP induction is indicated by arrowheads. (A) Top. Tracings show EPSC increase during LTP. The plot shows increase in both AMPA and NMDA currents (the dashed line shows EPSC amplitude in control recordings). Middle. Intrinsic excitability increases during LTP. Tracings show that less current is needed to reach firing threshold, and that this is later reduced. The plot shows the time course of the change in current needed to activate action potentials from rest (the dashed line indicates control recordings). Bottom. As a result of EPSC and excitability changes, synaptic excitation is increased. EPSPs, which were initially subthreshold, determine spike activation during LTP. The plot reports the probability of firing. (B) Presynaptic mf current increase during LTP. The time course of changes parallels that of postsynaptic modifications. Adapted with permission from Armano et al. (2000), Hansel et al. (2001), and Maffei et al. (2003).

et al., 2000). LTP was also prevented by intra- or extracellular blockage of PKC (D'Angelo et al., 1999; Maffei et al., 2002), a kinase requiring DAG and Ca^{2+} for activation. Mf–GrC LTP induction resembles therefore a general scheme in which high-frequency stimulation activates ionotropic and metabotropic glutamate receptors initiating a Ca^{2+}-dependent signaling cascade (Bliss and Collingridge, 1993).

Another process that is likely to occur during induction is NO production. The NO producing enzyme, nitric oxide synthase (NOS), is expressed in GrCs (Bredt et al., 1990; Garthwaite et al., 1998). During LTP induction, NO rises in the nanomolar

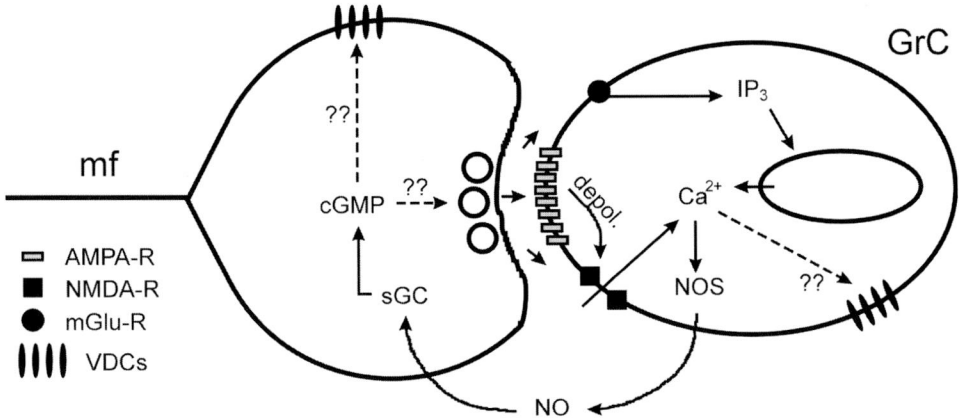

Fig. 4. Schematic illustration of present understanding of mf–GrC LTP induction and expression. Induction occurs postsynaptically through NMDA and mGlu receptor activation. These trigger an intracellular Ca^{2+} increase determining NO release. NO regulates presynaptic functions including terminal excitability and neurotransmitter release. The specific ionic channels and intracellular mechanisms involved in LTP expression remain largely unknown (question marks).

range, reaching concentrations sufficient to fully activate soluble guanylil cyclase, the major effector of the NO pathway (Maffei et al., 2003). Thus, NO is a potential candidate for signal back-propagation from post- to presynaptic site (see below).

LTP expression: synaptic conductance changes

The origin of synaptic conductance changes during long-term synaptic plasticity has been the object of a hot debate mostly focused on hippocampal LTP (Bliss and Collingridge, 1993; Edwards, 1995; Malenka and Nicoll, 1999). At the mf–GrC synapse, the fact that both the AMPA and NMDA currents are potentiated could indicate either a common presynaptic mechanism or a simultaneous postsynaptic change in both receptor types. Nonetheless, NMDA is usually stronger and develops more slowly than AMPA current potentiation. This would suggest that postsynaptic modulation also plays a role (D'Angelo et al., 1999; Fig. 3A). A presynaptic mechanism of expression has been suggested in a recent investigation (Sola et al., 2004). The demonstration is based on several points. First, during LTP, EPSC coefficient of variation (CV), failures and paired-pulse ratio (PPR) decreased. Similar changes were observed by raising neurotransmitter release (high Ca^{2+}/Mg^{2+}), while the opposite occurred by decreasing the release (low Ca^{2+}/Mg^{2+}, Cl-adenosine). No changes followed postsynaptic modifications (different holding potential), while only CV and failures decreased by raising the number of active synapses. LTP was occluded by raising release probability and was observed in the spillover-dependent component of AMPA EPSCs and in NMDA EPSCs. Finally, during LTP, miniature synaptic currents did not change their amplitude or variability but increased their frequency. Binomial analysis explained EPSC changes through an increased release probability.

As in mf–GrC LTP, an increased neurotransmitter release was proposed to occur at other central synapses in situ (Malinow and Tsien, 1990; Schulz et al., 1994; Kullmann et al., 1996; Gasparini et al., 2000) and in neuronal cell cultures (Bekkers and Stevens, 1990; Malgaroli et al., 1995). Coupling between postsynaptic NMDA receptor-dependent induction and presynaptic expression may be provided by NO (Arancio et al., 1996). Indeed, blocking NOS, scavenging NO, or blocking sGC prevented mf–GrC LTP, while NO donors induced it (Maffei et al., 2003). The present observation does not exclude that, in different functional or developmental conditions, LTP expression might change. For instance, silent synapse awakening characterizes the developmental process of the cerebellar mf–GrC relay, leading from purely NMDA to mixed AMPA-NMDA EPSCs (D'Angelo et al., 1993; Losi et al., 2002). Moreover, NMDA receptor stimulation leads to CREB activation in granule cells (Monti et al.,

2002), and may therefore prime postsynaptic gene expression and protein synthesis in later LTP phases.

LTP expression: changes in granule cell intrinsic excitability

In addition to showing an increased synaptic conductance, during LTP, GrC show an increased intrinsic excitability (Armano et al., 2000; Fig. 3A). The existence of excitability changes during LTP was indicated previously in the hippocampus, by Bliss and Lomo (1973), who called this LTP component E-S potentiation (or potentiation of EPSP-spike coupling). GrC E-S potentiation can be identified in the presence of blockers of inhibitory transmission and reflects therefore changes in intrinsic excitability. GrC E-S potentiation consists of an increased input resistance in the subthreshold region and of a spike threshold decrease. E-S potentiation is induced by NMDA receptor activation but is less sensitive to membrane depolarization than potentiation of synaptic conductance. Thus, a protracted weak stimulation may be able to enhance GrC excitability, and E-S potentiation would assume a homeostatic effect. Clearly, E-S potentiation is all postsynaptic, although the channels involved remain speculative. A reduction in voltage-dependent K^+ currents and an increase in persistent Na^+ currents may be involved.

LTP expression: changes in mossy fiber intrinsic excitability

During mf–GrC LTP, modifications can be revealed in the presynaptic terminal current measured by using loose-patch recording techniques (Maffei et al., 2002; Fig. 3B). During LTP, the presynaptic current increases with the same time course of LTP. The presynaptic current increase is prevented by the same factors preventing LTP, including block of NMDA receptors, mGlu receptors, or PKC, as well as by postsynaptic inhibition through Golgi cells. Moreover, the presynaptic current increase is prevented by NOS inhibitors, NO scavengers, and sGC inhibitors, and is mimicked by the NO donor DEA-NO (Maffei et al., 2003). Thus, it seems that postsynaptic induction can cause presynaptic expression through NO production.

The role of presynaptic current changes is unknown at present. The Ca^{2+}-dependent component of the presynaptic terminal current increased during LTP, and LTP as well as the presynaptic current change could be occluded by application of the K^+ channel blocker TEA (Maffei et al., 2002). Thus, presynaptic current changes may reflect regulation of excitation dynamics during repetitive stimulation and may be linked to the release process through a regulation of Ca^{2+} influx. It is currently unknown whether there is a mechanistic relationship between the increased neurotransmitter release revealed by EPSC quantal analysis and the presynaptic terminal current change. Moreover, it is unknown whether NO could determine LTP by raising presynaptic terminal excitability or by regulating the release machinery.

LTP impairment in mutant mice

If LTP is of functional significance, then it should be possible to impair LTP and motor learning in mice bearing mutation in molecules linked to the LTP mechanism. To this aim we have used mice with a mutation in the NMDA receptor, in particular a C-terminal truncation of the NR2A and NR2C C-terminal chains (Rossi et al., 2002). Due to the late expression of these subunits, the ontogenetic process is unaltered. Moreover, NR2A and NR2C show a selective co-expression in the cerebellar GrCs. Thus, NR2A-$C^{\Delta C/\Delta C}$ mice provide a rather target- and time-specific mutation. Interestingly, mutated NMDA channels show a reduced open probability but their density in the plasma membrane appears normal. In NR2A-$C^{\Delta C/\Delta C}$ mice LTP is impaired, and mice show impaired motor learning in the thin-rod test (Sprengel et al., 1998; Georg Kohr, unpublished observation). It seems therefore that mf–GrC LTP plays a role in motor coordination and learning. Investigation of other mutants is currently being performed in order to extend this observation.

Functional consequences of mossy fiber–granule cell LTP

The cerebellum is thought to operate in feed-forward mode anticipating the corrections needed to regulate complex sequences of movements (Ghez and Thach, 2003). As every feed-forward device, the cerebellum needs to store information to be used in a predictive manner. No surprise therefore that the cerebellar circuitry expresses mechanisms for learning and memory. As proposed by Marr (1969), a major form of plasticity occurs at the pf-PC synapse, allowing heterosynaptic depression when a motor error is detected (pf-PC LTD). Instruction or error signals are conveyed by cfs, so that pf-PC synapses relaying relevant sensorimotor signals are persistently depressed. The experimental observation of pf-PC LTD, which is thought to provide the biological counterpart of theoretical predictions, obviously has a huge impact on the present understanding of cerebellar functions (reviewed by Ito, 2001). Nonetheless, the discovery of mf–GrC LTP, together with various forms of plasticity at other cerebellar synapses, suggests that the classical concept of cerebellar learning needs to be extended (Hansel et al., 2001; see also DeSchutter, 1997; Llinas et al., 1997).

By being induced, omosynaptic activity, mf–GrC LTP implements a modality of unsupervised learning. By being dependent on high-frequency mf activity and postsynaptic depolarization, mf–GrC LTP is associative, implementing a process of coincidence detection. The functional consequences of mf–GrC LTP depend on several factors including the molecular and cellular mechanisms involved, the spatial distribution of plasticity, local network activity (primarily related to endogenous rhythms and synaptic inhibition), and long-range modulation (primarily related to cholinergic, serotoninergic and noradrenergic innervation of the cerebellum). Although understanding of these mechanisms is far from being complete, some preliminary considerations can be advanced.

(1) The enhanced release probability during mf–GrC LTP was associated with enhanced short-term synaptic depression, suggesting that mf–GrC LTP may play an important role in regulating cerebellar dynamics (Hansel et al., 2001; Carey and Lisberger, 2002). Moreover, since EPSC variability decreased, LTP should improve the reliability of neurotransmission. The postsynaptic increase in intrinsic excitability (Armano et al., 2000) is suited for determining the number of active GrCs minimizing redundancy and optimizing sparse representation of mf activity (Schweighofer et al., 2000; Philipona et al., 2003). It may also represent a homeostatic process compensating for low activation in certain GrCs.

(2) A critical step in understanding LTP functional implications is that of defining the *learning rules*. These include BCM and the STDP at cortical synapses (Siöström and Nelson, 2002). What we know is that a single 100-ms train is not usually sufficient for inducing LTP, while the train has to be repeated at least four times to be effective. LTD can be observed when the induction is weakened by pharmacological manipulations (such as intracellular Ca^{2+} chelation, NMDA receptor blockage, synaptic inhibition), although its natural induction paradigms have still to be understood. STDP should be considered in relation to oscillations generating in the cerebellar network (Pellerin and Lamarre, 1997; Hartmann and Bower, 1998), so that inputs in phase with the oscillation may be able to induce LTP, while those falling in anti-phase would be inefficient or even cause LTD. These aspects are currently under investigation in our laboratory by using a combination of electrophysiological and imaging recordings (Gall et al., 2004).

(3) NO diffusion may determine spatial spread of LTP. By diffusing transcellularly, NO released from neighboring GrCs may sum up exerting a collective control on the mf terminal. In turn, membrane depolarization needed to unblock NMDA receptors and release NO should follow synchronous discharge in several mfs (see Armano et al., 2000). Thus, the NO signal may be generated depending on the effective number and location of active GrCs, influencing temporospatial processing of mf discharge and sensorimotor control by the cerebellum.

Conclusions

Although several aspects remain to be investigated, mf–GrC LTP provides a wide substrate for information storage in the cerebellum. In the rat cerebellum, there are 10^{11} GrCs and four times as many mf–GrC synapses. Mathematical models have suggested that mf–GrC LTP improves mutual information transfer and regulates codon representation (Schweighofer et al., 2000; Philipona et al., 2003). Moreover, a preliminary investigation suggests a role in regulating neurotransmission dynamics (Nieus et al., 2004). These observations challenge the simple view of spatial pattern separation proposed by Marr. The potential consequences of mf–GrC LTP need to be further investigated and confronted with computational models of the cerebellar network.

Abbreviations

EPSC	excitatory postsynaptic current
EPSP	excitatory postsynaptic potential
LTP	long-term potentiation
GrC	granule cell
GoC	Golgi cell
PC	Purkinje cell
LC	Lugaro cell
SC	stellate cell
BC	basket cell
UBC	unipolar brush cell
mf	mossy fiber
pf	parallel fiber
cf	climbing fiber

Acknowledgments

Supported by European Community grants IST-2001-35271 and QLG3-CT-2001-02256, by MIUR and INFM of Italy.

References

Albus, J.S. (1971) A theory of cerebellar function. Math. Biosci., 10: 25–61.

Arancio, O., Kiebler, M., Lee, C.J., Lev-Ram, V., Tsien, R.Y., Kandel, E. and Hawkins, R.D. (1996) Nitric oxide acts directly in the presynaptic terminal to produce long-term potentiation in cultured hippocampal neurons. Cell, 87: 1025–1035.

Armano, S., Rossi, P., Taglietti, V. and D'Angelo, E. (2000) Long-term potentiation of intrinsic excitability at the mossy fiber–granule cell synapse of rat cerebellum. J. Neurosci., 20: 5208–5216.

Bekkers, J.M. and Stevens, C.F. (1990) Presynaptic mechanism for long-term potentiation in the hippocampus. Nature, 346: 724–729.

Bredt, D.S., Hwang, P.M. and Snyder, S.H. (1990) Localization of nitric oxide synthase indicating a neural role for nitric oxide. Nature, 347: 768–770.

Bliss, T.V.P. and Collingridge, G.L. (1993) A synaptic model of memory: long-term potentiation in the hippocampus. Nature, 361: 31–39.

Carey, M.R. and Lisberger, S.G. (2002) Embarassed, but not depressed: eye opening lessons for cerebellar learning. Neuron, 35: 223–226.

Chatala, L., Brickely, S., Cull-Candy, S. and Farrant, M. (2003) Maturation of EPSCs and intrinsic membrane properties enhances precision at a cerebellar synapse. J. Neurosci., 23: 6074–6085.

D'Angelo, E., De Filippi, G., Rossi, P. and Taglietti, V. (1995) Synaptic excitation of individual rat cerebellar granule cells in situ: evidence for the role of NMDA receptors. J. Physiol. (Lond.), 484: 397–413.

D'Angelo, E., Rossi, P., Armano, S. and Taglietti, V. (1999) Evidence for NMDA and mGlu receptor-dependent long-term potentiation of mossy fibre–granule cell transmission in rat cerebellum. J. Neurophysiol., 81: 277–287.

D'Angelo, E., Rossi, P. and Taglietti, V. (1993) Different proportions of N-methyl-D-aspartate and non-N-methyl-D-aspartate receptor currents at the mossy fibre–granule cell synapse of developing rat cerebellum. Neuroscience, 53: 121–130.

D'Angelo, E., Sola, E., Prestori, F., Rossi, P. and Nieus, T. (2003) Evidence for increased neurotransmitter release during cerebellar mossy fiber–granule cell LTP. Neurosci. Soc. Abs., 690.7.

DeSchutter, E. (1997) A new functional role for cerebellar long-term depression. Prog. Brain Res., 114: 529–542.

Dieudonne, S. and Damoulin, A. (2000) Serotonin-driven long-range inhibitory connections in the cerebellar cortex. J. Neurosci., 20: 1837–1848.

DiGregorio, D.A. and Silver, R.A. (2002) Spillover of glutamate onto synaptic AMPA receptors enhances fast transmission at a cerebellar synapse. Neuron, 35: 521–533.

Eccles, J.C., Ito, M. and Szentagothai, J. (1967) The Cerebellum as a Neuronal Machine. Springer Verlag, Berlin.

Edwards, F.A. (1995) Anatomy and electrophysiology of fast central synapses lead to a structural model for long-term potentiation. Physiol. Rev., 75: 759–787.

Gall, D., Roussel, C., Nieus, T., Cheron, G., Servasi, L., D'Angelo, E. and Schiffmann, S.N. (2004) Role of calcium binding proteins in the control of cerebellar granule cell neuronal excitability: experimental and modeling studies. Progr. Brain. Res.

Gall, D., Prestori, F., Roussel, C., Sola, E., Forti, L., Rossi, P., Schiffmann, S.N. and D'Angelo, E. (2004) Calcium dynamics and plasticity at the mossy fiber/cerebellar granule cell.

Garthwaite, J., Charles, S.L. and Chess-Williams, R. (1998) Endothelium-derived relaxing factor release on activation of NMDA receptors suggests role as intercellular messenger in the brain. Nature, 336: 385–388.

Gasparini, S., Saviane, C., Voronin, L.L. and Cherubini, E. (2000) Silent synapses in the developing hippocampus: lack of functional AMPA receptors or low probability of glutamate release? Proc. Nat. Acad. Sci. (USA), 97: 9741–9746.

Ghez, C. and Thach, W.T. (2003) The cerebellum. In: Kandel, E., Schwartz, J.H. and Jessel, T.M. (Eds.), Principles of Neural Science. Appleton and Lange, Norwalk, pp. 626–646.

Hansel, C., Linden, D.J. and D'Angelo, E. (2001) Beyond parallel fiber LTD: the diversity of synaptic and non-synaptic plasticity in the cerebellum. Nature Neurosci., 4: 467–475.

Hartmann, M.J. and Bower, J.M. (1998) Oscillatory activity in cerebellar hemispheres on unrestrained rats. J. Neurophysiol., 80: 1598–1604.

Hawkins, R.D., Kandel, E.R. and Siegelbaum, S.A. (1993) Learning to modulate transmitter release: themes and variations in synaptic plasticity. Annu. Rev. Neurosci., 16: 625–665.

Ito, M. (1984) The Cerebellum and Neural Control. Raven Press, New York.

Ito, M. (2001) Cerebellar long-term depression: characterization, signal transduction, and functional roles. Physiol. Rev., 81: 1143–1195.

Jakab, J. and Hamori, J. (1988) Quantitative morphology and synaptology of cerebellar glomeruli in the rat. Anat. Embriol., 179: 81–88.

Kase, M., Miller, D.C. and Noda, H. (1980) Discharges of Purkinje cells and mossy fibres in the cerebellar vermis of the monkey during saccadic eye movements and fixation. J. Physiol. (Lond.), 300: 539–555.

Kullmann, D.M., Erdemli, G. and Asztely, F. (1996) LTP of AMPA and NMDA receptor-mediated signals: evidence for presynaptic expression and extrasynaptic glutamate spillover. Neuron, 17: 461–474.

Llinas, R., Lang, E.J. and Welsh, J.P. (1997) The cerebellum, LTD, and memory: alternative views. Learn. Mem., 3: 445–455.

Losi, G., Prybylowski, K., Fu, Z., Luo, J.H. and Vicini, S. (2002) Silent synapses in developing cerebellar granule neurons. J. Neurophysiol., 87: 1263–1270.

Maffei, A., Prestori, F., Rossi, P., Taglietti, V. and D'Angelo, E. (2002) Presynaptic current changes at the mossy fiber-granule cell synapse of cerebellum during LTP. J. Neurophysiol., 88: 627–638.

Maffei, A., Prestori, F., Shibuki, K., Rossi, P., Taglietti, V. and D'Angelo, E. (2003) NO enhances presynaptic currents during cerebellar mossy fiber-granule cell LTP. J. Neurophysiol., 90: 2478–2483.

Malenka, R.C. and Nicoll, R.A. (1999) Long-term potentiation — A decade of progress? Science, 285: 1870–1874.

Malgaroli, A., Ting, A.E., Wendland, B., Bergamaschi, A., Villa, A. and Tsien, R.W. (1995) Presynaptic component of long-term potentiation visualized at individual hippocampal synapses. Science, 268: 1624–1628.

Malinow, R. and Tsien, R.W. (1990) Presynaptic enhancement shown by whole-cell recording of long-term potentiation in hippocampal slices. Nature, 346: 177–180.

Marr, D. (1969) A theory of the cerebellar cortex. J. Physiol. (Lond.), 202: 437–470.

Monti, B., Marri, L. and Contestabile, A. (2002) NMDA receptor-dependent CREB activation in survival of cerebellar granule cells during in vivo and in vitro development. Eur. J. Neurosci., 16: 1490–1498.

Nieus, T., Sola, E. and D'Angelo, E. (2004) Experimental and modelling investigation of synaptic dynamics in cerebellar granule cells. Europ. J. Neurosci., A119.2.

Nunzi, M.G., Birnstiel, S., Bhattacharyya, B.J., Slater, N.T. and Mugnaini, E. (2001) Unipolar brush cells form a glutamatergic projection system within the mouse cerebellar cortex. J. Comp. Neurol., 434: 329–341.

Pellerin, P.-P. and Lamarre, Y. (1997) Local field potential oscillations in primate cerebellar cortex during voluntary movement. J. Neurophysiol., 78: 3502–3507.

Petralia, R.S., Wang, Y.-X. and Wenthold, R.J. (2002) NMDA receptors and PSD-95 are found in attachment plaques in cerebellar granular layer glomeruli. Eur. J. Neurosci., 15: 583–587.

Philipona, D., Dognin, E. and Coenen, O.J.-M.D. (2003) A model of the granular layer of the cerebellum. Comp. Neurosci., to appear.

Rossi, P., D'Angelo, E. and Taglietti, V. (1996) Differential long-lasting potentiation of the NMDA and non-NMDA synaptic currents induced by metabotropic and NMDA receptor coactivation in cerebellar granule cells. Eur. J. Neurosci., 8: 1182–1189.

Rossi, P., Sola, E., Taglietti, V., Borchardt, T., Steigerwald, F., Utvik, K., Ottersen, O.P., Kohr, G. and D'Angelo, E. (2002) Cerebellar synaptic excitation and plasticity require proper NMDA receptor positioning and density in granule cells. J. Neurosci., 22: 9687–9697.

Schulz, P.E., Cook, E.P. and Johnston, D. (1994) Changes in paired-pulse facilitation suggest presynaptic involvement in long-term potentiation. J. Neurosci., 14: 5325–5337.

Schweighofer, N., Doya, K. and Lay, F. (2000) Unsupervised learning of granule cell sparse codes enhances cerebellar adaptive control. Neuroscience, 103: 35–50.

Silver, R.A., Cull-Candy, S.G. and Takahashi, T. (1996) Non-NMDA glutamate receptor occupancy and open probability at a rat cerebellar synapse with single and multiple releases sites. J. Physiol. (Lond.), 494: 231–250.

Siöström, P.J. and Nelson, S.B. (2002) Spike timing, calcium signals and synaptic plasticity. Curr. Opin. Neurobiol., 12: 305–314.

Sola, E., Prestori, F., Rossi, P., Taglietti, V. and D'Angelo, E. (2004) Increased neurotransmitter release during long-term potentiation at mossy fibre-granule cell synapses in rat cerebellum. J. Physiol., 557:843–861.

Sprengel, R., Suchanek, B., Amico, C., Brusa, R., Burnashev, N., Rozov, A., Hvalby, V., Jensen, N., Paulsen, O., Andersen, P., Kim, J.J., Thompson, R.F., Sun, W., Webster, L.C., Grant, S.G.N., Eilers, J., Konnerth, A., Li, J., McNamara, J.O. and Seeburg, P.H. (1998) Importance of the intracellular domain of NR2 subunits for NMDA receptor function in vivo. Cell, 92: 279–289.

Xu-Friedman, M.A. and Regher, W.G. (2003) Ultrastructural contributions to desensitization at cerebellar mossy fiber to granule cell synapses. J. Neurosci., 23: 2182–2192.

CHAPTER 8

Climbing fiber synaptic plasticity and modifications in Purkinje cell excitability

Matthew T. Schmolesky, Chris I. De Zeeuw and Christian Hansel*

Department of Neuroscience, Erasmus Medical Center, P.O. Box 1738, 3000 DR Rotterdam, The Netherlands

Keywords: cerebellum; climbing fiber; long term depression; Purkinje cell; complex spike; afterhyperpolarization; afterdepolarization; intrinsic excitability; calcium-activated potassium current

Introduction

The cerebellum is a brain area that is critical for the fine adjustment of motor output and for the formation of several types of motor memory. The Purkinje cell (PC) is the only type of neuron that projects out of the cerebellar cortex and, as such, is considered to be central for cerebellar processing. Purkinje cells receive two types of excitatory input: the parallel fibers rising from the granule cells of the cerebellar cortex and the climbing fibers rising from neurons of the inferior olive nucleus found in the brainstem. In the adult mammal, the dendritic tree of each PC is innervated by hundreds of thousands of parallel fibers (PFs) and one single climbing fiber (CF). A vast body of research has been dedicated to the study of long term depression (LTD) at the parallel fiber–Purkinje cell synapse. The physiological and molecular requirements for PF–LTD induction have been well characterized, as have the molecular cascades leading to PF–LTD expression (Bear and Linden, 2001; Ito, 2001). Furthermore, the importance of PF–LTD for the formation of several types of motor memory, such as adaptation of the vestibulo-ocular reflex and associative eyeblink conditioning, has been convincingly demonstrated (De Zeeuw et al., 1998; Koekkoek et al., 2003). In contrast to the wealth of information currently available about PF–LTD, we have only recently discovered and begun to characterize long term depression at the climbing fiber–PC synapse (Hansel and Linden, 2000; Schmolesky et al., 2002; Shen et al., 2002; Weber et al., 2003).

Although each PC is innervated by only one climbing fiber, this input is disproportionately strong as it forms more than 1500 high-probability release synapses that are near-simultaneously active (Dittman and Regehr, 1998; Silver et al., 1998; Strata and Rossi, 1998; Foster et al., 2002). Under physiological conditions, excitation of the climbing fiber leads to the synchronous activation of ionotropic (AMPA) and type-1 metabotropic (mGluR1) glutamate receptors. Subsequently, Na^+ influx depolarizes the PC with the consequence that voltage gated calcium channels (VGCCs) are activated and conduct Ca^{2+}

---

*Corresponding author. Tel.: +31-10-4087404; Fax: +31-10-4089459; E-mail: c.hansel@erasmusmc.nl

DOI: 10.1016/S0079-6123(04)48008-X

into the neuron. At the same time, the binding of glutamate to mGluR1s leads to the synthesis of the second messengers inositol-trisphosphate (IP$_3$) and diacylglycerol (DAG). IP$_3$ causes the liberation of calcium from the IP$_3$-mediated internal stores. This calcium combines with that flowing through VGCCs and with DAG to elevate protein kinase C (PKC) activity. The rise in postsynaptic [Ca^{2+}]$_i$ levels also initiates calcium-activated potassium (K$_{Ca^{2+}}$) conductances. The sum total of these events is a complex spike that, during whole-cell current clamp recording from single PCs in the brain slice preparation, appears as a single fast component (primarily due to Na$^+$ influx), one or more slow components (due to Ca^{2+} and Na$^+$ influx), and a slow afterhyperpolarization (AHP; due to K$^+$ efflux) (see Schmolesky et al., 2002).

Complex spikes fire spontaneously in awake behaving animals at a rate of $\sim$1 Hz, but can reach rates of up to 10 Hz for very short periods of time (Simpson et al., 1996). Hansel and Linden (2000) found that tetanization of the climbing fiber at 5 Hz for 30 s was sufficient to cause LTD at the CF–PC synapse. This form of synaptic plasticity is expressed as a $\sim$20% decrease in the excitatory postsynaptic current (EPSC) amplitude. As in the case of PF–LTD, expression of CF–LTD is postsynaptic (Shen et al., 2002) and is dependent upon a rise in the postsynaptic calcium levels, activation of type 1 metabotropic glutamate receptors (mGluR1), and activation of PKC. However, CF–LTD is unique in that its expression is not restricted to the reduction of the AMPA-R generated EPSC. Instead, the second, slow component of the complex spike is also substantially reduced (Hansel and Linden, 2000) as is the CF-evoked PC dendritic Ca^{2+} signal (Weber et al., 2003). Neither the success rate of evoking complex spikes nor the fast Na$^+$-dependent component of the complex spike is affected by CF–LTD induction.

The goal of this study was to extend our findings on long term plasticity at the CF–PC synapse. More specifically, we sought to confirm preliminary evidence that CF–LTD results not only in a reduction of the second spike, but also in a suppression of the complex spike AHP (Schmolesky et al., 2002). Particular attention was paid to the slow AHP, following the complex spike for two reasons. First, the fact that the dendritic Ca^{2+} signals are depressed by CF–LTD suggested to us that the Ca^{2+}-activated K$^+$ conductances presumed to underlie the AHP might also be affected. Second, we are greatly interested in the functional implications of CF–LTD and a reduction in the AHP is likely to have consequences for PC dendritic integration and axonal firing patterns.

Materials and methods

Slice preparation and electrophysiology

Parasagittal slices of the cerebellar vermis (200 μm thick) were prepared from postnatal day 18–28 Sprague–Dawley rats by using a vibratome (LeicaVT1000S). The dissection and slicing were conducted in ice cold artificial cerebrospinal fluid (ACSF) containing (in mM): 124 NaCl, 5 KCl, 1.25 Na$_2$HPO$_4$, 2 MgSO$_4$, 2 CaCl$_2$, 26 NaHCO$_3$, and 10 D-glucose bubbled with 95% O$_2$/5% CO$_2$. The slices were maintained in room temperature ACSF for 1–6 h before recording. The slices were transferred to a submerged recording chamber and perfused at room-temperature with ACSF supplemented with either 20 μM bicuculline methiodide or picrotoxin to block γ-aminobutyric acid (GABA) type A receptors (flow rate 1–3 ml/min). Visually guided whole-cell patch clamp recordings were conducted using a Zeiss Axioskop FS with IR-DIC optics, a Hamamatsu C7500 CCD camera attached to a Sony monitor, and an EPC-9 amplifier (HEKA Electronics, Lambrecht/Pfalz, Germany). All recordings were conducted with borosilicate capillaries pulled to provide electrode resistances of 2–4 MΩ and tip sizes of 1–2 μm. The patch pipettes were filled with a solution containing (in mM): 9 KCl, 10 KOH, 120 K gluconate, 3.48 MgCl$_2$, 10 HEPES, 4 NaCl, 4 Na$_2$-ATP, 0.4 Na$_3$-GTP, and 17.5 sucrose (pH adjusted to 7.25; osmolarity = 305 ± 5 mOsm). For experiments including BAPTA in the pipette solution, K gluconate was reduced to 67 mM and both K$_4$ BAPTA (20 mM) and CaCl$_2$ (2 mM) were added. Before seal formation, the liquid junction potential was compensated using standard procedures. Currents were filtered at 3 kHz, digitized at 8 kHz and acquired using PULSE software. Small negative currents were passed to move the PC membrane potential into the range of −75 to −65 mV to prevent spontaneous spike activity. Electrical stimulation of a climbing fiber was conducted using a stimulus isolation unit attached to a standard pipette filled with ACSF and placed in the granule cell layer. The test responses were evoked at 0.05 Hz with $\sim 3.0\,\mu A$ square waves pulses of 0.5 ms duration. The induction of CF–LTD was carried out by tetanizing the climbing fiber at 5 Hz for 30 s using the same stimulus. Only cells that were innervated by one CF input, as indicated by a single step in the CF input–output relation, were used. All drugs were purchased from Sigma.

Data inclusion criteria and analysis

All data analysis was conducted using PulseFit (HEKA), IGOR and SPSS software. To monitor the complex spike waveform modifications, the peak amplitudes of spike components were measured relative to a preceding trough or the prestimulus resting membrane potential, chosen for each cell to maximize stability during the baseline period. Voltage integrals of the slow spike region were measured by summing the area under the curve starting at the trough separating spike 1 from spike 2 and ending 10 ms later (Fig. 1A–D). The amplitude of the complex spike afterhyperpolarization was determined by subtracting the resting membrane potential from the minimum voltage recorded in the 50–500 ms time period following the stimulus (Fig. 2A). The amplitude of the afterdepolarization was measured relative to the resting membrane potential and analyzed at 50 and 100 ms following the stimulus onset (Fig. 3A). For all measures, the data for three sweeps collected per min of the recording were averaged and normalized to the 5 min baseline. Short term recordings from a large number of PCs revealed that the initial AHP amplitude is quite variable. Data for this study were not included from any cell where the AHP amplitude was < 1 mV at the beginning of the recording. Short negative current pulses were delivered at the end of each sweep cycle (1.5 s following synaptic stimulation) and the resulting voltage response was taken as a measure of input resistance. Data were included from cells that demonstrated stability in all response parameters for at least a 5 min baseline period and stability in resting membrane potential ($\Delta < \pm 5\,mV$) and input resistance ($\Delta < \pm 15\%$) for ~ 30 min or more. A mixed design two-way ANOVA with one between group factor (tetanus vs. control condition) and one within group factor (time) was used to assess statistically significant differences in the baseline and post-baseline periods for each measure. F-values were Greenhouse–Geisser corrected. Planned comparisons were carried out using an independent or paired-sample t-tests, as appropriate, with the Bonferroni correction applied.

Results

CF–LTD is accompanied by an attenuation of the complex spike afterhyperpolarization

In our initial study, we found that the second, partly Ca^{2+} dependent component of the complex spike was reduced by the same climbing fiber tetanization protocol that had induced LTD of the CF EPSC (Hansel and Linden, 2000). The main goal of the present study was to extend this work by replicating the experiment, analyzing the early components in greater detail, and recording increased sweep durations to capture the slow afterdepolarization (ADP) and afterhyperpolarization (AHP). As before, we found that the early components of the complex spike were stable over a long period of time in the control condition (Fig. 1E, F; $n = 8$). Also as before, tetanization resulted in a substantial reduction of the second spike amplitude in a subset (8/13) of cells ($86 \pm 2.0\%$, $t = 25$ min, $n = 8$; Fig. 1B). The remaining five cells in the tetanus condition did not demonstrate a spike 2 amplitude decrease ($105 \pm 3.5\%$, $t = 25$ min, $n = 5$) but did show a substantial reduction in the voltage integral (area under the curve) of the 10 ms period, following a spike 2 initiation (example shown in Fig. 1C, D). Thus, the voltage integral of the 'slow spikes' was analyzed for all cells and compared across conditions. The tetanization resulted in a significant reduction of the slow spike integral compared to baseline ($90 \pm 1.6\%$, $t = 25$ min, $n = 12$) and compared to the control condition ($p < 0.05$) but had no effect upon the spike 1 amplitude ($103 \pm 3.2\%$, $t = 25$ min,

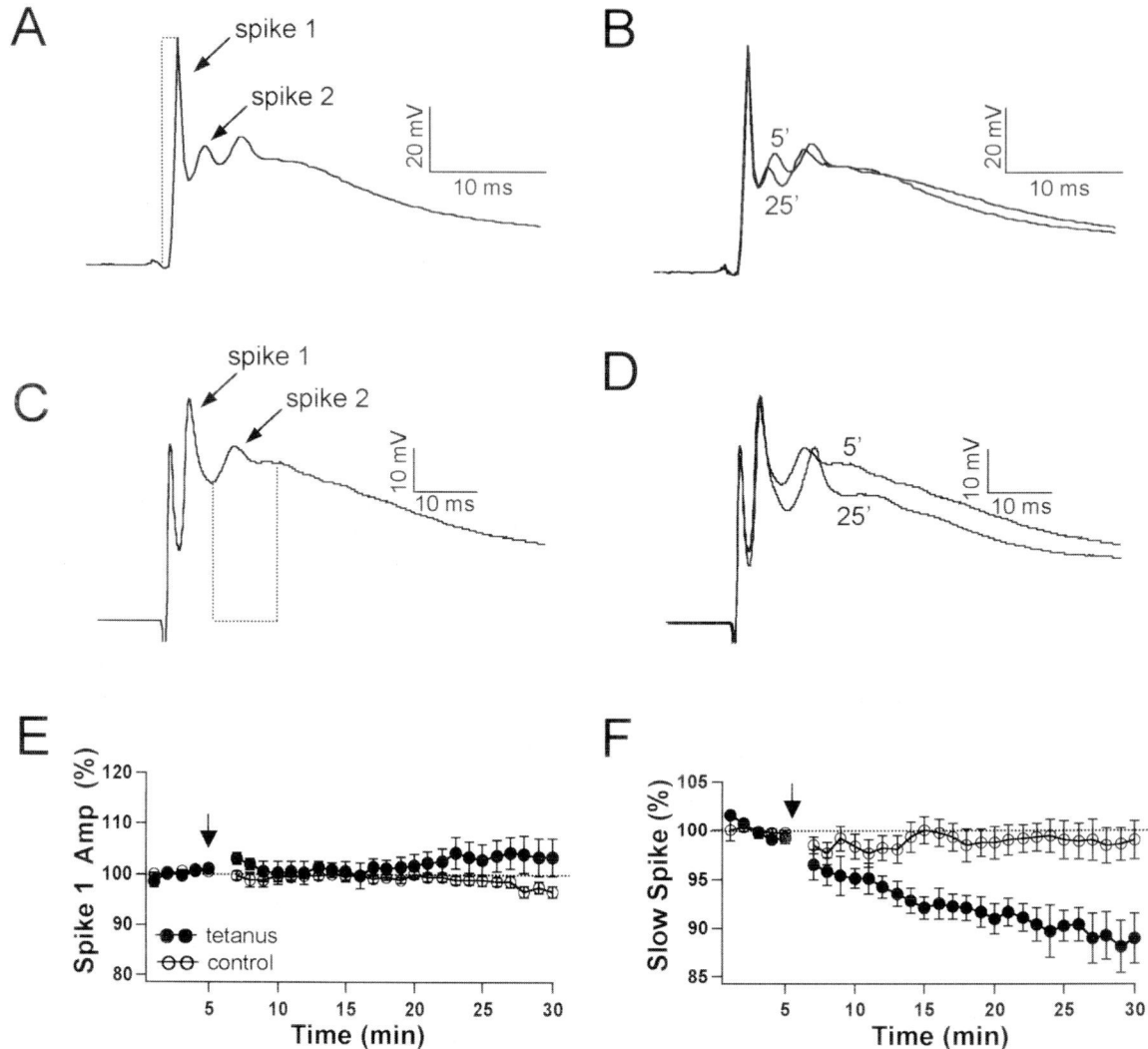

Fig. 1. Excitation of the climbing fiber evokes a complex spike in the Purkinje cell. A complex spike recorded from the PC soma demonstrates one fast Na^+ spike (spike 1) and one to four slower spikes atop a depolarization plateau (see Schmolesky et al., 2002). In some cases, stimulation also activated the PC axon and resulted in a retrograde that precedes the complex spike (example in C). Representative complex spike traces are shown from two cells recorded at the fifth min of the baseline (A, C) and at 20 min post-tetanus (B, D). All spike amplitudes are measured as the difference from the peak to the preceding trough or resting membrane potential (A, dashed line). Slow spike integrals are measured by summing the area under the curve of the 10 ms region following the onset of spike 2 (C, dashed line). Stimulus artifacts have been truncated or digitally reduced. Tetanization of the climbing fiber at 5Hz for 30 s (arrowhead) results in an attenuation of the spike 2 amplitude (example in B) and/or slow spike integral (F; $n = 12$) compared to baseline or to control ($n = 8$). Tetanization has no effect upon the spike1 amplitude (E; $n = 13$).

$n = 13$). From this, we can conclude that CF–LTD may depress the magnitude of the depolarization plateau, the slow spikes riding atop it or, potentially, both.

Long term recordings of the complex spike AHP indicate that this parameter was also stable throughout the recording period in a subset of control cells (3 of 8; Fig. 2C). On average, the AHP

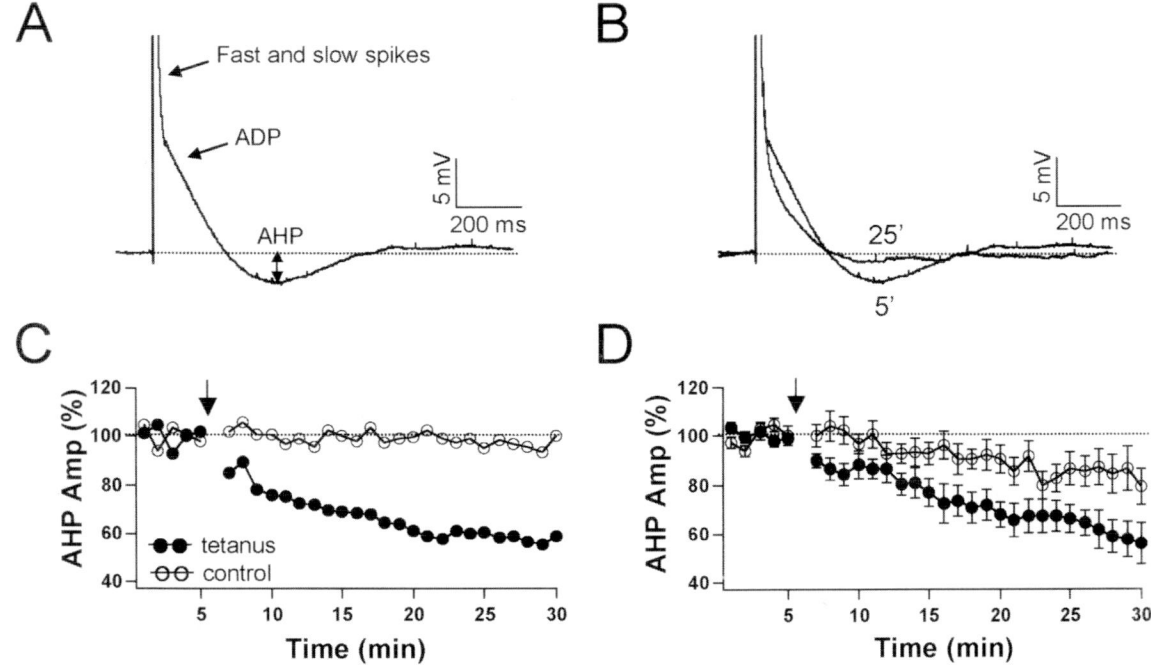

Fig. 2. (A, B) Subsequent to the fast and slow spikes, a complex spike is composed of a slow afterdepolarization (ADP) and afterhyperpolarization (AHP). The complex spike AHP for an individual PC in the control condition could be held stable for over 30 min (C). On average, the control AHP demonstrated a slow decline over time (D). However, climbing fiber tetanization resulted in a significant reduction in the complex spike AHP ($n = 13$) compared to baseline or to control ($n = 8$).

stability could be retained for 5–10 min but then steadily declined at a slow rate ($86 \pm 6.8\%$, $t = 25$ min, $n = 8$) (Fig. 2D). This was true despite the fact that the complex spike amplitudes/integrals, resting membrane potential, holding current, and input resistance all remained stable. Attempts to further stabilize the AHP in the control condition failed. Filling the patch electrodes with a methylsulfate based internal solution reported to improve K conductance stability (Velumian et al., 1997) did not influence the AHP decline ($n = 3$, data not shown). The GABA$_A$ receptor blocker we initially used, bicuculline methiodide, has been reported to inhibit SK ('small' conductance potassium) channels (Johnson and Seutin, 1997). However, a replacement of bicuculline with another common GABA$_A$ antagonist, picrotoxin, also did not yield any effect upon the AHP decline. Nonetheless, CF tetanization resulted in a significant reduction in the complex spike AHP compared to baseline ($66 \pm 5.9\%$, $t = 25$ min, $n = 13$, $p < 0.05$) or to the control condition (Fig. 2D). The difference in the AHP amplitude between conditions reached $\sim 20\%$ within 10 min after tetanization and was evident for as long as the recordings could be maintained (> 1 h).

The complex spike afterdepolarization is not affected by CF–LTD

Under our recording conditions, the complex spikes repolarize at a slow rate. Afterdepolarizations typically spanned a ~ 150 ms period following the last slow spike (Fig. 3A; also see Hashimoto and Kano, 1998). Similar to the fast and slow spikes, the ADP is presumed to reflect mixed Ca^{2+} and Na^+ conductances. Thus, one logical prediction was that the ADP, like the early spikes, would remain stable in the control condition and be significantly depressed

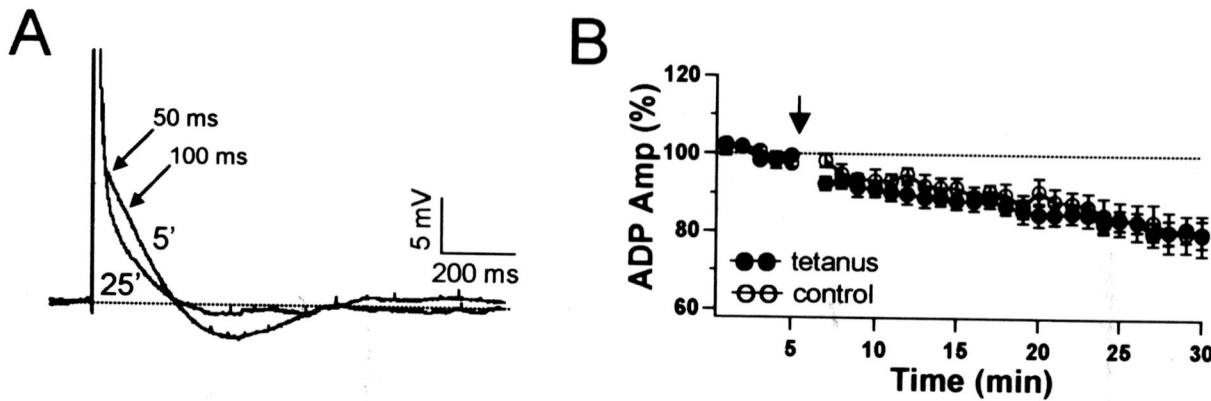

Fig. 3. (A) The afterdepolarization (ADP) was measured at 50 and 100 ms following stimulus onset. (B) The ADP measured at 100 ms shows a significant downward trend over time in both conditions, whereas the differences between conditions proved to be insignificant.

following tetanization. In fact, the opposite was observed: the ADP showed no significant differences between conditions ($p > 0.05$) and declined steadily over the recording period, regardless of the measurement point. At 50 and 100 ms following climbing fiber stimulation, the ADP in the control condition was reduced to $89 \pm 3.2\%$ and $84 \pm 4.3\%$, respectively ($t = 25$ min, $n = 8$). The same measures in the tetanus condition came to $86 \pm 1.4\%$ and $83 \pm 2.4\%$, respectively ($t = 25$ min, $n = 11$) (Fig. 3A, B).

The results described above suggest that the complex spike AHP amplitude could be determined by the magnitude of both the slow spikes and the ADP. Scatterplots of the values for these measures across the 30 min recording period for individual cells (Fig. 4A–D) and averaged by condition (Fig. 4E, F) provide graphic evidence for these relationships. In the tetanus condition, there is a strong positive correlation between the slow spike and AHP for all cells (Fig. 4A, E) and between the ADP and AHP for most (Fig. 4B, F). In contrast, while the ADP and AHP also show a strong positive correlation for most cells in the control condition (Fig. 4D, F), the correlations between the slow spike and the AHP are evenly split between a positive and a negative relationship (Fig. 4C, E). The correlation values provide evidence of association and not causality. However, the temporal order of the early spikes, ADP and AHP suggest that the steady decline of the ADP could be directly responsible for the AHP decline in the control condition. Assuming this to be true, we should expect that the ADP decline accounts for approximately half of the AHP reduction seen in the tetanus condition, with the balance directly attributable to the depression of the slow spikes. This explanation is also consistent with the notion that the early spikes and the ADP reflect Ca^{2+} and Na^+ currents, while the AHP reflects the action of Ca^{2+}-activated K^+ channels that may be voltage sensitive (BK) or not (SK). We sought to test this explanation and directly demonstrated the calcium dependence of the complex spike AHP through the use of a calcium chelator.

The complex spike afterhyperpolarization is mediated by a Ca^{2+}-sensitive K^+ current

As predicted, the inclusion of the calcium chelator BAPTA (20 mM) in the patch pipette internal saline caused a dramatic reduction in the complex spike AHP ($29 \pm 4.4\%$ relative to baseline values, $t = 25$ min, $n = 7$; Fig. 5B, D). This AHP suppression is not the result of impaired postsynaptic activity or complex spike generation, as the amplitude of both spike 1 and spike 2 are unaffected by this treatment ($101 \pm 4.1\%$ and $100 \pm 4.3\%$, respectively, $t = 25$ min, $n = 7$; Fig. 5C) and the probability of evoking

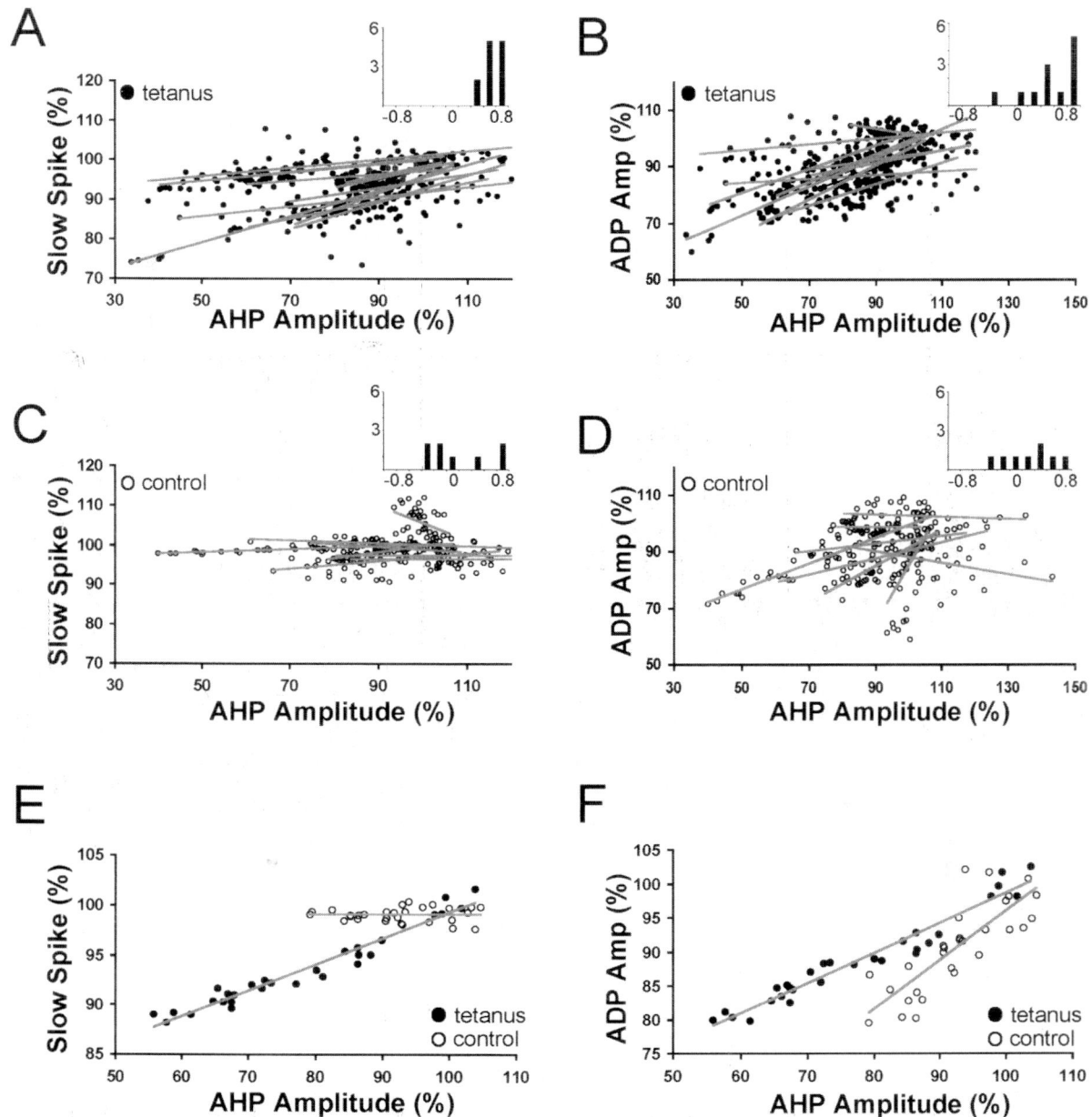

Fig. 4. (A–D) Scatterplots of the slow spike integrals, AHP amplitudes, and ADP amplitudes taken over the 30 min recording period for individual cells with trend lines superimposed. Insets indicate the number of cells (y axis) with a given correlation value (x axis; 0 indicates r values of 0 to 0.2; 0.8 indicates r values of 0.8 to 1.0). The open and closed captions represent data from the control ($n = 8$) and tetanus ($n = 11$) conditions, respectively. (E–F) Scatterplots of the average slow spike integrals, AHP amplitudes, and ADP amplitudes taken over the 30 min recording period with trend lines superimposed. Conventions as in A–D.

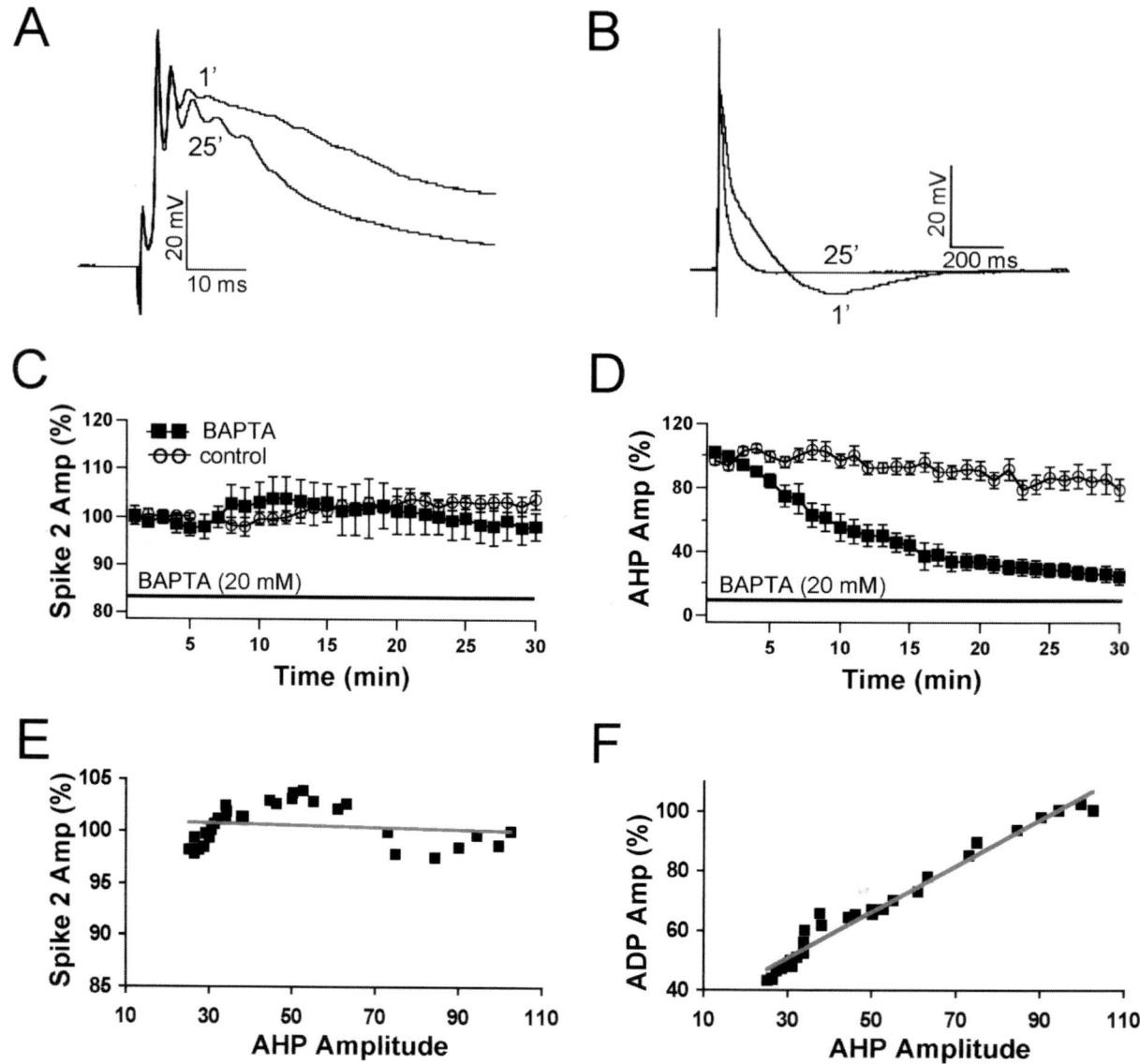

Fig. 5. Inclusion of a calcium chelator (20 mM BAPTA) into the patch pipette results in a swift, selective reduction in the complex spike AHP (B, D) and ADP ($n = 7$). Spike 1 and spike 2 amplitudes are not affected by the BAPTA inclusion (A, C). Scatterplots of the average values taken over the 30 min recording period reveal no relationship between the spike 2 amplitude and AHP (E) but demonstrate a strong positive correlation between the ADP and AHP (F). Trend lines are superimposed.

a complex spike remained near 100%. However, calcium chelation did result in a dramatic reduction of the complex spike ADP measured 100 ms following climbing fiber activation ($48 \pm 9.0\%$, $t = 25$ min, $n = 7$; Fig. 5F). These data suggest that the AMPA-mediated depolarization, fast Na^+ spike, and initial Ca^{2+} influx remain intact after dialysis of the cell with BAPTA, but that calcium chelation prevents slower Ca^{2+}-mediated depolarization (i.e., the ADP), and activation of $K_{Ca^{2+}}$ channels.

Discussion

The long term modification of synaptic strength following an associative event is a favored biological mechanism for memory formation. This phenomenon, termed synaptic plasticity, has been extensively documented in multiple brain areas including the hippocampus, the cortex and the cerebellum and, in several cases, has been directly linked to a specific form of memory. Long term depression at the parallel fiber–Purkinje cell (PF–PC) synapse is thought to be necessary for at least two forms of motor learning: associative eyeblink conditioning and vestibulo-ocular reflex adaptation (Ito et al., 1982; De Zeeuw et al., 1998; Ito, 2001; Koekkoek et al., 2003; for contrasting views see De Schutter, 1995; Llinas et al., 1997). The parallel fiber LTD under physiological conditions is associative, wherein the CF and PF inputs must be coactivated for LTD induction to occur. It has long been believed that the single CF innervating each PC provides massive and invariant excitation. However, we now know that activation of the climbing fiber alone at a moderate frequency for a short duration leads to LTD of the CF–PC synapse (Hansel and Linden, 2000; Hansel et al., 2001). The primary goal of this study was to further explore the physiological consequences of CF–LTD at the cellular level. This goal was accomplished by using whole-cell patch clamp recording in the rat cerebellar brain slice preparation.

Our previous and current investigations collectively reveal four aspects of CF–LTD. That is, the single CF-LTD induction protocol results in the selective reduction of the CF-evoked: (a) EPSC amplitude, (b) dendritic Ca^{2+} transient, (c) slow spike amplitude or integral, and (d) complex spike AHP. The possibility exists that these phenomena are derived in a simultaneous but independent fashion. For instance, direct modification or internalization of the AMPA receptors, voltage gated calcium channels (VGCCs) and K^+ channels (Ca^{2+}-sensitive or otherwise) could instantiate the first, second and fourth aspects of CF–LTD, respectively. While the evidence to date is insufficient to discount this possibility, the data and the parsimony favor the serial determination of these four aspects. In this model, CF-LTD induction initially leads to AMPA-R internalization and the attenuation of subsequent CF-evoked EPSCs. Decreased local dendritic depolarizations following climbing fiber activation will translate into reduced Ca^{2+} influx through VGCCs which is, in turn, reflected in the altered complex spike waveform (i.e., reduced slow spike). The reduced Ca^{2+} influx also results in less current through $K_{Ca^{2+}}$ channels and, thus, directly causes the impaired AHP. The evidence for this serial determinant model is manifold.

CF-LTD at the EPSC level is postsynaptic (Hansel and Linden, 2000; Shen et al., 2002) and at least three of the key molecular requirements are identical for the induction of PF- and CF-LTD at the EPSC level. These facts suggest that AMPA-R internalization occurs in CF-LTD as has been proven for PF-LTD (Wang and Linden, 2000). LTD of the EPSC, slow spike, and Ca^{2+} transient is blocked by the PKC antagonist chelerythrine (Hansel and Linden, 2000; Weber et al., 2003). Furthermore, the same concentration of NBQX that provides a $\sim 20\%$ reduction in the CF-EPSC also selectively reduces slow complex spike components (Weber et al., 2003), suggesting a causal link between these two factors. Finally, the fact that the complex spike AHP is dependent upon internal Ca^{2+} concentration is consistent with the notion that LTD of CF-evoked Ca^{2+} transients dictates the AHP reduction (Fig. 2) in part or in total.

Regarding the specific ion conductances underlying the complex spike AHP, it has long been assumed that Ca^{2+}-activated K^+ channels are involved. Climbing fiber activation of a PC results in considerable Ca^{2+} influx (Llinas and Sugimori, 1980a,b) and K^+ efflux (Bruggencate et al., 1976). In situ hybridization and immunochemistry have revealed that both the SK and the BK ('small' and 'big' conductance potassium) channels are present in the PC membrane (Knaus et al., 1996; Chang et al., 1997; Stocker and Pedarzani, 2000). Furthermore, a number of electrophysiological experiments have characterized currents in the PC that match known SK and BK characteristics (Gruol et al., 1991; Jacquin and Gruol, 1999; Womack and Khodakhah, 2002, 2003; Edgerton and Reinhart, 2003). However, these studies have not directly tested the Ca^{2+}-dependence or the pharmacological profile of the complex spike AHP. The current data directly demonstrate that the complex spike AHP is calcium dependent: inclusion of the calcium chelator BAPTA in the patch pipette causes a swift and near-complete suppression of the complex spike AHP while leaving the early spike components unaffected (Fig. 5). Both SK and BK channels have been reported to be tightly coupled or co-localized with VGCCs in different neurons (Vergara et al., 1998; Hallworth et al., 2003). Such tight coupling, if it exists in the PCs, would have implications for the cellular expression and specificity of CF-LTD (see below). The BK channels have been specifically implicated in controlling the AHP seen after calcium spikes (Edgerton and Reinhart, 2003) and may, therefore, prove to underlie the complex spike AHP as well. Future studies will examine the pharmacological profile of the complex spike slow AHP to determine its dependence upon the BK and/or SK channels.

Calcium chelation also suppresses the complex spike ADP (Fig. 3) indicating that it too is a Ca^{2+}-dependent phenomenon. The ~20% reduction of the CF-evoked Ca^{2+} transient by CF-LTD (Weber et al., 2003) is apparently not sufficient to significantly impact the ADP compared to control, though a slight trend in that direction is observed (Fig. 3). We have found that the ADP and AHP are correlated in all three of the cell groups described here (Figs. 3 and 5). In this context, it is worth noting that paired pulse CF activation with short inter-stimulus intervals (<500 ms) leads to reduced early spikes, ADP and AHP in the second complex spike, but has no effect on the fast Na^+ spike (Hashimoto and Kano, 1998). The ADP amplitude itself may contribute to the AHP amplitude as BK channels are not only Ca^{2+}-sensitive, but also voltage-sensitive.

Modification of synaptic and intrinsic excitability

Synaptic plasticity is often activity-dependent, bi-directional, and input-specific. That is, a particular pattern of input activity is required to induce plasticity, the strength of an individual synapse can be either potentiated or depressed, and changes can be made at one synapse, independent of those adjacent or distant to it. These properties are ideal for memory formation. However, synaptic plasticity is not the only cellular mechanism underlying the memory formation. An additional, nonexclusive candidate is that of modifying neuronal intrinsic excitability (Zhang and Linden, 2003).

All neurons express voltage-gated ion channels on their dendritic and axosomatic membrane. If the number, composition or conformation of these channels is substantially altered the responsiveness of the neuron to ongoing electrochemical events (i.e., its intrinsic excitability) would be expected to change in kind (Bernard and Johnston, 2003). Thus, it is possible that a particular pattern of input activity could cause a change in the intrinsic excitability of a neuron rather than, or in addition to, a change in synaptic strength per se. For instance, a mossy fiber stimulation at different frequencies is sufficient to induce changes in the intrinsic excitability in either cerebellar granule cells (Armano et al., 2000) or neurons of the deep cerebellar nuclei (Aizenman and Linden, 2000).

In practice, experience-dependent alterations in neuronal excitability are often due to a modification of K^+ currents. In hippocampal CA1 pyramidal neurons, the coincidence of backpropagating spikes and synaptic events may cause both LTP and altered dendritic excitability, with the latter due to changes in A-type K^+ currents (Johnston et al., 2003). The excitability of hippocampal pyramidal cells may also be modified as the result of trace eyeblink conditioning. Here the postburst AHP following somatic current injection is reduced, as is the Ca^{2+}-sensitive K current I_{AHP} (Disterhoft et al., 1988; Coulter et al., 1989; de Jonge et al., 1990; Sanchez-Andres and Alkon, 1991). Pyramidal neurons in the olfactory cortex from trained rats demonstrate reduced postburst AHP for several days after odor-discrimination learning, and this reduction is also due to decreased current through $K_{Ca^{2+}}$ channels (Saar et al., 1998). In the cerebellum, intradendritic PC current injection reveals a K^+ channel-mediated transient hyperpolarization that is specifically reduced following delay eyeblink conditioning in rabbits (Schreurs et al., 1998). Additionally, the deletion of the HCN1 channel, which mediates an important K^+ conductance in PCs, interferes with motor learning and memory (Nolan et al., 2003). Finally, the neurons in the vestibular nucleus show long-term increases in excitability, following brief periods of inhibitory synaptic stimulation. This change is thought to be due to decreases in $[Ca^{2+}]_i$ and, thereby, $K_{Ca^{2+}}$ current and is hypothesized to influence motor learning in the vestibulo-ocular reflex (Nelson et al., 2003). It is important to note that, in cases such as those listed above, modified K^+ currents may be due to changes in a K^+ channel itself (i.e., intrinsic excitability), another voltage-gated channel that influences the K^+ currents (also intrinsic excitability), or synaptic efficacy that indirectly influences the K^+ currents (synaptic, but not intrinsic, excitability).

Is the Purkinje cell complex spike AHP reduction described here the result of altered synaptic or intrinsic excitability? Based on the serial determinance model for CF–LTD proposed herein, we hypothesize that the $K_{Ca^{2+}}$ channels are not directly modified, that depressed CF–PC synaptic efficacy leads indirectly to reduced $K_{Ca^{2+}}$ transients, and that the reduced AHPs observed are specific to CF-evoked events. Thus, fast and/or slow K^+ currents evoked by non-CF synaptic excitation or somatic current injection are expected to be unaffected by CF–LTD induction. Ongoing studies have been designed to test this prediction.

What is the functional significance of long-term depression at the climbing fiber to Purkinje cell synapse? It should be clear from the examples given above that learning-associated declines in the potassium-mediated membrane hyperpolarization are common. While it is possible that CF–LTD could provide a substrate for adaptive learning (for instance, when an unconditional stimulus is repetitively presented), this form of plasticity is more likely to affect motor behavior and learning by modulating ongoing cerebellar activity and other forms of synaptic plasticity.

Under natural conditions in the intact animal, beyond the climbing fiber input, the Purkinje cells receive synaptic inputs from granule, stellate and Golgi cells and recursive inputs from their own axon collaterals (Eccles et al., 1966b, 1967; De Zeeuw and Berrebi, 1995). Even in the absence of synaptic inputs, the PCs may demonstrate a high rate of spontaneous activity (Häusser and Clark, 1997; Raman and Bean, 1997, 1999). We have suggested that the $K_{Ca^{2+}}$ depression seen after CF–LTD induction will be restricted to CF-evoked events. However, the CF-evoked Ca^{2+} transient spreads throughout large regions of the PC dendritic tree and may elevate $[Ca^{2+}]_i$ for over 500 ms (Ross and Werman, 1987; Knöpfel et al., 1991; Konnerth et al., 1992; Miyakawa et al., 1992; Weber et al., 2003). Furthermore, the duration of the complex spike AHP itself may be on the order of 50 ms as measured in vivo (Eccles et al., 1966a; Murphy and Sabah, 1970; Bloedel and Roberts, 1971). Thus, even if non-CF synaptic inputs and intrinsically generated phenomena are not directly impacted by CF–LTD induction, each could be substantially altered if and when they occur shortly after a complex spike.

In vivo extracellular recordings have shown that a single spontaneous or stimulus-evoked complex spike is followed by a pause in ongoing simple spike activity (Simpson et al., 1996; Schmolesky et al., 2002). One of the favored candidate mechanisms for this 'complex spike pause' is that the hyperpolarized state of the membrane potential during the complex spike AHP creates a refractory period for simple spike firing (McDevitt et al., 1982; Rawson and Tilokskulchai, 1982). Thus, by reducing the amplitude of the complex spike AHP, CF–LTD could alter the duration of the complex spike pause and, thereby, affect PC throughput. Another interesting feature of the complex spike pause is that upon its cessation simple spike firing may resume at frequencies equal to, less than, or greater than that seen before the pause (Sato et al., 1992). It is conceivable that alterations in the complex spike AHP amplitude could influence not only the pause duration, but the postpause firing frequency as well.

Conclusions

We have reported previously that CF–LTD results in a long term depression of CF-evoked Ca^{2+} transients in PC dendrites (Weber et al., 2003). As a result, CF–LTD could also prove significant for Ca^{2+}-sensitive developmental processes, neuroprotection, and/or shifting the probability of plasticity at other synapses onto the PC (Hansel et al., 2001; Schmolesky et al., 2002; Weber et al., 2003).

In conclusion, the current data provide a critical replication of the CF–LTD induction experiments characterized at the complex spike level. We have also demonstrated, for the first time, that CF–LTD induction is accompanied by a substantial reduction in the slow AHP and that this AHP is Ca^{2+}-dependent. This study provides a better understanding of plasticity in one of the most important synapses in the cerebellum.

Acknowledgments

The authors wish to thank M.P.H. Coesmans and M.M. De Ruiter for critically reading the manuscript. This work was supported by an EUR fellowship (M.T.S.), the Dutch Neurofederation (C.D.Z., M.T.S., C.H.), NWO (C.D.Z.), NWO-VIDI (C.H.), and the Royal Dutch Academy of Sciences (C.H.).

References

Aizenman, C.D. and Linden, D.J. (2000) Rapid, synaptically driven increases in the intrinsic excitability of cerebellar deep nuclear neurons. Nat. Neurosci., 3: 109–111.

Armano, S., Rossi, P., Taglietti, V. and D'Angelo, E. (2000) Long-term potentiation of intrinsic excitability at the mossy fiber-granule cell synapse of rat cerebellum. J. Neurosci., 20: 5208–5216.

Bear, M.F. and Linden, D.J. (2001) The mechanisms and meaning of long-term synaptic depression in the mammalian brain. In: Cowan, W.M., Sudhof, T.C. and Stevens, C.F. (Eds.), Synapses. Johns Hopkins University Press, Baltimore, pp. 455–517.

Bernard, C. and Johnston, D. (2003) Distance-dependent modifiable threshold for action potential back-propagation in hippocampal dendrites. J. Neurophysiol., 90: 1807–1816.

Bloedel, J.R. and Roberts, W.J. (1971) Action of climbing fibers in cerebellar cortex of the cat. J. Neurophysiol., 34: 17–31.

Bruggencate, G.T., Nicholson, C. and Stockle, H. (1976) Climbing fiber evoked potassium release in cat cerebellum. Pflugers Arch., 367: 107–109.

Chang, C.P., Dworetzky, S.I., Wang, J. and Goldstein, M.E. (1997) Differential expression of the alpha and beta subunits of the large-conductance calcium-activated potassium channel: implication for channel diversity. Brain. Res. Mol. Brain Res., 45: 33–40.

Coulter, D.A., Lo Turco, J.J., Kubota, M., Disterhoft, J.F., Moore, J.W. and Alkon, D.L. (1989) Classical conditioning reduces amplitude and duration of calcium-dependent afterhyperpolarization in rabbit hippocampal pyramidal cells. J. Neurophysiol., 61: 971–981.

de Jonge, M.C., Black, J., Deyo, R.A. and Disterhoft, J.F. (1990) Learning-induced afterhyperpolarization reductions in hippocampus are specific for cell type and potassium conductance. Exp. Brain Res., 80: 456–462.

De Schutter, E. (1995) Cerebellar long-term depression might normalize excitation of Purkinje cells: a hypothesis. Trends Neurosci., 18: 291–295..

De Zeeuw, C.I. and Berrebi, A.S. (1995) Postsynaptic targets of Purkinje cell terminals in the cerebellar and vestibular nuclei of the rat. Eur. J. Neurosci., 7: 2322–2333.

De Zeeuw, C.I., Hansel, C., Bian, F., Koekkoek, S.K., van Alphen, A.M., Linden, D.J. and Oberdick, J. (1998) Expression of a protein kinase C inhibitor in Purkinje cells blocks cerebellar LTD and adaptation of the vestibulo-ocular reflex. Neuron, 20: 495–508.

Disterhoft, J.F., Golden, D.T., Read, H.L., Coulter, D.A. and Alkon, D.L. (1988) AHP reductions in rabbit hippocampal neurons during conditioning correlate with acquisition of the learned response. Brain Res., 462: 118–125.

Dittman, J.S. and Regehr, W.G. (1998) Calcium dependence and recovery kinetics of presynaptic depression at the climbing fiber to Purkinje cell synapse. J. Neurosci., 18: 6147–6162.

Eccles, J.C., Llinas, R. and Sasaki, K. (1966a) The excitatory synaptic action of climbing fibres on the Purkinje cells of the cerebellum. J. Physiol., 182: 268–296.

Eccles, J.C., Llinas, R. and Sasaki, K. (1966b) Parallel fibre stimulation and the responses induced thereby in the Purkinje cells of the cerebellum. Exp. Brain Res., 1: 17–39.

Eccles, J.C., Sasaki, K. and Strata, P. (1967) A comparison of the inhibitory actions of golgi cells and of basket cells. Exp. Brain Res., 3: 81–94.

Edgerton, J.R. and Reinhart, P.H. (2003) Distinct contributions of small and large conductance Ca^{2+}-activated K^+ channels to rat Purkinje neuron function. J. Physiol., 548: 53–69.

Foster, K.A., Kreitzer, A.C. and Regehr, W.G. (2002) Interaction of postsynaptic receptor saturation with presynaptic mechanisms produces a reliable synapse. Neuron, 36: 1115–1126.

Gruol, D.L., Jacquin, T. and Yool, A.J. (1991) Single-channel K^+ currents recorded from the somatic and dendritic regions of cerebellar Purkinje neurons in culture. J. Neurosci., 11: 1002–1015.

Hallworth, N.E., Wilson, C.J. and Bevan, M.D. (2003) Apamin-sensitive small conductance calcium-activated potassium channels, through their selective coupling to voltage-gated calcium channels, are critical determinants of the precision, pace, and pattern of action potential generation in rat subthalamic nucleus neurons in vitro. J. Neurosci., 23: 7525–7542.

Hansel, C. and Linden, D.J. (2000) Long-term depression of the cerebellar climbing fiber-Purkinje neuron synapse. Neuron, 26: 473–482.

Hansel, C., Linden, D.J. and D'Angelo, E. (2001) Beyond parallel fiber LTD: the diversity of synaptic and non-synaptic plasticity in the cerebellum. Nat. Neurosci., 4: 467–475.

Hashimoto, K. and Kano, M. (1998) Presynaptic origin of paired-pulse depression at climbing fibre-Purkinje cell synapses in the rat cerebellum. J. Physiol., 506: 391–405.

Häusser, M. and Clark, B.A. (1997) Tonic synaptic inhibition modulates neuronal output pattern and spatiotemporal synaptic integration. Neuron, 19: 665–678.

Ito, M. (2001) Cerebellar long-term depression: characterization, signal transduction, and functional roles. Physiol. Rev., 81: 1143–1195.

Ito, M., Sakurai, M. and Tongroach, P. (1982) Climbing fibre induced depression of both mossy fibre responsiveness and glutamate sensitivity of cerebellar Purkinje cells. J. Physiol., 324: 113–134.

Jacquin, T.D. and Gruol, D.L. (1999) Ca^{2+} regulation of a large conductance K^+ channel in cultured rat cerebellar Purkinje neurons. Eur. J. Neurosci., 11: 735–739.

Johnson, S.W. and Seutin, V. (1997) Bicuculline methiodide potentiates NMDA-dependent burst firing in rat dopamine neurons by blocking apamin-sensitive Ca^{2+}-activated K^+ currents. Neurosci. Lett., 231: 13–16.

Johnston, D., Christie, B.R., Frick, A., Gray, R., Hoffman, D.A., Schexnayder, L.K., Watanabe, S. and Yuan, L.L. (2003) Active dendrites, potassium channels and synaptic plasticity. Philos. Trans. R. Soc. Lond. B Biol. Sci., 358: 667–674.

Knaus, H.G., Schwarzer, C., Koch, R.O., Eberhart, A., Kaczorowski, G.J., Glossmann, H., Wunder, F., Pongs, O., Garcia, M.L. and Sperk, G. (1996) Distribution of high-conductance Ca(2+)-activated K+ channels in rat brain: targeting to axons and nerve terminals. J. Neurosci., 16: 955–963.

Knöpfel, T., Vranesic, I., Staub, C. and Gähwiler, B.H. (1991) Climbing fibre responses in olivo-cerebellar slice cultures. II. Dynamics of cytosolic calcium in Purkinje cells. Eur. J. Neurosci., 3: 343–348.

Koekkoek, S.K., Hulscher, H.C., Dortland, B.R., Hensbroek, R.A., Elgersma, Y., Ruigrok, T.J. and De Zeeuw, C.I. (2003) Cerebellar LTD and learning-dependent timing of conditioned eyelid responses. Science, 301: 1736–1739.

Konnerth, A., Dreessen, J. and Augustine, G.J. (1992) Brief dendritic calcium signals initiate long-lasting synaptic depression in cerebellar Purkinje cells. Proc. Natl. Acad. Sci. USA, 89: 7051–7055.

Llinas, R. and Sugimori, M. (1980a) Electrophysiological properties of in vitro Purkinje cell dendrites in mammalian cerebellar slices. J. Physiol., 305: 197–213.

Llinas, R. and Sugimori, M. (1980b) Electrophysiological properties of in vitro Purkinje cell somata in mammalian cerebellar slices. J. Physiol., 305: 171–195.

Llinas, R., Lang, E.J. and Welsh, J.P. (1997) The cerebellum, LTD, and memory: alternative views. Learn. Mem., 3:445–455.

McDevitt, C.J., Ebner, T.J. and Bloedel, J.R. (1982) The changes in Purkinje cell simple spike activity following spontaneous climbing fiber inputs. Brain Res., 237: 484–491.

Miyakawa, H., Lev-Ram, V., Lasser-Ross, N. and Ross, W.N. (1992) Calcium transients evoked by climbing fiber and parallel fiber synaptic inputs in guinea pig cerebellar Purkinje neurons. J. Neurophysiol., 68: 1178–1189.

Murphy, J.T. and Sabah, N.H. (1970) The inhibitory effect of climbing fiber activation on cerebellar purkinje cells. Brain Res., 19: 486–490.

Nelson, A.B., Krispel, C.M., Sekirnjak, C. and du Lac, S. (2003) Long-lasting increases in intrinsic excitability triggered by inhibition. Neuron, 40: 609–620.

Nolan, M.F., Malleret, G., Lee, K.H., Gibbs, E., Dudman, J.T., Santoro, B., Yin, D., Thompson, R.F., Siegelbaum, S.A., Kandel, E.R. and Morozov, A. (2003) The hyperpolarization-activated HCN1 channel is important for motor learning and neuronal integration by cerebellar Purkinje cells. Cell, 115: 551–564.

Raman, I.M. and Bean, B.P. (1997) Resurgent sodium current and action potential formation in dissociated cerebellar Purkinje neurons. J. Neurosci., 17: 4517–4526.

Raman, I.M. and Bean, B.P. (1999) Ionic currents underlying spontaneous action potentials in isolated cerebellar Purkinje neurons. J. Neurosci., 19: 1663–1674.

Rawson, J.A. and Tilokskulchai, K. (1982) Climbing fibre modification of cerebellar Purkinje cell responses to parallel fibre inputs. Brain Res., 237: 492–497.

Ross, W.N. and Werman, R. (1987) Mapping calcium transients in the dendrites of Purkinje cells from the guinea-pig cerebellum in vitro. J. Physiol., 389: 319–336.

Saar, D., Grossman, Y. and Barkai, E. (1998) Reduced afterhyperpolarization in rat piriform cortex pyramidal neurons is associated with increased learning capability during operant conditioning. Eur. J. Neurosci., 10: 1518–1523.

Sanchez-Andres, J.V. and Alkon, D.L. (1991) Voltage-clamp analysis of the effects of classical conditioning on the hippocampus. J. Neurophysiol., 65: 796–807.

Sato, Y., Miura, A., Fushiki, H. and Kawasaki, T. (1992) Short-term modulation of cerebellar Purkinje cell activity after spontaneous climbing fiber input. J. Neurophysiol., 68: 2051–2062.

Schmolesky, M.T., Weber, J.T., De Zeeuw, C.I. and Hansel, C. (2002) The making of a complex spike: ionic composition and plasticity. Ann. N. Y. Acad. Sci., 978: 359–390.

Schreurs, B.G., Gusev, P.A., Tomsic, D., Alkon, D.L. and Shi, T. (1998) Intracellular correlates of acquisition and long-term memory of classical conditioning in Purkinje cell dendrites in slices of rabbit cerebellar lobule HVI. J. Neurosci., 18: 5498–5507.

Shen, Y., Hansel, C. and Linden, D.J. (2002) Glutamate release during LTD at cerebellar climbing fiber-Purkinje cell synapses. Nat. Neurosci., 5: 725–726.

Silver, R.A., Momiyama, A. and Cull-Candy, S.G. (1998) Locus of frequency-dependent depression identified with multiple-probability fluctuation analysis at rat climbing fibre-Purkinje cell synapses. J. Physiol., 510: 881–902.

Simpson, J.J., Wylie, D.R. and De Zeeuw, C.I. (1996) On climbing fiber signals and their consequence(s). Behav. Brain Sci., 19: 384–398.

Stocker, M. and Pedarzani, P. (2000) Differential distribution of three Ca(2+)-activated K(+) channel subunits, SK1, SK2, and SK3, in the adult rat central nervous system. Mol. Cell Neurosci., 15: 476–493.

Strata, P. and Rossi, F. (1998) Plasticity of the olivocerebellar pathway. Trends Neurosci., 21: 407–413.

Velumian, A.A., Zhang, L., Pennefather, P. and Carlen, P.L. (1997) Reversible inhibition of I_K, I_{AHP}, I_h and I_{Ca} currents by internally applied gluconate in rat hippocampal pyramidal neurones. Pflugers Arch., 433: 343–350.

Vergara, C., Latorre, R., Marrion, N.V. and Adelman, J.P. (1998) Calcium-activated potassium channels. Curr. Opin. Neurobiol., 8: 321–329.

Wang, Y.T. and Linden, D.J. (2000) Expression of cerebellar long-term depression requires postsynaptic clathrin-mediated endocytosis. Neuron, 25: 635–647.

Weber, J.T., De Zeeuw, C.I., Linden, D.J. and Hansel, C. (2003) Long-term depression of climbing fiber-evoked calcium transients in Purkinje cell dendrites. Proc. Natl. Acad. Sci. USA, 100: 2878–2883.

Womack, M.D. and Khodakhah, K. (2002) Characterization of large conductance Ca^{2+}-activated K^+ channels in cerebellar Purkinje neurons. Eur. J. Neurosci., 16: 1214–1222.

Womack, M.D. and Khodakhah, K. (2003) Somatic and dendritic small-conductance calcium-activated potassium channels regulate the output of cerebellar purkinje neurons. J. Neurosci., 23: 2600–2607.

Zhang, W. and Linden, D.J. (2003) The other side of the engram: experience-driven changes in neuronal intrinsic excitability. Nat. Rev. Neurosci., 4: 885–900.

CHAPTER 9

Bases and implications of learning in the cerebellum — adaptive control and internal model mechanism

Masao Ito*

RIKEN Brain Science Institute, Wako, Saitama 351-0198, Japan

Keywords: cerebellum; learning; internal model; mental model; LTD; automatism

Abstract: The cerebellum has a fine compartment structure, which represents various discrete functions. Evolutionarily old, medial parts of the cerebellum are involved in the adaptive control of vairous brainstem and spinal cord functions, and long-term depression (LTD) plays a key role in this adaptive control. To extend these views to evolutionarily newer, lateral parts of the cerebellum involved in cerebral cortical functions such as voluntary movement, perception and language, the internal model hypothesis is instrumental. This hypothesis explains not only the characteristic cerebellar symptom, dysmetria, but also a number of otherwise unexplainable phenomena displayed in movements and mental actions.

Introduction

Investigations of the cerebellum have been greatly advanced during the past century, and several lines of basic knowledge accumulated thus, to provide the basis for understanding the neuronal mechanisms and functional roles of the cerebellum. First, the neuroanatomical studies revealed that the cerebellum has a unique geometrical architecture represented by the checkered board map shown in Fig. 1. It is divided into the middle longitudinal area, vermis, and the intermediate and lateral parts of the hemisphere, which are further subdivided into 13 major longitudinal zones (A in the middle and B, C1, C2, C3, D1 and D2 on the right and left sides) (Groenewegen et al., 1979). At the vermis, grooves running along the right–left axis divide the cerebellum into ten lobules I–X, which continue to the hemisphere as lobules HII–HX (Larsell and Yansen, 1972). While certain divisions are missing in the posterior part of the B, C1 and C3 zones, and there are additional zones such as x in the anterior vermis (Fig. 1), the cerebellum contains more than one hundred divisions.

Second, the cerebellar divisions have uniformly structured neuronal circuits and hence are presumed to operate by the same local circuit principles. The elaborate network structures of the cerebellar cortex composed of the Purkinje, basket, stellate, Golgi and granule cells, and mossy and climbing fiber afferents have been analyzed extensively. More recently, Lugaro cells (see Lane and Axelrod, 2002), unipolar brush cells (see Dino et al., 1999) and serotonergic, noradrenergic, dopaminergic and acetylcholinergic afferents (see Schweighofer et al., 2004) have also been recognized as unique elements of the network. While various types of synaptic plasticity have been located in the network as

*Tel.: +81-48467-6984; Fax: +81-48476-6975;
E-mail: masao@brain.riken.jp

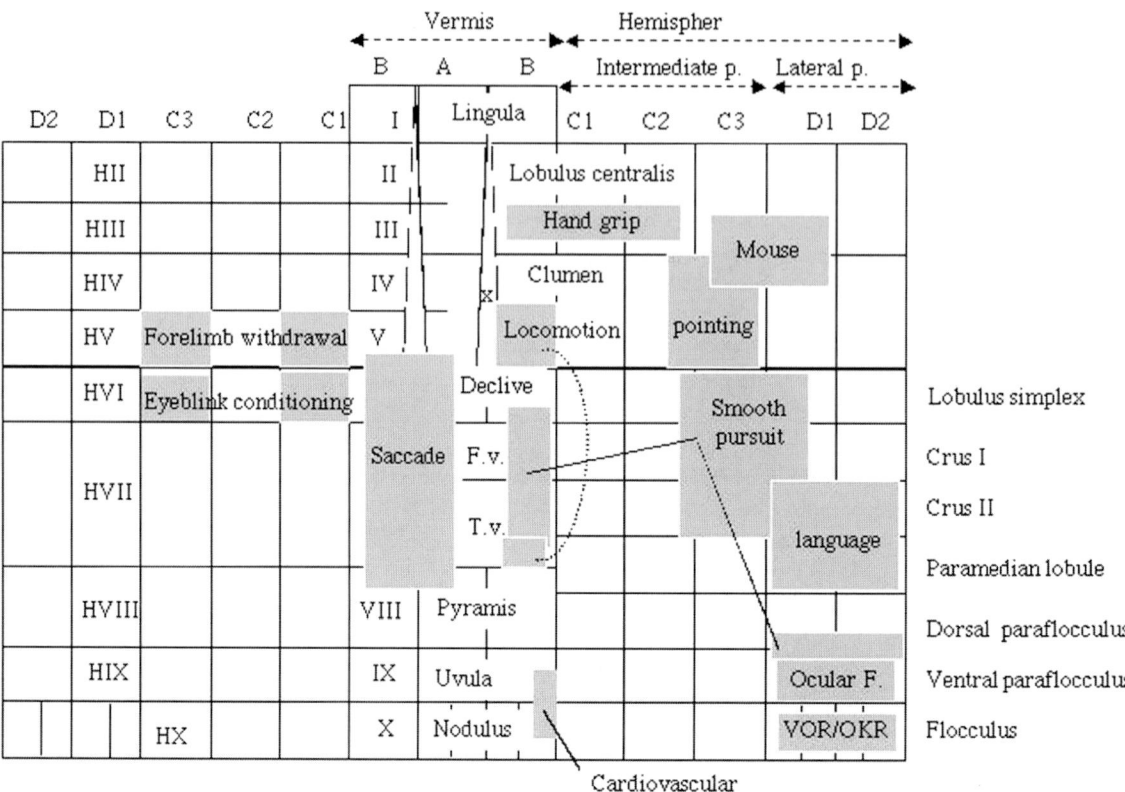

Fig. 1. Surface map of the cerebellar cortex. Ordinate, lobule numbers. Abscissa, zone address. x, extra zone. Anatomical names of lobules (Jansen and Brodal, 1954) are also indicated. Note that there is an extra zone x in the anterior vermis, that D zones of the flocculus are divided to four zones, that the A and B zones of the nodulus are also divided to four zones, and that certain divisions are missing (in humans, C1 and C2 zones in Crus I and II and dorsal and ventral paraflocculus and flocculus, Voogd, 2003). Dark areas represent approximate locations of the indicated functions. Multiply represented areas are connected by lines. (References: cardiovascular control, see the legend of Fig. 3; VOR/OKR, Ito, 1998; ocular following, Shidara et al., 1993; saccade, Noda and Fujikado, 1987; Barash et al., 1999; smooth pursuit, Ron and Robinson, 1973; Suh et al., 2000; locomotion, Yanagihara and Udo, 1994; Apps and Lee, 1999; eye blink conditioning, Attwell et al., 2001; forelimb withdrawal, Ekerot et al., 1997; hand pointing, Kitazawa et al., 1998; computer mouse manipulation, Imamizu et al., 2000; hand grip, Kawato et al., 2003; language, Petersen et al., 1989.)

possible memory elements (Hansel et al., 2001), the long-term depression (LTD) is unique being induced by the convergence of signals from both granule cell axons and climbing fibers to a Purkinje cell. The signal transduction underlying LTD has recently been analyzed extensively (see Ito, 2001, 2002). Earlier, the author proposed a cerebellar cortico-nuclear microcomplex (hereafter referred to as 'microcomplex') as a functional unit capable of learning with LTD as a major memory process (Ito, 1984). A microcomplex is composed of a microzone of the cerebellar cortex interconnected to a small group of cerebellar nuclear neurons (or vestibular nuclear neurons) and a small group of inferior olive neurons that project climbing fiber afferents. A microcomplex modifies the relationship between the input from mossy fiber afferents and the output from nuclear neurons in response to error signals conveyed by the climbing fibers.

Third, these microcomplexes are connected to different extracerebellar structures, and hence are involved in diverse functions. Since a microcomplex has an area of 10 mm^2 in the cerebellar cortex, and since the entire cortex has an area of 50,000 mm^2, a human cerebellum may contain 5000 microcomplexes (Ito, 1984). A prototype of a cerebellar control has

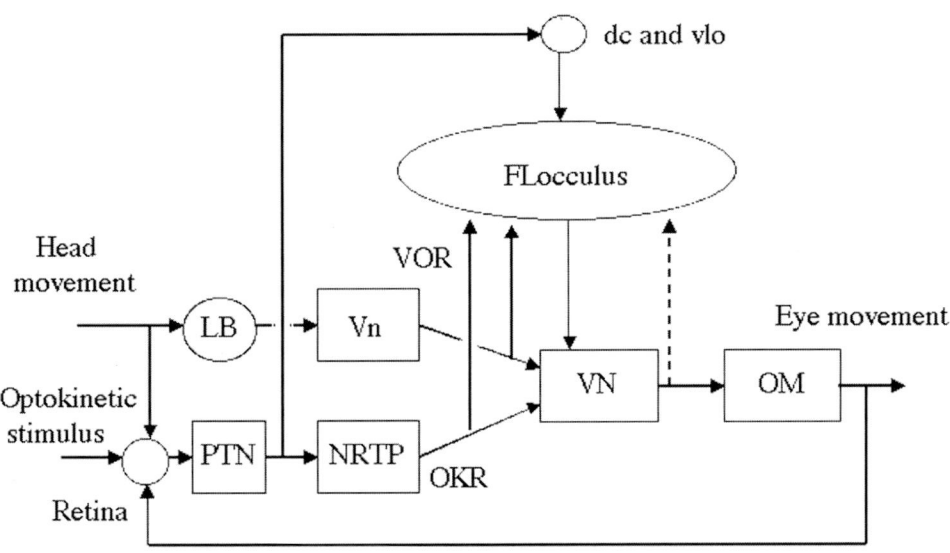

Fig. 2. Adaptive control system structure for ocular reflexes. VOR and OKR are combined in the manner of the two-degrees-of-freedom control system, which is attached with the flocculus as an adaptive control center. LB, labyrinth. Vn, vestibular nerve. VN, vestibular nucleus. OM, oculomotor neurons. PTN, pretectal nucleus. NRTP, nucleus reticularis tegmenti pontis. dc, dorsal cap of the inferior olive. vlo, ventrolateral outgrowth of the inferior olive. (References: see Ito, 1998.)

been developed from the manner in which the vermis, including the A and B zones, and the intermediate part of the hemisphere including the C1–C3 zones are connected to the spinal cord and the brainstem. Figure 2 illustrates the involvement of the flocculus in the two-degrees-of-freedom control of the eye movements in response to the head movements and optokinetic stimulation. A relatively small number of specific functions have so far been localized in the cerebellar surface, as illustrated in Fig. 1. Locomotion is represented in the vermis of lobule V (Yanagihara and Udo, 1994) and the C1 and C3 zones of the rostral paramedian lobule (Apps and Lee, 1999), and the saccadic eye movements in the oculomotor vermis covering lobules VI–VIII (Noda and Fujikado, 1987; Barash et al., 1999). A major site involved in the eye-blink conditioning is the C1 and C3 zones of lobule HVI (Attwell et al., 2001). The scheme of adaptive control applies well to the evolutionarily old, medial part of the cerebellum connected to the brainstem and the spinal cord (Ito, 1984; Barlow, 2002). In particular, the vestibuloocular reflex adaptation (Nagao and Ito, 1991; De Zeeuw et al., 1998), eyeblink conditioning (Bao et al., 1998) and locomotion (Ichise et al., 2000) are now widely used as model systems for testing the cerebellar learning function under various pharmacological and genetic manipulations of the cerebellar neuronal circuits.

Fourth, during the course of evolution in vertebrates beyond birds toward humans, the cerebellar hemisphere expanded markedly in association with the development of the cerebral cortex. The cerebellar hemisphere is thus involved in voluntary movements such as pointing with a hand (Kitazawa et al., 1998), visuomotor tracking with a computer mouse (Imamizu et al., 2000) and the hand-grip control (Kawato et al., 2003) (Fig. 1). Furthermore, the posterolateral part of the hemisphere represents language (Petersen et al., 1989; Fiez et al., 1992; Gebhart et al., 2002). In this chapter, the author presents the effort to expand the adaptive control system concepts thus far developed in studies of the evolutionarily old, medial parts of the cerebellum involved in the brainstem and spinal cord functions to the evolutionarily newer, lateral parts of the cerebellum involved in the cerebral cortical functions. To this effort, the internal model mechanism is instrumental. It is still largely hypothetical, but it receives increasing support from neuroanatomy and imaging analyses of human brains and computational studies in robotics. The internal model mechanism explains a number of seemingly unexplainable observations in movements as well as mental actions, covering a broad range of central nervous system functions, and its verification will be a major target for future cerebellar research.

Forward model for voluntary movement

The intermediate part of the hemisphere including the C1–C3 zones, is connected not only to the brainstem and the spinal cord via the projection from the interpositus nucleus to the red nucleus and other nuclei, but also to the cerebral primary motor cortex (hereafter referred to as the motor cortex) via the projection from the interpositus nucleus to the thalamus. Since the motor cortex projects to the cerebellum via certain brainstem nuclei (pontine nuclei, nucleus reticularis tegmenti pontis and lateral reticular nucleus), the well-known cerebrocerebellar communication loop is formed through the intermediate part of the cerebellar hemisphere and the motor cortex (Fig. 3). Since this loop connection does not match the adaptive control system scheme shown in Fig. 2, as the author previously suggested, the intermediate part of the cerebellar hemisphere provides an internal model (forward model type) simulating a motor apparatus (including related segmental circuits) that is controlled in a learnt voluntary movement (Ito, 1970, 1984). The control system scheme shown in Fig. 4, thus consists of the motor cortex (controller), the motor apparatus (controlled object), and a forward model of the motor apparatus in the cerebellum. A forward model can be regarded as a predictor of the consequences of motor commands, similar to the Smith Predictor adopted in the chemical processing systems (Miall et al., 1993).

Such a forward model can be formed in the cerebellum owing to the aforementioned learning capability inherent to the cerebellar microcomplex. When common input signals drive both a microcomplex and a neuronal circuit, and when the difference in their outputs is returned to the microcomplex via the climbing fibers, the input–output relationship of the microcomplex will gradually be modified due to the induction of LTD until it closely mimics the other neuronal circuit (Fig. 5). Thus, a microcomplex is capable of forming a forward model by copying the signal transfer characteristics of another system running in parallel. The control system scheme for the voluntary movements shown in Fig. 4 incorporates this mechanism. The actual movements of the motor apparatus are observed through a sensory system and compared with the predicted movements by the forward model, probably through the inhibitory connection from the cerebellar nuclei to the inferior olive (see Ito, 2001).

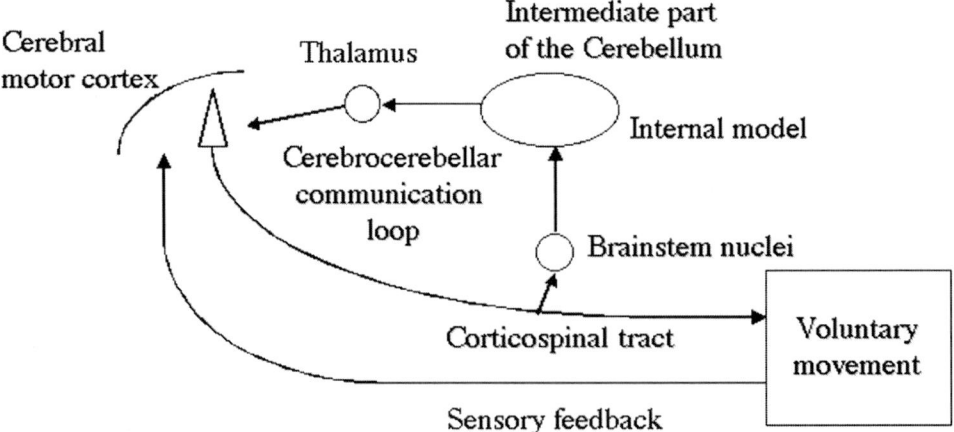

Fig. 3. Cerebrocerebellar communication loop. It shows how the loop suggests an internal model for voluntary movement control. (References: Ito, 1970, 1984.)

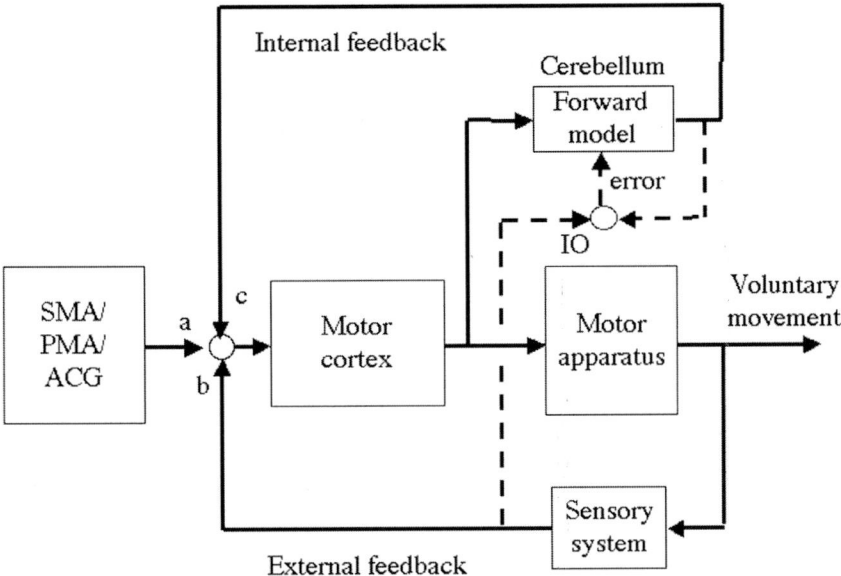

Fig. 4. Forward model control system. a, Instruction signal. Note that the output of the forward model (c) functions in two ways. First, c replaces the external feedback b so that the system operates even if the external feedback is not available. Second, c cancels the effect of external feedback b disruptive to the operation of the system. Broken lines indicate how error signals are derived via the inferior olive (IO) from the sensory system monitoring the voluntary movement and the cerebellum.

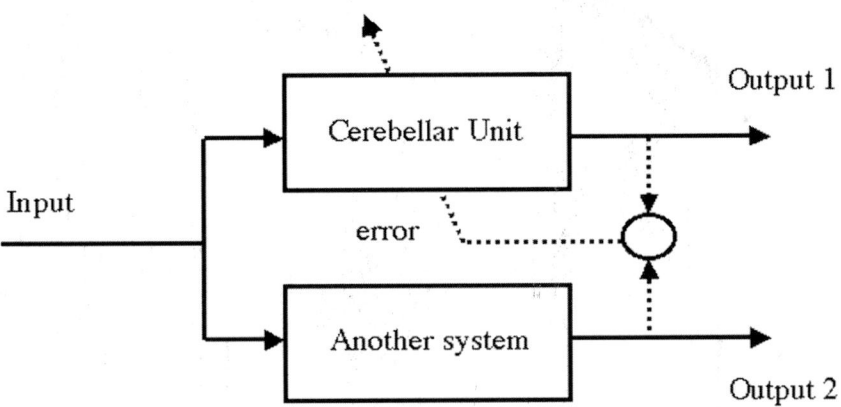

Fig. 5. The formation of an internal model in a cerebellar miccomplex. It shows how a microcomplex copies the input–output relationship of another system.

The internal feedback, through a forward model of the cerebellum, explains how we eventually learn to move in an increasingly more accurate and smooth manner, even with our eyes closed, through repeated practice. When we voluntarily perform a movement, the motor cortex generates command signals and sends them to a motor apparatus. Simultaneously, the motor cortex sends impulses to the intermediate part of the cerebellar hemisphere, which in turn, sends back impulses to the motor cortex, via the thalamus. Owing to this loop connection, the command signals generated by the motor cortex return via the cerebellum to the motor cortex before the actual results of the performed movement are fed back to the motor cortex through a sensory system (Fig. 4). As long as the forward model simulates the motor apparatus closely, the motor cortex should be able to perform precise control of the motor apparatus by referring to the prediction given by the forward model, even without referring to the consequences of the actual movement through external feedback via the sensory system. The accuracy of the prediction by the forward model can be maintained by the above-mentioned learning mechanism for the microcomplex, which corrects the difference in behaviors between the model and the actual motor apparatus. This hypothesis explains how we can perform an intended movement precisely even with our eyes closed, or even when the movement is too fast to be controlled by a visual feedback. For example, a skilled batter who can strike a pitched ball that approaches him within 0.2 s must predict the course of the ball accurately by simulation, using the forward model.

A similar explanation applies to the finger-to-nose or finger-to-finger test commonly performed in neurological clinics. These tests reveal a symptom characteristic of cerebellar disorders, that is, dysmetria, due to the lack of prediction using the forward model. The slow initiation of movement, another symptom characteristic of the cerebellar damage (Holmes, 1939), can likewise be explained to be due to the dependence of the movement execution on visual feedback in the absence of a predicting forward model. The cause of stuttering is unknown, but it is tempting to speculate that it arises from a failure of a forward model that normally guides the speech–motor activity predictively. At a speed sufficiently slow for effective auditory feedback, a stuttering subject can speak smoothly. A stuttering subject can also speak fluently in choral reading, not in solo reading, of syllables, presumably because hearing colleagues' speaking replaces the inefficient internal feedback. A PET study revealed that, in stuttering subjects, the right cerebellum is markedly activated as compared with the control subjects (Fox et al., 2000). This observation can be interpreted as representing an unconscious effort to form an efficient forward model in the cerebellum that enables fluent utterances in persons who stutter.

Sensory cancellation by prediction

In the above-mentioned mode of a forward model control, internal feedback replaces external feedback and makes the system perform effectively even if the external feedback is unavailable (c replaces b in Fig. 4). In another mode, internal feedback predicts a sensory consequence of a movement and acts to cancel it, which otherwise induces a sensation that disrupts the execution of the movement (c cancels b in Fig. 4). A clear example has been observed in fish cerebellum-like tissues, which generate signals to cancel the disruptive signals caused by swimming to the lateral line organs (Bell et al., 1997). This second mode of the forward model control is equivalent to the efference copy mechanism proposed earlier by von Holst (1954). It explains our experience of a self-generated tactile stimulus being perceived as less ticklish than the same stimulus applied externally (Blakemore et al., 1998). This sensory cancellation may apply to the disruptive, unpleasant effects of a self-generated stimulus that would produce error signals to shape a forward model for effective cancellation. However, the innocuous or pleasant effects would be left out since these will not give rise to error signals to shape a forward model.

Automatization by the inverse model

After a long practice in motor learning, we become so skillful that we no longer need to be conscious about an executed movement. It often happens that, after a grueling match, Sumo wrestlers remember very little about how they have performed, which suggests that they performed largely automatically. Such skill automatization can be explained, at least in part, based on the forward model (Fig. 4), which enables us to perform a precise movement without paying attention to the consequences of an actual movement. However, the inverse model control by the cerebellum as developed by Kawato et al. (1987) implies a possibility of developing a higher degree of skill automatization.

The most lateral part of the hemisphere, including the D1–D2 zones, is connected to the motor cortex in parallel, not in a loop. Based on this connection, Kawato et al. (1987) suggested that these cerebellar

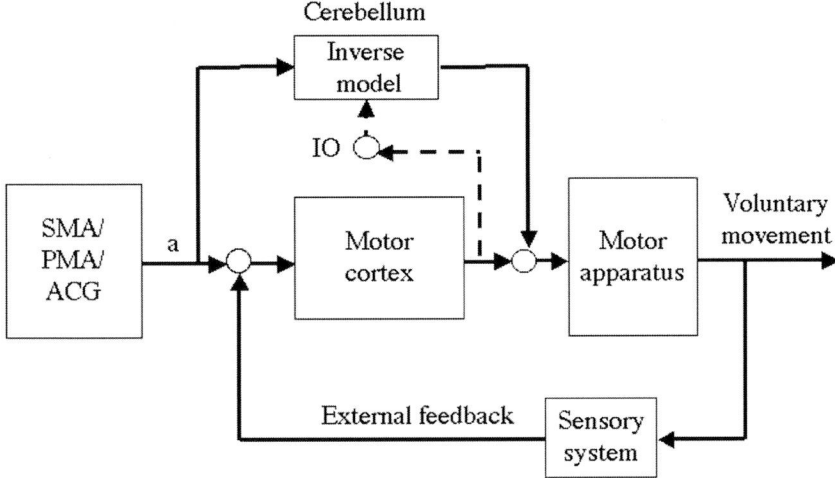

Fig. 6. Inverse model control system. (References: Kawato et al., 1987; Ito, 2001.)

areas provide an inverse model of a motor apparatus, which replaces the controller action of the motor cortex in a learned voluntary movement (Fig. 4). In contrast to the cerebrocerebellar communication loop in the forward model control system (Fig. 4), the inverse model control system is characterized by the convergence of cerebral and cerebellar outputs to a motor apparatus (Fig. 6). Such a convergence as this occurs in the rubrospinal or the reticulospinal tract neurons in the brainstem, or even in certain segmental neurons in the cervical cords to which the corticospinal and cerebellar nucleofugal axons may converge. Kawato et al. (1987) proposed that an inverse model is formed by the feedback learning mechanism in which the output of the motor cortex provides the climbing fiber error signals, as shown in Fig. 6. Darlot et al. (1996), however, suggest that a forward model formed in the cerebellar cortex is converted to an inverse model through computation in neuronal circuits involving cerebellar nuclei.

It is generally thought that the cerebellum is not involved in consciousness, because even a large deficit of the cerebellum does not interfere with consciousness. Hence, the author assumes that events in the cerebellum are not reflected in consciousness in which the cerebral cortex has been generally involved. The operation of a cerebellar internal model, either forward or inverse, will not come up to the level of unconsciousness. An inverse model in the cerebellum bypasses the motor cortex and performs movement control without the feeling of movements. The inverse model may thus offer an explanation of the automatism, such as that observed in cases of brain damage, hypnosis, and spiritualism, which accompanies no apparent sense of conscious 'will' associated with the observed movement (Wegner, 2002).

Source of instruction signals for a voluntary movement

To complete the control system scheme shown in Figs. 4 and 6, sources should be defined for instruction signals to the motor cortex and the cerebellum, which designate the content of the movement to be performed, for example, a desired trajectory of an arm movement. These instruction signals should come from the cortical areas devoted to motor planning and preparation, directly from the anterior cingulate gyrus (ACG), supplementary motor area (SMA) and premotor area (PMA), and indirectly from pre-SMA via SMA. Eccles (1994) considered SMA as a mediator of the force of mind representing conscious will. Electric stimulation of SMA performed in patients induced movements of limbs and also certain sensations such as a preliminary sensation of 'urge' to perform a movement or anticipation that a movement was about to occur were evoked (Fried et al., 1991). A stimulation of the ventral bank of the anterior cingulate sulcus evoked in a patient an irresistible urge to grasp something, resulting in exploratory eye movements and a wandering arm (Kremer et al., 2001). These are in contrast to the fact that an electric stimulation of the motor cortex induces a mere feeling of movements. An fMRI imaging study revealed that self-paced thumb movements caused cerebral activation to spread from the anterior cingulate gyrus through the SMA and PMA to the primary motor and sensory cortices (Hulsmann et al., 2003). This cascade in the cerebral cortex occurred temporally in parallel to the cerebellar activations propagating from lateral to medial through the cerebellum. These observations support the view that instruction signals for voluntary movement arise from the ACG, the SMA and the PMA as shown in Figs. 4 and 6.

Internal model for mental action

The most lateral part of the hemisphere (D1 and D2 zones) is connected to the cerebral association cortex. In humans these zones expand markedly and have been suggested to be involved not only in motor but also in certain cognitive functions such as language (Leiner et al., 1986). In contrast to the sensory and the motor cortices that receive stimuli from the external environment and send outputs to the external environment, the association cortex contains a self-sufficient internal loop in which the anterior part (prefrontal cortex) acts on the posterior part (temporoparietal cortex) and vice versa. Here, we find the structural basis of our mental action such as thought, because the association cortex operates through the loop in isolation from the external environment. Current neuroscience knowledge suggests that the prefrontal cortex performs an executive role in initiating a thought and is likely to be associated with the will to think (see Fuster, 2000).

In expanding the adaptive control theory to mental actions, the author takes a key psychological concept of mental models that represent parts of the mind. This concept originated from Craik (1943) statement that the mind constructs 'small-scale models' of reality that it uses to anticipate events, to reason, and to explain. When we think, the object to be controlled is a mental model, more concretely, a visual image or an abstract concept or idea. The author assumes that this psychological concept has a counterpart in the brain. Mental models are formed in the neuronal circuit of the cerebral cortex by associating various pieces of information received through the sensory cortex or arising from other areas of the cerebral cortex, and stored in the temporoparietal cortex that constitutes the internal environment of the brain. The prefrontal cortex acts as an executive cortex and manipulates the mental models represented in the internal environment, and this is thinking.

Based on a formalistic analogy between movement and thought, the author views a thought to be the control of a mental model, just as a voluntary movement is the control of a motor apparatus (Ito, 1991, 1993). In movements and thoughts, the objects to be controlled are very different from each other in nature, but they are equivalent to each other as control objects. Thus, a forward model control system for voluntary movements in Fig. 4 applies to the forward model control of thought as schematically illustrated in Fig. 7. It illustrates that, while the executive cortex manipulates a mental model, the cerebellum copies it to form a forward model and thereafter, the executive cortex can perform the thought by manipulating the internal model. The inverse model control system for voluntary movements in Fig. 6 can also apply to the inverse control of thought. If the forward and inverse model controls are combined, an interesting possibility emerges that the cerebellum conducts the entire process of thinking (Fig. 8), which will not come up to the level of consciousness. This may explain our daily experience that, after repeated trials of learning, a correct answer pops out readily without a conscious effort.

A contribution of the cerebellum to the cognitive function is typically seen in its involvement in language. A study using a noun-to-verb conversion task for testing language (Petersen et al., 1989) revealed activation of the anterior cingulate gyrus, prefrontal cortex (area 44), Wernicke's area and contralateral part of the cerebellar cortex (Petersen et al., 1989), corresponding well to structures shown in Figs. 7 and 8. This noun-to-verb conversion test

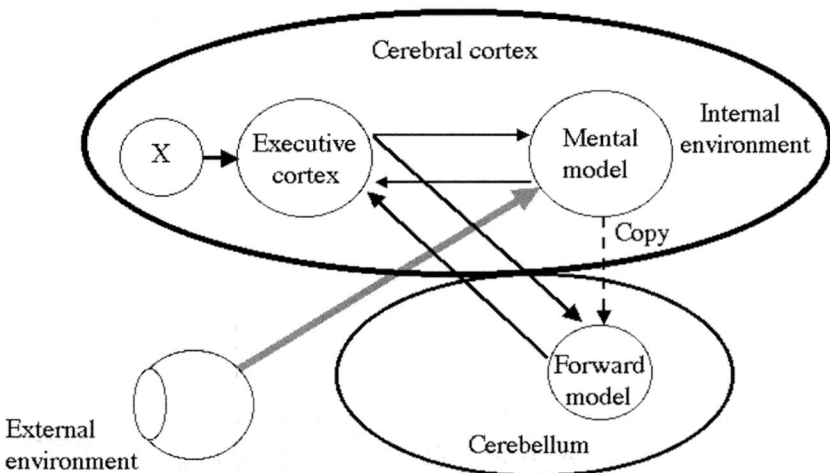

Fig. 7. Schematic diagram for the forward model control of mental action. X, anterior cingulate gyrus and probably some other, unidentified cortical areas, which gives rise to instruction signals for the executive cortex.

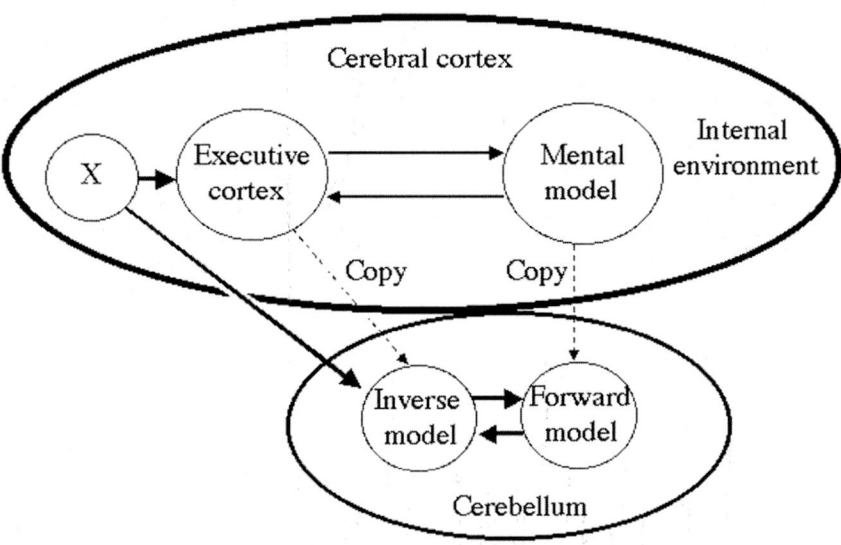

Fig. 8. Schematic digram showing the possibility of channeling mental actions through the cerebellum. Both the executive cortex and mental model are copied in the cerebellum.

paradigm raised a question whether it examines 'mental movement,' but a later study using other tasks confirmed that a deficit in language is of cognitive nature (Gebhart et al., 2002). Developmental dyslexia is a disorder in children who fail to attain the language skills of reading, writing, and spelling despite conventional classroom capability. Even though the developmental disturbance in cerebral visuomotor functions has been proposed to be the cause of dyslexia (Stein and Walsh, 1997), a deficit in cerebellar function has also been proposed to be a major cause of dyslexia (Nicolson et al., 2001). It is tempting to speculate that the guidance of fast reading by a forward model that functions for grapheme-to-phoneme conversion (Proverbio et al., 2004) is impaired in dyslexia.

The above-discussed thought mechanism involving mental model in the temporoparietal cortex and an internal model in the cerebellum may be extended to the possible brain mechanism for our understanding what goes on in the external environment. Children at 3–4 years of age are able to guess how other children think and feel, as revealed by the 'Theory of Mind' tasks (Siegal and Varley, 2002). A delay in the formation of this capability is specific to autism and Asperger syndrome, in which patients cannot guess how others feel so that they tend to develop antisocial behavior. A shrinkage of Purkinje cells by 24% in the average cross-sectional areas (Fatemi et al., 2002) and nicotinic receptor abnormalities in both the parietal and the cerebellar cortices (Martin-Ruiz et al., 2004) have been reported in autistic patients. Dysfunction of mental models in the parietal cortex and its copies in the cerebellum due to these abnormalities is a likely cause of autism.

The development of mental models of other people may imply the establishment of one's own identity. The prefrontal executive cortex continuously acting on the mental models in the cerebral cortex and readily receiving feedback from these would be the basis of our awareness of the self. It may be like watching ourselves in a mirror and mental models may act as a mirror that reflects the self. An impairment of this internal feedback would cause thoughts to be isolated from the sense of will normally associated with them, and this can be the cause for the auditory hallucination characteristic to schizophrenic patients (Blakemore et al., 2000). It is unlikely that internal models in the cerebellum contribute to the awareness of self because of its irrelevance to consciousness, but these must form large unconscious parts of our mental life.

Internal models for motor actions

Hitherto, we have discussed the possible mechanisms of voluntary movements and thoughts, separately. However, various motor actions that we daily perform integrate both movements and thoughts. One expects that such a motor action involves not only an internal model of a motor apparatus in the cerebellum (Figs. 4 and 6) but also a movement-related mental model in the association cortex and its internal model in the cerebellum (Figs. 7 and 8). A brain imaging study demonstrated the representation of a movement-related mental model in the left posterior parietal cortex, which was activated by imagining auditory-cued hand movements (Gerardin et al., 2000). This situation explains the involvement of both the parietal association cortex and the cerebellum in certain motor actions as recently reviewed by Blakemore and Sirigu (2003).

For example, the delusion of alien control, which schizophrenic patients often exhibit, makes them misattribute self-generated motor actions to externally generated motor actions. Blakemore et al. (2003) explain this passivity phenomenon as arising from the inefficient operation of a forward model; if internal feedback through a forward model does not cancel the sensation induced by self-generated actions, the latter will reach the level of consciousness similarly to that induced by external stimuli. In this situation, one cannot distinguish the self-generated actions from those generated externally. The alien control seen in hypnosis may likewise arise from suppression of the operation of a forward model under the influences of signals arising from the prefrontal cortex that is known to be activated when specific suggestions are made to hypnotized subjects. A brain imaging study revealed that the active movement attributed to an external source resulted in significantly higher activations in both the parietal cortex and the cerebellum than an identical active movements correctly attributed to the self (Blakemore et al., 2003). Blakemore and Sirigu (2003) interpreted their observations as indicating that both the parietal cortex and the cerebellum provide internal models of different nature and speculated that, while a cerebellar internal model makes rapid predictions about the sensory consequences of self-generated movement at a very low level of movement execution without being available to awareness, a model in the parietal cortex may address more cognitive aspects of movements and can be available to conscious awareness.

Another example involving the parietal cortex is the phantom limb, a sensation that a missing limb following amputation is still present and moves (Frith et al., 2000). This sensation is represented in the parietal cortex (Ramachandran and Hirsten, 1998), and may arise from a mental model of the limb if it is maintained after the loss of the limb. An internal model in the cerebellum may also be maintained, but it would not contribute to the phantom limb sensation, because cerebellar events will not be available to self-awareness.

A mental model is in fact an internal model of the cerebral cortex, which is formed within the temporo-parietal association cortex, and to which the executive cortex refers to when it acts on the external environment. However, in order to avoid confusion, the author maintains the term of mental models as distinct from internal models in the cerebellum. The author proposes the following steps for developing an integrated motor action by learning (Fig. 9). First, the executive cortex acts on a motor system to perform a motor action relying on external feedback. Second, while this step continues, the parietal cortex creates a mental model of the motor system by means of a cortical learning mechanism. Internal feedback through the mental model replaces or cancels the effect of external feedback from the performed motor action, and hence the motor action is executed quickly relying on the internal feedback from the mental model. Third, the mental model is further copied by the cerebellum to form a forward model. Thence, the executive cortex generates the motor action relying neither upon the external feedback nor upon the internal feedback from the mental model, and therefore, the operation of the executive cortex will become less conscious and more automatic. To explain the entire brain mechanism for learning motor actions, a sequential combination of a conscious cerebral process for creating a mental model and an unconscious cerebellar process for copying the created mental model to an internal model is proposed here.

This proposed dual learning mechanism for motor actions is consistent with the common observation on Alzheimer's disease (AD). AD patients exhibit serious impairment of memory and intellect, yet they

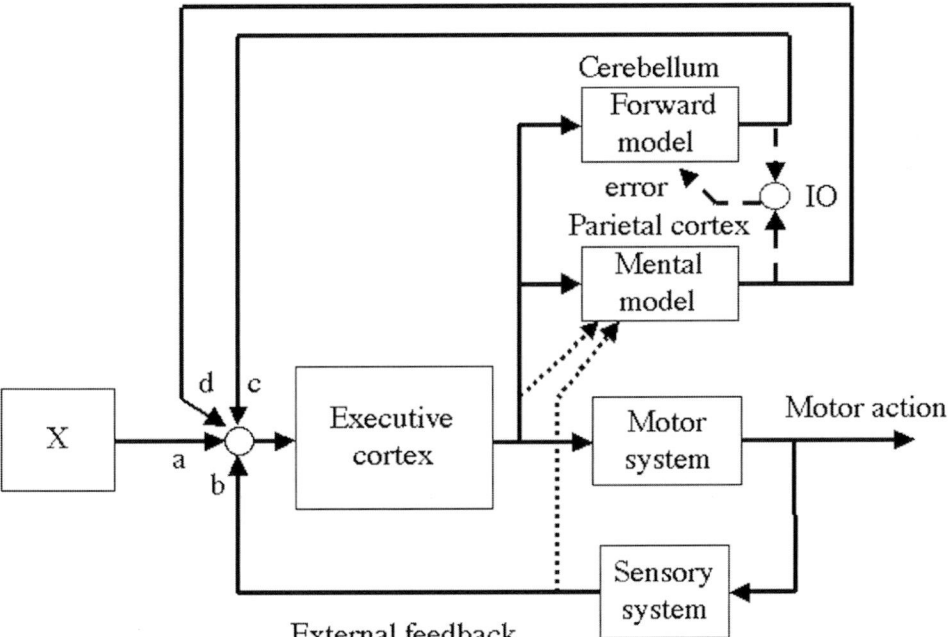

Fig. 9. Block diagram for a motor action control system. Motor system includes the motor cortex, motor apparatus and internal models indicated in Figs. 4 and 6. Dotted lines indicate signal flow that may be utilized in forming mental models in the parietal cortex. lines. d, internal feedback from the mental model. Other symbols are defined in the legends for other figures.

retain nurtured patterns of daily life. This may suggest that in AD patients, whereas the mental model function in the cerebral cortex is lost, the internal model function in the cerebellum is preserved. In fact, AD patients are characterized by the regional impairment of cerebral glucose metabolism in neocortical association areas including the posterior cingulate, temporoparietal, and frontal association cortices, whereas the primary visual and sensorimotor cortices, basal ganglia, and cerebellum are relatively well preserved (Herholz, 2003).

Perspectives

The internal model hypothesis of the cerebellum is based on a combination of three lines of approaches to solving problems pertaining the brain and mind: (1) neuroscience of neuronal circuits including synaptic plasticity, (2) system neuroscience based on modern control theories, and (3) mental model in psychology. It broadly covers problems of cognition and conation and, in particular, the mechanisms of physical and mental exercises that automate our behavior. It does not directly cover emotion that represents another important aspect of the mind, but we may detect a linkage in which emotion motivates positively or negatively the formation of an internal model. As it covers a large part of the brain–mind relationship, one would expect a tremendous impact of the internal model hypothesis on our understanding of the brain–mind relationship and the application of this hypothesis to childcare and education, treatment of mental diseases, control engineering, and information technology.

These perspectives would set forth three major targets of future cerebellar research. First, beyond the synaptic plasticity that is followed for 1–3 h in general, longer-term or even permanent memory mechanisms in the cerebellum are to be explored. While we are still remote from a stage in which we understand such permanent memory mechanisms of the cerebellum, we presently follow the general hypothesis that new memories expressed in synaptic plasticity are initially labile and sensitive to disruption before becoming permanently stored in the wiring of the brain. However, on reactivation, consolidated memories return to a labile, sensitive state where it can be modified or even erased, and hence its storage in structural changes of synapses is doubtful (Nader, 2003). Dynamic chemical processes may be a more likely site of memory storage (Arshavsky, 2003)

Second, even though neuronal network models and concepts of adaptive control and internal model have successfully been applied to interpret roles of the cerebellum in various functions, functions of large areas of the cerebellar surface map remain unidentified. In particular, connections between the D1 and D2 zones and the cerebral association cortex in humans should play important roles in mental actions, but our knowledge of these connections is still relatively meager (Voogd, 2003). Numerous microcomplexes, packed in the cerebellum could be combined to form multiple unit system that displays superior capabilities of control. Wolpert and Kawato (1998) developed multiple pairs of tightly coupled inverse (controller) and forward (predictor) models. During learning and use, these forward models determine the contribution of each inverse model's output to the final motor command. This architecture can simultaneously learn the multiple inverse models necessary for control as well as how to select the inverse models appropriate for a given environment. One would also imagine a situation in which multiple control systems, with either inverse or forward models, operate in cascade in parallel with sequential operations of the cerebral association cortices. It is of particular interest to know how the control systems are organized in stages upstream of those shown in Figs. 4 and 6. Steady efforts to fill gaps in the functional map of the cerebellum would lead to the discovery of new neuronal circuit mechanisms and new control system principles in the cerebellum.

Third, the formalistic analogy is a powerful method of finding an essential relationship between two seemingly heterogeneous systems such as movement and thought. At a next step, however, computational modeling, as has been done for the analogy between the biological motor systems and machine control systems must substantiate the analogy between movements and cognitive functions. Such computational modeling of thought still remains difficult because the encoding of mental models, either verbally or nonverbally, in the neuronal circuit of the cerebral cortex has yet been unsuccessful. The substantiation of mental models in physical terms will be essential in future neuroscience research.

Conclusions

The cerebellum has a fine compartment structure, which represents various discrete functions. Evolutionarily old, medial parts of the cerebellum is involved in the adaptive control of various brainstem and spinal cord functions, and long-term depression plays a key role in this adaptive control. To extend these views to evolutionarily newer, lateral parts of the cerebellum involved in cerebral cortical functions such as voluntary movement, perception and language, the internal model hypothesis is instrumental. This hypothesis explains not only the characteristic cerebellar symptom, dysmetria, but also a number of otherwise unexplainable phenomena displayed in movements and mental actions.

References

Arshavsky, Y.I. (2003) Cellular and network properties in the functioning in the nervous system; from central patter generators to cognition. Brain Res. Rev., 41: 229–267.

Attwell, P.J., Rahman, S. and Yeo, C.H. (2001) Acquisition of eyeblink conditioning is critically dependent on normal function in cerebellar cortical lobule HVI. J. Neurosci., 21: 5715–5722.

Apps, R. and Lee, S. (1999) Gating of transmission in climbing fibre paths to cerebellar cortical C1 and C3 zones in the rostral paramedian lobule during locomotion in the cat. J. Physiol. (Lond.), 516: 875–883. Comment in: J. Physiol., (Lond.) (1999) 516: 629.

Bao, S., Chen, L., Qiao, X., Knusel, B. and Thompson, R.F. (1998) Impaired eye-blink conditioning in waggler, a mutant mouse with cerebellar BDNF deficiency. Learn. Mem., 5: 355–364.

Barash, S., Melikyan, A., Sivakov, A., Zhang, M., Glickstein, M. and Their, P. (1999) Saccadic dysmetria and adaptation after lesions of the cerebellar cortex. J. Neurosci., 19: 10931–10939.

Barlow, J.S. (2002) The Cerebellum and Adaptive Control. Cambridge Univ. Press, Cambridge, England, p. 340.

Bell, C., Bodznick, D., Montgomery, J. and Bastian, J. (1997) The generation and subtraction of sensory expectations within cerebellum-like structures. Brain Behav. Evol., 50 Suppl 1: 17–31.

Blakemore, S.J., Wolpert, D.M. and Frith, C.D. (1998) Central cancellation of self-produced tickle sensation. Nat. Neurosci., 1: 635–640.

Blakemore, S.J., Smith, J., Steel, R., Johnstone, C.E. and Frith, C.D. (2000) The perception of self-produced sensory stimuli in patients with auditory hallucinations and passivity experiences: evidence for a breakdown in self-monitoring. Psychol. Med., 30: 1131–1139.

Blakemore, S.J., Oakley, D.A. and Frith, C.D. (2003) Delusions of alien control in the normal brain. Neuropsychologia, 41: 1058–1067.

Blakemore, S.J. and Sirigu, A. (2003) Action prediction in the cerebellum and in the parietal lobe. Exp. Brain Res., 153: 239–245.

Craik, K.J.W. (1943) The Nature of Explanation. Cambridge University Press, Cambridge, England, p. 143.

Darlot, C., Zupan, L., Etard, O., Denise, P. and Maruani, A. (1996) Computation of inverse dynamics for the control of movements. Biol. Cybern., 75: 173–186.

De Zeeuw, C.I., Hansel, C., Bian, F., Koekkoek, S.K., van Alphen, A.M., Linden, D.J. and Oberdick, J. (1998) Expression of a protein kinase C inhibitor in Purkinje cells blocks cerebellar LTD and adaptation of the vestibulo-ocular reflex. Neuron, 20: 495–508.

Dino, N.R., Willard, F.H. and Mugnani, E. (1999) Distribution of unipolar brush cells and other calrectin immunoreactive components in the mammalian cerebellar cortex. J. Neurocytol., 28: 99–123.

Eccles, J.C. (1994) How the Self Controls its Brain? Springer-Verlag, Berlin, Heidelberg., p. 197.

Ekerot, C.F., Garwicz, M. and Jorntell, H. (1997) The control of forelimb movements by intermediate cerebellum. Prog. Brain Res., 114: 423–429.

Fatemi, S.H., Halt, A.R., Realmuto, G., Earle, J., Kist, D.A., Thuras, P. and Merz, A. (2002) Purkinje cell size is reduced in cerebellum of patients with autism. Cell. Mol. Neurobiol., 22: 171–175.

Fiez, J.A., Petersen, S.E., Cheney, M.K. and Raichle, M.E. (1992) Impaired non-motor learning and error detection associated with cerebellar damage. A single case study. Brain, 115: 155–178.

Fox, P.T., Ingham, R.J., Ingham, J.C., Zamarripa, F., Xiong, J.-H. and Lancaster, J.L. (2000) Brain correlates of stuttering and syllable production. A PET performance-correlation analysis. Brain, 123: 1985–2004.

Fuster, J.M. (2000) Executive frontal functions. Exp. Brain Res., 133: 66–70.

Fried, I., Katz, A., McCarthy, G., Sass, K.J., Williamson, P., Spencer, S.S. and Spencer, D.D. (1991) Functional organization of human supplementary motor cortex studied by electrical stimulation. J. Neurosci., 11: 3656–3666.

Frith, S.D., Blakemore, S.J. and Wolpert, D.M. (2000) Abnormalities in the awareness and control of action. Phil. Trans. R. Soc. Lond. B., 355: 1771–1788.

Gebhart, A.L., Petersen, S.E. and Thach, W.T. (2002) Role of the posterolateral cerebellum in language. Ann. NY Acad. Sci., 978: 318–333.

Gerardin, E., Sirigu, A., Lehericy, S., Poline, J.B., Gaymard, B., Marsault, C., Agid, Y. and Le Bihan, D. (2000) Partially Overlapping Neural Networks for Real and Imagined Hand Movements. Cerebral Cortex, 10: 1093–1104.

Groenewegen, H.J., Voogd, J. and Freedman, S.L. (1979) The parasagittal zonation within the olivocerebellar projection. II. Climbing fiber distribution in the intermediate and hemispheric parts of cat cerebellum. J. Comp. Neurol., 183: 551–601.

Hansel, C., Linden, D.J. and D'Angelo, E. (2001) Beyond parallel fiber LTD: the diversity of synaptic and non-synaptic plasticity in the cerebellum. Nat. Neurosci., 4: 467–475.

Herholz, K. (2003) PET studies in dementia. Ann Nucl Med., 17: 79–89.

Holmes, G. (1939) The cerebellum of man. Brain, 62: 1–30.

Hulsmann, E., Erb, M. and Grodd, W. (2003) From will to action: sequential cerebellar contributions to voluntary movement. Neuroimage, 20: 1485–1492.

Imamizu, H., Miyauchi, S., Tamada, T., Sasaki, Y., Takino, R., Putz, B., Yoshioka, T. and Kawato, M. (2000) Human cerebellar activity reflecting an acquired internal model of a new tool. Nature, 403: 192–195.

Ichise, T., Kano, M., Hashimoto, K., Yanagihara, D., Nakao, K., Shigemoto, R., Katsuki, M. and Aiba, A. (2000) mGluR1 in cerebellar Purkinje cells essential for long-term depression, synapse elimination, and motor coordination. Science, 288: 1832–1835.

Ito, M. (1970) Neurophysiological basis of the cerebellar motor control system. Intern. J. Neurol., 7: 162–176.

Ito, M. (1984) The Cerebellum and Neural Control. Raven Press, New York, p. 580.

Ito, M. (1991) Neural control as a major aspect of high-order brain function. In: Eccles, J.C. and Creutzfeldt, O. (Eds.), The Principles of Design and Operation of the Brain, Exp. Brain Res. Suppl., Vol. 20. Springer-Verlag, pp. 281–292.

Ito, M. (1993) Movement and thought: identical control mechanisms by the cerebellum. Trends Neurosci., 16: 448–450.

Ito, M. (1998) Cerebellar learning in the vestibulo-ocular reflex. Trends Cogn. Sci., 2: 313–321.

Ito, M. (2001) Cerebellar long-term depression — characterization, signal transduction and functional roles. Physiol. Rev., 81: 1143–1195.

Ito, M. (2002) The molecular organization of cerebellar long-term depression. Nat. Rev. Neurosci., 3: 896–902.

Jansen, J. and Brodal, A. (1954) Aspects of Cerebellar Anatomy. Forlagt Johan Grundt Tanum, Oslo.

Kawato, M., Furukawa, K. and Suzuki, R. (1987) A hierarchical neural-network model for control and learning of voluntary movement. Biol. Cybern., 57: 169–185.

Kawato, M., Kuroda, T., Imazumi, H., Nakano, E., Miyauchi, S. and Yoshioka, T. (2003) Progr. Brain Res., 142: 171–188.

Kitazawa, S., Kimura, T. and Yin, P.B. (1998) Cerebellar complex spikes encode both destinations and errors in arm movements. Nature, 392: 494–497.

Kremer, S., Chassagnon, S., Hoffmann, D., Benabid, A.L. and Kahane, P. (2001) The Cingulate hidden hand. J. Neurol. Neurosurg. Psychiatry, 70: 264–265.

Lane, J. and Axelrod, H. (2002) Extending the cerebellar Lugaro cell class. Neuroscience, 115: 363–374.

Larsell, O. and Yansen, J. (1972) The comparative anatomy and histology of the cerebellum. The human Cerebellum, Cerebellar Connections, and the Cerebellar Cortex. Univ. Minnesota Press, Minneapolis.

Leiner, H.C., Leiner, A.L. and Dow, R.S. (1986) Does the cerebellum contribute to mental skills? Behav. Neurosci., 100: 443–454.

Martin-Ruiz, C.M., Lee, M., Perry, R.H., Baumann, M., Court, J.A. and Perry, E.K. (2004) Molecular analysis of nicotinic receptor expression in autism. Mol. Brain Res., 123: 81–90.

Miall, R.C., Weir, D.J., Wolpert, D.M. and Stein, J.F. (1993) Is the Cerebellum a Smith Predictor? J. Mot. Behav., 25: 203–216.

Nader, K. (2003) Memory traces unbound. Trends Neurosci., 26: 65–72.

Nagao, S. and Ito, M. (1991) Subdural application of hemoglobin to the cerebellum blocks vestibuloocular reflex adaptation. Neuroreport, 2: 193–196.

Nicolson, R.I., Fawcett, A.J. and Dean, P. (2001) Developmental dyslexia; The cerebellar deficit hypothesis. Trends Neurosci., 24: 508–511.

Noda, H. and Fujikado, T. (1987) Topography of the oculomotor area of the cerebellar vermis in macaques as determined by microstimulation. J. Neurophysiol., 58: 359–378.

Petersen, S.E., Fox, P.T., Posner, M.I., Mintun, M. and Raichle, M.E. (1989) Positron emission tomographic studies of the procxessing of single-words. J. Cogn. Neurosci., 1: 153–170.

Proverbio, A.M., Vecchi, L. and Zani, A (2004) From Orthography to Phonetics: Measures of Grapheme-to-Phoneme Conversion Mechanisms in Reading. J. Cogn. Neurosci., 16: 301–317.

Ramachandran, V.S. and Hirsten, W. (1998) The perception of phantom limbs. Brain, 121: 1603–1630.

Ron, S. and Robinson, A. (1973) Eye movements evoked by cerebellar stimulation in the alert monkey. J. Neurophysiol., 36: 1004–1022.

Schweighofer, N., Doya, K. and Kuroda, S. (2004) Cerebellar aminergic neuromodulation: towards a functional understanding. Brain Res. Rev., 44: 103–116.

Shidara, M., Kawano, K., Gomi, H. and Kawato, M. (1993) Inverse-dynamics model eye movement control by Purkinje cells in the cerebellum. Nature, 365: 50–52.

Siegal, M. and Varley, R. (2002) Neural systems involved in 'theory of mind'. Nat. Rev. Neurosci., 3: 463–471.

Stein, J. and Walsh, V. (1997) To see but not to read; the magnocellular theory of dyslexia. Trends Neurosci., 20: 147–152.

Suh, M., Leung, H.C. and Kettner, R.E. (2000) Cerebellar flocculus and ventral paraflocculus Purkinje cell activity during predictive and visually driven pursuit in monkey. J. Neurophysiol., 84: 1835–1850.

von Holst, E. (1954) Relations between the central nervous system and peripheral organs. Br. J. Anim. Behav. 2: 89–94.

Voogd, J. (2003) The human cerebellum. J. Chem. Neuroanat., 26: 243–252.

Wegner, D.M. (2002) The Illusion of Conscious Will. MIT Press, Cambridge, Mass., p. 405.

Wolpert, D.M. and Kawato, M. (1998) Multiple paired forward and inverse models for motor control. Neural Netw., 11: 1317–1329.

Yanagihara, D. and Udo, M. (1994) Climbing fiber responses in cerebellar vermal Purkinje cells during perturbed locomotion in decerebrate cats. Neurosci. Res., 19: 245–248.

SECTION IV

Imaging of Cerebellar Activity

CHAPTER 10

Synaptic transmission and long-term depression in Purkinje cells in an in vitro block preparation of the cerebellum isolated from neonatal rats

Akiko Arata* and Masao Ito

Laboratory for Memory and Learning, Brain Science Institute, RIKEN, Hirosawa 2-1, Wako, Saitama 351-0198, Japan

Keywords: cerebellum; in vitro preparation; inferior olive; parallel fibers; optical imaging; LTD; neonatal rat

Introduction

The cerebellum contains intricate neuronal circuits consisting of several types of neuron and fiber. The major components of cerebellum are granule cells and unipolar brush cells as excitatory neurons; Purkinje, basket, stellate, Golgi and Lugaro cells as inhibitory neurons; and mossy fibers arising from numerous precerebellar nuclei, climbing fibers (CFs) from the inferior olive (IO) and certain aminergic fibers. Cerebellar long-term depression (LTD) is a major type of synaptic plasticity expressed in the unique structure of the cerebellar cortex described by Cajal (1911), that is, the convergence of synaptic inputs from parallel fibers (PFs, axons of granule cells) and CFs in Purkinje cell dendrites. LTD is the persistent depression of synaptic transmission from PFs to a Purkinje cell, and is induced when these PFs are activated conjunctively with the CF that forms a synaptic contact with the same Purkinje cell. LTD is considered to represent a major memory process underlying the error-driven learning in the cerebellum (for review, see Ito, 2001).

LTD was first discovered in in vivo preparations (Ito and Kano, 1982; Ito et al., 1982; Ekerot and Kano, 1985), but more recently, in vitro slice preparations of the cerebellum (Sakurai, 1987; Crepel and Krupa, 1988; Hartell, 1994; Lev-Ram et al., 1997) and tissue-cultured Purkinje cells (Linden et al., 1991; Launey et al., 2004) are widely used in the analyses of cellular and molecular mechanisms of LTD (for review, see Ito, 2001, 2002).

However, these preparations lack fiber connections to and from the cerebellar cortex with other structures such as the inferior olive, the pontine nuclei of cerebellar nuclei, and retain only a part of the neuronal circuits normally present in the cerebellum. Moreover, the severance of fibers may modify synaptic transmission even at remote intact synapses, as has been demonstrated in the inhibitory synapses that Purkinje cells supply to cerebellar nuclear neurons. The transmission efficacy across these synapses substantially attenuates in one hour following the lesion of the inferior olive (Ito et al., 1979; Karachot et al., 1987). To overcome these

*Corresponding author. Tel.: +81-48-462-1111 ext. 6444; Fax +81-48-462-4646; E-mail: ako@brain.riken.jp

shortcomings, we developed a unique block preparation of the cerebellum isolated from neonatal rats; the isolated block contains the pons and the medulla oblongata (Arata and Ito, 2001, 2003, 2004). The connections of the cerebellar cortex with the inferior olive, the pontine nuclei and the cerebellar nuclei are preserved in this preparation. Block preparations derived from postnatal days 0 to 8 (P0–P8) rats retain the basal electrical activity under perfused in vitro conditions. Block preparations derived later than P8 do not survive under in vitro conditions, partly because the increased size of blocks from older rats impedes the sufficient diffusion of oxygen into deep tissues and also because the tolerance to hypoxia is exhibited only in neonatal animals (Singer, 1999). We demonstrated, using electrophysiological techniques, that, in block preparations at P5–P8, Purkinje cells responded to the stimulation of CFs and PFs, and also exhibited LTD. We also demonstrated these responses and LTD induction using optical imaging techniques (Cohen and Yarom, 1999; Gao et al., 2003). When using optical imaging, the block preparations provide a convenient material for the pharmacological investigation of cerebellar circuits. Another advantage of using the block preparation is that we can follow the functional development of cerebellar neuronal circuits during the first postnatal week. In this article, we present the results recently obtained using the block preparations.

Block preparation

Under deep ether-anesthesia, the cerebellum was dissected out in a block containing the pons and the medulla from P0–P8 rats. The block was placed in a recording chamber, with the cut surface of the medulla exposed (Fig. 1), and perfused with modified Krebs solution equilibrated with 95% O_2 and 5% CO_2 at 26–27°C. Since the lateral portion of the cerebellum (flocculus and paraflocculus) and the most posterior part of the vermis (lobules IX (uvula) and X (nodulus)) develop faster than other areas in the cerebellum (Altman and Bayer, 1985), we paid particular attention to these areas.

Of the neurons recorded in the lateral portion of the cerebellum, the Purkinje cells were identified by their characteristic responses to IO stimulation (see below) and also by their location and characteristic dendritic morphology revealed by the infusion of Lucifer Yellow through a patch pipette.

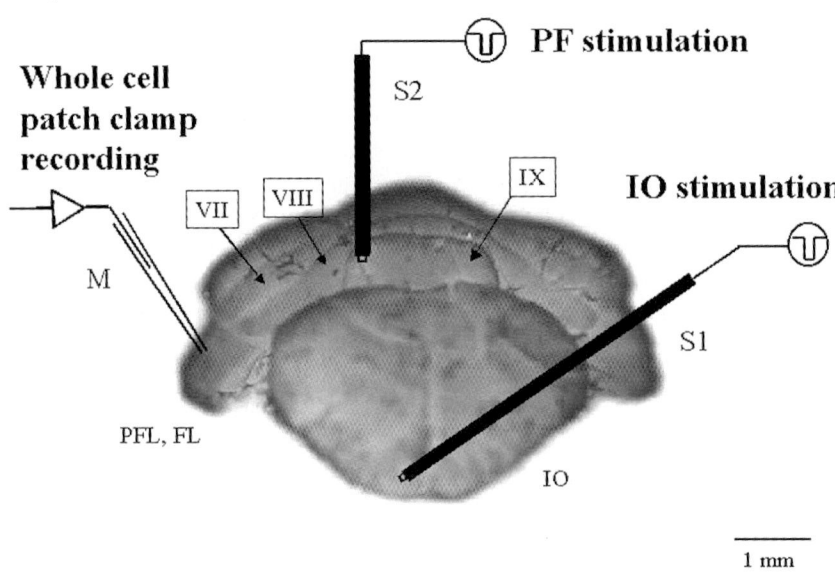

Fig. 1. Experimental arrangements. Viewing a block preparation from the side of the medulla oblongata. PFL, paraflocculus. FL, flocculus. VII–IX, lobules. S1, S2, concentric stimulation electrode. M, glass pipette.

Electrical activity and LTD

The whole-cell-clamped Purkinje cells exhibited resting potentials and membrane resistance in a range similar to those reported in the cerebellar slices. At P5, the stimulation of CFs at the IO evoked complex responses consisting of Ca^{2+} spikes and underlying large excitatory postsynaptic potentials (EPSPs) in Purkinje cells, as characteristic of CF-mediated responses, with a latency of 20–30 ms. The large EPSPs were isolated from Ca^{2+} spikes by hyperpolarizing the membrane slightly (Fig. 2A). Purkinje cells recorded in the flocculus or paraflocculus frequently responded to the IO stimulation. The stimulation of PFs at the edge of the posterior vermis ipsilateral to the recording site induced relatively small graded EPSPs in Purkinje cells with a latency of 5–10 ms (Fig. 2B).

At P0–P3, PF-mediated EPSPs were often missing. At P5 onward, sampled Purkinje cells frequently received both CF- and PF-mediated EPSPs. In them, when conjunctive IO and posterior vermis stimulation was performed at 1 Hz for 5 min (protocol adopted by Karachot et al., 1994), the CF-mediated EPSPs did not change (Fig. 2C), but the PF-mediated EPSPs often underwent persistent depression (Fig. 2D). Figure 2E illustrates the average depression curve for 15 P5 Purkinje cells, among which the extent of depression varied from cell to cell. As measured 35–44 min after the onset of conjunctive stimulation, the extent of reduction in PF-mediated EPSPs was 25.8% on the average ($\pm$ 2.3% S.E., $n = 15$). The depression is similar to the LTD observed in adult cerebellar slices, except that in the adult rat cerebellar slices, the extent of depression progressively increased with time over the period of 1 h (Karachot et al., 1994), while in P5 block preparations, a marked short-term depression lasting for 10 min superposes on LTD (Fig. 2E). The rate of the occurrence of LTD was as low as 17% at P4, and then increased through P5–P8 up to 57% (Fig. 3).

Optical images

A cerebellum–pons–medulla block preparation was stained with a voltage-sensitive dye (Di-4 ANEPPS or Di-2 ANEPEQ, 0.2 mg/ml) for 30–50 min. In the perfusion chamber, fluorescence changes corresponding to membrane potential changes induced by the stimulation of IO and posterior vermis were observed by optical recording using a MiCAM01 system (Brainvision Inc., Tokyo, Japan; see also Tominaga et al., 2000).

The IO stimulation induced responses in the lateral portion of the cerebellum (Fig. 4A). The spatial pattern of IO-stimulation-evoked excitation is consistent with the anatomically demonstrated olivocerebellar projection from the medial accessory olive and principal olive in adult rats (Ruigrok et al., 1992; Sugihara et al., 2000). In contrast, the stimulation of the PFs at the lateral edge of the posterior vermis induced a long band of PF-mediated responses (Fig. 4C). This band is presumably induced by the propagation of impulses along the long PFs prevailing in the growing molecular layer during the first postnatal week (Liljelund and Levine, 1998). These optical responses, illustrated in Fig. 4, as well as in later figures (Figs. 5 and 8) lasted for 300–500 ms. The fact that the duration of the optical responses is much longer than that of the EPSPs (Fig. 3A and B) suggests that optical responses involve more than EPSPs generated in the Purkinje cells (see below and Fig. 8).

The band of the PF-mediated excitation reached the lateral portion of the cerebellum where CF-mediated responses predominated, while the medial part of the cerebellum did not show CF-mediated responses (compare Fig. 4A with C). We therefore compared the events occurring in these two parts of the cerebellum in testing the effect of conjunctive stimulation of IO and the posterior vermis. The conjunctive stimulation only slightly reduced the PF-mediated responses around the site of posterior vermis. In contrast, it induced a pronounced depression in the lateral portion of the cerebellum where the CF- and PF-mediated responses overlapped each other (Fig. 4D, compare the area pointed by an arrow with the corresponding area in C). The depression lasted over an hour corresponding to the LTD observed electrophysiologically (Fig. 2C). LTD was observed in responses evoked by the stimulation at lobules VII, VIII, or IX. We confirmed that conjunction did not affect the responses induced by IO stimulation alone (Fig. 4B, compare with A). We also confirmed that a repetitive stimulation of the posterior vermis alone at 1 Hz for 5 min induced

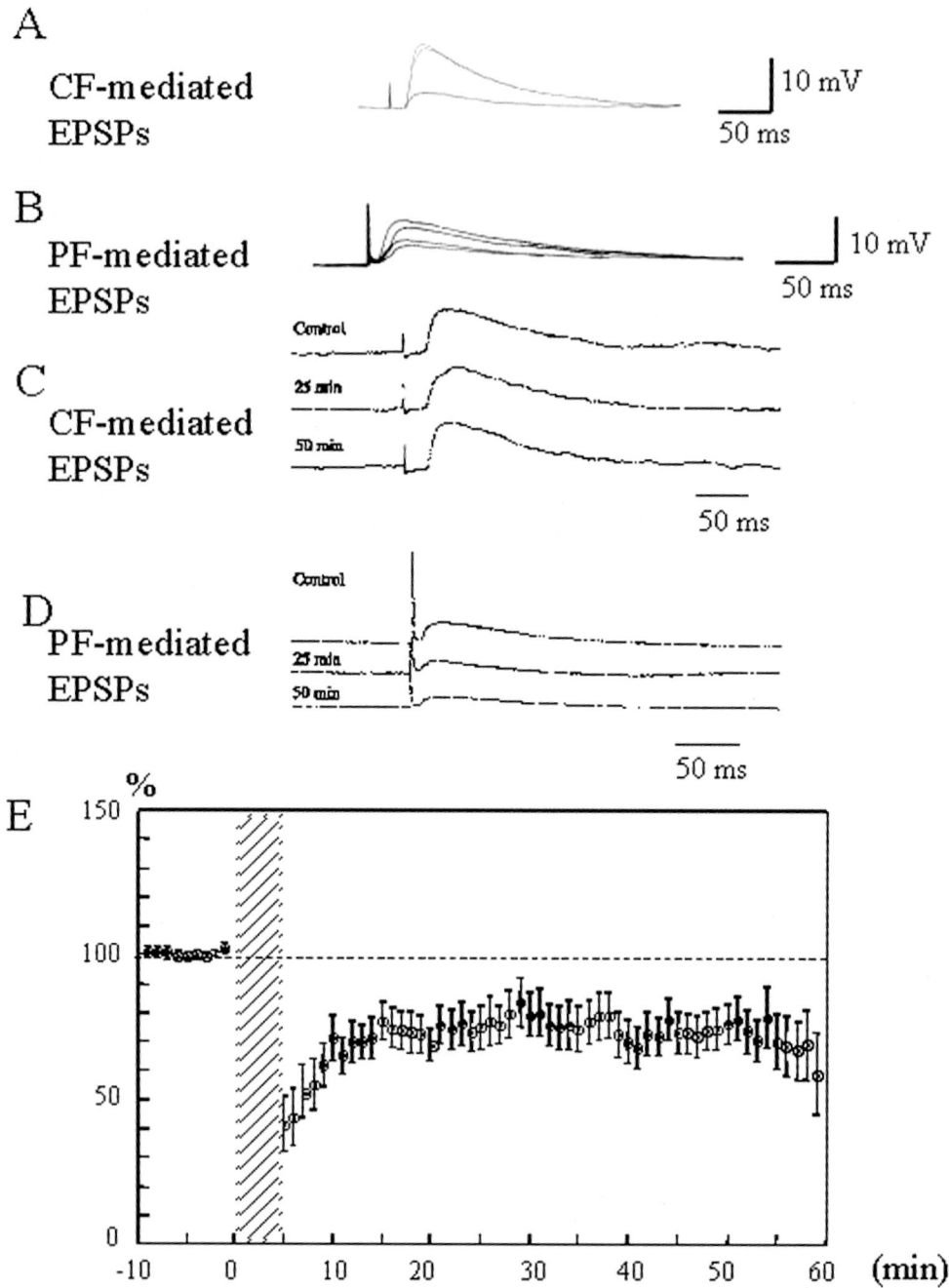

Fig. 2. Synaptic transmission and LTD. A, large all-or-none EPSP evoked by IO stimulation. B, graded EPSP evoked by posterior vermis stimulation. C, IO-stimulation-evoked EPSP before and 25 and 50 min after conjunction. D, posterior vermis-stimulation-evoked EPSP shown similarly to C. E plots against time the mean peak amplitude of posterior vermis-evoked EPSPs in 15 P5 Purkinje cells. Obliquely shaded column indicates conjunctive stimulation for 5 min. Broken line shows the amplitude during 5 min period before conjunction, that is taken as 100%. Vertical bars, S.E. (From Arata and Ito, 2004.)

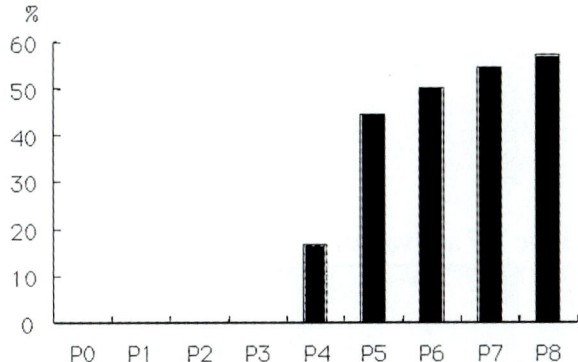

Fig. 3. Development of LTD. Ordinate, rate of occurrence of LTD among 17–40 Purkinje cells examined in each stage of postnatal development.

no more than a very slight reduction of the PF-mediated responses (Fig. 5).

The nature of the optical responses in the cerebellar cortex has been examined in the following experiments. The responses mediated by CFs and PFs were abolished in the presence of low-Ca^{2+} and high-Mg^{2+} solution that blocks synaptic transmission (Fig. 6), indicating that these responses are postsynaptic to PFs and CFs. The extent of PF-mediated responses increased in the presence of 10 μM $GABA_A$ antagonist, bicuculline (A. Arata, unpublished observation). This observation also supports the view that the PF-mediated responses are postsynaptic to the CFs or PFs. CF-mediated responses were completely

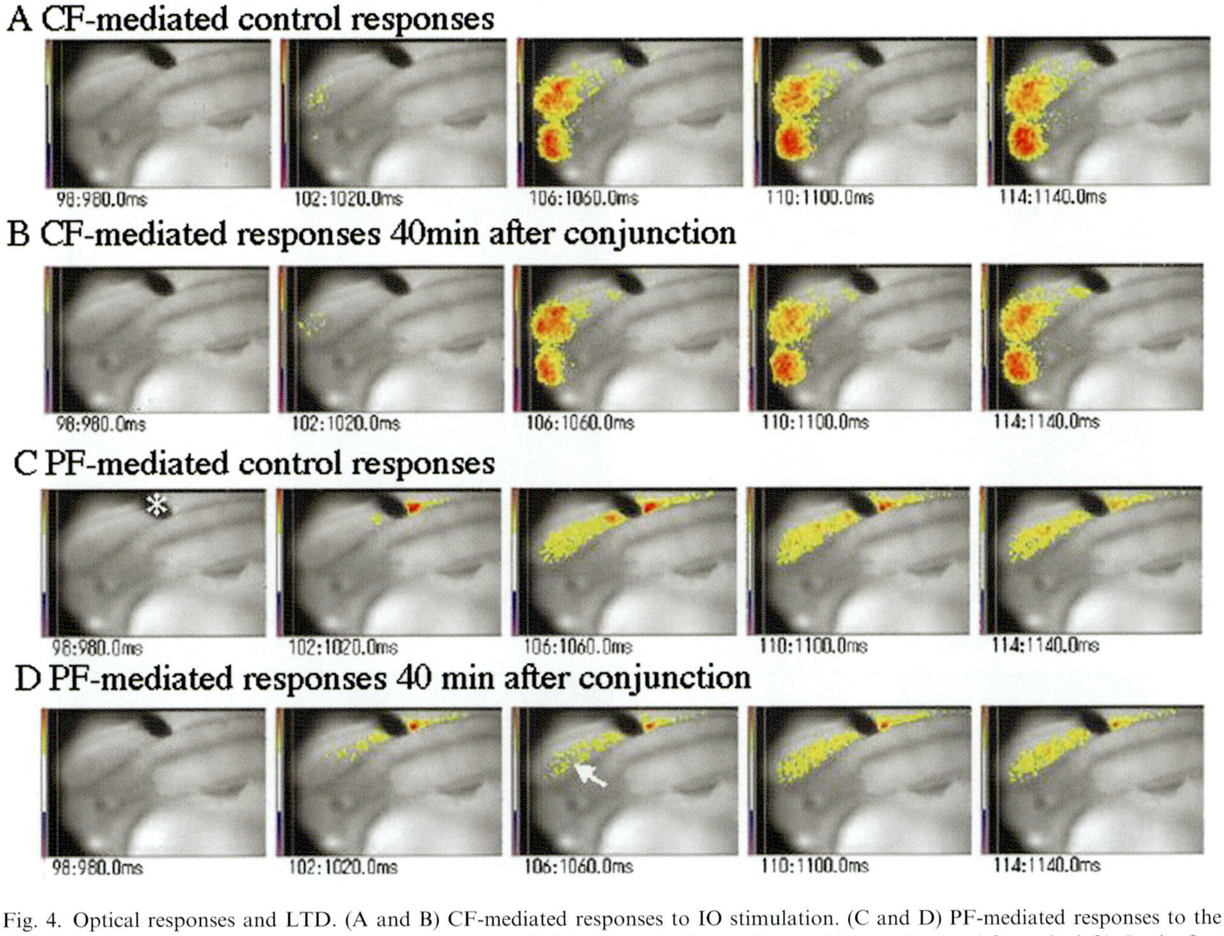

A CF-mediated control responses

B CF-mediated responses 40min after conjunction

C PF-mediated control responses

D PF-mediated responses 40 min after conjunction

Fig. 4. Optical responses and LTD. (A and B) CF-mediated responses to IO stimulation. (C and D) PF-mediated responses to the posteior vermis stimulation. Stimulation pulses were applied between the first and second images (counted from the left). In the first image of C indicates the stimulating electrode (S2 in Fig. 1.). A and C, control responses. B and D, responses 40 min after conjunction. From the left to the right, images are shown at 40 ms intervals. An arrow in the third image of D points to the area where PF-mediated responses were markedly depressed as compared with the third image of C.

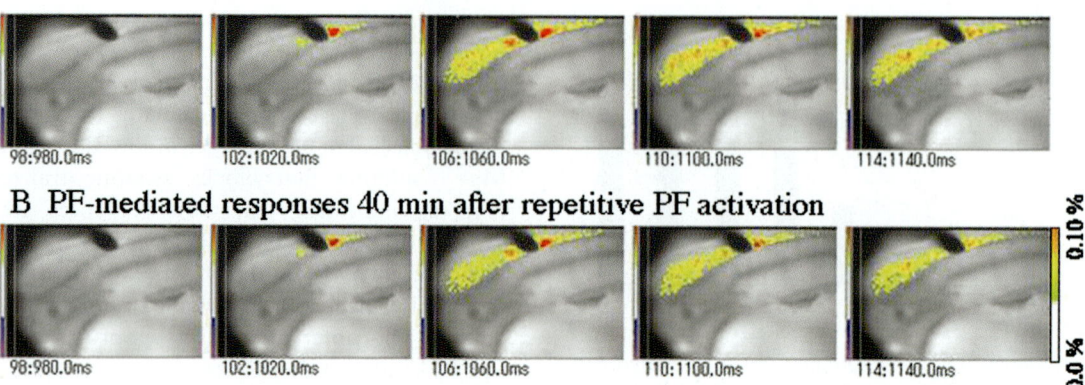

Fig. 5. Effect of repetitive stimulation of PFs on the PF-mediated optical responses. (A) Control responses to PF stimulation at the lateral edge of the posterior vermis. (B) Same as A, but recorded 40 min after repetitive stimulation of PFs at 1 Hz for 5 min.

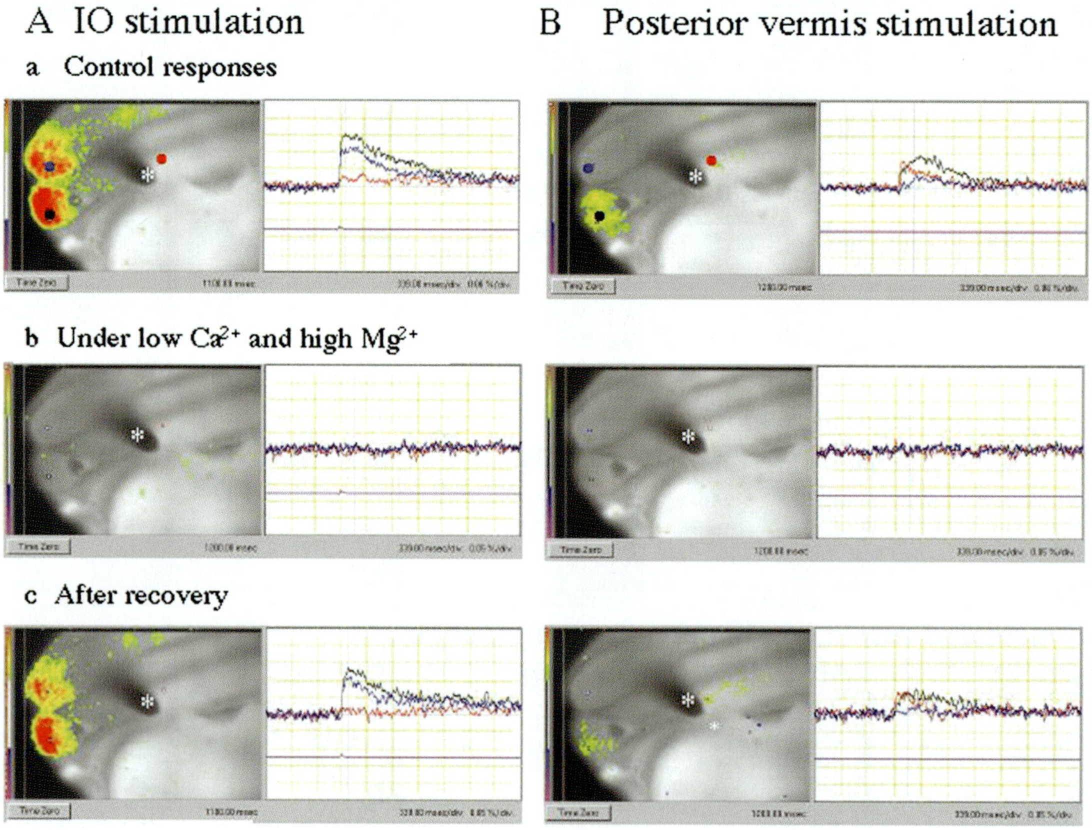

Fig. 6. Optical images in the presence of low-Ca^{2+} and high Mg^{2+} concentrations. (A) IO stimulation. (B) Posterior vermis stimulation. a, control responses. b, in the presence of 0.2 mM Ca^{2+} and 5 mM Mg^{2+}. c, recovery in the standard perfusate. Curves attached to images indicate the time course of the responses recorded at three sites. Black and green lines, at the two foci of excitation in the lateral portion of the cerebellum. Red line, at the site medial to the S2 stimulating electrode. The three stimulated sites are indicated by the same colors.

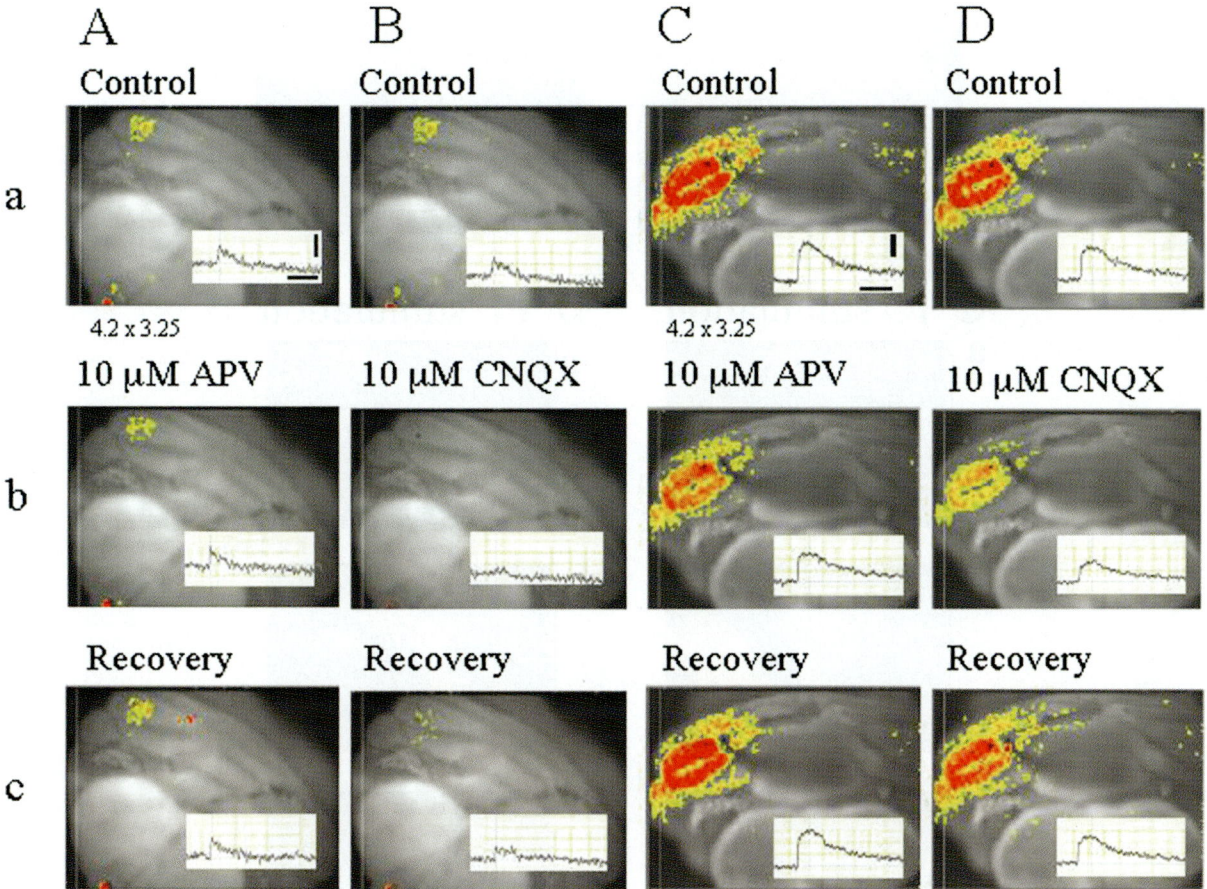

Fig. 7. Glutamate receptors mediating optical responses. (A and B) IO-stimulation-evoked responses (20 ms after stimulation). (C and D) posterior vermis-stimulation-evoked responses (20 ms after stimulation). A, control responses. B, during application of glutamate receptor antagonists as indicated. C, recovery after washout. Inset curves show the time course of the optical responses recorded at one site in the focus of excitation. Horizontal scale, 340 ms. Vertical scale, 0.05% in Δ F/F (F: intensity of fluorescence). (From Arata and Ito, 2004)

blocked by the non-NMDA-receptor antagonist CNQX (10 μM), and not affected by the NMDA-receptor antagonist APV (10 μM) (Fig. 7A and B). In contrast, PF-mediated responses were partially blocked by either 10 μM APV or 10 μM CNQX (Fig. 7C and D). These results indicate that the IO-stimulation-evoked responses are mediated solely by the AMPA receptors, while both the AMPA and the NMDA receptors mediate posterior vermis-stimulation-induced responses. The observations of a cut surface of the cerebellar lobule in the lateral portion of the cerebellum (Fig. 8A and B) also support the postsynaptic localization of CF- or PF-mediated responses because the early responses, corresponding to the EPSPs in Purkinje cells, distributed over the Purkinje cell layer and molecular layer, suggest an activation of the Purkinje, basket, and stellate cells (Fig. 8Ca, Da). However, the late responses in the cut surface spread to granule cell layer, and lasted for more that 140 ms (Fig. 8Cb, Cc, Db, Dc). One may suppose that the late optical responses reflect excitation of not only Purkinje, basket, and stellate cells but also Golgi cells or Lugaro cells located in the granule cell layer. It is also possible that granule cells are activated via a neuronal circuit involving the cerebellar nuclei; nucleocortical mossy fiber projections may increase if Purkinje cell inhibition on nuclear neurons is reduced as a consequence of a

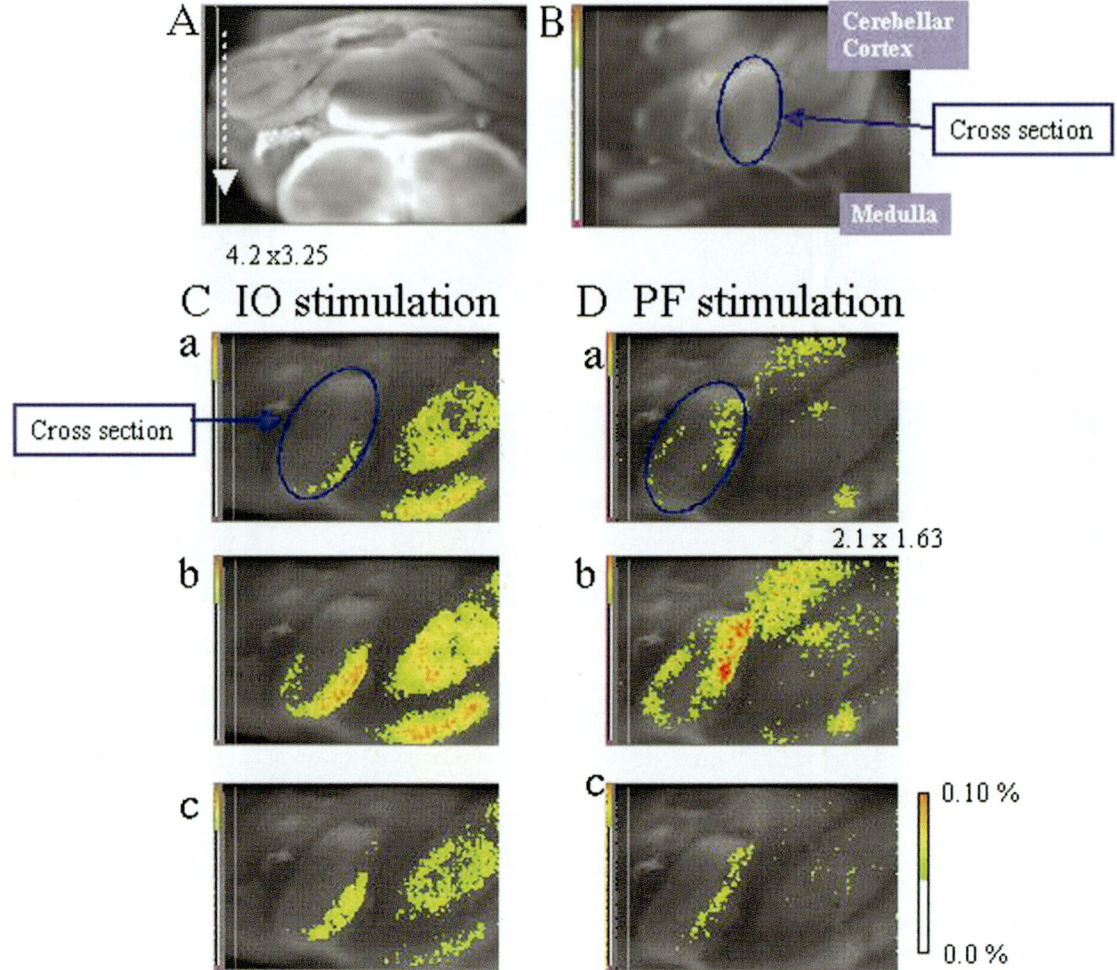

Fig. 8. Localization of the optical responses in the laminar profile of the cerebellar cortex. (A) Photograph of a block preparation, in which a lobule was cut in the lateral portion of the cerebellum, as indicated by a broken arrow. (B) The same preparation as in A, but photographed facing the cut surface, which is enclosed within a frame. In C and D, the cut surface is shown at a higher magnification and in Ca and Da is outlined at its pial surface. C, responses to IO stimulation (Ca, 10 ms; Cb, 80 ms; Cc, 140 ms after stimulation). D, responses to posterior vermis stimulation recorded at the same timing as Ca–c). Note that the responses in the cut surface were initially localized along Purkinje cell layers (Ca and Da), but then they spread to granule cell layers (Cb and Db). The responses finally retracted to Purkinje cell layer (Cc, Dc). Da and Db shows that responses observed along the cut lobule are continuous to its cut surface.

complex neuronal circuit mechanism involving cortical interneurons. If this occurs, all types of neurons in the cerebellar cortex will be excited via nucleocortical mossy fiber projections. Such a long-lasting excitation of the granule cells mediated by neuronal circuits, as assumed here, has not been recognized in studies with cerebellar slices in which, however, corticonuclear as well as nucleocortical projections are severed.

Development of neuronal circuits

The olivocerebellar fibers are present in the white matter of the cerebellum at birth (Sotelo et al., 1984). The present electrophysiological and optical imaging studies revealed that CF and PF inputs to Purkinje cells become functional by P5 (Fig. 3). In the early postnatal stage, each Purkinje cell receives multiple branches of CFs, whereas in adult animals, each

Purkinje cell receives only one CF. The peak of the multiple innervation of Purkinje cells by CFs is attained at P5 (Crepel et al., 1976, 1981; Mariani and Changeux, 1981). In mice, the extent of innervation by multiple CFs increases from P3 to P7 (Hashimoto and Kano, 2003). Multiple CF inputs to a Purkinje cell probably exist in the presently studied P5 rats. In this respect, CFs in P5 rats are still immature, even though CF stimulation induces large EPSPs and LTD.

AMPA receptor immunolabeling (with anti-GluR2/3 and GluR2 antibodies) was clearly observed at P2 and P5 and was stronger in the CF synapses than in the PF synapses (Zhao et al., 1998). A few PF synapses were observed in the vermis of lobules V and VI at P5 (Zhao et al., 1998), but our data showed that the rate of LTD induction in AMPA receptors in the lateral portion of the cerebellum significantly increased at P5 (Fig. 3). At P5, optical imaging revealed that PFs link the lateral edge of the posterior vermis and the lateral portion of cerebellar hemisphere. At P7–P8, PF-mediated optical responses were more localized near the stimulated sites.

Behavioral development

The nodulus, uvula, flocculus, and paraflocculus show earlier morphological growth during the postnatal stages of the rat cerebellum (Wassef et al., 1992), which corresponds to the early development of the vestibulo-ocular reflex function (De Zeeuw et al., 1998). We examined the motor behavior of neonatal rats and found that the abilities to roll over, to turn around the head, to move forward and to balance the body (pivoting) developed around P5 stage (Fig. 9, A. Arata, unpublished observation). Therefore, LTD observed at P4–P8 may play functional roles in motor control even in neonatal stages.

Conclusions

The cerebellum–pons–medulla block preparation provides material suitable for investigating the neuronal circuitry of the cerebellum and its early postnatal development. A similar block preparation has been used for studying the neuronal mechanisms of respiration (Onimaru and Homma, 2003). We

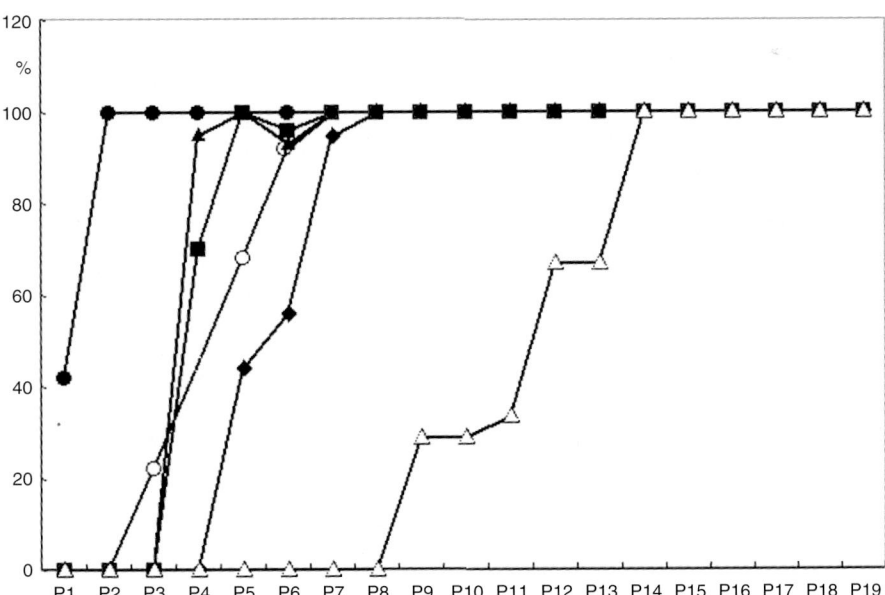

Fig. 9. Postnatal development of motor behavior in rats. Motor behavior was observed at P1–P19 in 31 neonatal rats. Ordinate, the rate of finding the rats exhibiting various motor behaviors. ●, rolling over. ▲, turning the head around. ■, pivoting. ○, moving forward. ◆, locomotion. △, eye blinking.

observed, in the P5 block preparation, that Purkinje cells are already functionally mature in terms of synaptic transmission and synaptic plasticity, yet they undergo rapid changes in their morphology and molecular mechanisms. A typical example of such a molecular change is that, whereas LTD induction requires protein phosphatase 2A in Purkinje cells cultured for 22–35 days (Launey et al., 2004), protein phosphatase 1 plays a major role in Purkinje cells cultured for 12–15 days (Eto et al., 2002). Hence, LTD required for cerebellar function appears to be supported by molecular mechanisms, which dynamically change with time during postnatal development.

References

Altman, J. and Bayer, S.A. (1985) Embryonic development of the rat cerebellum. III. Regional differences in the time of origin, migration, and setting of Purkinje cell. J. Comp. Neurol., 231: 42–65.

Arata, A. and Ito, M. (2001) Synaptic responses of Purkinje cell to stimulation of the inferior olive and parallel fibers in the cerebellum-pons-medulla preparation from neonatal rat. Neurosci. Res., 25: S118.

Arata, A. and Ito, M. (2003) Spatio-temporal pattern of optical responses induced by stimulation of the inferior olive and parallel fibers in the cerebellum-pons-medulla preparation from neonatal rat. Neurosci. Res., 27: S192.

Arata, A. and Ito, M. (2004) Purkinje cell functions in the *in vitro* cerebellum isolated from neonatal rats in the block with the pons and medulla. *Neurosci. Res.* (in press).

Cajal, Ramon y. (1911). In: Azoulay, L. (Ed.), Histologie du système nerveux de l'homme et des vertebrés. II. French edition. Inst. Ramon y Cajal, Madrid, 1955.

Cohen, D. and Yarom, Y. (1999) Optical measurements of synchronized activity in isolated mammalian cerebellum. Neuroscience, 94: 859–866.

Crepel, F., Delhaye-Bouchaud, N. and Dupont, J.L. (1981) Fate of the multiple innervation of cerebellar Purkinje cells by climbing fibers in immature control, X-irradiated and hypothyroid rats. Dev. Brain Res., 1: 59–71.

Crepel, F. and Krupa, M. (1988) Activation of protein kinase C induces a long-term depression of glutamate sensitivity of cerebellar Purkinje cells. An in vitro study. Brain Res., 458: 397–401.

Crepel, F., Mariani, J. and Delhaye-Bouchaud, N. (1976) Evidence for a multiple innervation of Purkinje cells by climbing fibers in the immature rat cerebellum. J. Neurobiol., 7: 567–578.

De Zeeuw, C.I., Hansel, C., Bian, F., Koekkoek, S.K., van Alphen, A.M., Linden, D.J. and Oberdick, J. (1998) Expression of a protein kinase C inhibitor in Purkinje cells blocks cerebellar LTD and adaptation of the vestibule-ocular reflex. Neuron, 20: 495–508.

Ekerot, C.F. and Kano, M. (1985) Long-term depression of parallel fibre synapses following stimulation of climbing fibers. Brain Res., 342: 357–360.

Eto, M., Bock, R., Brautigan, D.L. and Linden, D.J. (2002) Cerebellar long-term synaptic depression requires PKC-mediated activation of CPI-17, a myosin/moesin phosphatase inhibitor. Neuron, 36: 1145–1158.

Gao, W., Dunbar, R.L., Chen, G., Reinert, K.C., Oberdick, J. and Ebner, T.J (2003) Optical imaging of long-term depression in the mouse cerebellar cortex in vivo. J. Neurosci., 23: 1859–1866.

Hartell, N.A. (1994) Induction of cerebellar long-term depression requires activation of glutamate metabotropic receptors. NeuroReport, 5: 913–916.

Hashimoto, K. and Kano, M. (2003) Functional differentiation of multiple climbing fiber inputs during synapse elimination in the developing cerebellum. Neuron, 38: 785–796.

Ito, M. (2001) Cerebellar long-term depression-characterization, signal transduction and functional roles. Physiol. Rev., 81: 1143–1195.

Ito, M. (2002) The molecular organization of cerebellar long-term depression. Nat. Rev. Neurosci., 3: 896–902.

Ito, M. and Kano, M. (1982) Long-lasting depression of parallel fiber-Purkinje cell transmission induced by conjunctive stimulation of parallel fibers and climbing fibers in the cerebellar cortex. Neurosci. Lett., 33: 253–258.

Ito, M., Nisimaru, N. and Shibuki, K. (1979) Destruction of inferior olive induces rapid depression in synaptic action of cerebellar Purkinje cells. Nature, 277: 568–569.

Ito, M., Sakurai, M. and Tongroach, P. (1982) Climbing fibre induced depression of both mossy fibre responsiveness and glutamate sensitivity of cerebellar Purkinje cells. J. Physiol. (Lond.), 324: 113–134.

Karachot, L., Ito, M. and Kanai, Y. (1987) Long-term effects of 3-acetylpyridine-induced destruction of cerebellar climbing fibers on Purkinje cell inhibition of vestibulospinal tract cells of the rat. Exp. Brain Res., 66: 229–246.

Karachot, L., Kado, R.T. and Ito, M. (1994) Stimulus parameters for induction of long-term depression in in vitro rat Purkinje cells. Neurosci. Res., 21: 161–168.

Launey, T., Endo, S., Sakai, R., Harano, J. and Ito, M. (2004) Protein phosphatase 2A inhibition induces cerebellar long-term depression and declustering of synaptic AMPA receptor. Proc. Natl. Acad. Sci. USA, 101: 676–681.

Lev-Ram, V., Jjang, T., Wood, J., Lawrence, D.S. and Tsien, R.Y. (1997) Synergies and coincidence requirements between NO, cGMP, and Ca^{2+} in the induction of cerebellar long-term depression. Neuron, 18: 1025–1038.

Liljelund, P. and Levine, J.M. (1998) Dynamic behavior of the ends of growing parallel fibers in early postnatal rat cerebellum. J. Neurobiol., 36: 91–194.

Linden, D.J., Dickinson, M.H., Smeyne, M. and Connor, J.A. (1991) A long-term depression of AMPA currents in cultured cerebellar Purkinje neurons. Neuron, 7: 81–89.

Mariani, J. and Changeux, J.P. (1981) Ontogenesis of olivocerebellar relationships. I. Studies by intracellular recordings of the multiple innervation of Purkinje cells by climbing fibers in the developing rat cerebellum. J. Neurosci., 1: 696–702.

Onimaru, H. and Homma, I. (2003) A novel functional neuron group for respiratory rhythm generation in the ventral medulla. J. Neurosci., 23: 1478–1486.

Ruigrok, T.J., Osse, R.J. and Voogd, J. (1992) Organization of inferior olivary projections to the flocculus and ventral paraflocculus of the rat cerebellum. J. Comp. Neurol., 316: 129–150.

Sakurai, M. (1987) Synaptic modification of parallel fibre-Purkinje cell transmission in *in vitro* guinea-pig cerebellar slice. J. Physiol. (Lond.), 394: 463–480.

Singer, D. (1999) Neonatal tolerance to hypoxia: a comparative-physiological approach. Comp. Biochem. Physiol. A Mol. Integr. Physiol., 123: 221–234.

Sotelo, C., Bourrat, F. and Triller, A. (1984) Postnatal development of the inferior olivary complex in the rat. II. Topographic organization of the immature olivocerebellar projection. J. Comp. Neurol., 222: 177–199.

Sugihara, I., Bailly, Y. and Mariani, J. (2000) Olivocerebellar climbing fibers in the franuloprival cerebellum: morphological study of individual axonal projections in the X-irradiated rat. J. Neurosci., 20: 3745–3760.

Tominaga, T., Tominaga, Y., Yamada, H., Matsumoto, G. and Ichikawa, M. (2000) Quantification of optical responses with electrophysiological responses in neural activities of Di-4-ANEPPS stained rat hippocampal slices. J. Neurosci. Methods, 102: 11–23.

Wassef, M., Cholley, B., Heizmann, C.W. and Sotelo, C. (1992) Development of the olivocerebellar projection in the rat: II. Matching the developmental compartmentations of the cerebellum and inferior olive through the projection map. J. Comp. Neurol., 323: 537–550.

Zhao, H.M., Wenthold, R.J. and Petralia, R.S. (1998) Glutamate receptor targeting to synaptic populations on Purkinje cells is developmentally regulated. J. Neurosci., 18: 5517–5528.

CHAPTER 11

Optical imaging of cerebellar functional architectures: parallel fiber beams, parasagittal bands and spreading acidification

Timothy J. Ebner*, Gang Chen, Wangcai Gao and Kenneth Reinert

Department of Neuroscience, University of Minnesota, 2001 Sixth Street SE, 421 LRB, Minneapolis, MN 55455, USA

Keywords: Purkinje cell; episodic ataxia; autofluorescence; neutral red; Kv1 potassium channels

Introduction

An intriguing feature of the cerebellar cortex is its highly ordered circuitry and architectures (Eccles et al., 1967). Organized into three distinct layers, the cerebellar cortical circuitry is noted for its almost 'crystalline' structure and this leads to the concept of the cerebellum as a computational machine. The afferent and efferent systems to and from the cerebellar cortex are also notable for their striking topographies (Eccles et al., 1967). Understanding the function of this stereotypic circuitry and neuronal architectures is needed to unravel the nature of the neural computations performed by the cerebellar cortex.

The parasagittal zonation of the cerebellar cortex and its afferent and efferent projections is an organizing principle. Voogd was the first to demonstrate that Purkinje cell axons are organized into parasagittal zones (Voogd, 1967). The olivocerebellar projection is also organized into parasagittal zones similar to those of the corticonuclear projection (Brodal and Kawamura, 1980; Voogd and Bigare, 1980; Bloedel and Courville, 1981). Histochemical and immunocytochemical markers have revealed the parasagittal compartmentalization at the molecular level (Hawkes et al., 1992). The functional nature of this parasagittal zonation has been confirmed in electrophysiological studies (Ekerot and Larson, 1980; Llinas and Sasaki, 1989; Welsh et al., 1995).

A second major architecture is the parallel fibers, the molecular layer extension of granule cell axons. Running for several millimeters along a folium, parallel fibers synapse on and excite Purkinje cells and cerebellar interneurons. The parallel fiber–Purkinje cell circuit and parasagittal zones are nearly orthogonal, with numerous theories of cerebellar function focused on the interactions between these two systems (Ebner and Bloedel, 1987; Hansel et al., 2001; Ito, 2001). Although providing 100–200,000 inputs to Purkinje cells, it still remains unclear as to under what conditions parallel fibers are activated (Bower, 1997).

One of the challenges in studying these cellular architectures is the need to monitor the activity of

*Corresponding author. Tel.: +1 612-626-2205; Fax: +1 612-626-9201; E-mail: ebner001@umn.edu

neurons spatially. Developments in optical imaging have made it possible to image neuronal activity with high spatial resolution (Lieke et al., 1989; Ebner and Chen, 1995). This chapter reviews our efforts to develop and use optical imaging methodologies in the cerebellar cortex in vivo. The review also emphasizes how optical imaging has proven to be valuable in studying the interactions between the parallel fiber and climbing fiber systems, monitoring synaptic plasticity at the circuitry level, and observing novel phenomena such as spreading acidification and depression.

Neutral red imaging

Neutral red imaging has been used by our laboratory to map cerebellar cortical neuronal activity in vivo (Chen et al., 1998), exploiting the known close coupling between pH changes and neuronal activation (Chesler, 2003). Activating parallel fibers using a brief stimulus train in the rat (Chen et al., 1998; Chen et al., 2001) or mouse cerebellar cortex (Gao et al., 2003a; Dunbar et al., 2004) evokes an optical response consisting of a longitudinal beam running parallel to the long axis of the folium (Fig. 1A). The optical signal is biphasic, consisting of a beam of increased fluorescence (acidic shift) that returns to baseline in approximately 60–80 s, followed by a beam of decreased fluorescence (alkaline shift) for up to 120 s (Fig. 1B). A major advantage of the neutral red optical signal is that its amplitude is 5–20 times larger (1–5% $\Delta F/F$) than the epifluorescence signals (0.1–0.5% $\Delta F/F$) obtained with voltage sensitive dyes (Cohen et al., 1978; Ebner and Chen, 1995) (0.1–0.5% $\Delta F/F$) or the reflectance change (0.1–0.5% $\Delta R/R$) from the hemodynamic intrinsic signal (Lieke et al., 1989; Frostig et al., 1990).

The neutral red based optical signal has several useful properties for monitoring spatial patterns of neuronal activation in the cerebellar cortex. First, the optical signal evoked by parallel fiber activation is monotonically and linearly related to the simulation amplitude and frequency (Fig. 1). Second, the majority of the optical signal (80–85%) is postsynaptic (i.e., Purkinje cells and interneurons) with contributions from both ionotropic and metabotropic glutamate receptors (Chen et al., 2001; Gao et al., 2003a; Dunbar et al., 2004). Similarly, the optical response evoked by stimulation of the inferior olive or the periphery reflects postsynaptic activation (Chen et al., 1996; Dunbar et al., 2004). Therefore, neutral red optical signals reflect intrinsic cerebellar cortical activity and not afferent input. Third, the spatial resolution obtainable using neutral red is excellent (Fig. 1A). The lowest amplitude of surface stimulation (100 μA) evokes an optical beam 40 μm in width, considerably smaller than the ~300 μm width of a Purkinje cell dendritic tree (Eccles et al., 1967). Therefore, the optical signal reflects activation of a small fraction of the dendritic tree. Field potential recordings confirm that the neutral red signal is tightly correlated to the underlying neuronal activation (Chen et al., 1998; Hanson et al., 2000; Gao et al., 2003a; Dunbar et al., 2004).

Initial efforts were focused on understanding the source(s) of the neutral red optical signal. The results show that the optical signals originate from pH

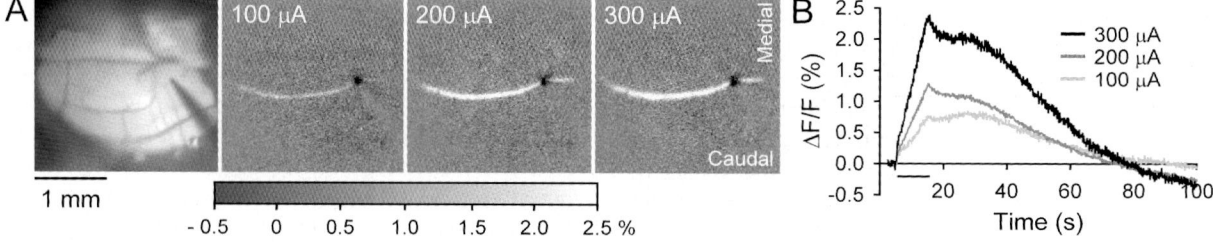

Fig. 1. (A) Examples of the optical response evoked by surface stimulation on Crus IIA in the mouse cerebellar cortex. Shown are 'raw' subtraction maps in which stimulation amplitude is varied. The stimulus train was 10 s in duration, and consisted of 200 μs pulses. The stimulation frequency was kept constant (10 Hz). Location of the surface stimulating electrode can be appreciated in the images. (B) Temporal profiles of the $\Delta F/F$ (%) for the optical responses shown in A.

changes associated with neuronal activation (Chen et al., 1998, 2001). Neutral red is highly pH sensitive and selective, with its fluorescence modulated inversely with pH change. Superfusion of the cerebellar cortex with solutions of varying pH produces the predicted changes in fluorescence, even in the presence of multiple ion channel and receptor blockers. Neutral red fluorescence is independent of a wide range of Ca^{2+}, Na^+, or K^+ concentrations. The optical signal is not due to changes in the extracellular space or cell volume changes. A large number of findings are consistent with the pH shifts being primarily intracellular in origin and primarily neuronal (Chen et al., 1998).

We have used neutral red to map the spatial patterns of activation (Crus I and II) in response to peripheral inputs or stimulation of the contralateral inferior olive (Chen et al., 1996; Hanson et al., 2000; Gao et al., 2003a; Dunbar et al., 2004). Electrical or natural stimulation of the vibrissae area of the ipsilateral face evokes parasagittal bands when using ketamine/xylazine anesthesia. The optical bands are in alignment with the parasagittal zones delineated by immunostaining with anti-zebrin II. Hypothesized to be due to climbing fiber activation of Purkinje cells, the parasagittal bands show a preferred frequency of face stimulation (6–10 Hz), consistent with the inherent rhythmicity of the inferior olivary neurons (Llinas and Yarom, 1986). Figure 2 shows the parasagittal bands evoked by natural stimulation of the upper lip. Contralateral inferior olivary (CIO) stimulation evokes parasagittal bands with nearly identical spatial and frequency tuning characteristics (Hanson et al., 2000; Dunbar et al., 2004). Lidocaine injection into the inferior olive blocks the parasagittal bands evoked from the periphery and leads to the conclusion that the bands are primarily due to activation of Purkinje cells by climbing fibers.

Therefore, neutral red imaging provides a dramatic visualization that the cerebellar cortex is activated in parasagittal zones, the functional organization reflecting the anatomical organization.

Autofluorescence imaging

More recently, we described an intrinsic autofluorescence signal in the cerebellar cortex that reflects the coupling between neuronal activity and mitochondrial metabolism (Reinert et al., 2004). As described over 40 years ago, neuronal activity leads to changes in the oxidation/reduction of flavoproteins and nicotinamide adenine dinucleotide (NADH) (Chance et al., 1962; Duchen, 1992). Autofluorescence of flavoproteins and NADH has been used as an indirect measure of neuronal activity in isolated cell cultures and brain slice but only to a limited extent in vivo. Recently, there has been renewed interest in using autofluorescence to monitor neuronal activity in vitro and in vivo (Shibuki et al., 2003; Shuttleworth et al., 2003). As described in this review, autofluorescence based on the oxidation/reduction of

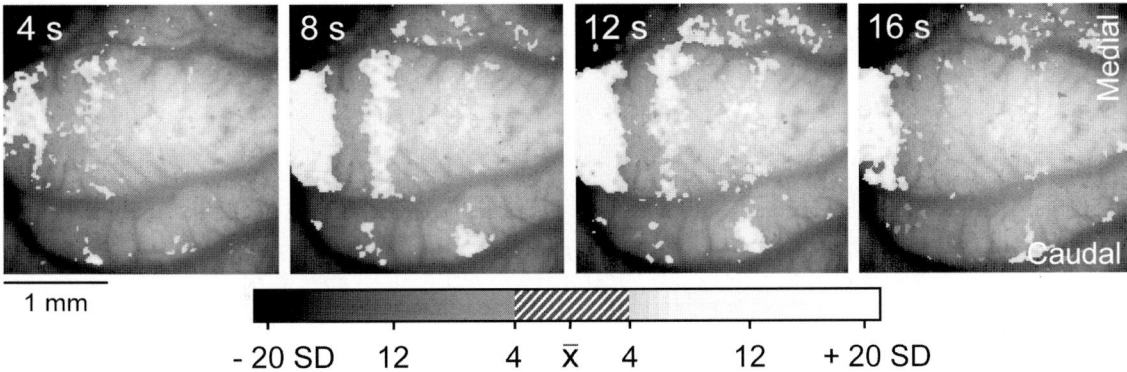

Fig. 2. Optical responses in Crus I and II to mechanical stimulation of the upper lip area evoked with a mechanical vibrator (square wave at 8 Hz for 20 s, 1.0 mm peak-to-peak displacement). The statistically significant optical responses at 4 time points throughout the stimulation are shown as indicated in the grey scale Z-score bar.

flavoproteins in the cerebellar cortex provides a robust, activity-dependent signal.

Imaging in the unstained cerebellar cortex at flavoprotein wavelengths (420–490 nm excitation and 510–570 nm emission) surface stimulation evokes a transverse beam of optical activity consisting of a large amplitude increase in fluorescence (peak ~1.5 ΔF/F %) followed by a longer duration decrease (Fig. 3A–B). Hypothesized to be the result of oxidation and subsequent reduction of flavoproteins, this autofluorescence is markedly reduced by blocking mitochondrial respiration with sodium cyanide (NACN) or selective inactivation of flavoproteins with diphenyleneiodonium (DPI) (Reinert et al., 2004). The reduction in the autofluorescence signal was accomplished without altering the integrity of presynaptic and postsynaptic components of the electrophysiological response after carefully defining the appropriate dose and time window of NACN or DPI. The changes in reflectance at these excitation and emission wavelengths are an order of magnitude smaller than the flavoprotein autofluorescence, which is also independent of hemodynamic changes. Furthermore, the wavelength selectivity and fluorescence nature of this signal demonstrate that it is not the hemodynamic intrinsic signal (Frostig et al., 1990). Similar conclusions were reached for the flavoprotein autofluorescence signal monitored in the rat somatosensory cortex (Shibuki et al., 2003).

Flavoprotein autofluorescence has excellent signal-to-noise characteristics and temporal resolution (Reinert et al., 2004). The time course of the initial increase in fluorescence is tightly coupled to the duration of the stimulation. The signal can be detected with a single surface stimulation pulse and has

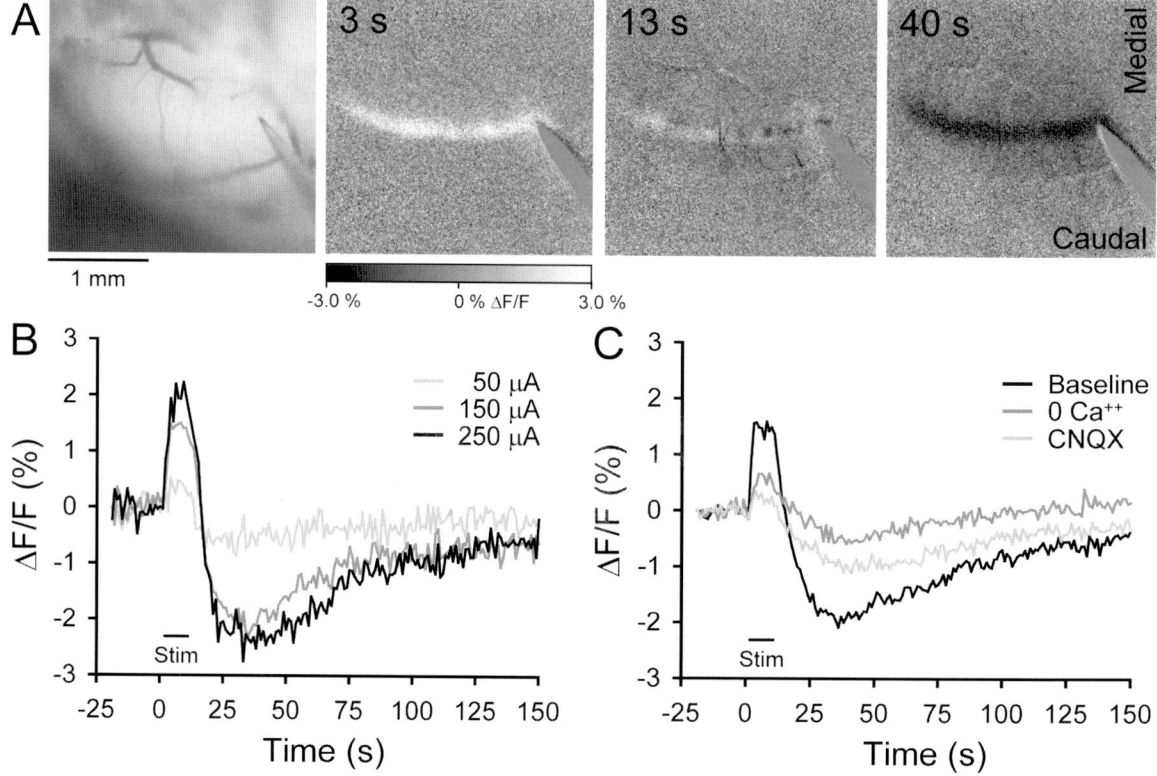

Fig. 3. (A) Light and dark phase of the autofluorescence signal evoked by surface stimulation (200 μA, 100 μs pulses at 10 Hz for 10 s). Subtraction images at three different time points are shown. (B) Temporal profile of the optical response evoked at different stimulation amplitudes. Stimulation consisted of a 10 s train of 200 μs pulses at 10 Hz. (C) Temporal profiles of the optical response evoked in normal Ringer's, zero Ca^{2+} Ringer's, and in 50 μM CNQX.

a latency as short as 50 ms. A brief train of stimuli (2 or 3 pulses at 100 Hz) produces a large optical response (~1.0% $\Delta F/F$). The signal amplitude is linearly related to the amplitude (Fig. 3B), and frequency of surface stimulation. Using low Ca^{2+} Ringer's or blocking ionotropic glutamate receptors with CNQX shows that a majority of the flavoprotein autofluorescence is due to synaptic transmission and activation of the postsynaptic targets of the parallel fibers (Fig. 3C), similar to neutral red based signal (Chen et al., 2001; Gao et al., 2003a). Therefore, the on-beam optical signal response provides a measure of parallel fiber–Purkinje cell synaptic activity.

An important property of flavoprotein autofluorescence is the ability to monitor excitation and inhibition. Surface stimulation evokes both an on-beam excitation response and off-beam inhibition, the latter due to activation of molecular layer interneurons (Eccles et al., 1967). The off-beam inhibition became readily apparent when urethane anesthesia was used and the responses over a small number of trials averaged. As shown in Fig. 4A, surface stimulation evokes an off-beam decrease in fluorescence in addition to the on-beam increase in fluorescence.

The off-beam decrease in fluorescence is highly structured, consisting of parasagittal bands of approximately 300–500 μm wide in the transverse plane and 400–700 μm long in the parasagittal plane (Fig. 4A). Furthermore, the off-beam response is opposed by troughs (relative decreases in the intensity of the signal) along the on-beam response, also consistent with the activation of molecular layer inhibitory interneurons and subsequent inhibition of on-beam Purkinje cells. The net effect is a non-uniform, on-beam response in which the relative minima along the beam are in alignment with the off-beam decrease in fluorescence (Fig. 4B).

Three findings demonstrate that the off-beam bands of decreased fluorescence are inhibitory. First, the off-beam response was eliminated by application of the $GABA_A$ blocker bicuculline and the on-beam response is enhanced (Fig. 5). Second, field potential recordings in the molecular layer are consistent with the decreased fluorescence reflecting synaptic inhibition. Third, the off-beam bands are activated by very low amplitude PF stimulation, below the threshold required to activate the on-beam response. This is consistent with recent findings that stellate cells are activated by parallel fibers at lower stimulation amplitudes than are Purkinje cells (Carter and Regehr, 2002). The decrease in fluorescence off-beam and in the intensity on-beam likely reflect the inhibitory actions of stellate and basket cell activation (Eccles et al., 1967). Therefore, the off-beam inhibitory responses do not appear to be uniform, as conventional wisdom dictates.

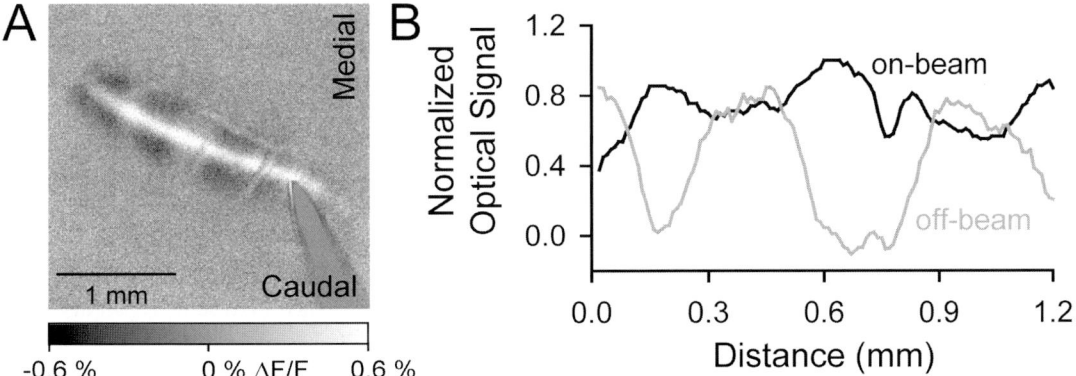

Fig. 4. Off-beam response observed with autofluorescence imaging. (A) Average of 5 surface stimulation trials (200 μA, 100 μs pulse at 10 Hz for 10 s) reveals parasagittal regions of decreased fluorescence lateral to the on-beam. Note that the on-beam response is non-uniform. (B) Normalized intensity profiles measured on-beam and off-beam show the alternating pattern of relative increase and decrease. Normalization involved scaling the maximum of both intensity profiles to 1. The off-beam responses on both sides of the beam were averaged. Image in A was transformed into grey scales based on $\Delta F/F$ as indicated by the grey scale bar.

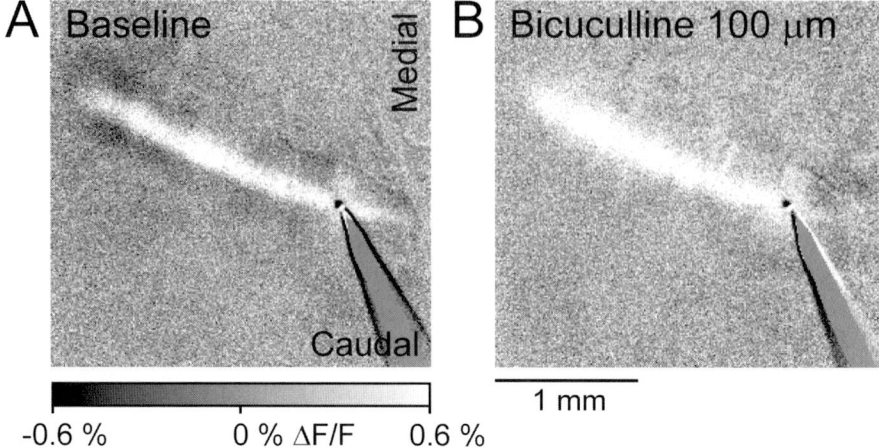

Fig. 5. Off-beam response is blocked by bicuculline. (A) Surface stimulation (200 μA, 100 μs pulses at 10 Hz for 10 s) evoked on-beam and off-beam fluorescence responses. (B) Bicuculline (100 μM) abolished the off-beam response and increased the on-beam response.

Therefore, flavoprotein imaging has many useful properties for monitoring neuronal activation. As for neutral red, the autofluorescence signal has a large amplitude, linear dependence on stimulus parameters, short latency and results primarily from postsynaptic activity. Temporal dynamics of the initial increase in fluorescence is excellent. The autofluorescence signal can detect the activation produced by a single stimulus and can monitor both excitation and inhibition. Due to the signal's intrinsic nature, there are no concerns over toxicity or pharmacological actions of exogenous optical dyes. The flavoprotein signal is larger and has better spatial and temporal resolution than the hemodynamic intrinsic optical signal (Frostig et al., 1990; Malonek and Grinvald, 1996).

Optical imaging of long term depression

Conjunctive stimulation of climbing fiber and parallel fiber inputs results in LTD at the parallel fiber–Purkinje cell synapse (Ito and Kano, 1982). The properties and cellular and molecular mechanisms of parallel fiber–Purkinje cell LTD, including demonstrating its postsynaptic nature, have been documented extensively in vitro. The signaling cascade involves activation of group 1 metabotropic glutamate receptors (mGluR$_1$) (Aiba et al., 1994; Ichise et al., 2000) and protein kinase C (PKC) (Crepel and Krupa, 1988; Linden and Connor, 1991). The nitric oxide–cyclic GMP–protein kinase G cascade is also a component of LTD induction (Shibuki and Okada, 1991) and maybe downstream of mGluR$_1$ activation (Lev-Ram et al., 1997). These signaling cascades result in clathrin-mediated internalization of postsynaptic AMPA receptors (Wang and Linden, 2000). To understand the role LTD plays in cerebellar function including motor learning will require characterizing LTD in vivo, with the ultimate goal to understand LTD in the awake, behaving animal. While studied primarily in vitro using culture and slice preparations (Sakurai, 1987; Crepel and Krupa, 1988; Linden et al., 1991; Hartell, 1994; Lev-Ram et al., 1997), PF–PC LTD has been shown to exist in decerebrate animals (Ito et al., 1982; Ekerot and Kano, 1985).

Using neutral red imaging, we have been able to optically map and characterize in vivo LTD of parallel fiber evoked responses following conjunctive stimulation (Gao et al., 2003a). At the site of the intersection of the parallel fiber-evoked beam and the climbing fiber induced parasagittal band, the subsequent responses to parallel fiber stimulation are depressed for at least 60 min. The depression is spatially specific, requires the conjunction of both

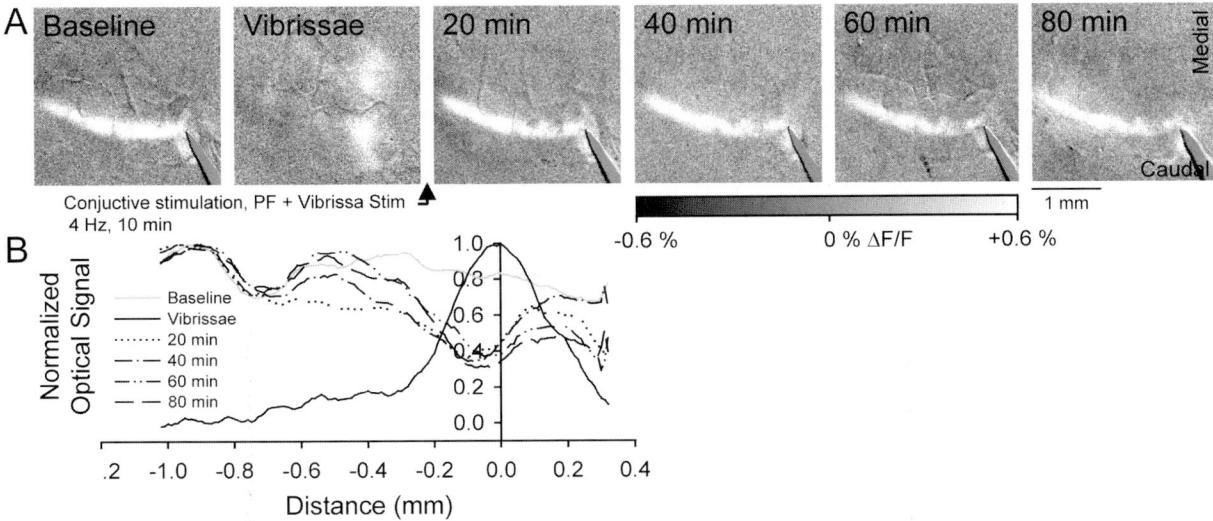

Fig. 6. Conjunction of peripheral input with parallel fiber activation evokes LTD. (A) Series of images depicts the on-beam response before and after conjunctive stimulation with vibrissae stimulation (20 V, 300 μs pulses at 4 Hz for 10 min). Vibrissae stimulation evoked a response consisting of a parasagittal band. (B) Plots of the intensity profiles along the beam for vibrissae stimulation and PF stimulation at different times. Decrease in on-beam following conjunctive stimulation spatially overlaps with parasagittal band activated by vibrissae stimulation.

inputs, and is blocked by metabotropic glutamate receptor antagonists. The depression does not occur in the PKC inhibitor transgenic mouse, which expresses a pseudosubstrate of PKC in Purkinje cells (De Zeeuw et al., 1998). Climbing fiber activation alone did not produce any depression. Conversely, LTD occurs in the $mGluR_4$ knock-out mouse, which is known to have intact LTD (Pekhletski et al., 1996). In addition to providing the first visualization of parallel fiber–Purkinje cell LTD in the cerebellar cortex, this study demonstrates the spatial specificity of LTD and its dependence on $mGluR_1$ and PKC in vivo.

The critical question is whether peripheral inputs can evoke comparable long-term changes in parallel fiber–Purkinje cell synapses. To test this question we have been examining the changes in the cerebellar cortex including evoking parallel fiber–Purkinje cell LTD by the conjunction of parallel fiber and peripheral inputs. These studies have used flavoprotein autofluorescence imaging and normal Ringer's (i.e., no GABA blockers) (Gao et al., 2003b). Conjunctive stimulation of peripheral and parallel fiber inputs can evoke LTD-like changes in the subsequent responses to PF stimulation. Stimulation of the contralateral vibrissae evokes a parasagittal response (Fig. 6A), that when applied conjunctively with parallel fiber stimulation results in an LTD-like reduction in the beam that persists for over 1 h. The depression is spatially specific, occurring at the intersection of the peripheral evoked parasagittal bands and parallel fiber evoked on-beam response (Fig. 6B). Presumably, the vibrissae stimulation activated climbing fiber input that when paired with parallel fiber stimulation resulted in a parallel fiber–Purkinje cell LTD. Our previous studies demonstrated that the parasagittal bands are primarily due to postsynaptic activation of Purkinje cells by CF input (Hanson et al., 2000; Dunbar et al., 2004). This is the first demonstration that when peripheral inputs are paired with parallel fiber stimulation, the parallel fiber–Purkinje cell synaptic strength is modified.

Spreading acidification and depression

While monitoring the neutral red optical responses evoked by surface stimulation we observed an unexpected phenomenon. If the surface stimulation is sufficiently intense, the evoked optical beam began to 'spread' across the folium (Fig. 7A). The

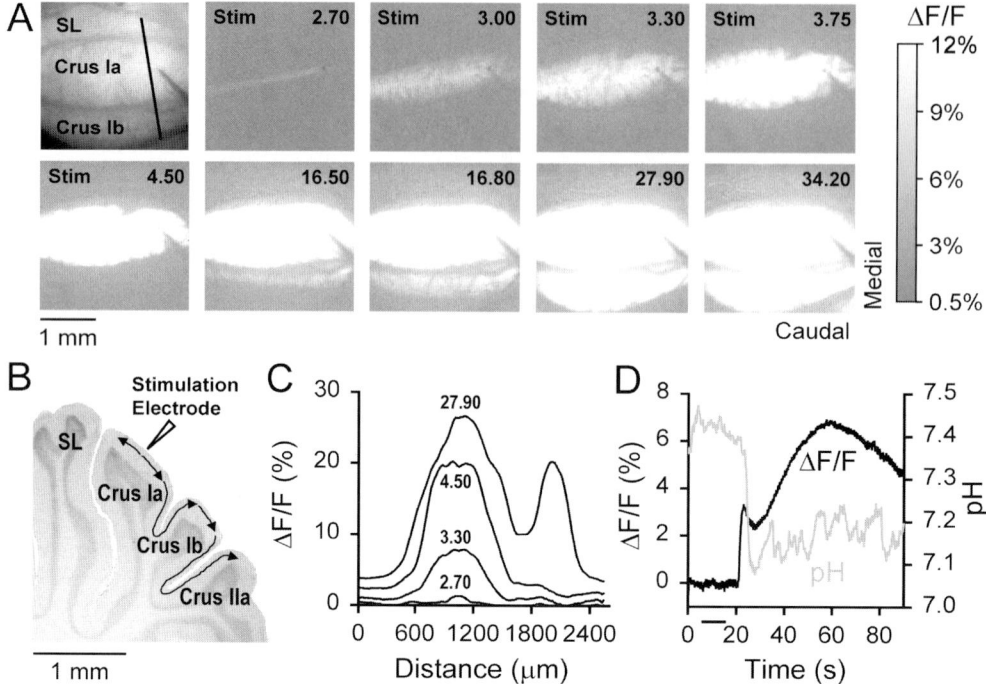

Fig. 7. (A) Series of optical images illustrating SAD in the rat cerebellar cortex evoked by surface stimulation with a brief train of pulses (150 μs, 150 μA at 10 Hz for 10 s). The images were created by superimposing the increase in fluorescence (ΔF/F %) on the background surface image. The first image in the series is a background image showing the folia imaged (SL = simplex), the position of the stimulation electrode, and a black line indicating the location of a section perpendicular to the evoked beam. Different time intervals are shown to emphasize different aspects of the propagation. Numbers in the upper right hand corner are the times in seconds relative to the onset of stimulation. In the upper left corner, 'Stim' denotes that the surface stimulation was ongoing. (B) Parasagittal section (along the line indicated in the first image of A) illustrating the presumed propagation pathway perpendicular to the optical beam. (C) ΔF/F as function of distance along the propagation pathway (line in A) at four time points (in seconds) in the evolution of the spread. (D) The relation between the increased fluorescence (ΔF/F, dark line) and extracellular pH (gray line) in a different animal. Extracellular pH was monitored with a pH sensitive electrode. Horizontal bar on the time axis denotes the period of surface stimulation.

propagation of the signal is characterized by a 10–15 fold increase in fluorescence (i.e., large acidic shift Fig. 7D) and the spread of the optical signal outlasts the time of surface stimulation by minutes. Propagating at speeds reaching 1100 μm/s, the signal spreads anteriorly and posteriorly (parasagittally) over multiple folia (Fig. 7B). The spread is initiated nearly simultaneously along the beam of activated parallel fibers evoked by the surface stimulation and travels perpendicular to the beam. This geometry is maintained as the optical signal spreads to neighboring folia (Fig. 7A and C).

The large increase in the fluorescence in SAD primarily reflects an intracellular acidic shift (Chen et al., 1998). In addition, SAD is associated with an extracellular acidification (Fig. 7D), consistent with the observation that neuronal activity results in both intracellular and extracellular acidification (Chesler, 2003). The fluorescence changes during SAD usually last for 60–90 s and have a refractory period as short as 90 s (Chen et al., 1999). The temporal profile of the increased fluorescence has two phases, an initial rapid increase over a few hundred milliseconds to a first peak followed by a slower increase over tens of seconds to a second peak (Fig. 7D). The latter phase is consistent with a regenerative process underlying SAD in which after a threshold is reached the acidification is actively reinforced.

The spread of the optical signal is accompanied by a transient suppression of the cerebellar cortical

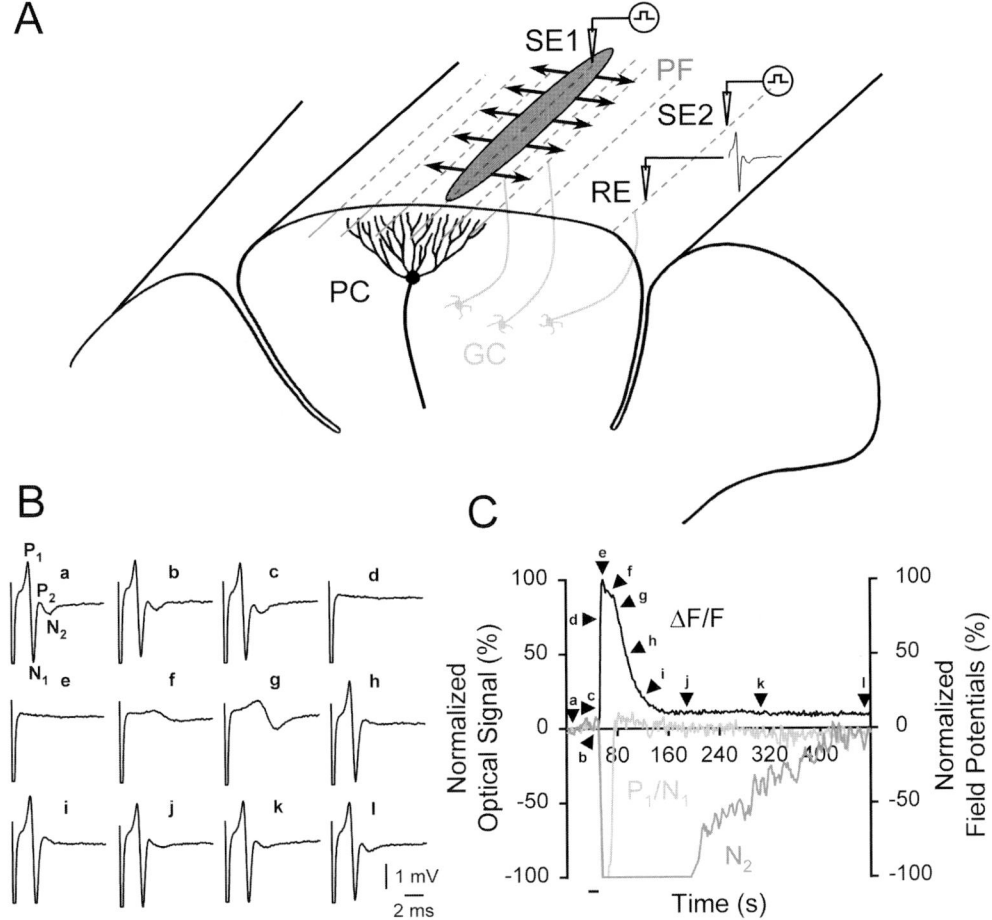

Fig. 8. (A) Schematic drawing of the cerebellar cortex shows the granule cell (gray)–parallel fiber (dashed lines)–Purkinje cell (black) circuit and the experimental setup used to monitor cortical excitability relative to SAD. Stimulation electrode 1 (SE1) was used to evoke an optical beam and SAD (wide strip along the long axis of the folium represents the optical beam and the small arrows indicate the propagation of SAD). Field potentials evoked by the second stimulation electrode (SE2) in the same folium but lateral to the beam evoked by SE1 were monitored by an extracellular recording electrode (RE). (B) Field potentials evoked by SE2 to assay cortical excitability during SAD. Field potentials have the characteristic waveform consisting of a parallel fiber volley ($P_1/N_1/P_2$) followed by the post-synaptic response (N_2). Small letters on the field potentials indicate the time relative to the changes in fluorescence (C) initiated by SE1. (C) Time course of the normalized optical signal ($\Delta F/F$ in black) and field potentials (P_1/N_1 is the parallel fiber volley in lightest gray and N_2 is the post-synaptic response in medium gray). Arrows and corresponding letters on the graph indicate the recording time of the field potentials shown in B. SAD was associated with a loss of field potentials and subsequent recovery. All experimental data from the rat.

circuitry (Fig. 8). Therefore, we have named this phenomenon spreading acidification and depression (SAD). The experimental design used to examine the integrity of the parallel fiber–Purkinje cell circuitry during SAD is shown in Fig. 8A. SAD was initiated with a surface stimulation electrode (SE1). A second stimulation electrode (SE2) and a microelectrode (RE) was used to assess the cerebellar excitability lateral to the site of SAD initiation. A dramatic suppression of the pre- and postsynaptic responses occurs as the increased fluorescence spreads across the folium (Fig. 8B and C). The onset of the suppression in the pre- and postsynaptic responses is tightly coupled to the initial increase in fluorescence (Fig. 8C). The depression in the postsynaptic component generally outlasts the increase in fluorescence

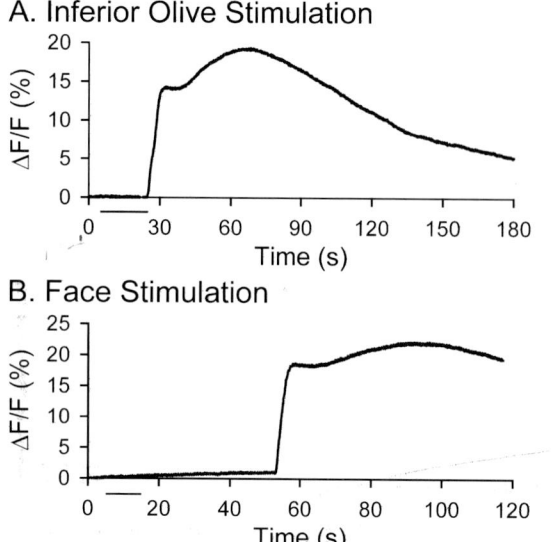

Fig. 9. Climbing fiber and peripheral input evoke SAD. (A) Stimulation of the contralateral inferior olive evoked SAD in Crus II. (B) Stimulation of the ipsilateral face, evoked SAD. Plots of $\Delta F/F$ show the SAD in response to inferior olive (10 Hz for 10 s at 200 µA) or face stimulation (bipolar wires in vibrissae, 10 Hz for 10 s at 20 V). Face stimulation was below threshold for evoking muscle twitch/movements. Data from the rat.

and may be involved with the self-termination of SAD.

Inferior olive stimulation and peripheral stimulation can initiate SAD. In Fig. 9A contralateral inferior olivary stimulation evokes SAD that spreads parasagittally as found for surface stimulation. Also ipsilateral face stimulation can evoke SAD (Fig. 9B). These observations are important, demonstrating that central projections and afferent pathways to the cerebellar cortex can activate SAD.

Spreading acidification differs from other forms of propagated activity found in the CNS. SAD's high speed of propagation is at least an order of magnitude greater than classical spreading depression of Leao (1944), calcium waves (Basarsky et al., 1998; Newman, 2001), or ATP dependent spread (Guthrie et al., 1999). A large number of other characteristics differentiate SAD from classical spreading depression or its variants (Ebner and Chen, 2003). Therefore, SAD appears to be a unique type of propagating activity.

Multiple factors are likely to contribute to the spread of the acidification and depression (Chen et al., 2001). First, hyperexcitability of the cerebellar circuitry plays a role. The probability of evoking SAD increases with greater frequency and/or amplitude stimulation, suggesting that strong activation of the parallel fiber–Purkinje cell circuit is important and that triggering SAD requires reaching a threshold. Second, SAD is initiated nearly simultaneously along the activated beam and propagates uniformly away from the beam suggesting that the highly ordered cerebellar circuitry plays a role. Local depolarization of the neurons or glia is not sufficient to generate SAD, since TTX completely blocks SAD. Third, activation of the postsynaptic targets of parallel fibers plays a role. Ca^{2+} free Ringer's solution blocks SAD and AMPA and/or metabotropic glutamate receptor antagonists increase the threshold for SAD. Therefore, both pre- and postsynaptic circuitry contribute to SAD.

One possible candidate to explain the parasagittal spread is the molecular layer interneurons, including basket cells and deep stellate cells, with their 'off-beam' inhibitory projections that extend several hundred microns perpendicularly to the parallel fibers (Eccles et al., 1967; Mugnaini, 1972; Palay and Chan-Palay, 1974). Potentially these inhibitory projections play a role in the fast propagation speed of SAD and the accompanying suppression in cortical excitability. However, GABA antagonists lower the threshold for evoking SAD (Chen et al., 2001), inconsistent with the hypothesis that inhibitory interneurons play a role. Furthermore, preliminary results suggest the threshold and properties of SAD are normal in the D2-cyclin null mouse ($-/-$), which lacks stellate cells (Huard et al., 1999). Therefore, the off-beam inhibitory projections of molecular layer interneurons are highly unlikely to be involved in the propagation of SAD.

Potassium channels, episodic ataxia type 1 and SAD

Increased excitability of the parallel fiber–Purkinje cell circuitry leads to SAD. Voltage gated potassium channels including the Kv1 family, play a major role in controlling the excitability of neurons (Hille, 1992).

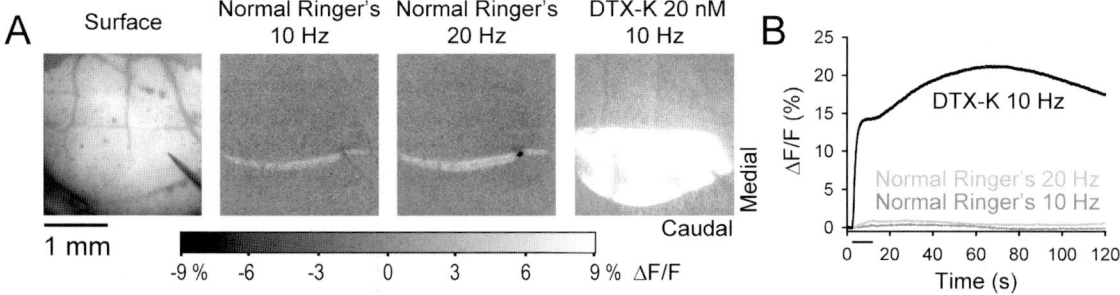

Fig. 10. (A) In normal Ringer's stimulation at 10 and 20 Hz (10 s at 200 µA) failed to evoke SAD. After the superfusion of DTX-K (20 nM), stimulation at 10 Hz evoked SAD in the rat. (B) Plot of the change in ΔF/F for the three conditions shown in A. Bar beneath the x-axis denotes surface stimulation period.

Members of the Kv1 family of delayed-rectifier K^+ channels are highly expressed in the cerebellar cortex (Wang et al., 1994; Veh et al., 1995). Therefore, these channels are logical candidates for involvement in SAD.

Preliminary findings implicate Kv1.1 potassium channels in SAD. Specific blockers of voltage gated potassium channels produce a dramatic decrease in the threshold for evoking SAD. Concentrations as low as 20 nM of the Kv1.1 blocker, DTX-K (dendrotoxin K) (Robertson et al., 1996) are effective at evoking the spread (Fig. 10). At these concentrations, DTX-K primarily blocks Kv1.1 and to a lesser degree Kv1.2 α-subunits. In addition, SAD with DTX-K was readily evoked using stimulation amplitudes as low as 100 µA. Stimulation at this amplitude rarely evoked SAD without DTX-K in several hundred animals tested (Chen et al., 2001). Also in the presence of DTX-K, SAD can occur spontaneously and can be initiated from those spontaneously occurring optical beams. These observations support our early concept that hyperexcitability is critical to SAD and reveal a link between SAD and Kv1.1 potassium channels.

Episodic ataxia type 1 (EA1) is an ion channelopathy characterized by two major symptoms: transient attacks of generalized cerebellar dysfunction and persistent, interictal motor unit activity (i.e., myokymia) (Brandt and Strupp, 1997). Exertion, stress or startle can trigger the transient attacks of cerebellar dysfunction. A dominantly inherited disorder EA1 is associated with a number of different mutations of the gene (KCNA1 on 12 p) that codes for human Kv1.1 (Browne et al., 1994; Kullmann et al., 2001). The myokymia is attributed to hyperexcitability in the preterminal region of motor axons (Zhou et al., 1998), where Kv1.1 potassium channels are enriched (Wang et al., 1994; Veh et al., 1995). A major question is what produces the transient cerebellar dysfunction.

The KCNA1 mutations in EA1 alter a variety of the properties of the Kv1.1 channels that impair axonal repolarization. These include reduction in peak current amplitude, increase in voltage threshold and changes in the activation and deactivation rates (Eunson et al., 2000; Kullmann et al., 2001). The net effect would be to increase neuronal excitability. Some EA1 families have an epilepsy phenotype, consistent with this concept.

How does this hyperexcitability in the cerebellar circuitry translate into transient dysfunction? The most prominent hypothesis is focused on the high density of Kv1.1 potassium channels at the basket cell axon terminals in the pinceau (Southan and Robertson, 1998; Zhang et al., 1999). The resulting hyperexcitability of the basket cell terminal arborization leads to increased GABA release that is hypothesized to interfere with cerebellar cortical function (Southan and Robertson, 1998; Zhang et al., 1999). However, this hypothesis does not explain the transient nature of the cerebellar disorder (Kullmann et al., 2001; Ebner and Chen, 2003).

We have proposed that the episodic cerebellar symptoms of EA1 are due to evoking SAD in the cerebellar cortex (Ebner and Chen, 2003). In this hypothesis, the central tenet is that the altered Kv1.1 potassium channels increase cerebellar cortical excitability leading to a greater likelihood of evoking SAD. The increased excitability would not be limited to basket cell axon terminals, but include granule cells/parallel fibers and Purkinje cells (Veh et al., 1995; Wang et al., 1994). This increased neuronal excitability allows stress, startle, or exertion to evoke SAD. The rapid propagation of SAD across the cortical surface produces a transient but profound depression in cerebellar cortical excitability, that results in a brief period of ataxia.

This hypothesis is supported by several parallels between EA1 and SAD. First, the attacks in EA1 are generally short lasting (seconds to minutes) consistent with the time course of SAD (Kullmann et al., 2001). Spreading acidification and the associated depression of cortical excitability lasts seconds to minutes and both are transient events (Chen et al., 1999, 2001). Second, EA1 appears to be initiated by a triggering event associated with increased activation or excitability of cerebellar neurons. A threshold needs to be reached and startle, stress, or exercise can be viewed as stimuli that contribute to reaching the required threshold. SAD also requires a triggering event, such as sufficiently intense parallel fiber, olivary, or peripheral stimulation and is associated with increased excitability in the cerebellar cortex (Chen et al., 2001). Lastly, both EA1 and SAD are coupled through alterations in the Kv1.1 potassium channels. The mutations in Kv1.1 channels lead to abnormal repolarization after action potential generation and blocking Kv1.1 channels lowers the threshold for SAD.

Spreading acidification and depression in the cerebellar cortex is a newly described propagating activity in the CNS. Much is still to be learned about SAD including the mechanism of spread and the nature of the regenerative process. The parallels between SAD and EA1 are intriguing. If the cerebellar dysfunction in EA1 is due to SAD, this could potentially open avenues for new therapies based on controling SAD and provide insights in the mechanisms of other episodic ataxias. While this link to EA1 suggests that SAD is a pathophysiological process, it remains to be explored whether SAD plays a role in normal cerebellar information processing.

References

Aiba, A., Kano, M., Chen, C., Stanton, M.E., Fox, G.D., Herrup, K., Zwingman, T.A. and Tonegawa, S. (1994) Deficient cerebellar long-term depression and impaired motor learning in mGluR1 mutant mice. Cell, 79: 377–388.

Basarsky, T.A., Duffy, S.N., Andrew, R.D. and MacVicar, B.A. (1998) Imaging spreading depression and associated intracellular calcium waves in brain slices. J. Neurosci., 18: 7189–7199.

Bloedel, J.R. and Courville, J. (1981) A review of cerebellar afferent systems. In: Brooks, V.B. (Ed.), Handbook of Physiology, Sect. 1, The Nervous System, Motor Control, Part 2, Vol. II. Williams and Wilkins, Baltimore, pp. 735–830.

Bower, J.M. (1997) Is the cerebellar sensory for motor's sake, or motor for sensory's sake: the view from the whiskers of a rat? Prog. Brain Res., 114: 483–516.

Brandt, T. and Strupp, M. (1997) Episodic ataxia type 1 and 2 (familial periodic ataxia/vertigo). Audiol. Neurootol., 2: 373–383.

Brodal, A. and Kawamura, K. (1980) Olivocerebellar projection: a review. Adv. Anat. Embryol. Cell. Biol., 64: 1–140.

Browne, D.L., Gancher, S.T., Nutt, J.G., Brunt, E.R., Smith, E.A., Kramer, P. and Litt, M. (1994) Episodic ataxia/myokymia syndrome is associated with point mutations in the human potassium channel gene, KCNA1. Nat. Genet., 8: 136–140.

Carter, A.G. and Regehr, W.G. (2002) Quantal events shape cerebellar interneuron firing. Nat. Neurosci., 5: 1309–1318.

Chance, B., Cohen, P., Jobsis, F.F. and Schoener, B. (1962) Intracellular oxidation-reduction states *in vivo*. Science, 137: 499–508.

Chen, G., Dunbar, R.L., Gao, W. and Ebner, T.J. (2001) Role of calcium, glutamate neurotransmission, and nitric oxide in spreading acidification and depression in the cerebellar cortex. J. Neurosci., 21: 9877–9887.

Chen, G., Hanson, C.L., Dunbar, R.L. and Ebner, T.J. (1999) Novel form of spreading acidification and depression in the cerebellar cortex demonstrated by neutral red optical imaging. J. Neurophysiol., 81: 1992–1998.

Chen, G., Hanson, C.L. and Ebner, T.J. (1996) Functional parasagittal compartments in the rat cerebellar cortex: An *in vivo* optical imaging study using neutral red. J. Neurophysiol., 76: 4169–4174.

Chen, G., Hanson, C.L. and Ebner, T.J. (1998) Optical responses evoked by cerebellar surface stimulation *in vivo* using neutral red. Neuroscience, 84: 645–668.

Chesler, M. (2003) Regulation and modulation of pH in the brain. Physiol. Rev., 83: 1183–1221.

Cohen, L.B., Salzberg, B.M. and Grinvald, A. (1978) Optical methods for monitoring neuron activity. Annu. Rev. Neurosci., 1: 171–182.

Crepel, F. and Krupa, M. (1988) Activation of protein kinase C induces a long-term depression of glutamate sensitivity of

cerebellar Purkinje cells. An in vitro study. Brain Res., 458: 397–401.
De Zeeuw, C.I., Hansel, C., Bian, F., Koekkoek, S.K., van Alphen, A.M., Linden, D.J. and Oberdick, J. (1998) Expression of a protein kinase C inhibitor in Purkinje cells blocks cerebellar LTD and adaptation of the vestibulo-ocular reflex. Neuron, 20: 495–508.
Duchen, M.R. (1998) Ca^{2+}-dependent changes in the mitochondrial energetics in single dissociated mouse sensory neurons. Biochem. J., 283 (Pt 1): 41–50.
Dunbar, R.L., Chen, G., Gao, W., Reinert, K., Feddersen, R.M., Ebner, T.J. (2004) Imaging parallel fiber and climbing fiber responses and their short-term interactions in the mouse cerebellar cortex in vivo. Neuroscience, 126: 213–227.
Ebner, T.J. and Bloedel, J.R. (1987) Climbing fiber afferent system: intrinsic properties and role in cerebellar information processing. In: King, J.S. (Ed.), New Concepts in Cerebellar Neurobiology. Alan R. Liss, Inc., New York, pp. 371–386.
Ebner, T.J. and Chen, G. (1995) Use of voltage-sensitive dyes and optical recordings in the central nervous system. Prog. Neurobiol., 46: 463–506.
Ebner, T.J. and Chen, G. (2003) Spreading acidification and depression in the cerebellar cortex. Neuroscientist, 9: 37–45.
Eccles, J.C., Ito, M. and Szentagothai, J. (1967) The Cerebellum as a Neuronal Machine. Springer-Verlag, Berlin.
Ekerot, C.-F. and Kano, M. (1985) Long-term depression of parallel fibre synapses following stimulation of climbing fibers. Brain Res., 342: 357–360.
Ekerot, C.-F. and Larson, B. (1980) Termination in overlapping sagittal zones in cerebellar anterior lobe of mossy and climbing fiber paths activated from dorsal funiculus. Exp. Brain Res., 38: 163–172.
Eunson, L.H., Rea, R., Zuberi, S.M., Youroukos, S., Panayiotopoulos, C.P., Liguori, R., Avoni, P., McWilliam, R.C., Stephenson, J.B., Hanna, M.G., Kullmann, D.M. and Spauschus, A. (2000) Clinical, genetic, and expression studies of mutations in the potassium channel gene KCNA1 reveal new phenotypic variability. Ann. Neurol., 48: 647–656.
Frostig, R.D., Lieke, E.E., Ts'o, D.Y. and Grinvald, A. (1990) Cortical functional architecture and local coupling between neuronal activity and the microcirculation revealed by in vivo high-resolution optical imaging of intrinsic signals. Proc. Natl. Acad. Sci. USA, 87: 6082–6086.
Gao, W., Dunbar, R.L., Chen, G., Reinert, K.C., Oberdick, J. and Ebner, T.J. (2003a) Optical imaging of long-term depression in the mouse cerebellar cortex in vivo. J. Neurosci., 23: 1859–1866.
Gao, W., Reinert, K., Chen, G. and Ebner, T.J. (2003b) Peripheral stimulation induced long-term depression like phenomenon in the cerebellum in vivo: an autofluorescence imaging study. Soc. Neurosci. Abstr., 75: 18.
Guthrie, P.B., Knappenberger, J., Segal, M., Bennett, M.V., Charles, A.C. and Kater, S.B. (1999) ATP released from astrocytes mediates glial calcium waves. J. Neurosci., 19: 520–528.
Hansel, C., Linden, D.J. and D'Angelo, E. (2001) Beyond parallel fiber LTD: the diversity of synaptic and non-synaptic plasticity in the cerebellum. Nat. Neurosci., 4: 467–475.
Hanson, C.L., Chen, G. and Ebner, T.J. (2000) Role of climbing fibers in determining the spatial patterns of activation in the cerebellar cortex to peripheral stimulation: an optical imaging study. Neuroscience, 96: 317–331.
Hartell, N.A. (1994) Induction of cerebellar long-term depression requires activation of glutamate metabotropic receptors. NeuroReport, 5: 913–916.
Hawkes, R., Brochu, G., Dore, L., Graval, C. and Leclerc, N. (1992) Zebrins: molecular markers of compartmentation in the cerebellum. In: Llinas, R. and Sotelo, C. (Eds.), The Cerebellum Revisited. Springer-Verlag, New York, pp. 22–55.
Hille, B. (1992) Ionic Channels of Excitable Membranes. Sinauer, Sunderland, MA.
Huard, J.M., Forster, C.C., Carter, M.L., Sicinski, P. and Ross, M.E. (1999) Cerebellar histogenesis is disturbed in mice lacking cyclin D2. Development, 126: 1927–1935.
Ichise, T., Kano, M., Hashimoto, K., Yanagihara, D., Nakao, K., Shigemoto, R., Katsuki, M. and Aiba, A. (2000) mGluR1 in cerebellar Purkinje cells essential for long-term depression, synapse elimination, and motor coordination. Science, 288: 1832–1835.
Ito, M. (2001) Cerebellar long-term depression: characterization, signal transduction, and functional roles. Physiol. Rev., 81: 1143–1195.
Ito, M. and Kano, M. (1982) Long-lasting depression of parallel fiber–Purkinje cell transmission induced by conjunctive stimulation of parallel fibers and climbing fibers in the cerebellar cortex. Neurosci. Lett., 33: 253–258.
Ito, M., Sakurai, M. and Tongroach, P. (1982) Climbing fibre induced depression of both mossy fibre responsiveness and glutamate sensitivity of cerebellar Purkinje cells. J. Physiol., 324: 113–134.
Kullmann, D.M., Rea, R., Spauschus, A. and Jouvenceau, A. (2001) The inherited episodic ataxias: how well do we understand the disease mechanisms. Neuroscientist, 7: 80–88.
Leao, A.A.P. (1944) Spreading depression of activity in the cerebral cortex. J. Neurophysiol., 7: 359–390.
Lev-Ram, V., Jiang, T., Wood, J., Lawrence, D.S. and Tsien, R.Y. (1997) Synergies and coincidence requirements between NO, cGMP, and Ca2+ in the induction of cerebellar long-term depression. Neuron, 18: 1025–1038.
Lieke, E.E., Frostig, R.D., Arieli, A., Ts'o, D.Y., Hildesheim, R. and Grinvald, A. (1989) Optical imaging of cortical activity: real-time imaging using extrinsic dye-signals and high resolution imaging based on slow intrinsic-signals. Annu. Rev. Physiol., 51: 543–559.
Linden, D.J. and Connor, J.A. (1991) Participation of postsynaptic PKC in cerebellar long-term depression in culture. Science, 254: 1656–1659.

Linden, D.J., Dickinson, M.H., Smeyne, M. and Connor, J.A. (1991) A long-term depression of AMPA currents in cultured cerebellar Purkinje neurons. Neuron, 7: 81–89.

Llinas, R. and Sasaki, K. (1989) The functional organization of the olivo-cerebellar system as examined by multiple Purkinje cell recordings. Eur. J. Neurosci., 1: 587–602.

Llinas, R. and Yarom, Y. (1986) Oscillatory properties of guinea-pig inferior olivary neurones and their pharmacological modulation: an *in vitro* study. J. Physiol., 376: 163–182.

Malonek, D. and Grinvald, A. (1996) Interactions between electrical activity and cortical microcirculation revealed by imaging spectroscopy: implications for functional brain mapping. Science, 272: 551–554.

Mugnaini, E. (1972) The histology and cytology of the cerebellar cortex. In: Larsell, O. (Ed.), The Comparative Anatomy and Histology of the Cerebellum, The Human Cerebellum, Cerebellar Connections and Cerebellar Cortex. University of Minnesota Press, Minneapolis, pp. 201–265.

Newman, E.A. (2001) Propagation of intercellular calcium waves in retinal astrocytes and Muller cells. J. Neurosci., 21: 2215–2223.

Palay, S.L. and Chan-Palay, V. (1974) Cerebellar Cortex. Springer-Verlag, New York.

Pekhletski, R., Gerlai, R., Overstreet, L.S., Huang, X.P., Agopyan, N., Slater, N.T., Abramow-Newerly, W., Roder, J.C. and Hampson, D.R. (1996) Impaired cerebellar synaptic plasticity and motor performance in mice lacking the mGluR4 subtype of metabotropic glutamate receptor. J. Neurosci., 16: 6364–6373.

Reinert, K.C., Dunbar, R.L., Gao, W., Chen, G. and Ebner, T.J. (2004) Flavoprotein autofluorescence imaging of neuronal activation in the cerebellar cortex *in vivo*. J. Neurophysiol, 92: 199–211.

Robertson, B., Owen, D., Stow, J., Butler, C. and Newland, C. (1996) Novel effects of dendrotoxin homologues on subtypes of mammalian Kv1 potassium channels expressed in Xenopus oocytes. FEBS Lett., 383: 26–30.

Sakurai, M. (1987) Synaptic modification of parallel fibre-Purkinje cell transmission in *in vitro* guinea-pig cerebellar slices. J. Physiol. [Lond.], 394: 463–480.

Shibuki, K., Hishida, R., Murakami, H., Kudoh, M., Kawaguchi, T., Watanabe, M., Watanabe, S., Kouuchi, T. and Tanaka, R. (2003) Dynamic imaging of somatosensory cortical activity in the rat visualized by flavoprotein autofluorescence. J. Physiol., 549: 919–927.

Shibuki, K. and Okada, D. (1991) Endogenous nitric oxide release required for long-term synaptic depression in the cerebellum. Nature, 349: 326–328.

Shuttleworth, C.W., Brennan, A.M. and Connor, J.A. (2003) NAD(P)H fluorescence imaging of postsynaptic neuronal activation in murine hippocampal slices. J. Neurosci., 23: 3196–3208.

Southan, A.P. and Robertson, B. (1998) Modulation of inhibitory post-synaptic currents (IPSCs) in mouse cerebellar Purkinje and basket cells by snake and scorpion toxin K^+ channel blockers. Br. J. Pharmacol., 125: 1375–1381.

Veh, R.W., Lichtinghagen, R., Sewing, S., Wunder, F., Grumbach, I.M. and Pongs, O. (1995) Immunohistochemical localization of five members of the Kv1 channel subunits: contrasting subcellular locations and neuron-specific co-localizations in rat brain. Eur. J. Neurosci., 7: 2189–2205.

Voogd, J. (1967) Comparative aspects of the structure and fibre connections of the mammalian cerebellum. Prog. Brain Res., 25: 94–135.

Voogd, J. and Bigare, F. (1980) Topographical distribution of olivary and corticonuclear fibers in the cerebellum. A review. In: Courville, J., deMontigny, C. and Lamarre, Y. (Eds.), The Inferior Olivary Nucleus. Raven, New York, pp. 207–234.

Wang, H., Kunkel, D.D., Schwartzkroin, P.A. and Tempel, B.L. (1994) Localization of Kv1.1 and Kv1.2, two K channel proteins, to synaptic terminals, somata, and dendrites in the mouse brain. J. Neurosci., 14: 4588–4599.

Wang, Y.T. and Linden, D.J. (2000) Expression of cerebellar long-term depression requires postsynaptic clathrin-mediated endocytosis. Neuron, 25: 635–647.

Welsh, J.P., Lang, E.J., Sugihara, I. and Llinas, R. (1995) Dynamic organization of motor control within the olivocerebellar system. Nature, 374: 453–457.

Zhang, C.L., Messing, A. and Chiu, S.Y. (1999) Specific alteration of spontaneous GABAergic inhibition in cerebellar purkinje cells in mice lacking the potassium channel Kv1.1. J. Neurosci., 19: 2852–2864.

Zhou, L., Zhang, C.L., Messing, A. and Chiu, S.Y. (1998) Temperature-sensitive neuromuscular transmission in Kv1.1 null mice: role of potassium channels under the myelin sheath in young nerves. J. Neurosci., 18: 7200–7215.

CHAPTER 12

Imaging cerebellum activity in real time with magnetoencephalographic data

Andreas A. Ioannides* and Peter B.C. Fenwick

Laboratory for Human Brain Dynamics, Brain Science Institute (BSI), RIKEN, 2-1 Hirosawa, Wako-shi, Saitama, 351-0198, Japan

Keywords: MEG; MFT; single trial analysis; schizophrenia; cerebellum; eye movement

Abstract: The cerebellum has traditionally been associated with motor movements but recent studies suggest its involvement with fine timing, sensory analysis and cognition. Much of the new data comes from neuroimaging techniques such as fMRI and PET, which have high spatial resolution and show that for even simple stimuli many cerebellar and cortical areas are involved. We use examples from recent studies to demonstrate that magnetic field tomograhy (MFT) offers a new and powerful tool for studying cerebellar function through real time localization of cortical, brainstem and cerebellar activations over timescales ranging from a fraction of a millisecond to seconds, minutes and hours. The examples include demonstration of cerebellar activations along well-established anatomical pathways during saccades and the visualization of the ascending medullar volley after median nerve stimulation. MFT analysis of single trial MEG signals elicited by the presentation of faces in emotion and object recognition tasks, show changes in cerebellar activation between schhizophrenics and normal subjects in agreement with proposals for disturbed cerebellar function in schizophrenia. The ability of MFT to identify cerebellar, brainstem and cortical activations in real time can add new insights about dynamics of brain activity to the recent findings about cerebellar function from PET and fMRI.

Introduction

Neuroimaging techniques relying on hemodynamic flow, namely positron emission tomography (PET) and functional magnetic resonance imaging (fMRI) have rekindled the old debate about cerebellar function by demonstrating that the cerebellum is activated in a wide range of tasks. Although these techniques can delineate increases and decreases of activity in each task, their temporal resolution is limited by their reliance on hemodynamic flow. In the case of fMRI, the temporal resolution barely approaches one second, while in the case of PET it is of the order of a few minutes. This is too slow to decipher what the cerebellum is doing. Although advocates of different roles for the cerebellum hold passionate and apparently diverse views about its role, they agree about the cerebellum's critical role in fine timing, where the relevant timescale is milliseconds, about three orders of magnitude faster than the hemodynamic techniques can handle. To move forward from the understanding gained from recent PET and fMRI studies, a technique is required that can deliver similar spatial accuracy but with much improved temporal resolution. We suggest magnetoencephalography (MEG) to be potentially such a technique and we provide as evidence examples from our recent results.

*Corresponding author. Tel.: +81 48 467 9730; Fax: +81 48 467 9731; E-mail: ioannides@postman.riken.go.jp

MEG deals with the detection and interpretation of the minute magnetic field generated by the electrical activity of neurons in the brain. The same neuronal populations generate the MEG signal and the more familiar measurements of the electrical potential on the scalp, the electroencephalogram (EEG). The complementary aspects of the electrical current density in the brain give rise to the MEG and the EEG signal, but once a model for the generators has been introduced these contributions become less distinct. In general, however, the identification of the location of the generators is much easier and can be achieved more accurately with MEG than with EEG using simple models for the conductivity of the surrounding medium. Magnetoencephalography is not the most accurate method for studying any one property of neuronal function in isolation: other techniques can be used to study activity in the brain and the cerebellum with better spatial or temporal resolution, or with better specificity in terms of anatomy, neurochemistry, or connectivity. The unique aspect of MEG is its ability to provide information about cerebellar function with sufficient detail in both the temporal and the spatial domains and to do this in the context of brainstem and cortical activations. Specifically, MEG can provide a millisecond-by-millisecond record of activity with spatial resolution well below one centimeter throughout the cerebellum, brainstem, and cortex for a range of tasks. These include (limited) motor, perceptual, and cognitive tasks. We suggest that MEG can provide a much-needed tool to explore how activity in well-circumscribed parts of the cerebellum relates to activity in the rest of the brain and brainstem. These relationships can be probed over a very wide range of timescales, from a fraction of a millisecond (fine timing) to seconds (preparation for (eye) movement) and to minutes and hours (habituation and learning).

In proposing MEG as a method for studying the cerebellum, we must first show that cerebellar activity contributes to the MEG signal and that such activity can be reliably disentangled from other generators of the MEG signal. As early as 1987, the MEG signal elicited by electrical stimulation of isolated turtle cerebellum was recorded with a single channel (Okada et al., 1987). Evidence for the contribution of human cerebellar generators to the MEG signal had to wait for the introduction of a full head system. The first reports were in association with visually guided horizontal saccades (Jousmaki et al., 1996) and somatosensory stimulation (Tesche and Karhu, 1997). These and the few more recent reports of cerebellar activity extracted from the MEG data relied on heavy averaging of evoked responses to somatosensory stimuli either real (Hashimoto et al., 2003) or omitted (Tesche and Karhu, 2000). In making the case for MEG as a tool for the study of cerebellar function we contrast these earlier approaches with tomographic analysis of MEG single trial data. We believe that this is essential in the case of the cerebellum where plasticity and habituation are likely to be extremely important. We will clarify the distinction between the tomographic single trial analysis and the analysis of average signals in later sections dealing explicitly with the MEG methodology and recent applications.

Cerebellum function

For many years, the cerebellum has been regarded by neurophysiologists and neurologists as an organ serving motor function. During a neurological examination, it is routine to ascribe certain deficits of the coordination of movement to disorders of the cerebellum. Neurophysiological monitoring of cerebellar output nuclei suggests a predominant influence on motor structures. That cerebellar activity is associated with movement is also supported by a significant body of evidence from both functional neuroimaging in humans and electrophysiological recordings in primates.

There is, however, evidence pointing in a different direction. To begin with, the cerebellum first evolved in the context of a sensory filtering function and embryologically it is derived from sensory and not motor tissue. Added to this, the floccular regions of the cerebellum that are specifically involved in eye movement control are phylogenetically older than the regions which it has been suggested are involved in movement control (Welsh and Llinas, 1997).

More recently the role of the cerebellum has been questioned and a wider function suggested for it (Ito, 1993), ranging from classical conditioning

(Thompson, 1988) to sensory and cognitive tasks (Parsons et al., 1997). Some have seen it involved in the control of attention (Allen et al., 1997) while others suggest it to be used for perceptual timing (Keele and Ivry, 1990). A recent study looked at the ability of the cerebellum to discriminate temporal information in a nonverbal auditory memory task (Mathiak et al., 2004). The subjects had to decide whether pauses between tones were short or long and which had been the longest. This task is analogous to verbal working memory tasks. They suggested that the storage of precise temporal structures relies on a cerebellar prefrontal loop. This follows from an earlier suggestion by the same group that the cerebellum monitors future movements and reduces sensory inflow to the central nervous system from the active areas (Leube et al., 2003). They studied fMRI activation in subjects who could see their hand movements on a video screen with different time delays. There was a positive correlation between temporal delay and activation in the right posterior superior temporal cortex and a differential activation in the cerebellum. They concluded that conscious detection of small temporal deviations might be based on signals generated in the cerebellum which provide fine-grained temporal information. In 1997, Bower suggested that the cerebellum is specifically involved in monitoring and adjusting the acquisition of most of the sensory data on which the rest of the nervous system depends. He went on to suggest that the cerebellum is not by itself responsible for any particular behaviorally related function but facilitates the functioning of other structures. So he concluded that the cerebellum might be useful but not necessary, for brain function as a whole. He suggested that the computations performed by the cortical cerebellar circuitry are such that the dendritic state of the Purkinje cells reflect sensory information arising from a large number of different tactile surfaces distributed over a wide extent of the cerebellar cortex. This information in turn modulates information from specific areas. Thus the theory suggests that the cerebellum fine-tunes sensory information within the central nervous system.

There is now a significant body of research suggesting a relationship between disturbances of cerebellar functioning and schizophrenia and other disorders. There are also a number of neurotransmitter and structural studies which suggest abnormalities of neurotransmitter activity and atrophy in the cerebellum. Szeszko et al. (2003a) studied the relationship between the structure of the cerebellum and the neuropsychological functioning in schizophrenia. They found that larger cerebellar volume significantly correlated with better global cognitive functioning in healthy subjects, but not among patients with schizophrenia. They concluded that the cerebellum plays a role in higher cognitive functions in healthy individuals, but that the association between cerebellar size and function was absent in patients experiencing a first episode of schizophrenia. Okugawa et al. (2003) studied cerebellar volume in chronic schizophrenia. They showed that men as well as women with schizophrenia had significantly smaller total vermis volume and smaller vermian sub regions than did the healthy subjects.

Szeszko et al. (2003b), using fMRI, showed that regional cerebellar volumes did not differ between controls, but that among male patients with first episode schizophrenia there were greater reversals in measures of cerebellar asymmetry and they concluded that aberrant neuro-developmental processes involving the metencephalon were involved in the pathophysiology of schizophrenia. Keller et al. (2003), in an MRI study of cerebellar volume in childhood-onset schizophrenia, showed significant progressive loss of cerebellar volume during adolescence, consistent with previously reported decreases in total cerebral and cortical grey matter. They commented that this loss appears secondary to a generalized process.

Andreasen and colleagues suggested that schizophrenia may be associated with dysfunctional prefrontal–thalamic–cerebellar circuitry (Andreasen et al., 1996) and further demonstrated that the cerebellum plays a role in conscious episodic memory retrieval in the absence of any motor output (Andreasen et al., 1999). More recently, PET was used to study schizophrenic patients' response to pleasant and unpleasant visual stimuli. It was found that when patients consciously evaluated unpleasant images they did not activate the phylogenetically older fear–danger recognition circuit. Patients showed reduced activation in other areas, like the thalamus and the cerebellum (Paradiso et al., 2003).

Methods

MEG: early days

The MEG has always been recognized as a method for describing fast changes of activity, but in many people's minds it was considered to be a technique with low spatial resolution, partly because of the non-uniqueness of the underlying mathematical inverse problem but mainly for historical reasons (Ioannides, 1994). For about 20 years after its introduction in the late 1960s and early 1970s, the MEG hardware allowed measurements first with just one and then with a few channels. In order to make a map of the distribution of magnetic fields around a suspected generator site, it was necessary to repeat the experiment many times with a simple and stereotyped stimulus. After a heroic effort (by both experimenters and subjects) a reproducible evoked response could be obtained, first in one location and after many repetitions in enough locations to make a reasonable map. Obviously, such a methodology allows the use of simple stimuli only. A rather indirect consequence of the early limitations was that the generators most likely to contribute to the measurements, i.e., the ones which were most likely to produce a reproducible activation, corresponded to focal activity from well-circumscribed cortical regions, usually at or close to primary sensory areas. Such generators produce a clear and easy-to-spot dipolar pattern and are of course, describable by point like models for the generators, or equivalent current dipoles (ECD).

MEG: whole head coverage

The introduction of a helmet like MEG system capable of capturing in a snapshot the entire magnetic field around the head, is qualitatively a new kind of measurement. Researchers however continued to use the old techniques, because they were easy to use and were easily available. In addition, the trial by trial patterns recorded by the helmet-like MEG devices appeared to be too variable and were easy to interpret as being noisy. Averaging the responses of many trials reduces this 'noise' to the comforting and familiar dipolar patterns. Even today, the recording of a good MEG signal is often equated with stimulus designs leading to clear dipolar patterns. This standard practice for the MEG and EEG data analysis was reinforced strongly by PET and fMRI. An image of activity from a snapshot of PET or fMRI data is extremely noisy; hence reliable and reproducible brain activations can only be obtained after statistical processing of many such images. The widely used statistical parametric maps (SPM) show responses in well-defined areas and they are conceptually very compatible with the ECD activations. This 'signal plus noise' alliance of models is now in full swing with many studies focusing on the combination of hemodynamic methods to fix the location for ECD analysis of average MEG and/or EEG data.

Problems with the classical view

It is important to stress that the map of the magnetic field around the head is captured by modern hardware instantaneously, because it propagates from the place of generation to the sensors at the speed of light. Nowadays it is a routine to sample such instantaneous MEG maps millisecond-by-millisecond or faster. In addition, a combination of hardware and software techniques can largely eliminate all external contributions from outside the room and much of the uninteresting (for the moment) activity from the rest of the body (e.g., muscles, heart, and eyes). What is left is a clean signal with very large dynamic range that is generated entirely by brain activity. Rather than investigating what information might be contained in this signal, an explanation was given for the variability justifying the more familiar but limited analysis techniques: it was assumed that the variability was caused by random electrical activity in the brain, aptly labeled 'brain noise,' possibly related to uninteresting activity from other brain areas. This interpretation of variability is unlikely to be correct for practical as well as theoretical reasons. Put simply random coherent electrical activity, enough to produce the measured magnetic field, would be uneconomical and it is unlikely that it would have survived in the game of evolution. Although evidence against the signal plus noise model was available for many years within the MEG and EEG communities, the real appreciation came from outside. It was the advent of event-related fMRI designs that convinced the wider research community that variability was real and possibly a part of the normal functioning of the brain.

Magnetic field tomography

In 1989, we had introduced the magnetic field tomography (MFT) (Ioannides et al., 1989, 1990) and since then used it essentially in the same form. Specific descriptions for MFT can be found elsewhere, e.g., implementation details in Ioannides (1994) and the mathematical background in terms of lead field expansions in Taylor et al. (1999). The MFT solutions obtained from the unaveraged MEG data showed new aspects of brain function that were incompatible with the 'signal plus noise' view of brain function. The complex patterns of activity related to background rhythms (Liu and Ioannides, 1996, Liu et al., 1996, Laskaris et al., 2003) and dynamics related directly to the stimulus or its perception (Liu et al., 1996, 1999, 2003) that do not follow stimulus onset with identical delay each time. It became clear that important information was washed away by averaging and ECD analysis with apparently very little or no reduction of noise (Ioannides, 2001). The millisecond-by-millisecond, single trial tomographic MFT solutions produces a huge volume of data. To ease the resulting information overload, post-MFT analysis was introduced in the form of standard statistical analysis (Liu et al., 1999; Ioannides, 2001), pattern analysis (Laskaris and Ioannides, 2001, 2002) and information theoretic methods (Ioannides et al., 2000). We summarize next the two key results that established the validity of tomographic analysis and paved the way for moving with confidence away from the classical methods of analysis of MEG data.

How well can MEG localize?

People have addressed the question of localization of MEG data by studying how well one can identify one or more point like generators from the MEG signal they generate. In some studies, the input MEG signal is generated by a forward model computer calculation, while in others the average MEG for some easy-to-deliver stimulus is used. The problem in these approaches is that the activity in the brain is not likely to be anything like a simple point source. To address the question of localization of MEG in a meaningful way, we compared time-dependent statistical parametric mapping (SPM) images derived from our MFT solutions with similar images derived from fMRI data using identical stimuli and the same four subjects. The results showed remarkable agreement for the activation of the primary visual cortex, V1. The post-MFT SPMs around 40 ms after stimulus onset were within a few millimeters of the fMRI SPMs (Moradi et al., 2003). Within a few milliseconds the post-MFT SPMs show widespread activity in extrastriate areas. A new wave of V1 activity was seen again just after 50 ms leading to the well-known peak at 70 ms. The excellent localization demonstrated for superficial activity in V1 cannot be generalized to the entire brain because the MEG sensitivity is reduced with depth and brain areas like the cerebellum and brainstem are not well covered by the sensor array of modern MEG devices. We have used the knowledge of the eye movement system to test localization in the brainstem where we have demonstrated spatial localization of the gaze centers across the midline (Ioannides et al., 2004a). In this and many of our recent studies very robust results were obtained, that were directly comparable to results from fMRI and PET, by combining statistically significant activations across subjects after transformation to a common Talairach space (Talairach and Tournoux, 1988)

Signal plus noise?

The application of pattern analysis and information theoretic methods allowed us to quantify and explore the variability in single trials. These methods revealed a complex but well-ordered and highly organized network of interactions between areas. This organization is not apparent at the level of the signal where variability appears random. The single trial regional activation timecourses appear less random, but they are still difficult to interpret in isolation, especially, if a boring stimulus is repeated many times.

We used the data from the fMRI/MEG experiment to classify the single trial responses according to the dominant patterns of the MEG signal. This classification scheme revealed that the variability in single trials relates to activity in polymodal areas, possibly expressed as background rhythmic activity. The activity in the polymodal areas is coupled to the activity in V1, first preceding the evoked response and then modified by it. This result provides some of the strongest evidence yet against the 'signal plus noise' model and hence averaging (Laskaris et al., 2003).

MEG studies of the cerebellum

Studies using averaged MEG signals

Hari and colleagues used the 122-channel MEG probe to study visually guided horizontal saccades. Using separate averages for saccades to the left and the right, they first estimated from a subset of channels the location and direction of ECD in the posterior parietal cortex (PPC) and the cerebellum (posterior vermis). The time course of activations for these two sources was then estimated from the average signal. The cerebellar source activity began 30 ms before and peaked about 170 ms after the saccade onset (Jousmaki et al., 1996). Significantly, saccades in darkness showed less suppression for cerebellar than PPC sources and a delay of 14 ms in the cerebellar activation.

Tesche and Karhu, in a series of papers, used a similar procedure and the same MEG hardware to study cerebellar contributions following somatosensory stimulation (Tesche and Karhu, 1997) or its omission in a run of stimuli (Tesche and Karhu, 2000). Their analysis identified activity in the relevant cerebellar source areas 13–19 ms after median nerve stimulation delivered with ISI of 500 ms. In the second study, intermittent electrical stimulation of the finger and the median nerve elicited a stimulus-locked cerebellar response with oscillatory components at 6–12 Hz and 25–35 Hz. The random stimulus omissions produced sustained oscillatory activity with cerebellar initiation before the next overt stimulus. Recently Hashimoto and colleagues recorded the MEG signal for right median nerve stimulation (Hashimoto et al., 2003). The authors used a 'vector beam former' technique to identify a four-component cerebellar response from the average MEG signal of 10,000 trials!

Studies using a single trial MEG data

In our early single trial MFT analysis, activations were often identified in the cerebellum. In these early studies, however, it was important first to demonstrate that the tomographic analysis of single trials was consistent with the large body of MEG studies, hence, rather simple sensory stimuli were used to emphasize reproducible activity derived from the average signal in primary sensory areas. We show the following results from a recent single trial MFT analysis using stimuli and tasks similar to the ones used in the MEG experiments described earlier to highlight the consistency with the standard results and to contrast them with the new information that becomes accessible by the new analysis.

Figure 1 shows an example of the detail that can be extracted from post-MFT analysis of 50 single trial solutions for median nerve stimulation. The figure shows the SPM foci obtained by contrasting the distribution of instantaneous (single timeslice, single trial) current density for a fixed poststimulus latency with the corresponding pre-stimulus distribution. A separate SPM map is obtained for each latency (1.6 ms apart). The figure shows the sensory volleys ascending through the brain stem starting at 10.4 ms following median nerve stimulation of the arm. The spreading of the volley to the cerebellum is maximum between 12 and 13.6 ms. The spread in time reflects the duration of the volley and variability across single trials. The SPM maps cover the entire cortex, the brainstem, and the cerebellum. Focal, strong, and highly statistically significant activation in SI (not shown) is identified a few milliseconds after the activation sequence displayed in the figure.

We have recently completed an eye movement study with three subjects comparing saccades to the left, right, up, and down executed after an auditory cue (OCS) or self-paced (SPS) (Ioannides et al., 2004a). The same three subjects were studied while they slept. The MEG signal was recorded for the entire night, so that data would be available from all sleep stages, including the rapid eye movement (REM) sleep. Our published data emphasized the activity in the gaze centers and the contrasted activity over time for OCS and SPS with that during REM saccades (REMS). MFT analysis was used to obtain

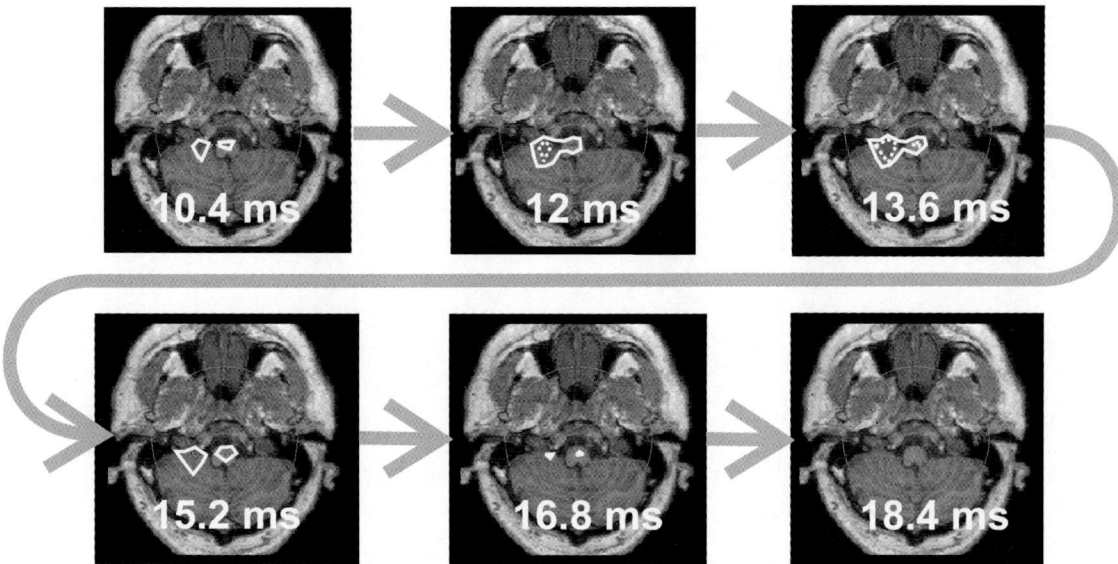

Fig. 1. SPM maps showing the ascending volleys as they traverse the brain stem at the level of the cuneate nucleus. Derived from single trials following somatosensory stimulation. The contours show regions with statistically significant difference in activation in the post-stimulus period compared to the activity in the prestimulus period (solid contour $p < 10^{(-5)}$ and dashed contours $p < 10^{(-7)}$).

tomographic descriptions of activity for each of about 15 single trials in each subject, in each of the OCS, SPS (in a given direction), and REMS (mainly leftwards). Averaging of the MFT solutions for saccades in the same direction produced consistent current flows at widespread cortical, brainstem, and cerebellar sites, including activity similar to that reported by the earlier MEG study of visually guided eye movements (Jousmaki et al., 1996). The activity identified from the MFT analysis of the average signal was, however, dwarfed by nontimelocked activity in the same areas which started many hundreds of milliseconds before and lasted for many hundreds of milliseconds after saccade onset. By computing the mutual information (MI) between the time-lagged segments of regional activations and/or the electrooculogram (EOG) a complex mechanism for the eye movement control was unveiled, revealing linked activity between brain areas with very long time delays (Ioannides et al., 2004a). In the cerebellum rapid and strong changes were identified as the trials progressed. Over longer timescales however, a more stable patterns prevail even for the cerebellum. Figure 2 shows SPMs obtained from statistical comparisons between distributions before or during saccades with distributions recorded while the subject was resting prior to sleep with eyes closed (ECW). For these comparisons, we pooled together the single trial current density values within 500 ms long latency windows into one distribution for each active condition (saccades to the left). Each active distribution was contrasted with the distribution obtained by pooling together the single trial current density values from identical 500 ms long latency windows from the control condition. The SPMs showed a decrease in the left cerebellum before and during saccades for both the OCS and SPS conditions. The SPMs between the REMS and ECW showed a more complex pattern of both increases and decreases in activity but most of these corresponded to state differences between the ECW and REM conditions, highlighting the problems associated with the choice of baseline in such comparisons.

Cerebellum activations in schizophrenia

Our single trial MFT analysis shows how the brain responds to an external stimulus. We have studied in some detail the brain activations elicited by faces showing the standard emotional expressions in a facial emotion recognition task and contrasted these

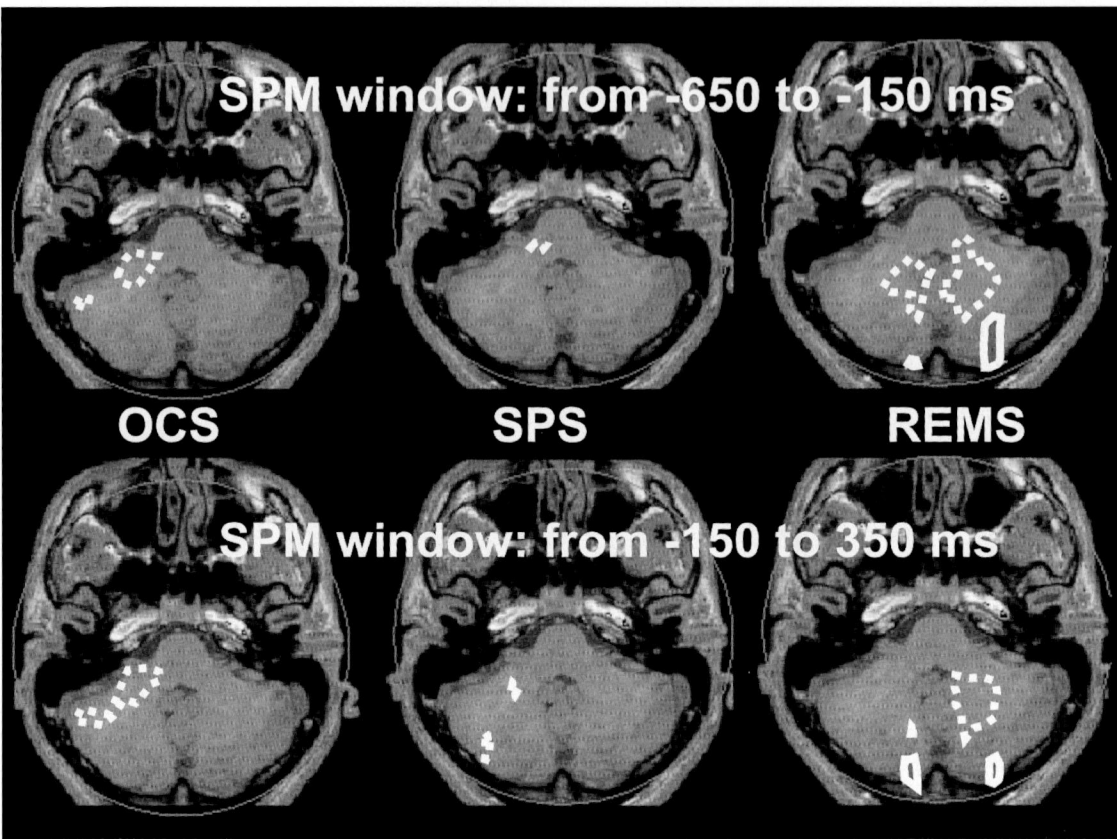

Fig. 2. SPM maps showing the statistically significant changes in activity common to all three subjects (after transformation to a common Talairach space) when 500 ms long windows before (top) and during (bottom) saccades to the left were compared with the activity in the control condition (eyes closed awake). The comparisons shown are for saccades cued on auditory tone (OCS, left), self paced (SPS, middle) and during REM sleep (right). Over the 500 ms only for REMS to the left an increase in activity is identified in the active condition (solid contors). The major change for saccades in awake condition is a decrease in activity on the ipsilateral cerebellum.

with brain activations elicited by images of neutral faces in an object recognition task. We studied the responses from normal subjects (Liu et al., 1999) and from people with schizophrenia (Streit et al., 2001). In the normal subjects the interactions between brain areas grouped together into stages of processing which follow each other in sequence. The activity from any one area contributes to these different stages and primary sensory cortices even take part in the down stream processing (Liu et al., 2003). The evoked response captured by averaging can be thought of as an extreme case of stereotyped stage of processing. The arrival of a stimulus introduces (in at least some of the trials) some new activity and/or resets the background rhythms, thus initiating the sequence of stages for processing. The responses from the schizophrenia group showed no such clear stages of processing (Ioannides et al., 2004b). In agreement with earlier studies, we found that the average MEG signal of schizophrenic subjects was weaker and that the MFT activity derived from the average MEG signal was also weaker in a network of cortical areas (Streit et al., 2001). Our ongoing examination of the MFT solutions from single trial MEG data provided a more sophisticated explanation. The single trial response was not lower in people with schizophrenia. On the contrary, some of the early responses were stronger (Ioannides et al., 2004b). The reduction observed in the average signal is the result of a higher variability; individual responses

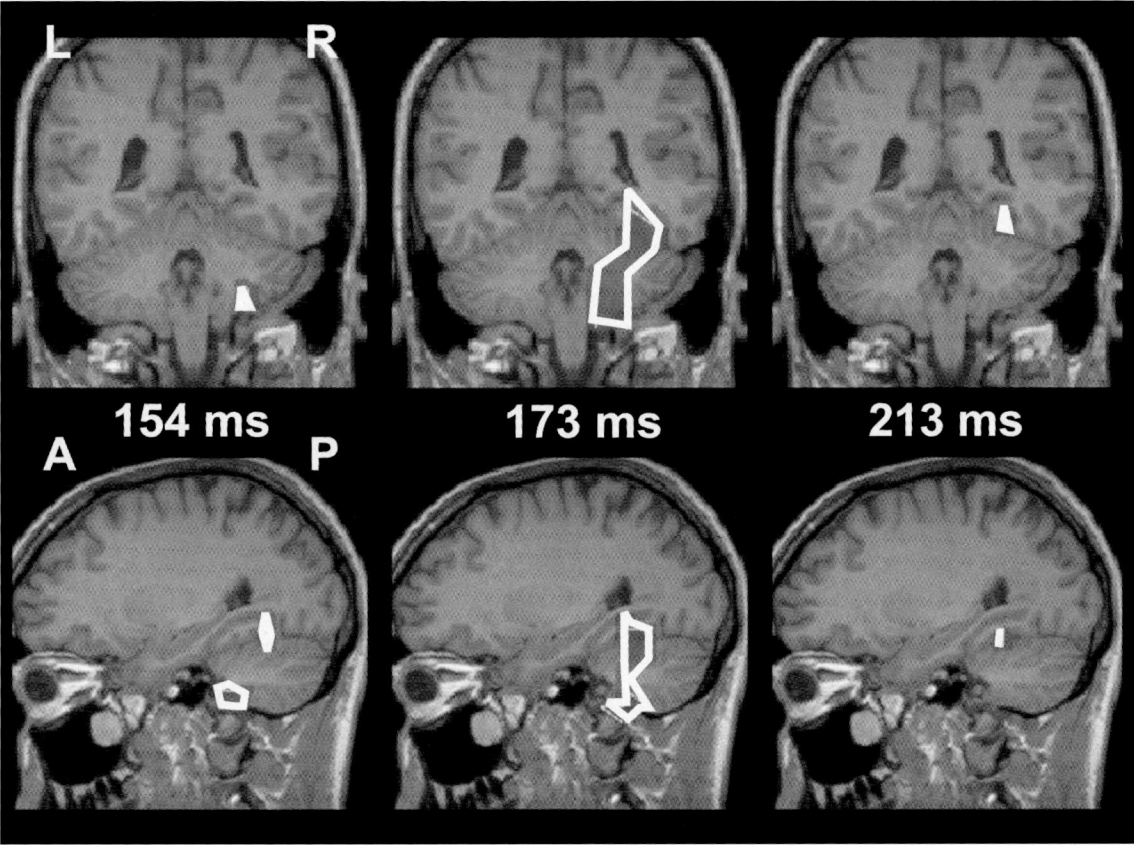

Fig. 3. SPM maps showing the statistically significant changes in activity common to all four normal subjects (after transformation to a common Talairach space) using windows 20 ms long in the contrast between post-stimulus and pre-stimulus activity in the facial expression recognition task.

lack coordination and do not organize themselves into well-defined stages of proccessing. Given the considerable literature implicating the cerebellum in schizophrenia we studied in detail the single trial activity from the cerebellum and brainstem. The key results are summarized in the two SPM maps in Figs. 3 and 4. Figure 3 shows, after transformation to the Talairach space (Talairach and Tournoux, 1988), the statistically significant activations for all the normal controls when compared with the corresponding distributions from the pre-stimulus period. The distribution of values for the current density modulus of the single trial MFT solutions was in a 20 ms window centered at the displayed latency. Significant increases in activity were identified in the right cerebellum and right fusiform face area (FFA). The two cerebellum foci and their Talairach coordinates (TC: x, y, z) in millimeters, were in the inferior lateral aspect (TC: 21, −45, −37) and in the superior part (TC: 26, −47, −9), directly inferior to the FFA. No cerebellum activity was identified in this comparison for schizophrenic subjects. The SPM comparisons were also made between the distribution of values for the current density modulus of the single trial MFT solutions for the emotional facial expression recognition task and an identical window with neutral faces in an object recognition task. This comparison showed common statistically significant increase of activity in the emotion recognition task for the schizophrenia subjects in the prestimulus period, as shown in Figure 4. This activation was in the left inferior lateral cerebellum (TC: −33, −59, −38). No difference could be identified in this comparison in the responses of the normal subjects.

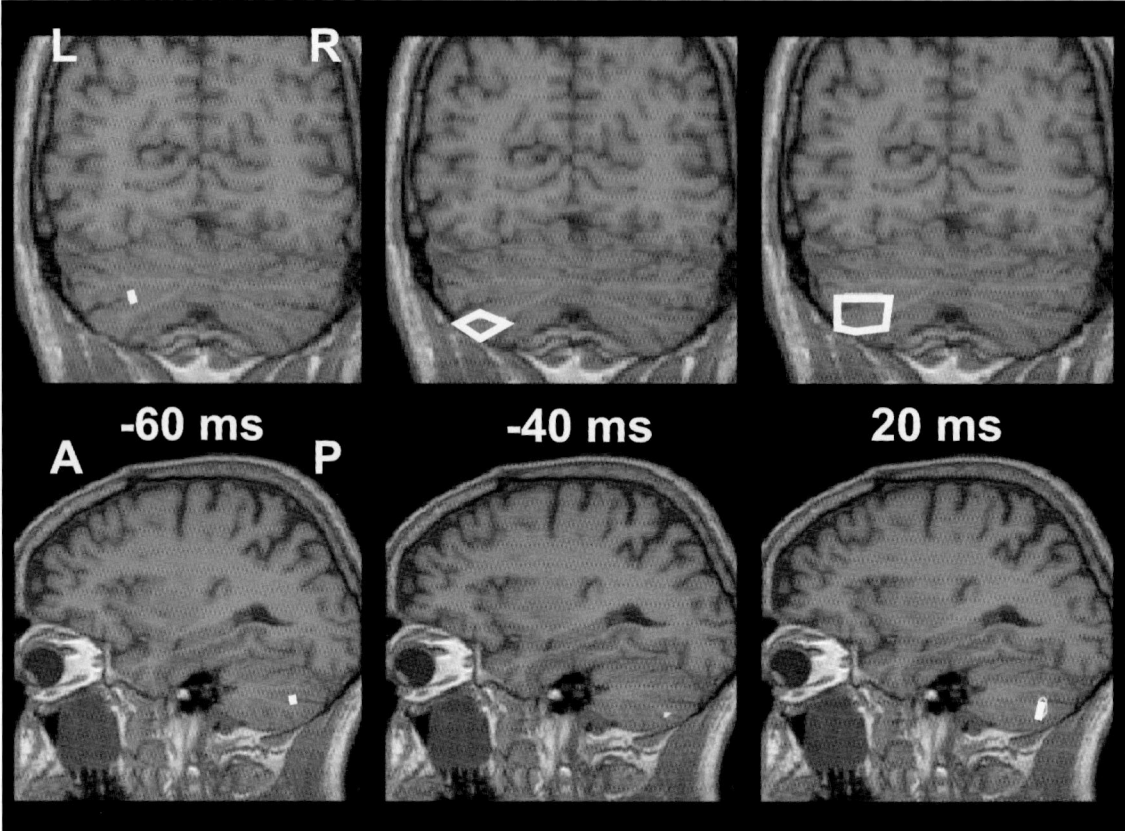

Fig. 4. SPM maps showing the statistically significant changes in activity common to at least 3 of 4 schizophrenia subjects (after transformation to a common Talairach space) using windows 20 ms long in the contrast between the facial expression recognition task and the object recognition task (neutral faces).

Conclusions

The cerebellum has been a conceptual battle ground of ideas about how the brain functions. Recent neuroimaging studies have opened up old debates and provided surprising new results about the contribution of the cerebellum to a bewildering range of tasks. And yet patients with cerebellar damage appear to recover sufficiently to perform these tasks almost as well as normal subjects. Deficits do however persist and these deficits become evident when fine timing is involved. In this paper, we have introduced the MEG methodology, first in its traditional form using heavy averaging and point-like models for the generators, and then in the more demanding tomographic analysis of single trial data. Studies using the traditional methods have demonstrated that cerebellar activity can be identified, even with MEG hardware with planar gradiometers, which are primarily sensitive to nearby sources. We have finally presented results from the single trial MFT analysis showing highly specific activations in well defined brainstem and cerebellar regions. The results for cerebellar activity in the object and facial expression recognition task are worth reiterating. The processing of emotional expression in normal subjects showed clear cerebellar activity around 150 ms after the image onset as compared to the prestimulus activity. No such activation was observed in patients, who instead, showed increased activity in the prestimulus period when the facial expression recognition task was compared to the task of recognizing objects, including faces (with neutral facial expression).

The TC of the activated focus in the left cerebellum almost coincides with the center of gravity of the cerebellar activations in an attentional task without motor contamination (Allen et al., 1997). Taken together, the presence and absence of activations in the control and the schizophrenia pools of subjects in the respective comparisons between tasks and between post- and pre-stimulus periods provide support for the idea that one aspect of schizophrenia is the inability to manage attentional control.

In summary, the results presented here provide good evidence for the promise of the MEG single trial tomographic analysis in studying the function of the cerebellum.

Abbreviations

EEG	electroencephalogram
EOG	electroocculogram
ECD	equivalent current dipoles
ECW	eyes closed waking
fMRI	functional magnetic resonance imaging
FFA	fusiform face area
ISI	inter-stimulus-interval
MFT	magnetic field tomography
MEG	magnetoencephalography
MI	mutual information
MRI	magnetic resonance imaging
OCS	on command saccade
PET	positron emission tomography
PPC	posterior parietal cortex
REM	rapid eye movement
REMS	REM saccades
SPS	self-paced saccade
SPM	statistical parametric map
TC	Talairach coordinates

Acknowledgments

We thank the members of the Laboratory for Human Brain Dynamics and especially Dr Vahe Poghosyan for his help with data for the analysis. We wish to dedicate the work to the memory of Dr Marcus Streit, who introduced one of us (AAI) to the study of schizophrenia with facial emotion recognition tasks.

References

Allen, G., Buxton, B.B., Wong, E.C. and Courchesne, E. (1997) Attentional activation of the cerebellum independent of motor involvement. Science, 275: 1940–1943.

Andreasen, N.C., O'Leary, D.S., Cizadlo, T., Arndt, S., Rezai, K., Ponto, L.L., Watkins, G.L. and Hichwa, R. (1996) Schizophrenia and cognitive dysmetria: a positron-emission tomography study of dysfunctional prefrontal-thalamic-cerebellar circuitry. Proc. Natl. Acad. Sci. USA, 93: 9985–9990.

Andreasen, N.C., O'Leary, D.S., Paradiso, S., Cizadlo, T., Arndt, S., Watkins, G.L., Ponto, L.L. and Hichwa, R. (1999) The cerebellum plays a role in conscious episodic memory retrieval. Human Brain Mapp., 8: 226–234.

Bower, J.M. (1997) Is the cerebellum sensory for motor's sake, or motor for sensory's sake? The view from the whiskers of a rat. Prog. Brain Res., 114: 483–516.

Hashimoto, I., Kimura, T., Tanosaki, M., Iguchi, Y. and Sekihara, K. (2003) Muscle afferent inputs from the hand activate human cerebellum sequentially through parallel and climbing fiber systems. Clin. Neurophysiol., 114: 2107–2117.

Ioannides, A.A. (1994) Estimates brain activity using magnetic field tomography and large-scale communication within the brain. In: Ho, M.W., Popp, F.A. and Warnke, U. (Eds.), Bioelectrodynamics and Biocommunication. World Scientific, Singapore, pp. 319–353.

Ioannides, A.A. (2001) Real time human brain function: observations and inferences from single trial analysis of magnetoencephalographic signals. Clinical EEG, 32: 98–111.

Ioannides, A.A., Bolton, J.P.R., Hasson, R. and Clarke, C.J.S. (1989) Localised and distributed source solutions for the biomagnetic inverse problem II. In: Williamson, S.J., Hoke, M., Stroink, G. and Kotani, M. (Eds.), Advances in Biomagnetism. Plenum Press, New York, pp. 591–594.

Ioannides, A.A., Bolton, J.P.R. and Clarke, C.J.S. (1990) Continuous probabilistic solutions to the biomagnetic inverse problem. Inverse Problem, 6: 523–542.

Ioannides, A.A., Liu, L.C., Kwapien, J., Drozdz, S. and Streit, M. (2000) Coupling of regional activations in a human brain during an object and face affect recognition task. Human Brain Mapp., 11: 79–92.

Ioannides, A.A., Corsi-Cabrera, M., Fenwick, P.B.C., Rio Portilla, Y., Laskaris, N., Khurshudyan, A., Theofilou, D., Shibata, T., Uchida, S., Nakabayashi, T. and Kostopoulos, G.K. (2004a) MEG tomography of human cortex and brainstem activity in waking and REM sleep saccades. Cerebral Cortex, 14: 56–72.

Ioannides, A.A., Poghosyan, V., Dammer, J. and Streit, M. (2004b) Real time neural activity and connectivity in healthy individuals and schizophrenic patients. NeuroImage (In press).

Ito, M. (1993) Movement and thought: identical control mechanisms by the cerebellum. Trends Neurosci., 16: 448–450.

Jousmaki, V., Hamalainen, M. and Hari, R. (1996) Magnetic source imaging during a visually guided task. Neuroreport, 7: 2961–2964.

Keele, S. and Ivry, R. (1990) Does the cerebellum provide a common computation for diverse tasks: a timing hypothesis. Ann. N.Y. Acad. Sci., 608: 179–211.

Keller, A., Castellanos, F.X., Vaituzis, A.C., Jeffries, N.O., Giedd, J.N. and Rapoport, J.L. (2003) Progressive loss of cerebellar volume in childhood-onset schizophrenia. A. J. Psychiatry, 160: 128–133.

Laskaris, N. and Ioannides, A.A. (2001) Exploratory data analysis of evoked response single trials based on minimal spanning tree. Clin. Neurophysiol., 112: 698–712.

Laskaris, N. and Ioannides, A.A. (2002) Semantic geodesic maps: a unifying geometrical approach for studying the structure and dynamics of Single trials evoked responses. Clin. Neurophysiol., 113: 1209–1226.

Laskaris, N.A., Liu, L.C. and Ioannides, A.A. (2003) Single-trial variability in early visual neuromagnetic responses: an explorative study based on the regional activation contributing to the N70m peak. Neuroimage, 20: 765–783.

Leube, D.T., Knoblich, G., Erb, M., Grodd, W., Bartels, M. and Kircher, T. (2003) The neural correlates of perceiving one's own movements. Neuroimage, 20: 2084–2090.

Liu, L.C. and Ioannides, A.A. (1996) A correlation study of averaged and single trial MEG signals: The average describes multiple histories each in a different set of single trials. Brain Topogr., 8: 385–396.

Liu, L.C., Ioannides, A.A. and Streit, M. (1999) Single trial analysis of neurophysiological correlates of the recognition of complex objects and facial expressions of emotion. Brain Topogr., 11: 291–303.

Liu, L.C., Fenwick, P.B.C., Laskaris, N.A., Schellens, M., Poghosyan, V., Shibata, T. and Ioannides, A.A. (2003) The human primary somatosensory cortex response contains components related to stimulus frequency and perception in a frequency discrimination task. Neuroscience, 121: 141–154.

Liu, M.J., Fenwick, P.B.C., Lumsden, J., Lever, C., Stephan, K.M. and Ioannides, A.A. (1996) Averaged and Single-trial analysis of cortical activation sequences in movement preparation, initiation and inhibition. Human Brain Mapp., 4: 254–264.

Mathiak, K., Hertrich, I., Grodd, W. and Akermann, H. (2004) Discrimination of temporal information at the cerebellum: functional magnetic resonance imaging of non-verbal auditory memory. Neuroimage, 21: 154–162.

Moradi, F., Liu, L.C., Cheng, K., Waggoner, R.A., Tanaka, K. and Ioannides, A.A. (2003) Consistent and precise localization of brain activity in human primary visual cortex by MEG and fMRI. NeuroImage, 18: 595–609.

Okada, Y.C., Lauritzen, M. and Nicholson, C. (1987) Magnetic field associated with neural activities in an isolated cerebellum. Brain Res., 412: 151–155.

Okugawa, G., Sedvall, G.C. and Agartz, I. (2003) Smaller cerebellar vermis but not hemisphere volumes in patients with chronic schizophrenia. A. J. Psychiatry, 160: 1614–1617.

Paradiso, S., Andreasen, N.C., Crespo-Facorro, B., O'Leary, D.S., Watkins, G.L., Boles Ponto, L. and Hichwa, R. (2003) Emotions in unmedicated patients with schizophrenia during evaluation with positron emission tomography. A. J. Psychiatry, 160: 1775–1783.

Parsons, L.M., Bower, J.M., Gao, J.H., Xiong, J., Li, J. and Fox, P.T. (1997) Lateral cerebellar hemispheres actively support sensory acquisition and discrimination rather than motor control. Learning and Memory, 4: 49–62.

Streit, M., Ioannides, A.A., Sinnemann, T., Woelwer, W., Dammers, J., Zilles, K. and Gaebel, W. (2001) Disturbed facial affect recognition in patients with schizophrenia associated with hypoactivity in distributed brain regions: a magnetoencephalographic study. A. J. Psychiatry, 158: 1429–1436.

Szeszko, P.R., Gunning-Dixon, F., Goldman, R.S., Bates, J., Ashtari, M., Snyder, P.J., Lieberman, J.A. and Bilder, R. (2003a) Lack of normal association between cerebellar volume and neuropsychological functions in first-episode schizophrenia. A. J. Psychiatry, 160: 1884–1887.

Szeszko, P.R., Gunning-Dixon, F., Ashtari, M., Snyder, P.J., Lieberman, J.A. and Bilder, R. (2003b) Reversed cerebellar asymmetry in men with first episode schizophrenia. Biol. Psychiatry, 53: 450–459.

Talairach, J. and Tournoux, P. (1988) Co-planar stereotaxic atlas of the human brain: 3-dimensional proportional system: an approach to cerebral imaging. Thieme, New York.

Taylor, J.G., Ioannides, A.A. and Mueller-Gaerdner, H.-W. (1999) Mathematical expansion of lead field expansions. IEEE Trans. Med. Imag., 18: 151–163.

Tesche, C.D. and Karhu, J. (1997) Somatosensory evoked magnetic fields arising from sources in the human cerebellum. Brain Res., 744: 23–31.

Tesche, C.D. and Karhu, J. (2000) Anticipatory cerebellar responses during somatosensory somatosensory omission in man. Human Brain Mapp., 9: 119–142.

Thompson, R.F. (1988) The neural basis of basic associative learning of discrete behavioral responses. Trends Neurosci., 11: 152–155.

Welsh, J.P. and Llinas, R. (1997) Some organising principles for the control of movements based on olivocerebellar physiology. Prog. Brain Res., 114: 449–461.

CHAPTER 13

The cerebellum in the cerebro-cerebellar network for the control of eye and hand movements — an fMRI study

M.F. Nitschke[1,*], T. Arp[1], G. Stavrou[1], C. Erdmann[1,2] and W. Heide[1,4]

[1]Department of Neurology, Medical University of Lübeck, Germany
[2]Department of Radiology, Medical University of Lübeck, Germany
[3]Department of Neuroradiology, Medical University of Lübeck, Germany
[4]Department of Neurology, Municipal Hospital Celle, Germany

Keywords: cerebellum; sequential movements; saccades; memory guided; fMRI

Abstract: The coordination of optical information and manipulation of objects in space by eye and hand movements is controlled by a cerebro-cerebellar network. The differential influence of prefrontal, motor, or parietal areas in combination with cerebellar areas, especially within the posterior hemispheres, on the control of eye and hand movements is not very well defined. Using fMRI we investigated the functional representation of isolated or combined eye and hand movements within the cerebellum and the impact of differential cognitive preload on the activation patterns. Each task consisted of the performance of saccades or hand movements triggered by a cue presented on a screen in front of the scanner. Saccades were tested for visually guided saccades, triple step saccades, and for visuospatial memory. Sequential finger opposition movements were tested for predictive and nonpredictive movements. Combined and isolated eye-hand reaching movements were tested toward a target presented in 5 different horizontal positions. Visually guided saccades activated the cerebellar vermis lobuli VI–VII, triple step saccades, including visuospatial memorization, in addition the cerebellar hemispheres lobuli VII–VIII. Sequential finger movements and reaching movements activated a cerebellar network consisting of the lobuli IV–VI, the vermis, and the lobuli VII–VIII with broader areas and additional regions especially within the lobus VII for more complex movements. The combined in contrast to the isolated performance of eye and hand movements demonstrated specialized activation foci within the cerebellar vermis and posterior hemispheres. We could demonstrate a differential representation of eye and hand movements within the cerebellum. Additional "cognitive" preload within a given task leads to additional activation of the posterior cerebellar hemispheres, with a subspecialization corresponding to premotor and parietal area connections.

Introduction

The coordination of optical information and manipulation of objects in space by eye and hand movements is controlled by a cerebro-cerebellar network (Nobre et al., 2000; Culham and Kanwisher, 2001). The differential influence of cortical areas like the prefrontal, motor, or parietal areas in cooperation with the cerebellum on the control of isolated and combined eye and hand movements in humans until recently could not be examined in the functional context of ongoing movements. Using the newly developed technique of functional Magnetic Resonance Imaging (fMRI),

*Corresponding author. Tel.: +49-451-500 2492; Fax: +49-451-500 2489; E-mail: nitschke_m@neuro.mu-luebeck.de

we investigated the representation of isolated or combined eye and hand movements within the cerebellum and the impact of a differential cognitive preload on the activation patterns.

Prediction of desired movement sequences and adaptation to new internally generated movement and externally acquired sensory parameters seem to be processed in part by the cerebellum. It is hypothesized that the cerebellum provides an internal model of what is to be moved to predict a proper movement sequence as a feedforward motor command, modulated by error signals conveyed by climbing fibers to the respective mossy fiber — Purkinje cell–parallel fiber system (Kawato, 1999; Ito, 2000). The integration of the human cerebellum into motor performance could be demonstrated previously by PET and fMRI studies. The lobuli IV–V of the ipsilateral anterior cerebellum were activated by simple hand or finger movements (Fox et al., 1985; Grafton et al., 1993; Nitschke et al., 1996; Lotze et al., 1999), but the activation pattern also comprised the corresponding contralateral anterior lobe and medial parts of the posterior hemispheres, lobuli VII–VIII (Desmond et al., 1997; Luft et al., 1998; Nitschke et al., 1998; Grodd et al., 2001). Anatomical data in the monkey propose a functional segregation within the cerebellar hemispheres, as the motor cortical areas project to the anterior lobe, prefrontal areas to lobuli VI–VII and parietal areas to lobuli VII–VIII (Brodal and Bjaalie, 1992; Middleton and Strick, 1994, 1997, 2001; see also Fig. 1). Accordingly, increased activation of the posterior lobes of the cerebellum could be detected by performance of increasing complex sequential movements or in the phase of movement preparation (Sadato et al., 1996; Catalan et al., 1998; Cui et al., 2000). Our first experiment looked at the modulation of cerebellar activation by predictive and nonpredictive externally triggered sequential finger movements at constant frequencies to keep the number of performed movements constant. We assumed that a nonpredictive task with presentation of unpredictable movement trajectories should result in an increased error correction and information processing for movement planning and execution.

For the control of saccadic eye movements, anatomical and electrophysiological experiments in primates as well as clinical data demonstrate a role

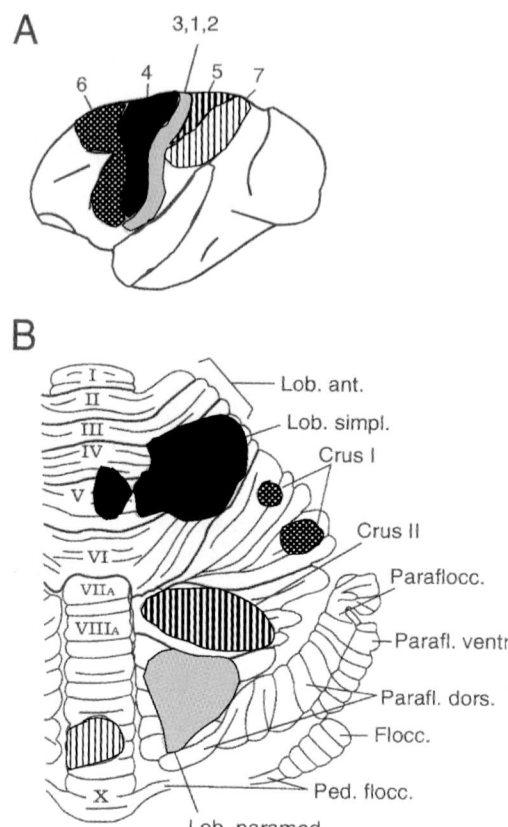

Fig. 1. Schematical drawing of cerebro-ponto-cerebellar connections in the monkey (modified after Colin et al., 2002) demonstrating a subspecification within the cerebellum. A: Numbers represent Brodmann areas. B: Cerebellar regions are matched to the cerebral areas.

especially of the posterior vermis in controlling the accuracy and timing of saccades (Büttner, 1999; Thier et al., 2000). Some PET and fMRI studies demonstrated activation of the vermis during saccades (Sweeney et al., 1996; Dejardin et al., 1998; Desmurget et al., 1998). In accordance with the control of sequential finger movements, the control of eye movements could involve premotor functions and information about the visual space. In the second experiment we investigated the functional significance of the postulated cerebro-cerebellar connections by looking for saccade-specific activation in the vermis and for memory related activation in the hemispheres especially lobuli VI–VIII, which are connected with the parieto-prefrontal network. In a task with no memory, load subjects had to perform externally triggered visually guided reflexive saccades. In a task with a moderate memory load they had to generate a sequence of saccades to the remembered locations of three peripheral targets that had been flashed in rapid succession. To further enhance the memory load and to dissociate the saccade-related components of this task from its cognitive components (e.g., working memory), we compared the triple-step task with a pure visuospatial working memory task that required memorization of the same sequence of three targets for at least 2.5 s, but without actually performing saccades. We expected saccade-specific activation in the posterior vermis and activation of the hemispheres in the memory conditions.

In the third experiment we investigated the activation patterns during the performance of isolated in comparison to the combined execution of goal directed, eye and hand movements towards horizontally presented targets. Besides the definition and confirmation of the executive areas for finger or eye movements during isolated performance, we looked for activation foci, which are selectively involved in the processing of the coordination of goal directed eye–hand movements.

Methods

Altogether we examined 16 volunteers at 1.5-T in different experimental settings. None of the subjects had a current or past history of neurological or ophthalmological disorders and all were normal on examination. The studies were approved by the local Ethics Committee. All subjects gave written informed consent prior to examination after the nature of the experimental procedures was explained. Each task consisted of the performance of saccades or hand movements or the combination triggered by a cue presented on a screen in front of the scanner.

Motor sequences

Seven right handed male and female subjects (25–35 years) were studied. Functional MRI was performed using echoplanar sequences (Siemens Symphony, Erlangen, Germany; standard head coil; TR/TE = 6 s/81 ms, matrix 81×128, slice thickness 4 mm) in contiguous sagittal slices. The protocols consisted of seven cycles alternating between activation (16 s) and control conditions (16 s). Each task consisted of sequential finger to thumb opposition movements, either in predictive (repeatedly 2, 3, 4, and 5) or non-predictive (randomized) fashion. The movement sequence was displayed on the monitor. The finger to move was coded by a number (index finger 2, middle finger 3, ring finger 4, and little finger 5). The frequency of the movement was kept constant at 1 Hz. The actual performance within the MR-scanner was controlled by recording the movements with SMD-switches fixed to each finger (for details see Nitschke et al., 2003).

Saccades

Six right handed male and female subjects (25–35 years) were studied. We performed functional imaging using echoplanar sequences (Siemens Vision, Erlangen, Germany; standard headcoil; TR/TE/flip angle = 5 s/66 ms/90°, slice thickness 4 mm) of the brain in 30 contiguous horizontal slices parallel to the intercommissural line covering the whole brain. The protocols consisted of six cycles of activation (25 s) and a control condition (25 s) in block design. Visual stimuli were presented with a red laserpoint. The task designs (A, B, and C) were measured in all six volunteers, the task sequence was pseudorandomized to counterbalance possible task specific influences. Task D was added to the task protocol for three volunteers. Task A: Visually guided saccades (VG) versus central fixation (Fix): Peripheral targets with different horizontal eccentricities (5–10°) were presented successively for a pseudorandomized time interval of 1200–1800 ms to trigger 16 saccades balanced throughout one activation period (25 s). The fixation point alternated between the left and the right visual hemifield, equally distributed within the whole sample of trials in a pseudorandomized order. The instruction was to continuously track the target. The control condition (Fix) was fixation of the stationary laserpoint positioned in the center of the field of view. Task B: triple-step saccades (TR) versus Fix: After the presentation of a central fixation point for 1500 ms, three successively flashed laser targets were presented with different horizontal eccentricities (5–10°)

alternating between the left and the right visual hemifield analogous to A. Presentation times were 400, 300, and 200 ms for the three targets that had to be memorized while subjects kept on fixating the central fixation point. After a delay of 750 ms (memorization time) the fixation point disappeared as the go-signal to perform saccades to the memorized locations of the three targets in the presented order in darkness. After 3100 ms, the central fixation point reappeared to start the next trial. Altogether, one triple-step trial lasted for 6250 ms with 16 saccades in one activation period (25 s, four triple-step trials). Task C: TR as the active condition was contrasted with VG as the high baseline condition. Task D: TR was contrasted with a visuospatial working memory task (WM). WM consisted of the flashed presentation and memorization of visual targets in TR sequence. After 1500 ms of central fixation, the triple-step sequence was presented and the volunteers had to memorize the target positions. After another 2000 ms of central fixation (memorization time) a new target was presented that had to be compared with the three targets presented before. This target had to be acquired by a single visually guided saccade only when its location was identical to one of the three targets, which happened randomly in 50% of the trials. Four of these trials were performed during one activation period (25 s). Performance was controlled by infrared reflection oculography outside the scanner prior to the experiments. In addition, a qualitative assessment of the subjects' global eye movement performance (presence and direction of saccades) was obtained by electro-oculographic recordings during an fMRI measurement. For further details see also Nitschke et al. (2004).

Eye hand coordination

We investigated 16 volunteers (Siemens Symphony, EPI-Sequences, whole brain, TR 6 s, TE 81 ms, matrix 81 × 128, slice thickness 5.0 mm). Each task consisted of the performance of saccades or reaching movements towards a target presented on a screen in front of the scanner. The target, a computer generated laser point, was pseudorandomly presented in five different horizontal positions (20° left and right visual field). The experimental paradigm consisted of performance of isolated or combined fixation of the laser point and reaching movements towards the target with their right hand. Saccade performance was controlled with an MR-compatible infrared-oculography (Cambridge Research Systems) during the scan.

Data analysis

Statistical image analysis was performed using Matlab (Mathworks Inc., Natiek, MA, USA) and the statistical parametric mapping package (SPM 96b and 99, Wellcome Department of Cognitive Neurology, London, UK; Friston et al., 1995, 1997). The imaging data were corrected for head movements and signal intensity variation and transformed into a standard stereotactic space (Tailarach and Tournoux, 1988). During spatial normalization, pixels were smoothed and the effects of global volume activity and time were removed as confounds, using linear regression and sine/cosine functions. The contrast between task and control conditions was thresholded (saccades: $p < 0.05$ (uncorrected); sequential finger and combined eye-hand movements: $p < 0.01$ (corrected)) for each voxel using the delayed box-car reference function of the SPM software. For visualization, color-coded quantitative maps of positive contrasts were superimposed on corresponding T1-weighted templates. The anatomical nomenclature was based on the three-dimensional atlas of the human cerebellum in proportional stereotactic space by Schmahmann et al. (1999). For 3-D visualization in Fig. 2 we used the BrainVoyager software (Brain Innovation B.V., Netherlands).

Results

Motor sequences

For the predictive sequence, the main activation focus was located within the lobuli IV–VI ipsilateral to the movements and additional foci could be detected within the contralateral lobuli IV–VI, the vermis, the ipsilateral lobuli VIIB–VIII, and a small spot within the contralateral crus I. Nonpredictive (randomized) finger opposition compared to the predictive task showed activation of a broader area

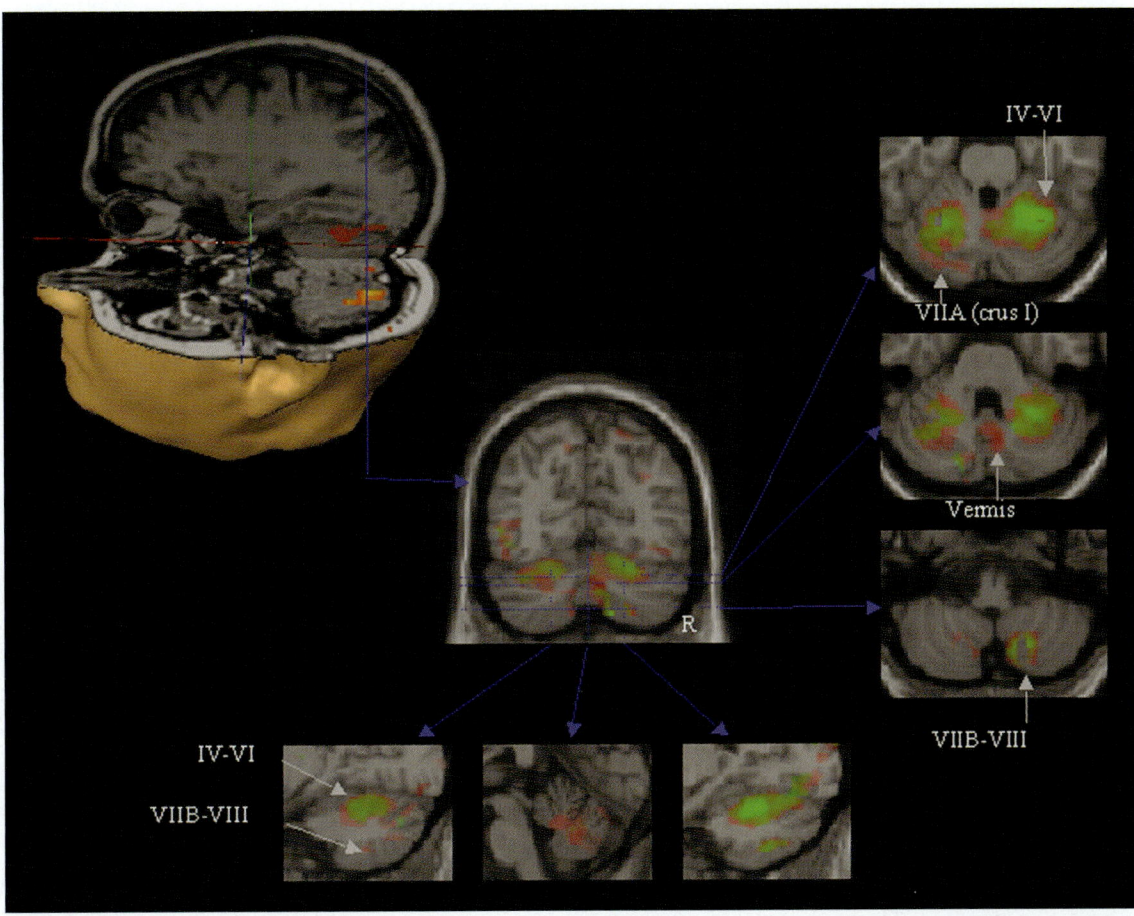

Fig. 2. 3-D map of cerebellar activation by non-predictive sequential finger movements (upper left, visualized with Brainvoyager) and comparison of predictive and non-predictive sequential finger movements. Activation foci of predictive movements (green) were located within the ipsilateral and contralateral lobuli IV–VI, vermis lobulus VI, ipsilateral lobuli VIIB–VIII, and the contralateral crus I. Non-predictive finger opposition (red) compared to the predictive task showed activation of a broader area within the anterior cerebellum (Lobuli IV–VI ipsilateral and especially contralateral), the vermis, and the ipsilateral lobuli VIIB–VIII. Additional activation foci were found in the contralateral left lobulus crus I and the contralateral lobuli VIIB–VIII.

within the anterior cerebellum (lobuli IV–VI ipsilateral and especially contralateral), the vermis, and the ipsilateral lobuli VIIB–VIII. Additional activation foci were found in the contralateral left lobulus crus I and the contralateral lobuli VIIB–VIII (Fig. 2). The number of performed movements were equal for the predictive and nonpredictive paradigm. The performance control revealed an increased error rate (thumb opposition movements to the wrong finger) for the nonpredictive movement task. The error rate for the predictive task was less than 5%, whereas nonpredictive movements revealed ca. 10% errors (for further details see also Nitschke et al., 2003).

Saccades

Visually guided saccades with central fixation as control (task A) activated the cerebellar vermis lobuli VI–VII and paravermal regions (Fig. 3A). Group analysis (task B) of the contrast between the triple-step task (TR) and fixation resulted in the predominant activation of the left cerebellar hemisphere corresponding to lobule VI–crus I. In addition to experiment A + B, where central fixation served as control, we used a differential task design with TR as active condition and VG or WM as control to look at different oculomotor and cognitive components of

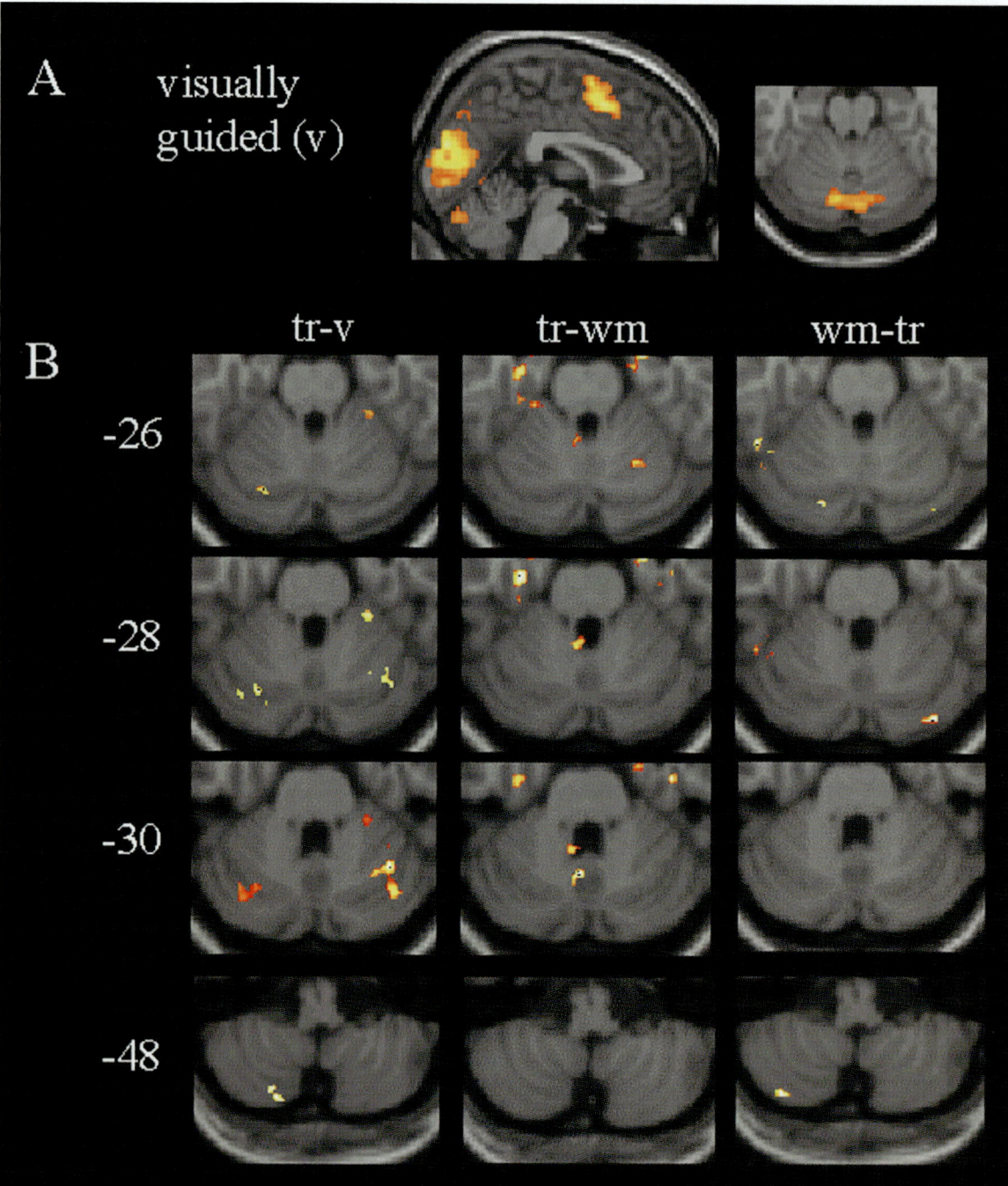

Fig. 3. Group analysis data of visually guided saccades (v) and of a differential task design. Triple step saccades (tr) with V as control predominantly activated areas within the lobuli VI–crus I and the lower part of the left cerebellar hemisphere (left column) corresponding to the lobuli VIIB–VIII. TR with a test of visuospatial working memory (wm) as control (tr-wm) revealed activation within the vermal/left paravermal region (middle column), whereas the reverse condition (wm-tr) demonstrated predominant hemispheric activation of areas within lobulus crus I (right column).

the triple-step task (Fig. 3B). The contrast between TR and VG (Fig. 3B, left column) revealed activation of both cerebellar hemispheres within the lobule VI–crus I (Z-coordinates of −26 to −30), identical to the area activated during TR with central fixation as control. Another area was activated within the lower part of the left cerebellar hemisphere located within the lobule VIIB–VIII (Z −48). To further separate the oculomotor from the visuospatial, attentional, and working memory components of the triple-step task we compared the triple-step task with the visuospatial working memory task (WM) in three volunteers. TR as contrasted to WM, demonstrated predominant activation within the posterior vermis. The reverse comparison (WM as active condition, TR as control) activated areas within the hemispheric lobule VI–crus I and the lower part of the left cerebellar hemispheres lobule VIIB–VIII (Fig. 3B, middle and right column) (for further details see also Nitschke et al., 2004).

Eye hand coordination

Isolated visually guided saccadic eye movements predominantly activated the saccade related areas predominantly in the vermis and paravermis lobuli VI–VII and an additional area bilaterally in the lobule VI on both sides. Reaching movements with the right hand towards the fixation point activated areas related to the known movement related areas in the cerebellum including the lobuli IV–VI on both sides, the lobuli crus I, VIIB–VIII, and within the posterior vermis lobuli IV–VII (Fig. 4A). We could detect an overlap of activation within the vermis and within the lobule VI on both sides performing the isolated movements. The combined, in contrast to the isolated, performance of eye and hand movements demonstrated additional activation foci within the cerebellar vermis and the posterior hemispheres. The activation within the vermis seems to correlate to predominant information processing for saccadic eye movements as it was evident by looking at the signal intensity time course, whereas the activation foci within a region corresponding to the location of the right dentate nucleus correlates to the processing of reaching movements (Fig. 4B).

Figure 5 schematically displays the activation foci within the Talairach space and overlaid onto a flattened surface of the cerebellum. The pattern of activation foci by sequential finger movements define a cerebellar network that is modulated by the cognitive preload. The tasks with a dominant oculomotor component (visually guided saccades, triple-step vs. visuospatial working memory) are preferentially oriented along the midline, whereas the tasks with a cognitive preload localize within the lateral hemispheres.

Discussion

The maps of cerebellar activations represent a full integration of the cerebellum into the calculation of motor and oculomotor processes. We could demonstrate activation of probably 'primary' cerebellar areas that are primarily involved in the execution of movements. The activation patterns by sequential finger movements and by goal directed hand movements define a network of cerebellar areas within the anterior and posterior lobe. The performance of the predictive, relatively automatic, sequences activated primarily the ipsilateral lobuli IV–VI, a smaller area within the corresponding contralateral anterior lobe, and a focus within the ipsilateral lobuli VIIB–VIII representing a cerebellar network that is typically activated by more simple finger or hand movements (Luft et al., 1998; Grodd et al., 2001). If one compares this network with previous physiological data, it becomes evident that different electrophysiological studies in animals revealed different organizational patterns for the representation of sensory information within the cerebellum. A direct stimulation of the sensory cortex and measurement of the evoked potentials over the cerebellar cortex, demonstrated a representation of the body surface within the cerebellum with the hand area located to the region of Larsell lobuli H IV–V reflecting the cerebro-ponto-cerebellar input signals (Snider and Eldred, 1952). Physiological studies, which examined cerebellar neuronal responses after stimulation of peripheral nerves, demonstrated sagitally oriented longitudinal zones in the anterior lobe (Ekerot and Larson, 1982; Voogd and Glickstein, 1998). Natural tactile stimulation instead of electrical stimulation of peripheral nerves demonstrated a patch- or mosaic-like pattern for mapping of climbing fiber and mossy

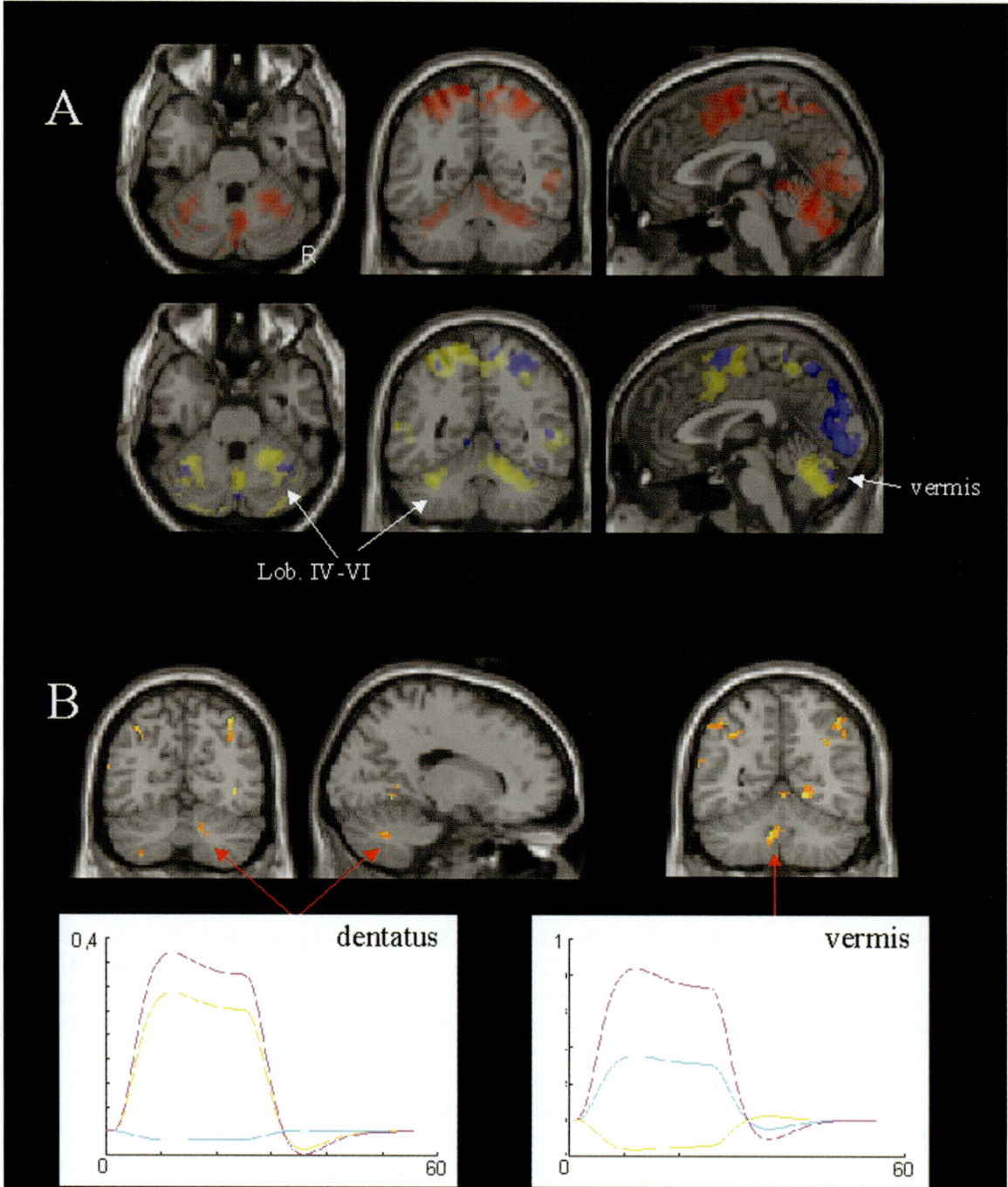

Fig. 4. A: cerebellar activations during combined (red), saccades (blue), and hand movements (yellow). B: areas that are specifically activated during the combined performance of eye and hand movements (combined performance versus isolated performance, masking): in the cerebellum: vermis and an area corresponding to the ipsilateral right dentate nucleus. Signal intensities in several areas of interest demonstrate different predominance. Combined performance of eye and hand movements is printed in violet. Hand movements (yellow) are predominantly processed in the dentate nucleus, eye movements (blue) in the cerebellar vermis.

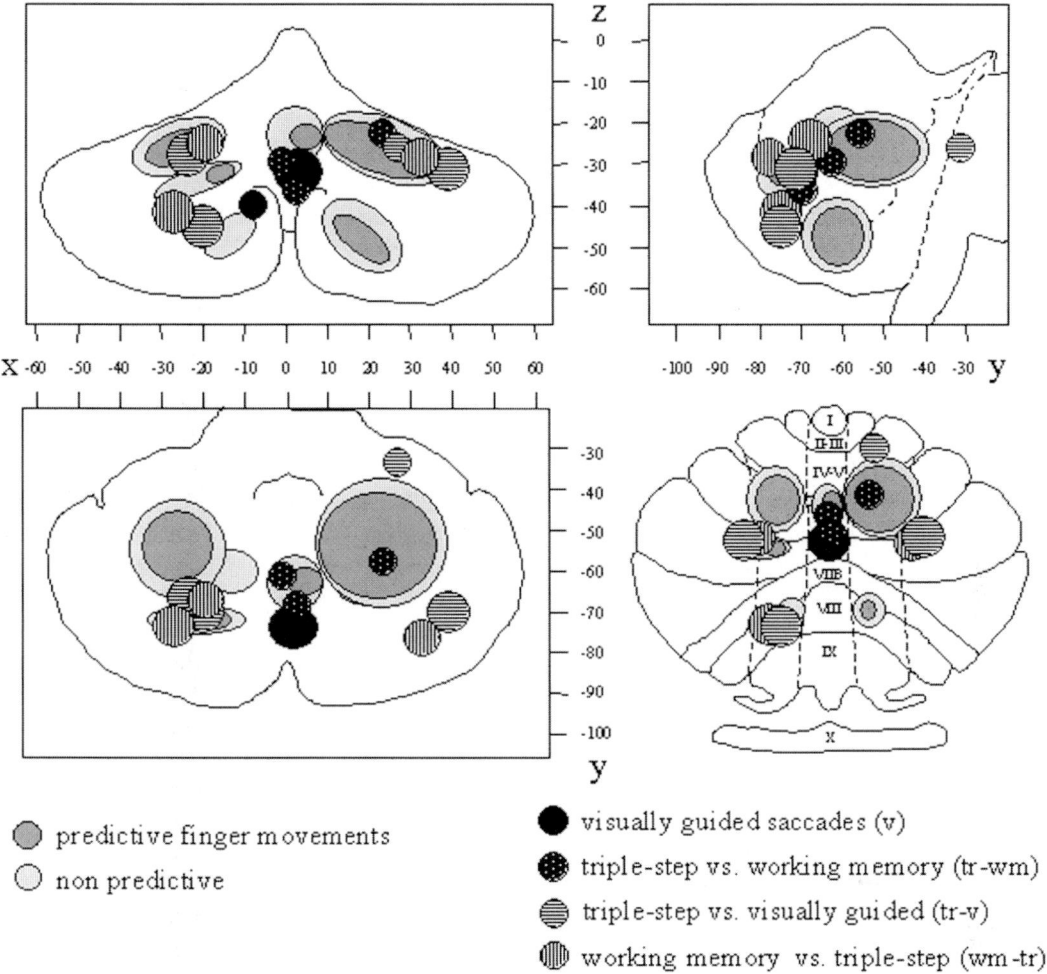

Fig. 5. Summary of the activation foci that are schematically displayed within the Talairach space for the coronal (upper left), sagittal (upper right), and axial (lower left) orientation. Outer limits of the Cerebellum at the specific projections are outlined and related to the x, y, and z coordinates. An additional projection onto an anatomical map of the unfolded cerebellum is provided (right lower corner). The pattern of activation foci by sequential finger movements define a cerebellar network that is modulated by the cognitive preload. The tasks with a dominant oculomotor component (visually guided saccades, triple-step vs. visuospatial working memory) are preferentially oriented along the midline, whereas the tasks with a cognitive preload (triple-step vs. visually guided saccades, visuospatial working memory) localize within the lateral hemispheres.

fiber responses (Miles and Wiesendanger, 1975; Kassel et al., 1984; Shambes et al., 1987). Although these studies revealed somewhat contradictory results, partly based on the different techniques and stimulation methods with different anatomical resolutions, these studies measured climbing fiber or mossy fiber responses after tactile stimulation of the hand or after electrical stimulation of peripheral nerves and revealed a converging forelimb representation more posterior and lateral within the anterior lobe corresponding to the lobuli IV–VI and to a lesser degree for the posterior lobe lobuli VII–VIII. It seems likely that these representation maps just reflect different qualities of sensory or motor information that are integrated and processed within the relevant cerebellar lobuli in the context of voluntary movements in contrast to the more artificial setting in the electropyhsiological studies.

We could also confirm the involvement of the posterior vermis in the execution of saccades corresponding to earlier electrophysiological and clinical data (Fuchs et al., 1993; Desmurget et al., 1998; Barash et al., 1999). Compared to some previous functional MRI and PET studies (Fox et al., 1985; Sweeney et al., 1996; Petit et al., 1997; Dejardin et al., 1998), we could map the activation more precisely to the vermal lobuli VI–VII corresponding to a more recent fMRI study that also located visual guided saccade activity within these lobuli (Hayakawa et al., 2002; Stephan et al., 2002).

Apart from the activation of the saccade and hand movement related cerebro-cerebellar network during isolated performance, we could demonstrate additional activation foci that are specific for the combined execution of eye and hand movements. These areas within the cortex and the posterior cerebellar hemispheres exert a specific role in the coordination of eye and hand movements during performance of actual reaching movements. Analysis of signal intensities in regions of interest demonstrate predominance of M1 and the dentate nucleus for hand movements and of the vermis for eye movements.

Another important observation of these studies was that the change in the 'cognitive' preload during constant motor performance changed the activation patterns within the cerebellum. Performance of the nonpredictive motor sequences showed an extended activation in the ipsilateral and especially the contralateral anterior lobe, the vermis, and the ipsilateral lobulus VIIIA + B. One possible explanation focuses on the fact that the nonpredictive task produced an error rate about twice that of the predictive task. The cerebellum modifies its output signals with unexpected sensory input suggesting a role for error correction (Kawato, 1999; Dreher and Grafman, 2002). This causes an increased firing of the climbing fibers that project directly to the Purkinje cells involved in movement execution (Thach, 1996). The increased activation within these lobuli during our nonpredictive task could relate to this process of error correction in relation to an unpredictable stimulus with a fast changing actual sensory status of the system that is primarily measured by the cerebellum (Brown and Bower, 2002). Recent fMRI studies also demonstrated an increased activation during the initial error-intensive phase of learning new finger movements within the anterior lobe, the vermis, and the contralateral hemisphere (Toni et al., 1998). In accordance with this interpretation, an increase in activation especially within the lobuli IV–VI could be detected by the acquisition of an internal model during motor learning. The cerebellar activation in the initial learning phase was proportional to the error signal that accompanied the acquisition of the new model and was significantly larger than the representative area for the subsequently acquired specific model (Imamizu et al., 2000). A sufficient error correction enables smooth motor performance that includes online control of movements and another or additional aspect of the function of the cerebellum is to measure the timing of movements (Ivry, 2000; Thier et al., 2000) possibly via specific mossy fiber sequences to which a parallel fiber is tuned 'sculpting' a motor sequence by a succession of well timed inhibitory volleys (Braitenberg et al., 1997). Timing related activation changes within the lobuli IV–VI could be detected during memory timed movements (Kawashima et al., 2000). Other imaging studies could demonstrate activation of the cerebellum and especially of the posterior hemispheres during tasks that involved time measurements (Jüptner et al., 1996; Tracy et al., 2000; Dreher and Grafman, 2002) and additional vermal activation could be demonstrated for irregular, in contrast to regular, finger tapping (Lutz et al., 2000). Exact timing is a prerequisite for correct error correction and is, therefore, probably just part of the task to enable smooth and correct movements.

Corresponding to the hand and finger movements data, a higher cognitive load during the oculomotor tasks demonstrated additional activation within the posterior cerebellar hemispheres lobuli VI–crus I and VIIB–VIII by sequences of memory guided saccades in the triple-step task and the visuospatial working memory task, which includes the cueing of visuospatial attention, the suppression of visual reflexive saccades, and the use of visuospatial working memory (Heide et al., 2001). A recent fMRI study also demonstrated activation of the lateral cerebellum by performance of an oculomotor delayed response task in normal volunteers (Luna et al., 2002). Some other oculomotor studies reported activation of the cerebellar hemispheres by the performance of self-initiated saccades in the vermis and the posterior hemispheres (Ellermann et al., 1998; Law et al., 1998; Leigh and Zee, 1999). The authors interpreted the hemispheric activation as related to the initiation of saccades. Another study that tested optokinetic stimulation and saccades interpreted activation of the vermis as reflecting oculomotor performance and activation of the cerebellar hemispheres as reflecting the processing of visuospatial attention (Dieterich et al., 2000).

This modulation of cerebellar activation depending on the context of movement, apart from controlling ongoing movements, can additionally be explained by the cerebral cortical input. Anatomical studies demonstrated connections from premotor or parietal areas that reach the cerebellar hemispheres, especially the lobuli VII–VIII (Glickstein, 2000; Middleton and Strick, 2001). At least in monkeys, premotor areas primarily project to the lobuli VI and VII, whereas parietal areas project to the lobuli VIIB–VIII (Brodal and Bjaalie, 1992, see also Fig. 1). Several imaging data support the hypothesis that the cerebellum, and especially the lobulus VII and the vermis, are involved in the movement preparation phase (Sadato et al., 1996; Cui et al., 2000). Another study investigated cerebellar movement execution paced by an auditory stimulus with or without a regular or irregular random tone omission. Motor execution alone, irrespective of regular tone omission, activated the anterior lobe, lobuli IV–V, and demonstrated the executive role for these lobuli in nearly automatic off-line processing dependent on an internal time-keeping system without relevant sensory feedback mechanisms. Random tone omission, however, additionally activated the lateral hemispheres localized to the lobulus VIIA (crus I) (Sakai et al., 1998). The interpretation for this was that subjects had to abruptly stop and adjust the tapping sequence online to the randomly missing tone sequence. Our first paradigm also contained random finger sequences that demanded online readjustment to a nonpredictive stimulus with integration of visual and sensory feedback and the additional activation in our study fits this interpretation. A corresponding activation within the lobulus VII (crus I) was detected by Imamizu et al. (2000). Switching between different newly learned visuomotor sequences activated the lobulus crus I probably representing the cerebellar adaptation to different motor-subsets. The lobulus VII in addition seems to be involved in the coordination and integration of sensory input and motor output, as activation within this lobulus was significantly enhanced during coordinated performance of eye and hand movements in contrast to eye or hand movements alone (Miall et al., 2000).

Another additional aspect is the participation of the cerebellum in attentional processes. This could be demonstrated by the performance of a visual attention task that resulted in activation of the lobulus VIIA (crus I) (Allen et al., 1997). The combination of a movement task and a visual attention task resulted in bilateral activation of the cerebellar hemispheres, lobuli VI–VII, for the attentional task alone and the combination of movement and attention activated the vermis in addition to pure motor performance (Indovina and Sanes, 2001). Another study demonstrated activation of the cerebellar hemispheres, lobuli VII, during shifting visual attention (Le et al., 1998). However, it has recently been suggested that attentional activation might be secondary to the coordination of motor responses (Bischoff-Grethe et al., 2002). In our study, nonpredictive performance and visuospatial memorization during the saccade task also required increased attention to the stimulation protocol and especially the activation in lobulus VII might be explained in part due to attentional processing. The participation in attentional aspects seems reasonable in the context that the cerebellum needs to predict, prepare, and eventually immediately adjust movement sequences.

The influence of the lobuli VIIB and VIII on the movement pattern is still not very well defined, but the aforementioned anatomical connections suggest the processing of parietal functions possibly in the context of spatially oriented movement planning. The increased activation in the ipsilateral lobulus VIIB–VIII together with the additional activation within the corresponding contralateral side in our study could reflect online updating of spatial coordinates during the nonpredictive performance or directing saccades to memorized location in the visual space.

However, the specific role of the cerebellum during 'cognitive' or mnemonic information processing and its mechanisms are still under debate (Thier et al., 1999). Lesion data in monkeys showed that the cerebellum is not critical for spatial working memory, though it may contribute to the preparation of responses (Nixon and Passingham, 1999). In contradiction, another study demonstrated cognitive impairment of set-shifting and working memory in cerebellar patients with lesions within the posterior cerebellar hemispheres and the vermis (Schmahmann and Sherman, 1998). In a recent fMRI study on attention and eye movements, the posterior vermis was exclusively activated by the execution of saccadic eye movements, whereas the cerebellar hemispheres were activated by covert shifts of attention, a 'pure' cognitive task, to peripheral visual stimuli (Corbetta et al., 1998). Accordingly, patients with damage to the posterior hemispheres and the vermis (lobuli VI + VII) are impaired in covertly orienting visuospatial attention (Townsend et al., 1999). In our study, covert shifts of attention were crucial for performing the triple-step and visuospatial working memory tasks and thus may have contributed to activation of the cerebellar hemispheres, besides the influence of working memory.

Conclusion

We could demonstrate a differential representation of eye and hand movements within the cerebellum. Whereas motor execution seems to be represented predominantly within the anterior lobe or the vermis, additional 'cognitive' preload leads to additional activation including the posterior cerebellar hemispheres, with a subspecialization corresponding to premotor and parietal area connections.

References

Allen, G., Buxton, R.B., Wong, E.C. and Courchesne, E. (1997) Attentional activation of the cerebellum independent of motor involvement. Science, 275: 1940–1943.

Barash, S., Melikyan, A., Sivakov, A., Zhang, M., Glickstein, M. and Thier, P. (1999) Saccadic dysmetria and adaptation after lesions of the cerebellar cortex. J. Neurosci., 19: 10931–10939.

Bischoff-Grethe, A., Ivry, R.B. and Grafton, S.T. (2002) Cerebellar involvement in response reassignment rather than attention. J. Neurosci., 22: 546–553.

Braitenberg, V., Heck, D. and Sultan, F. (1997) The detection and generation of sequences as a key to cerebellar function: experiments and theory. Bev. Brain Sci., 20: 229–277.

Brodal, P. and Bjaalie, J.G. (1992) Organization of the pontine nuclei. Neurosci. Res., 13: 83–118.

Brown, I.E. and Bower, J.M. (2002) The influence of somatosensory cortex on climbing fiber responses in the lateral hemispheres of the rat cerebellum after peripheral tactile stimulation. J. Neurosci., 22: 6819–6829.

Büttner, U. (1999) Eye movements deficits in cerebellar disease. In: Becker, U. (Ed.), Current Oculomotor Research. Plenum Press, New York, pp. 383–389.

Catalan, M.J., Honda, M., Weeks, R.A., Cohen, L.G. and Hallett, M. (1998) The functional neuroanatomy of simple and complex sequential finger movements: a PET study. Brain, 121: 253–264.

Colin, F., Ris, L. and Godeaux, E. (2002) Neuroanatomy of the cerebellum. In: Manto, M.U. and Pandolfo, M. (Eds.), The Cerebellum and its Disorders. Cambridge University Press, p. 21.

Corbetta, M., Akbudak, E., Conturo, T.E., Snyder, A.Z., Ollinger, J.M., Drury, H.A., Linenweber, M.R., Petersen, S.E., Raichle, M.E., Van-Essen, D.C. and Shulman, G.L. (1998) A common network of functional areas for attention and eye movements. Neuron, 21: 761–773.

Cui, S.Z., Li, E.Z., Zang, Y.F., Wenig, X.C., Ivry, R. and Wang, J.J. (2000) Both sides of human cerebellum involved in preparation and execution of sequential movements. Neuroreport, 11: 3849–3853.

Culham, J.C. and Kanwisher, N.G. (2001) Neuroimaging of cognitive functions in human parietal cortex. Curr. Opin. Neurobiol., 11: 157–163.

Dejardin, S., Dubois, S., Bodart, J.M., Schiltz, C., Delinte, A., Michel, C., Roucoux, A. and Crommelinck, M. (1998) PET study of human voluntary saccadic eye movements in darkness: effect of task repetition on the activation pattern. Eur. J. Neurosci., 10: 2328–2336.

Desmond, J.E., Gabrieli, J.D., Wagner, A.D., Ginier, B.L. and Glover, G.H. (1997) Lobular patterns of cerebellar activation in verbal working-memory and finger-tapping tasks as revealed by functional MRI. J. Neurosci., 17: 9675–9685.

Desmurget, M., Pelisson, D., Urquizar, C., Prablanc, C., Alexander, G.E. and Grafton, S.T. (1998) Functional anatomy of saccadic adaptation in humans. Nat. Neurosci., 1: 524–528.

Dieterich, M., Bucher, S.F., Seelos, K.C. and Brandt, T. (2000) Cerebellar activation during optokinetic stimulation and saccades. Neurology, 54: 148–155.

Dreher, J.C. and Grafman, J. (2002) The roles of the cerebellum and basal ganglia in timing and error prediction. Eur. J. Neurosci., 16: 1609–1619.

Ekerot, C.F. and Larson, B. (1982) Branching of olivary axons to innervate pairs of saggital zones in the cerebellar anterior lobe of the cat. Exp. Brain. Res., 48: 185–198.

Ellermann, J.M., Siegal, J.D., Strupp, J.P., Ebner, T.J. and Ugurbil, K. (1998) Activation of visuomotor systems during visually guided movements: a functional MRI study. J. Magn. Reson., 131: 272–285.

Fox, P.T., Fox, J.M., Raichle, M.E. and Burde, R.M. (1985) The role of cerebral cortex in the generation of voluntary saccades: a positron emission tomography study. J. Neurophysiol., 54: 348–369.

Friston, K.J., Holmes, A.P., Poline, J.B., Grasby, P.J., Williams, S.C.R., Frackowiak, R.S.J. and Turner, R. (1995) Analysis of fMRI time-series revisited. NeuroImage, 2: 5–53.

Friston, K.J., Ashburner, J., Poline, J.B., Frith, C.D., Heather, J.D. and Frackowiak, R.S.J. (1997) Spatial realignment and normalization of images. Hum. Brain Mapp., 2: 165–189.

Fuchs, A.F., Robinson, F.R. and Straube, A. (1993) Role of the caudal fastigial nucleus in saccade generation. I. Neuronal discharge patterns. J. Neurophysiol., 70: 1723–1740.

Glickstein, M. (2000) How are visual areas of the brain connected to motor areas for the sensory guidance of movement? Trends Neurosci., 23: 613–617.

Grafton, S.T., Woods, R.P. and Mazziotta, J.C. (1993) Within-arm somatotopy in human motor areas determined by positron emission tomography imaging of cerebral blood flow. Exp. Brain Res., 95: 172–176.

Grodd, W., Hülsmann, E., Lotze, M., Wildgruber, D. and Erb, M. (2001) Sensorimotor mapping of the human cerebellum: fMRI evidence of somatotopic organization. Hum. Brain Mapp., 13: 55–71.

Hayakawa, Y., Nakajima, T., Takagi, M., Fukuhara, N. and Abe, H. (2002) Human cerebellar activation in relation to saccadic eye movements: a functional magnetic resonance imaging study. Ophthalmologica, 216: 399–405.

Heide, W., Binkofski, F., Seitz, R.J., Posse, S., Nitschke, M.F., Freund, H.J. and Kömpf, D. (2001) Activation of fronto-parietal cotrices during memorized triple-step sequences of saccadic eye movements: An fMRI study. Eur. J. Neurosci., 13: 1177–1189.

Imamizu, H., Miyauchi, S., Tamada, T., Sasaki, Y., Takino, R., Putz, B., Yoshioka, T. and Kawato, M. (2000) Human cerebellar activity reflecting an acquired internal model of a new tool. Nature, 403: 192–195.

Indovina, I. and Sanes, J.N. (2001) Combined visual attention and finger movement effects on human brain representations. Exp. Brain Res., 140: 265–279.

Ito, M. (2000) Neurobiology: internal model visualised. Nature, 403: 153–154.

Ivry, R. (2000) Exploring the role of the cerebellum in sensory anticipation and timing: commentary on Tesche and Karhu. Hum. Brain Mapp., 9: 115–118.

Jüptner, M., Rijntjes, M., Weiller, C., Faiss, J.H., Timmann, D., Müller, S.P. and Diener, H.C. (1996) Localization of a cerebellar timing process using PET. Neurology, 47: 306–307.

Kassel, J., Shambes, G.M. and Welker, W. (1984) Fractured cutaneous projections to the granule cell layer of the posterior cerebellar hemisphere of the domestic cat. J. Comp. Neurol., 225: 458–468.

Kawashima, R., Okuda, J., Umetsu, A., Sugiura, M., Inoue, K., Suzuki, K., Tabuchi, M., Tsukiura, T., Narayan, S.L., Nagasaka, T., Yanagawa, I., Fujii, T., Takahashi, S., Fukuda, H. and Yamadori, A. (2000) Human cerebellum plays an important role in memory-timed finger movement: an fMRI study. J. Neurophysiol., 83: 1079–1087.

Kawato, M. (1999) Internal models for motor control and trajectory planning. Curr. Opin. Neurobiol., 9: 718–727.

Law, I., Svarer, C., Rostrup, E. and Paulson, O.B. (1998) Parieto-occipital cortex activation during self-generated eye movements in the dark. Brain, 121: 2189–2200.

Le, T.H., Pardo, J.V. and Hu, X. (1998) 4 T-fMRI study of nonspatial shifting of selective attention: cerebellar and parietal contributions. J. Neurophysiol., 79: 1535–1548.

Leigh, J.R. and Zee, D.S. (1999) The Neurology of eye movements; 3rd edition. Oxford University Press, New York, 121 p.

Lotze, M., Montoya, P., Erb, M., Hulsmann, E., Flor, H., Klose, U., Birbaumer, N. and Grodd, W. (1999) Activation of cortical and cerebellar motor areas during executed and imagined hand movements: an fMRI study. J. Cogn. Neurosci., 11: 491–501.

Luft, A.R., Skalej, M., Stefanou, A., Klose, U. and Voigt, K. (1998) Comparing motion- and imagery-related activation in the human cerebellum: a functional MRI study. Hum. Brain Mapp., 6: 105–113.

Luna, B., Minshew, N.J., Garver, K.E., Lazar, N.A., Thulborn, K.R., Eddy, W.F. and Sweeney, J.A. (2002) Neocortical system abnormalities in autism. An fMRI study of spatial working memory. Neurology, 59: 834–840.

Lutz, K., Specht, K., Shah, N.J. and Jancke, L. (2000) Tapping movements according to regular and irregular visual timing signals investigated with fMRI. Neuroreport, 11: 1301–1306.

Miall, R.C., Imamizu, H. and Miyauchi, S. (2000) Activation of the cerebellum in co-ordinated eye and hand tracking movements: an fMRI study. Exp. Brain Res., 135: 22–33.

Middleton, F.A. and Strick, P.L. (1994) Anatomical evidence for cerebellar and basal ganglia involvement in higher cognitive function. Science, 266: 458–461.

Middleton, F.A. and Strick, P.L. (1997) Dentate output channels: motor and cognitive components. Prog. Brain Res., 114: 553–566.

Middleton, F.A. and Strick, P.L. (2001) Cerebellar projections to the prefrontal cortex of the primate. J. Neurosci., 21: 700–712.

Miles, T.S. and Wiesendanger, M. (1975) Organization of climbing fibre projections to the cerebellar cortex from trigeminal cutaneous afferents and from the SI face area of the cerebral cortex in the cat. J. Physiol. (Lond.), 245: 409–424.

Nitschke, M.F., Kleinschmidt, A., Wessel, K. and Frahm, J. (1996) Somatotopic motor representation in the human anterior cerebellum. A high resolution functional MRI study. Brain, 119: 1023–1029.

Nitschke, M.F., Hahn, C., Melchert, U.H., Handels, H. and Wessel, K. (1998) Activation of the human anterior cerebellum by finger movements and sensory finger stimulation: A functional magnetic resonance imaging study. J. Neuroimaging, 8: 127–131.

Nitschke, M.F., Stavrou, G., Melchert, U.H., Erdmann, C., Kreuder, F., Petersen, D., Heide, W. and Kömpf, D. (2003) Modulation of cerebellar activation during predictive and non predictive finger movements. The Cerebellum, 2: 233–240.

Nitschke, M.F., Binkofski, F., Buccino, G., Posse, S., Erdmann, C., Kömpf, D., Seitz, R.J. and Heide, W. (2004) Activation of cerebellar hemispheres in spatial memorization of saccadic eye movements — a fMRI study. Hum. Brain Mapp., 22: 155–164.

Nixon, P.D. and Passingham, R.E. (1999) The cerebellum and cognition: cerebellar lesions do not impair spatial working memory or visual associative learning in monkeys. Eur. J. Neurosci., 11: 4070–4080.

Nobre, A.C., Gitelman, D.R., Dias, E.C. and Mesulam, M.M. (2000) Covert visual spatial orienting and saccades: overlapping neural systems. Neuroimage, 11: 210–221.

Petit, L., Clark, V.P., Ingeholm, J. and Haxby, J.V. (1997) Dissociation of saccade-related and pursuit-related activation in human frontal eye fields as revealed by functional fMRI. J. Neurophysiol., 77: 3386–3390.

Sadato, N., Campbell, G., Ibanez, V., Deiber, M.P. and Hallett, M. (1996) Complexity affects regional cerebral blood flow change during sequential finger movements. J. Neurosci., 16: 2693–2700.

Sakai, K., Takino, R., Hikosaka, O., Miyauchi, S., Sasaki, Y., Putz, B. and Fujimaki, N. (1998) Separate cerebellar areas for motor control. Neuroreport, 13: 2359–2363.

Schmahmann, J.D. and Sherman, J.C. (1998) The cerebellar cognitive affective syndrome. Brain, 121: 561–579.

Schmahmann, J.D., Doyon, J., McDonald, D., Holmes, C., Lavoie, K., Hurwitz, A.S., Kabani, N., Toga, A., Evans, A. and Petrides, M. (1999) Three-dimensional MRI atlas of the human cerebellum in proportional stereotaxic space. NeuroImage, 10: 233–260.

Shambes, G.M., Gibson, J.M. and Welker, W.I. (1987) Fractured somatotopy in granular cell tactile areas in rat cerebellar hemispheres revealed by micromapping. Brain. Behav. Evol., 15: 94–140.

Snider, R.S. and Eldred, E. (1952) Cerebro-cerebellar relationships in the monkey. J. Neurophysiol., 15: 27–40.

Stephan, T., Mascolo, A., Yousry, T.A., Bense, S., Brandt, T. and Dieterich, M. (2002) Changes in cerebellar activation pattern during two successive sequences of saccades. Hum. Brain Mapp., 16: 63–70.

Sweeney, J.A., Mintun, M.A., Kwee, S., Wiseman, M.B., Brown, D.L., Rosenberg, D.R. and Carl, J.R. (1996) Positron emission tomography study of voluntary saccadic eye movements and spatial working memory. J. Neurophysiol., 75: 454–468.

Tailarach, J. and Tournoux, P. (1988) Co-Planar Stereotactic Atlas of the Human Brain. Thieme Medical Publishers, Stuttgart, New York, pp. 1–122.

Thach, W.T. (1996) On the specific role of the cerebellum in motor learning and cognition: clues from PET activation and lesion studies in man. Behav. Brain Sci., 19: 411–431.

Thier, P., Haarmeier, T., Treue, S. and Barash, S. (1999) Absence of a common functional denominator of visual disturbances in cerebellar disease. Brain, 122: 2133–2146.

Thier, P., Dicke, P.W., Haas, R. and Barash, S. (2000) Encoding of movement time by populations of cerebellar Purkinje cells. Nature, 405: 72–76.

Toni, I., Krams, M., Turner, R. and Passingham, R.E. (1998) The time course of changes during motor sequence learning. A whole brain fMRI study. Neuroimage, 8: 50–61.

Townsend, J., Courchesne, E., Covington, J., Westerfield, M., Harris, N.S., Lyden, P., Lowry, T.P. and Press, G.A. (1999) Spatial attention deficits in patients with acquired or developmental cerebellar abnormality. J. Neurosci., 19: 5632–5643.

Tracy, J.I., Faro, S.H., Mohamed, F.B., Pinsk, M. and Pinus, A. (2000) Functional localization of a 'Time Keeper' function separate from attentional resources and task strategy. Neuroimage, 11: 228–242.

Voogd, J. and Glickstein, M. (1998) The anatomy of the cerebellum. Trends Neurosci., 21: 370–375.

SECTION V

Oscillations and Synchrony in Cerebellar Cortex and Inferior Olive

Fast oscillation in the cerebellar cortex of calcium binding protein-deficient mice: a new sensorimotor arrest rhythm

Guy Cheron[1,2,*], Laurent Servais[1,3], Bernard Dan[1,2], David Gall[3], Céline Roussel and Serge N. Schiffmann[3]

[1]Laboratory of Neurophysiology, Université Mons-Hainaut, Mons, Belgium
[2]Laboratory of Movement Biomechanics CP168, Université Libre de Bruxelles, 1070 Brussels, Belgium
[3]Laboratory of Neurophysiology CP601, Université Libre de Bruxelles, 1070 Brussels, Belgium

Keywords: Purkinje cells; fast oscillation; cerebellum; calcium-binding protein

Abstract: Fast oscillations (> 100 Hz) may serve physiological roles when regulated properly. They may also appear in pathological conditions. In cerebellum, 160 Hz oscillation emerge in mice lacking calbindin and/or calretinin, two proteins devoted to calcium buffering in Purkinje and granule cells, respectively. Here, we review the pharmacological and spatiotemporal properties of this fast cerebellar oscillation and the related Purkinje cell firing behaviour in alert mice. We show that this oscillation is highly synchronized along the parallel fiber beam and reversibly inhibited by gap junctions, $GABA_A$ and NMDA receptors blockers. Cutaneous stimulation of the whisker region transiently suppressed the oscillation which shows in some aspects similarities with cerebral "resting" rhythmic activities of wakefulness arresting to sensory or motor information such as alpha and mu rhythms. The Purkinje cells of these mutants present an increased simple spike firing rate, rhythmicity and synchronicity, and a decreased complex spike duration and subsequent pause. Both simple and complex spikes may be tightly phase-locked with the oscillation. Contrastingly, on slice recordings, the intrinsic membrane properties of Purkinje cell are similar in wild type mice and in mice lacking calbindin. The role played by this fast cerebellar oscillation in the emergence of ataxia is yet to be solved.

Introduction

Hippocampal oscillations at 200 Hz were recorded in normal rats during sleep or at rest (Buzsáki et al., 1992; Ylinen et al., 1995; Chrobak et al., 2000), indicating that high frequency oscillations may serve physiological roles when regulated properly. In this context, theoretical studies (Hopfield and Brody, 2001) have demonstrated that the precise degree of synchronization depends on how close together the firing frequencies are, the strength of coupling between the neurons, and the number of neurons involved. Details of synapse time response and membrane time constants can strongly affect the computation of synchronizing neurons (Traub and Bibbig, 2000; Whittington and Traub, 2003). This also means that the transition between the normal regime (selective synchronization in response to specific patterns) and the pathological state ('epileptic' synchronization in response to any pattern) seems to have the nature of a phase transition (Hopfield and Brody, 2001).

*Corresponding author. Tel.: +3226502187;
Fax: +3265373566; E-mail: gcheron@ulb.ac.be

In the cerebellum, the firing dynamics of the Purkinje cells (PC), the sole output neurons of the cerebellar cortex, is important for the generation of precise timing signals involved in the motor coordination function of the cerebellum (Welsh et al., 1995). The dynamics of intracellular Ca^{2+} fluxes into the PC is crucial for the regulation of their firing pattern. For example, when the P/Q-type calcium channels were blocked in vitro, the spontaneous tonic firing of a PC was replaced by a bursting mode followed by a cessation of firing (Womack and Khodakhah, 2003). Voltage-gated calcium channels, intrinsic neuronal Ca^{2+} conductances and intracellular Ca^{2+} homeostasis are considered to be crucial partners in the emergence of network activity, including oscillations (Llinas, 1988; Huguenard, 1996; McCormick and Bal, 1997).

Among the various mechanisms involved in the Ca^{2+} homeostasis, calcium-binding proteins such as calretinin (CR), calbindin D28k (CB), and parvalbumin (PV) are important modulators of intracellular Ca^{2+} dynamics and contribute to shaping pre- and postsynaptic signaling (Llano et al., 1994; Airaksinen et al., 1997; Caillard et al., 2000; Edmonds et al., 2000). For example, PV-containing neurons such as cerebellar basket and stellate cells are always able to fire in burst at high rate (Eccles et al., 1966). In PV knock-out mice, the paired-pulse depression at the level of the GABAergic synapse between the molecular interneuron and the PC is replaced by a paired-pulse facilitation (Caillard et al., 2000). In CB knockout, mice specific alterations in the PC dendritic compartment have been revealed by the detection of increased synaptically evoked Ca^{2+} transients (Airaksinen et al., 1997; Barski et al., 2003). Mice with inactivated Cb or Cr genes ($Cb^{-/-}$ and $Cr^{-/-}$ mice) have impaired motor coordination, but no apparent changes in the overall structure of the cerebellar cortex (Airaksinen et al., 1997; Schiffmann et al., 1999; Cheron et al., 2000). Although, $Cr^{-/-}$ PCs recorded in vitro show no changes in somatic cellular physiology (Schiffmann et al., 1999), in an alert animal the PCs fire simple spikes (SS) at a dramatically increased rate, probably reflecting the enhanced excitability of their afferent granule cells (Gall et al., 2003). Together with the reduced duration of the complex spike (CS) and the related pause in SS firing (Schiffmann et al., 1999), these changes suggest influences of the gene defect at the circuit level.

This review will focus on the effect of null mutations of Cr or Cb genes on neural network activity, recorded with multiple electrodes neurons and local field potentials (LFPs) from the cerebellum of $Cr^{-/-}$, $Cb^{-/-}$ as well as double knock-out $Cb^{-/-}Cr^{-/-}$ mice. In particular, we describe the basic elements of the emergence of spontaneous high-frequency (~ 160 Hz) oscillations recently described in the cerebellum of these mutant mice (Cheron et al., 2004).

Altered Purkinje cell firing leads to the emergence of 160 Hz oscillation in alert mutant mice

Spontaneous LFP oscillations (LFPOs) were found throughout the explored cerebellar regions (vermis, uvula, and nodulus) of alert mutant mice (Fig. 1A). They appeared as spindle-shaped episodes of oscillations (maximal amplitude 0.54 ± 0.23 mV) with a mean rate of occurrence of 4.5 ± 1.7 episodes/s. The LFPOs were recorded in 6 out of 7 $Cr^{-/-}$ mice, in 11 out of 12 $Cb^{-/-}$ and in all 24 $Cr^{-/-}Cb^{-/-}$ mice, with frequencies of 167.8 ± 36.0 Hz, 164.5 ± 36.0 Hz, and 164.9 ± 45.3 Hz, respectively. The oscillation indices were significantly higher in $Cr^{-/-}Cb^{-/-}$ (13.9 ± 6.2) than in $Cr^{-/-}$ (6.4 ± 1.8) and $Cb^{-/-}$ (6.7 ± 4.1) mice ($p < 0.001$), indicating the reinforcement of the oscillating state in double knock-out animals. Fast LFPOs were not recorded in WT mice ($n = 10$).

Spontaneous firing constitutes an efficient way of representing different functional states in a neural network (Hausser and Monsivais, 2003; Nelson et al., 2003). In the absence of synaptic inputs, the PCs are not quiescent but intrinsically generate action potentials at a high rate (Häusser and Clark, 1997). Voltage-activated sodium channels are the main inward current necessary for the maintenance of tonic action potential firing (Raman and Bean, 1999). An intrinsic trimodal pattern of activity (tonic firing, bursting, and silent modes) has been described (Llinás and Sugimori, 1980; Swensen and Bean, 2003; Womack and Khodakhah, 2003). In vivo, despite the presence of high levels of spontaneous inhibitory synaptic activity (Eccles et al., 1966; Häusser and Clark, 1997) the PC typically fires tonically at an irregular rate SS around 50 Hz in different studied

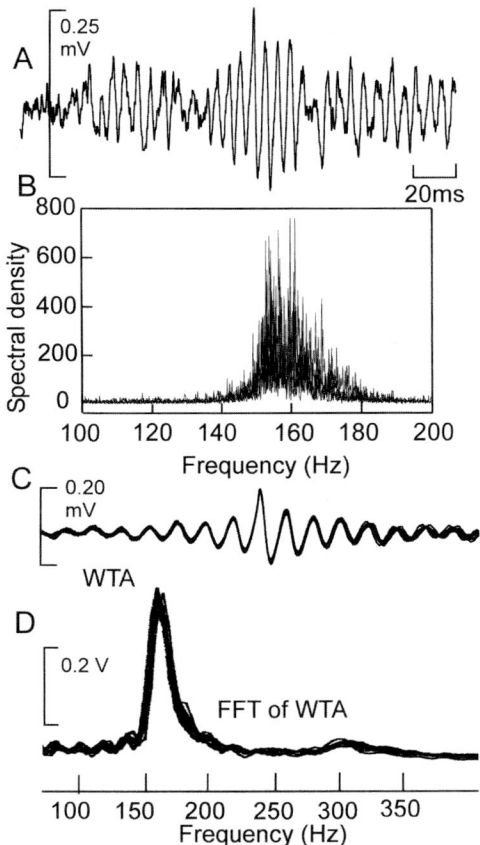

Fig. 1. Emergence of high-frequency oscillations in the cerebellum of $Cr^{-/-}Cb^{-/-}$ mice. (A) Sample records of LFPOs $Cr^{-/-}Cb^{-/-}$ mice. (B) Superimposition of spectral analysis (Fast Fourier Transform) of the raw LFPO performed every each minute during 10 min of continuous recording. (C) Superimposition of the wave triggered averaging (WTA) of the LFPO performed in the same way as in B. (D) Superimposition of the spectral density of the WTA presented in C.

Table 1. Purkinje cell firing behavior in alert WT, $Cb^{-/-}$ and $Cb^{-/-}Cr^{-/-}$ mice

Parameter	WT $n=35$ cells	$Cb^{-/-}$ $n=78$ cells	$Cb^{-/-}Cr^{-/-}$ $n=150$ cells
SSf, Hz	41.7 ± 17.5	86.3 ± 45.3*	80.4 ± 44.0*
CSf, Hz	0.42 ± 0.12	0.50 ± 0.22	0.50 ± 0.23
CSd, ms	12.3 ± 3.8	6.8 ± 2.2*	6.9 ± 2.4*
Pause, ms	24.3 ± 8.0	9.4 ± 6.8*	6.4 ± 6.3*

CSd, complex spike duration; CSf complex spike firing rate; SSf, simple spike firing rate; WT, wild-type. *$p < 0.00001$ as compared to age-matched, wild-type animals; one-way ANOVA. (From Cheron et al. (2004).)

change in the rhythmicity of SSs in $Cb^{-/-}Cr^{-/-}$ compared to WT mice. One-sided peak counts of the autocorrelogram were 0.8 ± 0.1 ($n = 36$) and 6.6 ± 1.5 ($n = 7$) for WT and $Cb^{-/-}Cr^{-/-}$ mice, respectively ($p < 0.0001$) (Fig. 2A, B). A rhythm index (RI) was also significantly higher ($p < 0.0001$) in mutant mice (0.37 ± 0.28) than in WT (0.03 ± 0.02).

In simultaneous recordings of LFPOs and single PCs, the SS (Fig. 3) and CS (Fig. 4) discharges were phase-locked to the LFPOs. The SS discharges occurred at the depth of the LFP wave (Fig. 3D) whereas the CS discharge appeared during the LFP ascending phase (Fig. 4D). In contrast, simultaneous recordings of LFPs and Golgi cell never demonstrated phase locking between LFP waves and Golgi cell spikes. In this case, an increased Golgi cell firing rate was correlated with a significant reduction of concomitant LFPOs.

Spatiotemporal mapping of the 160 Hz oscillation and synchrony of PCs along the parallel fiber beam

The spatiotemporal mapping of the oscillation is facilitated by cerebellar architecture, which is based on repeats of an anatomically well defined circuit. The LFPO amplitude increased when the electrode approached the PC layer, reaching its maximum just beneath the PC bodies. In all instances, the oscillations recorded in the deep PC layer of the nodulus disappeared abruptly when the electrode tip reached the IV ventricle or the medulla (Cheron et al., 2004).

Spatial coherence in the frontal plane (i.e., along a PF beam) vs. the sagittal plane was measured with multiple electrodes. The LFPOs were synchronized at sites along the frontal plane (Fig. 5A) separated by

species (Lisberger and Fuch, 1974; Ghelarducci et al., 1975; Cheron et al., 1997, 2000; Goossens et al., 2001), though it may occasionally show bursting patterns as well as periods of complete silence in SS firing.

The spontaneous activity of 263 PCs in $Cb^{-/-}$, $Cb^{-/-}Cr^{-/-}$, and WT alert mice were recorded (Cheron et al., 2004). As for $Cr^{-/-}$ the firing rate of spontaneous SS was much higher in both mutants than in WT animals while the durations of CSs, and of the pauses in SS firing following CSs, were significantly reduced (Table 1). Conversely, the spontaneous rate of the CS was not statistically different. We also demonstrated a highly significant

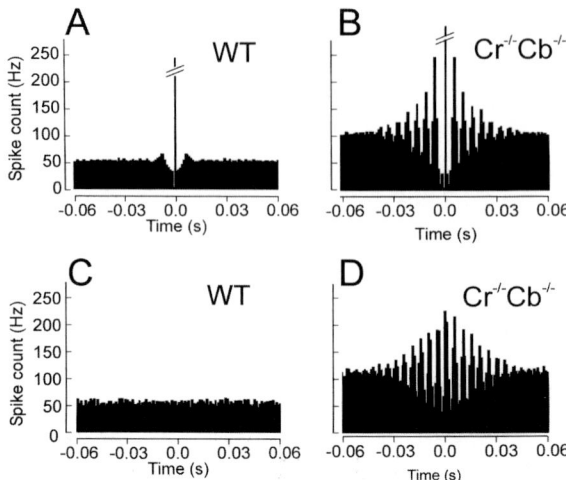

Fig. 2. PC rhythmicity in alert wild-type (WT) and in Cb$^{-/-}$Cr$^{-/-}$ mice. (A,B) SS autocorrelograms for one representative WT (A) and one Cb$^{-/-}$Cr$^{-/-}$ PC (B) demonstrating increased rhythmicity in the mutant. (C,D) Crosscorrelograms of SSs from PC pairs multi-recorded along a PF beam (0.5 mm apart) in WT (C) and Cb$^{-/-}$Cr$^{-/-}$ mice (D), demonstrating PC synchronization in the mutant. (Reprinted with permission from Cheron et al. (2004).)

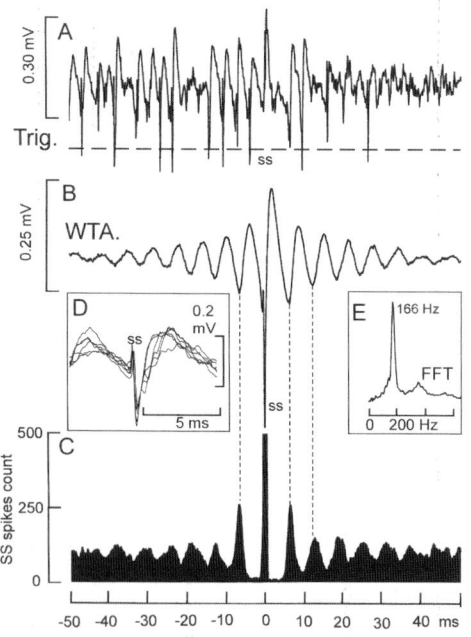

Fig. 3. Temporal relationships between LFPOs and PC simple spike (SS) firing. (A) Simultaneous recordings by the same electrode of an isolated PC and a 166 Hz LFPO. In this case the trigger (Trig) was adjusted on the SSs. (B) WTA of the LFPO, using PC SSs as trigger (n = 1000). (C) SS autocorrelogram, with central peak truncated (lower part), has the same rhythmicity as the WTA trace. (D) Superimposition of SS single traces (n = 6). (E) Spectral density of the WTA records. (Modified from Cheron et al. (2004).)

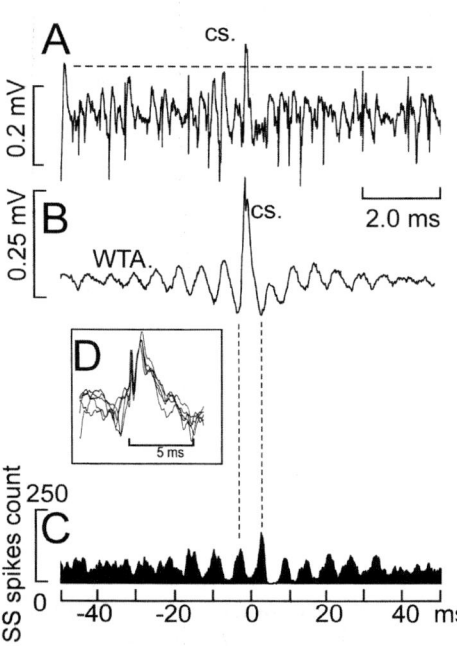

Fig. 4. Temporal relationships between LFPOs and PC complex spike (CS) firing firing. (A) The same procedure as in Fig. 3 was performed with the trigger adjusted on the CSs of the same PC. (B) WTA shows the presence of 166 Hz oscillation around the CS (middle part). (C) SS crosscorrelogram has the same rhythmicity as the WTA trace. (D) Superimposition of unaveraged traces (n = 6) confirmed the recurrent occurrence of the CS in the ascending phase of the LFP oscillation. (Modified from Cheron et al. (2004).)

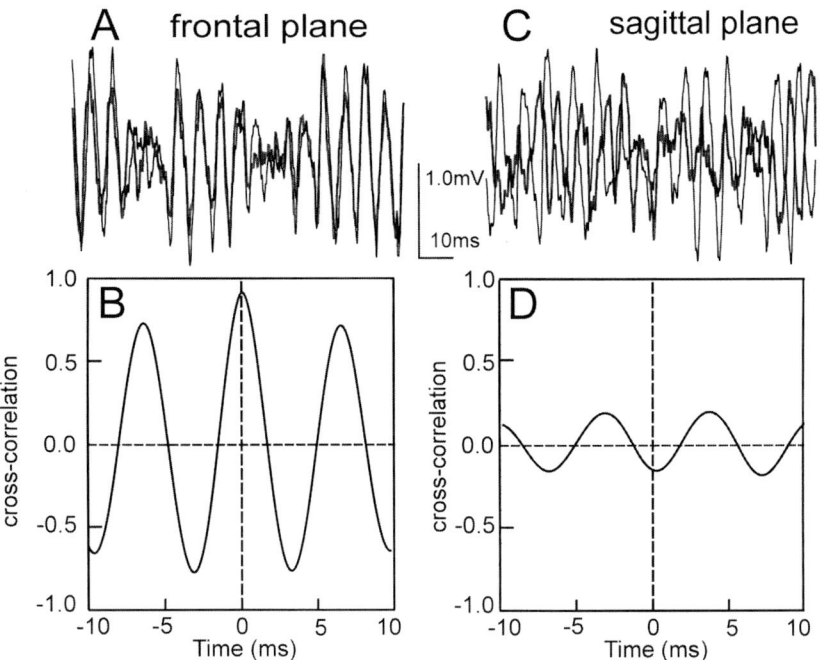

Fig. 5. Spatial coherence of LFPOs. Simultaneous recordings from electrode pairs aligned along the longitudinal (frontal plane) (A) or rostro-caudal (sagittal) axis of a folium (C). Cross-correlation analysis of the LFPO traces recorded in the frontal (B) or in the sagittal (D) plane. Note the synchronization of the LFPO in the frontal plane (A–B) and the absence of synchronization in the sagittal plane (C–D).

up to 2 mm (cross-correlation coefficient 0.91 ± 0.05, $n=5$) (Fig. 5B). In contrast, paired sites aligned in the sagittal plane (Fig. 5C) did not show synchronization (-0.03 ± 0.10, $n=5$) (Fig. 5D). In order to verify if LFPO synchronization along the PF beam is correlated to the synchronous PC firing, we recorded 33 PC pairs during LFPOs in 4 $Cb^{-/-}Cr^{-/-}$ mice. All mutant PC pairs showed a significant central peak (synchrony index (SI) 0.24 ± 0.13) on their cross-correlogram and high-frequency oscillations with a number of side peaks of 6.3 ± 1.3 (RI 3.14 ± 1.45) (Fig. 2D). Conversely, in WT mice ($n=48$ pairs), coronal cross-correlograms were flat in all but three pairs (SI 0.02 ± 0.05) and high-frequency oscillations were always absent (RI 0.02 ± 0.09) (Fig. 2C).

The intrinsic membrane properties of $Cb^{-/-}$ Purkinje cell somata remain unchanged

To determine whether the altered in vivo activity resulted from an increased PC excitability, the intrinsic electroresponsiveness of $Cb^{-/-}$ PCs was studied in vitro (Fig. 6) We measured the SS firing rate (Fig. 6A) at increasing intensities of injected current in slices from 5 WT and 5 $Cb^{-/-}$, mice using the perforated-patch whole-cell configuration. The slope of the linear part of these current–frequency plots (Fig. 6B, C), which reflects the intrinsic somatic excitability, was not significantly different between WT (0.207 ± 0.024 Hz pA^{-1}) and $Cb^{-/-}$ mice (0.188 ± 0.045 Hz pA^{-1}). In addition, we did not observe significant modifications in action potential half-width, (0.55 ± 0.04 ms vs. 0.59 ± 0.07 ms) (Fig 6D), resting potential (-64.8 ± 2.2 mV vs. -55.8 ± 4.1 mV) and input resistance (158.5 ± 24.5 $M\Omega$ vs. 212.5 ± 47.9 $M\Omega$).

The effect of peripheral stimulation and motor responses on 160 Hz oscillation

Given the involvement of the PCs in the sensorimotor processing (Bower and Woolston, 1983), we

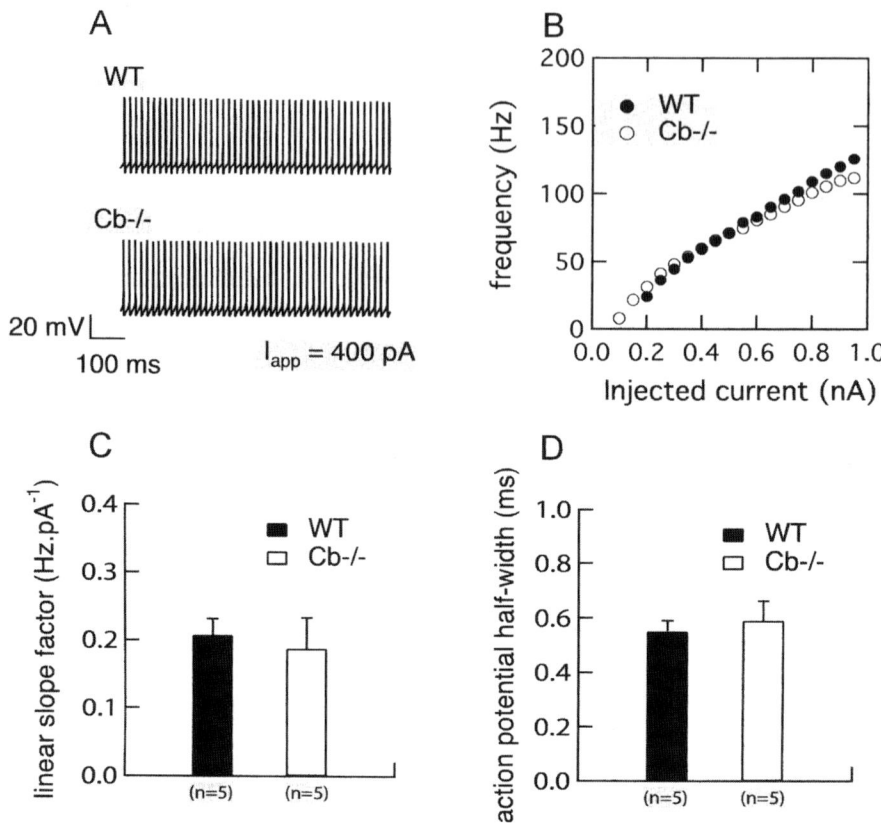

Fig. 6. Evaluation of the intrinsic excitability of a cerebellar PC in wild type and $Cb^{-/-}$ mice. Shown in (A) is the repetitive spiking obtained by the injection of a 400 pA depolarizing current in a WT PC (A, top) and in a $Cb^{-/-}$ purkinje cell with similar threshold current for action potential generation (A, bottom). No difference is seen between the $Cb^{-/-}$ PC compared to the WT cell for the same injected current from a holding potential of −80 mV. Because the frequency remained stable throughout the stimulation, average frequency was measured over the total time of current injection (1 s) at increasing intensities. From this data current-frequency plots could be constructed (B). In the studied interval of injected current intensities, PC showed a linear encoding of stimulus intensity. The slope of the linear part of current-frequency plots was used as a measure of the intrinsic PC excitability. Histograms report slope factor of the current-frequency plots for WT vs. $Cb^{-/-}$ (C) and corresponding durations of action potentials at half-amplitude (D) evaluated at the threshold potential where fast repetitive spiking is obtained. In the perforated patch configuration, $Cb^{-/-}$ PC ($n = 5$) do not show a significantly increased excitability or altered action potentials duration compared to WT ($n = 5$).

examined the effect on LFPOs and PC SS firing of cutaneous stimulation of the whisker region. Using microelectrodes separated by 250 μm Eckhorn system (Eckhorn and Thomas, 1993), we simultaneously recorded the LFPOs and a single unit PC activity in Crus II, a lobule controlling oro-facio-lingual movements (Bower and Woolston, 1983), in conjunction with EMG activity of the face muscles (Fig. 7). Interestingly, the LFPO was suppressed in response to an air puff on the whisker region, whether or not this stimulus caused a facial movement, but only if the simultaneously recorded PC also responded to the stimulus (Fig. 7A, B). Moreover, the LFPO was also suppressed during the spontaneous muscle activity documented by EMG, but again only if the PC SS firing was also modulated (compare first and second EMG burst in Fig. 7C). Individual PCs ($n = 5$) demonstrated the classical PC responses to cutaneous stimulation (Fig. 7A, B) characterized by firing rate excitation (Fig. 7E) or inhibition (Fig. 7F) or by transient responses as sequences of inhibition–excitation or excitation–inhibition (Bower and Woolston, 1983). At stimulus frequencies greater than 0.5 Hz, habituation occurred and the LFPO

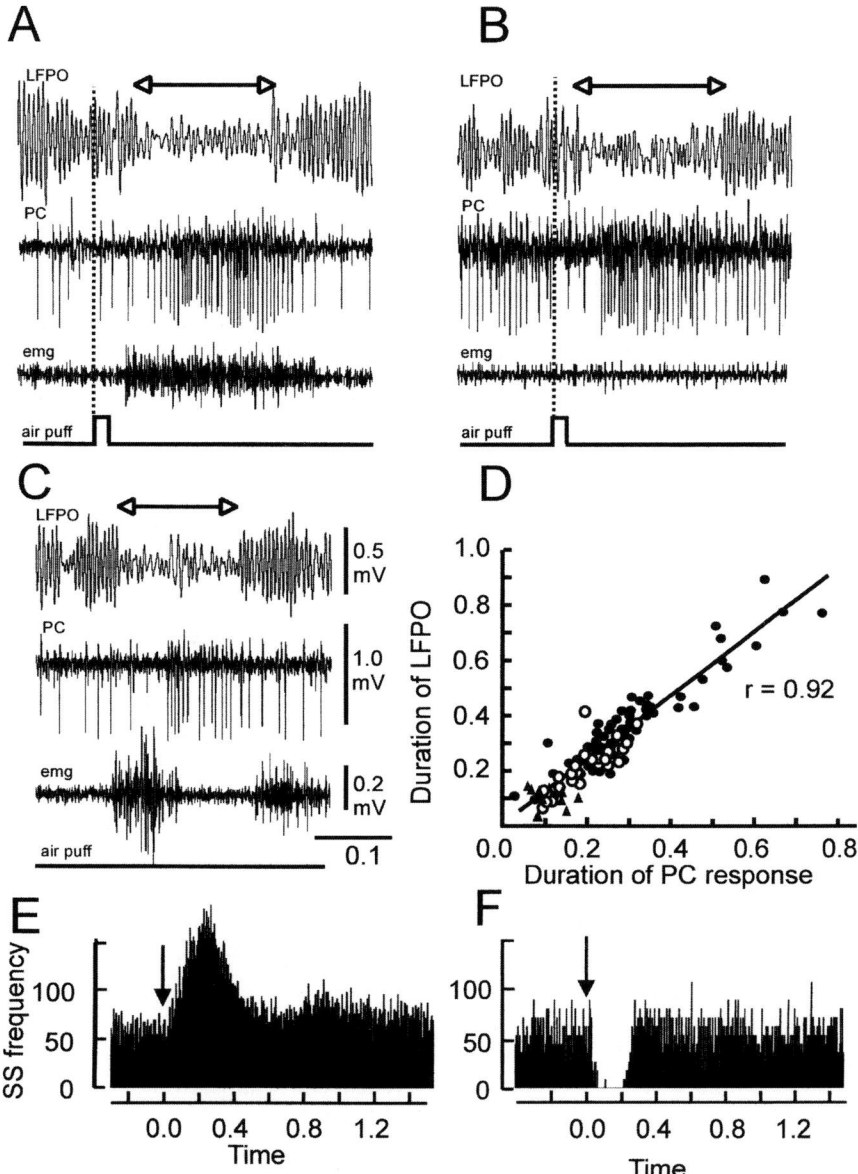

Fig. 7. Local field potential oscillation (LFPO), PC firing and EMG responses during cutaneous stimulation of the whisker region. (A, B) An air puff is produced and evokes suppression of LFPO, PC firing response and EMG burst (only in A). (C) Spontaneous EMG activities accompanied by LFPO inhibition only when concomitant PC transient responses were recorded. Horizontal arrows indicate the period of LFPO inhibition. (D) Scatter plot of durations of LFPO inhibition corresponding to tactile stimulation without motor response (triangles), with evoked motor activity (closed circles) and during spontaneous motor activity (open circles) and durations of PC firing modifications. Note the strong linear relationship between duration of PC response and of LFPO inhibition. LFPO recording is filtered with a low pass digital filter (200 Hz). (E, F) Peri-event histograms of SS frequency for two representative PCs excited (E) or inhibited (F) by air puff stimulation (33 and 22 trials in E and F, respectively; 5 ms bins). Arrows indicate stimulus onset.

suppression or PC modulation were no longer observed (data not shown).

The LFPO was inhibited during 110 ± 42 ms ($n = 13$), 377 ± 241 ms ($n = 66$), 217 ± 79 ms ($n = 26$) by tactile stimulation without motor response (Fig. 7D, triangles), with evoked motor response (Fig. 7D, closed circles) and during spontaneous motor activity (Fig. 7D, open circles), respectively. The duration of the inhibition period of the LFPO was significantly longer ($p < 0.001$) when inhibition was induced by an evoked motor response and, importantly, was highly correlated ($r = 0.92$) to the duration of the transient responses of the PC (Fig. 7D). Mean onset latencies of LFPO inhibition, of PC firing rate modulation and of EMG response were of 33 ± 21, 43 ± 22, and 33 ± 12 ms, respectively. In some cases, an EMG burst was not recorded despite the presence of LFPO inhibition and the related PC response (Fig.7B).

Pharmacological properties of 160 Hz oscillation

To examine which types of synapses contributed to, or regulated, the LFPOs, we applied blockers of GABA$_A$ and NMDA receptors and of gap junctions (Cheron et al., 2004).

Figure 8 illustrated the temporal evolution of the 160 Hz LFPO after carbenoxolone microinjections performed in two Cb$^{-/-}$Cr$^{-/-}$ mice in comparison to microinjection of saline. The amplitude of the power spectrum of the WTA greatly decreased in the few minutes after the injection of the gap junction blocker. In these experiments, the 160 Hz LFPO completely disappeared 3 min after carbenoxolone injection (Fig. 8C). The time of LFPO recovery depended the injected volume. In half portion of the case, we recorded a post-effect increase in the power amplitude of the LFPO (Fig. 8A, C). In contrast, the LFPO were unaffected by injection of saline (Fig. 8B). We may summarize our pharmacological investigation by the fact that microinjection of bicuculline ($n = 12$) or SR95531 ($n = 4$), APV ($n = 9$) and carbenoxolone (Draguhn et al., 1998) ($n = 10$) reversibly reduced the power of the LFPOs, down to $35.0 \pm 17.8\%$, $45.2 \pm 17.7\%$, $48.3 \pm 20\%$, and $9.5 \pm 10.5\%$ of pre-injection values, respectively. We ruled out non-specific actions of carbenoxolone on PC firing ($n = 5$ in 2) by simultaneously recording LFPs and spikes from single PCs (using separate electrodes) before, during, and after microinjection of carbenoxolone. One minute after the injection, the LFPO disappeared while the PC firing remained unchanged.

Discussion

Purkinje cells as the generators of the 160 Hz oscillation

The depth recording profile, the synchronization between 160 Hz oscillation and both the SS and CS, and the LFPO synchronization along the PF beam, strongly suggest that PC populations appear to be the major generators of the 160 Hz oscillation. The highly stereotyped arrangement of closely apposed PCs and their orientation as well as their physiological functioning as an open field make PC assemblies ideal candidates for generating coherent LFPs when their activity is boosted and synchronized. Gradients in synaptic arrangements, with inhibitory synapses mostly located more proximally than parallel fiber synapses, may act as a dipole, generating the largest power close to the plane of PC bodies.

Involvement of the 'mixed synapses' of the molecular interneurons

Molecular interneurons have been shown to be interconnected through coupled electrical and chemical synapses (Sotelo and Llinas, 1972; Mann-Metzer and Yarom, 1999). These composite 'mixed synapse' works as single functional units (Mamiya et al., 2003). The inhibition of 160 Hz LFPO by microinjections of gabazine and carbenoxolone may be explained by the respective actions of these two pharmacological agents on the 'mixed synapses' of the molecular interneurons. These 'mixed synapses' of molecular interneurons share many features with interneurons in the neocortex and are considered to be functional units which act to synchronize the firing of neurons (Galarreta and Hestrin, 1999; Tamas et al., 2000) or to induce neuronal rhythmogenesis (Galarreta and Hestrin, 2001a,b). It was demonstrated in the pyloric circuit of the spiny

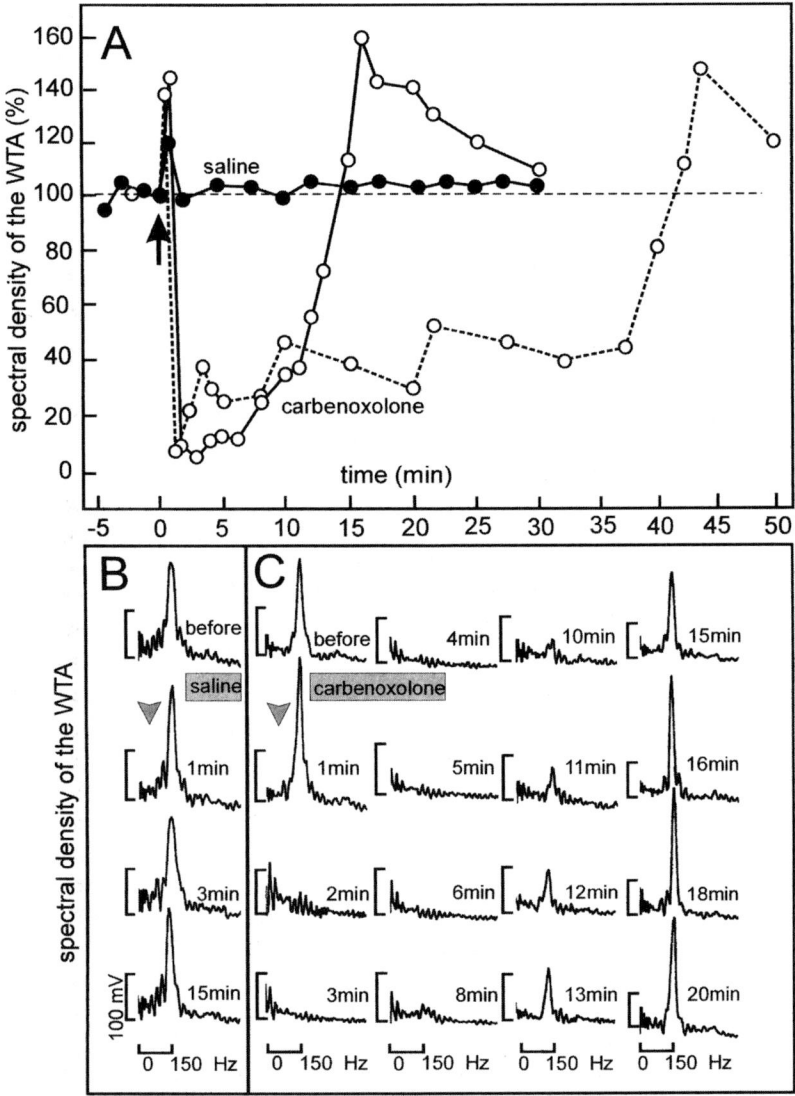

Fig. 8. Effects of carbenoxolone microinjections on the 160Hz LFPO. (A) Temporal evolution of the maximal amplitude of WTA power of the oscillation during two different experiments (carbenoxolone, ○) compared to a microinjection of saline (●). Note the presence of a post-effect increase in the power amplitude of the LFPO. (B) Fast Fourier Transform spectrum of the WTA 160 Hz LFPO before and after (1, 3, and 15 min) microinjection of saline. (C) WTA 160 Hz LFPO spectrum before and after (from 1 to 20 min) microinjection of carbenoxolone.

lobster that during an ongoing rhythm the electrical component of the synapse is dominant (Mamiya et al., 2003). There are at least two possible sites where gap junctions would be able to 'sharpen' the 160 Hz LFPO: at the level of stellate and basket cells (Sotelo and Llinas, 1972; Mann-Metzer and Yarom, 1999) and at the level of the axonal plexus of PC collaterals.

Several subtypes of connexins (connexin 36 (Teubner et al., 2000 but see Meier et al., 2002), connexin 43 (Simburger et al., 1997) and connexin 47 (Teubner et al., 2001)) are expressed in different neuronal populations in the cerebellum including molecular interneurons and PCs. However, expression of connexins and hence presence of gap junction in the

PCs has recently been called into question (Odermatt et al., 2003).

The hypothetical link between the absence of Cr and Cb and the activation of a gap junction-coupled-network producing the 160 Hz oscillation could be explained by modifications in the intracellular Ca^{2+} homeostasis in PC (Airaksinen et al., 1997; Schiffmann et al., 1999). Indeed, gating of gap junctions could be modified by increased intracellular Ca^{2+} either through activation of Ca^{2+}/calmoduline kinase II and phosphorylation of connexins (De Pina-Benabou et al., 2001) or by the direct induction of conformational changes of the channel proteins (Lazrak and Peracchia, 1993).

In the molecular layer, stellate cells form a densely connected network through fast $GABA_A$ receptor synapses (Kondo and Marty, 1998; Carter and Regehr, 2002) as well as through dendrodendritic gap junctions (Mann-Metzer and Yarom, 1999). This network provides the dominant inhibitory input to PCs through fast GABA-ergic synapses (Puia et al., 1994). In addition, parallel fibers make powerful synapses on stellate cells (Carter and Regehr, 2002) favoring synchronization of the oscillations along the folial axis (Isope et al., 2002). Moreover, the recent finding that inhibitory and excitatory GABA synapses coexist in cerebellar interneuron network (Chavas and Marty, 2003) may provide an additional mechanism whereby PC collaterals projecting on molecular interneurons would trigger a rhythmic inhibition–excitation sequence in the coupled interneuron network. However, as proposed for the hippocampus a more complex mechanism could be involved (Traub et al., 2003a, b). Experimental and theoretical studies demonstrated that the collective behavior of a gap-junctional coupling between axons of pyramidal cells result in high-frequency field oscillation (>100 Hz). This fast oscillation occurs as a consequence of random activity within the axonal plexus and is uncovered when all chemical synapses are blocked. It was proposed that ectopic spike generation at a low rate (0.05 to 1 per second) is able to sustain such high-frequency oscillation (Traub et al., 2003a,b). The interplay between fast (>100 Hz) and gamma oscillations seems to be governed by a dual role of the $GABA_A$ receptor. Nonsomatic $GABA_A$ receptor activation enhances this collective oscillation whereas perisomatic $GABA_A$ receptors, that are activated by interneuron input, phase this random activity at gamma rhythm (Traub et al., 2003a, b).

The study of the spontaneous firing of the Golgi cells which is not locked to the 160 Hz oscillation can exclude this inhibitory neurons as possible generator. Moreover, the LFPOs disappeared immediately, when an increased Golgi cell activity suppressed granule cells (Cheron et al., 2004), reducing the PF input to PCs, or when the NMDA-receptor-mediated mossy fiber input to granule cells (D'Angelo et al., 1995) was blocked. Finally, because the 160 Hz oscillation is a network phenomenon, we cannot exclude that other types of neurons than PC and stellate also participate to the rhythm, such as basket cells, whose axons are electrically interconnected in the 'pinceau' ensheathing the initial axon segment of PCs, or Lugaro cells, whose myelinated axons can span 2 mm along a folium (Dieudonne and Dumoulin, 2000).

Common set of physiological alterations despite the specific distribution of Cr and Cb in the cerebellar cortex

In spite of a distinct, cell-specific expression of Cr and Cb in the cerebellar cortex (Résibois and Rogers, 1992) inactivation of either gene gave rise to a common set of physiological alterations reinforced in the double knockout construct. In the $Cr^{-/-}$ mice, the increased PC firing rate, and the concomitant 160 Hz oscillation, most likely result from the increased intrinsic excitability of the afferent granule cells (Gall et al., 2003). In addition, the PCs of the $Cr^{-/-}$ mice have been demonstrated to exhibit a paradoxical calretinin-immunoreactivity probably related to an increase in intracellular Ca^{2+} concentration (Schiffmann et al., 1999).

Although we failed to find in the $Cb^{-/-}$ mice some changes in somatic intrinsic PC excitability, specific alterations in the PC dendritic compartment have been revealed by the detection of increased synaptically evoked Ca^{2+} transients (Airaksinen et al., 1997; Barski et al., 2003). Moreover, the absence of Cb resulted in an increase of various spine parameters of the PC, which were interpreted as a morphological compensation for the lack of soluble calcium buffer in the PC cytoplasm (Vecellio et al., 2000).

Physiological or pathological nature of the 160 Hz oscillation

In normal animals the spontaneous SS frequency of PCs is characterized by an irregular firing rate with a weak rhythmicity (Ebner and Bloedel, 1981; Goossens et al., 2001; Schwarz and Welsh, 2001; present results). In contrast, in the $Cr^{-/-}Cb^{-/-}$ mice, SS rhythmicity was highly amplified and was synchronized along the PF beam. As for cortical neurons (Gerstein and Mandelbrot, 1964; Shadlen and Newsome, 1998), we may infer that in normal cerebellum, the balanced excitation and inhibition is one of the mechanisms behind the irregular firing of the PC. Consequently, the presence of a very rhythmic PC firing in the mutant mice may be interpreted as a disruption of the balance between the excitatory and inhibitory drive of the PC. As demonstrated in the cerebral cortex (Singer, 1999), the enhancement of the discharges rate, rhythmicity, and synchrony are the basic requirements for the emergence of oscillatory patterning.

The central question concerning the exact nature and the role played by the 160 Hz oscillation is yet to be solved. It could appear as a pathological feature, perhaps of dysfunction of the PC–molecular interneurons dynamics, or represent an adaptive process to overcome deficits in sensorimotor coordination. It has been proposed that spontaneous activity of synchronized oscillations may be useful for the adjustment of the synaptic strength, not only during development (Katz and Shatz, 1996) but in the mature brain as well (McCormick, 1999). Oscillations have been reported in brain areas such as neocortex and hippocampus in different frequency ranges. Synchronization of oscillation in the gamma range (30–70 Hz) between different cortical areas is supposed to underlie the binding of several features into a single perceptual entity (Gray et al., 1989; Singer, 1993). Oscillations may also reflect pathological states such as those recorded in seizure (Steriade et al., 1998) or in Parkinson's disease. In this later state, basal ganglia do exhibit very strong beta-band oscillation reflecting an abnormal synchrony in basal ganglia network-level activity preventing normal circuit function (Williams et al., 2002).

The properties of the present oscillation show in some aspects similarities with cerebral 'resting' rhythmic activities of wakefulness arresting to sensory or motor information, such as alpha and mu rhythms. Mu oscillations most often occur during a pre-movement period and cease around movement onset (Donoghue et al., 1998; Ohara et al., 2000; Pfurtscheller et al., 2003). The decrease in mu oscillation roughly coincides with an increase in the gamma oscillation above 30 Hz (Pfurtscheller et al., 2003) and with the appearance of firing rate modulation coupled to the motor action (Donoghue et al., 1998). Cortical gamma oscillation with different frequency bands has been documented in different areas and their possible functions remain largely debated. For some authors, the gamma oscillating activity in the motor system is involved in resetting the descending motor commands needed for changes in motor state (Baker et al., 1999). For others, they represent a neural correlate of attention during demanding sensorimotor behaviors (Murthy and Fetz, 1996). In this context, it was proposed (Fetz et al., 2000) that one functional role could be a global long term potentiation of synaptic interactions underlying increased attention and motor learning. As we have mentioned above, the 160-Hz oscillation in our null-mutant mice may be considered as a pathological feature resulting from the deletion and leading to ataxia or as a compensatory mechanism allowing the cerebellum to work despite abnormal PC firing. In the latter view, the 160-Hz LFPO synchrony could act as a spatiotemporal filter sharpening the action of selected rostro-caudal modules of the cerebellum (Voogd and Glickstein, 1998). The focal mossy fiber input related to the cutaneous stimulation probably overcome the generalized state of synchronous activation of PC populations leading to silent or bursty mode capable of affecting their targets in cerebellar nuclei. A similar mechanism has been proposed by Courtemanche et al. (2003) for the beta-band oscillation of the monkey striatum modules. Whether or not 160 Hz cerebellar oscillation leads to ataxia, its emergence in different mice models targeting different cellular elements of the cerebellar cortex (Cheron et al., 2003, 2004; Servais et al., 2003) outline the ability of this structure to generate fast oscillations.

Acknowledgments

The authors wish to thank S. Schurmans and M. Meyer for providing $Cr^{-/-}$ and $Cb^{-/-}$ mice, respectively. M. Escudero, F. Colin, V. Seutin, and R. Traub for their helpful discussion, and M.P. Dufief for in vivo electrophysiology, and H. Nguyen and M. Bourguet for their technical assistance. We thank Sanofi (Paris) for providing us the SR95531. This work was sponsored by the Fonds National de la Recherche Scientifique (Belgium) and research funds of Université Libre de Bruxelles (ULB) and Université de Mons-Hainaut, Belgium. L. Servais is supported by a grant from the Fondation Erasme (ULB).

References

Airaksinen, M.S., Eilers, J., Garaschuk, O., Thoenen, H., Konnerth, A. and Meyer, M. (1997) Ataxia and altered dendritic calcium signaling in mice carrying a targeted null mutation of the calbindin D28k gene. Proc. Natl. Acad. Sci. USA, 94: 1488–1493.

Baker, S.N., Kilner, J.M., Pinches, E.M. and Lemon, R.N. (1999) The role of synchrony and oscillations in the motor output. Exp. Brain Res., 128: 109–117.

Barski, J.J., Hartmann, J., Rose, C.R., Hoebeek, F., Morl, K., Noll-Hussong, M., De Zeeuw, C.I., Konnerth, A. and Meyer, M. (2003) Calbindin in cerebellar Purkinje cells is a critical determinant of the precision of motor coordination. J. Neurosci., 23: 3469–3477.

Bower, J.M. and Woolston, D.C. (1983) Congruence of spatial organization of tactile projections to granule cell and Purkinje cell layers of cerebellar hemispheres of the albino rat: vertical organization of cerebellar cortex. J. Neurophysiol., 49: 745–766.

Buzsáki, G., Horváth, Z., Urioste, R., Hetke, J. and Wise, K. (1992) High-frequency network oscillation in the hippocampus. Science, 256: 1025–1027.

Caillard, O., Moreno, H., Schwaller, B., Llano, I., Celio, M.R. and Marty, A. (2000) Role of the calcium-binding protein parvalbumin in short-term synaptic plasticity. Proc. Natl. Acad. Sci. USA, 97: 13372–13377.

Carter, A.G. and Regehr, W.G. (2002) Quantal events shape cerebellar interneuron firing. Nat. Neurosci., 5: 1309–1318.

Chavas, J. and Marty, A. (2003) Coexistence of excitatory and inhibitory GABA synapses in the cerebellar interneuron network. J. Neurosci., 23: 2019–2031.

Cheron, G., Dufief, M.P., Gerrits, N.M., Draye, J.P. and Godaux, E. (1997) Behavioural analysis of Purkinje cell output from the horizontal zone of the cat flocculus. Prog. Brain Res., 114: 347–356.

Cheron, G., Schurmans, S., Lohof, A., D'alcantara, P., Meyer, M., Draye, J.P., Parmentier, M. and Schiffmann, S. (2000) Firing behaviour of Purkinje cells and sensori-motor coordination in calretinin knockout mice. Prog. Brain Res., 124: 299–308.

Cheron, G., Gall, D., Servais, L., Dan, Maex, R. and Schiffmann, S. (2004) High-frequency oscillations in cerebellar cortex induced by inactivation of calcium binding protein genes. J. Neurosci., 24: 434–441.

Cheron, G., Servais, L., Wagstaff, J., Dan, B. (2003) High frequency oscillation in cerebellum of mice with inactivated maternally-inherited *UBE3A* gene. Proceedings of Symposium in honour of Constantino Sotelo's lifetime work. Creating Coordination in the cerebellum, p. 37.

Chrobak, J.J., Lorincz, A. and Buzsaki, G. (2000) Physiological patterns in the hippocampo-entorhinal cortex system. Hippocampus, 10: 457–465.

Courtemanche, R., Fujii, N. and Graybiel, A.M. (2003) Synchronous, focally modulated beta-band oscillations characterize local field potential activity in the striatum of awake behaving monkeys. J. Neurosci., 23: 11741–11752.

D'Angelo, E., De Filippi, G., Rossi, P. and Taglietti, V. (1995) Synaptic excitation of individual rat cerebellar granule cells in situ: evidence for the role of NMDA receptors. J. Physiol. (Lond.), 484: 397–413.

De Pina-Benabou, M.H., Srinivas, M., Spray, D.C. and Scemes, E. (2001) Calmodulin kinase pathway mediates the K^+-induced increase in Gap junctional communication between mouse spinal cord astrocytes. J. Neurosci., 21: 6635–6643.

Dieudonne, S. and Dumoulin, A. (2000) Serotonin-driven long-range inhibitory connections in the cerebellar cortex. J. Neurosci., 20: 1837–1848.

Donoghue, J.P., Sanes, J.N., Hatsopoulos, N.G. and Gaal, G. (1998) Neural discharge and local field potential oscillations in primate motor cortex during voluntary movements. J. Neurophysiol., 79: 159–173.

Draguhn, A., Traub, R.D., Schmitz, D. and Jefferys, J.G.R. (1998) Electrical coupling underlies high-frequency oscillations in the hippocampus in vitro. Nature, 394: 189–192.

Ebner, T.J. and Bloedel, J.R. (1981) Temporal patterning in simple spike discharge of Purkinje cells and its relationship to climbing fiber activity. J. Neurophysiol., 45: 933–947.

Eccles, J.C., Llinas, R. and Sasaki, K. (1966) The inhibitory interneurones within the cerebellar cortex. Exp. Brain Res., 1: 1–16.

Eckhorn, R. and Thomas, U. (1993) A new method for the insertion of multiple microprobes into neural and muscular tissue, including fibre electrodes, fine wires, needles and microsensors. J. Neurosci. Methods, 49: 175–179.

Edmonds, B., Reyes, R., Schwaller, B. and Roberts, W.M. (2000) Calretinin modifies presynaptic calcium signaling in frog saccular hair cells. Nat. Neurosci., 3: 786–790.

Fetz, E.E., Chen, D., Murthy, V.N. and Matsuura, M. (2000) Synaptic interactions mediating synchrony and oscillations in primate sensorimotor cortex. J. Physiol. (Paris), 94: 323–331.

Galarreta, M. and Hestrin, S. (1999) A network of fast-spiking cells in the neocortex connected by electrical synapses. Nature, 402: 72–75.

Galarreta, M. and Hestrin, S. (2001a) Spike transmission and synchrony detection in networks of GABAergic interneurons. Science, 292: 2295–2299.

Galarreta, M. and Hestrin, S. (2001b) Electrical synapses between GABA-releasing interneurons. Nat. Rev. Neurosci., 2: 425–433.

Gall, D., Roussel, C., Susa, I., d'Angelo, E., Rossi, P., Bearzatto, B., Galas, M.-C., Blum, D., Schurmans, S. and Schiffmann, S.N. (2003) Altered neuronal excitability in cerebellar granule cells of mice laking calretinin. J. Neurosci., 23: 9320–9327.

Gerstein, G.l. and Mandelbrot, B. (1964) Random walk models for the spike activity of a single neuron. Biophys. J., 71: 41–68.

Ghelarducci, B., Ito, M. and Yagi, N. (1975) Impulse discharges from flocculus Purkinje cells of alert rabbits during visual stimulation combined with horizontal head rotation. Brain Res., 87: 66–72.

Goossens, J., Daniel, H., Rancillac, A., van der Steen, J., Oberdick, J., Crepel, F., De Zeeuw, C.I. and Frens, M.A. (2001) Expression of protein kinase C inhibitor blocks cerebellar long-term depression without affecting Purkinje cell excitability in alert mice. J. Neurosci., 21: 5813–5823.

Gray, C.M., Koenig, P., Engel, A.K. and Singer, W. (1989) Stimulus-specific neuronal oscillations in cat visual cortex exhibit inter-columnar synchronization which reflects global stimulus properties. Nature, 338: 334–337.

Häusser, M. and Clark, B.A. (1997) Tonic synaptic inhibition modulates neuronal output pattern and spatiotemporal synaptic integration. Neuron, 19: 665–678.

Häusser, M. and Monsivais, P. (2003) Less means more: inhibition of spontaneous firing triggers persistent increases in excitability. Neuron, 40(3): 449–451.

Hopfield, J.J. and Brody, C.D. (2001) What is a moment? Transient synchrony as a collective mechanism for spatiotemporal integration. Proc. Natl. Acad. Sci. USA, 98: 1282–1287.

Huguenard, J.R. (1996) Low-threshold calcium currents in central nervous system neurons. Annu. Rev. Physiol., 58: 329–348.

Katz, L.C. and Shatz, C.J. (1996) Synaptic activity and the construction of cortical circuits. Science, 274: 1133–1138.

Kondo, S. and Marty, A. (1998) Synaptic currents at individual connections among stellate cells in rat cerebellar slices. J. Physiol. (Lond.), 509: 221–232.

Isope, P., Dieudonne, S. and Barbour, B. (2002) Temporal organization of activity in the cerebellar cortex: a manifesto for synchrony. Ann. N. Y. Acad. Sci., 978: 164–174.

Llano, I., Di Polo, R. and Marty, A. (1994) Calcium-induced calcium release in cerebellar Purkinje cells. Neuron, 12: 663–673.

Lazrak, A. and Peracchia, C. (1993) Gap junction gating sensitivity to physiological internal calcium regardless of pH in Novikoff hepatoma cells. Biophys. J., 65: 2002–2012.

Llinás, R.R. and Sugimori, M. (1980) Electrophysiological properties of in vitro Purkinje cell somata in mammalian cerebellar slices. J. Physiol. (Lond.), 305: 171–195.

Llinas, R.R. (1988) The intrinsic electrophysiological properties of mammalian neurons: insights into central nervous system function. Science, 242: 1654–1664.

Lisberger, S.G. and Fuchs, A.F. (1974) Response of flocculus Purkinje cells to adequate vestibular stimulation in the alert monkey: fixation vs. compensatory eye movements. Brain Res., 69: 347–353.

Mann-Metzer, P. and Yarom, Y. (1999) Electronic coupling interacts with intrinsic properties to generate synchronized activity in cerebellar networks of inhibitory interneurons. J. Neurosci., 19: 3298–3306.

Mamiya, A., Manor, Y. and Nadim, F. (2003) Short-term dynamics of a mixed chemical and electrical synapse in a rhythmic network. J. Neurosci., 23: 9557–9564.

McCormick, D.A. (1999) Spontaneous activity: signal or noise? Science, 23: 541–543.

McCormick, D.A. and Bal, T. (1997) Sleep and arousal: thalamocortical mechanisms. Annu. Rev. Neurosci., 20: 185–215.

Meier, C., Petrasch-Parwez, E., Habbes, H.W., Teubner, B., Guldenagel, M., Degen, J., Sohl, G., Willecke, K. and Dermietzel, R. (2002) Immunohistochemical detection of the neuronal connexin36 in the mouse central nervous system in comparison to connexin36-deficient tissues. Histochem. Cell. Biol., 117: 461–471.

Murthy, V.N. and Fetz, E.E. (1996) Oscillatory activity in sensorimotor cortex of awake monkeys: synchronization of local field potentials and relation to behavior. J. Neurophysiol., 76: 3949–3967.

Nelson, A.B., Krispel, C.M., Sekirnjak, C. and du Lac, S. (2003) Long-lasting increases in intrinsic excitability triggered by inhibition. Neuron, 40: 609–620.

Odermatt, B., Wellershaus, K., Wallraff, A., Seifert, G., Degen, J., Euwens, C., Fuss, B., Bussow, H., Schilling, K., Steinhauser, C. and Willecke, K. (2003) Connexin 47 (Cx47)-deficient mice with enhanced green fluorescent protein reporter gene reveal predominant oligodendrocytic expression of Cx47 and display vacuolized myelin in the CNS. J. Neurosci., 23: 4549–4559.

Ohara, S., Ikeda, A., Kunieda, T., Yazawa, S., Baba, K., Nagamine, T., Taki, W., Hashimoto, N., Mihara, T. and Shibasaki, H. (2000) Movement-related change of electrocorticographic activity in human supplementary motor area proper. Brain, 123: 1203–1215.

Pfurtscheller, G., Graimann, B., Huggins, J.E., Levine, S.P. and Schuh, L.A. (2003) Spatiotemporal patterns of beta desynchronization and gamma synchronization in corticographic data during self-paced movement. Clin. Neurophysiol., 114: 1226–1236.

Puia, G., Costa, E. and Vicini, S. (1994) Functional diversity of GABA-activated Cl− currents in Purkinje versus granule neurons in rat cerebellar slices. Neuron, 12: 117–126.

Raman, I.M. and Bean, B.P. (1999) Ionic currents underlying spontaneous action potentials in isolated cerebellar Purkinje neurons. J. Neurosci., 19: 1663–1674.

Résibois, A. and Rogers, J.H. (1992) Calretinin in rat brain: an immunohistochemical study. Neuroscience, 46: 101–134.

Schwarz, C. and Welsh, J.P. (2001) Dynamic modulation of mossy fiber system throughput by inferior olive synchrony: a multielectrode study of cerebellar cortex activated by motor cortex. J. Neurophysiol., 86: 2489–2504.

Schiffmann, S.N., Cheron, G., Lohof, A., d'Alcantara, P., Meyer, M., Parmentier, M. and Schurmans, S. (1999) Impaired motor coordination and Purkinje cell excitability in mice lacking calretinin. Proc. Natl. Acad. Sci. USA, 96: 5257–5262.

Servais, L. De Saedeleer, C., Schwaller, B., Schiffmann, Sn. and Cheron, G. (2003). Emergence of a double high-frequency oscillation in the cerebellum of mice lacking parvalbumin and calbindin. Proceedings of Symposium in honour of Constantino Sotelo's lifetime work. Creating Coordination in the cerebellum, p. 62.

Simburger, E., Stang, A., Kremer, M. and Dermietzel, R. (1997) Expression of connexin43 mRNA in adult rodent brain. Histochem. Cell. Biol., 107: 127–137.

Shadlen, M.N. and Newsome, W.T. (1998) The variable discharge of cortical neurons: implications for connectivity, computation, and information coding. J. Neurosci., 18: 3870–3896.

Singer, W. (1993) Synchronization of cortical activity and its putative role in information processing and learning. Annu. Rev. Physiol., 55: 349–374.

Singer, W. (1999) Neuronal synchrony: a versatile code for the definition of relations. Neuron, 24: 31–65, and references therein.

Sotelo, C. and Llinas, R. (1972) Specialized membrane junctions between neurons in the vertebrate cerebellar cortex. J. Cell. Biol., 53: 271–289.

Steriade, M., Amzica, F., Neckelmann, D. and Timofeev, I. (1998) Spike-wave complexes and fast components of cortically generated seizures. II. Extra- and intracellular patterns. J. Neurophysiol., 80: 1456–1479.

Swensen, A.M. and Bean, B.P. (2003) Ionic mechanisms of burst firing in dissociated Purkinje neurons. J. Neurosci., 23: 9650–9663.

Tamas, G., Buhl, E.H., Lorincz, A. and Somogyi, P. (2000) Proximally targeted GABAergic synapses and gap junctions synchronize cortical interneurons. Nat. Neurosci., 3: 366–371.

Teubner, B., Dengen, J., Sohl, G., Guldenagel, M., Bukauskas, F.F., Trexler, E.B., Verselis, V.K., De Zeeuw, C.I., Lee, C.G., Kozak, C.A., Petrasch-Parwez, E., Dermietzel, R. and Willecke, K. (2000) Functional expression of the murine connexin 36 gene coding for a neuron-specific gap junctional protein. J. Membrane Biol., 176: 249–262.

Teubner, B., Odermatt, B., Guldenagel, M., Sohl, G., Degen, J., Bukauskas, F., Kronengold, J., Verselis, V.K., Jung, Y.T., Kozak, C.A., Schilling, K. and Willecke, K. (2001) Functional expression of the new gap junction gene connexin 47 transcribed in mouse brain and spinal cord neurons. J. Neurosci., 21: 1117–1126.

Traub, R.D. and Bibbig, A. (2000) A model of high-frequency ripples in the hippocampus based on synaptic coupling plus axon-axon gap junctions between pyramidal neurons. J. Neurosci., 20: 2086–2093.

Traub, R.D., Cunningham, M.O., Gloveli, T., LeBeau, F.E., Bibbig, A., Buhl, E.H. and Whittington, M.A. (2003a) GABA-enhanced collective behavior in neuronal axons underlies persistent gamma-frequency oscillations. Proc. Natl. Acad. Sci. USA, 100: 11047–11052.

Traub, R.D., Pais, I., Bibbig, A., LeBeau, F.E., Buhl, E.H., Hormuzdi, S.G., Monyer, H. and Whittington, M.A. (2003b) Contrasting roles of axonal (pyramidal cell) and dendritic (interneuron) electrical coupling in the generation of neuronal network oscillations. Proc. Natl. Acad. Sci. USA, 100: 1370–1374.

Vecellio, M., Schwaller, B., Meyer, M., Hunziker, W. and Celio, M.R. (2000) Alterations in Purkinje cell spines of calbindin D-28 k and parvalbumin knock-out mice. Eur. J. Neurosci., 12: 945–954.

Voogd, J. and Glickstein, M. (1998) The anatomy of the cerebellum. Trends Neurosci., 21: 370–375.

Welsh, J.P., Lang, E.J., Suglhara, I. and Llinás, R.R. (1995) Dynamic organization of motor control within the olivocerebellar system. Nature, 374: 453–457.

Whittington, M.A. and Traub, R.D. (2003) Interneuron Diversity series: Inhibitory interneurons and network oscillations in vitro. Trends Neurosci., 26: 676–682.

Williams, D., Tijssen, M., Van Bruggen, G., Bosch, A., Insola, A., Di Lazzaro, V., Mazzone, P., Oliviero, A., Quartarone, A., Speelman, H. and Brown, P. (2002) Dopamine-dependent changes in the functional connectivity between basal ganglia and cerebral cortex in humans. Brain, 125: 1558–1569.

Womack, M.D. and Khodakhah, K. (2003) Somatic and dendritic small-conductance calcium-activated potassium channels regulate the output of cerebellar purkinje neurons. J. Neurosci., 23: 2600–2607.

Ylinen, A., Bragin, A., Nadasdy, Z., Jando, G., Szabo, I., Sik, A. and Buzsaki, G. (1995) Sharp wave-associated high-frequency oscillation (200 Hz) in the intact hippocampus: network and intracellular mechanisms. J. Neurosci., 15: 30–46.

CHAPTER 15

Oscillations in the cerebellar cortex: a prediction of their frequency bands

Reinoud Maex and Erik De Schutter*

Laboratory of Theoretical Neurobiology, Born-Bunge Foundation, University of Antwerp, B-2610 Antwerp, Belgium

Keywords: synchrony; oscillation; resonance; feedback; inhibition; inhibitory postsynaptic current; dynamics; delay

Abstract: Local recurrent connections endow the cerebellar cortex with an intrinsic dynamics. We performed computer simulations to predict the frequency bands of the oscillations that will most likely emerge. Feedback inhibition from the Golgi to the granule cells induced 10–50 Hz oscillations, the period at resonance being approximately equal to four times the maximum conduction delay generated along the parallel-fiber connections from granule to Golgi cells. In the molecular layer, the interneurons tended to induce fast oscillations (100–250 Hz), having a period equal to about four times the delay over their reciprocal synaptic connections. Finally, although the presence of lateral inhibition among the Purkinje cells has not been firmly established, reciprocal Purkinje-cell synapses are predicted to transform the cerebellar cortex into a potential temporal integrator.

Introduction

Because of its fast responsiveness and the temporal precision with which it operates, the cerebellar cortex is often described as a pure feed forward circuit conveying the mossy-fiber input over granule cells and along parallel fibers to Purkinje cells, which send their axons to the deep cerebellar (or vestibular) nuclei. Plasticity of the synapses connecting the consecutive stages, guided by supervised and unsupervised learning rules, continually adapts the circuit (for reviews see Hansel et al., 2001; Ito, 2001; Barlow, 2002). Inhibitory interneurons likewise sharpen the input–output relationship and control its gain.

Nevertheless, all inhibitory neurons within the cerebellar cortex also make recurrent connections (Palay and Chan-Palay, 1974). Whatever their function, the resulting feedback endows the circuit with intrinsic dynamics, the simplest manifestation of which would be the emergence of oscillatory activity. Although one of the functions of the cerebellum is likely to dampen mechanical oscillations (Thach et al., 1992), over the years oscillations in a variety of frequency bands have been observed in different experimental settings, suggesting that the cerebellum is also a rhythm generator (Pellerin and Lamarre, 1997; Baçar, 1998; Hartmann and Bower, 1998; Tesche and Karhu, 2000; Courtemanche et al., 2002; Isope et al., 2002).

We recently reported that inhibitory circuits have a resonance frequency which is determined by the average delay time required for one inhibitory neuron to inhibit a nearby inhibitory neuron (Maex and De Schutter, 2003). More particularly, in purely inhibitory networks, the oscillation period at resonance is equal to about four times the average latency to the onset of the evoked inhibitory postsynaptic currents

*Corresponding author. Universiteitsplein 1, B-2610 Antwerpen, Belgium. Tel.: +32 3 820 26 44; Fax: +32 3 820 26 69; E-mail: erik@bbf.uia.ac.be

(IPSCs). We use this rule in combination with computer simulations to predict the frequency bands of the oscillations likely to arise in the granular layer (through recurrent inhibition from Golgi to granule cells), in the molecular layer (through reciprocal inhibition among the interneurons), and in the Purkinje cell layer (through Purkinje axon collaterals). We do not consider oscillations resulting from reverberation in compound circuits comprising the inferior olive, cerebellar nuclei, thalamus, or neocortex (see Kistler et al., 2000).

Recurrent inhibition from Golgi to granule cells

The dynamics of the Golgi-granule cell feedback loop determines the temporal pattern of parallel-fiber input that Purkinje cells receive from granule cells. In principle, the granule cells could produce synchronous oscillations, fire completely asynchronously, or burst. For each state, we discuss which parameters are critical for its generation, its possible functional significance, and supporting experimental data. Thereafter we examine in which frequency band oscillations presumably emerge.

The asynchronous state

In the asynchronous state, the population of Golgi cells fires incoherently, as do the granule cells. For this state to arise in the granular layer, pairs of Golgi cells along the folial axis must be prevented from firing synchronously to the common excitatory input they receive from parallel fibers (Vos et al., 1999). In models, such a desynchronization could be most readily achieved by increasing the strength of the monosynaptic excitation that Golgi cells receive from mossy fibers (Maex and De Schutter, 1998a). Indeed, mossy fibers diverge over shorter distances within a folium, presumably make fewer but stronger synapses, and fire less coherently than do parallel fibers (Dieudonné, 1998b). Excitation through stronger synapses also deteriorated the spike rhythmicity of single model Golgi cells (Maex and De Schutter, 1998b). In addition, pairs of granule cells lying within the same Golgi axonal arbor must be prevented from synchronizing their spike emission by randomization of the strength and time-course of the synapses they receive from the common Golgi afferent.

The functional advantage of the asynchronous state is a maximization of the information content of granule cell spikes (Buonomano and Mauk, 1994). If each granule cell would fire at a unique interval after or during the course of stimulation, stimulus time would be mapped on the granule cell population. Long-term depression of the synapses originating from the granule cells selectively active during a particular stimulus phase would then endow the Purkinje cells with the capacity of recognizing temporal stimulus patterns (Steuber and De Schutter, 2002). The usefulness of this temporal coding was demonstrated in semi-realistic simulations of a conditioned response (Buonomano and Mauk, 1994) and a motor coordination task (Schweighofer et al., 1998).

The bursting state

In an alternative state, the granular layer is mostly silent except for bursts of granule-cell spikes evoked, for instance, by sensory stimulation. In models, this behavior could be reproduced by locally applying a bursting mossy-fiber input (Franck et al., 2001). The resulting bursts of the model granule cells were not aborted by inhibition from parallel-fiber activated Golgi cells provided that the patch of active granule cells was small and mossy-fiber activation of Golgi cells weak (Finch and Augustine, 1998; De Schutter and Bjaalie, 2001). Bursts evoked by peripheral stimuli were recently recorded in granule cells in anesthetized rats (Chadderton et al., 2003), and extracellularly from presumed granule cells in anesthetized or decerebrated cats (Eccles et al., 1971). Intrinsic membrane currents through non-inactivating Na^+ and K^+ channels contribute to the bursting behavior of granule cells in vitro (D'Angelo et al., 2001). Bursting of granule cells can enhance the reliability and plasticity of transmission at the parallel-fiber synapses (Casado et al., 2002).

The synchronous state

From theoretical considerations, the delayed feedback from Golgi to granule cells was predicted to induce synchronous oscillations (Maex and De Schutter,

1998a) or traveling waves (Roberts, 1997). As far as we know, most if not all network models of the granular layer exhibited (wanted or unwanted) oscillations at one or another stage of their development. In the next section, we predict that the granular layer has a characteristic frequency restricted to the 10–50 Hz frequency band, the exact value depending on the conduction speed of parallel fibers.

Independently, experimentalists observed that Golgi and granule cells are, more than the average neuron is, intrinsic oscillators. Potent I_h channels, which rapidly activate during hyperpolarization, induce rebound spikes in rat Golgi cells (Dieudonné, 1998a; Forti et al., 2003). The optimal frequency of alternating hyper- and depolarization is not known, but the firing pattern of a simplified model Golgi cell was most regular at about 11 Hz during parallel-fiber stimulation (Maex et al., 2000). The firing rate of rat granule cells, on the other hand, is maximal at a burst frequency of 3–12 Hz (D'Angelo et al., 2001). This resonance frequency is expected to increase by a factor of two after correction for the 30°C recording temperature.

It is unclear until now how the cerebellum could benefit from synchrony in its granular layer. Not only would synchronous oscillations send regular volleys of parallel-fiber spikes to the Purkinje cells, they would also determine which patterns of mossy-fiber input most easily pass (Kistler et al., 2000). Another conspicuous difference between the synchronous and asynchronous states concerns their effect on gain control. The firing rate of a granule cell depends on the degree of synchrony in the activation of the ten inhibitory synapses it receives on average (Jakab and Hámori, 1988). If the synapses originate from different Golgi afferents, then the summed inhibitory postsynaptic potentials (IPSPs) readily saturate in the synchronous state, because the reversal potential for the $GABA_A$ receptor current is close to the resting membrane potential. In contrast, asynchronously induced IPSPs do not saturate and hold the granule cell in a state of sustained suppression. Figure 1 demonstrates how increasing the number of Golgi afferents hardly affected the firing rate of granule cells in a synchronous network, whereas the increased convergence of Golgi to granule cells suppressed the granule cells in an asynchronous network (Maex and De Schutter, 1998a).

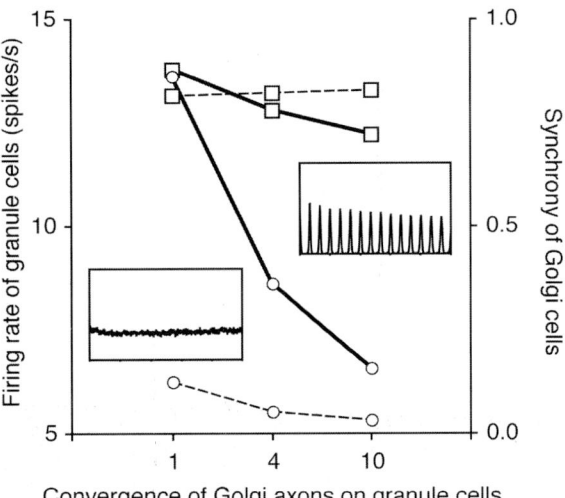

Fig. 1. The firing rate of granule cells is lower in an asynchronous network than in a synchronous network. We varied the degree of convergence from Golgi to granule cells in a synchronous (rectangles) and an asynchronous (circles) model of the granular layer. More particularly, each of the 5355 granule cells received inhibition from either only the closest of the 30 Golgi cells, or from the most nearby 4 or 10 Golgi cells, as indicated on the horizontal axis. The latter two connection schemes implement the innervation of each of the four granule dendrites, versus each of the ten inhibitory synapses, by a different Golgi cell. Individual synaptic strengths were normalized over the synaptic number, so that the mean synaptic current into each granule cell remained approximately constant. The two networks differed only in the number of monosynaptic connections from the mossy fibers to the Golgi cells. In the asynchronous network, each Golgi cell received synapses from a separate, contiguous set of 18 out of the 540 mossy fibers. Increasing the arborization of mossy fibers such that each Golgi cell received 108 (proportionally weaker) synapses sufficed to synchronize the network. The peak conductance of the mossy-fiber activated AMPA channel of Golgi cells was set to 60% of the conductance of the parallel-fiber activated AMPA channel. Solid lines: average firing rate of granule cells. Dashed lines: synchronization index of Golgi cell population. Insets: one-sided autocorrelation histograms of the Golgi cell population (range 0–500 ms) in the asynchronous (left) and synchronous (right) network.

Oscillations in, or close to, the expected 10–50 Hz band were recorded in the granular layer of unrestrained, alert rats (Hartmann and Bower, 1998) and monkeys (Pellerin and Lamarre, 1997; Courtemanche et al., 2002) and, locked to sensory stimulation, over the cerebellar surface of humans (Tesche and Karhu, 2000).

The characteristic frequency

Assuming that synchronous oscillations are able to arise in the granular layer as a resonance phenomenon of the delayed feedback from Golgi to granule cells, we assessed their optimum frequency in a detailed computer model (Maex and De Schutter, 1998a).

The critical parameters, determining the speed of feedback, were updated as follows. The excitatory postsynaptic currents which granule cells evoke in Golgi cells had a minimum latency of 0.5 ms, in addition to a delay proportional to distance due to spike propagation along the parallel fibers (length: 2.5 mm in either direction; conduction speed: 0.1, 0.3, or 0.5 m/s). The IPSCs evoked by Golgi cells in granule cells had a default latency of 0.5 ms. These IPSCs rose with a time-constant of 0.29 ms, and decayed following the sum of a fast and slow exponential with time constants of 4.1 and 22.5 ms, and relative strengths of 40% and 60%, respectively (reversal potential −75 mV; values as reported in Mitchell and Silver, 2003). Their peak conductance was set to 3.1 nS (Mitchell and Silver, 2003) in pure feedback models lacking monosynaptic mossy-fiber excitation of the Golgi cells, and to 1.55 nS in versions of the model in which the Golgi cells also mediated feed forward inhibition from the mossy fibers to the granule cells. A 300 pS membrane conductance was added to simulate tonic inhibition in the granule cells (Hamann et al., 2002).

After randomization of all neurons and synapses, the model granular layer was excited uniformly and continually to a level determined by the average mossy-fiber firing rate. At each excitation level, the degree of synchronous oscillatory activity in the granular layer was quantified from the spike time histograms of the entire Golgi and granule cell populations, using a synchrony index (Maex and De Schutter, 1998a; this index is equivalent to peak power divided by the power at zero frequency).

Figure 2 illustrates that, both in networks with and without feed forward inhibition, power was maximal at a particular level of excitation. This excitation level, and the corresponding oscillation frequency, increased with the value of the parallel-fiber conduction speed. We predicted the optimum frequency from the average delay time d it would take for one Golgi cell to inhibit or completely disfacilitate another Golgi cell. (Golgi cells are not monosynaptically connected but can be considered to inhibit each other disynaptically over granule cells.) This delay time d comprises the delay time required for a Golgi spike to inhibit a granule cell (the default 0.5 ms synaptic latency) plus the interval between inhibition of the granule cell and disfacilitation of all its efferent Golgi cells, the latter being equal to the sum of the maximal propagation delay generated along parallel fibers and the 0.5 ms synaptic latency (see above). The vertical lines in Fig. 2 are drawn at frequencies equal to $1(4d)$, and give an approximation of the preferred network frequency for each parallel-fiber conduction speed. It must be noted that to each delay time d, there corresponds an optimal decay time-constant for the IPSC (Maex and De Schutter, 2003). The use of the same IPSC kinetics across the different values of the parallel-fiber conduction speed can therefore explain deviations from the predicted resonance frequencies.

In conclusion, the frequency of the oscillations in a slice preparation will critically depend on the average length of the preserved parallel fibers, and hence on the orientation of their plane of section (see also Maex and De Schutter, 1999, for the effect of varying the parallel-fiber length). Likewise, the oscillation frequency will increase with the parallel-fiber conduction speed, and hence with the recording temperature. The reported speeds vary from less than 0.2 m/s to 0.7 m/s, but we note that the present network model was able to reproduce the in vivo responses of rat Golgi cells to peripheral cutaneous stimuli when a 0.3 m/s conduction speed was used (Volny-Luraghi et al., 2002, and discussion therein).

Reciprocal inhibition between the interneurons of the molecular layer

The inhibitory interneurons of the molecular layer (stellate and basket cells) are connected through chemical and electrical synapses, probably forming local networks of synchronized activity (Palay and Chan-Palay, 1974; Kondo and Marty, 1998; Mann-Metzer and Yarom, 1999, 2000).

In a computational study, purely inhibitory networks were found to have a resonance frequency

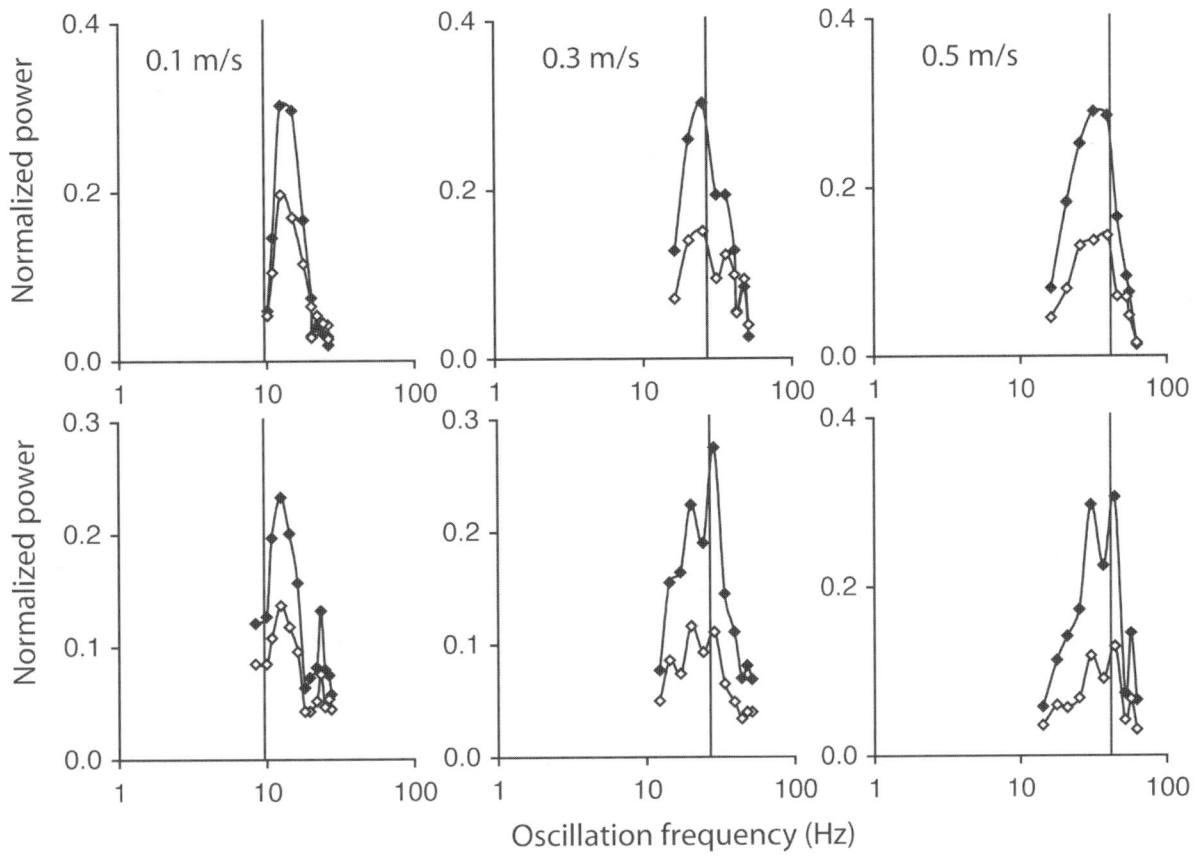

Fig. 2. The granular layer has a characteristic frequency determined by the parallel-fiber conduction speed. The panels show tuning curves obtained from simulations of networks with (upper panels) and without (lower panels) monosynaptic mossy-fiber excitation of Golgi cells, using a conduction speed for parallel fibers of 0.1 m/s (left), 0.3 m/s (middle) or 0.5 m/s (right). Each data point plots the oscillation frequency (horizontal axis) and the corresponding power (vertical axis), obtained at a particular level of excitation, in the population of Golgi cells (closed symbols) and granule cells (open symbols). The tuning curves were constructed by varying the average firing rate of mossy fibers by a factor of 1.41 between simulations. The vertical lines indicate frequencies of 9.6 Hz (speed 0.1 m/s), 26.8 Hz (0.3 m/s) and 41.7 Hz (0.5 m/s).

close to $1/(4d)$, with d the average delay time or latency of the evoked IPSCs (Maex and De Schutter, 2003). Electrotonic coupling through gap junctions improved network synchrony but did not affect the resonance frequency.

The confinement of the interneuron axons within the molecular layer, where they spread less than 300 μm (Sultan and Bower, 1998), the short latency of the IPSCs evoked in paired recordings (mean 1.7 ms at 20–22°C in Kondo and Marty, 1998), and the absence of excitatory neurons capable of polysynaptically interconnecting the interneurons, all suggest that the average connection delay within the interneuron network is small. Average delay times of 2.5 and 1 ms are predicted to induce resonance at 100 and 250 Hz, respectively. The fast time-course of the IPSCs (3.2 ms half-decay time at 34–35°C in Carter and Regehr, 2002) is near optimal to sustain oscillations in this frequency band.

Local field potential oscillations at 160 Hz, recorded in the cerebellar cortex of alert transgenic mice deficient for the Ca^{2+}-binding proteins calbindin and/or calretinin, were attributed to the interneuron network (Cheron et al., 2004). Older studies, discussed in Baçar (1998) and Isope et al. (2002), also reported cerebellar oscillations at frequencies of 180 Hz and beyond.

Recurrent Purkinje axon collaterals

A conspicuous finding of the study of Cheron et al. (2004) was that individual Purkinje cells, recorded from during the oscillations, produced regular trains of simple spikes with a rate close to the oscillation frequency (120–160 Hz). This contrasts with the high-frequency oscillations observed in hippocampus and neocortex, which during their occurrence typically suppressed the firing of simultaneously recorded pyramidal neurons (see Whittington and Traub, 2003).

Although an inhibitory network can generate a population rhythm of a frequency an order of magnitude greater than the mean firing rate of its component neurons (Brunel and Wang, 2003), the high discharge rate of the Purkinje cells in conjunction with the depth profile of field potentials reaching their peak amplitude close to the Purkinje cell bodies (Cheron et al., 2004) suggest that the inhibitory Purkinje cells might play an active role in rhythm generation, rather than being the passive recorders of synchronous IPSCs through the interneurons' synapses.

Recurrent axon collaterals make synapses on neighboring Purkinje cells (Palay and Chan-Palay, 1974) and were proposed to synchronize stripes of Purkinje cells (Hawkes and Leclerc, 1989). The recurrent axon is myelinated, has varicosities confined within 300 μm of the parent cell (O'Donoghue and Bishop, 1990) and forms synapses on somata or proximal dendrites, providing in this way a substrate for fast mutual inhibition. In support of a high resonance frequency for the Purkinje cell network is the observation that spontaneous IPSCs in Purkinje cells (although putatively induced by interneurons) are fast decaying as opposed to the typically slow time-course of the Golgi–granule cell IPSC (Puia et al., 1994). However, quantitative predictions require a physiological characterization of the Purkinje–Purkinje cell synapse, and a measurement of the latency and decay time of the IPSCs evoked in paired recordings. As Purkinje collaterals were demonstrated to inhibit also basket cells (O'Donoghue et al., 1989), it cannot be excluded that fast rhythms are generated by mixed networks containing several types of inhibitory neurons.

Finally, it is worth mentioning that the mere presence of mutual inhibition between the Purkinje cells would have a profound effect on the overall dynamics of the cerebellar cortex. In conjunction with the high spontaneous firing rate of the Purkinje cells and the off-beam inhibition which they receive from stellate cells, the cortex would resemble a circuit for temporal integration designed by Robinson and collaborators (Cannon et al., 1983; Robinson, 1989). Trying to explain temporal integration in the oculomotor system, these authors demonstrated that mutual inhibition increases the effective time-constant of a couple of neurons that receive complementary (push–pull) inputs, enabling them to almost mathematically integrate their inputs over time. Hence Purkinje–Purkinje cell synapses, if present, would extend the time-scale on which the cerebellar cortex is able to operate.

Conclusion

Recurrent inhibition from Golgi to granule cells induces 10–50 Hz oscillations in models of the granular layer, the resonance frequency being determined primarily by the length and conduction speed of the parallel fibers. In the molecular layer, reciprocal synapses between the interneurons are predicted to induce oscillations with a period length about four times the mean synaptic latency (100–250 Hz). Depending also on the strength and time-course of the poorly characterized synapses between Purkinje cells, similar frequencies can be generated in the Purkinje cell layer. Several rhythms observed in the cerebellum in vivo fall within these frequency bands. Although all rhythms generated in the cerebellar cortex, due to the nature of its neurons, probably depend on inhibitory synaptic interactions, a definitive elucidation of their mechanisms requires controlled in vitro experiments. Such experiments are also needed to resolve the discordance with the hippocampus and neocortex, where interneuron networks produce gamma-frequency oscillations (40 Hz) and where ripples (200 Hz) are accounted for on the basis of axoaxonal gap junctions between the pyramidal neurons (Whittington and Traub, 2003).

Abbreviations

IPSC   inhibitory postsynaptic current
IPSP   inhibitory postsynaptic potential

Acknowledgments

Mike Wijnants provided much appreciated computer assistance. This work was supported by grants from the FWO (Flanders), the IUAP (Belgium) and the European Union.

References

Baçar, E. (1998) Brain Function and Oscillations. Springer, Berlin.

Barlow, J.S (2002) The Cerebellum and Adaptive Control. Cambridge University Press, Cambridge, England.

Brunel, N. and Wang, X.J. (2003) What determines the frequency of fast network oscillations with irregular neural discharges? I. Synaptic dynamics and excitation-inhibition balance. J. Neurophysiol., 90: 415–430.

Buonomano, D.V. and Mauk, M.D. (1994) Neural network model of the cerebellum: temporal discrimination and the timing of motor responses. Neural Comput., 6: 38–55.

Cannon, S.C., Robinson, D.A. and Shamma, S. (1983) A proposed neural network for the integrator of the oculomotor system. Biol. Cybern., 49: 127–136.

Carter, A.G. and Regehr, W.G. (2002) Quantal events shape cerebellar interneuron firing. Nat. Neurosci., 5: 1309–1318.

Casado, M., Isope, P. and Ascher, P. (2002) Involvement of presynaptic N-methyl-D-aspartate receptors in cerebellar long-term depression. Neuron, 33: 123–130.

Chadderton, P.T., Margrie, T.W. and Hausser, M. (2003) Sensory stimuli evoke bursting in cerebellar granule cells *in vivo*. Program No. 75.17. Abstract Viewer/Itinerary Planner. Society for Neuroscience, Washington, DC. Online.

Cheron, G., Gall, D., Servais, L., Dan, B., Maex, R. and Schiffmann, S.N. (2004) Inactivation of calcium-binding protein genes induces 160 Hz oscillations in the cerebellar cortex of alert mice. J. Neurosci., 24: 434–441.

Courtemanche, R., Pellerin, J.-P. and Lamarre, Y. (2002) Local field potential oscillations in primate cerebellar cortex: modulation during active and passive expectancy. J. Neurophysiol., 88: 771–782.

D'Angelo, E., Nieus, T., Maffei, A., Armano, S., Rossi, P., Taglietti, V., Fontana, A. and Naldi, G. (2001) Theta-frequency bursting and resonance in cerebellar granule cells: experimental evidence and modeling of a slow K^+-dependent mechanism. J. Neurosci., 21: 759–770.

De Schutter, E. and Bjaalie, J.G. (2001) Coding in the granular layer of the cerebellum. Prog. Brain Res., 130: 279–296.

Dieudonné, S. (1998a) Submillisecond kinetics and low efficacy of parallel fibre-Golgi cell synaptic currents in the rat cerebellum. J. Physiol. (Lond.), 510: 845–866.

Dieudonné, S. (1998b) Étude fonctionnelle de deux interneurones inhibiteurs du cortex cérébelleux: les cellules de Lugaro et de Golgi. Thèse de doctorat de l'université Pierre et Marie Curie – Paris VI. (Unpublished doctoral thesis).

Eccles, J.C., Faber, D.S., Murphy, J.T., Sabah, N.H. and Taborikova, H. (1971) Afferent volleys in limb nerves influencing impulse discharges in cerebellar cortex. I. In mossy fibers and granule cells. Exp. Brain Res., 13: 15–35.

Finch, E.A. and Augustine, G.J. (1998) Local calcium signalling by inositol-1,4,5-trisphosphate in Purkinje cell dendrites. Nature, 396: 753–756.

Forti, L., Mapelli, J. and D'Angelo, E. (2003) Membrane mechanisms underlying intrinsic tonic firing of rat cerebellar Golgi cells. Program No. 171.13. Abstract Viewer/Itinerary Planner. Society for Neuroscience, Washington, DC. Online.

Franck, P., Maex, R. and De Schutter, E. (2001) Synchronization between patches of local excitation in a cerebellar granular layer model. Neurocomputing, 38–40: 595–599.

Hamann, M., Rossi, D.J. and Attwell, D. (2002) Tonic and spillover inhibition of granule cells control information flow through cerebellar cortex. Neuron, 33: 625–633.

Hansel, C., Linden, D.J. and D'Angelo, E. (2001) Beyond parallel fiber LTD: the diversity of synaptic and non-synaptic plasticity in the cerebellum. Nat. Neurosci., 4: 467–475.

Hartmann, M.J. and Bower, J.M. (1998) Oscillatory activity in the cerebellar hemispheres of unrestrained rats. J. Neurophysiol., 80: 1598–1604.

Hawkes, R. and Leclerc, N. (1989) Purkinje cell axon collateral distributions reflect the chemical compartmentation of the rat cerebellar cortex. Brain Res., 476: 279–290.

Isope, P., Dieudonné, S. and Barbour, B. (2002) Temporal organization of activity in the cerebellar cortex: a manifesto for synchrony. Ann. N.Y. Acad. Sci., 978: 164–174.

Ito, M. (2001) Cerebellar long-term depression: characterization, signal transduction, and functional roles. Physiol. Rev., 81: 1143–1195.

Jakab, R.L. and Hámori, J. (1988) Quantitative morphology and synaptology of cerebellar glomeruli in the rat. Anat. Embryol., 179: 81–88.

Kistler, W.M., van Hemmen, J.L. and De Zeeuw, C.I. (2000) Time window control: a model for cerebellar function based on synchronization, reverberation, and time slicing. Prog. Brain Res., 124: 275–297.

Kondo, S. and Marty, A. (1998) Synaptic currents at individual connections among stellate cells in rat cerebellar slices. J. Physiol. (Lond.), 509: 221–232.

Maex, R. and De Schutter, E. (1998a) Synchronization of Golgi and granule cell firing in a detailed network model of the cerebellar granule cell layer. J. Neurophysiol., 80: 2521–2537.

Maex, R. and De Schutter, E. (1998b) The critical synaptic number for rhythmogenesis and synchronization in a network model of the cerebellar granular layer. In: Niklasson, L., Boden, M. and Ziemke, T. (Eds.), ICANN 98 Proceedings of the 8th International Conference on Artificial Neural Networks. Springer-Verlag, London, pp. 361–366.

Maex, R. and De Schutter, E. (1999) An optimal connection radius for long-range synchronization. ICANN 99 Proceedings of the 9th International Conference on Artificial Neural Networks. IEE (Conference Publication No. 470), London, pp. 557–562.

Maex, R. and De Schutter, E. (2003) Resonant synchronization in heterogeneous networks of inhibitory neurons. J. Neurosci., 23: 10503–10514.

Maex, R., Vos, B.P. and De Schutter, E. (2000) Weak common parallel fibre synapses explain the loose synchrony observed between rat cerebellar Golgi cells. J. Physiol. (Lond.), 523: 175–192.

Mann-Metzer, P. and Yarom, Y. (1999) Electrotonic coupling interacts with intrinsic properties to generate synchronized activity in cerebellar networks of inhibitory interneurons. J. Neurosci., 19: 3298–3306.

Mann-Metzer, P. and Yarom, Y. (2000) Electrotonic coupling synchronizes interneuron activity in the cerebellar cortex. Prog. Brain Res., 124: 115–122.

Mitchell, S.J. and Silver, R.A. (2003) Shunting inhibition modulates neuronal gain during synaptic excitation. Neuron, 38: 433–445.

O'Donoghue, D.L. and Bishop, G.A. (1990) A quantitative analysis of the distribution of Purkinje cell axonal collaterals in different zones of the cat's cerebellum: an intracellular HRP study. Exp. Brain Res., 80: 63–71.

O'Donoghue, D.L., King, J.S. and Bishop, G.A. (1989) Physiological and anatomical studies of the interactions between Purkinje cells and basket cells in the cat's cerebellar cortex: evidence for a unitary relationship. J. Neurosci., 9: 2141–2150.

Palay, S.L. and Chan-Palay, V. (1974) Cerebellar Cortex. Springer-Verlag, New York.

Pellerin, J.-P. and Lamarre, Y. (1997) Local field potential oscillations in primate cerebellar cortex during voluntary movement. J. Neurophysiol., 78: 3502–3507.

Puia, G., Costa, E. and Vicini, S. (1994) Functional diversity of GABA-activated Cl^- currents in Purkinje versus granule neurons in rat cerebellar slices. Neuron, 12: 117–126.

Roberts, P.D. (1997) Stochastic recruitment in parallel fiber activity patterns. Beh. Brain Sci., 20: 263–264.

Robinson, D.A. (1989) Integrating with neurons. Annu. Rev. Neurosci., 12: 33–45.

Schweighofer, N., Spoelstra, J., Arbib, M.A. and Kawato, M. (1998) Role of the cerebellum in reaching movements. II. A neural model of the intermediate cerebellum. Eur. J. Neurosci., 10: 95–105.

Steuber, V. and De Schutter, E. (2002) Rank order decoding of temporal parallel fibre input patterns in a complex Purkinje cell model. Neurocomputing, 44: 183–188.

Sultan, F. and Bower, J.M. (1998) Quantitative Golgi study of the rat cerebellar molecular layer interneurons using principal component analysis. J. Comp. Neurol., 293: 353–373.

Tesche, C.D. and Karhu, J.J.T. (2000) Anticipatory cerebellar responses during somatosensory omission in man. Hum. Brain Mapp., 9: 119–142.

Thach, W.T., Kane, S.A., Mink, J.W. and Goodkin, H.P (1992) Cerebellar output: multiple maps and modes of control in movement coordination. In: Llinás, R. and Sotelo, C. (Eds.), The Cerebellum Revisited. Springer-Verlag, New York, pp. 283–300.

Volny-Luraghi, A., Maex, R., Vos, B.P. and De Schutter, E. (2002) Peripheral stimuli excite coronal beams of Golgi cells in rat cerebellar cortex. Neuroscience, 113: 363–373.

Vos, B.P., Maex, R., Volny-Luraghi, A. and De Schutter, E. (1999) Parallel fibers synchronize spontaneous activity in cerebellar Golgi cells. J. Neurosci., 19: RC6.

Whittington, M.A. and Traub, R.D. (2003) Interneuron diversity series: inhibitory interneurons and network oscillations in vitro. Trends Neurosci., 26: 676–682.

CHAPTER 16

Gap junctions synchronize synaptic input rather than spike output of olivary neurons

W.M. Kistler and C.I. De Zeeuw*

Department of Neuroscience, Erasmus MC, P.O. Box 1738, 3000 DR Rotterdam, The Netherlands

Keywords: cerebellum; complex spikes; Connexin 36; coupling; climbing fibers

Abstract: Electronic coupling in the inferior olive is supposed to underlie the synchrony of complex spike activities of Purkinje cells in the cerebellar cortex. Here we show a computational model which suggests that the olivary gap junctions may synchronize the input rather than the neuronal output. As such, coupling may influence the absolute moment in time of the complex spike activity rather than their synchrony.

Introduction

Neurons in the inferior olivary (IO) nucleus of the brain stem are densely coupled by dendritic gap junctions (Sotelo et al., 1974). Traditionally, these gap junctions are thought to synchronize action potentials among neighboring olivary cells; a mechanism that would lead to synchronous complex spikes in the cerebellar cortex. Purkinje cells from a common parasagittal zone do indeed show a tendency to fire complex spikes synchronously (Bell and Kawasaki, 1972; Llinás and Sasaki, 1989; Sasaki et al., 1989; Lang et al., 1996) recent experimental results, however, cast doubts on the involvement of olivary gap junctions in complex spike synchronization (Kistler et al., 2002). As it has already been stated 30 years ago in the original paper by Bell and Kawasaki (1972), complex spikes are usually only roughly synchronized with 'broad symmetric peaks (about 50 ms wide)' in the cross-correlograms. Obviously, fast electrotonic coupling is not required to synchronize complex spikes with a rather poor precision of a few tens of milliseconds. Moreover, the absence of functional gap junctions in connexin36 null-mutant mice does not affect the degree of complex spike synchrony (Kistler et al., 2002). Apart from the rough form of complex spike synchrony, exceptional cases have been observed where the complex spike onsets in two different Purkinje cells differ by less than a millisecond and where every complex spike in one cell is accompanied by a complex spike in the other cell (Bell and Kawasaki, 1972; Kistler et al., 2002). This effect is most likely due to an innervation of both the Purkinje cells by climbing fiber collaterals from the same IO neuron and is thus independent of gap junctional coupling.

An indirect evidence against the role of olivary gap junctions in synchronizing complex spikes is provided by the morphology of the inferior olive. Olivary gap junctions are predominantly located on the dendritic spines far from the soma where they would have the greatest efficiency in synchronizing the activity of two coupled cells. Since the spines are also a major target of the GABAergic projections from the deep cerebellar nuclei (De Zeeuw et al., 1989), it has been suggested that synaptic activity can

*Corresponding author. Tel.: +31(10)4087299;
Fax: +31(10)4089459; E-mail: c.dezeeuw@erasmusmc.nl

modulate the strength of the electrotonic coupling by shunting the transjunctional current and thus 'sculpturing' dynamical clusters of IO neurons as they are required for motor coordination (Llinás, 1991; Welsh et al., 1995; Lang et al., 1999). This suggestion conflicts with the observation that the electrical conductance of olivary gap junctions is exceptionally small (Srinivas et al., 1999) and that even in the unshunted situation, the coupling coefficient of two IO neurons is well below 5% (Devor and Yarom, 2002).

As an alternative to the classical role of olivary gap junctions, we suggest a radically different paradigm where gap junctions synchronize the synaptic input to the neighboring IO neurons rather than their output spikes. This interpretation is supported by several observations. First, gap junctions directly couple spines of different IO neurons rather than their somata. The coupling coefficient between the spine heads is therefore much larger than the coupling coefficient of the somata. Second, dendritic spines are the very first step in the integration of synaptic input in IO neurons. It is thus very likely that the gap junctions play a major role in synaptic integration. Third, the connexin36 null-mutants do not show any obvious behavioral phenotype in motor performance (Güldenagel et al., 2001; Kistler et al., 2002). While it is difficult to envision any compensatory mechanisms that can substitute electrotonically induced synchronization of action potentials, there are ample possibilities to compensate the effects of missing gap junctions for synaptic integration such as a raised number of collaterals or synaptic transmission sites.

In the following, we present a computational model of electrotonically coupled olivary neurons based on a two-compartment model developed by Schweighofer et al. (1999). We show that active dendritic spines can exhibit an all-or-none response that can be conveyed via gap junctions to the neighboring IO neurons. The synaptic terminals and dendritic spines are closely packed together in glomerular structures. The glomeruli form functional units in the sense that all the spines within a glomerulus are excited collectively in an all-or-none fashion due to the gap junctional coupling. Our simulations indicate that the synaptic input is integrated *locally* at the glomeruli before activity from all glomeruli is integrated at the soma. This has interesting implications on how basic computations are performed in the inferior olive, for example with respect to the reliability of the synaptic transmission.

Methods

Simulations are based on the active membrane model of olivary neurons developed by Schweighofer et al. (1999). The original model consists of two compartments, one compartment for the soma containing fast sodium channels, low-threshold calcium conductances, a delayed-rectifying potassium current and the h-current; and the other compartment representing the dendritic tree with high-threshold calcium and calcium activated potassium channels. In addition, all compartments carry passive leak conductances. We have extended this model by additional compartments for the spine heads, the spine necks and the dendritic shaft the spines are attached to. The spine neck and the dendritic shaft were described as a passive membrane. The spine head was equipped with fast sodium channels (Na = 140), high-threshold calcium channels ($g_{Ca\_h} = 4.0$), and calcium-activated potassium channels ($g_{K\_Ca} = 35$; all conductances in mS/cm^2). The kinetic properties of these channels were the same as in the original model. All compartments had a cylindrical shape with the following dimensions (length × diameter): Dendritic shaft (200 μm × 1 m), spine neck (3 μm × 0.2 μm), spine head (1 μm × 1 μm). These values have been chosen so as to mimic morphological data (De Zeeuw et al., 1989, 1990). The synaptic input to the spine head was described by time-dependent conductances where we have assumed — for want of experimental data — typical values for rise and decay time of the postsynaptic currents: 1 ms/10 ms (GABA), 0.09 ms/1.5 ms (AMPA), and 3 ms/40 ms (NMDA). GABA and AMPA receptor gated channels had a reversal potential of −75 mV and 0 mV, respectively. The magnesium block of NMDA receptor gated ion channels was taken into account via a nonlinear dependency of the synaptic current on the membrane potential, $I \propto [1 + 0.336 \exp(0.062\, v)]^{-1}$ (Gabbiani et al., 1994). The gap junctions connected the spine heads from different neurons and had a conductance of 1 nS (Srinivas et al., 1999).

The simulation of several, electrotonically coupled IO neurons requires the solution of high-dimensional nonlinear differential equations. This has been done with Mathematica (Wolfram, 1991) and custom made C-programs based on the Bulirsch–Stoer method adapted for stiff differential equations (Press et al., 1992) running on a Linux PC.

Results

Olivary neurons are known to have extraordinarily long and thin dendritic spines (De Zeeuw et al., 1998). The electrotonic coupling of two IO cells is thus not only restricted by the low conductance of the gap junction but also by the cytosolic resistance of the spine neck. The typical values for the length (0.5–6 μm) and diameter (~0.2 μm) of a spine neck (De Zeeuw et al., 1989, 1990) and specific resistance of the cytosol (400 Ω cm) give a resistance of the spine neck of 0.06–0.8 GΩ. The total resistance between two IO cells is of the order of 0.7–8 GΩ (Devor and Yarom, 2002). Thus, the cytosolic resistance of the spine necks contributes substantially to the overall electrical resistance between two electronically coupled IO neurons.

In order to study the electrical properties of the coupled IO cells, we have built an active membrane model with separate compartments for the somata, dendrites, spine necks, and heads. We determined the coupling coefficient (ratio of pre- and post-junctional voltage response) of two IO cells by simulating the somatic voltage response to current injection into the soma of the pre-junctional cell. In keeping with the simultaneous double patch recordings in slices (Devor and Yarom, 2002) the (somatic) coupling coefficients (cc) were rather low. Cells coupled by a single gap junction had $cc_{soma} = 0.01$; cells coupled via several spines had an accordingly larger coupling coefficient. We also determined the coupling coefficients for the spine heads that are directly coupled by a gap junction. To this end we 'injected' a constant current into one spine head and simulated the voltage response at the post-junctional spine. As expected, the coupling coefficients for the spine heads ($cc_{spine} = 0.40$) are more than one order of magnitude larger than those for the somata because the current does not have to travel all the way through dendrites and spine necks; cf. Fig. 1A. Even a weak electrotonic coupling can thus have a huge effect on the two spine

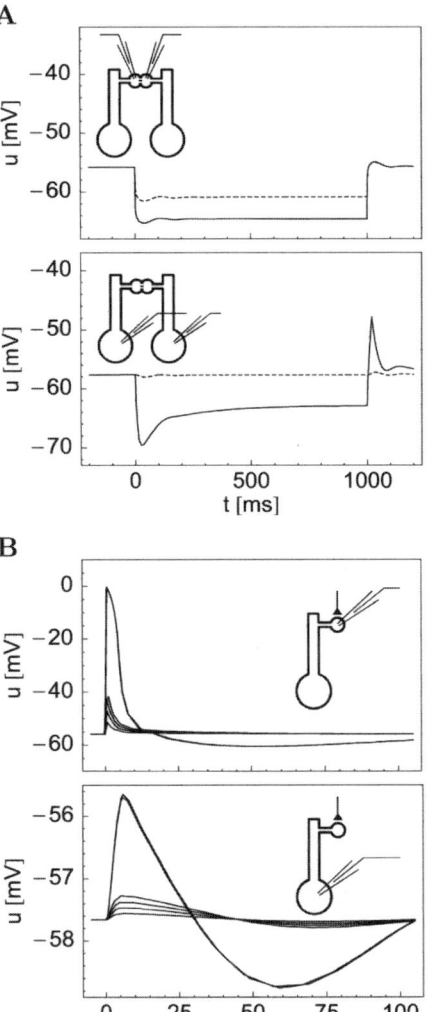

Fig. 1. (A) Gap junctional coupling of spines in the IO results in a much tighter coupling of the spine heads than of the somata. The upper panel shows the simulated voltage response of the pre- (lower trace) and the post-junctional (upper trace) spine head to current injection into the pre-junctional spine. The coupling coefficient in this case is cc = 0.40. The lower panel shows somatic membrane potentials in a similar simulation where current is injected into one of the somata. The coupling coefficient of the somata amounts to only cc = 0.01. The insets indicate the configuration of the electrodes. (B) Local all-or-none response of spines to excitatory synaptic input. The panels show the membrane potential at the spine head (upper panel) and at the soma (lower panel) in response to the activation of a single excitatory synapse at time $t = 0$. Different traces correspond to different amplitudes of the postsynaptic current. The insets indicate the configuration of the synapse and the recording electrode.

heads though it hardly affects the electrically well separated somata.

It is well-established that the dendrites carry voltage-dependent ion channels that endow them with a rich repertoire of nonlinear properties. In particular, it has been shown that the spines of the olivary neurons, pyramidal cells, and Purkinje cells contain voltage-gated calcium channels that lead to local all-or-none responses to the excitatory synaptic input (Llinás and Yarom, 1981; Eilers et al., 1995; Wang et al., 2000), a behavior that has been predicted on purely theoretical grounds more than 25 years ago (Jack et al., 1975); see also (Segev and Rall, 1988). We were able to reproduce this phenomenon in our model by equipping the membrane of the spine heads with fast sodium channels and the high-threshold calcium current as described in Methods section. The resulting behavior is illustrated in Fig. 1B. We have simulated the response of the neuron to the activation of an excitatory synapse located at the spine head. Activation of a weak synapse (low peak postsynaptic current) leads to a subthreshold response that has hardly any effect on the membrane potential at the soma. A small increase in the amplitude of the postsynaptic current, however, triggers a full-blown calcium spike in the spine and gives rise to a post-synaptic potential of a few millivolts at the soma. The active electrical properties of the spine head membrane amplify the tiny current that flows through the synaptically activated ion channels so as to generate a macroscopic postsynaptic potential at the soma.

Interestingly, the all-or-none response at the spines cannot only be triggered by direct synaptic input but also be triggered indirectly through a gap junction. Figure 2 shows results from a simulation of three IO cells that are mutually coupled via spines that are part of a common glomerulus. We have studied the neuronal response to the synaptic input to only *one* of the three spines within the glomerulus. The neuron that is directly innervated shows the typical all-or-none behavior, depending on the amplitude of the postsynaptic current as described above. As long as the amplitude of the postsynaptic current stays subthreshold, there is only a small and brief response in the spine, which hardly affects the soma and the post-junctional cells. However, as soon as the threshold is exceeded, a local calcium spike is generated and the transjunctional current triggers a

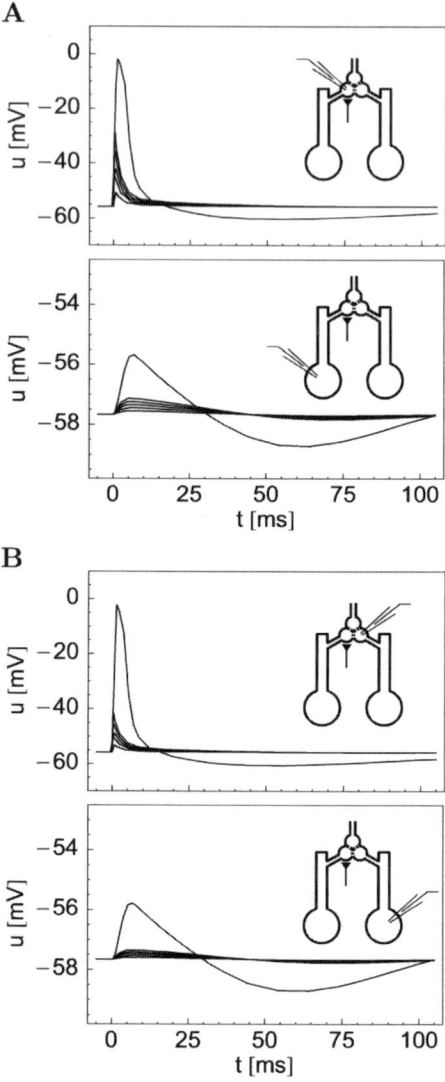

Fig. 2. Glomeruli form functional units. Simulation of three mutually coupled IO neurons. (A) Synaptic input to one of the three spines in the glomerulus results in an all-or-none response depending on the amplitude of the postsynaptic current (similar plots as in Fig. 1B). (B) Due to the electrotonic coupling the same all-or-none response is triggered simultaneously in all post-junctional spines (upper panel). The somatic potential (lower panel) is indistinguishable from the postsynaptic potential of the directly innervated cell.

similar response in all post-junctional spines as well (Fig. 2B). Spines within a glomerulus behave as a functional unit in that they respond collectively to the excitatory synaptic input even if only a few of them are directly innervated. Synaptic input to one of them is conveyed to all the postglomerular neurons and the somatic potentials are indistinguishable from direct synaptically evoked potentials (lower panels in Fig. 2).

The fact that olivary glomeruli form the functional units has important implications on how the synaptic input to IO neurons is integrated. Our simulations indicate that synaptic integration consists of two steps. In the first step, the synaptic input is integrated *locally* at the glomeruli. In the second step, the synaptically activated glomeruli are integrated at the soma and an action potential is either fired or not fired. Figure 3 shows the results from a simulation of an IO neuron that is coupled to six other IO neurons via three different glomeruli. The excitatory synaptic input to one of the glomeruli triggers the all-or-none response in all of its spines and, consequently, a unitary response at the soma. The somatic potentials that are generated by different glomeruli sum up linearly (upper panel in Fig. 3A). However, if the same number of spikes impinges onto a single glomerulus, no summation takes place because the all-or-none response is triggered only in this glomerulus, irrespective of the number of activated synapses (lower panel in Fig. 3A).

The fact that the synaptic input is integrated locally at the glomeruli can be seen most clearly in the simulations shown in Fig. 3B. Both panels show the somatic membrane potential that results from the simultaneous activation of one excitatory and one inhibitory synapse. In the upper panel, both the synapses are located on two different spines within the *same* glomerulus. In this case excitatory and inhibitory inputs cancel each other and no all-or-none response is triggered; the soma remains at the resting potential. In the lower panel, the excitatory and inhibitory input impinge on the two spines from two different glomeruli. The excitatory input triggers the all-or-none response in all the spines of the corresponding glomerulus; the inhibitory input, on the other hand, does not have any effect on the glomerulus ('silent inhibition') nor on the soma because of the large cytosolic resistance of the spine neck. Altogether, there is an excitatory postsynaptic potential at the soma. In summary, it clearly makes a difference whether synaptic input reaches the neuron via a common glomerulus or whether the input is distributed over several distinct glomeruli — a phenomenon that is incompatible with the simplistic

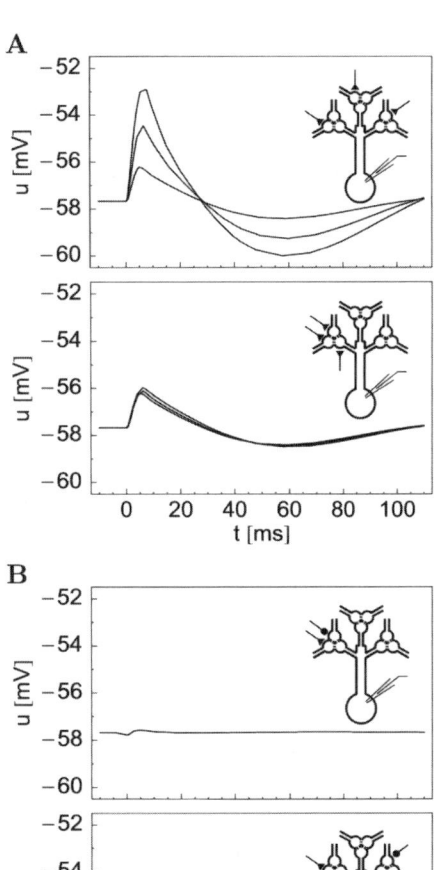

Fig. 3. Synaptic integration in IO neurons. (A) Integration of excitatory synaptic input depends on the position of the synapses. The upper panel shows somatic potentials generated by one, two, and three (traces) presynaptic spikes that impinge simultaneously on three different glomeruli. Individual EPSPs sum up linearly. If the same spikes impinge onto the same glomerulus virtually no summation takes place (lower panel). (B) Integration of excitatory and inhibitory synaptic input depends on the position of the synapses. In the upper panel an excitatory and an inhibitory synapse within the same glomerulus are activated simultaneously so that their effects cancel each other. In the lower panel excitatory and inhibitory synapses are located in two different glomeruli. In this case excitatory and inhibitory input does not cancel and there is a clear EPSP at the soma. Insets indicate the configuration of the cell, the recording electrode, and excitatory (triangular symbols) and inhibitory (circular symbols) synapses.

idea that synaptic input is collected by the dendritic tree and integrated at the soma; see also Rapp et al. (1996); Segev and Rall (1998).

Discussion

We have shown that gap junctions can play a major role in the integration of synaptic input in the inferior olive. This conclusion is based on two prerequisites. First, the gap junctions couple spine heads more tightly than the corresponding somata. Given the morphology of olivary neurons there is no doubt that this is actually the case. Second, the excitable spines exhibit an all-or-none response to the synaptic input. While this has been studied extensively in the computational models (Jack et al., 1975; Miller et al., 1985; Perkel and Perkel, 1985; Segev and Rall, 1988, 1998) there is no direct experimental evidence so far that this is the case in olivary neurons. Sodium and calcium imaging experiments in the hippocampal pyramidal cells and cerebellar Purkinje cells (Denk et al., 1995; Eilers et al., 1995; Yuste and Denk, 1995; Wang et al., 2000; Rose and Konnerth, 2001), however, indicate that the dendritic spines indeed produce local super-linear responses to synaptic stimulation. We think that it is quite natural to assume that this holds true for the olivary cells as well, because in these cells, the high-threshold calcium current is actually located on the dendrites (Llinás and Yarom, 1981).

The parameters in the simulations have been chosen so as to meet the two conditions for electrotonic coupling of spines and their excitability. We found that there is a wide range of parameters that produce the desired behavior; especially different combinations of spine morphology and ion channel densities result in the very same form of all-or-none responses at the spine (Perkel and Perkel, 1985; Segev and Rall, 1988). The conclusions drawn from our simulations are thus generic, in the sense that, they do not depend on a particular set of parameters but only on the excitability of spines and the fact that spines from different cells can be tightly coupled by gap junctions, whereas different spines from the same cell are electrically well-separated by the large resistance of their spine necks.

One of the principal findings of the present study is that the excitatory postsynaptic potentials in the olivary neurons can be elicited indirectly through gap junctions and synaptic input to post-junctional spines. The activation of the local all-or-none response in one of the post-junctional spines gives rise to a current through the gap junction. The nonlinear properties of the spine membrane amplify this current so as to trigger the very same all-or-none response in the pre-junctional spine and the corresponding potential at the soma. Hence, glomeruli form functional units in the sense that excitatory synaptic input to some of its spines can trigger unitary somatic potentials in all post-glomerular cells collectively.

Dendritic spines from 5 to 8 different IO neurons and about the same number of both the GABAergic and non-GABAergic fibers intermingle in glomerular structures (De Zeeuw et al., 1990). Each fiber makes several synaptic contacts with different spines within the glomerulus. The redundancy of multiple synaptic innervation and mutual gap junctional coupling within a glomerulus results in a substantial gain in synaptic transmission reliability. Suppose that there are six different spines that are mutually coupled by gap junctions and that all of them receive excitatory synaptic input from the same presynaptic cell, even if each individual synapse has a failure probability of 50%, the overall probability that transmission fails is only $0.5^6 = 0.016$, because a successful transmission across a single synapse suffices to trigger the somatic potential in all post-glomerular neurons. The reliability of synaptic transmission may well be an issue in the olivo-cerebellar system given the close-to-perfect transmission at the next stage, the climbing fiber-Purkinje cell synapse [see Kistler and De Zeeuw (2002) for a functional interpretation]. On the other hand, an indirect triggering of somatic potentials via gap junctions can substitute direct synaptic innervation in a situation where for geometric reasons a synaptic contact is not possible. It has been reasoned that spines are used to maximize the number of potential synapses (Swindale, 1981; Stepanyants et al., 2002). The indirect triggering of somatic potentials via gap junctions may be a mechanism to boost the effective synaptic connectivity within the already pretty crowded glomeruli while economizing neurotransmitter and synaptic channels.

The collective all-or-none type of response of olivary glomeruli clearly affects the way the synaptic input is integrated. We have shown that this is a two-step process where the synaptic input is first integrated in each glomerulus individually before the all-or-none responses of all glomeruli are integrated at the soma. This may be a mechanism to accurately subtract the excitatory and inhibitory input; a goal that would be difficult to achieve, if the excitatory and inhibitory synapses were spread over the whole dendritic tree. A similar function has been proposed for the synaptic glomeruli of the cerebellar cortex where excitatory mossy fiber synapses and inhibitory Golgi cell terminals meet at the dendritic claws of granule cells (Kistler and van Hernmen, 1999; Kistler et al., 2000).

The new view of gap junctional function put forward by the present study may also help to solve the puzzle that connexin36 knockout mice, i.e., mice without functional neuronal gap junctions, have no obvious behavioral phenotype (Güldenagel et al., 2001; Kistler et al., 2002). It is difficult to envision compensatory mechanisms that can generate electrotonically mediated synchronization of action potentials other than gap junctions. The above described role of gap junctions in synaptic integration, however, can at least partially be taken over by altered synaptic innervation patterns in the glomeruli. A raised number of synapses within each glomerulus and/or an increase in the number of afferent collaterals can lead to a collective activation of all glomerular spines even if they are not coupled by gap junctions. The primary effect of the absence of gap junctions is thus not a reduction of complex spike synchrony or firing rate but rather a deterioration of the reliability with which a given Purkinje cell is firing a complex spike in response to sensory input. This is particularly interesting in the context of the experimentally observed long-latency responses following electrical stimulation of the mesodiencephalic junction or the motor cortex (Ruigrok and Voogd, 1995; Schwarz and Welsh, 2001). It has been suggested that these long-latency responses are due to a reverberating activity in the olivo-cerebellar loop and might play an important role in establishing a dynamical short-term memory (Kistler and De Zeeuw, 2003). It has also been shown that the capacity of the dynamical short-term memory depends critically on the reliability of synaptic transmission within the reverberating loop (Kistler and De Zeeuw, 2002). From these observations we would expect an impaired performance of the connexin36 knock-out mice in a classical delay-conditioning paradigm. The forthcoming experiments will have to show if these predictions hold valid.

Conclusions

Among the functions commonly attributed to gap junctions is the fast and accurate synchronization of action potentials within a population of electrotonically coupled neurons. In particular, the high density of gap junctions in the inferior olive (IO) is thought to underlie synchronous complex spike responses in cerebellar Purkinje cells. Recent experiments in knock-out mice that lack functional gap junctions, however, have questioned this interpretation. In the present study we thus develop a radically different view of gap junctional function. Based on the morphology — primarily the co-location of synaptic terminals and gap junctions on dendritic spines — we suggest that olivary gap junctions play a major role in the integration of a synaptic input rather than in the synchronization of output spikes. We show by means of a computational model (i) that the dendritic spines can be tightly coupled by gap junctions while the corresponding somata are electrically well-separated and (ii) that the spines equipped with voltage-dependent ion channels can generate local 'micro spikes' in an all-or-none fashion that can be triggered either directly by the synaptic input or, indirectly by the synaptic input to a neighboring spine attached by a gap junction. At the soma, the directly and indirectly triggered potentials are indistinguishable. Due to the dense gap junctional coupling glomeruli form the functional units, in the sense that all spines are collectively activated even by synaptic input to only some of them. This has interesting implications for the integration of the synaptic input in IO cells and for reliability of the synaptic transmission.

Acknowledgments

Part of this work was finished at the Interdisciplinary Center for Neural Computation of the Hebrew University Jerusalem, Israel, and it is my (W.M.K.) pleasure to thank Idan Segev and Yosef Yarom for their hospitality and inspiring discussions. The work was supported financially by the ARI program of the EC and a NWO pioneer grant to C.I. De Zeeuw.

References

Bell, C.C. and Kawasaki, T. (1972) Relations among climbing fiber responses of nearby Purkinje cells. J. Neurophysiol., 35: 155–169.

De Zeeuw, C.I., Holstege, J.C., Ruigrok, T.J. and Voogd, J. (1989) Ultrastructural study of the gabaergic, cerebellar, and mesodiencephalic innervation of the cat medial accessory olive: anterograde tracing combined with immunocytochemistry. J. Comp. Neurol., 284: 12–35.

De Zeeuw, C.I., Ruigrok, T.J., Holstege, J.C. and Jansen, J.V. (1990) Intracellular labeling of neurons in the medial accessory olive of the cat: 11. Ultrastructure of dendritic spines and their GABAergic innervation. J. Comp. Neurol., 300: 478–494.

De Zeeuw, C.I., Simpson, J.I., Hoogenraad, C.C., Galjart, N., Koekkoek, S.K.E. and Ruigrok, T.J.H. (1998) Microcircuitry and function of the inferior olive. Trends Neurosci., 21: 391–400.

Denk, W., Sugimori, M. and Llinás, R. (1995) Two types of calcium response limited to single spines in cerebellar Purkinje cells. Proc. Natl. Acad. Sci. USA, 92: 8279–8282.

Devor, A. and Yarom, Y. (2002) Electrotonic coupling in the inferior olivary nucleus revealed by simultaneous double patch recordings. J. Neurophysiol., 87: 3048–3058.

Eilers, J., Augustine, G.J. and Konnerth, A. (1995) Subthreshold synaptic signalling in fine dendrites and spines of cerebellar Purkinje neurons. Nature, 373: 155–158.

Gabbiani, F., Midtgaard, J. and Knoepfl, T. (1994) Synaptic integration in a model of cerebellar granule cells. J. Neurophysiol., 72: 999–1009.

Güldenagel, M., Ammermüller, J., Feugenspan, A., Teubner, B., Degen, J., Söhl, G., Willecke, K. and Weiler, R. (2001) Visual transmission deficits in mice with targeted disruption of the gap junction gene connexin36. J. Neurosci., 21: 6036–6044.

Jack, J.J.B., Noble, D. and Tsien, R.W. (1975) Electric current flow in excitable cells. Clarendon Press.

Kistler, W.M., De Zeeuw. C.I. (2002) Dynamical working memory and timed responses: The role of reverberating loops in the olivo-cerebellar system. Neural Comput. 14: 2597–2626.

Kistler, W.M. and De Zeeuw, C.I. (2003) Time windows and everberating loops: A reverse-engineering approach to cerebellar function. The Cerebellum, 2: 44–54.

Kistler, W.M. and van Hermmen, J.L. (1999) Delayed reverberation through time windows as a key to cerebellar function. Biol. Cybern., 81: 373–380.

Kistler, W.M., van Hermmen, J.L. and De Zeeuw, C.I. (2000) Time window control: A model for cerebellar function based on synchronization, reverberation, and time slicing. Progr. Brain Res., 124: 275–297.

Kistler, W.M., De Jeu, M.T., Elgersma, Y., Van Der Giessen, R.S., Hensbroek, R., Luo, C., Koekkoek, S.K.E., Hoogenraad, C.C., Hamers, F.P., Gueldenagel, M., Sohl, C., Willecke, K., and De Zeeuw, C.I. (2002) Analysis of Cx36 knockout does not support the tenet that olivary gap junctions are required for complex spike synchronization and normal motor performance. Ann. NY. Acad. Sci., 978: 391–404.

Lang, E.J., Sugihara, I. and Llinás, R. (1996) GABAergic modulation of complex spike activity by the cerebellar nucleoolivary pathway in rat. J. Neurophysiol., 76: 255–275.

Lang, E.J., Sugihara, I., Welsh, J.P. and Llinás, R. (1999) Patterns of spontaneous Purkinje cell complex spike activity in the awake rat. J. Neurosci., 19: 2728–2739.

Llinás, R. (1991) The noncontinuous nature of movement execution. In: Humphrey, D.R. and Freund, H.J. (Eds.), Motor Control: Concepts and Issues. Wiley, New York, pp. 223–242.

Llinás, R. and Sasaki, K. (1989) The functional organization of the olivo-cerebellar system as examined by multiple Purkinje cell recordings. Eur. J. Neurosci., 1: 587–602.

Llinás, R. and Yarom, Y. (1981) Properties and distribution of ionic conductances generating electroresponsiveness of mammalian inferior olivary neurones *in vitro*. J. Physiol. Lond., 315: 569–584.

Miller, J.P., Rall, W. and Rinzel, J. (1985) Synaptic amplification by active membrane in dendritic spines. Brain Res., 325: 325–330.

Perkel, D.H. and Perkel, D.J. (1985) Dendritic spines: role of active membrane in modulating synaptic efficacy. Brain Res., 325: 331–335.

Press, W.H., Teukolsky, S.A., Vetterling, W.T. and Flannery, B.P. (1992) Numerical Recipes in C, second edition. Cambridge University Press.

Rapp, M., Yarom, Y. and Segev, I. (1996) Modeling back propagating action potential in weakly excitable dendrites of neocortical pyramidal cells. Proc. Natl. Acad. Sci. USA, 93: 11985–11990.

Rose, C.R. and Konnerth, A. (2001) NMDA receptor-mediated Na+ signals in spines and dendrites. J. Neurosci., Jun 15; 21(12): 4207–4214.

Ruigrok, T.J.H. and Voogd, J.Cerebellar influence on olivaXy excitability in the cat Eur. J. Neurosci., 7: 679–693.

Sasaki, K., Bower, J.M. and Llinas, R. (1989) Multiple Purkinje cell recording in rodent cerebellar cortex. Eur. J. Neurosci., 1: 572–586.

Schwarz, C. and Welsh, J.P. (2001) Dynamic modulation of mossy fiber system throughput by inferior olive synchrony: a multielectrode study of cerebellar cortex activated by motor cortex. J. Neurophysiol., 86: 2489–2504.

Schweighofer, N., Doya, K. and Kawato, M. (1999) Electrophysiological properties of inferior olive neurons: A compartmental model. J. Neurophysiol., 82: 804–817.

Segev, I. and Rall, W. (1988) Computational study of an excitable dendritic spine. J. Neurophysiol., 60: 499–523.

Segev, I. and Rall, W. (1998) Excitable dendrites and spines: earlier theoretical insights elucidate recent direct observations. Trends Neurosci., 21: 453–460.

Sotelo, C., Llinás, R. and Baker, R. (1974) Structural study of inferior olivary nucleus of the cat: Morphological correlations of electrotonic coupling. J. Neurophysiol., 37: 541–559.

Srinivas, M., Rozental, R., Kojima, T., Dermietzel, R., Mehler, M., Condorellia, D.F., Kessler, J.A. and Spray, D.C. (1999) Functional properties of channels formed by the neuronal gap junction protein connexin36. J. Neurosci., 19: 9848–9855.

Stepanyants, A., Hof, P.R. and Chklovskii, D.B. (2002) Geometry and structural plasticity of synaptic connectivity. Neurod, 34: 275–288.

Swindale, N.V. (1981) Dendritic spines only connect. Trends Neurosci., 4: 240–241.

Wang, S.S., Denk, W. and Hausser, M. (2000) Coincidence detection in single dendritic spines mediated by calcium release. Nat. Neurosci., 3: 1266–1273.

Welsh, J.P., Lang, E.J., Sugihara, I. and Llinás, R. (1995) Dynamic organization of motor control within the olivocerebellar system. Nature, 374: 453–457.

Wolfram, S. (1991) Mathematica: a system for doing mathematics by computer, second edition. Addison-Wesley Publishing Company.

Yuste, R. and Denk, W. (1995) Dendritic spines as basic functional units of neuronal integration. Nature, 375: 682–684.

SECTION VI

Cerebellar Motor Control

CHAPTER 17

Is the cerebellum ready for navigation?

Laure Rondi-Reig* and Eric Burguière

Laboratoire de Physiologie de la Perception et de l'Action, UMR CNRS 7124, 11 place Marcellin Berthelot, Collège de France, 75005 Paris, France

Keywords: cerebellum; navigation; declarative learning; procedural learning; LTD

Abstract: Spatial navigation required the acquisition of at least two complementary processes: the organization of the spatial representation of the environment (declarative learning) and the acquisition of a motor behaviour adapted to the specific context (procedural learning). The potential role of the cerebellum in spatial navigation is part of the debate concerning its role in cognitive function. Experiments ranging from cerebellar patients to animal models have indicated that cerebellar damage affects the processing of spatial information. The main unresolved issue concern the interpretation of these deficits. Is the cerebellum involved in both declarative and procedural components of navigation? Could all deficits in navigation paradigms be interpreted by a deficit in a motor-dependant process? The purpose of this review is to examine different results coming from anatomical data, experimental paradigms and models in order to give a critical answer to this question.

The cerebellum has first been considered to contribute to motor control and coordination (Holmes, 1939). Since then, many other functions have been attributed to the cerebellum, such as motor learning (Thompson and Kim, 1996; Thach, 1998) or cognitive functions (Leiner et al., 1993) including spatial navigation (Lalonde, 1997; Petrosini et al., 1998; Rondi-Reig et al., 2002). However, the role of the cerebellum in cognitive function is still a matter of strong debate. Why is the road to prove the role of the cerebellum in non-motor functions so long and difficult? A recurrent problem is the difficulty in dissociating motor deficits per se from cognitive impairment in paradigms in which the measured variable is behavior. Is there clear evidence in favor of the involvement of the cerebellum in cognition and more particularly in navigation? By contrast, could all cognitive dysfunctions be explained by a deficit in a common (motor-dependant?) process for which cerebellar patients or animal models display a deficiency? This review attempts to give a critical, although not exhaustive, overview of different results coming from anatomical data, experimental paradigms and models showing a possible implication of the cerebellum in navigation.

Spatial navigation: a dual-process function

Spatial navigation is a cognitive function that can be defined as a self-controlled movement in space toward a non-visible goal. This function requires the integration of both internal (vestibular, proprioceptive, or kinesthetic) and external (visual, olfactory, or auditory) sensory–motor information. Although the first observations concerning the potential role of the cerebellum in cognitive function were made in humans, the putative role of the cerebellum in navigation essentially comes from experiments with animal models presenting a cerebellar impairment. Traditional paradigms used are mazes in which the

*Corresponding author. Tel.: +331-44-27-14-30; Fax: +331-44-27-13-82; E-mail: laure.rondi-reig@college-de-france.fr

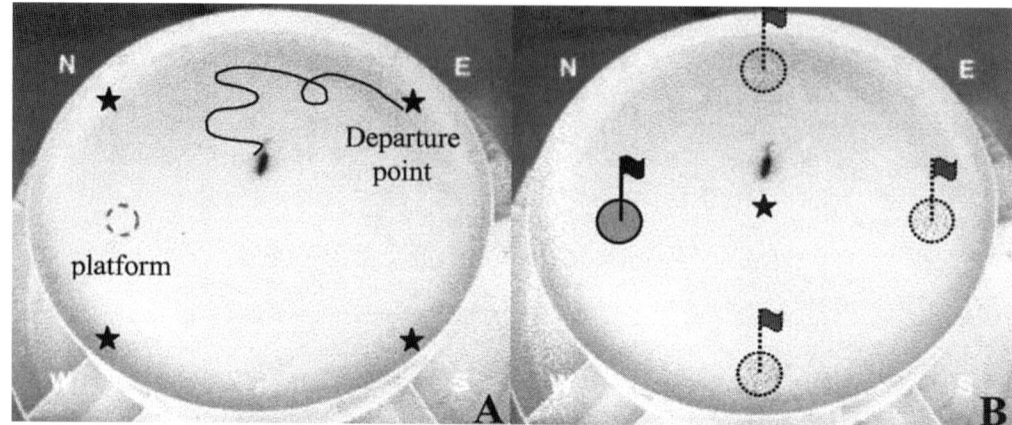

Fig. 1. The Morris water maze paradigm. (A) During the non-visible platform task, animals are required to find a submerged and non-visible platform (see dashed circle) located in a fixed position from random-departure points (indicated by stars). In order to solve this task, animals can use extra-maze cues located in the environment. (B) During the visible platform task the platform was cued with a proximal cue (represented by flags) is placed on the top of the submerged platform, therefore indicated the exact position of the platform to the animal. The visible platform positioned in one of four possible locations (north, east, west, south) as indicated on the figure.

animal has to localize the relative position of a non-visible goal and then to orient itself toward this goal as in the classical Morris water maze (Morris et al., 1984) (see Fig. 1A). Spatial navigation can therefore be defined as a dual-process function that requires the acquisition of at least two complementary processes: the organization of the spatial representation of the environment (declarative component) and the acquisition of a motor behavior adapted to the specific context (procedural learning). An important question currently under debate concerns the exact role of the cerebellum in these processes.

Patients with cerebellar lesions show unexpected cognitive deficits

In humans, a review of current literature concerning cerebellar patients shows a very heterogeneous range of propositions for cerebellum functions in cognitive processes. Reports of impairments range from visuo-spatial recall or serial reaction time task (Botez et al., 1989; Gomez-Beldarrain et al., 1998; Drepper et al., 1999) to higher cognitive levels such as judging the timing of events (Ivry, 1997), anticipatory planning (Grafman et al., 1992), or verbal fluency tasks (Hubrich-Ungureanu et al., 2002).

A number of neuropsychological studies have indicated that cerebellar damage affects the processing of spatial information. Friedreich's ataxia and

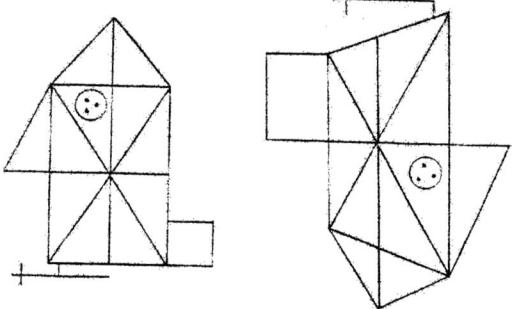

Fig. 2. The Rey Complex Figure Test is a widely used neuropsychological test for visuo-constructive skills and visual memory. This drawing and visual memory test examines the ability to construct a complex figure and remember it for later recall. The patient's methods of procedure as well as his specific copying errors are analyzed. It measures memory as well as visual-motor organization.

olivo-ponto-cerebellar atrophy (OPCA) patients, mostly characterized by atrophy of afferent cerebellar fibers or by damaged olivary and pontine nuclei and cerebellum respectively, were evaluated with a battery of neuropsychological tests. The Rey complex figure test evaluates visuo-spatial functioning. In this task, patients are either asked to copy a complex figure or to draw the figure from memory (see an example of the drawing in Fig. 2). Different components can be evaluated: the accuracy, the location (quantitative evaluation) and the organization (qualitative evaluation) of the drawings. Cerebellar patients evaluated with the copy version were both quantitatively and qualitatively impaired in this complex task (Botez-Marquard and Botez, 1992), suggesting a deficit in visuo-spatial organization.

Neurological and neuropsychological evaluation of other patients with diseases confined to the cerebellum also revealed that 'cerebellar cognitive affective syndrome' included visual–spatial impairment (Schmahmann and Sherman, 1997) in both versions of the Rey complex figure test (copy and memory), and was reflected in the disorganization of the sequential approach to the drawings and the conceptualization of the figures. The patients were also tested in the Hooper task, which tests the individual's ability to organize visual stimuli. The test consists of multiple line drawings, each showing a common object (such as an apple or a ball) that has been cut into several pieces. The pieces are scattered on the page like parts of a puzzle. The patient's task was to tell what the object would be if the pieces were put back together correctly. Patients were not significantly impaired in this task, suggesting an absence of visual organization deficit per se in cerebellar patients. A difference, however, between these two tasks is the verbal response required in the Hooper task and the motor response in the Rey complex figure test. This distinction might support the alternative explanation that impairment in the Rey test might also come from the motor component of the task. An argument in favor of the cognitive hypothesis, however, comes from the fact that the deficit was observed regardless of the severity of the dysmetria (i.e., the lack of coordination of movement), suggesting that the observed impairment could not be explained by difficulties with motor control. Some patients also showed simultagnosia, i.e., typically could see individual objects but were unable to recognize a visual scene as a whole.

Clinical observations of cerebellar patients have given accumulative evidence of impaired performances in a number of cognitive tasks. However, the interpretation of these deficits as a true cognitive impairment remains unclear. Taken together, these clinical observations suggest, more than they demonstrate, a possible visuo-spatial impairment in cerebellar patients.

The mutant cerebellar mouse as a first animal model: progress and limitation

Cerebellar mutant mouse studies first reported a possible implication of the cerebellum in navigation (see Table 1) (Lalonde, 1997). These mutant mice all performed very poorly in the traditional paradigm of the Morris water maze when required to find a hidden platform from random departure locations (Fig. 1A).

However, the significant ataxia in these models complicates the interpretation of a role of the cerebellum in the spatial learning process per se. In order to overcome this problem, different approaches have been used. First, the presence of the ataxic syndrome in cerebellar mutant was one of the reasons for using a water paradigm, as the motor responses required during swimming are less affected by

Table 1. Spatial impairments in cerebellar animals models

	Type of impairment	Water maze: hidden platform	Visible platform	Motor deficit	Cerebellar cortex impairment	Extracerebellar cortex impairment
Lurcher	mutant	Yes	Yes	Yes	Granule, Purkinje	Inferior olive
Staggerer	mutant	Yes	Yes	Yes	Granule, Purkinje	Inferior olive
Weaver	mutant	Yes	Yes	Yes	Granule	Substantia Nigra pars compacta
pcd	mutant	Yes	No	Yes	Purkinje	Inferior olive
nervous	mutant	Yes	No	Yes	Purkinje	Deep cerebellar nuclei
OX7	Immunotoxin	Yes	No	No locomotor activity	Purkinje	No

Weaver: in this case, the paradigm was an aquatic maze with walls. The animals presented an increased number of errors compared to controls and never used a direct path to the goal
Adapted from Lalonde et al. (1997).

cerebellar damage than those required during walking or running (Fortier et al., 1987). Second, the lack of a correlation between spatial and visuo-motor deficits could be achieved by either statistical analysis (Lalonde and Thifault, 1994) and/or the comparison of performance in paradigms allowing navigation toward a hidden or a visible goal (see Fig. 1).

Goodlett and collaborators first reported that Purkinje cell degeneration (pcd) mutant mice were deficient in the hidden goal paradigm but not in the visible paradigm, strengthening the idea of a role for the cerebellum in spatial learning (Goodlett et al., 1992). Gandhi and collaborators confirmed this previous report using OX7, an immunotoxin selective to Purkinje cells, and showed in addition that despite the lack of motor deficit in their model, spatial learning toward a hidden goal was still impaired (Gandhi et al., 2000).

Taken together, these results were the first demonstration of the involvement of the cerebellum, and even more specifically of Purkinje cells, in navigation. It was clear that this deficit was not due to motor impairment. A question however remained unsolved. As mentioned above, navigation is a dual process requiring both the organization of the spatial representation of the environment (declarative component) and the acquisition of a motor behavior adapted to the specific context (procedural learning). Is the cerebellum involved in both processes or only in the procedural component? If the cerebellum is involved in the procedural component, does that component also participate in declarative knowledge?

Declarative versus procedural learning

Investigations into the participation of the cerebellum in navigation raise the question of the characterization of its exact role in spatial tasks. Schenk and Morris defined declarative spatial memory as 'the representation of knowledge in a form which describes the position of the escape platform in relation to other cues of the environment' whereas procedural learning was defined as 'the representation of stimulus-responses necessary to guide the animal to the correct location' (Schenk and Morris, 1985). The studies mentioned above concerning cerebellar patients or cerebellar mouse models did not attempt to distinguish between cognitive versus procedural impairments. In recent years, authors have tried to distinguish exactly which processes computed by the cerebellum are implicated and essential for accurate navigation.

Using hemi-cerebellectomized rats in order to obtain a cerebellar model with partial deficit and no strong basal motor impairment, Petrosini and Molinari's group gave some answers. They examined the role of the cerebellum as an essential structure for the elaboration of a correct representation of the environment or as a player in the procedural component required during navigation. They tried to characterize the exact origin responsible for the poor performances observed in hemicerebellectomized rats relative to control animals during spatial navigation tasks, using various protocols, notably Morris water maze (MWM) and T-maze paradigms (Petrosini et al., 1996; Molinari et al., 1997; Mandolesi et al., 2001). In the MWM task, the authors measured latencies of hemicerebellectomized rats and observed the strategies used (direct finding, extended or restricted searching, circling) in order to reach a hidden platform from different departure locations (place phase) (see Fig. 1A). This task was done before or after the hemicerebellectomy, and before or after a cue phase in which the platform was visible (see Fig. 1B). Performances were different depending on the protocol used; hemicerebellectomized rats displayed a severe impairment if they were trained first in the place phase. However, if rats were trained before this place phase with a cue phase or if they were preoperatively trained, their deficit observed in the MWM were significantly reduced (see Fig. 3).

With these results, authors proposed that the cerebellum plays a crucial role during the acquisition, but not the retention, of a spatial task (Hilber et al., 1998; Petrosini et al., 1998). In addition, they proposed that the predominant role of the cerebellum was the organization of the motor behavior adopted to reach the target (procedural component) rather than in the elaboration of an internal map of the environment necessary to localize the target (Leggio et al., 1999). However, in another recent study, they proposed that the acquisition of the declarative spatial memory requires learning of a correct explorative behavior suggesting that the procedural impairment observed in cerebellar animals could

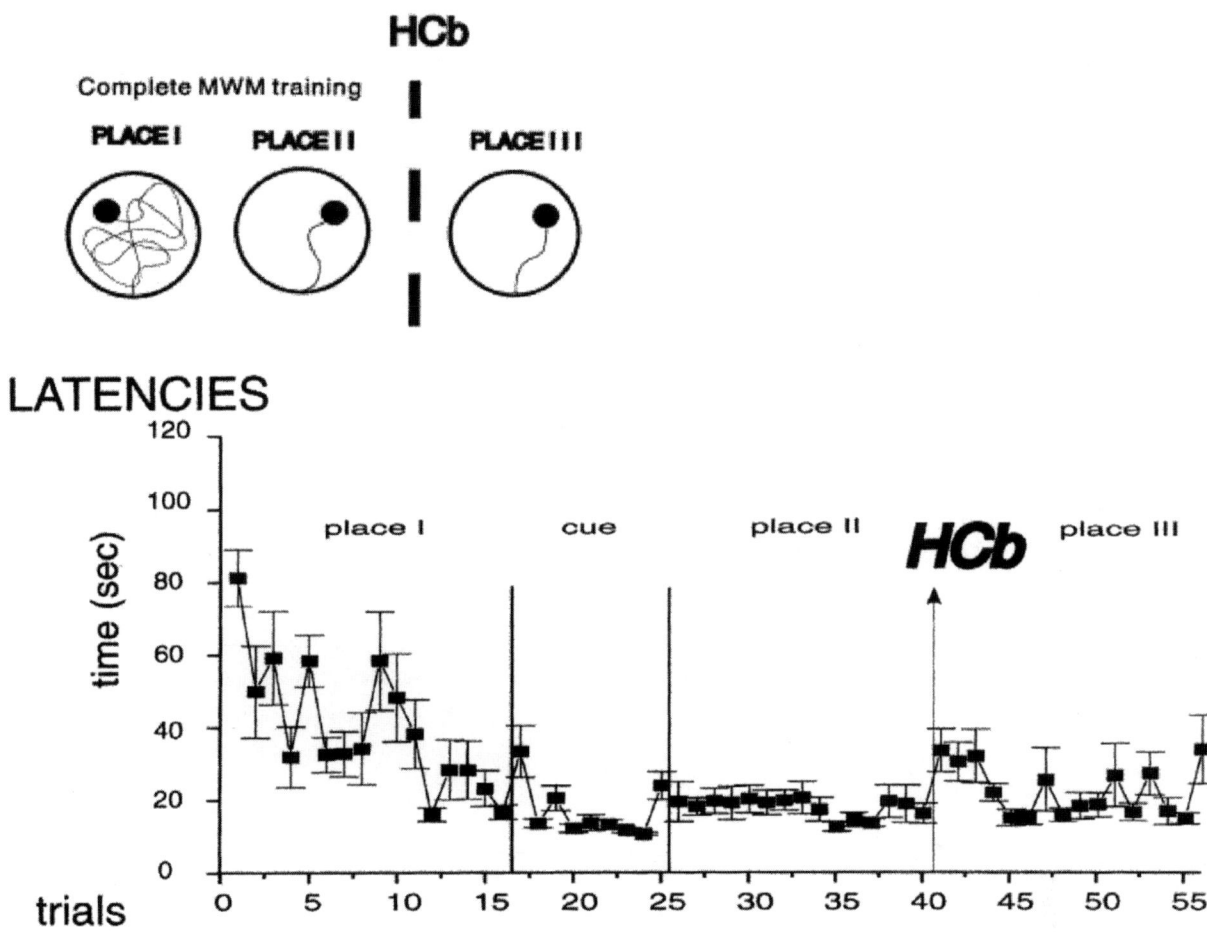

Fig. 3. To establish whether a cerebellar lesion would cause a deficit in spatial navigation, hemi-cerebellectomized rats (Hcbed) were tested in an MWM paradigm characterized by a first phase with the platform hidden in a pool position (place I phase), followed by a phase with a visible platform moved in a different pool position (cue phase), followed by a final phase with the platform hidden in the last position (place II phase). Mean escape latencies in MWM is represented. Rats were pre-operatively (place I, cue and place II) and post-operatively (place III) tested. Note the maintenance of the direct trajectory acquired preoperatively. Vertical arrows indicate hemicerebellectomy (HCb). Vertical bars indicate standard errors. [From Petrosini et al. (1996).]

then lead to a more cognitive impairment (Mandolesi et al., 2003).

Others studies have tried to dissociate motor and cognitive demands during spatial navigation (Gandhi et al., 2000; Martin et al., 2003). With pharmacological and genetic approaches, authors obtained specimens with reduced motor skills deficits but still significant impairment during spatial navigation tasks. These studies are in agreement with the proposition of a role for the cerebellum in spatial processing, in synergy with other brain areas (Petrosini et al., 1998). They also suggested that performances during navigation tasks were the consequence of both processes. One is a declarative component which allows learning where the target is and could involve the hippocampus and associated areas. The other is a procedural component necessary to develop an effective exploration behavior and could involve the cerebellum and other associated brain areas.

The anatomical substrate for the putative role of the cerebellum in navigation

The cerebellum receives multisensorial inputs

The anatomical connections of the cerebellum represent a good substrate for the putative role of the cerebellum in navigation. In mammals, the cerebellum receives proprioceptive, somatosensory, visual, and auditory information and projects to the tectum, the red nucleus, and the cerebral cortex via the thalamus. The cerebellum can be divided into three longitudinal zones (medial, intermediate, and lateral) based on the projection of the cerebellar cortex onto the three deep nuclei. These zones differ in the type of information that comes into them via mossy fibers and climbing fibers, the two main inputs to the cerebellar cortex. The medial zone is dominated by information from vestibular, somatosensory, visual, and auditory regions. The intermediate zone receives proprioceptive and somatosensory information from the spinal cord as well as information from the motor cortex via the ponto-cerebellar nuclei. The lateral zone receives information via the pontine nuclei from the motor cortex, the premotor, and motor cortex (see Fig. 4). These differences in sensory input suggest functional differences. Indeed, experiments using rats with lesions targeting the different parts of the cerebellum revealed the functional specificity of the lateral part (comprising the cerebellar hemispheres and dentate nucleus) compared to the midline part (comprising

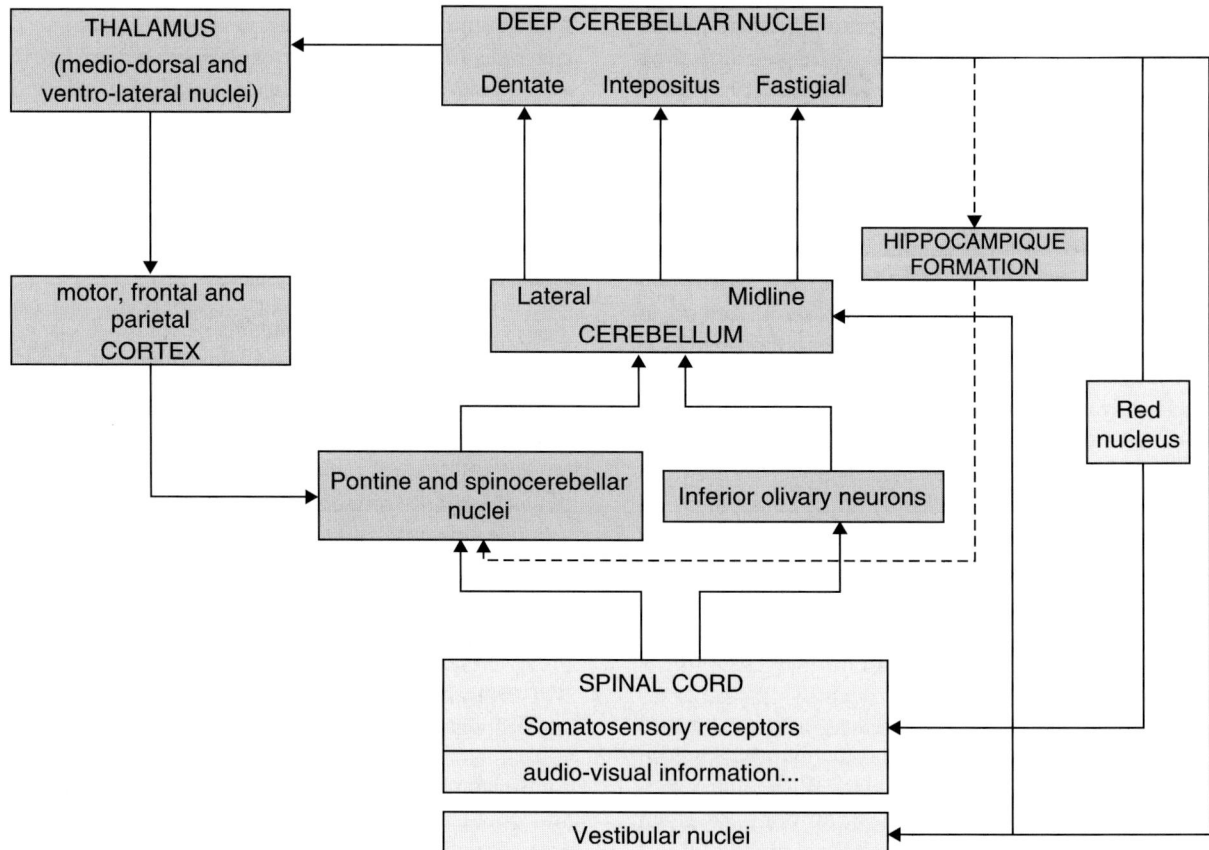

Fig. 4. Anatomical projections from and to the cerebellum. The cerebellum receives multi-sensorial inputs via pre-cerebellar nuclei. The different parts of the cerebellar cortex projects to three cerebellar nuclei and are connected to the cortical area via the thalamus and to sensori-motor related structures via the red nucleus.

the vermis and fastigial nucleus). Rats with midline lesions had disturbances in balance and equilibrium and presented visuo-motor functional deficits when assessed by the visible platform paradigm in the Morris water maze. Rats with lateral lesions had milder deficits and were deficient in spatial orientation (the hidden goal paradigm of the Morris water maze) but presented no visuo-motor deficit (visible platform paradigm) (Joyal et al., 1996). Lesions of the dentate nucleus alone did not affect equilibrium or balance performances but resulted in the same selective impairment in spatial orientation as the one observed with a combined lesion of the cerebellar hemispheres and dentate nucleus (Joyal et al., 2001). These results point towards a specific role of this lateral region in spatial orientation.

The cerebellum as a part of the spatial network

The outputs of the cerebellum also strongly suggest a role in a network involved in spatial orientation. It has long been thought that outputs of the three zones of the cerebellar cortex (and therefore deep nuclei) were also segregated, i.e., fastigial to vestibular and reticulospinal systems, interpositus to red nucleus and dentate (via the thalamus) to motor cortex. Although there may be predominant segregation of the projections along these lines, there is also a clear overlap (Bastian et al., 1999). Thus, vestibular, fastigial, interpositus, and dentate nuclei all project via the thalamus to the motor cortex and vestibular, fastigial, and interposed nuclei all project to the spinal cord (Asanuma et al., 1983a, b). In primates (and probably in humans), the cerebellar projections of the lateral part (neo-cerebellum) also expand to almost all parts of the associative cortex including the frontal and parietal association areas (Glickstein, 1993). Nothing is known yet about such projections in rodents. In addition, electrophysiology recordings in rats suggest reciprocal projections between the cerebellum and the hippocampus. Both climbing and mossy fibers pathways, which are the two main inputs of the cerebellar cortex were activated by fornix stimulation (the main efferent pathway of the hippocampus) (Saint-Cyr and Woodward, 1980). Conversely, stimulation of the fastigial nucleus (corresponding to the medial part) impaired neuronal responses in the septum and the hippocampus proper (Heath et al., 1978). It is not known whether these projections are direct or indirect.

The repetitive microcomplex and long term depression cellular mechanism

Ito was the first to propose the corticonuclear microcomplex as the functional module of the cerebellum (Ito, 2001). In this microcomplex, a cortical microzone is paired with a small distinct group of neurons in a deep cerebellar or vestibular nucleus. The mossy fiber (MF) afferents arising from various precerebellar nuclei supply excitatory synapses to granule cells in the microzone and also to the nuclear neurons via collaterals. Granule cell axons (called parallel fibers; PFs) then send excitatory projections to the Purkinje cells. Another set of inputs to the microcomplex is relayed by a small distinct group of inferior olive neurons, whose axons end as climbing fibers (CFs) on Purkinje cells and also supply excitatory synapses to the nuclear neurons via collaterals. According to Ito (2001), four basic notions constitute the operational principle of the microcomplex. First, MF signals to a microzone are relayed by granule cells and in turn excite Purkinje cells and other cortical neurons, eventually evoking simple spikes in Purkinje cells. Second, the MF signals drive the nuclear neurons, which generate output signals of the microcomplex under inhibitory influences of Purkinje cells. Third, CFs convey error signals to a microcomplex regarding the operation of the neural system that includes the microcomplex. The error signals are generated by various neuronal mechanisms in diverse pre-olivary structures. Fourth, CF error signals induce the cerebellar synapse plasticity, Long Term Depression (LTD), in the conjointly activated PF-Purkinje cells synapses and thereby modify the operation of the microcomplex until the error signals are minimized.

LTD has been proposed to be a crucial cellular mechanism underlying motor learning (Marr, 1969; Albus, 1971; Ito and Kano, 1982). A large amount of experimental data supporting this theory are available in studies of simple forms of motor learning such as the vestibulo-ocular reflex (VOR) or eyelid conditioning (Gilbert and Thach, 1977; McCormick and Thompson, 1984). The crucial role of LTD during motor learning has been well demonstrated with recent work using a new transgenic model called L7-PKCI (De Zeeuw et al., 1998). This model has a specific alteration of LTD at the Purkinje cell level and a deficit of motor learning was observed during VOR adaptation (Goossens et al., 2001; Van Alphen and De Zeeuw, 2002). A question of interest is now to see whether this cellular synaptic mechanism could be involved in other kinds of learning such as spatial learning. Indeed, the diversity of the deficits described after cerebellar impairment could appear to be in contradiction with the repetitive anatomical structure defined as microcomplex by Ito (2001) and found throughout the entire cerebellum. Although the potential role of the cerebellum in cognitive functions has been interpreted in many different ways, authors are currently trying to find a common process to all the deficits related to the cerebellum. The cerebellar LTD is one of the best cellular candidates.

The role of climbing fibers (CF) and parallel fibers (PF)

The first indirect evidence for a potential role of cerebellar input-dependent mechanisms in the process of spatial navigation comes from experimental lesions of these afferents. Our aim was to understand the specific and possible conjoint effect of a lesion of CF and PF inputs. Specific destruction of the inferior olivary neurons and their axons, the CFs, were made by i.p. administration of 3-acetylpyridine (3-AP) and niacinamide (Rondi-Reig et al., 2002). We used X-irradiation delivered to the cerebellum a few days after birth in order to prevent the genesis of granule cells and their axons, the PFs. The protocol of navigation we used was a 'fixed departure–fixed arrival' procedure in the water maze. Animals had to find a hidden platform located in a fixed position. This protocol minimized spatial processing; to solve, the task animals could use a single cue (rather than the configuration of cues in the room) or learn to orient their body directly to the platform. Our results demonstrated that the absence of the major cerebellar inputs, i.e., CF and PF, differentially impaired the acquisition of the non-visible platform task but not the visible platform task, suggesting a specific role of these inputs in spatial orientation rather than in visuo-motor orientation. A total lesion of PFs (with intact CFs) totally impaired the acquisition of the non-visible version of the water maze (Le Marec et al., 1997). This impairment was the same in the case of a double lesion (PF and CF). In contrast to control animals that, after a few trials of random exploration, learned to use an initial rotation followed by a direct trajectory toward the platform, rats lacking PFs were unable to learn to orient their bodies toward the non-visible platform. Instead, they adopted a circling behavior during the majority of the trials. These data were comparable to those obtained by Petrosini et al. (1996) in the 'fixed start–fixed arrival' procedure (corresponding to the first protocol they used in their study). Hemicerebellectomized rats had trouble developing efficient searching behaviors: they performed extended searching around the pool and peripheral circling before finding the platform. These effects were present in the acquisition phase (i.e., when efficient exploration is extremely important). The similarity of the results strongly suggests that deficits observed after a lesion of the cerebellum could be due to its disconnection via one of its major inputs, the PFs. An additional experiment showed that partial destruction of the parallel fiber inputs only delayed the acquisition of the spatial task (Le Marec et al., 1997) therefore suggesting that the number of granule cells (and thus PFs) is a key factor in normal spatial function.

In contrast, a total lesion of CFs alone only delayed the acquisition of this same task (Dahhaoui et al., 1992). The delayed but not impaired behavior observed with total CF destruction suggested a role of CF in the time-course of the acquisition, its presence increasing the rapidity of the process. The importance of an intact functional CF system was additionally revealed in the case of a partial lesion of PFs only (Rondi-Reig et al., 2002). A total lesion of CFs in combination with a partial lesion of PFs dramatically impaired the acquisition of the non-visible platform task (Rondi-Reig et al., 2002) whereas, as mentioned above, this acquisition was only delayed in the case of a partial lesion of PF and intact CF. All these results are summarized in Table 2.

The comparison of the two types of lesions gave interesting insight into the relative importance of

Table 2. Functional consequences of climbing fibres and parallel fibers lesions

	Non-visible task	Visible task (cued)
1. Controls	normal	normal
2. Total lesion of CF/Intact PF	delayed	(not done)
3. Intact CF/Partial lesion of PF	delayed	(not done)
4. Intact CF/Total lesion of PF	impaired	(not done)
5. Total lesion of CF/Partial lesion of PF (partially lesioned group)	impaired	normal
6. Total lesion of CF/Total lesion of PF (totally lesioned group)	impaired	normal

Delayed means that experimental animals learn the task and finally reach similar escape latencies than controls but need more training trials. Impaired means that experimental animals never learn the task and present significantly higher escape latencies compared to controls even at the end of the training.
From Rondi-Reig et al. (2002).

these two input systems. It seems that for this particular behavior, the presence of intact CFs can compensate for the absence of some of the PFs. It also suggests that the presence of PFs is crucial for the acquisition of such behavior, and that CFs play a role which seems minimal compared to that of the PFs. These results are consistent with the hypothesis of a possible role of a synaptic mechanism occurring at the level of the PFs. Indeed, the required presence of PFs for spatial learning and the importance of the presence of CFs when the number of PFs is decreased indicate a mechanism occurring at the PF level but also depending on CFs. Homo- and heterosynaptic LTD occurring at the synaptic level of PF could be good candidates.

Recent advances in conditional mutagenesis provide the required tool to address such a specific question. As mentioned above, L7-PKCI mice present a specific deficit of LTD at the synaptic level between PFs and Purkinje cells. In addition, this model presents no deficit in basic motor skills in contrast to all other cerebellar mutant models (Van Alphen and De Zeeuw, 2002). By testing these mutant mice for spatial navigation abilities, our first results revealed that they present spatial navigation deficit in the MWM (personal communication). We are currently characterizing these deficits in order to know if the procedural or declarative component is affected.

Discussion

Based on the accumulation of experimental data showing that cerebellar patients and animal models present cognitive deficits, what can one answer to the question: is the cerebellum ready for navigation? More than a decade after this question was first posed, one point is now recognized: the cerebellum must be added to the structural network involved in navigation. The anatomical projections of the cerebellum are consistent with this view. A crucial question nevertheless remains: how does the cerebellum participate? The role of the cerebellum in the procedural component of the navigation process is one aspect that has been the best documented experimentally. In contrast, the possible implication of the cerebellum in the declarative component of navigation remains unclear. Is the information processing in the cerebellum required for the acquisition of a spatial representation? Is the procedural impairment observed in cerebellar animals indirectly responsible for an impaired spatial representation? The possible bi-directional projections between the cerebellum and the hippocampus revealed by electric stimulation are intriguing. However, anatomical analyses to confirm such projections are lacking.

A computational model has suggested a complementary role of the cerebellum and the basal ganglia for the learning of sequential procedures (Hikosaka et al., 1999). Such a model is consistent with the parallel anatomical loops formed with cortical area, in particular the prefrontal cortex (Middleton and Strick, 2000, 2001), by the lateral part of the cerebellum on the one hand and the caudate nucleus-substantia nigra pathway on the other (see Fig. 5).

Recent studies have shown that brain areas other than the cerebellum contribute to different stages and

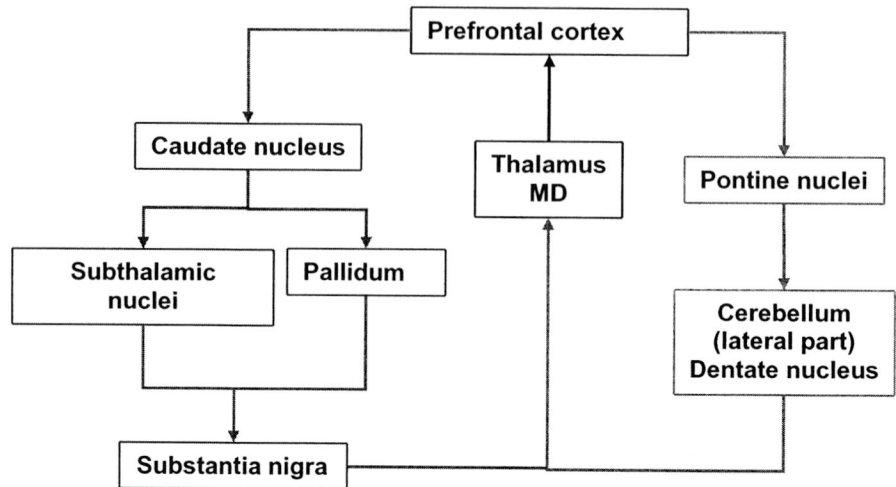

Fig. 5. Parallel anatomical loops of the lateral part of the cerebellum and of the basal ganglia with the prefrontal cortex.

aspects of procedural learning (Doya, 2000). On the basis of a series of studies using a sequence-learning task with trial-and-error in monkeys, Hikosaka and collaborators (Hikosaka et al., 1999) have proposed a hypothetical scheme in which a sequential procedure is acquired independently by two cortical systems, one based on basal ganglia and using spatial coordinates and the other based on the cerebellum using motor coordinates. These systems would be preferentially active in the early (basal ganglia) and late (cerebellum) stages of learning. Both the systems are supported by loop circuits formed with the basal ganglia and the cerebellum, the former for reward-based evaluation and the latter for processing of timing.

Another unanswered question concerns the potential mechanism(s) underlying the function of the cerebellum during navigation. As mentioned above, a number of apparently different functions have been attributed to the cerebellum, from basic motor skills to cognitive function, yet the internal anatomy of the cerebellum is highly homogenous. One possibility would be to consider that all these functions have a common process in which the cerebellum participates. This proposition is also consistent with a recent finding showing that protein kinase C-dependent long-term depression in Purkinje cells is necessary for learning-dependent timing (Koekkoek et al., 2003) as proposed in the model of Hikosaka et al. (1999).

We are currently trying to determine whether this process is also necessary for spatial learning.

Abbreviations

CF	climbing fibers
LTD	long term depression
MF	mossy fibers
MWM	Morris water maze
PF	parallel fibers
VOR	vestibulo-ocular reflex

Acknowledgments

The authors would like to thank Anne Lohof for reviewing the language of the paper. Work in the author's laboratory is supported by grants from the ACI 'Neurosciences Intégratives et Computationnelles' (NIC 0083) and the European Laboratory of Neurosciences and Action (LENA).

The authors also thank France Maloumian for her help in the preparation of the figures.

References

Albus, J. (1971) A theory of cerebellar function. Math Biosci., 10: 26–61.

Asanuma, C., Thach, W.T. and Jones, E.G. (1983a) Brainstem and spinal projections of the deep cerebellar nuclei in the monkey, with observations on the brainstem projections of the dorsal column nuclei. Brain Res., 286: 299–322.

Asanuma, C., Thach, W.T. and Jones, E.G. (1983b) Distribution of cerebellar terminations and their relation to other afferent terminations in the ventral lateral thalamic region of the monkey. Brain Res., 286: 237–265.

Bastian, A.J., Mugnaini, E. and Thach, W.T (1999) Cerebellum. In: Zigmund, M.J., Bloom, F.E., Landis, S.C., and Roberts, J.L. (Eds.), Fundamental Neuroscience. Squire IR, pp. 973–992.

Botez, M.I., Botez, T., Elie, R. and Attig, E. (1989) Role of the cerebellum in complex human behavior. Ital. J. Neurol. Sci., 10: 291–300.

Botez-Marquard, T. and Botez, M.I. (1992) Visual memory deficits after damage to the anterior commissure and right fornix. Arch Neurol., 49: 321–324.

Dahhaoui, M., Stelz, T. and Caston, J. (1992) Effects of lesion of the inferior olivary complex by 3-acetylpyridine on learning and memory in the rat. J. Comp. Physiol. [A], 171: 657–664.

De Zeeuw, C.I., Hansel, C., Bian, F., Koekkoek, S.K., van Alphen, A.M., Linden, D.J. and Oberdick, J. (1998) Expression of a protein kinase C inhibitor in Purkinje cells blocks cerebellar LTD and adaptation of the vestibulo-ocular reflex. Neuron, 20: 495–508.

Doya, K. (2000) Complementary roles of basal ganglia and cerebellum in learning and motor control. Curr. Opin. Neurobiol., 10: 732–739.

Drepper, J., Timmann, D., Kolb, F.P. and Diener, H.C (1999) Non-motor associative learning in patients with isolated degenerative cerebellar disease. Brain, 122(Pt 1): 87–97.

Fortier, P.A., Smith, A.M. and Rossignol, S. (1987) Locomotor deficits in the mutant mouse. Lurcher. Exp. Brain Res., 66: 271–286.

Gandhi, C.C., Kelly1, R.M., Wiley, R.G. and Walsh, T.J. (2000) Impaired acquisition of a Morris water maze task following selective destruction of cerebellar purkinje cells with OX7-saporin. Behav. Brain Res., 109: 37–47.

Gilbert, P.F. and Thach, W.T. (1977) Purkinje cell activity during motor learning. Brain Res., 128: 309–328.

Glickstein, M. (1993) Motor skills but not cognitive tasks. Trends Neurosci., 16: 450–451.

Gomez-Beldarrain, M., Garcia-Monco, J.C., Rubio, B. and Pascual-Leone, A. (1998) Effect of focal cerebellar lesions on procedural learning in the serial reaction time task. Exp. Brain Res., 120: 25–30.

Goodlett, C.R., Hamre, K.M. and West, J.R. (1992) Dissociation of spatial navigation and visual guidance performance in Purkinje cell degeneration (pcd) mutant mice. Behav. Brain Res., 47: 129–141.

Goossens, J., Daniel, H., Rancillac, A., van der, S.J., Oberdick, J., Crepel, F., De Zeeuw, C.I. and Frens, M.A. (2001) Expression of protein kinase C inhibitor blocks cerebellar long-term depression without affecting Purkinje cell excitability in alert mice. J. Neurosci., 21: 5813–5823.

Grafman, J., Litvan, I., Massaquoi, S., Stewart, M., Sirigu, A. and Hallett, M. (1992) Cognitive planning deficit in patients with cerebellar atrophy. Neurology, 42: 1493–1496.

Heath, R.G., Dempesy, C.W., Fontana, C.J. and Myers, W.A. (1978) Cerebellar stimulation: effects on septal region, hippocampus, and amygdala of cats and rats. Biol. Psychiatry, 13: 501–529.

Hikosaka, O., Nakahara, H., Rand, M.K., Sakai, K., Lu, X., Nakamura, K., Miyachi, S. and Doya, K. (1999) Parallel neural networks for learning sequential procedures. Trends Neurosci., 22: 464–471.

Hilber, P., Jouen, F., Delhaye-Bouchaud, N., Mariani, J. and Caston, J. (1998) Differential roles of cerebellar cortex and deep cerebellar nuclei in learning and retention of a spatial task: studies in intact and cerebellectomized lurcher mutant mice. Behav. Genet., 28: 299–308.

Holmes, G. (1939) The cerebellum of man. Brain, 62: 1–30.

Hubrich-Ungureanu, P., Kaemmerer, N., Henn, F.A. and Braus, D.F. (2002) Lateralized organization of the cerebellum in a silent verbal fluency task: a functional magnetic resonance imaging study in healthy volunteers. Neurosci. Lett., 319: 91–94.

Ito, M. (2001) Cerebellar long-term depression: characterization, signal transduction, and functional roles. Physiol. Rev., 81: 1143–1195.

Ito, M. and Kano, M. (1982) Long-lasting depression of parallel fiber-Purkinje cell transmission induced by conjunctive stimulation of parallel fibers and climbing fibers in the cerebellar cortex. Neurosci. Lett., 33: 253–258.

Ivry, R. (1997) Cerebellar timing systems. Int. Rev. Neurobiol., 41: 555–573.

Joyal, C.C., Meyer, C., Jacquart, G., Mahler, P., Caston, J. and Lalonde, R. (1996) Effects of midline and lateral cerebellar lesions on motor coordination and spatial orientation. Brain Res., 739: 1–11.

Joyal, C.C., Strazielle, C. and Lalonde, R. (2001) Effects of dentate nucleus lesions on spatial and postural sensorimotor learning in rats. Behav. Brain Res., 122: 131–137.

Koekkoek, S.K., Hulscher, H.C., Dortland, B.R., Hensbroek, R.A., Elgersma, Y., Ruigrok, T.J. and De Zeeuw, C.I. (2003) Cerebellar LTD and learning-dependent timing of conditioned eyelid responses. Science, 301: 1736–1739.

Lalonde, R. (1997) Visuospatial abilities. Int. Rev. Neurobiol., 41: 191–215.

Lalonde, R. and Thifault, S. (1994) Absence of an association between motor coordination and spatial orientation in lurcher mutant mice. Behav. Genet., 24: 497–501.

Le Marec, N., Dahhaoui, M., Stelz, T., Bakalian, A., Delhaye-Bouchaud, N., Caston, J. and Mariani, J. (1997) Effect of cerebellar granule cell depletion on spatial learning and memory and in an avoidance conditioning task: studies in postnatally X-irradiated rats. Brain Res. Dev. Brain Res., 99: 20–28.

Leggio, M.G., Neri, P., Graziano, A., Mandolesi, L., Molinari, M. and Petrosini, L. (1999) Cerebellar contribution to spatial event processing: characterization of procedural learning. Exp. Brain Res., 127: 1–11.

Leiner, H.C., Leiner, A.L. and Dow, R.S. (1993) Cognitive and language functions of the human cerebellum. Trends Neurosci, 16: 444–447.

Mandolesi, L., Leggio, M.G., Graziano, A., Neri, P. and Petrosini, L. (2001) Cerebellar contribution to spatial event processing: involvement in procedural and working memory components. Eur. J. Neurosci., 14: 2011–2022.

Mandolesi, L., Leggio, M.G., Spirito, F. and Petrosini, L. (2003) Cerebellar contribution to spatial event processing: do spatial procedures contribute to formation of spatial declarative knowledge? Eur. J. Neurosci., 18: 2618–2626.

Marr, D. (1969) A theory of cerebellar cortex. J. Physiol., 202: 437–470.

Martin, L.A., Goldowitz, D. and Mittleman, G. (2003) The cerebellum and spatial ability: dissection of motor and cognitive components with a mouse model system. Eur. J. Neurosci., 18: 2002–2010.

McCormick, D.A. and Thompson, R.F. (1984) Neuronal responses of the rabbit cerebellum during acquisition and performance of a classically conditioned nictitating membrane-eyelid response. J. Neurosci., 4: 2811–2822.

Middleton, F.A. and Strick, P.L. (2000) Basal ganglia and cerebellar loops: motor and cognitive circuits. Brain Res. Brain Res. Rev., 31: 236–250.

Middleton, F.A. and Strick, P.L. (2001) Cerebellar projections to the prefrontal cortex of the primate. J. Neurosci., 21: 700–712.

Molinari, M., Grammaldo, L.G. and Petrosini, L. (1997) Cerebellar contribution to spatial event processing: right/left discrimination abilities in rats. Eur. J. Neurosci., 9: 1986–1992.

Morris, R., Rawlins, J., Garrud, P. and O'Keefe, J. (1984) Place navigation impaired in rats with hippocampal lesions. Nature, 297(5868): 681–683.

Petrosini, L., Leggio, M.G. and Molinari, M. (1998) The cerebellum in the spatial problem solving: a co-star or a guest star? Prog. Neurobiol., 56: 191–210.

Petrosini, L., Molinari, M. and Dell'Anna, M.E. (1996) Cerebellar contribution to spatial event processing: Morris water maze and T-maze. Eur. J. Neurosci., 8: 1882–1896.

Rondi-Reig, L., Le Marec, N., Caston, J. and Mariani, J. (2002) The role of climbing and parallel fibers inputs to cerebellar cortex in navigation. Behav. Brain Res., 132: 11–18.

Saint-Cyr, J.A. and Woodward, D.J. (1980) Activation of mossy and climbing fiber pathways to the cerebellar cortex by stimulation of the fornix in the rat. Exp. Brain Res., 40: 1–12.

Schenk, F. and Morris, R.G. (1985) Dissociation between components of spatial memory in rats after recovery from the effects of retrohippocampal lesions. Exp. Brain Res., 58: 11–28.

Schmahmann, J.D. and Sherman, J.C. (1997) Cerebellar cognitive affective syndrome. Int. Rev. Neurobiol., 41: 433–440.

Thach, W.T. (1998) A role for the cerebellum in learning movement coordination. Neurobiol. Learn Mem., 70: 177–188.

Thompson, R.F. and Kim, J.J. (1996) Memory systems in the brain and localization of a memory. Proc. Natl. Acad. Sci. USA, 93: 13438–13444.

Van Alphen, A.M. and De Zeeuw, C.I. (2002) Cerebellar LTD facilitates but is not essential for long-term adaptation of the vestibulo-ocular reflex. Eur. J. Neurosci., 16: 486–490.

CHAPTER 18

The lateral cerebellum and visuomotor control

N.L. Cerminara[1], A.L. Edge[1], D.E. Marple-Horvat[2] and R. Apps[1,*]

[1]Department of Physiology, School of Medical Sciences, University Walk, University of Bristol, Bristol BS8 1TD, UK
[2]Institute for Biophysical and Clinical Research into Human Movement, Manchester Metropolitan University, Alsager, UK

Keywords: cerebellum; visual; visuomotor; climbing fiber; inferior olive

Abstract: The lateral cerebellum receives an abundance of visual input providing the link between visual and motor control centers. In this review we discuss experiments designed to increase our understanding of how visual inputs to the cerebellum are arranged in relation to the zonal organization of the cerebellar cortex, and how visual inputs are utilized to assist in the regulation of a visually guided movement. On the basis of anatomical and physiological characteristics our findings indicate that the medial-most folium in crus I of the cat lateral cerebellum can be subdivided into at least three functionally distinct zones; from lateral to medial along the length of the folium these correspond to zones D_1, lateral C_3 and C_2. Each zone displays clear differences in olivo-cortico-nuclear connectivity and in the anesthetized animal zones D_1 and C_2 both receive powerful visual inuts relayed via the climbing fiber system. Complementary experiments in awake behaving cats found that Purkinje cells located in the D_1 and D_2 zones of crus I exhibit changes in simple spike discharge time locked to target motion during a visually guided reaching task. These changes were unaffected by temporary visual denial of the target, raising the possibility that internally generated feedforward visuomotor control mechanisms are operating, in which a predictive model of the target's motion has been constructed by the CNS.

Introduction

Any investigation of how vision guides movement leads inevitably to the lateral cerebellum because this is where visual information is itself directed, via a massive projection that provides, probably, the main route linking visual to motor control centers (Robinson et al., 1984; Glickstein et al., 1985, 1994). The cerebellar hemispheres receive via the pontine nuclei powerful visual inputs from the parietal lobe extrastriate visual areas whose cells are particularly responsive to moving visual targets and to visual events (Zeki, 1974; Baker et al., 1976; Thier et al., 1988; reviewed in Glickstein, 1998). The importance of these inputs is betrayed by the profound deficit in visuomotor performance that results from their removal (Classen et al., 1995) or from any cerebellar injury (Holmes, 1939). Both anatomy and pathology therefore suggest a critical role for the lateral cerebellum in the visual guidance of movement.

The activity of the cells that forward information via the mossy fiber input to the lateral cerebellar cortex therefore suggests that the activity of single neurons in the cerebellar hemisphere should display at least passive signaling of visual events, and an online description of a moving target's motion. There is some evidence (though far from complete) that this is indeed the case (Chapman et al., 1986; Marple-Horvat and Stein, 1990; Mushiake and Strick, 1993; Marple-Horvat et al., 1998). In addition, if the target is familiar, consistent features of the way it moves might be built into an internal (cerebellar) model of its motion with predictive capacity (Miall et al., 1993; Miall, 1998).

*Corresponding author. Tel.: +44 (0) 117 928 7803; Fax: +44 (0) 117 928 8923; E-mail: r.apps@bristol.ac.uk

A straightforward way to distinguish between a passive online description of external events and an active operation of such an internal model is to temporarily deny visibility of the target. If cerebellar cortical neurons simply respond passively to a visual stimulus, then denying visibility of a target should lead to a loss of neuronal activity signaling features of its movement. On the other hand, if an internal model has been constructed, then when visibility is denied, the activity encoding features of a familiar target's movement might be actively maintained. One of the twin aims of the present study was to seek evidence for, and discriminate between, these two possibilities. The second aim was to identify the underlying neural circuitry involved. Interpretation of the activity patterns obtained in single unit recordings crucially benefits from a detailed knowledge of the input–output relations of the cells whose activity has been recorded. As a first step in studying lateral cerebellar contributions to visuomotor control, we have therefore charted the neuroanatomical connections of the climbing fiber and some of the major mossy fiber inputs and the Purkinje cell outputs of the cortical regions studied in our physiological experiments.

Neuroanatomical studies

A key anatomical feature of the cerebellar cortex is its division into a series of longitudinally oriented olivo-cortico-nuclear zones. Each is defined by its climbing fiber input from a specific subnucleus within the inferior olive, and by its Purkinje cell output to a specific domain within the cerebellar or vestibular nuclei (for reviews see Brodal and Kawamura, 1980; Voogd and Bigaré, 1980; Buisseret-Delmas and Angaut, 1993; Voogd and Glickstein, 1998). Proceeding from the midline laterally, each half of the cerebellar cortex can be divided into at least eight zones: termed A, X, and B in the vermis; C_1, C_2, and C_3 in the paravermis; and D_1 and D_2 in the lateral hemisphere. Each zone is typically 1 mm in mediolateral width but extends for many millimeters in the rostrocaudal plane, and some can be further subdivided into medial and the lateral subzones (e.g., medial and lateral C_3). While a considerable amount is now known about the functional organization of the vermal and paravermal zones (for recent reviews see Voogd and Ruigrok, 1997; Apps, 1999; Garwicz, 2000), by comparison, much less information is available about those present within the hemispheres, despite the fact that this cortical region forms the main body of the mammalian cerebellum. Nonetheless, it is clear that visual events such as a flash of light (Snider and Stowell, 1944; Fadiga and Pupilli, 1964; Marple-Horvat and Stein, 1990; Budanur et al., 1999) or appearance of a target, (Kitazawa et al., 1998) can be forwarded to the lateral cerebellum by the climbing fiber system (see also Maekawa and Natsui, 1973; Simpson and Alley, 1974; Simpson et al., 1996), suggesting that such information is important in lateral cerebellar contributions to motor control.

To gain a more comprehensive understanding of the relationship between functional localization and the presence of zones within the lateral cerebellar cortex, we have used a double label axonal tracing strategy in combination with electrophysiological mapping techniques to chart the input–output connections of crus I in cats.

Methods

In brief, the methods involve the use of aseptic techniques under general anesthesia (propofol, which has a barbiturate-like action) to deliver minute volumes (c. 100 nl) of bi-directional tracer material into different parts of crus I (the medial-most folium of the ansiform lobule), guided by physiological responses to electrical stimulation of different body parts: ipsilateral and contralateral forelimbs, ipsilateral periorbital region, and optic chiasm. The anterograde and retrograde tracers used in each injection were red or green fluorescently tagged dextran amines and beads (a 20% solution of Fluoro-Ruby combined with red beads or a 20% solution of Fluoro-Emerald combined with green beads).

After a 7-day survival period to allow axonal transport to occur, the animals are deeply re-anesthetized and fixed by transcardial perfusion. Brain tissue is then prepared for histological examination and sections scrutinized, using an epi-fluorescence microscope fitted to a custom-designed image analysis system (for further details see Edge et al., 2003). The locations within the pontine nuclei and the inferior olivary nucleus of retrogradely labeled cells (single and double-labeled with beads) are plotted to characterize the mossy fiber and climbing fiber inputs, respectively, to the functionally defined areas of the lateral cerebellum where injections were made. Similarly, the anterograde terminal labeling in the cerebellar nuclei (due to transport of the dextran amines) is plotted to chart the cortico-nuclear output of the Purkinje cell axons arising from the same injection sites (cf. Garwicz et al., 1996). Thus the mossy and climbing fiber inputs and cortico-nuclear outputs can be related to functional localization within the same animal.

Functional localization

The responses recorded in the medial-most folium of crus I evoked by an electrical stimulation of different body parts had features typical of field potentials generated by activity in olivo-cerebellar pathways (cf. Ekerot and Larson, 1979; Trott and Apps, 1991, 1993). These include (i) a highly characteristic waveform (the main component of the response was always a sharply rising positive deflection with a duration of about 5 ms), (ii) an onset latency of at least 9 ms, and (iii) considerable variability in response amplitude between successive stimuli. Since responses mediated by mossy fiber pathways are usually severely depressed under barbiturate anesthesia, we conclude that the responses were mainly, if not exclusively, climbing fiber in origin. The topographical distribution of the responses also indicated that the most medial folium of crus I can be subdivided into at least three distinct areas: each defined in terms of differences in amplitude and patterns of convergence of the responses evoked by stimulation of the different body sites tested.

Table 1 summarizes our findings and shows that the three cortical areas can be distinguished on the basis of their somatic and visual input. From caudal to rostral (equivalent to medial to lateral along the length of the folium because of the orientation of the folia in the ansiform lobule in the cat cerebellum), area 1 receives convergent input from all four body sites tested (ipsilateral and contralateral forelimbs, ipsilateral periorbital region, and optic chiasm); area 2 receives input mainly from the ipsilateral periorbital region; and area 3 receives input mainly from the optic chiasm (for further details see Edge et al., 2003). It should be emphasized, however, that each of the cortical areas defined in the present study by their differences in forelimb, face or visual climbing fiber responses may also receive inputs from other parts of the body not tested, and also are highly likely to receive sensory inputs relayed via the mossy fiber system. As mentioned earlier, the latter pathways were most probably suppressed by the anesthesia used in the present experiments.

Olivo-cortico-nuclear connections

Figure 1 shows the pooled retrograde cell labeling in the inferior olive obtained from individual cases in which injections were placed into approximately the center of each of the three electrophysiologically defined cortical areas in crus I. The pooled diagrams show the density of cell labeling at different transverse levels of the olive, with filled regions indicating regions of highest density. An inspection of Fig. 1 shows that each cortical area receives its olivo-cerebellar projection from a topographically distinct territory within the contralateral inferior olive. From medial to lateral along the length of the folium: area 1

Table 1. Summary of functional localization within crus I and relation to cortical zones as deduced in the present experiments

Body site	Area 1 (C_2 zone)	Area 2 (C_3 zone)	Area 3 (D_1 zone)
Ipsilateral forelimb	22–34 ms ($n=4$)	X ($n=5$)	X ($n=4$)
Contralateral forelimb	22–37 ms ($n=4$)	X ($n=5$)	X ($n=4$)
Ipsilateral periorbital region	21–24 ms ($n=2$)	11–15 ms ($n=4$)	X ($n=3$)
Optic chiasm	23–27 ms ($n=4$)	X ($n=5$)	13–16 ms ($n=4$)

The range of response onset latencies is given for each cortical area (each value is an average of four stimulus trials per experiment). Numbers in brackets indicate number of cats. X, indicates that electrical stimulation of a particular body site either failed to evoke a detectable response or produced only a weak response.

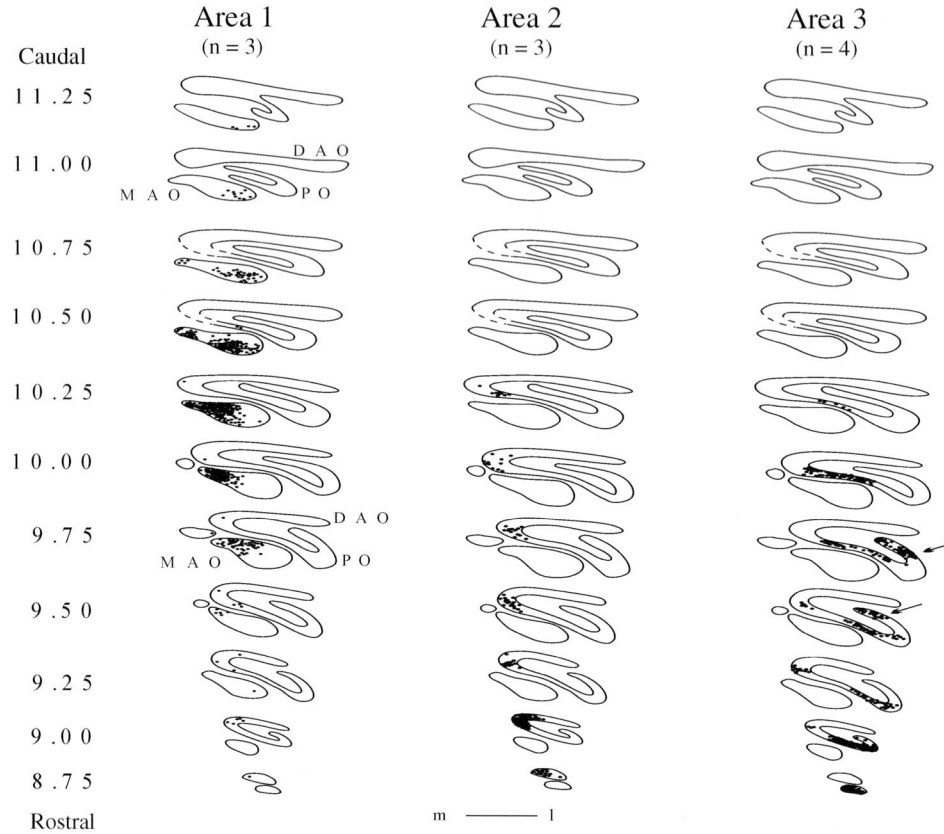

Fig. 1. Distribution of olive cell labeling after tracer injections into different cortical areas in crus I. Equally spaced transverse outlines of the inferior olive between AP levels 11.25 and 8.75, showing pooled retrograde cell labeling arising from injections into area 1, area 2 or area 3. Numbers in brackets refer to the number of individual cases pooled for each area. Each dot corresponds to one cell, except in regions of high density of labeling. Arrows indicate labeling in the dorsal lamella of PO arising from an injection that deviated lateral to area 3. DAO, dorsal accessory olive; l, lateral; m, medial; MAO, medial accessory olive; PO, principal olive. Scale bar = 1 mm. For further details see Edge et al. (2003).

receives its climbing fiber input from the rostral half of the medial accessory olive (rMAO); area 2 receives its input from the interface between the ventral lamella of the principal olive (vlPO) and the rostral part of the dorsal accessory olive (rDAO); and area 3 receives its input mainly from vlPO (for area 3 the additional cell labeling in the dorsal lamella indicated by arrows was obtained in one case in which the injection deviated lateral to area 3).

Overall the present experiments therefore support the zonal hypothesis of cerebellar compartmentation whereby different olivary regions provide climbing fiber input to different medio-lateral zones in the cortex. This is emphasized by the double tracer results. When, in the same animal, areas 1 and 3, or areas 2 and 3 were injected with two different tracers there was no overlap in the contralateral inferior olive between the territories occupied by cells single-labeled with red or green beads, and no double-labeled cells were found.

The one exception was in an experiment where tracer injections were made into areas 2 and 3, and in the olive a little overlap was present between the two single-labeled cell populations in the interface between vlPO and rDAO, and a single double-labeled cell was found in this region of overlap. However, the most likely explanation for these results is that some tracer material spread medially from the injection into area 3 to involve the neighboring area 2.

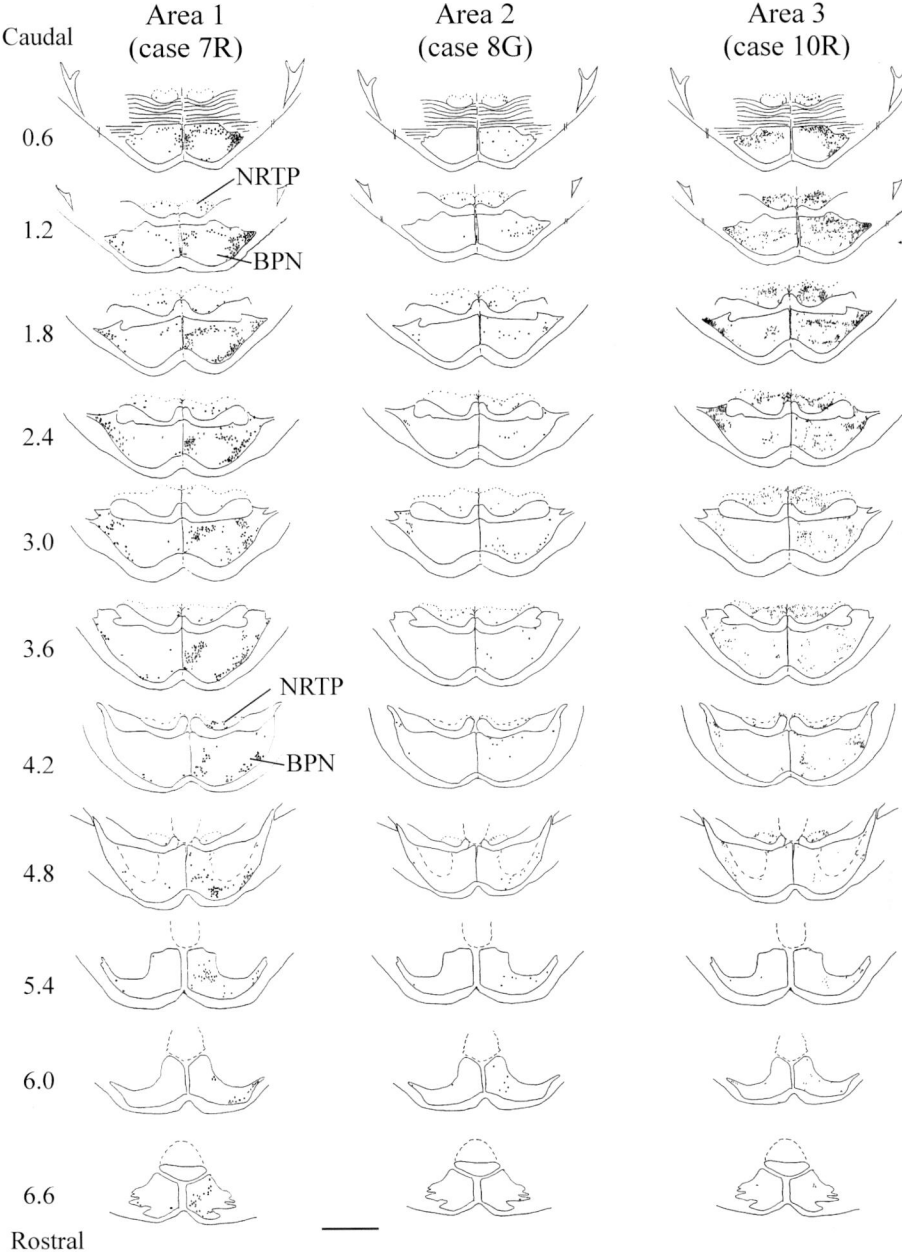

Fig. 2. Distribution of pontine cell labeling after tracer injections into different cortical areas in crus I. Equally spaced transverse outlines of the basal pontine nuclei (BPN) and nucleus reticularis pontis (NRTP) showing retrograde cell labeling in individual cases arising from injections into area 1, area 2, or area 3. Numbers to the left refer to the distance (in mm) of each level from the caudal pole of the pons. Each dot corresponds to one cell, except in regions of high density of labeling. Scale bar = 1 mm.

Figure 2 shows for representative cases the pattern of retrograde cell labeling in the pontine nuclei after injections centered on each of the three cortical areas. By contrast to the olivo-cerebellar results, no clear topographical differences are apparent. In each case, retrograde cell labeling was patchy and scattered throughout much of the rostro-caudal extent of nucleus reticularis tegmenti pontis (NRTP) and the basal pontine nuclei (BPN), mainly on the contralateral side. Double tracer experiments showed that cells single-labeled with red or green beads were mainly intermingled but very few double-labeled cells were found.

An inspection of Fig. 2 suggests, however, that one possible difference between cortical areas was that the cell labeling arising from injections into area 2 was less dense than that arising from injections into areas 1 or 3. Such differences may arise simply because of variations between cases in the size of the individual injection sites, and in the present experiments, the injection sites differed in cortical area by as much as a factor of 5 (for a detailed description of the methods we used to assess injection site size see Herrero et al. (2002)).

To take the differences in injection site size into account, we have therefore considered the data in terms of mean densities of projection. For each case, the total number of labeled cells in the pontine nuclei (counts on both sides of the brain in NRTP and BPN combined) has been normalized, i.e., the cell counts adjusted to a value per square millimeter of cortical area involved in the injection site, and the values for each area averaged.

The mean density of projection to area 1 was 4329 cells per mm^2; to area 2 was 1156 cells per mm^2; and to area 3 was 4182 cells per mm^2. Thus, on average, the density of the pontine projections to area 2 was substantially smaller than to areas 1 and 3 (by a factor of 3.7 and 3.6, respectively). Similar findings were obtained when NRTP and BPN were considered separately. However, the number of cases available for each area was small and the differences in the mean values were not statistically significant (one way ANOVA, $P > 0.05$), so no firm conclusions can be made at this stage.

Nonetheless, it is relevant to note that the territories occupied by labeled cells include parts of the pontine nuclei known to receive visual and/or sensorimotor projections from the cerebral cortex (for reviews and references see Brodal and Bjaalie, 1997; Glickstein, 1997). Of particular interest is the presence of cell labeling within dorsolateral parts of the BPN where cortico-pontine fibers originating from extrastriate visual areas of the cerebral cortex have been shown to terminate. These projections are thought to be especially concerned with the processing of visual motion and are likely therefore to play an important role in the visual guidance of movement (Glickstein, 1997).

By contrast to the pontine cell labeling, Fig. 3 shows that a rather clear topographical difference was evident between cortical areas in terms of their Purkinje cell cortico-nuclear projections. The data available for each area are pooled, showing that the cases in which an injection was centered on area 1 invariably resulted in anterograde terminal labeling within nucleus interpositus posterior (NIP), while for area 2 terminal labeling was located mainly within the transitional region between nucleus interpositus anterior (NIA) and dentate (medio-lateral levels 1.50–2.00). By comparison to area 2 the terminal labeling arising from injections into area 3 was centered farther laterally within dentate. These findings therefore support the olivo-cerebellar results by indicating that each electrophysiologically defined cortical area has a topographically distinct cortico-nuclear projection to the ipsilateral cerebellar nuclei. This is confirmed by the double tracer results which showed that there was no (or, at best, very little) overlap between the nuclear territories occupied by terminals labeled with Fluoro-Ruby or Fluoro-Emerald (for further details see Edge et al., 2003).

In summary, the present experiments show that in cats the medial-most folium of crus I can be subdivided into at least three functionally distinct areas which differ conspicuously in their olivo-cortico-nuclear connections, but less so in terms of their mossy fiber inputs from the pons. The latter finding should, however, be considered with some caution because the areas under study may differ, at least partially, with regard to the density of their pontine inputs, and a finer grain topography may be present that would require charting ponto-cerebellar projections to smaller areas of cortex than was the case in the present study (cf. the cortical patches of Shambes et al., 1978, see also Bower and Kassel, 1990).

With regard to the olivo-cortico-nuclear connections and physiological properties of area 1, these are consistent with the presence of a C_2 zone within the caudal part of crus I (corresponding to the medial aspect of the folium in terms of its long axis). Similarly, the anatomy and physiology of area 2 are consistent with the presence of a lateral C_3 subzone, while the olivo-cortico-nuclear connections

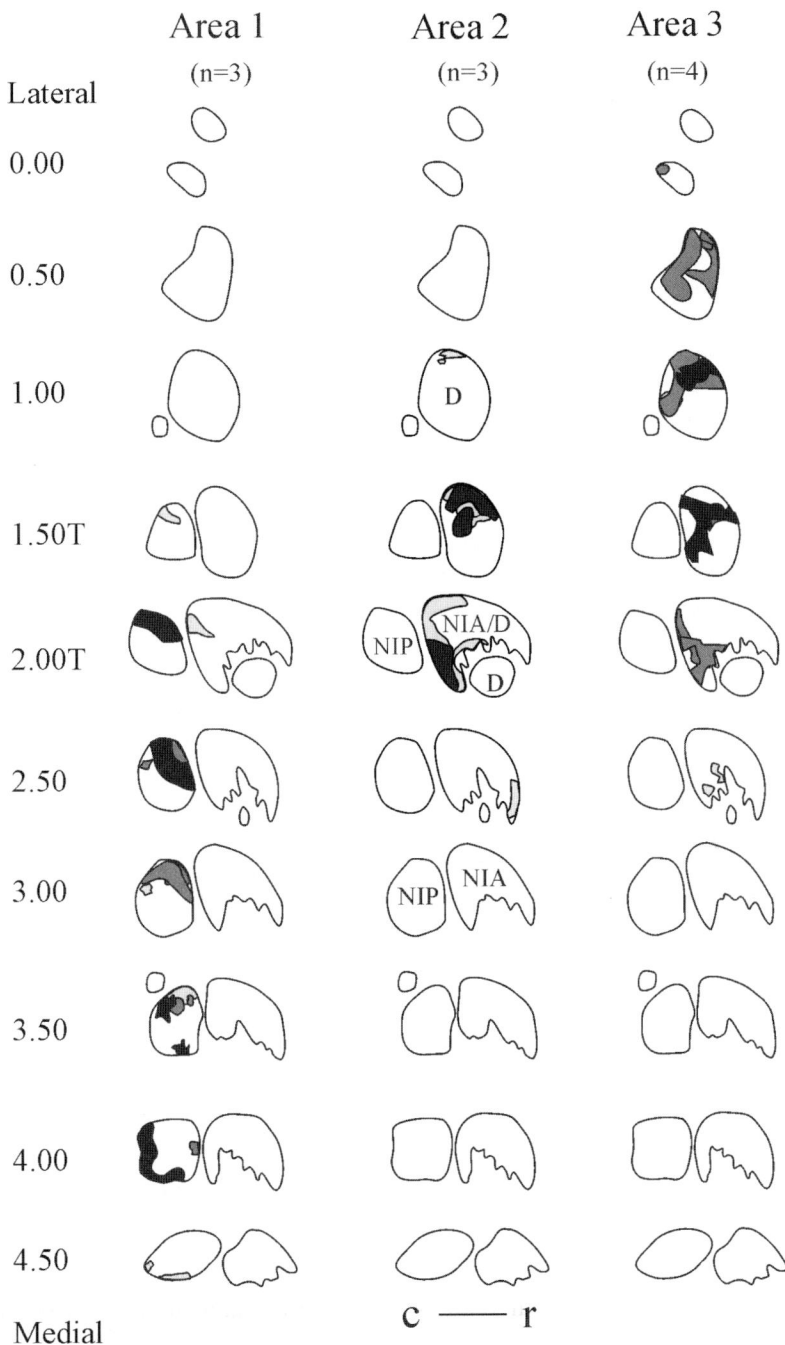

Fig. 3. Distribution of cerebellar nuclear terminal labeling after tracer injections into different cortical areas in crus I. Equally spaced sagittal outlines showing anterograde-labeled axon terminal fields arising from injections into area 1, area 2, or area 3. Same cases pooled as illustrated in Fig. 1. Numbers to the left refer to the distance (in mm) of each level from the lateral edge of the dentate nucleus. T, levels transitional between dentate and nucleus interpositus anterior. Density of terminal labeling is shown by three different levels of shading. Light gray indicates sparse labeling, dark gray moderate labeling and black, regions of dense labeling. c, caudal; D, dentate nucleus; NIA, nucleus interpositus anterior; NIP, nucleus interpositus posterior; r, rostral. Scale bar = 2 mm. For further details see Edge et al. (2003).

associated with rostral area 3 (corresponding to a lateral part of crus I in terms of the long axis of the folium), are consistent with the presence of a D_1 zone. Finally, the presence of retrograde cell labeling in the dorsal lamella of the principal olive, arising from an injection made lateral to area 3, suggests that a D_2 zone is also present in the rostral pole of crus I.

We also found that an electrical stimulation of the optic chiasm in the anesthetized cat consistently evoked substantial field potentials in both the C_2 and D_1 zones, implying that visual pathways are a potent source of climbing fiber input to these zones in crus I (for further discussion see Edge et al., 2003). Our results should therefore provide a useful frame of reference for future studies of lateral cerebellar contributions to visuomotor control and, in particular, have aided the interpretation of our on-going experiments in chronically instrumented cats in which we are recording neuronal activity patterns in crus I during performance of visually guided movements.

Neurophysiological studies

Two questions central to understanding how vision guides movement are firstly, what visual information is used; and secondly, when is it used? Regarding the first issue, behavioral studies involving restriction of the available visual information have shown that the bare minimum for successful performance is much less than a full uninterrupted view of the world (Prablanc et al., 1986; Weir et al., 1989; Hollands and Marple-Horvat, 1996). Indeed, as far as looking at, pointing to or stepping onto visual targets is concerned, the only requirement is for a visible target. The rest of the visual world, including the view of one's own body, can be removed with little effect on performance. Regarding the second issue, i.e., when visual information is required, experiments have shown that rendering the target temporarily and unpredictably invisible also has at some times little or no effect on performance, but at other times such a denial has a substantial effect (Becker and Fuchs, 1985; Laurent and Thomas, 1988; Hollands and Marple-Horvat, 1996).

Clearly, only certain visual information is required and then only at certain times during the planning and execution of visually guided movements. This implies that feedforward visuomotor control mechanisms are operating, which are robust in the face of denial of all visual information for several hundred milliseconds, including preferred inputs at the preferred times. There is some evidence that the ability to function effectively in this way is dependent upon the integrity of the lateral cerebellum (Miall et al., 1987).

Experimental protocols in awake animals that correlate neuronal activity patterns with visual denial at different times (and for different periods) during the planning and execution of movements are a direct way of identifying the times at which visual information is preferentially used and required.

Methods

We have trained cats to perform a visually guided reaching task and then, at an aseptic operation, chronically implanted a recording chamber over the lateral cerebellum (crus I) to permit access to this region of cortex with microelectrodes to record extracellularly the activity patterns of single cortical neurons whilst the animal performs the task. Figure 4 shows a schematic diagram of the experimental arrangements which are carried out in a light-proof room in the dark. The target to reach, consisting of a perspex tube containing a food reward, is dimly lit with a ring of LEDs (the only source of light to the cat). The tube is initially stationary and positioned 7 cm to the left of the animal at a comfortable reaching height. The tube then starts to move horizontally at a constant velocity rightwards across the animal's visual field. A 'go' signal (brightening of the ring of LEDs) is then given, which is a cue for the animal to make a reach with its forelimb ipsilateral to the cerebellar recording chamber to retrieve the food reward from the tube (which stops moving as soon as the animal's paw enters the tube).

At six different stages during the task (including in between trials when the target is stationary), the LEDs surrounding the tube are temporarily extinguished for either 200 or 300 ms ('denial' trials). During the performance of the task, cerebellar neuronal spike trains and electro-oculograms are recorded, the latter to monitor the left and rightward horizontal saccades. At the end of the recording

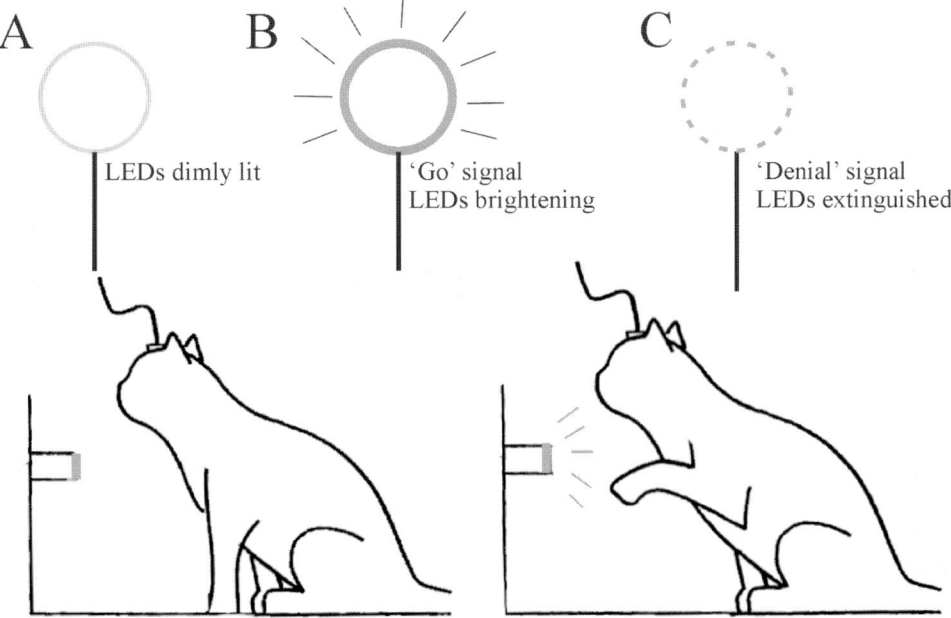

Fig. 4. Schematic diagram of chronic recording arrangements. (A) Cats are trained to perform a visually guided reaching task in the dark in which a tube, dimly lit by a ring of LEDs, moves horizontally in front of the animal. (B) Approximately 600 ms after the tube moves rightwards, the LEDs brighten to cue the animal to make a reach with its forelimb ipsilateral to the cerebellar recording, to retrieve a food reward from the tube. (C) At different times during the task the LEDs are extinguished for a period of 200 or 300 ms.

period (typically 8–12 weeks after the initial operation) fluorescent tracer material (red beads) is microinjected into the tips of the folia in the region of cerebellar cortex in which most of the microelectrode tracks were made during the recording sessions, to permit identification of the climbing fiber input to the cortical regions from which neuronal recordings have been obtained. As in the anatomical studies, the animals are kept for a further 7 days to allow transport of tracer to occur and then deeply re-anesthetized, fixed by transcardial perfusion and the cerebellum and brainstem removed for histological processing.

Neuronal activity patterns

To date, results have been obtained from a total of 92 lateral cerebellar cortical neurons in three cats. Of these neurons 68 have been positively identified as Purkinje cells by their cortical location, irregular discharge (cf. Armstrong and Rawson, 1979), and by the presence of complex spikes in their spike trains. The following section will focus on the simple spike activity of these cells in relation to (1) onset of target movement, (2) the denial signal, and (3) onset of eye movement. Since it was not possible to reliably discriminate all complex spike discharges of individual cells during performance of the task, these data have not been considered further here.

Overall, the simple spike activity of 31/68 (46%) of the Purkinje cells was modulated by the start of target movement. Most cells displayed an increase in discharge rate with an onset latency of 65 ± 37 ms ($n = 19$), while the remainder displayed a reduction in activity, with an onset latency of 76 ± 46 ms ($n = 12$). Figure 5 shows a peri-event discharge histogram for an example Purkinje cell which displayed a phasic increase in simple spike activity after onset of target movement. In this example the firing rate transiently increased from its pre-event levels of approximately 6 impulses per second to 14 impulses per second, about 60 ms after the start of tube motion. This change in firing rate was statistically significant, defined as a discharge frequency in at least two successive 5 ms bins more than two standard

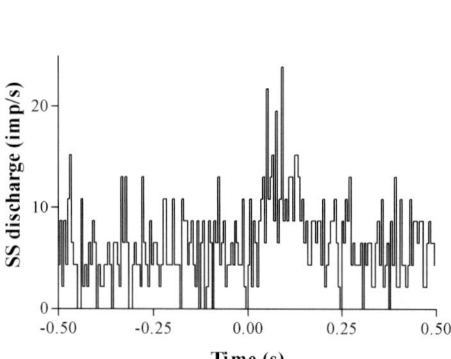

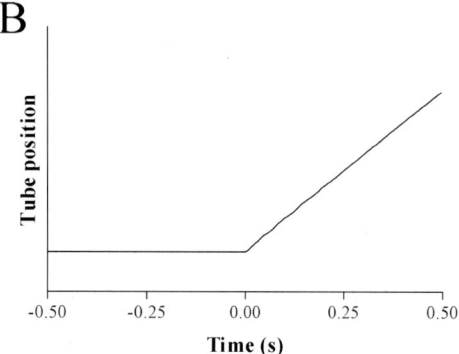

Fig. 5. Phasic modulation of Purkinje cell simple spike activity in crus I. (A) Peri-event time histogram in which the simple spike (SS) discharge of an individual Purkinje cell in crus I has been averaged in relation to time of onset of tube movement, which occurs at time zero. (B). In this example there was a statistically significant phasic increase in activity that was time locked to onset of tube movement. Average of 92 trials. Bin width = 5 ms.

deviations above the mean level of discharge in a 200 ms (40 bins) 'control' period immediately prior to the visual event (cf. Marple-Horvat et al., 1998). Thus, about half of the Purkinje cells we sampled in crus I were capable of responding to visual events associated with the forelimb reaching task under study.

In addition to the initial phasic modulation, the overwhelming majority (26/31, 84%) of the responsive Purkinje cells also displayed a significant difference in their tonic discharge rate while the target was moving, as compared to when it was stationary (Student's paired t-test, $P < 0.05$). This tonic change was either an increase (Fig. 6A) or a decrease (Fig. 6D) in simple spike activity after the onset of tube movement (Fig. 6B and 6E). Under the conditions of temporary visual denial (200 or 300 ms periods of extinguishing the target's LEDs, Fig. 6C and 6F, respectively) no significant change in discharge rate in any of the 26 tonically active cells was found.

To control for the possibility that the tonic discharge was due to saccadic eye movements during target motion, the activity of 19 Purkinje cells was studied in relation to the electro-oculogram (EOG). Figure 7 shows a peri-event time histogram of the cell displayed in Fig. 6A, but this time in relation to the onset of rightward horizontal eye movement. Typical of the sample as a whole, there is no statistically significant change in activity, suggesting that the tonic changes in firing rate during tube movement are not likely to be fully explained by concomitant horizontal eye movements.

Since the moving target had been encountered many times, a model of its movement could have been constructed by the CNS, and the tonic simple spike discharge that persisted during the temporary absence of the target is consistent with this notion. However, the site (or sites) where the internal model resides remains an open question, although the present data are certainly consistent with the possibility that this is in the lateral cerebellum.

Finally, Fig. 8 shows, for two of the chronic recording animals, the distribution of retrograde cell labeling in the contralateral inferior olive after an injection of red beads was made into the cerebellar cortex where the majority of the Purkinje cell recordings were obtained. In one case the cell labeling (indicated by dots) was located in the lateral bend of the principal olive, while in the second animal the labeling (indicated by crosses) was located in the dorsal lamella of the principal olive. On the basis of the findings obtained in our neuroanatomical experiments outlined above (see also Kotchabhakdi et al., 1978; Rosina and Provini, 1982), we conclude that in the first animal, the single unit recordings were probably made from a region of cortex that was transitional between the D_1 and D_2 zones, while in the second animal the recordings were probably made farther laterally, in the D_2 zone. With this in mind it is interesting to note that in terms of simple spike response patterns, animal 2 displayed a higher proportion of responsive cells to target movement than animal 1 (8/13, 62% and 13/27, 48%, respectively).

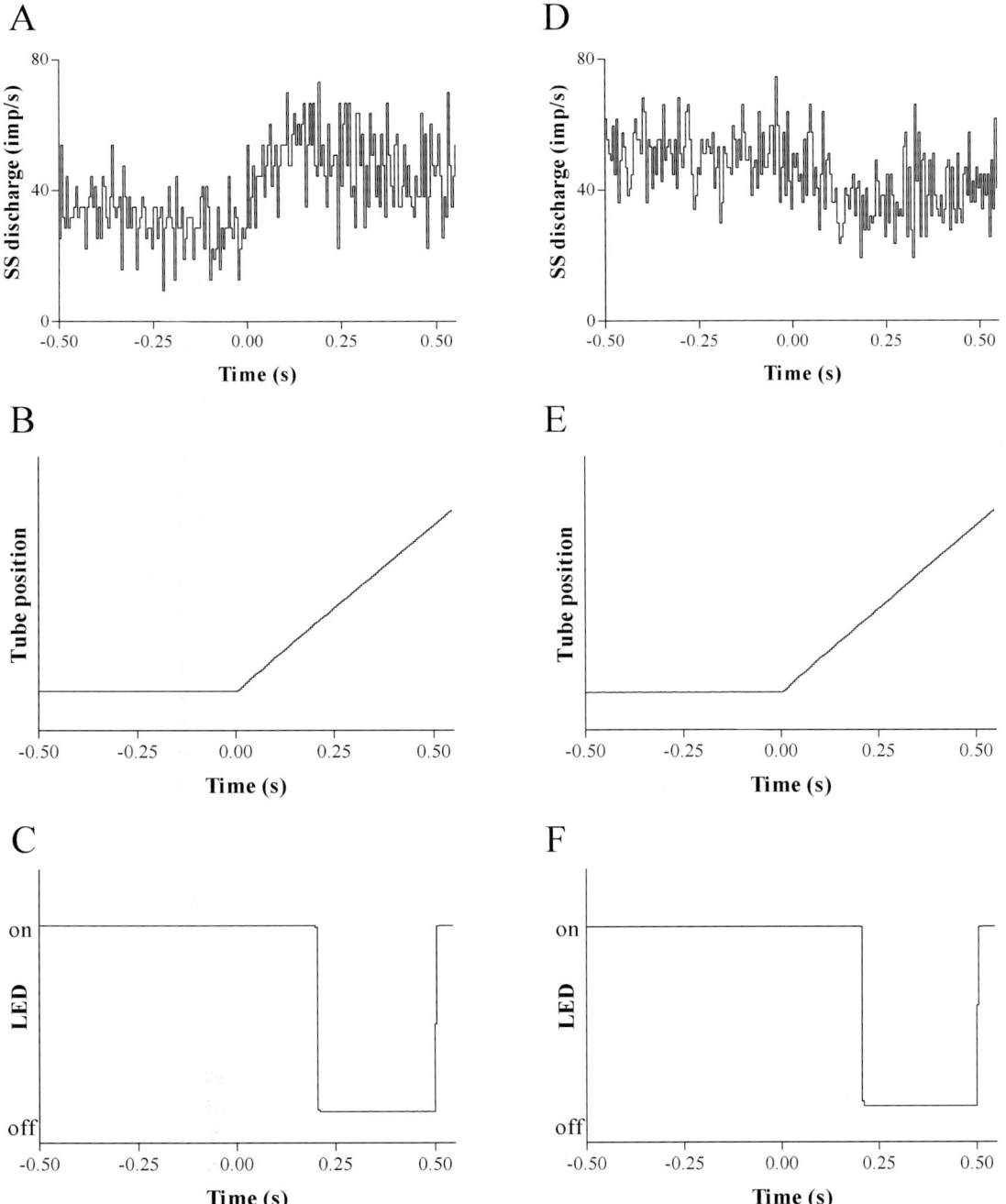

Fig. 6. Tonic changes in Purkinje cell simple spike activity in crus I. (A) Peri-event time histogram in which the simple spike (SS) discharge of an individual Purkinje cell in crus I has been averaged in relation to time of onset of tube movement, which was at time zero (B). In this example there was a statistically significant and sustained increase in activity which was unaffected by the target being temporarily absent from the cat's view (C). (D–F). Same as A–C but for a different Purkinje cell in crus I in which there was statistically significant and sustained decrease in simple spike activity after onset of tube movement. Average of 63 trials in A and 93 trials in D. Bin width = 5 ms.

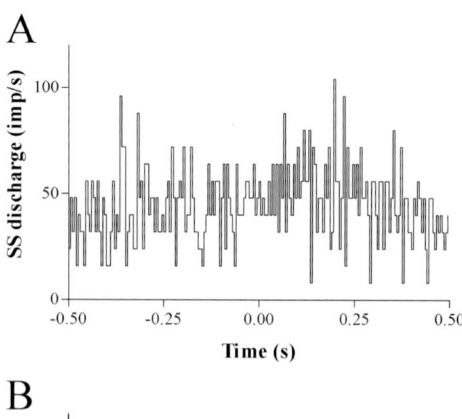

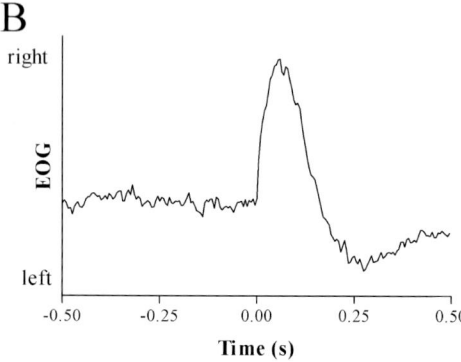

Fig. 7. Purkinje cell activity in relation to eye movements. (A) Peri-event time histogram in which the simple spike (SS) discharge of the same Purkinje cell illustrated in Fig. 6A has been averaged in relation to time of onset of rightward eye movement, which occurs at time zero (B). In this and all other cells tested there was no statistically significant change in activity related to onset of eye movement. Average of 25 trials. Bin width = 5 ms. EOG, electro-oculogram.

Conclusions

In this brief review we have outlined experiments in which we have investigated the anatomy and physiology of the lateral cerebellum in cats. We have found clear differences in olivo-cortico-nuclear connectivity and physiological properties, suggesting that the tips of the medial-most folium in crus I can be subdivided into at least three functionally distinct zones (from lateral to medial along the length of the folium: D_1, lateral C_3, and C_2), and that in the anesthetized animal zones D_1 and C_2 both receive a powerful visual input relayed via the climbing fiber system. The path(s) by which such signals are relayed from visual centers to this region of the cerebellar cortex via rostral parts of the olive remains to be

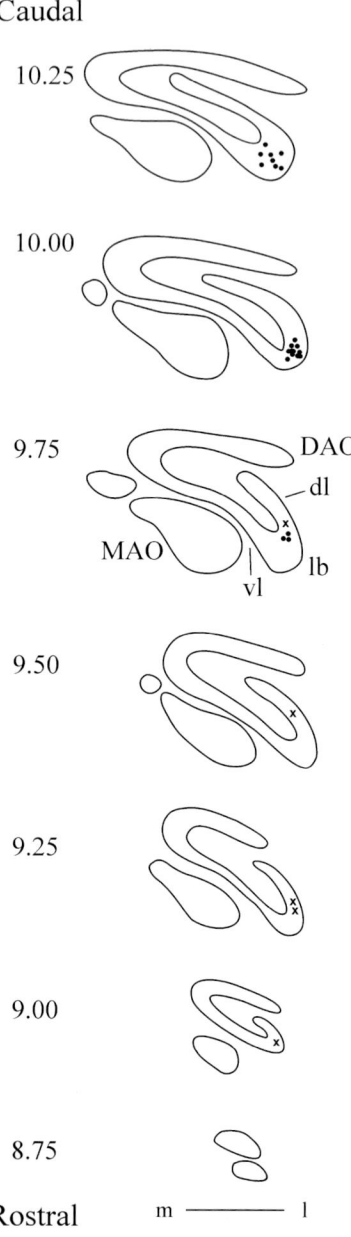

Fig. 8. Distribution of olive cell labeling after tracer injections into cortical areas studied in the awake cat. Equally spaced transverse outlines of the inferior olive between AP levels 10.25 and 8.75, showing pooled retrograde cell labeling arising from injections in two animals. Each symbol corresponds to one cell, (dots = animal 1, crosses = animal 2). DAO, dorsal accessory olive; dl, dorsal lamella of the principal olive; l, lateral; lb, lateral bend of the principal olive; m, medial; MAO, medial accessory olive; vl, ventral lamella of the principal olive. Scale bar = 1 mm.

determined, but presumably involves one or more intercalating synapses in the brainstem.

Our complementary experiments in awake behaving cats have found that Purkinje cells in crus I (probably located in the D_1 and D_2 zones) can exhibit changes in simple spike discharge time locked to the execution of a visually guided reaching task. These changes are unaffected by temporary visual denial of the target, raising the possibility that internally generated feedforward visuomotor control mechanisms are operating, in which a predictive model of the target's motion has been constructed by the CNS. It remains to be established if this internal model is located in the lateral cerebellum and/or resides upstream from the cells we recorded.

Acknowledgments

We thank Rachel Bissett, Clare Everard and Steve Gilbey for their expert technical assistance, and Professor David Armstrong for his help and guidance. A.E. was supported by a Wellcome Trust Prize Studentship and R.A. by an MRC Senior Research Fellowship. D.M.-H. is Reader in Motor Control, Manchester Metropolitan University, and a Visiting Fellow to the University of Bristol. Grant sponsors: BBSRC, MRC and the Wellcome Trust.

References

Apps, R. (1999) Gating of climbing fibre input to cerebellar cortical zones. Prog. Neurobiol., 57: 537–562.

Armstrong, D.M. and Rawson, J.A. (1979) Activity patterns of cerebellar cortical neurones and climbing fibre afferents in the awake cat. J. Physiol., 289: 425–448.

Baker, J., Gibson, A., Glickstein, M. and Stein, J. (1976) Visual cells in the pontine nuclei of the cat. J. Physiol., 255: 415–433.

Becker, W. and Fuchs, A.F. (1985) Prediction in the oculomotor system: smooth pursuit during transient disappearance of a visual target. Exp. Brain Res., 57: 562–575.

Bower, J.M. and Kassel, J. (1990) Variability in tactile projection patterns to cerebellar folia crus IIA of the Norway rat. J. Comp. Neurol., 302: 768–778.

Brodal, P. and Bjaalie, J.G. (1997) Salient anatomic features of the cortico-ponto-cerebellar pathway. Prog. Brain Res., 114: 227–249.

Brodal, A. and Kawamura, K. (1980) Olivocerebellar projection: a review. In: Brodal, A., Hild, W., Van Limborgh, J., Ortmann, R., Schiebler, T.H., Töndury, G. and Wolff, E. (Eds.), Adv. Anat. Embryol. Cell Biol, Vol. 64. Springer-Verlag, Berlin, Heidelberg, New York, pp. 1–140.

Budanur, O.E., Hollands, M.A. and Marple-Horvat, D.E. (1999) Precision signalling of visual events by complex spikes and simple spikes in hemispheral (crus I) Purkinje cells. Soc. Neurosci. Abs., 25(1): 149.8.

Buisseret-Delmas, C. and Angaut, P. (1993) The cerebellar olivo-corticonuclear connections in the rat. Prog. Neurobiol., 40: 63–87.

Chapman, C.E., Spidalieri, G. and Lamarre, Y. (1986) Activity of dentate neurons during arm movements triggered by visual, auditory, and somesthetic stimuli in the monkey. J. Neurophysiol., 55: 203–226.

Classen, J., Kunesch, E., Binkofski, F., Hilperath, F., Schlaug, G., Seitz, R.J., Glickstein, M. and Freund, H.J. (1995) Subcortical origin of visuomotor apraxia. Brain, 118: 1365–1374.

Edge, A.L., Marple-Horvat, D.E. and Apps, R. (2003) Lateral cerebellum: functional localization within crus I and correspondence to cortical zones. Eur. J. Neurosci., 18: 1468–1485.

Ekerot, C.-F. and Larson, B. (1979) The dorsal spino-olivocerebellar system in the cat. I. Functional organization and termination in the anterior lobe. Exp. Brain Res., 36: 201–217.

Fadiga, E. and Pupilli, G.C. (1964) Teleceptive components of the cerebellar function. Physiol. Rev., 44: 432–486.

Garwicz, M. (2000) Micro-organisation of cerebellar modules controlling forelimb movements. Prog. Brain Res., 124: 187–199.

Garwicz, M., Apps, R. and Trott, J.R. (1996) Micro-organization of olivocerebellar and corticonuclear connections of the paravermal cerebellum in the cat. Eur. J. Neurosci., 8: 2726–2738.

Glickstein, M. (1997) Mossy-fibre sensory input to the cerebellum. Prog. Brain Res., 114: 251–259.

Glickstein, M. (1998) Cerebellum and the sensory guidance of movement. In: Bock, G.R. and Good, J.A. (Eds.), Sensory Guidance of Movement. Wiley, Chichester, pp. 252–271.

Glickstein, M., May, J.G. 3rd and Mercier, B.E. (1985) Corticopontine projection in the macaque: the distribution of labelled cortical cells after large injections of horseradish peroxidase in the pontine nuclei. J. Comp. Neurol., 235: 343–359.

Glickstein, M., Gerrits, N., Kralj-Hans, I., Mercier, B., Stein, J. and Voogd, J. (1994) Visual pontocerebellar projections in the macaque. J. Comp. Neurol., 349: 51–72.

Herrero, L., Pardoe, J. and Apps, R. (2002) Pontine and lateral reticular projections to the C_1 zone in lobulus simplex and paramedian lobule of the rat cerebellar cortex. Cerebellum, 1: 185–199.

Hollands, M.A. and Marple-Horvat, D.E. (1996) Visually guided stepping under conditions of step cycle-related denial of visual information. Exp. Brain Res., 109: 343–356.

Holmes, G. (1939) The cerebellum of man. Brain, 62: 1–30.

Kitazawa, S., Kimura, T. and Yin, T.B. (1998) Cerebellar complex spikes encode both destinations and errors in arm movements. Nature, 392: 494–497.

Kotchabhakdi, N., Walberg, F. and Brodal, A. (1978) The olivocerebellar projection in the cat studied with the method of retrograde axonal transport of horseradish peroxidase VII. The projection to lobulus simplex, crus I and II. J. Comp. Neurol., 182: 293–313.

Laurent, M. and Thomas, J.A. (1988) The role of visual information in control of a constrained motor task. J. Mot. Behav., 20: 17–37.

Maekawa, K. and Natsui, T. (1973) Climbing fiber activation of Purkinje cells in rabbit's flocculus during light stimulation of the retina. Brain Res., 59: 417–420.

Marple-Horvat, D.E., Criado, J.M. and Armstrong, D.M. (1998) Neuronal activity in the lateral cerebellum of the cat related to visual stimuli at rest, visually guided step modification, and saccadic eye movements. J. Physiol., 506: 489–514.

Marple-Horvat, D.E. and Stein, J.F. (1990) Neuronal activity in the lateral cerebellum of trained monkeys, related to visual stimuli or to eye movements. J. Physiol., 428: 595–614.

Miall, R.C. (1998) The cerebellum, predictive control and motor coordination. In: Bock, G.R. and Good, J.A. (Eds.), Sensory Guidance of Movement. Wiley, Chichester, pp. 272–290.

Miall, R.C., Weir, D.J. and Stein, J.F. (1987) Visuo-motor tracking during reversible inactivation of the cerebellum. Exp. Brain Res., 65: 455–464.

Miall, R.C., Weir, D.J., Wolpert, D.M. and Stein, J.F. (1993) Is the cerebellum a smith predictor. J. Mot. Behav., 25: 203–216.

Miall, R.C. and Wolpert, D.M. (1996) Forward models for physiological motor control. Neural Net., 9: 1265–1279.

Mushiake, H. and Strick, P.L. (1993) Preferential activity of dentate neurons during limb movements guided by vision. J. Neurophysiol., 70: 2660–2664.

Prablanc, C., Pelisson, D. and Goodale, M.A. (1986) Visual control of reaching movements without vision of the limb I. Role of retinal feedback of target position in guiding the hand. Exp. Brain Res., 62: 293–302.

Robinson, F.R., Cohen, J.L., May, J., Sestokas, A.K. and Glickstein, M. (1984) Cerebellar targets of visual pontine cells in the cat. J. Comp. Neurol., 223: 471–482.

Rosina, A. and Provini, L. (1982) Longitudinal and topographical organization of the olivary projection to the cat ansiform lobule. Neuroscience, 7: 2657–2676.

Shambes, G.M., Gibson, J.M. and Welker, W. (1978) Fractured somatotopy in granule cell tactile areas of rat cerebellar hemispheres revealed by micromapping. Brain Behav. Evol., 15: 94–140.

Simpson, J.I. and Alley, K.E. (1974) Visual climbing fiber input to rabbit vestibulo-cerebellum: a source of direction-specific information. Brain Res., 82: 302–308.

Simpson, J.I., Wylie, D.R. and De Zeeuw, C.I. (1996) On climbing fiber signals and their consequence(s). Behav. Brain Sci., 19: 384–398.

Snider, R.S. and Stowell, A. (1944) Receiving areas of the tactile, auditory and visual systems in the cerebellum. J. Neurophysiol., 7: 331–357.

Thier, P., Koehler, W. and Buettner, U.W. (1988) Neuronal activity in the dorsolateral pontine nucleus of the alert monkey modulated by visual stimuli and eye movements. Exp. Brain Res., 70: 496–512.

Trott, J.R. and Apps, R. (1991) Lateral and medial subdivisions within the olivocerebellar zones of the paravermal cortex in lobule Vb/c of the cat anterior lobe. Exp. Brain Res., 87: 126–140.

Trott, J.R. and Apps, R. (1993) Zonal organization within the projection from the inferior olive to the rostral paramedian lobule of the cat cerebellum. Eur. J. Neurosci., 5: 162–173.

Voogd, J. and Bigaré, F. (1980) Topographical distribution of olivary and cortico nuclear fibers in the cerebellum: A review. In: Courville, J. (Ed.), The Inferior Olivary Nucleus. Raven Press, New York, pp. 207–234.

Voogd, J. and Glickstein, M. (1998) The anatomy of the cerebellum. Trends Neurosci., 21: 370–375.

Voogd, J. and Ruigrok, T.J. (1997) Transverse and longitudinal patterns in the mammalian cerebellum. Prog. Brain Res., 114: 21–37.

Weir, D.J., Stein, J.F. and Miall, R.C. (1989) Cues and control strategies in visually guided tracking. J. Mot. Behav., 21: 185–204.

Zeki, S.M. (1974) Functional organization of a visual area in the posterior bank of the superior temporal sulcus of the rhesus monkey. J. Physiol., 236: 549–573.

CHAPTER 19

Coupling of hand and foot voluntary oscillations in patients suffering cerebellar ataxia: different effect of lateral or medial lesions on coordination

Gabriella Cerri[1], Roberto Esposti[1], Marco Locatelli[2] and Paolo Cavallari[3],*

[1]*Istituto di Fisiologia Umana II, Università degli Studi, via Mangiagalli 32, 20133 Milan, Italy*
[2]*Dipartimento di Scienze Neurologiche, Università degli Studi, Ospedale Policlinico, via F. Sforza 35, I-20100 Milan, Italy*
[3]*Dipartimento di Medicina, Chirurgia e Odontoiatria, Università degli Studi, Ospedale S. Paolo, Via di Rudinì 8, I-20142 Milan Italy*

Keywords: associated movements; coordination; motor control; cerebellum; human

Abstract: Motor coordination has been investigated in seven ataxic patients who underwent surgery of the cerebellar hemisphere (4) or of the vermis–paravermis region (3). Subjects, tested ipsilaterally to the lesion, were asked to couple in-phase rhythmic oscillations of the prone hand and the ipsilateral foot for at least 10 s. The oscillation frequency, paced by a metronome, ranged 0.8–3 Hz. Hand and foot angular displacements were measured by a potentiometric technique; EMG from Extensor Carpi Radialis and Tibialis Anterior was recorded by surface electrodes. The phase-relations between the hand and foot movements, as well as between the onsets of motor commands, were calculated. For each of the limbs the frequency-response curve was estimated by plotting the mean phase values between the onset of the motor command and the onset of the related movement. The experiment was repeated with the same schedule after a strong artificial increase of the hand inertial momentum (15 g m^2).

In the unloaded condition, all patients failed to achieve a hand–foot synchrony (0°), the hand movement showing a net phase-lag. In four *hemispheric* and one *vermian* patients (group 1) this lag progressively grew with frequency up to 110°, in the other two *vermian* patients (group 2) the hand lag kept almost constant ($\approx 45°$). Group 1 subjects were unable to adequate the delay between the motor commands to the increase in frequency, as instead did group 2 subjects, although this was insufficient to produce movement synchrony.

Subjects reacted to hand loading with different strategies. In group 1, due to the net increase of hand inertia, movement synchrony required a strong advance of the hand motor command. Patients succeeded in this, but because of their inability to compensate for changes in frequency, they still produced a progressive lag between movements. In group 2, loading strongly increased the hand dynamic stiffness while it slightly lowered that of the foot, resulting in a rather small difference between mechanical properties of the limbs. Thus, compensation required only a slight anticipatory activation of the hand motor command. Patients failed to do so, however they were able to adjust the command delay to the required frequency and produced a constant hand lag. Their main motor handicap was found to to be the incapability of judging the hand lag as a lack of synchrony.

*Corresponding author. Tel.: +39 2 50315459; Fax: +39 2 50315455; E-mail: paolo.cavallari@unimi.it

DOI: 10.1016/S0079-6123(04)48019-4

These results seems to indicate that the cerebellum must be involved both in measuring the time difference between hand and foot movements and in weighting this delay in function of the oscillation frequency. These two processes may be confined to the vermis–paravermis region and to the hemisphere, respectively.

Introduction

It is evident that humans are able to organise limb movements in a large variety of associations. Some of these are performed very easily and naturally. For example, it has been shown that it is very easy to couple cyclic flexion-extensions of one hand to the same oscillation of the ipsilateral foot when the two segments move in-phase, i.e., in the same angular direction (Baldissera et al., 1982, 1991). In this particular task, the nervous system takes into account the different biomechanical properties of the two limbs and achieves a rather good synchrony by adopting two different strategies: either by anticipating activity in the hand movers (the 'heavier' segment, Baldissera et al., 2000) or by equalizing the mechanical properties of hand and foot (Baldissera and Cavallari, 2001). In adults, these two mechanisms seem to coexist, although each mechanism is utilized to a different degree by males and females. In children, the central structures controlling the hand and foot coupling seems to be immature before 10 years of age and the hand biomechanical characteristics shows age-related changes, while those of the foot remains constant (Cavallari et al., 2001).

In an attempt to identify the nervous structures governing the association of in-phase movements, it has been useful to test hand–foot coupling in subjects suffering well-identified lesions of the central nervous system. It has been observed that hemiplegic patients, performing in-phase coupling on the healthy side, were not able to reach the requested synchrony, since the hand movement progressively lagged the foot movement as the frequency increased (Baldissera et al., 1994). Also a total resection of the corpus callosum led to a progressive phase-advance of the foot movement both with the hand unloaded and with the hand loaded with a supplementary mass (Baldissera and Cavallari, 2002). Altogether these studies indicate that the elaboration of the afferent information controlling coupling of ipsilateral limbs needs the cooperation of both the cerebral hemispheres.

It has been proposed that coupling of in-phase hand and foot movements may be sustained by a single voluntary command integrated by the afferent signals generated by the ongoing movement (Baldissera et al., 1982, 1991, 1994, 2000). Thus, a *rhythm generator* should be coupled to a *neural controller* so as to compensate for the mechanical differences in the limbs and to maintain a stable phase-relation between them. The neural controller may operate either by de-synchronizing the central commands to hand and foot movers, or by changing the physical properties of the two limbs. Indeed, the task designed in this study requires an internal timing system able to process temporal parameters of the movement at different levels. An *intralimb* control is needed to regulate the agonist–antagonist recruitment in the function of the limb equilibrium position at rest (Baldissera et al., 2003) and an *interlimb* control is needed to gauge the hand and foot motor commands, accounting for the different mechanical properties of the moving segments and for the frequency of oscillation.

Given these premises, the aim of the present study was to evaluate the control of hand–foot in-phase coupling in subjects suffering from severe cerebellar ataxia. Cerebellum is essential for controlling motor coordination (Miall, 1998), motor learning (Ito, 2000) posture, and vestibular functions (Timmann and Horak, 2001; Barmack, 2003; Thach and Bastian, 2004), and it appears to be especially important when the tasks entail event timing (Ivry et al., 2002). In fact, presence of cerebellar lesions results in disruption of several forms of motor behavior that require (i) regulation of agonist–antagonist activity (Ivry, 1997); (ii) event timing during fast movements (Ivry et al., 1988; Bekkelund et al., 1999; Timmann et al., 1999; Thaut, 2003); (iii) temporal control of force

(Hore and Flament, 1986; Hore et al., 2002; Nowak et al., 2002). The well-known incoordination exhibited by patients with cerebellar lesions can thus be seen as a problem in controlling and regulating the temporal patterns of movement. The role of the cerebellum in temporal processing is well-known since 1902, when Babinski coined the term dysdiadachokinesia to describe the inability of cerebellar patients to perform movements requiring rapid alternate contractions of antagonist muscles. The inappropriate timing in muscle activation has been repeatedly claimed to be the cause of this motor inaccuracy (Hore et al., 1991; Diener et al., 1993). It being engaged in the control of temporal parameters of movement, the cerebellum seems to be crucial in coordinating simple movements in complex compound ones. Indeed, a coordination specific activity of the cerebellum has been demonstrated in actions involving a single joint movement (Berardelli et al., 1996), or multijoint movements in a single limb and in different limbs (Blouin et al., 2003; Ullén et al., 2003). A significant activation of the cerebellar areas processing afferent information from the spinocerebellar system and efferent information from cortical motor areas has been demonstrated during tasks requiring coordination of the arm and the fingers (Ramnani et al., 2001) and a frequency-dependent increase in blood flow has been observed in the intermediate and lateral portions of the anterior cerebellum during bimanual movements (Jancke et al., 1999). Particularly interesting is the imaging study by Debaere et al. (2001) showing that during cyclical isodirectional coordination of wrist and ankle movements, the activation of the cerebellum together with other brain areas (SMA, CMC, PMC, S1/M1) exceeded the sum of activation observed during isolated limb movement.

Methods

Seven subjects who underwent surgical treatment, because of vascular or neoplastic lesion of the cerebellum, were recruited from the 'Dipartimento di Scienze Neurologiche' of the Università degli Studi di Milano. Patients were informed about the test, and never compelled to take part in it; all of them gave informed consent to the experimental procedure. Experiments were also authorized by the medical staff and approved by the local ethical committee in accordance with the ethical standards laid down in the 1964 Declaration of Helsinki.

In all the subjects, a single well-identified lesion affected only one side of the cerebellum, being confined to the vermis–paravermis portion in three of them (BM, 52 years, male, teratoma; GD, 54 years, female, cavernous angioma; DDA, 18 years, male, pilocytic astrocytoma) and to the lateral part of the hemisphere in the other four patients (CA, 37 years, male, medulloblastoma; BA, 30 years, male, PICA territory infarct; PD, 53 years, female, metastasis of a pulmonary adenocarcinoma; ML, 51 years, female, metastasis of an unidentified neoplasia). The time period between the surgical treatment and the experiment varied, among subjects, from 1 month to 1 year, except for BM (2 years) and BA (3.5 years). Subjects were tested on the side of the body ipsilateral to the lesion.

To identify the anatomical location of the lesion and to reconstruct its extension, a TC or MRI scan was performed both before and after their surgery. All patients underwent a complete neurological test (including eventually EEG, EMG, somato-sensory, visual, and brainstem evoked potentials) and were found free from other neurological disease. At the time of the experiments all patients were able to stand and walk without assistance, although they displayed severe signs of gait ataxia, and exhibited a sufficiently good degree of dexterity in manipulation.

Experimental procedure

During the experimental session the subject sat on an armchair with the forearm resting in a horizontal position and the leg hanging. The hand was fixed in a prone position to a light frame rotating coaxial with the wrist joint, while the foot was lying on a platform pivoting on the same axis as the ankle joint. Wrist and ankle angular displacements were recorded by a potentiometric devices positioned coaxial to the center of rotation of the joint. Electromyogram (EMG) from Extensor Carpi Radialis (ECR) and Tibialis Anterior (TA) muscles was recorded by surface silver electrodes. Both goniometric and EMG signals were amplified, digitized (National Instruments AT-MIO16-L, sampling rate of 1000 Hz per channel), visualized, and stored on a computer. In the first part of the experiment, the subject was asked to couple cyclic flexion–extensions of the hand and the ipsilateral foot, paced by a metronome. This task was achieved by associating hand extension to foot dorsal flexion and hand flexion to foot plantar flexion, so that the two segments moved isodirectionally in-phase, in a parasagittal plane. The movement amplitude was spontaneously chosen by the subject in a range between 20 and 40°. The frequency of the oscillation was varied, in successive trials, in steps of 0.2 Hz starting from 0.8 Hz up to the subject's maximal frequency, each trial lasting at least 10 s. If during the test the subject felt fatigued, or a coupling error occurred, the trial was stopped and repeated after a short rest. Trials performed at frequencies lower than 0.8 Hz and for less than 10 s were not considered.

In the second part of the experiment, both the tasks were repeated, with the same schedule, after applying a mass of 3 kg concentrically to the rotation axis of the wrist, in order to increase the inertial momentum of the limb (15 g m^2).

Interlimb phase-relation between hand and foot movements

The phase-relation between hand and foot oscillations was automatically calculated from the angular position signal. For each movement cycle, the timing of the half-excursion points, where speed is maximal and the measurement error minimal, was determined. To reduce the influence of possible asymmetries of the movement cycles, the difference in time between each hand and foot cycle was calculated by comparing the midpoint of the hand period to the respective midpoint of the foot period.

The phase-relation between the hand and the foot movements was expressed in degrees, taking each hand period as a reference. For each trial the average phase-relation and standard error were computed. Phase 0° indicates that the task was executed with the required synchrony. Conventionally, positive values indicate a phase-advance of the hand cycle.

Interlimb phase-relation between ECR and TA EMG activation

The EMG signal from ECR and TA muscles was rectified and integrated (time constant 15 ms). The onset of each EMG burst was determined by automatically selecting the point where the EMG signal crossed a threshold voltage. The threshold value was visually chosen so as to detect the very beginning of the EMG bursts. Then periods between the successive EMG bursts were measured.

The mean phase-difference between hand and foot EMG bursts was calculated by comparing the midpoints of each ECR and TA period, and expressed in degrees. For each trial, the average phase-relation and its standard deviation were computed. Phase 0° indicates that ECR activated synchronously with TA. Conventionally, positive values indicate a phase-advance of the ECR on TA burst.

Frequency-response of the hand and the foot

For each limb, the EMG periods of a single trial were summed up and averaged over ten or more cycles. Movement periods underwent the same mathematical treatment. The mean phase-relation was then calculated between the onset of the EMG burst and the onset of the related movement. Mean phase values were plotted against the movement frequency to obtain frequency-response plots.

Phase-relation between the motor command and the ensuing movement

During voluntary oscillations of a limb, the EMG activity in antagonist muscles follows a cyclic half-wave pattern (Fig. 1A) that may be conceived as one single sine-wave motor command, split to antagonists according to the oscillation ambit and to the passive equilibrium position of the limb (Baldissera et al., 2003). This perspective brings the great advantage to consider each limb as a single-input (*motor command*) single-output (*movement*) system.

Under the assumption that each limb behaves like a torsional pendulum forced by a sinusoidal input, the limb frequency-response may be then viewed as the phase-difference between the movement and its motor command ($MC \to mov$), which can be easily derived by subtracting the phase-difference between the EMG activity and the motor command from the

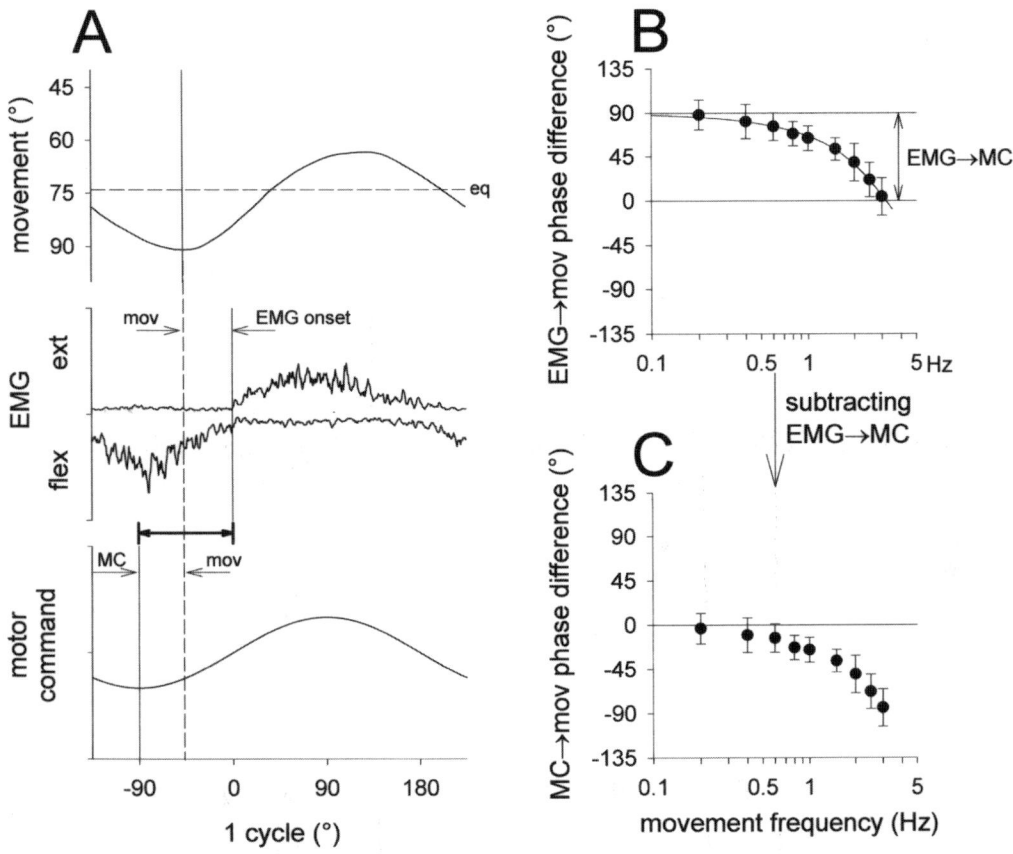

Fig. 1. Oscillation of a limb (A top graph) around its passive equilibrium position (*eq*, *dashed line*) is promoted by a pattern of alternated muscular activity in a couple of antagonist muscles (*ext* and *flex*, middle graph, *flex* trace inverted). Vertical solid lines set the onset of extension phase (*mov*) and the related muscular activity (*EMG onset*). Note that the *EMG onset* is delayed with respect to the movement onset, indicating that the first part of extension is driven only by elastic recoil. Phase-relations between *EMG onset* and movement are plotted in panel B against oscillation frequency. Positive values indicates delay of the EMG on movement. After inversion of *flex* trace, muscular activity may be conceived and visualized as one single sine-wave *motor command* (A, bottom graph). Vertical line sets the onset of the *motor command* extension phase (*MC*). Subtracting the phase-difference between *motor command* and *EMG onset* (*bold arrow*) from the phase between *EMG* and movements leads to the limb frequency response plotted in panel C. Data plotted in B and C are simulated. Solid line interpolating the data represents the fitting by the modified pendulum equation (see 'Methods' section).

$EMG \to mov$ data (Fig. 1B and 1C).

$$MC \to mov = EMG \to mov - EMG \to MC \quad (1)$$

The $EMG \to MC$ phase-difference has been estimated by fitting the $EMG \to mov$ data (see below) and extrapolating the phase value at zero frequency, where the biomechanical properties of the limb do not introduce phase-delays (i.e., where the $MC \to mov$ phase-difference is almost zero). Moreover, provided that oscillations are always performed into a fixed range, $EMG \to MC$ phase-difference is independent of the oscillation frequency.

Fitting procedure

The $EMG \to mov$ data were fitted with a modified pendulum equation (Baldissera et al., 2003) that takes into account both the frequency-dependent effect of the biomechanical properties and the frequency-independent $EMG \to MC$ contribution.

$$\Phi(f) = -\arctan\frac{\gamma f}{\pi(f_r^2 - f^2)} + y_0 \qquad (2)$$

This equation determines the value of the phase-difference $\Phi(f)$ at the oscillation frequency f as a function of the biomechanical properties of the limb, expressed by the resonance frequency f_r and the damping coefficient γ, and of the $EMG \to MC$ contribution y_0. Best fit was achieved using a least squared error procedure that led to an estimate of the mean f_r, γ, and y_0 values, with their respective SE. The quality of fitting was judged acceptable when R^2 coefficient was greater than 0.65.

The Gaussian distribution of f_r and γ parameters was ascertained by applying the Kolmogorov–Smirnov test for continuous data on the result of a Monte Carlo simulation (cf. Christopoulos, 1998) performed on Eq. (2). A direct comparison between couples of f_r or γ parameters belonging to different curves could then be performed by means of a Student's t analysis (cf. Motulsky, 1999) (see Fig. 1).

Phase-relation between motor commands to hand and to foot muscles

The phase-relation between motor commands driving the hand and foot ($MC_F \to MC_H$ curve) was derived from the phase-relation between the EMG onsets of the two limbs' prime movers ($EMG_H \to EMG_F$ curve), once the phase-differences between the EMG onset and the motor command in each limb has been estimated (see above).

$$MC_F \to MC_H = EMG_F \to EMG_H + \\ EMG_H \to MC_H - EMG_F \to MC_F \qquad (3)$$

The dependence of the phase-relation between motor commands on frequency was assessed by means of a linear regression.

Results

Hand free

All subjects with primary cerebellar lesions were able to perform in-phase oscillations of the hand and the ipsilateral foot up to frequencies of 2.2–2.8 Hz, i.e., always slightly lower than those attained by healthy subjects (maximal frequency 3.0–3.2 Hz, Baldissera et al., 2000).

When measuring the phase-relations between hand and foot movements, at the different frequencies (interlimb curves, see 'Methods' section), it was possible to classify the seven subjects into two categories of motor behavior: those who showed a frequency-dependent lag of the hand with respect to foot movement (CA, BA, PD, ML, DDA) and those who were able to maintain an almost constant phase-lag between hand and foot oscillations, despite the increase of movement frequency (BM, GD).

The *interlimb curves* measured in patients belonging to the first group are shown in Fig. 2B. Filled circles outline the frequency dependent phase-lag of the hand movement with respect of foot oscillation. The hand lagged the foot by about 30° at the lowest frequency and this lag increased to 90° above 2.6 Hz, indicating the difficulty of coordinating the two movements at this speed. Linear regression of the movement data gave a slope of $-36.9 \pm 4.5°\,\text{Hz}^{-1}$ ($p < 0.0001$) and an intercept not significantly different from zero. Also the phase between hand and foot motor commands (open circles) was frequency dependent and displayed a similar decay. Linear regression of these data showed a highly significant ($p < 0.0001$) dependence on the oscillation frequency, with a slope of $-30.0 \pm 4.6°\,\text{Hz}^{-1}$ and an intercept not significantly different from zero. To account for the progressive phase-lag, these subjects must have generated an almost constant time-delay between hand and foot activation. This is explicit in Fig. 2C, where the time difference between hand and foot motor commands is shown to be constant (about 81.84 ± 9.4 ms) all over the frequency range.

The phase-difference between the motor command and the ensuing movement was estimated at each frequency of oscillation (*frequency-response curves*, see 'Methods' section). In the first group of patients, the hand phase-response (Fig. 2A, filled diamonds)

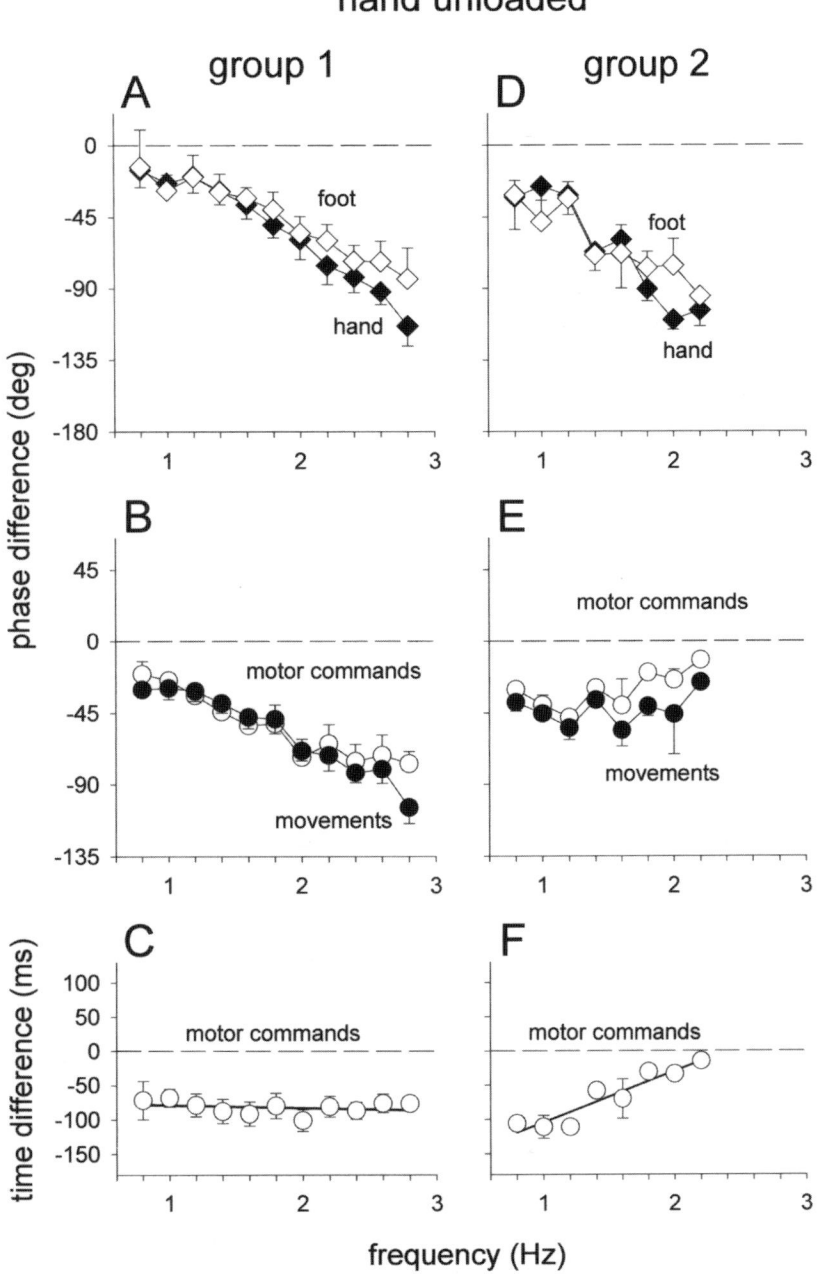

Fig. 2. In-phase coupling of hand and foot oscillations in two groups of ataxic patients, with the hand unloaded. A and D, frequency-responses of the hand (*filled diamonds*) and foot (*open diamonds*). Negative ordinates indicate a lag between movement and motor command. B and E, interlimb phase-relations, i.e., the phase-difference between hand and foot motor commands (*open circles*) and between hand and foot movements (*filled circles*). Dashed line indicates synchrony, negative values indicate that the hand *motor command*, or *movement*, lags that of the foot. *Open circles* in C and F display the difference between motor commands in the time domain, straight line depicts the linear regression. Each symbol plots the mean value (+1 SD) of 5 subjects in group 1 and 2 subjects in group 2.

shows that within 0.8–1.2 Hz the hand movement followed the motor command with an almost constant lag (about 20°). At higher frequencies this lag increased, reaching about 110° at 2.8 Hz. The foot curve (Fig. 2A, open diamonds) when superimposed on to that of the hand showed a synchrony up to 2 Hz, after which its decay was just less steep. At 2.8 Hz the phase-difference between motor commands and foot movement attained 84°. The segment biomechanical properties, estimated from raw experimental data, fitted with the modified pendulum equation (see 'Methods' section), were expressed in terms of resonance frequency (f_r, which depends on the ratio between elasticity and inertial momentum) and damping coefficient (γ, which depends on the ratio between viscosity and inertial momentum). The hand resonance frequency was 2.48 Hz and the relative damping factor had a value of 5.55 s^{-1}. These data were not significantly different from those calculated for the foot (f_r 2.9 Hz and γ 8.50 s^{-1}, respectively). Because of the similarity between the biomechanical properties of hand and foot, to produce synchronous movements it was expected that subjects just synchronized their muscle activity. This was not the case. In fact, these subjects could generate a fixed time-delay of about 81 ms between the hand and foot activation, which was responsible for the phase-shift of 30° between the hand and foot movements at 0.8 Hz and obviously resulted in a progressive increase of the movement phase-difference, as the frequency increased.

By contrast, subjects BM and GD, with lesions confined to the vermis–paravermis, were able to maintain an almost constant phase-lag between hand and foot oscillations despite the increase of movement frequency. The average phase-relation between hand and foot movements (Fig. 2E, filled circles) shows that the hand lagged the foot movement of about 38° at the lowest frequency and that this lag remained almost unchanged up to 2.0 Hz. On an average, all over the frequency range, the hand lagged the foot movement by 43 ± 9°. Linear regression of the movement data gave a slope of $-2.6 \pm 9.3°$ Hz^{-1} ($p < 0.7804$) and an intercept of $-41.7 \pm 13.7°$ indicating the independence of the phase-difference from the frequency of oscillation. Also the phase-difference between of the hand and foot motor commands (Fig. 2E, open circles) was almost constant up to 1.6 Hz (about 40°), and then slowly decreased to 12° at 2.2 Hz. Thus, these two subjects were sensitive to changes in frequency and were able to compensate for it. The course of this compensation is illustrated in Fig. 2F by the progressive reduction (from about 100 to 15 ms) of the delay between the onset of the hand and the foot motor command, as the frequency increases. Linear regression gave a slope of $74 \pm 10°$ Hz^{-1} ($p < 0.0001$) and an intercept of $-178.77°$.

As in the first group of patients, the mechanical properties of the hand and foot did not differ significantly from each other. This can be very well understood by the clear superimposition of the phase-response of the two limbs (Fig. 2D, filled and open diamonds). A rather constant phase-delay of the movement, with respect of the motor command, was found between 0.8 and 1.2 Hz (30 ± 4° for the hand and 37 ± 9° for the foot, respectively). Thereafter the two curves decayed almost parallel, being 104 ± 14° the delay of the hand movement at 2.2 Hz and 95° that of the foot at the same frequency. The hand resonance frequency and the relative damping factor (f_r 1.85 Hz and γ 4.69 s^{-1}, respectively) were not significantly different from those calculated for the foot (f_r 2.18 Hz and γ 9.95 s^{-1}, respectively). Due to the similarity between the biomechanical properties of the two segments, to synchronize the two oscillations a synchronous motor command was then needed. In fact, these subjects could generate a variable time-delay between hand and foot commands which could well compensate for the change in frequency, but this was not sufficient to keep a strict 0° phase value.

Hand loaded

In healthy subjects, due to the difference in the mechanical properties of the two limbs, an advance of the hand on foot activation is needed to produce almost synchronous movements (Baldissera et al., 2000). This advance is evidently enhanced when increasing the inertia of the hand (Baldissera and Cavallari, 2001). Also in ataxic patients this advance was present, but the level of compensation for the new condition was different in the two groups.

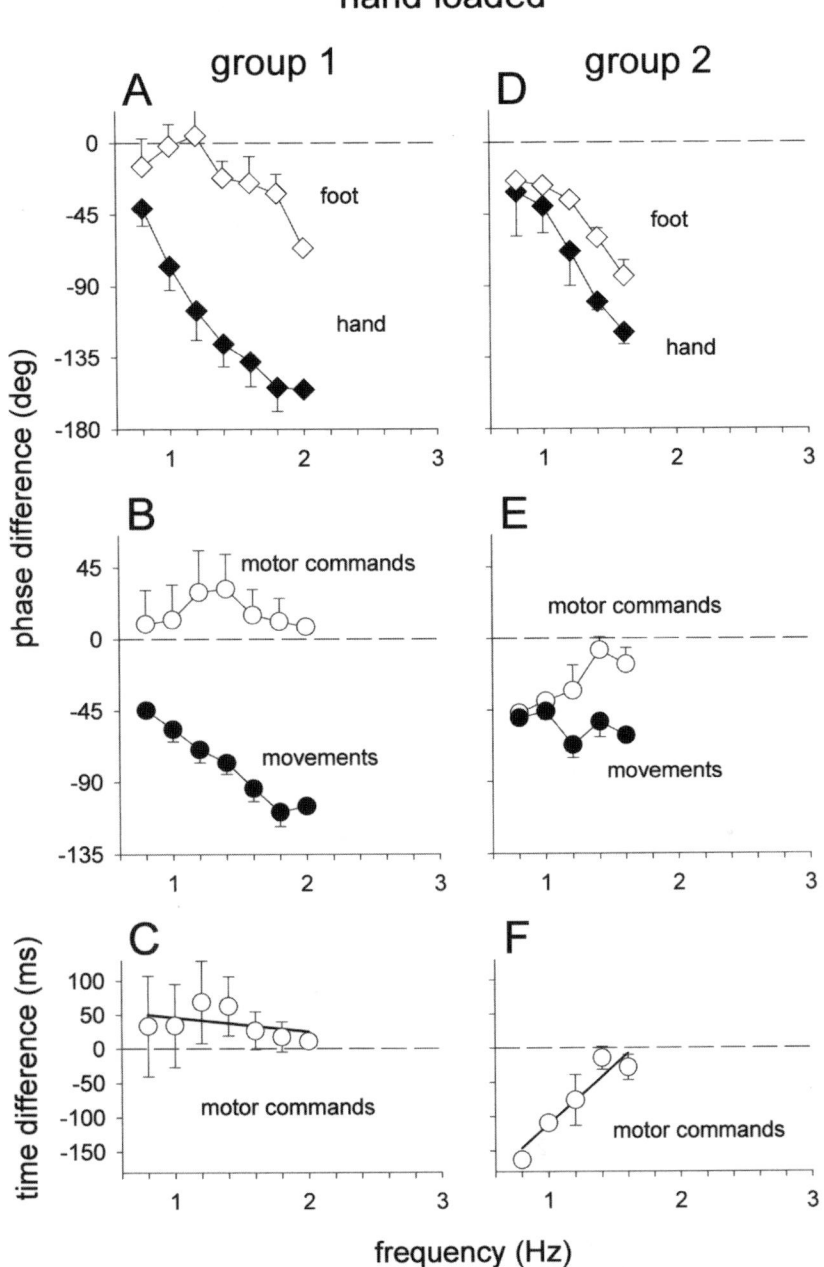

Fig. 3. In-phase coupling of hand and foot oscillations in two groups of ataxic patients, with an inertial load added to the hand. A and D, frequency-responses of the hand (*filled diamonds*) and foot (*open diamonds*). Negative ordinates indicate a lag between the movement and the motor command. B and E, interlimb phase-relations, i.e., the phase-difference between hand and foot motor commands (*open circles*) and between hand and foot movements (*filled circles*). Dashed line indicates synchrony, negative values indicate that the hand *motor command*, or *movement*, lags that of the foot. *Open circles* in C and F display the difference between motor commands in the time domain, straight line depicts the linear regression. Each symbol plots the mean value (+1 SD) of 3 subjects in group 1 and 2 subjects in group 2.

In group 1 patients (those unable to adapt to changes in frequency) the effect of adding an external load to the hand produced a dramatic worsening of their performance and for two of them (BA and ML) the task was in fact impossible. As shown in Fig. 3B (filled circles), the phase-difference between movements of the two limbs was largely higher than that observed with the hand unloaded: in the range between 0.8 and 2.0 Hz the movement curve quickly decayed from 45° to more than 110° and the maximal frequency was reduced to 1.8–2.0 Hz. Linear regression of the movement data showed an intercept not different from zero, and a slope of $-59.23 \pm 7.20°$ Hz^{-1}, a value significantly worse than that obtained with the hand unloaded ($p < 0.01$). This occurred despite the fact that hand motor command constantly advanced that of the foot (Fig. 3B, open circles), i.e., a clear compensation for the new mechanical properties occurred. In the new situation the hand advance was in average $16.8 \pm 9.7°$, without any dependence on the oscillation frequency. This advance is also confirmed by the course of the time difference between hand and foot motor commands depicted in Fig. 3C. On an average the foot lagged the hand command of about 36.4 ± 21.7 ms.

The hand frequency-response (Fig. 3A, filled diamonds) showed that the application of the load strongly and significantly ($p < 0.0001$) reduced its resonance frequency to 1.09 ± 0.18 Hz; the damping factor assuming a value of 2.13 s^{-1}. Also, the foot frequency-response (Fig. 3A, open diamonds) was slightly changed in its decay and resulted in a significantly different ($p < 0.02$) resonance frequency from that measured in the unloaded condition. The foot resonance frequency decreased to 2.09 ± 0.20 Hz and the damping factor assumed a value of 1.37 s^{-1}. After charging the hand, the resonance frequency of the two limbs became significantly different ($p < 0.01$). Given these premises, it can be concluded that although these patients could feel the changes in the biomechanical characteristics of the hand and partly compensate for it, the incapability to estimate a change in frequency strongly impaired their coordination ability.

In the second group of patients, those sensible to changes in frequency, applying load on the hand was almost ineffective in changing the phase-lag between movements, but efficiently reduced the maximal frequency of oscillation to 1.6 Hz. As shown in Fig. 3E (filled circles), the lag of the hand movement was rather constant (in average $55.3 \pm 8.5°$) and close to the average value measured in unloaded conditions. The load compensation is revealed by the hand lag in the motor command curve (Fig. 3E, open circles). Note that when frequency increases from 0.8 to 1.6 Hz the hand lag reduces from 47 to 16° and, consequently, the delay between the motor commands reduces from 160 to 30 ms (Fig. 3F). A significant ($p < 0.001$) dependence on the oscillation frequency was assessed by the linear regression, which gave a slope of $46.71 \pm 17.25°$ Hz^{-1} and an intercept of $-84.80 \pm 21.94°$.

In these subjects, hand loading resulted in a nonsignificant decrease of the resonance frequency from 1.85 to 1.35 Hz and in a change in the damping coefficient from 4.69 to 2.42 s^{-1}. Thus the hand curve became steeper than in unloaded condition, and the phase-lag between motor command and movement rapidly passed from 31° at 0.8 Hz to 100° at 1.6 Hz. The foot resonance frequency (1.62 Hz) and the relative damping factor (2.96 s^{-1}) were not significantly different from those calculated for the hand. At a difference from what happened in group 1 patients, loading induced only a marginal phase-difference between the hand and foot frequency responses. This difference, although not statistically significant, is however revealed by the changes in the motor commands both in the phase (Fig. 3E, open circles) and in the time domain (Fig. 3F). In this last case, it is quite apparent that the significant increase in the slope (from 74.5 to 173° Hz^{-1}, $p < 0.02$), has been induced by the compensation mechanism. Owing to the similarity between the biomechanical properties of the two segments, to synchronize the two oscillations, an almost synchronous motor command was then needed. The patients, however, failed to do so. In fact, they produced a constant hand movement lag of about 55°. Therefore, their main motor handicap resulted in an incapability to judge this hand lag as a lack of synchrony.

Discussion

The aim of the present study was to assess the role of cerebellum in coordinating isodirectional oscillations of the ipsilateral hand and foot.

The first observation is that all subjects with lesions affecting the cerebellum failed to achieve a synchrony between the two moving limbs, the hand movement showing a clear phase-lag with respect to the foot. Given the similarity between hand and foot biomechanical properties, to produce synchronous movements subjects were required only to synchronize their muscle activity. Note, instead, that even at 0.8 Hz, the lowest tested frequency, the lag was approximately identical in both groups (30° and 38°, respectively), and significantly higher than that observed in normal subjects (less than 15°, Baldissera et al., 2000). The seven patients studied were ranked in two categories of motor behavior: those in whom the hand phase-lag progressively increased with frequency (CA, BA, PD, ML, DDA) and those in whom the hand lag remained almost constant (BM, GD). Subjects of the first group adopted a motor program fully inadequate for the task. In fact, they displayed a time shift of about 80 ms between the hand and foot motor commands and they were not able to modify it with an increase in frequency. Subjects ranked in the second group were instead able to modify the delay between motor commands, however this was not sufficient to produce synchronous movements. This result seems to indicate that the cerebellum must be involved, although to different extents, both in estimating the time difference between hand and foot oscillatory movements and in weighting this delay in function of the oscillation frequency. The conversion of the delay into phase is, in fact, crucial to assess the quality of synchronism and eventually correct the timing between motor commands. The conversion from time-to-phase delay needs an estimate of the movement frequency (or period). Since all subjects voluntarily oscillated their limb, it seems to be reasonable to speculate that such an information is present at the cortical level and it would reach the cerebellum through the cortico-cerebellar pathway. This speculation is supported by the fact that group 1 patients, who produced a constant time-delay between hand and foot commands, abolishing this time-to-phase conversion, mainly suffered from lesions of the lateral part of the cerebellar hemispheres.

A comment would be pertinent regarding the ranking of the patients in the two groups. When analyzing the localization of the cerebellar lesions, it became apparent that the first group was mainly characterized by people suffering lesions confined to the lateral part of the cerebellar hemisphere, while the second group comprised people suffering lesions confined to the vermis–paravermis. The only exception was DDA whose MRI scan showed a major lesion of the vermis–paravermis, but displayed a motor behavior similar to the one adopted by group 1. Since it was not possible for us to detect the exact border of the surgical lesion, we are not in position to know if a lesion in the deep border had affected some cerebellar nuclei. Should this be the case, this subject would display a behavior that results from the impairment of both the hemisphere and paravermis. We are aware of the fact that this paper reports a very small number of cases, but the possible correlation of a specific functional behavior with a specific site of lesion in the cerebellum would be very suggestive. Although it is beyond the scope of this chapter to attribute different functions to medial and lateral cerebellum, it being our prime interest to assess the role of cerebellum in interlimb coordination, our findings are in agreement with several anatomical, clinical, and experimental observations. The vermis–paravermis is the only region of the cerebellum receiving direct spinal afferents (see Voogd (2003) for a review). In the vermis, somatosensory afferents from the head and proximal segments are implemented with vestibular, auditory, and visual signals, while the paravermis (intermediate) receives mainly somatosensory information from the limbs. This seems to be true also in humans in whom activation of the spinocerebellum following median nerve stimulation has been demonstrated by Tesche and Karhu (1997) and recently confirmed by the MEG study of Hashimoto et al. (2003). Projections from spinocerebellum mainly reach the brain stem regions controlling the descending systems. The lateral hemisphere, instead, receives projections mainly from the cerebral cortex (motor areas, frontal eye fields) and closes the loop projecting back to the same areas. Moreover, Ivry et al. (1988) demonstrated that, in man, the lateral regions of the cerebellum are required in the operation of the timing process, while the medial regions are involved in implementation and execution of motor responses, and Theoret et al. (2001) reported that repetitive transcranial magnetic stimulation differently affects the accuracy of finger tapping when applied to the lateral or the medial cerebellum.

Our results also demonstrate that whatever the kind of cerebellar lesion, the biomechanical properties of the limbs, and in particular those of the foot, were altered as compared to normal subjects. In particular, we show here that in all ataxic patients, the foot frequency-response in unloaded condition do not differ significantly from the hand frequency-response. In fact, in a subject belonging to group 1 the foot resonant frequency was decreased with respect to healthy subjects, thus causing the overlap between the hand and foot curves. In group 2 patients, a substantial decrease in the hand resonance frequency was accompanied by an even more accentuated reduction of that of the foot, which thus overlapped the hand curve. In contrast, in both male and female healthy subjects, the foot resonance frequency is always significantly higher than that of the hand (Baldissera and Cavallari, 2001). The more severe effect on the foot than on the hand biomechanical parameters may result from the fact that in a majority of the subjects, the lesion was localized in the posterior part of the cerebellum, which is known to mainly control the leg. This suspicion is also supported by the more severe ataxia in gait than in manipulation exhibited by the patients. Since the resonant frequency depends on the ratio between elasticity and inertia, and given that the cerebellar lesions are not expected to affect the mass of the limb, a possible explanation for this result may be that the inability to coordinate the agonist–antagonist activation may cause a decrease in the 'dynamic' elasticity during the oscillations. This result would be in keeping with the fact that cerebellum has a significant action in directing multijoint movement by controlling joint stiffness (Smith, 1993). In fact, cerebellar patients are known to exhibit slowed movement execution in terms of a reduced ratio of peak velocity to maximum amplitude (i.e., stiffness, Ackermann et al., 1997) and, indeed, one of the classical signs of cerebellar damage is the hypermetria due to decreased and delayed rate of rise of the antagonist EMG activity (Manto and Bosse, 2003).

To further explore the role of cerebellum in processing and compensating for the biomechanical properties of the moving limbs, the mechanical difference between the two limbs has been accentuated by applying an inertial load to the hand. It is, in fact, known that cerebellum is able to monitor the inertial load of the limbs: Purkinje cell activity is significantly modulated when grasping and lifting objects of different weight (Espinoza and Smith, 1990) and cerebellum seems to predict the dynamic interaction torques generated by different moving joints and compensate for them (Bastian et al., 1996). Moreover, neurologically silent cerebellar lesions may be unveiled by addition of a mass to the moving hand (Manto et al., 1995): in fact, facing an increased inertia of the moving limb, patients presenting a lesion of the lateral cerebellum are able to appropriately increase the intensity of the agonist, but are unable to adapt the intensity in the antagonist muscles.

The macroscopic effect of loading the hand was that under these conditions two subjects of the first group were unable to perform the task and that the maximal frequency of oscillation dropped by about 1 Hz in the other five. However, all subjects reacted to the increase in load, but again with different strategies in the two groups. In the first group the load significantly reduced the hand resonance frequency, so that it became significantly different from that of the foot. Given these premises, the movement synchrony would have now required a parallel advance of the hand motor command. Subjects succeeded in reversing the pattern of the motor command, i.e., displayed a phase-advance of the hand on the foot, however this was not sufficient to reach movement synchrony due to their inability to compensate for the change in frequency of oscillation. So, despite the sensible time advance of the hand command, the lack of delay-to-phase conversion produced a progressive phase-lag between movements.

Application of the load to the hand of subjects belonging to the second group lowered the resonance frequency of both limbs despite the fact that no load was applied to the foot. Anyway, the effect on the hand was dominant. It is apparent that group 2 patients reacted to loading by strongly rising the dynamic stiffness of the hand and by slightly lowering that of the foot. By observing the frequency responses of the two limbs, one may appreciate that in these patients the effect of hand loading did not increase much the differences between the biomechanical properties of the two segments, as it did in group 1 patients. Thus, to compensate for the load, a slight anticipatory activation of the hand motor command was expected, but actually the patients produced at all frequencies a lag of the hand command. However, due to the ability to transform delay into phase, these patients were able to adequate delay to frequency, thus producing an almost constant phase-lag of the hand movement. The fact that both in unloaded and in loaded conditions group 2 patients displayed an almost similar phase-lag between hand and foot movement (about 45°), suggests that their main motor handicap may relay on the incapability to be aware of judging this phase-delay as lack of synchrony, i.e., they seem unable to analyze and/or elaborate the timing from the kinaesthetic afferent signals.

It has been pointed out (in Introduction section) that kinaesthetic signals produced by the ongoing movement are critical in controlling the synchrony between the two oscillating segments (Baldissera et al., 1991, 1994; Baldissera and Cavallari, 2001). This means that the 'neural controller' producing synchrony between movements (i) has to be fed by afferent signals from the moving limbs, (ii) has to compute a signal error, and (iii) has to eventually correct the motor commands. In order to include the cerebellum as a part of the 'neural controller' for associated movements, it is thus necessary to define if and how the afferent signals related to the different kinematic parameters are encoded during cyclic limb movement in the discharge of cerebellar neurons. This is a very complex issue which is still unsolved, but several authors have shown that discharge properties of the cortical and nuclear cerebellar neurons is significantly modulated during forelimb and ankle passive cyclic movements (Perciavalle et al., 1998; Bosco et al., 2000). Recent evidence suggest that during voluntary arm movements, mossy fiber afferents to the cerebellar cortex provide information about movement direction and speed, and it has been proposed that the cerebellum uses these signals to control movement velocity during both step and tracking arm movements (see Ebner (1998) for a review). It seems also that the cerebellum plays an important role in regulating the stereotypic organization of complex goal-directed movements, including the temporal correlation among joint angle velocities (Milak et al., 1997).

A neural network model, which relies on kinematic parameters has been recently proposed (Contreras-Vidal et al., 1997). The model illustrates how a central pattern generator in the cortex and the basal ganglia, a neuromuscular force controller in spinal cord, and an adaptive cerebellum cooperate to reduce motor variability during multijoint arm movements using mono- and bi-articular muscles. Cerebellar learning modifies velocity commands to produce phasic antagonist bursts at interpositus nucleus cells whose feed-forward action overcomes inherent limitations of spinal feedback control of tracking. Excitation of alpha motoneuron pools, combined with inhibition of their Renshaw cells by the cerebellum, facilitate movement initiation, and optimal execution. Transcerebellar pathways are opened by learning through long-term depression (LTD) of parallel fiber–Purkinje cell synapses in response to conjunctive stimulation of parallel fibers and climbing fiber discharges that signal muscle stretch errors. The cerebellar circuitry also learns to control opponent muscles pairs, allowing cocontraction and reciprocal inhibition of muscles. Learning is stable, exhibits load compensation properties, and generalizes better across movement speeds if motoneuron pools obey the size principle. Simulated lesions of the cerebellar network reproduce symptoms of cerebellar disease, including sluggish movement onsets, poor execution of multijoint plans, and abnormally prolonged endpoint oscillations.

Conclusions

In conclusion, the integrity of the cerebellum is required to perform correct association of interlimb movements. From an analysis of our results, it is possible to speculate that different parts of the cerebellum contribute to different extents to the control of the performance: the lateral hemisphere seems more involved in generating the initial motor program, i.e., it accounts for the timing of the central commands to the limbs, adjusting the agonist–

antagonist activity within each limb and between the two limbs while the intermediate cerebellum (vermis–paravermis) may be more involved in monitoring the afferent signals of the ongoing movements, in error detection and correction, in order to adjust the central motor command to the actual limb position.

Acknowledgments

This study was supported by the 'Ministero della Università e della Ricerca' and the 'Università degli Studi di Milano.' We are indebted to Prof. Fausto Baldissera for helpful comments and suggestions and to Prof. Roberto Villani for clinical supervision.

References

Ackermann, H., Hertrich, I., Daum, I., Scharf, G. and Spieker, S. (1997) Kinematic analysis of articulatory movements in central motor disorders. Mov. Disord., 12(6): 1019–1027.

Babinski, J. (1902) Sur le rôle du cervelet dans les actes volitionnels necessitant une succession rapide de mouvements (Diadococinesie). Rev. Neurol. (Paris), 10: 1013–1015.

Baldissera, F., Cavallari, P. and Civaschi, P. (1982) Preferential coupling between voluntary movements of ipsilateral limbs. Neurosci. Lett., 34(1): 95–100.

Baldissera, F., Cavallari, P., Marini, G. and Tassone, G. (1991) Differential control of in-phase and anti-phase coupling of rhythmic movements of ipsilateral hand and foot. Exp. Brain Res., 83(2): 375–380.

Baldissera, F., Cavallari, P. and Tesio, L. (1994) Coordination of cyclic coupled movements of hand and foot in normal subjects and on the healthy side of hemiplegic patients. In: Swinnen S., Heuer H., Massion J. and Casaer P. (Eds.), Interlimb Coordination, Neural, Dynamical and Cognitive Constraints. Academic Press, San Diego, pp. 229–242.

Baldissera, F., Borroni, P. and Cavallari, P. (2000) Neural compensation for mechanical differences between hand and foot during coupled oscillations of the two segments. Exp. Brain Res., 133(2): 165–177.

Baldissera, F. and Cavallari, P. (2001) Neural compensation for mechanical loading of the hand during coupled oscillations of the hand and foot. Exp. Brain Res., 139(1): 18–29.

Baldissera, F. and Cavallari, P. (2002) Impairment in the control of coupled cyclic movements of ipsilateral hand and foot after total callosotomy. Acta Psychol., 110(2–3): 289–304.

Baldissera, F., Cavallari, P. and Esposti, R. (2003) Foot equilibrium position control partirion of voluntary command to antagonists during foot oscillation. Exp. Brain Res. (e-pub), DOI: 10.1007/s00221-003-1723-y.

Barmack, N.H. (2003) Central vestibular system: vestibular nuclei and posterior cerebellum. Brain Res. Bull., 60(5–6): 511–541.

Bastian, A.J., Martin, T.A., Keating, J.G. and Thach, W.T. (1996) Cerebellar ataxia: abnormal control of interaction torques across multiple joints. J. Neurophysiol., 76(1): 492–509.

Bekkelund, S.I., Pierre-Jerome, C., Winther, J. and Mellgren, S.I. (1999) Relationship between brain structure size and performing rapid limb movements. A quantitative magnetic resonance study. Eur. Neurol., 42(4): 185–189.

Berardelli, A., Hallett, M., Rothwell, J.C., Agostino, R., Manfredi, M., Thompson, P.D. and Marsden, C.D. (1996) Single-joint rapid arm movements in normal subjects and in patients with motor disorders. Brain, 119(2): 661–674.

Blouin, J.S., Bard, C. and Paillard, J. (2003) Contribution of the cerebellum to self initiated synchronized movements: a PET study. Exp. Brain Res. (e-pub), DOI: 10.1007/s00221-003-1709-9.

Bosco, G., Giaquinta, G., Valle, M.S., Caserta, C., Casabona, C., Casabona, A. and Perciavalle, V. (2000) Distribution of spinocerebellar Purkinje cell responses to passive forelimb movements in the rat. Eur. J. Neurosci., 12(11): 4063–4073.

Cavallari, P., Cerri, G. and Baldissera, F. (2001) Coordination of coupled hand and foot movements during childhood. Exp. Brain. Res., 141(3): 398–409.

Christopoulos, A. (1998) Assessing the distribution of parameters in models of ligand-receptors interaction: to log or not to log. Trends Pharmacol. Sci., 19: 351–357.

Contreras-Vidal, J.L., Grossberg, S. and Bullock, D. (1997) A neural model of cerebellar learning for arm movement control: cortico-spino-cerebellar dynamics. Learning Memory, 3(6): 475–502.

Debaere, F., Swinnen, S.P., Béatse, E., Sunaert, S., Van Hecke, P. and Duysens, J. (2001) Brain areas involved in interlimb coordination: a distributed network. NeuroImage, 14: 947–958.

Diener, H.C., Hore, J., Ivry, R. and Dichgans, J. (1993) Cerebellar disfunction of movement and perception. Can. J. Neurol. Sci., Suppl. 3: S62–S69.

Ebner, T.J. (1998) A role for the cerebellum in the control of limb movement velocity. Curr. Opin. Neurobiol., 8(6): 762–769.

Espinoza, E. and Smith, A.M. (1990) Purkinje cell simple spike activity during grasping and lifting objects of different textures and weights. J. Neurophysiol., 64(3): 698–714.

Hashimoto, I., Kimura, T., Tanosaki, M., Iguchi, Y. and Sekihara, K. (2003) Muscle afferent inputs from the hand activate human cerebellum sequentially through parallel and climbing fiber systems. Clin. Neurophysiol., 114(11): 2107–2117.

Hore, J. and Flament, D. (1986) Evidence that a disordered servo-like mechanism contributed to tremor in movements in cerebellar dysfunction. J. Neurophysiol., 56(1): 123–136.

Hore, J., Wild, B. and Diener, H.C. (1991) Cerebellar dysmetria at the elbow, wrist, and fingers. J. Neurophysiol., 65(3): 563–571.

Hore, J., Timmann, D. and Watts, S. (2002) Disorders in timing and force of finger opening in overarm throws made by cerebellar subjects. Ann. NY Acad. Sci., 978: 1–15.

Ito, M. (2000) Mechanisms of motor learning in the cerebellum. Brain Res., 886(1–2): 237–245.

Ivry, R. (1997) Cerebellar timing system. Int. Rev. Neurobiol., 41: 555–573.

Ivry, R.B., Keele, S.W. and Diener, H.C. (1988) Dissociation of the lateral and medial cerebellum in movement timing and in movement execution. Exp. Brain Res., 73(1): 167–180.

Ivry, R.B., Spencer, R.M., Zelaznik, H.N. and Diedrichsen, J. (2002) The cerebellum and event timing. Ann. NY Acad. Sci., 978: 302–317.

Jancke, L., Specht, K., Mirzazade, S. and Peters, M. (1999) The effect of finger-movement speed of the dominant and the subdominant hand on cerebellar activation: A functional magnetic resonance imaging study. NeuroImage, 9(5): 497–507.

Manto, M.U., Godaux, E. and Jacqui, J. (1995) Detection of silent cerebellar lesions by increasing the inertial load of the moving hand. Ann. Neurol., 37(3): 344–350.

Manto, M.U. and Bosse, P. (2003) A second mechanism of increase of cerebellar hypermetria in humans. J. Physiol., 547(3): 989–994.

Miall, R.C. (1998) The cerebellum, predictive control and motor coordination. Novartis Found. Symp., 218: 272–290.

Milak, M.S., Shimansky, Y., Bracha, V. and Bloedel, J.R. (1997) Effects of inactivating individual cerebellar nuclei on the performance and retention of an operantly conditioned forelimb movement. J. Neurophysiol., 78(2): 939–959.

Motulsky, H. (1999) Analyzing data with the GraphPad Prism. GraphPad Software Inc., San Diego.

Nowak, D.A., Hermsdorfer, J., Marquardt, C. and Fuchs, H.H. (2002) Grip and load force coupling during discrete vertical arm movements with a grasped object in cerebellar atrophy. Exp. Brain Res., 145: 28–39.

Perciavalle, V., Bosco, G. and Poppele, R.E. (1998) Spatial organization of proprioception in the cat spinocerebellum. Purkinje cell responses to passive foot rotation. Eur. J. Neurosci., 10(6): 1975–1985.

Ramnani, N., Toni, I., Passingham, R.E. and Haggard, P. (2001) The cerebellum and parietal cortex play a specific role in coordination: a PET study. NeuroImage, 14: 899–911.

Smith, A.M. (1993) Babinski and movement synergism. Rev. Neurol., 149(12): 764–770.

Tesche, C.D. and Karhu, J. (1997) Somatosensory evoked magnetic fields arising from sources in the human cerebellum. Brain Res., 744(1): 23–31.

Thaut, M.H. (2003) Neural basis of rhythmic timing networks in the human brain. Ann. NY Acad. Sci., 999: 364–373.

Thach, W.T. and Bastian, A.J. (2004) Role of the cerebellum in the control and adaptation of gait in health and disease. Prog. Brain Res., 143: 353–366.

Theoret, H., Haque, J. and Pasqual-Leone, A. (2001) Increased variability of paced finger tapping accuracy following repetitive magnetic stimulation of the cerebellum in humans. Neurosci. Lett., 306(1–2): 29–32.

Timmann, D., Watts, S. and Hore, J. (1999) Failure of cerebellar patients to time finger opening precisely causes ball high-low inaccuracy in overarm throws. J. Neurophysiol., 82(1): 103–114.

Timmann, D. and Horak, F.B. (2001) Perturbed step initiation in cerebellar subjects: 2. Modification of anticipatory postural adjustments. Exp. Brain Res., 141(1): 110–120.

UllénF., Forssberg, H. and Ehrsson, H. (2003) Neural networks for the coordination of the hands in time. J. Neurophysiol., 89: 1126–1135.

Voogd, J. (2003) The human cerebellum. J. Chem. Neuroanat., 26: 243–252.

Modulation of cutaneous reflexes in hindlimb muscles during locomotion in the freely walking rat: A model for studying cerebellar involvement in the adaptive control of reflexes during rhythmic movements

R. Bronsing, J. van der Burg and T.J.H. Ruigrok*

Department of Neuroscience, Erasmus MC Rotterdam, P.O. Box 1738, 3000DR Rotterdam, The Netherlands

Keywords: rhythmic movements; flexors; extensors; EMG; central pattern generator; cerebellum

Abstract: This study aims to demonstrate stepphase-dependent modulation in the gain of cutaneously triggered reflexes in the freely locomoting rat. Electromyographic recordings of biceps femoris (mainly involved in knee flexion) and gastrocnemius (mainly involved in ankle extension) muscles were continuously monitored during locomotion and cutaneous reflexes were induced by subcutaneously placed stimulation electrodes in the lateral malleolal region.

The results show that the reflex responses in both muscles during locomotion were generally reduced compared to reflexes induces in rest. For the biceps femoris reduction of reflex gain was highest during the stance phase whereas for the gastrocnemius the period of highest depression was found during the swing phase.

We conclude that stepphase-dependent modulation of peripheral reflexes can be measured in freely locomoting rats and generally concur with previous studies in cat and man that this type of modulation may be functionally important for maintaining and adjusting gait. Moreover, although the mechanism of inducing and maintaining this modulation is not fully known, it is now open to experimental investigation in rodents.

Introduction

Reflexes play a major role in the proper execution of ongoing motor behavior. For example, during walking, individuals depend heavily on their peripheral reflexes not only for adjusting to possible perturbations but also for initiating the correct sequence and duration of motor programs necessary for smooth walking (Duysens, 1977; Degtyarenko

---

*Corresponding author. Tel.: (+31) 10 408 7296; Fax: (+31) 10 408 9459; E-mail: t.Ruigrok@erasmusmc.nl

et al., 1998; McCrea, 2001). Although a spinal cord-based central program generator can provide the basic rhythm necessary for evoking a locomotor pattern, feedback from the body and limbs are used extensively and continuously to monitor and correct the resulting motor behavior in a reflexive way. In order to be able to make optimal use of these reflexes, it has become clear that they are not rigidly established but are adjusted to perform optimally in any given situation. In the case of locomotion it has become well established in both cat (Duysens, 1977; Duysens and Loeb, 1980; Drew and Rossignol, 1987;

Seki and Yamaguchi, 1997) and man (Stein and Capaday, 1988; Duysens et al., 1990; Duysens and Tax, 1993; Sinkjaer et al., 1996) that muscle reflexes induced by cutaneous stimulation are modulated with respect to the phase of the step cycle in which the stimulus was presented to the limb. Also, in other rhythmic movements such as hopping or chewing it has become apparent that reflexes show a phase-dependent modulation (van der Bilt et al., 1997; Hauglustaine et al., 2001; Hiraoka, 2001). In all these instances, it would appear that the reflexes are adjusted in such a way that they are of optimal use to the individual in that particular situation (Pearson et al., 1995).

Although it is not precisely known how the state-dependent modulation of reflexes is induced, there is some evidence that the corticospinal tract may be involved (Hayes et al., 1992; Iles and Pisini, 1992; Canu and Falempin, 1996). However, a cerebellar origin of reflex modulation can at least be suspected, since it is well known that cerebellar damage can have a profound effect on the execution of reflexes (Bloedel and Bracha, 1995). Also, the correcting influence of the cerebellum on the gain of the vestibulo-ocular reflex (VOR) has been well established (Ito, 1984; De Zeeuw et al., 1998; Ito, 1998). In this model, it is hypothesized that the cerebellum specifically functions by continuously adjusting the gain of the VOR on the basis of minimizing retinal slip. In analogy, it is attractive to assume that peripheral reflexes also, may be continuously monitored and adjusted by the cerebellum in a learning dependent way.

In order to further investigate the potential impact of the cerebellum on the control of reflex modulation during locomotion, we are developing a rat model (Ruigrok et al., 1996). The present study was aimed to show that step cycle dependent modulation of peripheral reflexes can be recorded in freely locomoting rats.

Experimental procedures

For these experiments a total of 14 male Wistar rats were used. The experiments were approved by the faculty's committee on animal experiments and all surgical procedures used adhered to the National Institute of Health guidelines.

Training

The rats initially underwent an intensive training to walk on a motor-driven rotating drum set at a speed of 10 m/min. Daily training sessions lasted at least 1 h for up to 10 days. After this training period, 10 rats were able to walk smoothly and continuously for at least one minute, which was followed by a 30 s to 1-min rest interval. After 10 of these 1-min walking periods, the rats were allowed to rest for up to 3 min. Some of these rats ($n = 3$) were also trained to walk at higher speeds (up to 20 m/min). When the rats were sufficiently trained, they were prepared for surgery in order to be able to chronically record muscle activity and deliver cutaneous stimuli.

Surgical procedure

Before surgery the rat was anaesthetized using halothane (5% initial, 0.5 to 2% maintenance in 0.5 L/min oxygen and 1.0 L/min nitrous oxide). Surgery on the rats consisted of implanting custom-made bipolar EMG electrodes, made of isolated stainless steel with a diameter 150 µm (Advent Research Materials Inc., Oxford, UK), in the posterior part biceps femoris (BF) and medial gastrocnemius (GC) muscles of the left hind limb of the rat. The individual wires were attached to the tip of a fine injection needle, which was pulled through the muscle for several millimeters. The isolation had been removed for approximately 2 mm near the tip of the electrode where the diameter had been enlarged by a knot covered by a drop of insulation and which was not pulled through the muscle fascia (see Fig. 1). The emerging part of the electrode was secured to the fascia of the muscle by a small suture. In a similar way, two custom-made stimulation electrodes (isolated stainless steel, diameter 150 µm) were implanted subcutaneously on the left hindlimb just lateral to the triceps surae tendon and within the innervation area of the sural nerve. With the aid of a long but thin injection needle the leads were pulled subcutaneously to the area that was exposed for identifying and implanting the muscles where they were further secured by suturing them to subcutaneous fascia. One stimulating wire was led posterior from the lateral malleolus, whereas the other passed it anteriorly.

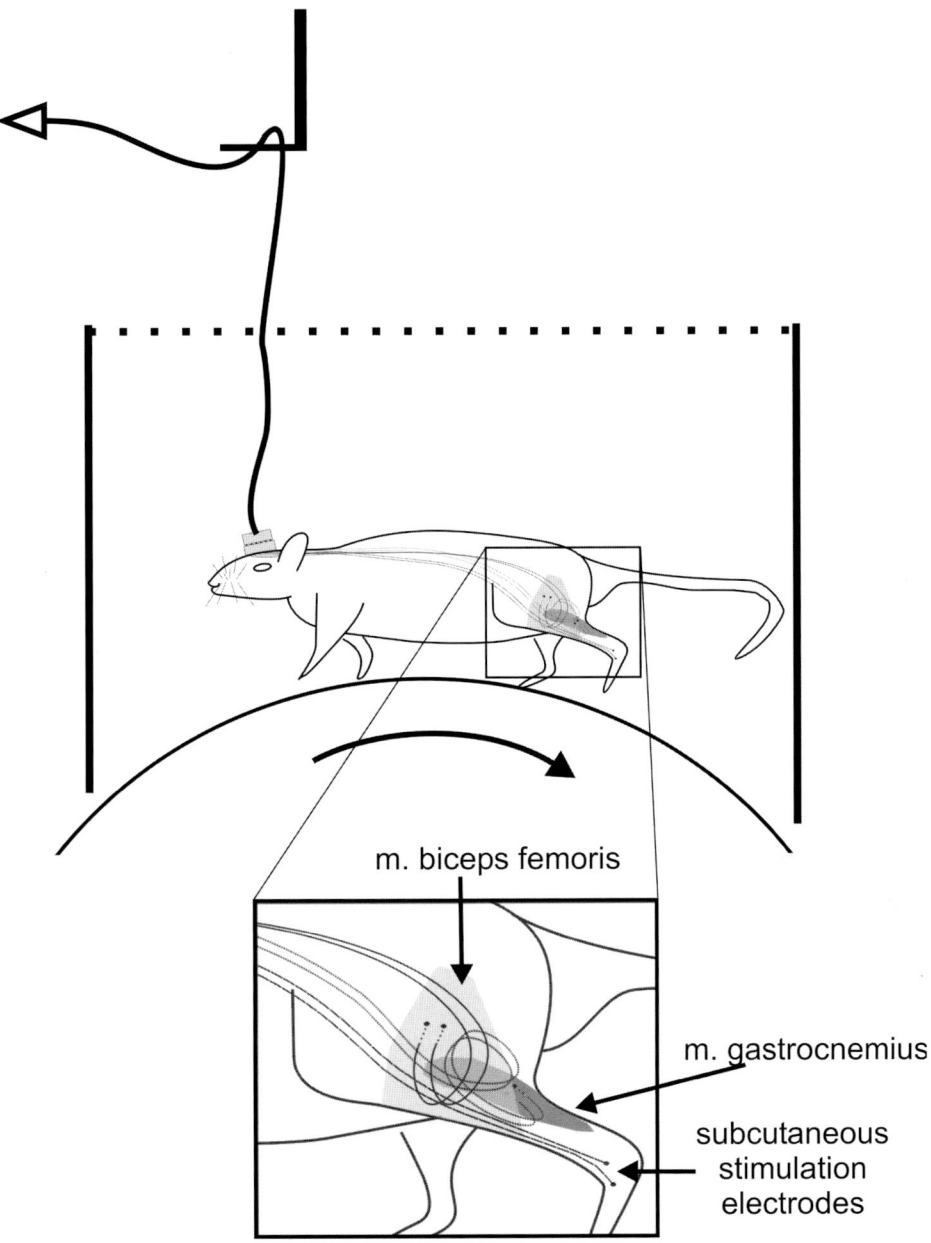

Fig. 1. Schematic diagram depicting the approximate position of the recording and stimulation electrodes (also see inset). The loops of the wires to the stimulation electrodes have not been shown for clarity. All wires were subcutaneously led to the head of the rat where the wires were connected to a connector that was mounted to the skull with dental cement and screws.

All wires, after providing sufficient slack by looping them in the exposed muscle region (Fig. 1), were led subcutaneously to the head where the wires were attached to a connector secured in a pedestal of dental cement that was fixed to the skull with screws.

Data acquisition

After a recovery phase of 2 to 3 days, the rats were subjected to daily test sessions consisting of up to 45 1-min intervals of treadmill walking. Usually, the rats would allow to be connected to the experimental set-up by inserting the connector plug into the pedestal socket. In some cases, very light halothane anaesthesia was necessary. All test sessions started with several minutes of normal walking. During these normal walking bouts no stimuli were induced and they served to obtain an average step and corresponding EMG pattern. Subsequently, the reflex threshold (T) was determined in a rest situation. Single stimuli typically lasted 0.1 ms and thresholds were usually found between 100 and 300 µA. Thresholds could change somewhat from day to day. Finally, the rats were subjected to a number of walking periods (stimulated walk) with randomly evoked stimuli with a stimulus intensity of 1.5 T unless indicated otherwise. These stimuli were administered to the rat using two Gaussian clocks programmed to generate stimuli with an average frequency of 0.4 Hz and with an interstimulus interval of at least 0.5 s. EMG signals were amplified using a Grass amplifier (on-line low-pass filtering at 10 KHz and high-pass filtering at 30 Hz, and including 50 Hz notch filtering) and individual traces were digitized at 10,000 Hz using a 1401 Plus^{TM} (CED, Cambridge, UK) and stored on disk for off-line analysis using the Spike2^{TM} data acquisition and analysis package (CED). Stimuli were generated using the stimulus sequencer of the 1401 Plus^{TM} (CED) that triggered a PG4000 digital pulse generator and isolated stimulation unit (SIU: both Neuro Data Instruments Corp, New York, USA).

Analysis

Off-line analysis of EMG data were performed with the Spike2^{TM} and MATLAB^{TM} software packages. First, EMG signals were rectified, low-pass filtered (FC = 1000 Hz, 101 kernel points) and any DC offset was removed. In Spike2^{TM}, a script was written to locate stimuli and export the stimuli and signals to the MATLAB^{TM} software package.

The first step in the analysis, after rectification and filtering, is the construction of an average step cycle EMG for the recorded muscles that will serve as the normal ongoing EMG activity during the step cycle. At least 35 and up to 100 step cycles were used to compute this average (Fig. 2). It was noted that the biceps EMG contained a very regular and sharply delineated burst of activity. From observation it was noted that this burst indicated the onset of the swing phase, which would correspond with the main function of this muscle, i.e., start of knee flexion (Pratt et al., 1991; Gillis and Biewener, 2001). Since the onset of this burst could be easily identified, it was chosen as the start of a step cycle. The step cycles were averaged and the step cycle length was normalized to 100% of the step cycle.

Responses to stimuli were extracted from the raw Spike2 data files, whilst omitting the first 6 to 10 ms in order to remove the stimulus artefacts. Subsequently, the individual reflex responses to delivered stimuli were corrected for normal ongoing activity by subtracting the averaged step cycle from the response EMG trace. The resulting response was numerically integrated over 150 ms using the trapezoid approximation algorithm for numerical integration (Zehr et al., 1997) and were plotted against the phase of the step cycle as a scatter plot. The y-axis of the scatterplots shows the size of the response relative to the maximum response in the experiment.

Results

Normal walking

Most of the trained rats ($n = 10$) were able to walk steadily on the rotating drum. This is exemplified by the EMG traces of the BF and GC muscles taken during normal walking of rat rb009 (Fig. 2A). In Fig. 2B, the distribution of the step cycle time of 98 consecutively executed steps of the same rat is shown. Note that the distribution is approximately Gaussian and, in this case, resulted in an average step cycle of 643 ± 156 ms. At a drum speed of 10 m/min, this would indicate an average step length of 107 mm. When calculated for 10 animals and including all usable records of each animal, the average step cycle duration was 710 ± 285 ms.

BF activity shows a conspicuous burst of activity that could be visually related to the onset of movement and reflected the period of knee flexion. Usually in between two bursts of activity a period with a lower level of activity could be registered which coincided with activity of the GC muscle. Since

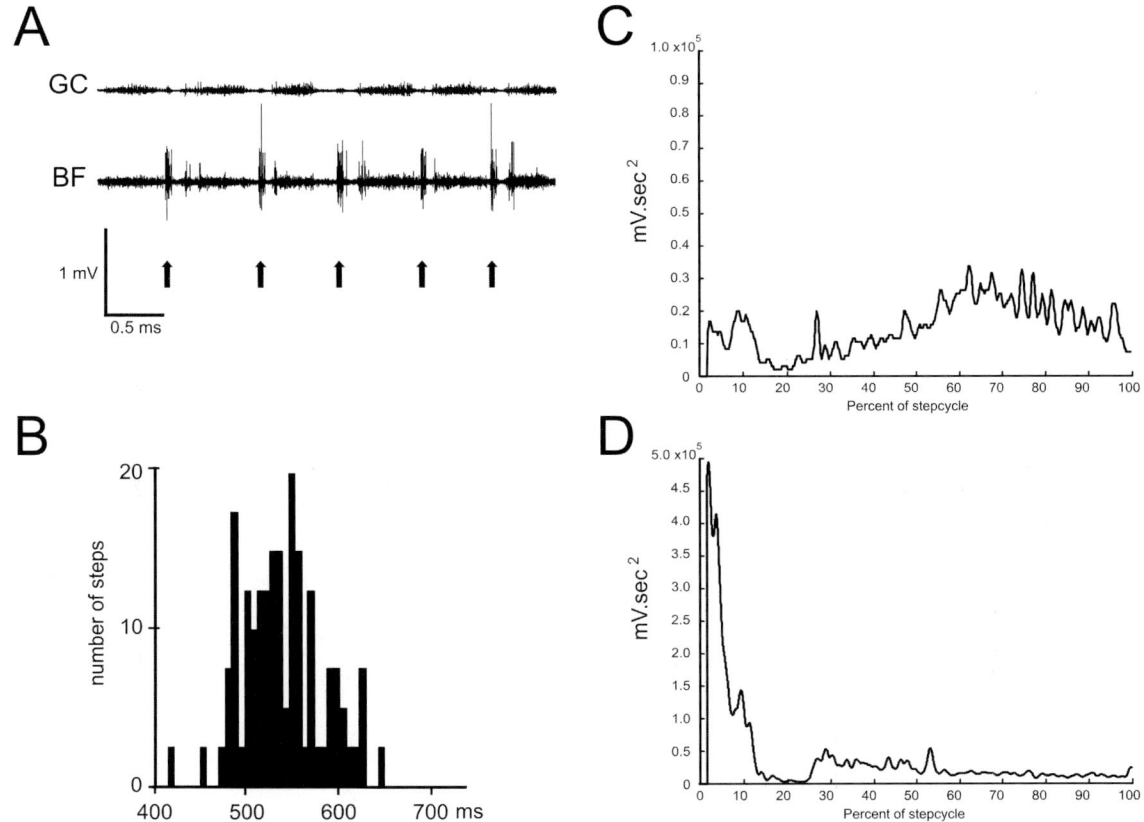

Fig. 2. A: Example of EMG traces for gastrocnemius muscle (top trace) and biceps femoris muscle (lower trace) under normal walking conditions at 10 m/min. The traces show a clear alternating pattern in activation. The arrows in the lower trace indicate the biceps burst, the onset of which is taken as the beginning of the step cycle. B: Histogram showing the distribution of step cycle length under normal walking conditions in the same rat. The distribution is approximately Gaussian (mean step cycle = 643 ± 156 ms, $n = 98$). C, D: Averaged and normalised EMG pattern (see experimental set up) of the gastrocnemius (C) and biceps femoris (D) muscles during a step cycle. The amplitude of the gastrocnemius muscle, averaged over 89 steps in this particular case is about 10 times lower than that of biceps femoris muscle. These averaged and normalized EMG values were used to correct the reflex response for normal, ongoing activity in the muscles.

the GC clearly is a muscle that functions as an ankle extensor, it shows its main activation when the animal is bearing its weight, i.e., during the stance phase. BF activity during this period may help in stabilizing the knee and hip joints. Averaging the BF and GC activity over at least 35 step cycles clearly shows the mean pattern of activity during an average step cycle (Fig. 2C, D).

Cutaneously induced reflexes

Suprathreshold stimulation of subcutaneously placed electrodes resulted in a short withdrawal movement of the limb. After a few of these stimuli were delivered the rat did not show any long latency reactions, although the twitch-like response of the limb was maintained. During walking, the rat did not show obvious signs of stress or unwillingness to perform the walking test while receiving the randomly delivered stimulations. The time of stimulation was registered with respect to the previously recorded BF burst. Figure 3A shows the EMG traces for both muscles (GC, upper trace; BF, lower trace) while the rat was randomly stimulated (strength 150 μA, 150 μs) at the indicated times (arrows). The onsets of the biceps burst, and therefore, by definition, the onsets of the step cycle are denoted by bullets. Note

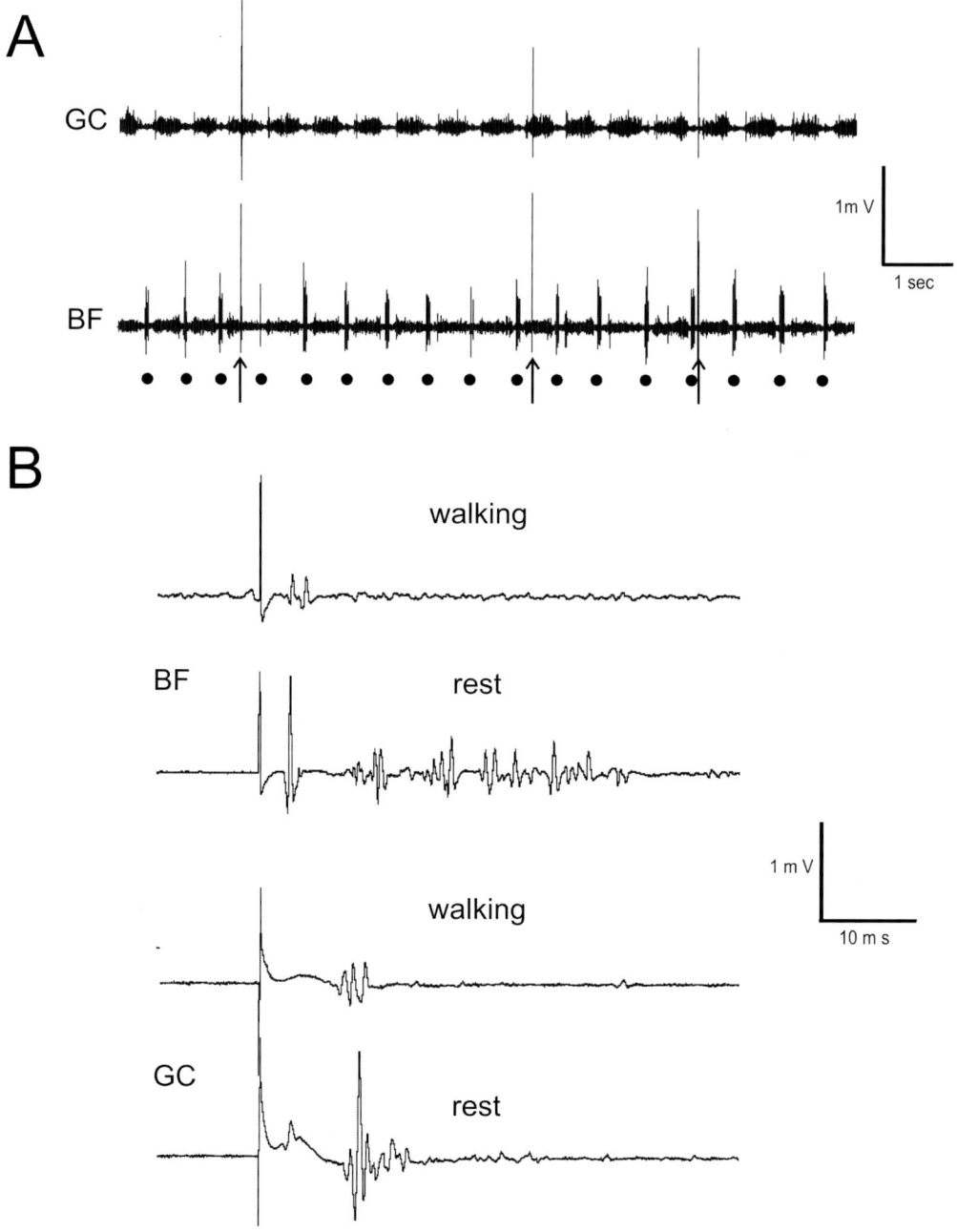

Fig. 3. A: Example of EMG traces for both gastrocnemius muscle (top trace: GC) and biceps femoris (BF) muscle while walking at 10 m/min under stimulus conditions (1.5 T). Arrows denote the occurrence of a stimulus. Note the stimulus artifacts in the EMG traces. B: Magnification of the first reflex shown in A (top and third trace for BF and GC, respectively) which was elicited in mid-stance, compared with a reflex at 1.5 T triggered at rest (second trace and bottom traces for BF and GC, respectively), indicating that, normally, reflexes elicited in both BF and GC are reduced under walking conditions when compared to rest situations.

that the stimulations resulted in a stimulus artefact, which is cut from the data before analysis, followed by a reflex response. However, the general rhythm and walking speed was not notably affected by the stimulus. Examples of reflex responses recorded at rest and during walking are shown in Fig. 3B, C. Note that reflexes are clearly reduced during walking.

Modulation of reflex response

After establishing the stimulation threshold and determining the average reflex response at rest using a 1.5x stimulation threshold (1.5 T), rats were tested for reflexes during several periods of walking. Figure 4 shows the size of the integrated reflex responses plotted against the relative moment the stimulus was delivered at in this particular step cycle. The EMG signals were integrated over 150 ms and corrected for the average muscle activity during the same period. All data points were obtained from several walking periods taken from the same test session. It is obvious that the response sizes are generally reduced compared to the reflexes recorded during rest. Moreover, in BF, the response is usually highest during or just after the BF burst, i.e., during the swing phase as compared to the stance phase. However, in GC the responses are highest during the stance phase, i.e., the phase in which the GC is most active. The responses during the swing phase can become smaller than the average normal, ongoing activity in the GC during the swing phase resulting in a negative value of the integration.

Effect of varying stimulation strength

By comparing Figs. 3 and 4, it can be noted that the periods demonstrating the highest level of muscle activity also tended to show the least depression of reflex activity. In order to see if this particular phase relation of the depression was in any way related to the method of analysis or to the intensity of stimulation, we have investigated the responses of the BF muscle using a graded series of stimulation intensities in all studied rats. As an indicator that our method of analysis did not induce inadvertent modulation of reflex strengths we have employed stimulus trigger sequences with zero stimulation strength (0 T). In the follow-up walking periods these were followed by 2 T and 3 T. As expected, and as is shown in Fig. 5 for rat Rb018, the subthreshold stimuli never resulted in an apparent step-cycle related modulation of responses. This justifies the conclusion that the method of calculation does not inherently introduce structure to the scatter plot. The scatter plots in Fig. 5B (2T) and 5C (3T) show that under these circumstances of increased stimulus strength also, the modulation of reflexes during walking is remarkably similar to the modulation under normal stimulation conditions (at 1.5 T).

Effect of varying walking speed

In order to investigate the effect of different walking speeds on the level and pattern of reflex modulation, rats were also required to walk at 20 m/min. At this speed, the basic gait pattern was quite similar to walking at 10 m/min (Gillis and Biewener, 2001). Figure 6A shows the reflex modulation in BF muscle during a normal experiment, whereas in Fig. 6B the modulation during the fast walk experiments is shown. It will be obvious that under both conditions, step phase dependent modulation of reflexes is present in this muscle but that the level of depression may show some minimal adaptations.

Reproducibility of modulation

Since some of these chronically prepared rats could be tested for several weeks, we have compared the pattern of modulation for the BF muscle at several consecutive days. Despite changes in stimulus thresholds the overall modulation pattern remained remarkably similar over these periods.

Figure 7 summarizes the data by showing the averaged data for both the BF ($n = 10$) and GC ($n = 6$) muscles as obtained in all rats. The step cycle was divided into 20 bins of 5% each and the rat averages within a bin are shown. The onset of BF burst at 0% is used as indicator of start of the swing phase. The graph clearly shows that in the biceps femoris muscle reflexes are less inhibited during the swing phase of the step cycle than during the stance phase. In contrast, the medial gastrocnemius muscle

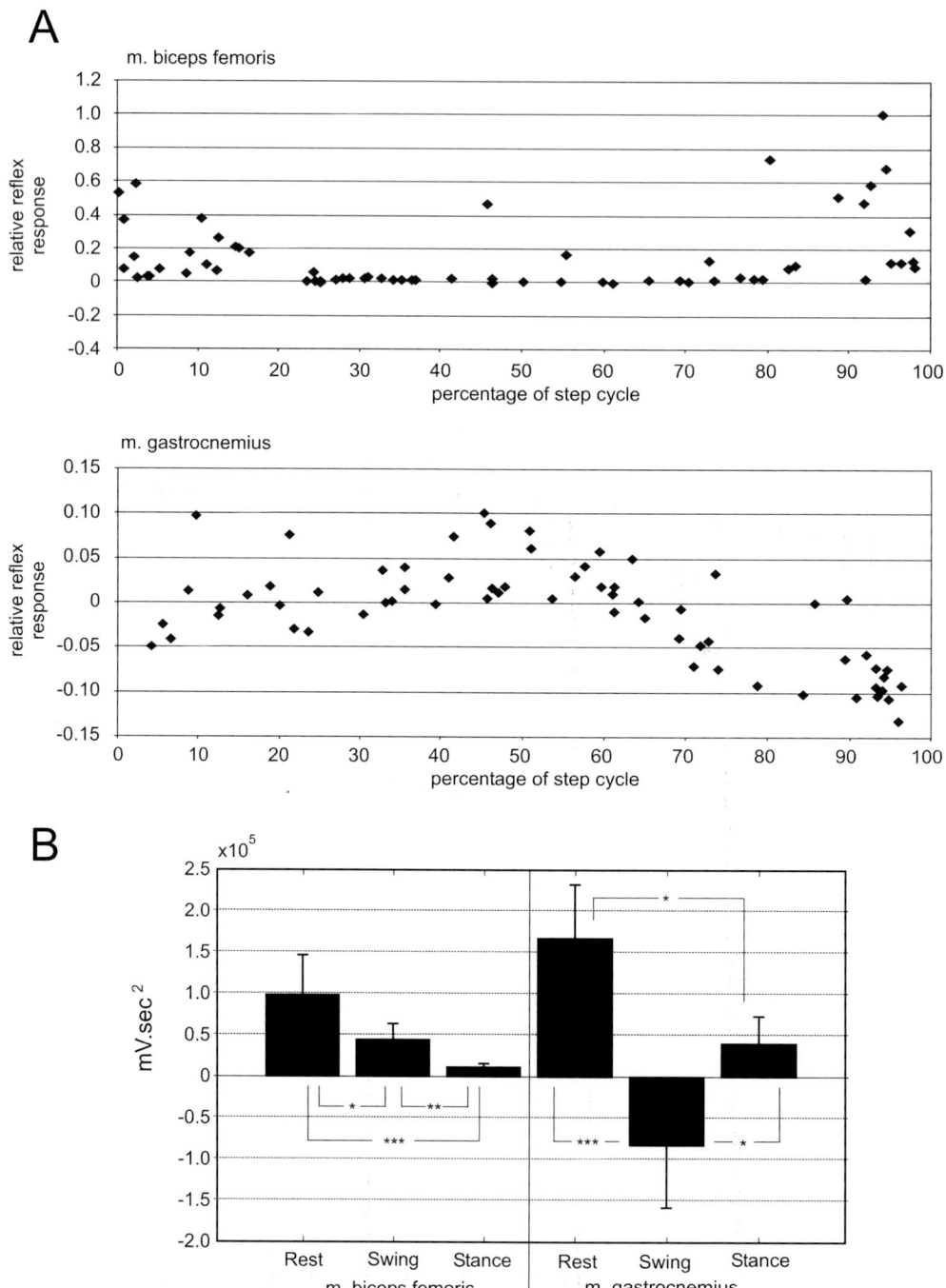

Fig. 4. (A) Scatter plots showing the modulation of cutaneous reflexes in both biceps femoris (top panel) and gastrocnemius muscles (bottom panel). Note that the trends in the modulation appear to be in counter phase with each other. Also note different scales on the response axes indicating a more severe reduction of reflexes in GC compared to BF. B. Pooled data showing average reflexes in BF and GC elicited under resting conditions and compared with averaged reflexes in swing and stance phases. Stars denote significance levels i.e., 0.05, 0.01, and 0.005, respectively. Error bars indicate standard deviation.

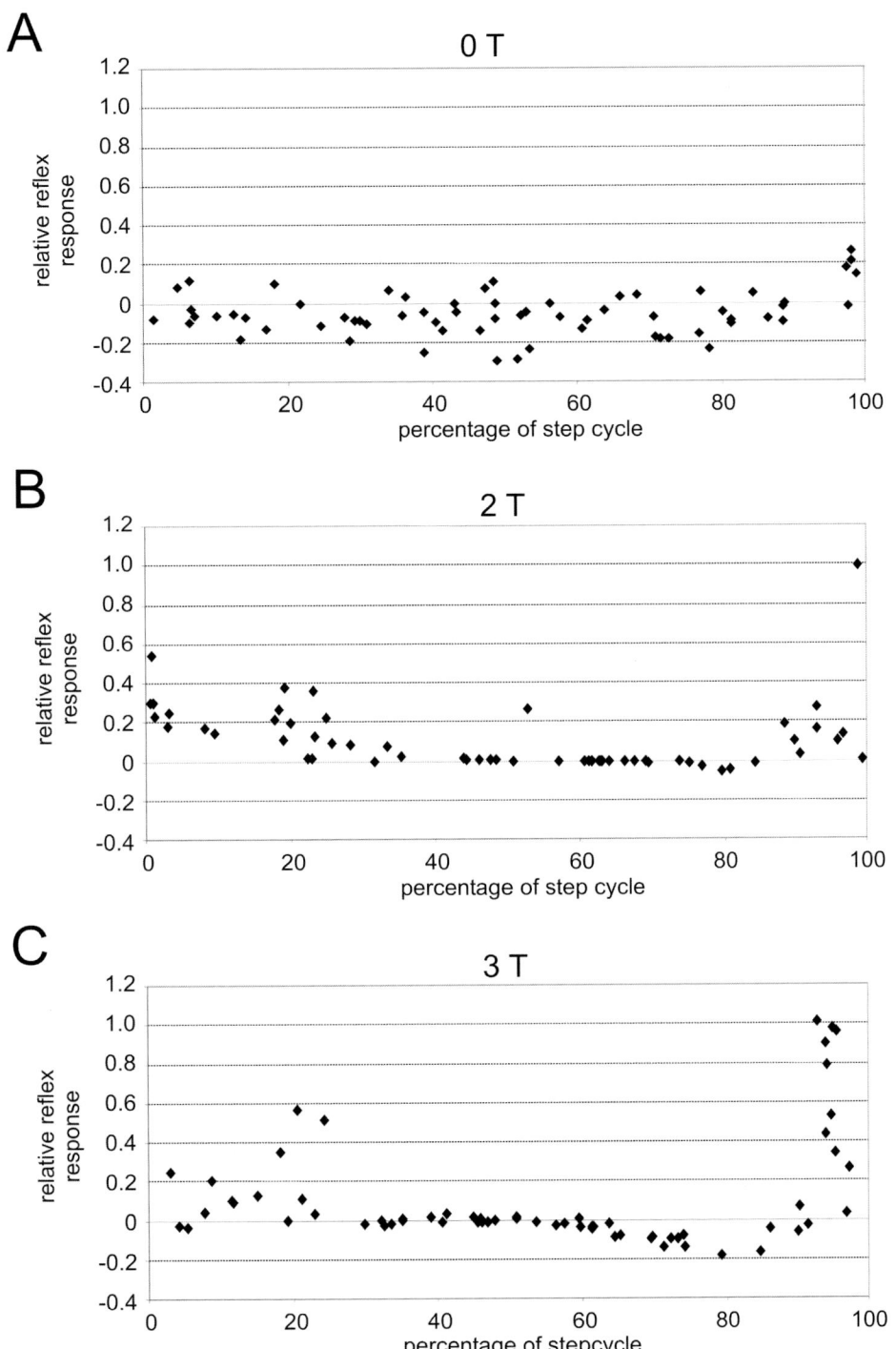

Fig. 5. Modulation of cutaneous reflexes in biceps femoris when stimulating at sub-threshold (0 T: A), 2 T (B) and 3 T (C) stimulation intensities. The responses are divided by the average response triggered under rest conditions using 2 T stimuli. Note that the modulation pattern is nearly identical in B and C, but cannot be discerned in A, indicating that the analysis process did not induce a systematic pattern.

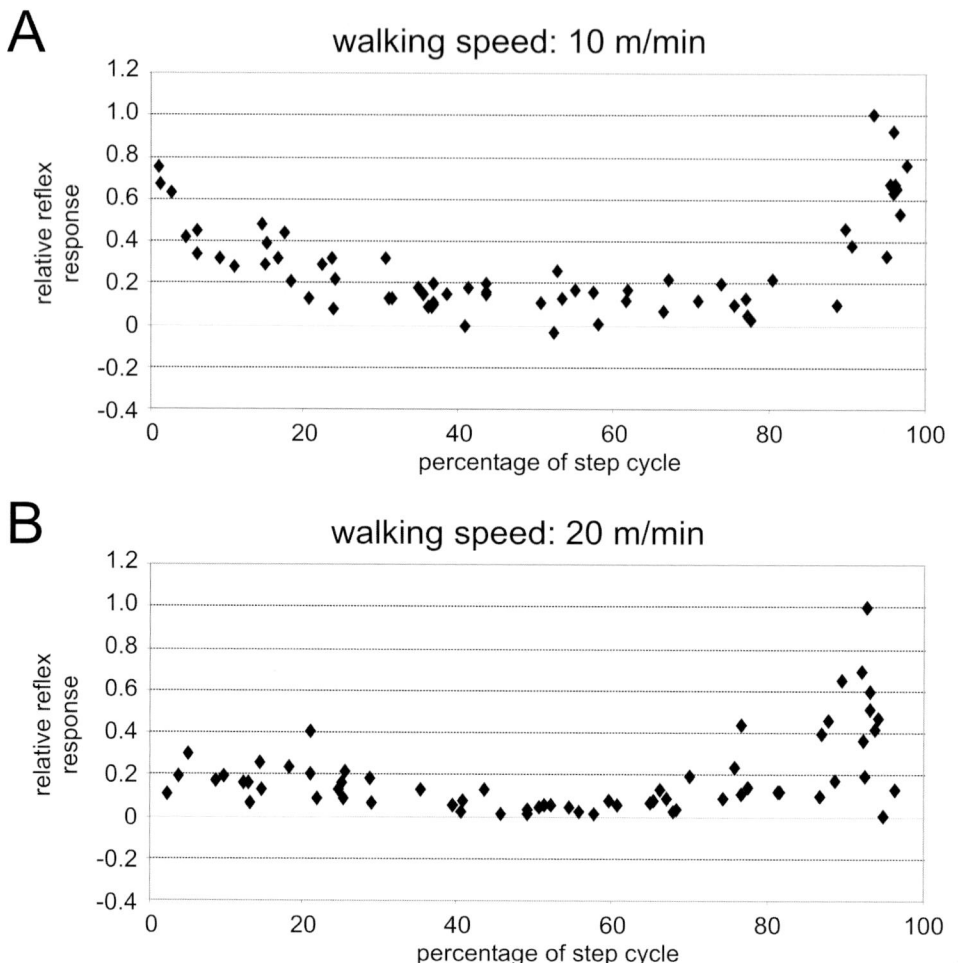

Fig. 6. Modulation of cutaneously induced reflexes in the biceps femoris obtained in the same session under normal walking (A: 10 m/min) and high speed walking (B: 20 m/min) conditions. Note that both conditions result in a similar modulation pattern (i.e., reflexes are less depressed during the swing phase of the step cycle than during the stance phase of the step cycle).

shows a strong depression and even reversal of reflex response (i.e., reduction of the normal EMG values) during or just prior to the swing phase. Note that GC reflex responses can best be measured during the stance phase but are still notably depressed compared to reflex responses triggered at rest.

Discussion

As far as we are aware this is the first report of step cycle phase-dependent modulation of peripheral reflexes recorded in the freely locomoting rat. Similar experiments have previously been conducted in man (Crenna and Frigo, 1984; Duysens et al., 1990; Hauglustaine et al., 2001) and in freely walking (Duysens and Loeb, 1980), decerebrated (Misiaszek et al., 2000), or fictitiously walking cats (Degtyarenko et al., 1998). In these and other studies, two conclusions were immediately apparent. First, it was noted that the cutaneously induced reflexes are generally inhibited during locomotion and, second, it was demonstrated conclusively that the gain of cutaneously induced reflexes was strongly dependent upon the phase of the step cycle in which they were triggered. These two main conclusions can also be

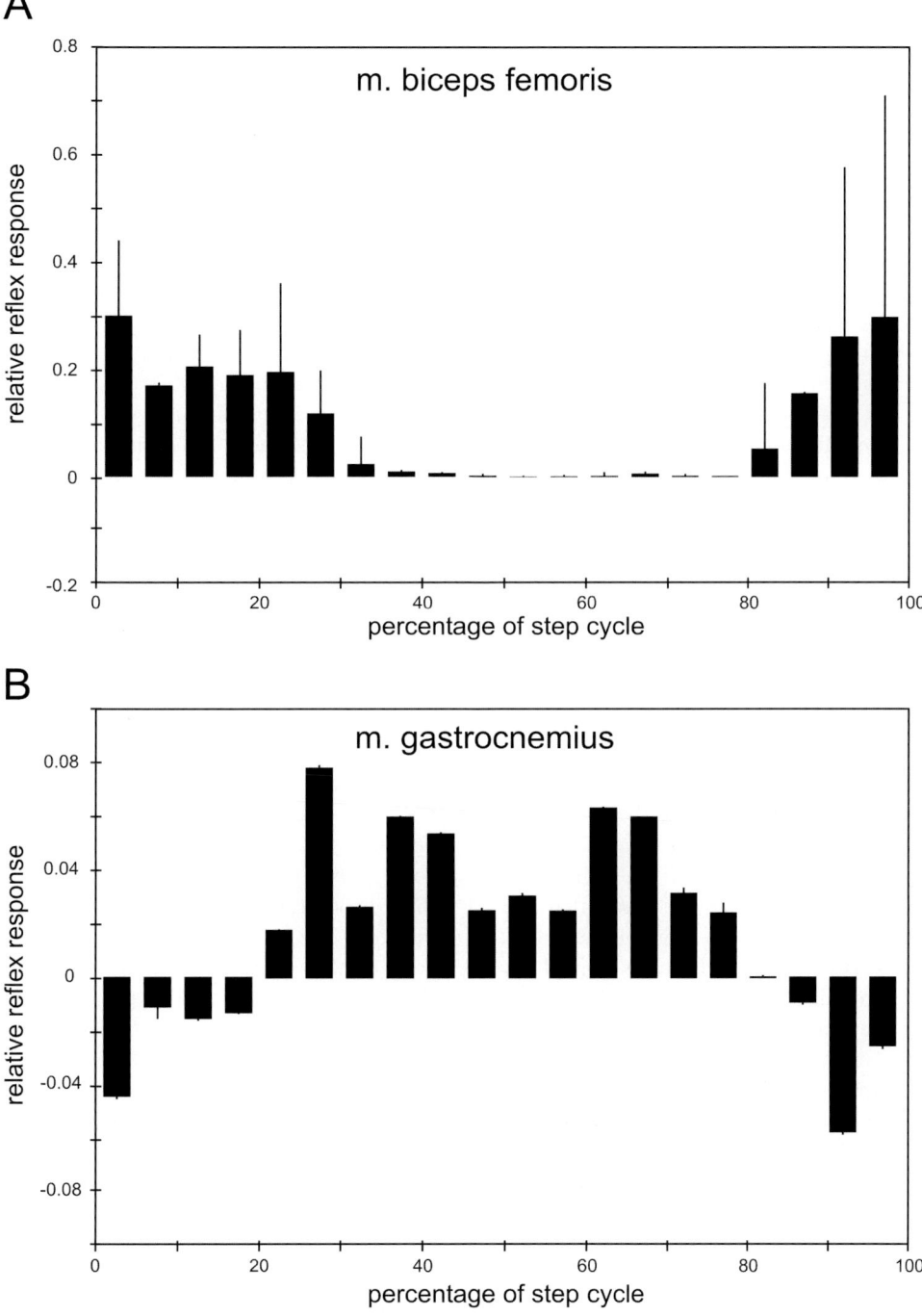

Fig. 7. Average of modulation pattern of cutaneous reflexes under normal stimulus and walking conditions in the biceps femoris (A: 10 rats) and gastrocnemius (B: 8 rats) muscles. Error bars depict standard error of mean.

made in our experiments, thus basically validating the currently used experimental set up.

Experimental technique and analysis

The use of rats as the experimental animal of choice to study structure and function of the nervous system has gained widespread acceptance in the literature as a low-cost alternative for cats. Obviously, however, size problems are inherent in the potential for conducting similar type of experiments as have been done in cats. In the present set up, we have recorded EMG activity of two hind limb muscles simultaneously for a period of up to a few weeks in conjunction with inducing cutaneous stimulation using a pair of permanently placed subcutaneous electrodes. Initial attempts to use a stimulation cuff that surrounded the sural nerve were aborted due to mechanical stretching of the nerve while placing the cuff and the high variability of the responses. However, with the currently used technique, the intended sensory activation of the sural nerve by the subcutaneously delivered stimulation could depend on the position of the two electrodes with respect to the skin and thus could be correlated to the executed movements during different phases of the step cycle. Hence, in case of such a correlation, changes in the size of a reflex response could result from a change in stimulation effect rather than be mediated by centrally induced changes in the gain of the reflex. In the cat, this problem has been examined by recording the stimulus-induced volley in the nerve and to see if any correlation could be observed between the size of the volley and phase of the step cycle (e.g., see Apps et al. (1990)). In our experimental set up, there was no objective way to study the sensory effect of the stimulation. However, circumstantial evidence suggests that our data are not severely corrupted by step-related movement of the stimulation electrodes. Firstly, it was noted that all rats showed the same type of modulation for the BF. It would seem unlikely that the electrode positions would always be less effective during the stance phase compared to the swing phase. Moreover, for the GC the reflex was least suppressed during the stance phase, suggesting that the reflex gain was optimal during swing for the BF, whereas it is optimal for the GC during stance. Also, it was noted that increasing the stimulus intensity did not affect the pattern of modulation, which suggests that the modulation is relatively independent of the intensity of the stimulus or the size of the stimulated area, but rather that the modulation pattern is really generated in the spinal cord. Finally, examination of the stimulation electrodes after termination of the experiments showed that they were firmly fixed within subcutaneous connective tissue from which they could not be easily dislocated.

Our method of data analysis largely follows those described in literature (Zehr et al., 1997). Moreover, using subthreshold stimuli, we have shown that the analytic procedure did not in itself introduce a modulation pattern.

Functional implications of reflex modulation

During locomotion, the posterior part of the bi-articulate BF muscle helps in knee flexion during the swing phase of the step cycle and stabilizes the knee during the extension phase of the step cycle. The average EMG of the BF muscle shows a number of features during different phases of the step cycle. The first major feature is the biceps burst, taken as the beginning of the step cycle, which from observation corresponds to knee flexion. Reflexes elicited during this period are least depressed and may serve to speed up knee flexion. Immediately after this burst of activity, there is a period where there is no activity in the BF muscle. During this period, stimuli still elicit strong responses. When a reflex is elicited during this time the rat lifts its foot and the stance phase cannot begin. This period corresponds to the time where the foot is replaced to the ground, so that the resulting reflex movement would tend to prevent the rat from putting its foot on the ground. During the stance phase of the step cycle, reflexes in the BF are most depressed. A strong response during this phase of the step cycle would cause the rat to lift its foot, whilst during the stance phase this foot is needed to help support the body weight. So, the strong depression of reflexes in BF during the stance phase seems to be functional in that the stance phase of the step cycle must reach near completion before the rat can lift that foot again in the next swing phase.

However, nearing the end of the stance phase the BF seems more susceptible to stimuli, so that at the end of the stance phase the rat will be able to swiftly lift its paw without this resulting in imbalance.

The reverse argument holds for GC muscle, which serves as an extensor muscle. During swing and late stance, this muscle shows a strong depression in reflex activity — even to the point of reflex reversal — suggesting that during this phase the hold that the GC muscle has on the ankle is relaxed. In contrast, during the stance phase, the reflexes in GC muscle are much less depressed which suggest that reflexes in GC during stance help stabilize the limb more, so that the rat is not inclined to lift its paw in mid-stance.

The basic form of step phase related modulation of BF and GC are generally in good agreement with findings in man and cat (Duysens and Loeb, 1980; Crenna and Frigo, 1984; Zehr et al., 1997).

Origin of reflex modulation during locomotion

The results show that the presently used experimental paradigm can be used to demonstrate the step cycle dependent modulation of peripheral reflexes in the rat. This opens up avenues to examine the origin of modulation using this rodent model. Basically, rhythmic muscle activations are suggested to be generated by a central pattern generator (CPG) located in the spinal cord (Guadagnoli et al., 2000; Ribotta et al., 2000; Burke et al., 2001; MacKay-Lyons, 2002). Since it has been shown that the CPG can activate and maintain locomotion and other rhythmic movements in spinal and/or deafferented preparations, it is thought to act rather independently from higher brain structures. However, it is obvious that locomotion in normal animals is much more refined and can be readily adapted or adjusted to deal with environmental demands. It is easy to understand how performance may be improved when the animal is able to make a variable, but nearly immediate use of cutaneous information while walking.

Available evidence implies that such a variable use of reflexes may take up different forms under different functional demands (Stein and Capaday, 1988; Dietz et al., 1994; Tinazzi and Zanette, 1998; Brooke et al., 2000; Kihara et al., 2000; Trimble et al., 2000), suggesting that the observed modulation of reflexes during rhythmical movements may be specifically adapted for a particular function. In analogy with the cerebellar involvement in the adaptation of the gain of the vestibulo-ocular reflex (Koekkoek et al., 1997; De Zeeuw et al., 1998; Ito, 1998; Lisberger, 1998), it stands to reason that the functional result of the modulation of peripheral reflexes may be constantly evaluated and, if necessary, adapted. Future research in our lab will therefore focus around the cerebellar role in inducing, maintaining, and adapting this reflex modulation.

Conclusions

This study demonstrates the step phase dependent modulation in the gain of cutaneously induced reflexes in walking rats. Electromyographic recordings of the posterior part of the m. biceps femoris (mainly involved in knee flexion) and the medial part of the m. gastrocnemius (mainly involved in ankle extension) were continuously monitored during locomotion and subcutaneously placed electrodes in the region distal to the lateral malleolus induced cutaneous reflexes.

The results show that during locomotion the reflex responses in both muscles were generally reduced compared to reflexes elicited at rest. For m. biceps femoris, reduction of reflex gain was highest during the stance phase whereas reflexes in the m. gastrocnemius were mostly depressed during the swing phase.

We conclude that step phase-dependent modulation of peripherally induced reflexes can be measured in walking rats and generally concur with previous studies in man and cat that this type of modulation may be functionally important for maintaining and adjusting gait. Although the mechanism of inducing and maintaining this modulation is not fully understood, we hypothesize, in analogy to the adaptive control of vestibulo-ocular reflexes, that the cerebellum will be actively involved in this process. We will use the present model to investigate this hypothesis in rodent locomotion.

Acknowledgments

This work was supported by ALW grant 809.37.007 (RB) and the Dutch ministry of Health, Welfare and Sport (JvdB and TJHR).

References

Apps, R., Lidierth, M. and Armstrong, D.M. (1990) Locomotion-related variations in excitability of spino-olivocerebellar paths to cat cerebellar cortical c2 zone. J. Physiol. (Lond.), 424: 487–512.

Bloedel, J.R. and Bracha, V. (1995) On the cerebellum, cutaneomuscular reflexes, movement control and the elusive engrams of memory. Behav. Brain Res., 68: 1–44.

Brooke, J.D., Peritore, G., Staines, W.R., McIlroy, W.E. and Nelson, A. (2000) Upper limb H reflexes and somatosensory evoked potentials modulated by movement. J. Electromyogr. Kinesiol., 10: 211–215.

Burke, R.E., Degtyarenko, A.M. and Simon, E.S. (2001) Patterns of locomotor drive to motoneurons and last-order interneurons: clues to the structure of the CPG. J. Neurophysiol., 86: 447–462.

Canu, M.H. and Falempin, M. (1996) Effect of hindlimb unloading on locomotor strategy during treadmill locomotion in the rat. Eur. J. Appl. Physiol., 74: 297–304.

Crenna, P. and Frigo, C. (1984) Evidence of phase-dependent nociceptive reflexes during locomotion in man. Exp. Neurol., 85: 336–345.

De Zeeuw, C.I., Hansel, C., Bian, F., Koekkoek, S.K.E., van Alphen, A., Linden, D.J. and Oberdick, J. (1998) Expression of a protein kinase C inhibitor in Purkinje cells blocks cerebellar long term potentiation and adaptation of the vestibulo-ocular reflex. Neuron, 20: 495–508.

Degtyarenko, A.M., Simon, E.S., Norden-Krichmar, T. and Burke, R.E. (1998) Modulation of oligosynaptic cutaneous and muscle afferent reflex pathways during fictive locomotion and scratching in the cat. J. Neurophysiol., 79: 447–463.

Dietz, V., Discher, M. and Trippel, M. (1994) Task-dependent modulation of short- and long-latency electromyographic responses in upper limb muscles. Electroencephalogr. Clin. Neurophysiol., 93: 49–56.

Drew, T. and Rossignol, S. (1987) A kinematic and electromyographic study of cutaneous reflexes evoked from the forelimb of unrestrained walking cats. J. Neurophysiol., 57: 1160–1184.

Duysens, J. (1977) Reflex control of locomotion as revealed by stimulation of cutaneous afferents in spontaneously walking premammillary cats. J. Neurophysiol., 40: 737–751.

Duysens, J. and Loeb, G.E. (1980) Modulation of ipsi- and contralateral reflex responses in unrestrained walking cats. J. Neurophysiol., 44: 1024–1037.

Duysens, J. and Tax, T. (1993) Interlimb reflexes during gait in cat and human. Interlimb Coordination: Neural Dynamical and Cognitive Contstraints. Academic Press, Inc., New York, pp. 97–126.

Duysens, J., Trippel, M., Horstmann, G.A. and Dietz, V. (1990) Gating and reversal of reflexes in ankle muscles during human walking. Exp. Brain Res., 82: 351–358.

Gillis, G.B. and Biewener, A.A. (2001) Hindlimb muscle function in relation to speed and gait: in vivo patterns of strain and activation in a hip and knee extensor of the rat (Rattus norvegicus). J. Exp. Biol., 204: 2717–2731.

Guadagnoli, M.A., Etnyre, B. and Rodrigue, M.L. (2000) A test of a dual central pattern generator hypothesis for subcortical control of locomotion. J. Electromyogr. Kinesiol., 10: 241–247.

Hauglustaine, S., Prokop, T., van Zwieten, K.J. and Duysens, J. (2001) Phase-dependent modulation of cutaneous reflexes of tibialis anterior muscle during hopping. Brain Res., 897: 180–183.

Hayes, K.C., Allatt, R.D., Wolfe, D.L., Kasai, T. and Hsieh, J. (1992) Reinforcement of subliminal flexion reflexes by transcranial magnetic stimulation of motor cortex in subjects with spinal cord injury. Electroencephalogr. Clin. Neurophysiol., 85: 102–109.

Hiraoka, K. (2001) Phase-dependent modulation of the soleus H-reflex during rhythmical arm swing in humans. Electromyogr. Clin. Neurophysiol., 41: 43–47.

Iles, J.F. and Pisini, J.V. (1992) Cortical modulation of transmission in spinal reflex pathways of man. J. Physiol., 455: 425–446.

Ito, M. (1984) The Cerebellum and Neural Control. Raven Press, New York.

Ito, M. (1998) Cerebellar learning in the vestibulo-ocular reflex. Trends Cogn. Sci., 2: 313–321.

Kihara, T., Matsuo, T., Sakamoto, M., Yasuda, Y., Yamamoto, Y. and Tanimura, T. (2000) Effects of prenatal aflatoxin B1 exposure on behaviors of rat offspring. Toxicol. Sci., 53: 392–399.

Koekkoek, S.K.E., van Alphen, A.M., van der Burg, J., Grosveld, F., Galjart, N. and De Zeeuw, C.I. (1997) Gain adaptation and phase dynamics of compensatory eye movements in mice. Genes and Function, 1: 175–190.

Lisberger, S.G. (1998) Physiologic basis for motor learning in the vestibulo-ocular reflex. Otolaryngol. Head Neck Surg., 119: 43–48.

MacKay-Lyons, M. (2002) Central pattern generation of locomotion: a review of the evidence. Phys. Ther., 82: 69–83.

McCrea, D.A. (2001) Spinal circuitry of sensorimotor control of locomotion. J. Physiol., 533: 41–50.

Misiaszek, J.E., de Serres, S.J., Stein, R.B., Jiang, W. and Pearson, K.G. (2000) Stretch and H reflexes in triceps surae are similar during tonic and rhythmic contractions in high decerebrate cats. J. Neurophysiol., 83: 1941–1950.

Pearson, C., Borovsky, B., Krueger, M., Curtis, R. and Ganz, E. (1995) Si(001) Step dynamics. Physic. Rev. Lett., 74: 2710–2713.

Pratt, C.A., Chanaud, C.M. and Loeb, G.E. (1991) Functionally complex muscles of the cat hindlimb. IV. Intramuscular distribution of movement command signals and cutaneous reflexes in broad, bifunctional thigh muscles. Exp. Brain Res., 85: 281–299.

Ribotta, M.G., Provencher, J., Feraboli-Lohnherr, D., Rossignol, S., Privat, A. and Orsal, D. (2000) Activation of locomotion in adult chronic spinal rats is achieved by transplantation of embryonic raphe cells reinnervating a precise lumbar level. J. Neurosci., 20: 5144–5152.

Ruigrok, T.J.H., Burg, Hvd. and Sabel-Goedknegt, E. (1996) Locomotion coincides with c-Fos expression in related areas of inferior olive and cerebellar nuclei in the rat. Neurosci. Lett., 214: 119–122.

Seki, K. and Yamaguchi, T. (1997) Cutaneous reflex activity of the cat forelimb during fictive locomotion. Brain Res., 753: 56–62.

Sinkjaer, T., Andersen, J.B. and Larsen, B. (1996) Soleus stretch reflex modulation during gait in humans. J. Neurophysiol., 76: 1112–1120.

Stein, R.B. and Capaday, C. (1988) The modulation of human reflexes during functional tasks. Trends Neurosci., 11: 328–332.

Tinazzi, M. and Zanette, G. (1998) Modulation of ipsilateral motor cortex in man during unimanual finger movements of different complexities. Neurosci. Lett., 244: 121–124.

Trimble, M.H., Du, P., Brunt, D. and Thompson, F.J. (2000) Modulation of triceps surae H-reflexes as a function of the reflex activation history during standing and stepping. Brain Res., 858: 274–283.

van der Bilt, A., Ottenhoff, F.A., van der Glas, H.W., Bosman, F. and Abbink, J.H. (1997) Modulation of the mandibular stretch reflex sensitivity during various phases of rhythmic open-close movements in humans. J. Dent. Res., 76: 839–847.

Zehr, E.P., Komiyama, T. and Stein, R.B. (1997) Cutaneous reflexes during human gait: electromyographic and kinematic responses to electrical stimulation. J Neurophysiol., 77: 3311–3325.

SECTION VII

Cerebellar Neuro-Anatomical Organization

CHAPTER 21

The basilar pontine nuclei and the nucleus reticularis tegmenti pontis subserve distinct cerebrocerebellar pathways

Federico Cicirata*, Maria Francesca Serapide, Rosalba Parenti, Maria Rosita Pantò, Agata Zappalà, Annalisa Nicotra and Deborah Cicero

Dipartimento di Scienze Fisiologiche, Università Catania, Viale A. Doria 6, 95125 Catania, Italy

Keywords: cerebellum; BPN; NRTP; rat; neuroanatomy

Abstract: Previous studies often considered the basilar pontine nuclei (BPN) and the nucleus reticularis tegmenti pontis (NRTP) as relays of a single cerebro-(ponto)-cerebellar pathway. Conversely, the different cortical afferences to the BPN and the NRTP, as well as the anatomical and functional features of the cerebellopetal projections from these pontine nuclei, support the different, and for some aspect, complementary arrangement of the cerebrocerebellar pathways relayed by the BPN or NRTP.

Both the BPN and the NRTP are innervated from the cerebral cortex, but with regional prevalence. The NRTP is principally innervated from motor or sensori-motor areas while the BPN are principally innervated from sensory, mainly teloceptive, and associative area. Projections from sensory-motor areas were also traced to the BPN.

The BPN and NRTP project to all parts of the cerebellar cortex with a similar pattern. In fact, from single areas of them projections were traced to set of sagittal stripes of the cerebellar cortex.

In variance to such analogies, the projections to the cerebellar nuclei differed between those traced from the NRTP and from BPN. In fact, BPN and NRTP have private terminal areas in the cerebellar nuclei with relatively little overlaps. The BPN innervated the lateroventral part of the nucleus lateralis and the caudoventral aspect of the nucleus interpositalis posterioris. The NRTP principally innervated the mediodorsal part of the nucleus lateralis, the nucleus interpositalis anterioris, the nucleus medialis. Since the single cerebellar nuclei have their specific targets in the extracerebellar brain areas, it follows that the BPN and the NRTP, passing through their cerebellar nuclei relays, are devoted to control different brain areas and thus likely to play different functional roles.

From single pontine regions (of both BPN and NRTP) projections were traced to the cerebellar cortex and to the cerebellar nuclei. In some cases these projections reached areas which are likely anatomically connected (by Purkinje axons). This pattern of the pontine projections was termed as coupled projection. In some other cases, the projections reached areas of the cerebellar cortex but not the nuclear regions innervated by them. We termed this as uncoupled projection. The existence of both coupled and uncoupled projections, open new vistas on the functional architecture of the pontocerebellar pathway. More in detail, this study showed the different quantitative and topographic distribution of the coupled and uncoupled projections visualized in the cerebellar projections from BPN and NRTP. All these evidences strongly support the anatomical and the functional differences that characterise the cerebrocerebellar pathways relayed by the BPN and the NRTP.

---

*Corresponding author. Tel.: (39) 095 333841; Fax: (39) 095 330645; E-mail: Cicirata@mbox.unict.it

Introduction

Cerebral cortex and cerebellum are computational entities. They transform a given input into a specific output that is then relayed to other parts of the brain. However, the cerebral cortex and the cerebellum are tightly interconnected by a large fiber system. It connects the cerebral and the cerebellar cortices reciprocally. The distribution of the onset latencies elicited by a motor behavior shows a broad overlap in the cerebral cortex and in the cerebellum (Thach, 1975; Fortier et al., 1993). This suggests that the cerebral cortex and the cerebellum process signals in conjunction, rather than passing them from one to the other in a strictly sequential manner. Therefore, the understanding of the function of each of the two structures will benefit greatly from studies on the communication between the two.

The cerebellum receives afferents from cerebral cortex by fiber systems that involve pontine nuclei as the major intercalated structure. These nuclei consist of neurons that terminate as the so-called mossy fibers in the granular cell layer of the cerebellar cortex. Pontine nuclei intercalated in the cerebrocerebellar communication system principally include the basilar pontine nuclei (BPN) and the reticulotegmental nucleus or nucleus reticularis tegmenti pontis (NRTP). We will use the term 'pontine nuclei' to refer to both BPN and NRTP.

NRTP was often considered along with BPN, thus grouping them into a single functional identity. Such a conclusion is not in agreement with some different features of the NRTP and BPN. In fact, NRTP is a specialized nucleus of the pontine reticular formation (Newman and Ginsberg, 1992) and differs from BPN in both ontogeny (Altman and Bayer, 1987) and afferent connectivity pattern.

These differences suggest that we consider the individual contribution of the BPN and the NRTP in the cerebrocerebellar communication system. In fact, it is possible that these pontine nuclei may be the relays of distinct cerebrocerebellar pathways that differ both anatomically and functionally, in variance to the current tendency to merge BPN and NRTP in a single pontine relay of a undifferentiated cerebroponto-cerebellar communication system. This hypothesis was investigated in this study, which used the rat as an experimental model.

Cytoarchitecture and cerebellopetal connecting fibers of the pontine nuclei

The BPN extend roughly 2 mm rostrocaudally. They are included between the interpeduncular nucleus and the trapezoid body. Four main subdivisions are generally recognized with respect to their position relative to the descending fibers of the cerebral peduncle (Mihailoff et al., 1981, Fig. 1B). The medial, ventral, and lateral subdivisions consist of rather tightly packed and homogeneously distributed neurons, and the peripeduncular nuclei immediately surround the peduncle. Various smaller subnuclei may also be identified. However, it is not easy to delineate borders between the various subdivisions. Golgi studies have indicated that, in general, four types of neurons may be distinguished (Mihailoff et al., 1981). The most common type is described as 'spineladen' and probably represents pontine projection neurons. Small bipolar and unipolar cells possess axons with collaterals within the pons. Small glutamic acid decarboxylase-positive, and therefore most likely GABA-ergic neurons have been demonstrated in the rat (Border and Mihailoff, 1985). BPN neurons project in to the medial cerebellar peduncle (mcp) of both sides. Most fibers cross the whole nuclear mass of the BPN to reach the contralateral mcp (Fig. 1C). Fewer fibers, principally those that arise from neurons lying near the ventral border, shift ventrally to the nuclear mass and run into the thin fiber layer that ventrally surrounds BPN (Fig. 1C). Relatively few fibers reach the ipsilateral mcp.

The NRTP is located dorsal to the medial lemniscus, along the midline. It appears to be continuous with the dorsomedial aspect of the pontine grey, especially at rostral level. Torigoe et al. (1986) described two cytoarchitectonically distinct portions of the NRTP. A central part (NRTPc) consists of rather tightly packed cells, whereas the pericentral part (NRTPp) is composed of loosely packed small neurons. The NRTPc is located dorsal to the medial lemniscus over the caudal two-thirds of the BPN. At some points it is contiguous with the dorsomedial BPN area (Mihailoff et al., 1981). Caudal to the BPN, the NRTP extends caudodorsally until it dissipates just rostral and ventral to the abducens nucleus. Cerebellopetal fibers from the two subdivisions of the NRTP follow different pathways. In fact, from the

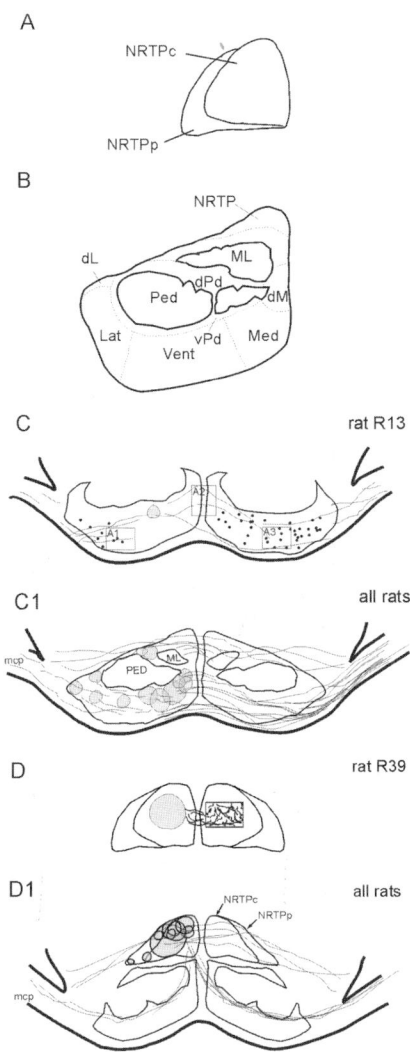

Fig. 1. A–B: Schematic delineation of the NRTP (A) and the BPN (B) of rat according to the nomenclature of Torigoe et al. (1986) and Mihailoff et al. (1981), respectively. C: Representative example of cell bodies (black dots) labeled bilaterally in the BPN after tracer deposition (gray area) segregated in a discrete area of the BPN of one side. Note also the fibers that radiate from the site of injection toward the medial cerebellar peduncle (mcp) of both sides. C1: Summary of the fibers from 12 injection sites involving this coronal section of the BPN. D: Injection segregated in the central part of the NRTP stained cell bodies in symmetric areas of the contralateral NRTP. D1: Summary of the fibers from 8 injection sites involving this coronal section of the NRTP. Note that projections to the mcp of both sides arise from the injections and follow two pathways, either ventrally, crossing the nuclear mass of the BPN, or dorsally, crossing the nuclear mass of the contralateral NRTP.

NRTPp most fibers were traced ventrally to the medial region of the BPN. There, they divided into two branches that respectively turned toward the mcp of both sides, passing through the nuclear mass of the BPN. Conversely, fibers from the NRTPc were principally traced to the contralateral mcp passing throughout the mass of the contralateral NRTPc (Fig. 1D).

The pathways followed by cerebellopetal pontine fibers open the way to three types of risks in the interpretation of the labeling of projecting fibers visualized after tracer deposition within one of the pontine nuclei. (i) The topographical contiguity of the two nuclei suggests the risk of the tracer spreading from one pontine nucleus to the other. This risk may be avoided by selecting for analysis rats with the core of the injections segregated within one single pontine nucleus. (ii) Some fibers of the NRTP reach the mcp passing through the nuclear mass of the BPN. Therefore, injections of tracer segregated within the BPN can also stain passing fibers from the NRTP. The risk is to attribute to the BPN what in reality is due to the NRTP. This risk was reduced by lowering the size of the injections and, more generally, the trauma. However, taking care does not eliminate the risk. Thus, it was essential to select for analysis only those animals that, in the histological examination, did not show cell bodies stained in the NRTP. Otherwise the animals were discarded and not considered further in the analysis. (iii) The fact that some fibers from the BPN and the NRTPc of one side reach the contralateral mcp passing through the nuclear mass of the contralateral homologous nucleus, gives rise to the possibility that injections segregated within one side could also stain projection fibers of the contralateral nucleus. This risk was real and occurred systematically in all of our animals, despite changing all possible parameters and the procedures of tracer injection. Therefore, in our experiments performed with tracer deposition within the pontine nucleus of one side it was not possible to analyse the laterality of the effects. With this aim specific experiments are currently in progress.

Afferents to the basilar pontine nuclei and nucleus reticularis tegmenti pontis

BPN afferents from the cerebral cortex arise from layer V neurons located throughout the entire ipsilateral cortex (Legg et al., 1989). However, there are clear regional differences in the relative contributions of each cortical area to the corticopontine system. Most fibers originate from the sensory motor and visual cortices. In addition, the primary auditory (rostral temporal) cortex, as well as cingulated, retrosplenial, and agranular insular cortices, also provide appreciable corticopontine projections. Relatively small contributions are derived from the caudal temporal cortex and from the perirhinal cortex (Burne et al., 1978b; Mihailoff et al., 1978, 1985; Wiesendanger and Wiesendanger, 1982a, b; Legg et al., 1989). A topographical pattern has been described within the corticopontine projections. Projections from the motor cortex predominantly terminate in the medial subdivision of the BPN, whereas the somatosensory and visual cerebral cortices project to more central and lateral areas, respectively (Wiesendanger and Wiesendanger, 1982b; Mihailoff et al., 1985). The visual corticopontine input terminates predominantly within the lateral third of the basal pontine nuclei with the exception of its lateral-most part. In addition, small patches of labeling are found rostromedially (Wiesendanger and Wiesendanger, 1982b). Auditory projections arising from the primary temporal cortex are rather weak and terminate ventrally within the lateral third of the nuclei. The anatomical descriptions of the corticopontine system suggest that there are major zones of convergence from various cortical regions (Wiesendanger and Wiesendanger, 1982b). Within the recipient areas of the somatosensory cortex, however, there appears to be less convergence. Nevertheless, electrophysiological data obtained in rat (Potter et al., 1978) have revealed that pontine neurons within a particular corticopontine cluster may respond to the stimulation of a number of different cortical sites. This convergence may be not only due to overlap of the corticopontine termination clusters, but also due to the tendency of the dendritic trees of the pontine neurons to invade neighboring corticopontine termination zones (Mihailoff et al., 1981).

NRTP afferents from the cerebral cortex stem from layer V neurons. By far the most dense projections to the NRTP are from the cingular cortex of both sides. This may, in the rat, be homologous to the frontal eye field of the cat and monkey which principally innervates the NRTPc. Considering other cortical afferents, projections were also traced principally to the NRTPp from the ipsilateral prefrontal cortex (Brodmann areas 8, 8a, 11, and 32) and the ipsilateral motor and somatosensory cortices (Brodmann areas 2, 4, 6, and 10), in particular (Torigoe et al., 1986).

Zonal projections to the cerebellar cortex from the BPN and NRTP

Some previous studies have reported that the climbing fibers are projected to the sagittal bands of the cerebellar cortex (CC; Ramòn y Cajal, 1909; Courville, 1975; Groenewegen and Voogd, 1976; Brodal and Walberg, 1977a, b; Groenewegen et al., 1979; Walberg and Brodal, 1979; Brodal and Kawamura, 1980; Buisseret-Delmas and Angaut, 1993) and that mossy fibers from different nuclei are projected to the sagittal stripes of the CC (see references in Tolbert et al., 1993). The zonal pattern of projections from these pre-cerebellar nuclei is assumed to be an expression of their anatomical and functional specificity. The projections of the pontine nuclei to the cerebellum reported by previous studies differed from this type of termination in the CC. In fact, it was generally assumed that the BPN and the NRTP project to the CC with a diffuse or aspecific pattern (For BPN see in: Kunzle, 1975; Brodal and Walberg, 1977a, b; Brodal, 1982; Brodal and Bjaalie, 1992. For NRTP see in: Hoddevik, 1978; Kawamura, and Hashikawa, 1981; Gerrits, and Voogd, 1986; Mihailoff, 1993), despite evidence reported in favor of the zonal arrangement of the BPN projections to discrete areas of the CC (Eisenman, 1981; Kawamura and Hashikawa, 1981; see also Fig. 5 of Mihailoff, 1993). The diffuse and aspecific projection pattern of the pontine projections to the cerebellum is not in agreement with the role of the cerebrocerebellar pathway as carrying cortical messages dedicated to the control of segregated motor effects. We therefore decided to investigate the pontine projections to the cerebellum in order to test whether they were arranged according to precise topographic pattern that is necessary to participate in the function played by the cerebro-cerebellar pathway. The projections to the CC were visualized after deposition of biotinylated dextrane amine (BDA) in the pontine nuclei of one side (Serapide et al., 1994, 2001, 2002a).

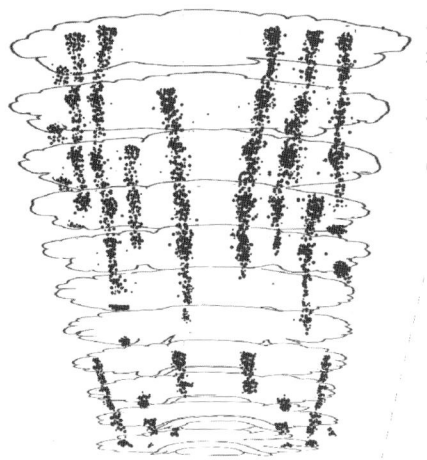

Fig. 2. 3D reconstruction of the zonal labeling found in the caudal two-thirds of the cerebellum of a rat injected in the nuclear mass of BPN. The reconstruction is rotated 60° about the x axis. The external borders of the cerebellar coronal sections, equally spaced at 180 μm, were drawn. The dots represent the labeled fiber terminals.

The main finding of this study was the visualization of projections from both the BPN (Fig. 2) and the NRTP to parasagittal stripes of the CC. The stripes of fiber terminals were sharply delimited on both sides by areas, interstripes, either virtually void of labeling or with a much lower density of labeling. The stripes extended along the rostrocaudal axis for some hundreds of micrometers, or more, and were usually a few micrometers in width. Figure 3 shows representative examples of stripes of projections in equally spaced horizontal sections of the cerebellum after tracer deposition within the BPN or NRTP. As may be noted, the stripes were usually located in symmetrical areas on the two sides of the cortex whereas stripes in different (non-symmetrical) areas were less frequently observed. The staining of BPN or NRTP visualized the degree of compartmentation of the various areas of the CC (see in Serapide et al., 2001, 2002a).

The zonal labeling of the CC was evidenced after injection of different types of tracer: ^{3}H-Leu, WGA-HRP and BDA. The different tracers showed similar zonation patterns of the CC. This evidence strongly supports the zonation pattern as an actual feature of the pontocerebellar projections. It was visualized only by injections of small size (injections of WGA-HRP or BDA that stained $66-16.80 \times 10^6$ μm^3 of pontine tissue), whereas large injections resulted in the diffuse labeling of areas of the CC (injections of ^{3}H-Leu or WGA-HRP that stained $360-113 \times 10^6$ μm^3 of pontine tissue). These findings strongly suggest that the zonal labeling is masked by diffuse labeling after large injections in the pontine nuclei. The liminal size was related to the type of tracer. The smallest injection volume (16.80×10^6 μm^3) effective in the visualization of cortical stripes was found for BDA, i.e., the more sensitive of the three tracers used in this study. The aggregation of fiber terminals into stripes was not visualized by a further reduction of the injection volume, whereas single rosettes or small patches of rosettes were still observed. These findings indicate that 16.80×10^6 μm^3 is the liminal volume of pontine tissue which must be injected to evidence the zonal projections.

The labeling in the CC was always present bilaterally, with a clear contralateral preponderance. This study showed two features of the BPN/NRTP projections to the CC that agree with previous findings (see references in Kawamura and Hashikawa, 1981): the divergent as well as the convergent pattern. An example of convergence was shown by the afferents that innervated the ventral paraflocculus from the lateral and the medial regions of the rostral two-thirds of the BPN, in agreement with the findings reported by Burne et al. (1978 a, b) and Eisenman and Noback (1980). The divergence was shown by pontine fibers that, immediately after entering the cerebellar white matter from mcp, divided into bundles directed to different cortical areas. In the immediate subcortical white matter they divided into various tufts of fibers that finally reached their cortical targets.

The comparison of the topographic distribution of the labeling in symmetric areas of the two sides showed that the projections were arranged according to a similar zonation pattern, independently from the number of stripes visualized on the two sides. Figure 4 shows a representative example of three stripes (i.e., the maximum number of stripes) labeled in the paraflocculus of both sides after injections in the medial (Fig. 4A) or in the lateral region of the BPN (Fig. 4B). The zones labeled on the two sides were completely, or at least largely, coincident. Figure 5 shows a representative example of a similar zonal pattern in the Crus II evidencing a different numbers

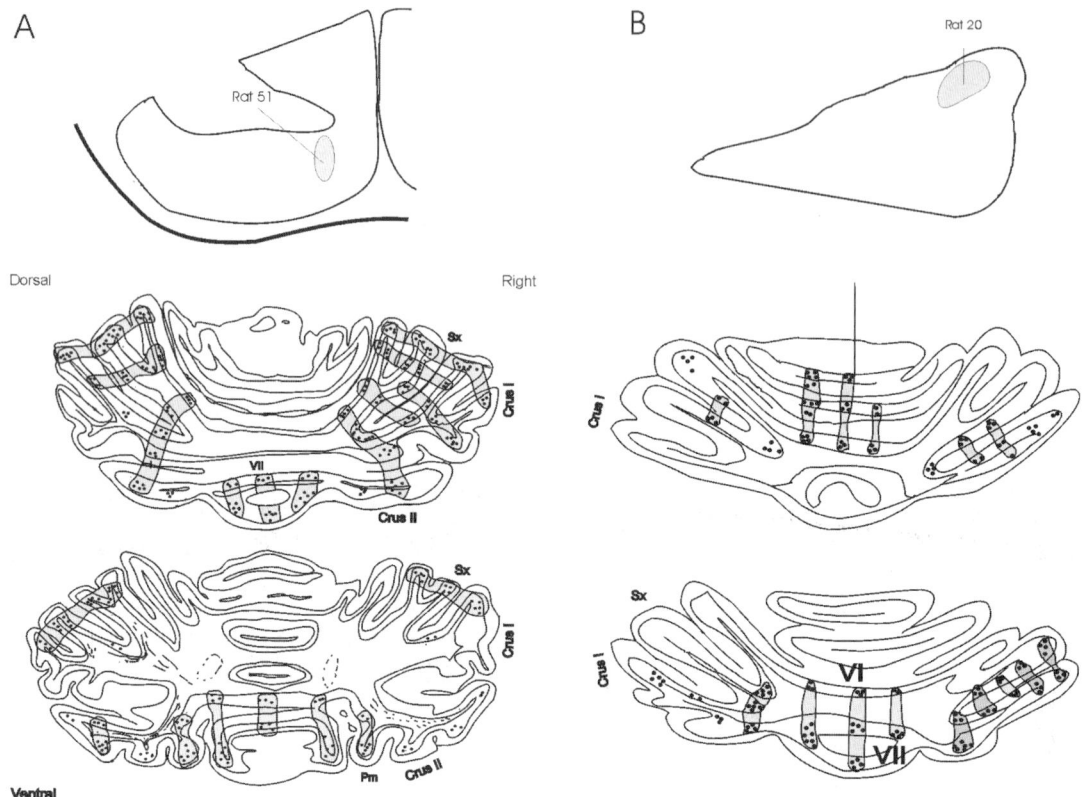

Fig. 3. Representative examples of the zonal pattern found in the CC after injection in the nuclear mass of the BPN of rat 51 (A), and the NRTP of rat 20 (B). The labeling was illustrated in two drawings of equally spaced horizontal sections of the cerebellar cortex. The pontine fiber terminals were symbolized by dots. Shaded areas show the stripes. Note that the stripes were usually located in symmetric areas of the two sides.

of stripes on the two sides. The digitized images of the two sides of the lobule were rotated and placed side by side. The stripes were numbered and arrows linked to similar areas of the two sides. Four stripes (corresponding to the full set of stripes) were labeled in the right Crus II (contralateral to the side of injection) whereas two stripes were labeled in the left Crus II. Stripes 2 and 3 labeled on the right largely correspond to the stripes labeled on the left. Thus, they are located according to a zonation pattern common to both sides.

Moreover, the single areas of different rats showed a similar zonation pattern after injections in different regions of the BPN. Figure 4 reports a representative example. In fact, after injections in the medial (Fig. 4A) or in the lateral region of the BPN (Fig. 4B) the three stripes labeled in the PFL of each side were largely coincident. Finally, Fig. 6 shows a representative example of complementary labeling (i.e., the concerned area of stripes labeled in one rat corresponds to the interstripe areas of the second rat) after injections in neighbouring regions of the BPN: three stripes were labeled in the right PFL of rat 27 (A1) whereas 2 stripes were labeled in the complementary areas of rat 37 (B1). The comparison of the zonal labeling in the two rats showed that the stripes of one rat largely corresponded to the interstripes of the other rat. Together, this evidence indicates that the projections of the pontine nuclei to the cerebellar cortex are arranged according to a fixed pattern specific for each cortical area, independent of the number of stripes labeled within it.

One main question is concerned with the degree of divergence of this pathway. In other words, is one region of the BPN connected to a set of stripes or to only one of them? The comparison of the size of the

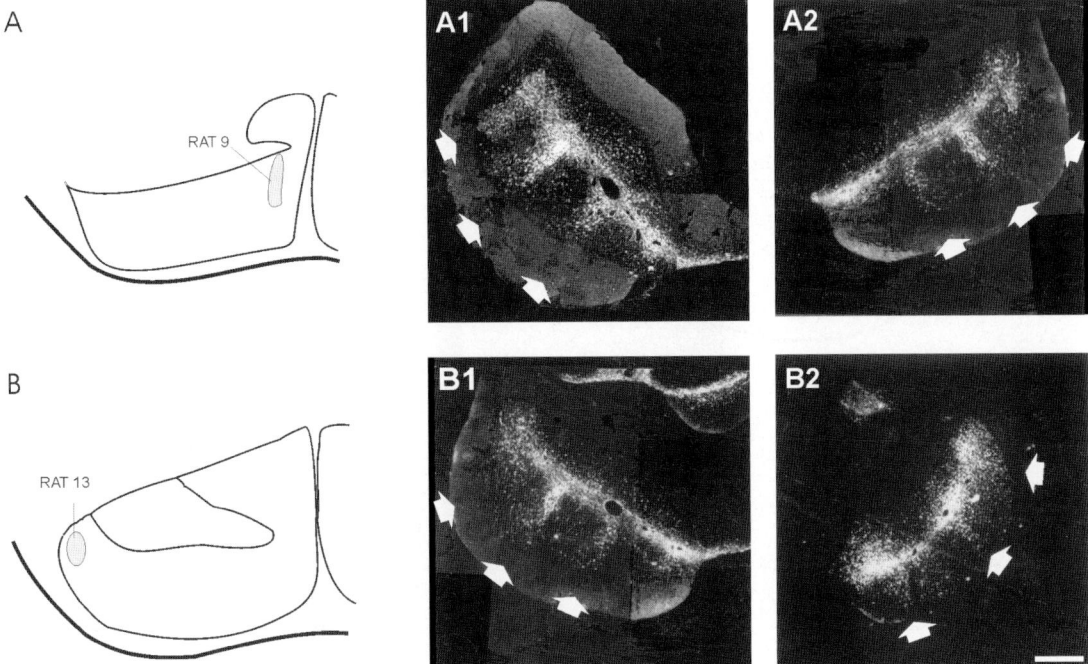

Fig. 4. Representative examples of reproducible zonal labeling in the paraflocculus after injection of tracer in two different regions of the BPN, rat 9 (A) and rat 13 (B). **A1–A2, B1–B2**: Darkfield photomicrographs of coronal sections of the paraflocculus contralateral (A2–B2), and ipsilateral (A1–B1) to the injection of ^{3}H-leu in the medial (A, rat 9) or in the lateral (B, rat 13) region of the central BPN. Note that three stripes (arrows) were labelled in both cases. Scale bar = 500 μm.

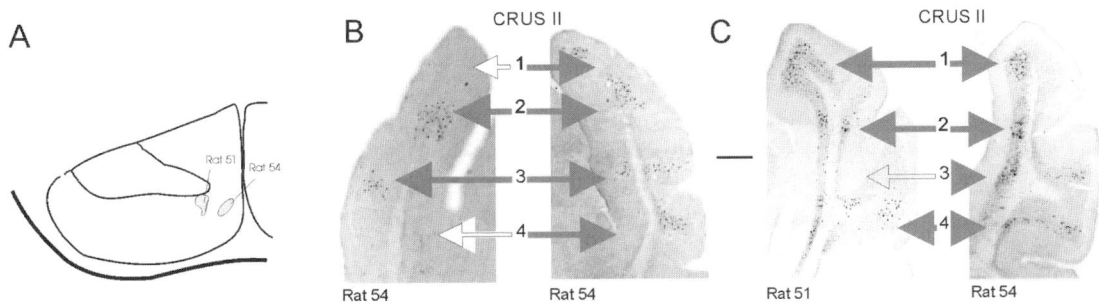

Fig. 5. Examples of the same fixed pattern of compartmentation of the Crus II, in two cases of injection (rats 54 and 51, A). B: Bright photomicrographs of horizontal sections of the Crus II of both sides, rotated and placed side by side, of rat 54. Four stripes were found in the right Crus II (contralateral to the side of injection) and 2 stripes in the left Crus II. Note that stripes 2 and 3 of the two sides were labeled in symmetric areas (double black arrows). C: bright photomicrographs of horizontal sections of the right Crus II (contralateral to the side of injection) of rats 51 and 54. The two Crura were put in parallel. Note that stripes 1, 2 and 4 of the two Crura were labeled in symmetric areas (double black arrows). Scale bar, 500 μm.

injections with the number of stripes in the CC showed that the number of stripes usually decreased in rats with progressively smaller size of injection sites and that at liminal injection volumes a set of few stripes were stained in various areas of the CC.

The divergent pattern is the anatomical basis for the specificity of the BPN projections to the CC. In fact, injections in different regions of the BPN labeled different sets of stripes. These sets were partially coincidental, while a complete overlap was never

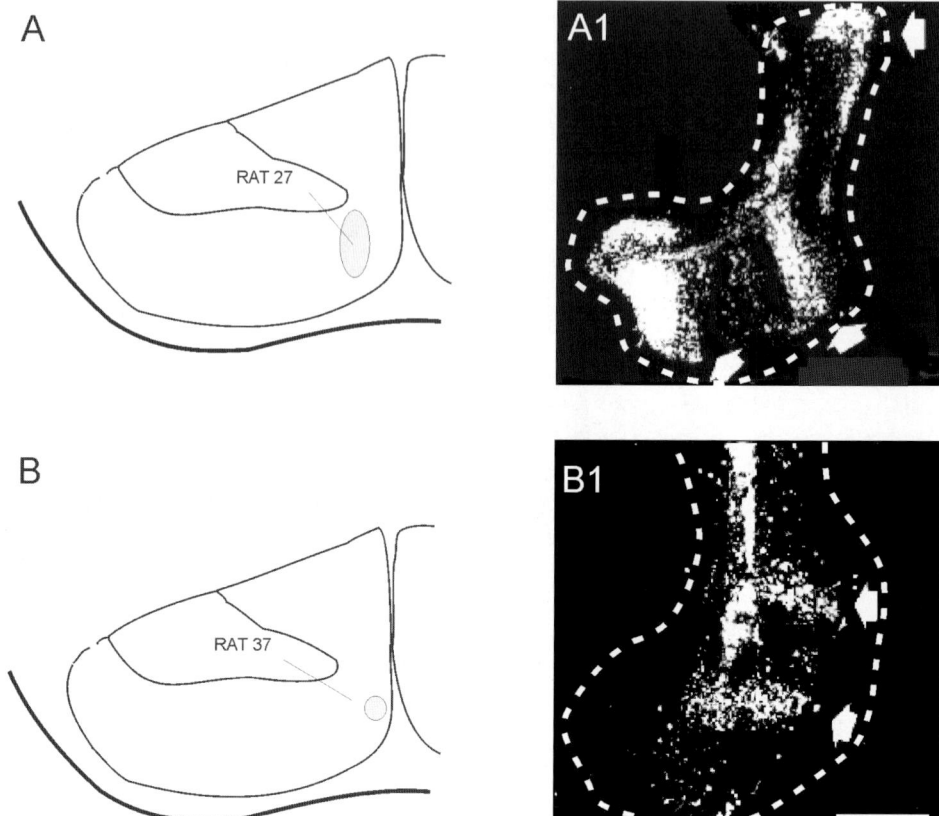

Fig. 6. Polarized light photomicrographs of horizontal sections of the paraflocculus contralateral to the injection of WGA-HRP in the BPN of rats 27 (A) and 37 (B), showing complementary zonal labeling. The injections concerned neighboring regions of the medial BPN of the two rats. Three and two stripes (arrows) were evidenced in rats 27 and 37 (A1, B1), respectively. Scale bar = 500 μm.

found. Thus, the specificity was given by the combination of the cortical stripes of granular cells projected upon from single regions of the BPN whereas the single stripes were usually associated with different sets in relation to the region of the BPN injected. These findings strongly support the idea that the BPN is at the origin of divergent projections to specific set of stripes of the CC.

The activation of longitudinal stripes of granular cells probably results in the activation, or preferential control, of overhanging stripes of Purkinje cells

The segregation of the pontine projections to sagittal stripes of granule cells raises the question whether this zonation pattern is lost in the successive transmission of pontine signals from granule to Purkinje cells (PCs). In fact, granule cells give rise to ascending axons that, at the level of the molecular layer dichotomized into parallel fibers that run for long distances, perpendicularly crossing the dendrite trees of PCs, with which they synapse.

The divergence of parallel fibers onto PCs is thought to underlie the coordination of multiple body parts by the cerebellum (Thach et al., 1992). The remarkable arrangement of parallel fibers and Purkinje cells means that many PCs along a parallel fiber 'beam' will share large fractions of their inputs. This fact gave rise to theories of cerebellar function based on the activation of beams of parallel fibers (Braitenberg and Atwood, 1958; Eccles, 1973). However, although Purkinje cell activity along the beams is demonstrated easily in vitro (Vranesic et al., 1994), the in vivo observations have proved to be more controversial (Bower and Woolston, 1983; Garwicz and Andersson, 1992; Cohen and Yarom, 1998), and PCs separated by more than ~100 μm do not show the correlation of their activity that would be expected from a common input (Bell and Grimm, 1969; Ebner and Bloedel, 1981). Moreover, PCs have well defined receptive fields and modalities (Bower and Woolston, 1983; Fushiki and Barmack, 1997; Ekerot and Jorntell, 2001), which may reflect the origin of the subjacent mossy fibers (Shambes et al., 1978), rather than the diffuse, generalized responsiveness that parallel fibers might be expected to produce. The usual explanation for these observations invokes granule cell "ascending axons" (Llinas, 1982), which run in the plane of the Purkinje cell dendrites and could potentially form strong multicontact synaptic connections PCs. Llinas (1982) proposed that with the restricted activation of Purkinje cells might reflect synapses made by granular cells axons as they ascend into the molecular layer past the Purkinje cell dendrite. These synapses were firstly described by Mugnaini (1972) and successively analysed by Gundappa-Suliur et al. (1999), who estimated that 20% of the granule cell synapses onto a Purkinje cell are actually made by the ascending segment, thus assuming that the two different regions of the granule cell axons may play different physiological roles in the cerebellar cortex. Recent electrophysiological investigations have shown different electrophysiological features in the synapses formed with PCs by both parallel fibers and granule cell ascending axons (Isope and Barbour, 2002).

In conclusion, despite the absence of conclusive evidences on the topographic segregation in the granule-PC projections, some evidence supports the idea that the activation of stripes of granule cells may result in the activation, or in preferential control, of the overhanging population of Purkinje cells.

Sagittal stripes of Purkinje cells innervate small areas of the cerebellar nuclei

Previous studies found that the functional units of the cerebellar nuclei (CN) were small regions that control

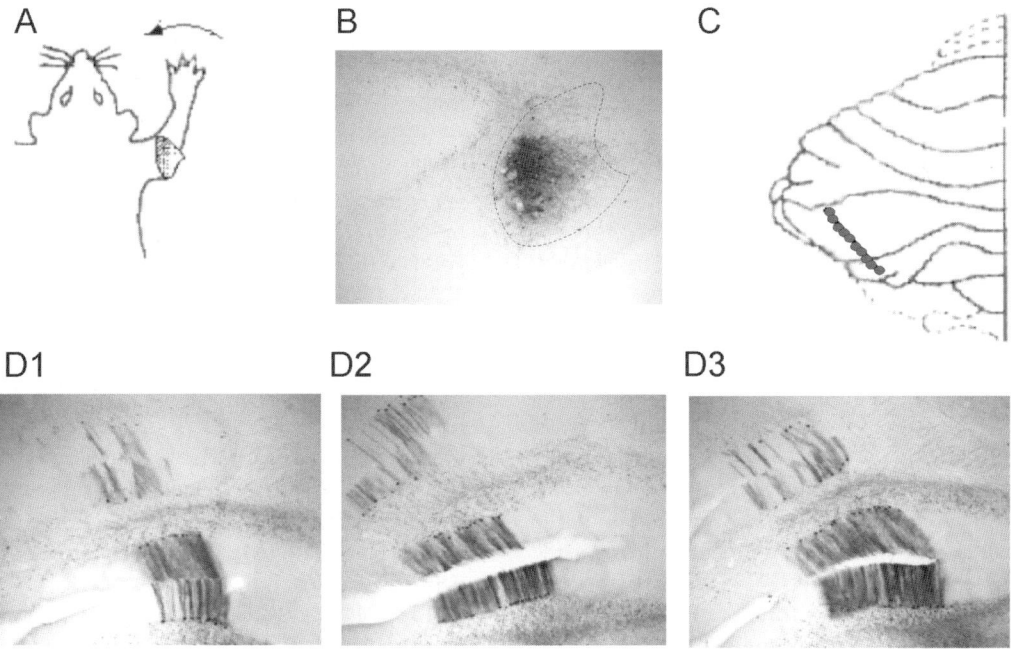

Fig. 7. Sagittal stripes of Purkinje cells innervate puntiform sites of the cerebellar nuclei. A: Motor effect (interesting the brachial biceps muscle) evoked by electrical stimulation of a puntiform site of NL nucleus. B: BDA injection in the same stimulated nuclear site. C: Dorsal surface of the cerebellum ipsilateral to the injection. A single sagittal stripe of stained PC was labeled. D1–D3: Rostrocaudal sequence of three coronal sections showing the PC labeled stripe, extending from the paramedian lobe to the CRUS II.

specific motor representations (see below). Thus, we planned a study to determine the topographic arrangement of the PCs that project to single motor representations in the CN.

Experimental procedure consisted of iontophoretically injecting BDA in progressively smaller areas of the CN, in order to determine the threshold volume of the injection able to evidence zonal staining of PCs. The injection sites were functionally identified by analyzing the motor effects elicited by their electrical stimulation. Thus, for each injection we identified the motor representation it concerned. The deposition of tracer in these nuclear sites labeled sagittal stripes of PCs of variable rostrocaudal extension. They usually extended rostrocaudally some hundreds of micrometers and were a few hundred micrometers wide. In most cases, the stained PCs were arranged in a single stripe (Fig. 7). It is worth noting that the stripes of PCs stained in the CC were reminiscent of the stripes stained at the granular layer by the pontine projections. The convergence of projections from sagittal stripes of PCs over small areas of the CN is largely confirmatory of the classical view that cortical projections (i.e., PC axons) converge over the CN with a radial pattern.

Projections from BPN and NRTP to the cerebellar nuclei

Our study on the pontine projections to the cerebellum also included the projections to the cerebellar nuclei (CN). Previous studies on the projections to the CN from the BPN (Eller and Chan-Paly, 1976; Brodal et al., 1986; Gerrits and Voogd, 1987; Mihailoff, 1993) and NRTP (Eller and Chan-Paly, 1976; McCrea et al., 1977; Gerrits and Voogd, 1987; Mihailoff, 1993) did not report significant differences in their projection patterns. They sustained largely diffuse projection patterns for both. On the other hand, the fact that the electrical stimulation of the mcp activates specific body segments through the cerebellum (Perciavalle et al., 1977) supports the idea of topographical precision in the pontine projections to the CN. This study was planned to analyse the projections from BPN and NRTP to the CN.

The experimental criterion was that of selecting for analysis only those animals with injection of BDA segregated within the BPN or the NRTP of one side.

As above, we excluded from the analysis any rats with tracer deposition within the BPN that showed cell bodies stained in the NRTP. Because tracer injections of both BPN and NRTP stained passing fibers from the homologous nucleus of the other side (Fig. 1), the unilateral injection of tracer in pontine nuclei always resulted in bilateral pontine staining.

The main result of the study was the evidence that the BPN and the NRTP innervated different regions of the CN.

BPN and NRTP project to private areas of the CNs

Fibers from the BPN were mainly traced to the NL and to a lesser degree to the ventrocaudal aspects of the nucleus interpositalis posterioris (NIP) of both sides, whereas no fibers were traced to the nucleus interpositalis anterioris (NIA) and nucleus medialis (NM, Fig. 8A). The BPN innervated the entire anteroposterior extension of the NL. The innervated region was arranged like a shell with lateral convexity in the central two thirds of the nucleus (Fig. 8A). The projections to the NL included both the magnocellular subdivision (MLm) and the parvo-cellular subdivision (or subnucleus lateralis parvocellularis, NLslp). A few fibers were traced to the dorsolateral hump (NLdlh). The target area of the NIP is restricted to the caudolateral region (Fig. 8A).

The question of whether the BPN is a homogeneous source of projections to the CNs was tested by comparing size and site of the injections in the BPN with the density of projections to the CN (see Table 1 of Parenti et al., 2002). The density of projections was not related to the size of injections but to the topographic localization of the injections within the BPN. The dorsal peripeduncular nucleus (dPd) was the main source of projections to the CN. Conspicuous projections were generally traced from the ventral (Vent) nucleus, ventral peripeduncular nucleus (vPd), lateral (Lat) and dorsolateral nuclei (dL). The medial (Med) was the source of a moderate density of projections to the CN.

Finally, the convergence from the pontine nuclei to single regions of the CN was studied (see Table 1 in Parenti et al., 2002). The NLm was projected upon from all parts of the BPN but principally from the Vent nucleus, both the Vent and the vPd nuclei and the dPd nucleus. The NLslp was principally projected

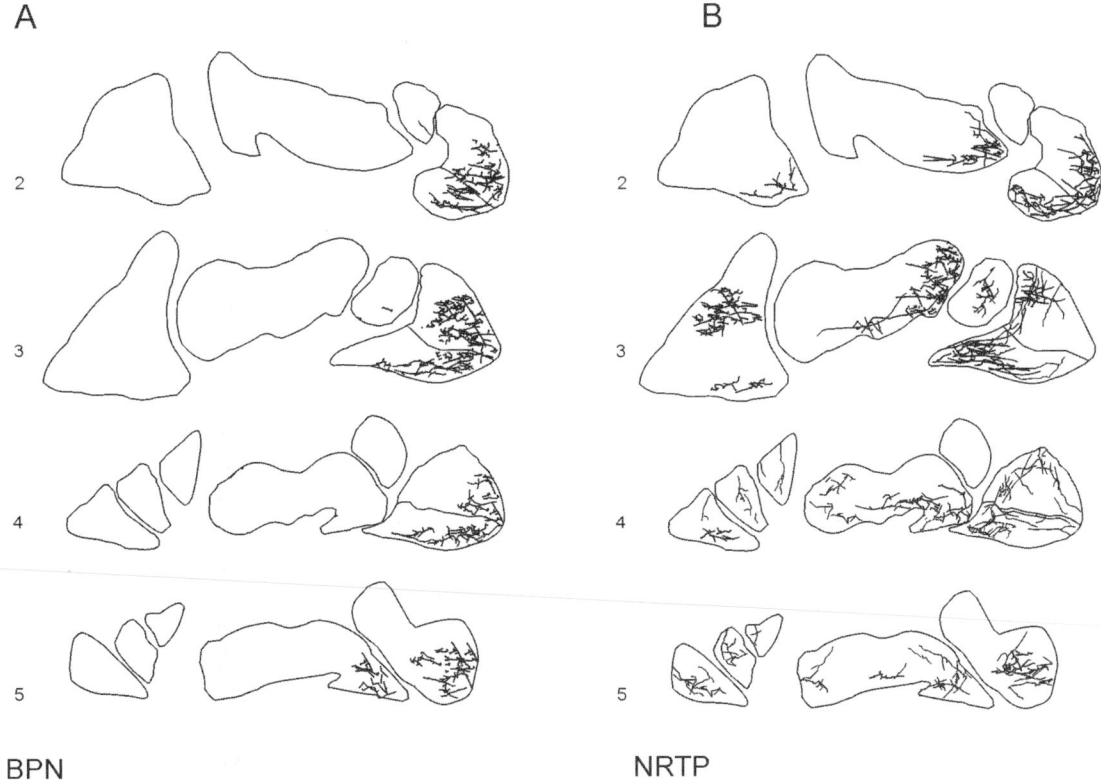

Fig. 8. Target areas in the cerebellar nuclei of 12 rats after injection in the BPN (A) and of 8 rats after injection in the NRTP (B).

upon from the Lat, the dL, the dPd nuclei and the Med. Moderate projections were traced to the NLslp from the vPd nucleus. The NIP was principally projected upon from the dPd, dL, Lat and Med nuclei. The NLdlh received sporadic projections only in rats injected in the Vent nucleus as well as in the vPd nucleus. These findings show that single regions of the CN are preferentially projected upon from specific regions of the BPN.

Fibers from the NRTP were traced to both NL, NIA, NIP and NM of both sides (Fig. 8B). A few fibers were also traced to the NLdlh. Projections to NL and NM were largely similar after staining of the two subregions of the NRTP, those to NIA/NIP were important only after staining of the NRTPp. Projections to the NL extended rostrocaudally for the entire extension of the nucleus and included both the NLm and the NLslp (Fig. 8B). The central part was principally traced to the medial side of the NL, a region roughly complementary to the target area of the BPN. Projections to the NM terminated in all divisions of the caudal part of the nucleus. Projections to the interposed nuclei principally concerned the lateral part of the NIA and the ventral part of the NIP.

The comparison of rats injected in the BPN and the NRTP showed that their terminal areas in the CN were basically complementary, with only discrete overlapping, principally in the rostral and caudal parts of the NL.

BPN and NRTP terminal fibers

The type of termination of the pontine fibers in the CN was studied. We assumed varicosized fibers to be sites of synapses (we referred to them as terminal fibers and smooth fibers as passing fibers). Both BPN and NRTP fibers were smooth in the cerebellar white matter. After penetration into the nuclear mass they bifurcated and, in the final sections, assumed a varicosized aspect. The configuration of these

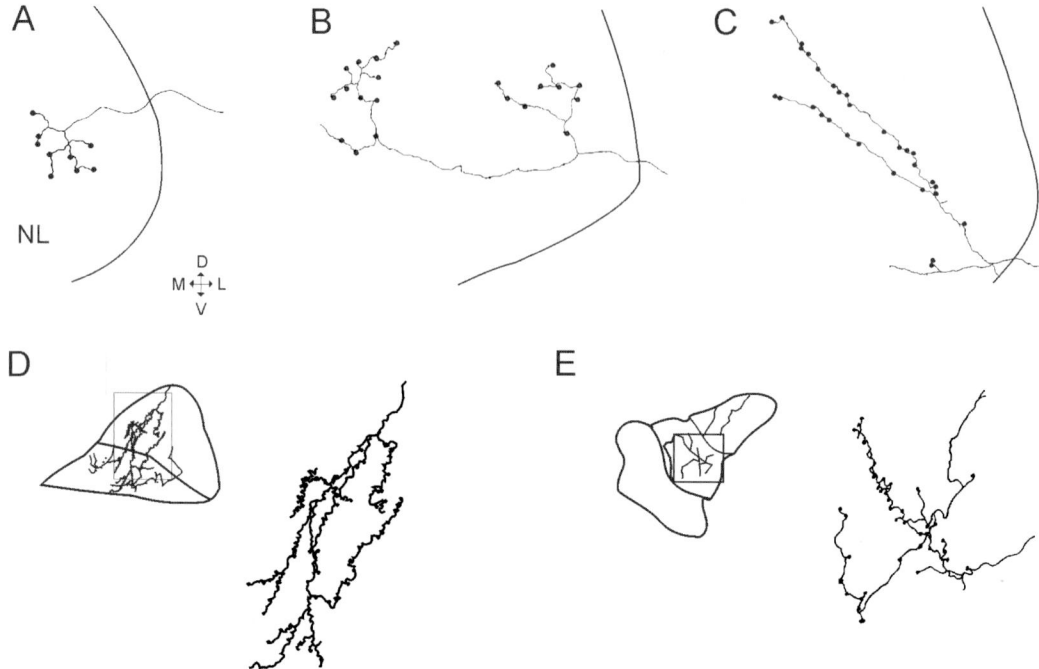

Fig. 9. Types of terminal fibers (distinguished from passing fibers by the presence of varicosities) in the cerebellar nuclei, projected from the BPN (A–C) and the NRTP (D–E), respectively.

varicosed (or terminal) fibers differed in relation to their BPN or NRTP origin.

Three types of BPN fiber terminations were found in our material. The most frequent termination consisted of parent, non-varicosed axons that bifurcated only once (Fig. 9A). Four to six short and thin varicosed fibers arose from the bifurcation. The collaterals usually radiated for a few tens of microns from the bifurcation point. Thus, the bush of collaterals concerned a small amount of nuclear mass. The second type consisted of parent, smooth axons, running across the nuclear mass and bifurcating two or three times along the parent axons (Fig. 9B). Each bifurcation process was similar to that reported previously. The third type of termination was found in relatively few cases. It consisted of parent axons which gave off two or three varicosed fibers that extended throughout the nuclear mass (Fig. 9C). The fiber diameters were around 0.4 µm at inter-varicosity level.

Only one type of termination of the NRTP fibers was found in the CN (Fig. 9D, E). This consisted of several bifurcating processes distributed at different levels along the axons. One or two collaterals usually originated at each bifurcation, and then further collateralised into thinner fibers. The diameter of the fibers ranged around 0.4–0.6 µm at inter-varicosity level. Terminal fibers evidenced a relatively small core, dense with varicosed fibers, plus (varicosed) collaterals that radiate in surrounding areas for a hundred microns or more, with some winding.

Single injections of tracer within the BPN or the NRTP usually evidenced a set of terminal fields in the CN (BPN: usually 2–4 in the NL and 1–2 in the NIP; NRTP: 1–2 in each CN: NL, NI, and NM). The comparison of the terminal fields of different rats showed that each field was usually projected upon from different pontine regions, while the complete overlap of the sets of terminal fields was never found. Therefore, the set of terminal fields is specific for each pontine region.

In conclusion, the study showed anatomical features which support relevant functional consequences. The terminal fibers visualized in single animals injected in few of the BPN or the NRTP concerned sites, each of small size, in the CN. This pattern of projection is apparently coherent with the possibility of activating discrete sets of nuclear neurons. The functional arrangement of the CN in motor representations (see below), suggests the possibility that the pattern of pontine projections to the CN may induce the activation of discrete sets of motor representations and hence the movements of specific body segments.

Coupled and uncoupled projections from BPN and NRTP to CC and CN

The projections to the CC and the CN from pontine nuclei evidenced in single animals showed a surprising and unexpected finding: the presence of uncoupled projections (i.e., which innervated the cortex, but not the corresponding regions of the CN), together with coupled projections (i.e., which innervated both the CC and the corresponding region of the CN), as inferred from the cartography of the corticonuclear projection pattern (Buisseret-Delmas and Angaut, 1993).

Coupled projections were traced from the BPN to the lateral and intermedial parts of the cerebellum (Fig. 10) and from the NRTP to both medial (vermis), intermedial and lateral parts (Fig. 10). This type of projection was assumed to be the standard pattern of afferent projections to the cerebellum, after the classic work of Andersson and Oscarsson (1978 a,b), who showed that in Deiter's nucleus the same neurons are excited by climbing fiber collaterals and indirectly inhibited by the same climbing fiber via Purkinje cells. Because of the inhibitory nature of the PCs, most theories of cerebellar function rest on the assumption that the main flow of information enters the CN via mossy fiber collaterals and leaves through cerebellar nuclear efferents (see e.g., Brooks and Thach, 1981; Llinas 1981; Ito, 1984). Thus, in a coupled projection pattern, the activity of the innervated PCs is probably associated with the movements elicited by the nuclear target.

Uncoupled projections were entirely evident in the projections from the BPN to the vermis (Fig. 10). In fact, most projections were always traced to the vermal cortex but none to the NM in any animals.

Moreover, some evidence supports the view of uncoupled projections as a usual mode of projections of both pontine nuclei to all over the CC. In fact, we found many more cortical stripes stained in the CC than terminal fields in the CN, in all rats injected either in the BPN or the NRTP. As the motor representations of the CN are innervated from cortical stripes of PCs in a ratio 1:1, the prevalence of cortical stripes over the terminal field stained in the CN of the animals studied suggests that the projections to the exceeding cortical stripes are uncoupled.

Because the activation of the innervated strips of PCs are probably concerned in the control of their nuclear targets and thus, of the motor activity represented there (see below), uncoupled projections are assumed to play a relevant role in the functional architecture of the cerebellum. This control could be involved in some putative functional activities intrinsic to the cerebellar machinery. Certainly, the fact that the pontine nuclei dispose of a double modality of control over cerebellar activity (i.e., coupled and uncoupled projections) seems more congruent with the elaborate control of the cerebellar output necessary to execute coordinated movements than the previous scheme which provided only excitatory drives to the CN, by coupled projections alone.

In a more general sense, the pools of PCs innervated by pontine projections can be divided into two main categories. The first consists of the pool of PCs that project to nuclear targets that are also directly innervated by pontine afferents. The activity of these PCs is probably associated with the movements elicited by the nuclear target output. The second category is composed by a pool of PCs that project to nuclear targets not directly innervated from the pontine nuclei. Therefore, the activation of these PCs is not associated with any motor effects. It is worth noting that these findings are in agreement with recent findings of Miller et al. (2002), who recorded the activity of PCs, primary motor cortical neurons and limb EMG signals while monkeys executed a sequential reaching and button pressing task. In this experimental procedure, they showed the activity of a group of PCs positively correlated with EMG activity, while the electrical activity of another group of PCs was negatively correlated.

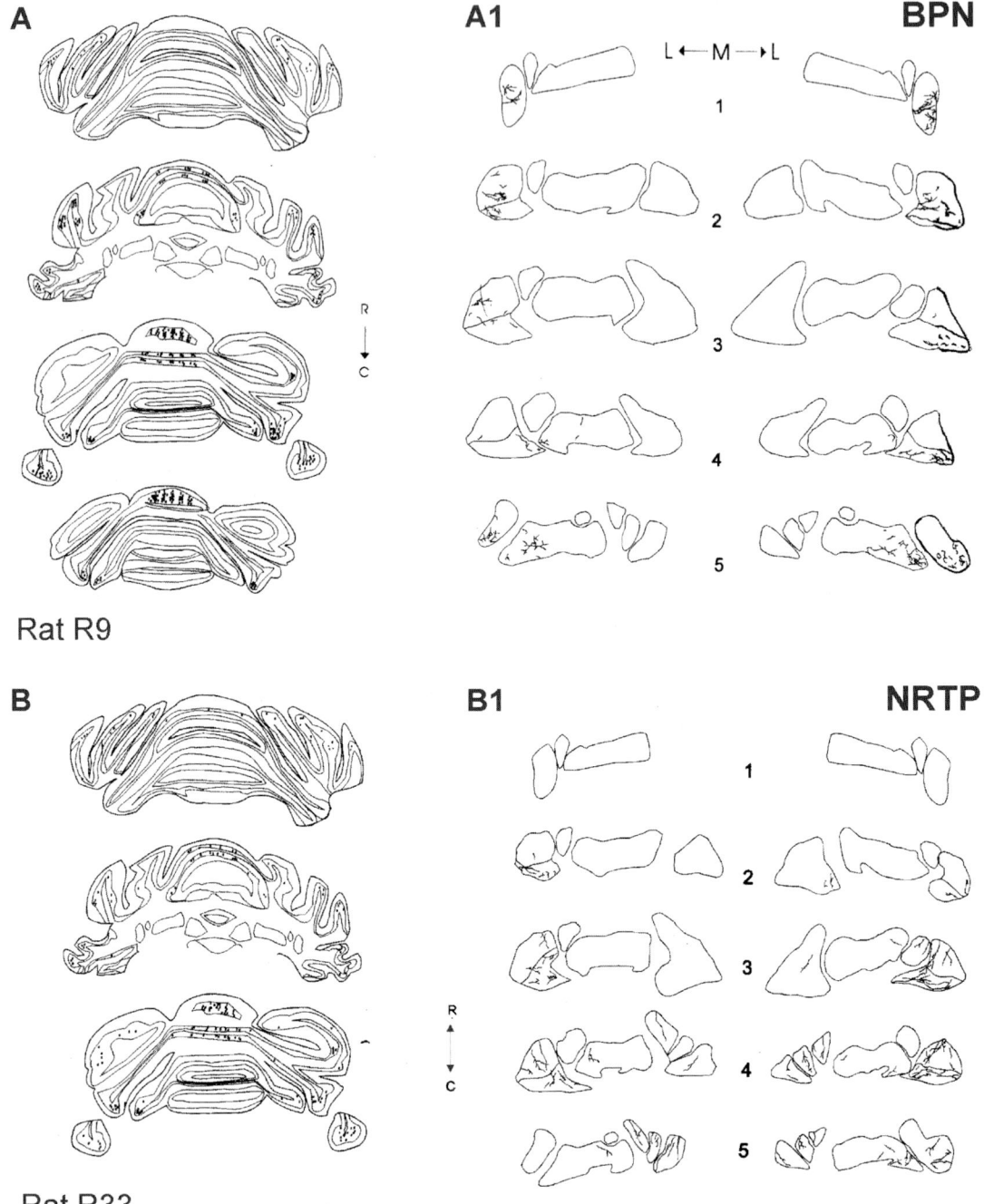

Fig. 10. Representative examples of coupled and uncoupled projections to CC and CN after injections of BDA in the BPN and NRTP. The labelling is illustrated in four drawings of equally spaced coronal sections of the CC (A, B) and in five drawings of rostrocaudal equally spaced coronal sections of the CN (A1, B1). Note projections from the BPN to the vermis of the cortex but none to the NM of the CN (uncoupled projections), while coupled projections were traced from the BPN to the lateral and intermedial parts of the CBL and from the NRTP to both medial, intermedial and lateral parts.

Functional organization of the CN in monkey and rat

As reported above, BPN and NRTP have private terminal areas in the CN, with only discrete overlapping. Moreover, they project to a set of nuclear parts (terminal fields) that are specific for each pontine region. Thus, the terminal fields visualized in the CN after the staining of the pontine nuclei seem to divide the CN into a set of zones. It is functionally relevant to determine whether a functional counterpart to the anatomical zonation of the CN exists as evidenced by the pontine projections. Despite the fact that this report specifically focuses on the rat as an experimental model, given the strict analogies between monkey and rat, we considered it useful to report the main findings provided by both species.

Monkey

Movements elicited by the stimulation of the cerebellar nuclei were studied in alert baboons chronically prepared (Rispal-Padel et al., 1980, 1981, 1982, 1983, 1985). Movements were filmed and the EMG of muscles were recorded. Two types of motor effects were observed (Fig. 11): (i) simple movements that concerned displacement of a limb segment; (ii) complex movements that involved distinct and frequently non-contiguous muscles were stereotyped and could not be dissociated. These movements are defined as motor synergies. Simple movements were due to activation of muscle of the involved segment in addition to the co-contraction of muscles of a nearby segment.

Somatotopic motor localization was evidenced in both the fastigial, interposed and dentate nuclei (Fig. 12). In the dentate nucleus, two subdivisions were distinguished: (i) an antero-medial region, corresponding to large cells or magnocellular subdivision, from which motor synergies can be elicited. Only complex movements were elicited from this nuclear subdivision. The movements concerned two medio-proximal joints of one or both limbs. Thus, this nuclear subdivision is involved either in the displacement of the forearm in space (when the moved joints concerned only this limb), or in postural adjustments (when they concerned only hindlimb), or posturo-kinetic integration (when they concerned both

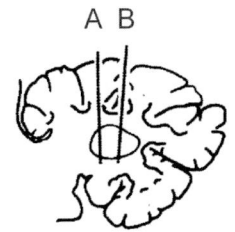

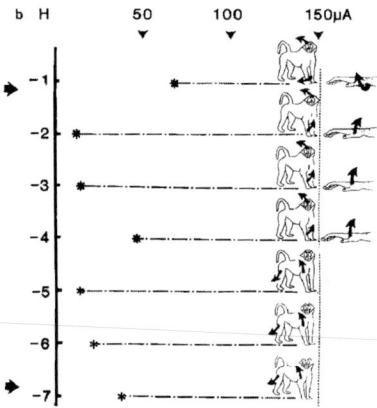

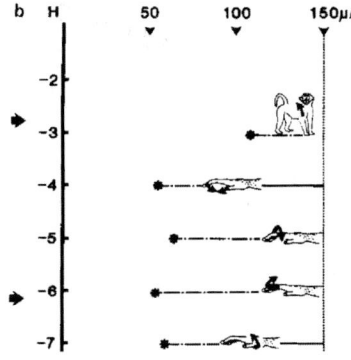

Fig. 11. Dentate control of movements. Upper: parasagittal section of the neocerebellum. Two tracks were placed at rostral (A) and caudal level (B) of the dentate nucleus. Central: A, diagram showing simple movements elicited by stimulation of the parvocellular region of the dentate nucleus in relation to the depths of the electrode tip (vertical coordinate at left indicated in mm) and to the threshold value of stimulation (horizontal coordinate). The boundaries of the nucleus are shown by arrows along the vertical scale. The stimulation threshold of each movement is represented by a star. Movements involving a single or few joints were induced from these dentate sites. Lower: B, diagram showing complex movements elicited by stimulation of the magnocellular region of the dentate nucleus. Same format as in A.

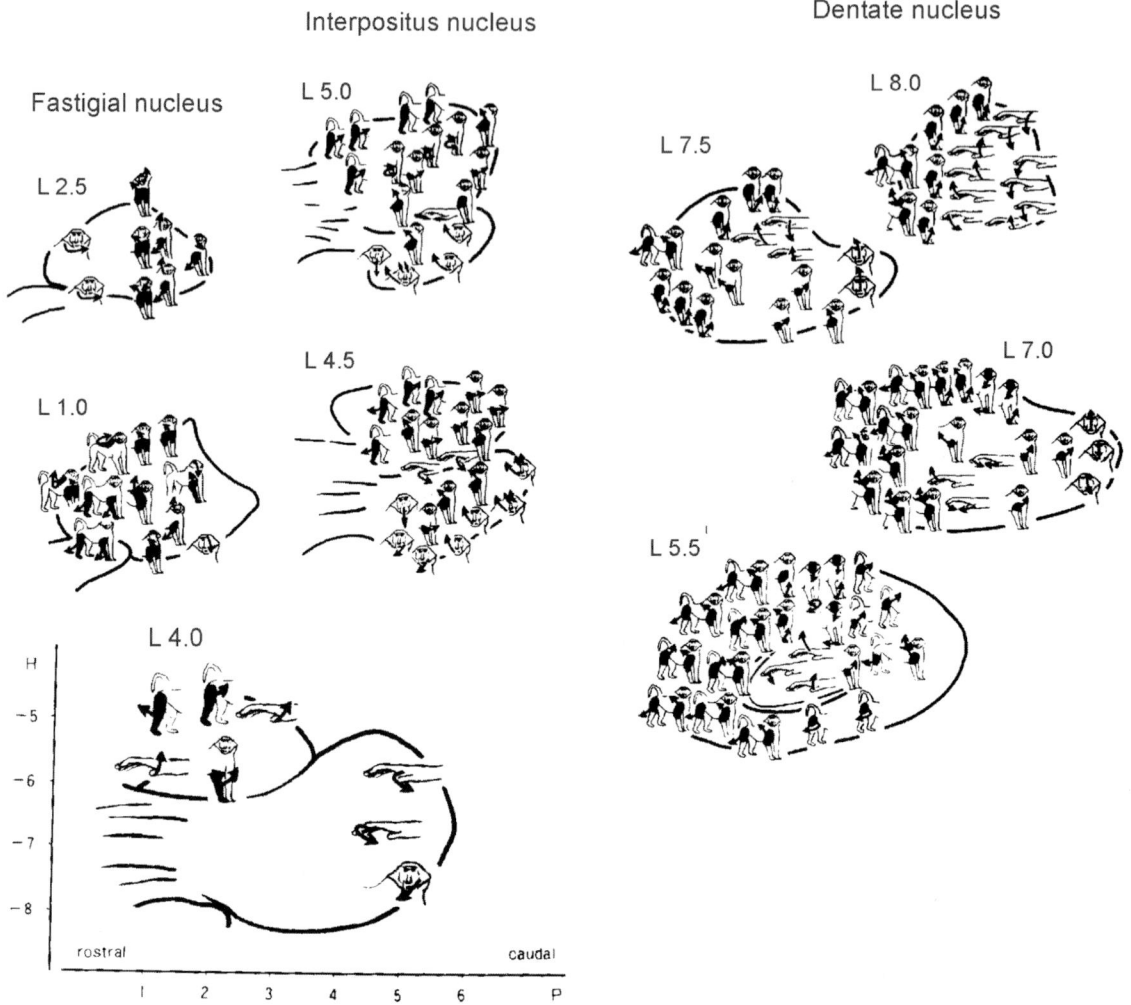

Fig. 12. Functional localization in baboon's cerebellar nuclei. Body regions where muscle contraction was induced are indicated on the stimulated regions. L, laterality, in mm.

limbs). (ii) The posterolateral region of the dentate, corresponding to small cell or parvocellular subdivision (the neodentatum of Demole), evoked only simple movements. These principally concerned either finger movements or eye rotations. Therefore, the parvocellular sub-division of the dentate is mainly concerned in either digital manipulation (skilled movements) or visual exploration of the environment.

Rat

The motor organization of the NL of the rat cerebellum was investigated by observing the motor effects of electrical microstimulations of the NL (Cicirata et al., 1992). The movements evoked by the NL mainly concerned forelimb and head segments. Only in a few cases were movements of hindlimb segments evoked. Motor effects were obtained according to a precise topographical pattern. This pattern delimited functional zones, or representations, within the NL, each zone being specifically related to a particular segment of the body (Fig. 13A). A few body segments were activated from single zones only (single representation) whereas some other body segments could be activated from different zones of the NL. Among them, the axio-proximal

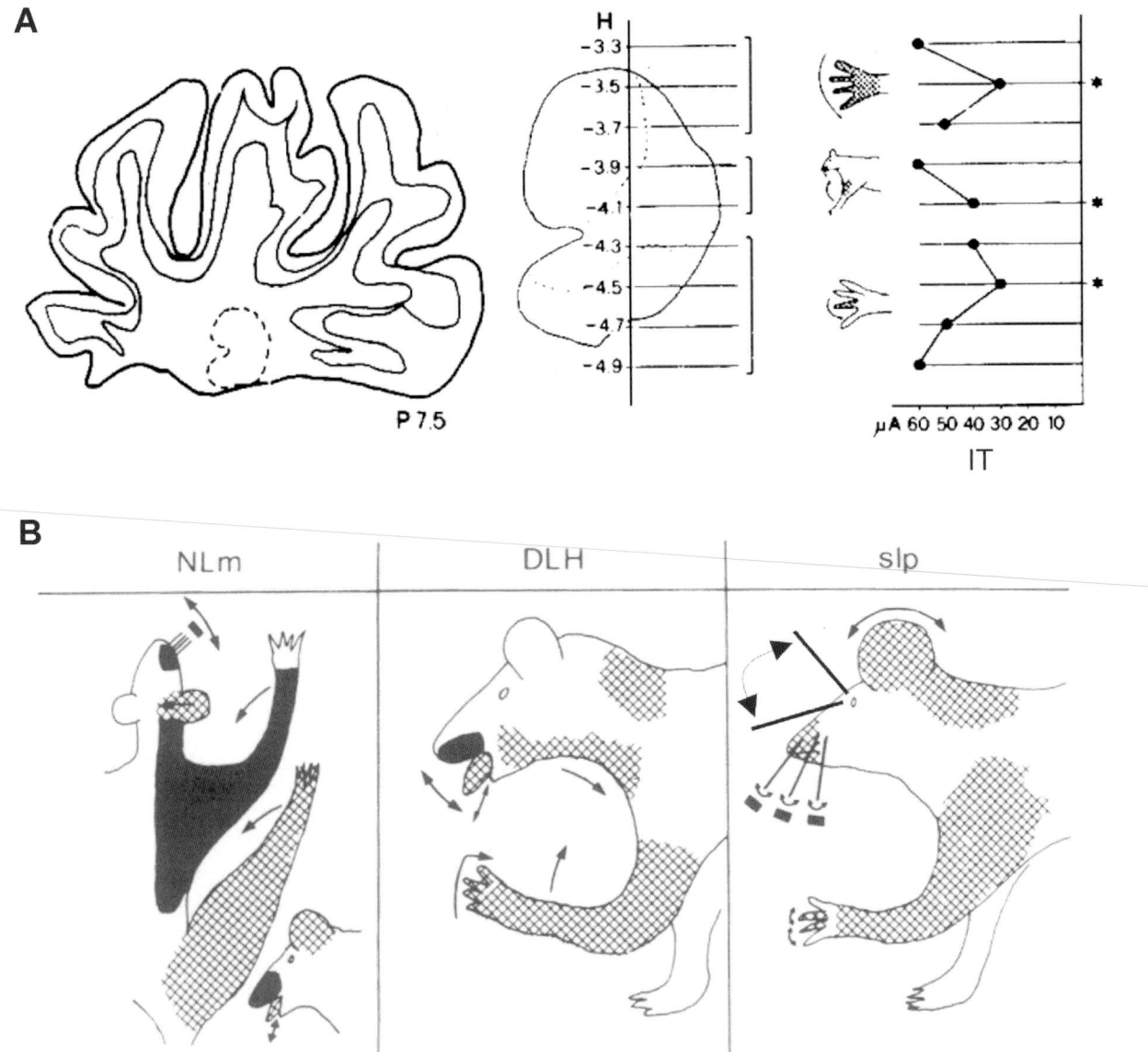

Fig. 13. (A) Example of microelectrode penetration through the NL. From left to right: parasagittal plane of penetration, magnification of the NL showing the electrode trajectory, motor effects and corresponding intensity thresholds (IT) of the electrical microstimulations. Below the cerebellar surface the descent of the electrode was stopped every 200 μm (reported as H). From top to bottom 3 representations are encountered: forelimb digits (as flexion of all digits), elbow (as flexion of forearm), forelimb digit (movement of single digit). Note that two forelimb digit representations were passed through by the electrode. They were topographically separated by the presence of a representation of a different body segment (elbow). Within each representation, the point with the lower IT (site) is identified by *. Only the sites (and not all the active points) were collected in the maps. (B) Variety of body segments activated (shaded areas) by the stimulation of the single NL regions. For each body segment the direction of the movements are indicated by arrows. The body segments most frequently activated by the stimulation of the single NL regions are blackened on the figurines.

body segments were activated in a similar way from all sites (multiple representation) whereas the distal body segments were differently activated from the various representation zones (specific representation). The multiple and specific representations were distributed between the 3 cytoarchitectonic subregions of the NL (MLm, NLslp, NLdlh) in such a way that the body segments were usually represented only once in each individual NL subdivision. Each NL subdivision included sets of representations concerning body segments characterized by a topographical continuity (e.g., the different segments of the forelimb in both NLdlh and NLslp). Thus, the individual NL subdivisions may bring into play coordinate plurisegmental muscular activities of the limbs and/or the head. The NLm controls movements of all the segments of the head and those of axio-proximal segments of both limbs (Fig. 13B). Thus it is principally concerned with the movements of the limbs or of the head in space. The NLdlh particularly controls movements of the head, including the oral region (frequent mouth movements), plus the proximal (neck) segment of the head. To a lesser degree, NLdlh also controls movements of the various segments of the forelimb, including synchronous flexion of all the digits (Fig. 13B). Thus, the NLdlh is principally involved in the control of integrated movements of forelimb/mouth (Fig. 13B), as occurs in feeding, nursing, and aggressive behavior. Finally, the NLslp is concerned with the motor control of distal segments of the head, such as motor ocular activity, displacement of individual vibrissae and rotation of the ear pinna, as well as of movements of individual digits of the forelimb (Fig. 13B). Thus, the NLslp is involved in the exploration of the environment by moving body segments (eyes, vibrissae, ear pinna) that are specifically devoted to this function in rat, as well as in digital manipulation (skilled movements), or in integrating both of these.

As reported, the functional organization of the CN of the two species shows a common pattern, with differences probably due to behavior specific to each species. In fact, both species showed a mosaic-like organization of the motor representations of the CN. The functional roles found for the magno- and parvocellular regions of the monkey dentate largely correspond to the functions found for the MLm and NLslp of the rat.

The comparison of the pontine projections with the functional arrangement of the CN leads to the question whether the mosaic-like arrangement of the motor representation identified by electrical stimulations of the CN coincides with the zonation pattern shown by pontine projections. For evident technical reasons it is quite impossible to give a formal answer to the question. Nevertheless, we think it is more likely that the two maps (functional and anatomical) are related than that they are reciprocally independent. In this hypothesis, the different terminal fields in the CN of projections from BPN and NRTP means that the motor representations in the CN are in a large part under the private control of the BPN or the NRTP.

Conclusions

The main findings of the study can be briefly commented. First of all, the anatomical dissection of the pontine projections to the cerebellum performed in this study showed that the stripes of projection to the CC and the terminal fields in the CN are the elementary units of the pontine projections respectively to CC and CN. In turn, electrophysiological investigations showed that the CN are arranged as a mosaic-like assembly of functional units, the motor representations. The identification of the unitary elements which are assembled together in a very complex system was very useful in our attempt to discern the machinery of the pontocerebellar pathway, otherwise irresolvable.

The second main finding of this study was the evidence that mossy fiber projections, at least those coming from pontine nuclei, terminated in the cerebellum with both coupled and uncoupled cortico/nuclear patterns. This opens new vistas into the functional architecture of the pontine projections to the cerebellum. In fact, the planning of the movement of a limb at the CN level theoretically demands a complex and differentiated control of the motor representations concerned with the moved limb. It means that while some motor representations are activated, some others could be modulated or wholly inhibited. The coexistence of coupled and uncoupled projections in single animals may represent the anatomical substrate for this functional pattern.

The initial question of this study was: can the BPN and the NRTP be considered as an unitary and undifferentiated relay of the cortico-ponto-cerebellar pathway, or do they subserve distinct pathways? All the steps of this study synergically support the different arrangement of the pathways passing through the BPN and the NRTP. In fact, they differ both in the afferent pattern from the cerebral cortex, in the types of termination within the CN and, principally in the segregation of their terminal fields in the CN. In previous studies we showed that the different CN, and even their subnuclear regions, have distinct terminal fields at the level of both the thalamus (Angaut et al., 1985) and the cerebral cortex (Cicirata et al., 1986). Therefore, the projection from BPN and NRTP to different areas of the CN involves the control of equally different areas of the cerebral cortex.

In the light of some of the new informations presented here, some aspects of the pontine projections to the cerebellum remain to be investigated. One particularly important unsolved question concerns the laterality of the pontine projections, or better, the anatomical and functional features that differentiate, if a difference exists, the projection that, from the pontine nucleus of one side, innervates the two sides of the cerebellum. In fact, the study performed by injecting a neuronal tracer within the pontine nucleus of one side always stained passing fibers from the homologous nucleus of the other side. Thus, the tracer injected into one side resulted in the staining of the nuclei of both sides. This technical limit is probably the reason for the indistinct bilateral projections to both sides of the cerebellum as reported by other studies also. Thus this study, and others performed with the same procedure, are inconclusive on this aspect. A specific line of investigation aimed at resolving the question is currently in progress in our laboratory. Preliminary results (Parenti et al., 2002; Serapide et al., 2002b) corroborate the hypothesis that the projections to the two sides are differently arranged.

Abbreviations

BPN	basilar pontine nuclei
BDA	biotine dextrane amine
CC	cerebellar cortex
CN	cerebellar nuclei
mcp	medial cerebellar peduncle
NI	nucleus interpositalis
NIA	nucleus interpositalis anterioris
NIP	nucleus interpositalis posterioris
NL	nucleus lateralis
NLm	nucleus lateralis, magnocellular subdivision
NLslp	nucleus lateralis, subnucleus lateralis parvocellularis
NLdlh	nucleus lateralis, dorsolateral hump
NM	nucleus medialis
NRTP	nucleus reticularis tegmenti pontis
NRTPc	nucleus reticularis tegmenti pontis, central part
NRTPp	nucleus reticularis tegmenti pontis, pericentral part
PCs	Purkinje cells

References

Altman, J. and Bayer, S.A (1987) Development of the precerebellar nuclei in the rat: IV. The anteriorprecerebellar extramural migratory stream and the nucleus reticularis tegmenti pontis and the basal pontine gray. J. Comp. Neurol., 257(4): 529–552.

Andersson, G. and Oscarsson, O. (1978a) Projections to lateral vestibular nucleus from cerebellar climbing fiber zones. Exp. Brain Res., 32(4): 549–564.

Andersson, G. and Oscarsson, O. (1978b) Climbing fiber microzones in cerebellar vermis and their projection to different groups of cells in the lateral vestibular nucleus. Exp. Brain Res., 32(4): 565–579.

Angaut, P., Cicirata, F. and Serapide, F. (1985) Topographic organization of the cerebellothalamic projections in the rat. An autoradiographic study. Neuroscience, 15(2): 389–401.

Bell, C.C. and Grimm, R.J. (1969) Discharge properties of Purkinje cells recorded on single and double microelectrodes. J. Neurophysiol., 32: 1044–1055.

Border, B.G. and Mihailoff, G.A (1985) GAD-immunoreactive neural elements in the basilar pontine nuclei and nucleus reticularis tegmenti pontis of the rat. I. Light microscopic studies. Exp. Brain Res., 59(3): 600–614.

Bower, J.M. and Woolston, D.C. (1983) Congruence of spatial organization of tactile projections to granule cell and Purkinje cell layers of cerebellar hemispheres of the albino rat: vertical organization of cerebellar cortex. J. Neurophysiol., 49: 745–766.

Braitenberg, V. and Atwood, R.P. (1958) Morphological observations on the cerebellar cortex. J. Comp. Neurol., 109: 1–34.

Brodal, A. and Walberg, F. (1977a) The olivocerebellar projection in the cat studied with the method of retrograde axonal transport of horseradish peroxidase. IV. The projection to the anterior lobe. J. Comp. Neurol., 172: 85–108.

Brodal, A. and Walberg, F. (1977b) The olivocerebellar projection in the cat studied with the method of retrograde axonal transport of horseradish peroxidase. VI. The projection onto longitudinal zones of the paramedian lobule. J. Comp. Neurol., 176: 281–294.

Brodal, A. and Kawamura, K. (1980) Olivocerebellar projection: a review. Adv. Anat. Embryol. Cell Biol., 64: 1–140.

Brodal, P. (1982) Further observations on the cerebellar projections from the pontine nuclei and the nucleus reticularis tegmenti pontis in the rhesus monkey. J. Comp. Neurol., 204: 44–55.

Brodal, P., Dietrichs, E. and Walberg, F. (1986) Do pontocerebellar mossy fibers give off collaterals to the cerebellar nuclei? An experimental study in the cat with implantation of crystalline HRP-WGA. Neurosci. Res., 4: 12–24.

Brodal, P. and Bjaalie, J.G. (1992) Organization of the pontine nuclei. Neurosci. Res., 13: 83–118.

Brooks, V.B. and Thach, W.T (1981) Cerebellar control of posture and movement. In: Brooks, V.B. (Ed.), Handbook of Physiology, Section 1, The nervous system, Motor control, part 2, Vol. II. Am. Physiol. Soc., Bethesda, Maryland, pp. 877–946.

Buisseret-Delmas, C. and Angaut, P. (1993) The cerebellar olivo-corticonuclear connections in the rat. Progr. Neurobiol., 40: 63–87.

Burne, R.A., Ericksson, M.A., Saint-Cyr, J.A. and Woodward, D.J. (1978a) The organization of the pontine projection to lateral cerebellar areas in the rat: dual zones in the pons. Brain Res., 139: 340–347.

Burne, R.A., Mihailoff, G.A. and Woodward, D.J (1978b) Visual corticopontine input to the paraflocculus: a combined autoradiographic and horseradish peroxidase study. Brain Res., 143(1): 139–146.

Cicirata, F., Angaut, P., Cioni, M., Serapide, M.F. and Papale, A. (1986) Functional organization of thalamic projections to the motor cortex. An anatomical and electrophysiological study in the rat. Neuroscience, 19(1): 81–99.

Cicirata, F., Angaut, P., Serapide, M.F., Pantò, M.R. and Nicotra, G. (1992) Multiple representation in the nucleus lateralis of the cerebellum: an electrophysiologic study in the rat. Exp. Brain Res., 89(2): 352–362.

Cohen, D. and Yarom, Y. (1998) Patches of synchronized activity in the cerebellar cortex evoked by mossy-fiber stimulation: questioning the role of parallel fibers. Proc. Natl. Acad. Sci. USA, 95: 15032–15036.

Courville, J. (1975) Distribution of olivocerebellar fibers demonstrated by radioautographic tracing method. Brain Res., 95: 253–263.

Ebner, T.J. and Bloedel, J.R. (1981) Correlation between activity of Purkinje cells and its modification by natural peripheral stimuli. J. Neurophysiol., 45: 948–961.

Eccles, J.C. (1973) The cerebellum as a computer: patterns in space and time. J. Physiol. (Lond.), 229: 1–32.

Eisenman, L.M. (1981) Pontocerebellar projections to the pyramis and copula pyramidis in the rat: evidence for a mediolateral topography. J. Comp. Neurol., 199: 77–86.

Eisenman, L.M. and Noback, C.R. (1980) The ponto-cerebellar projection in the rat: differential projections to sublobules of the uvula. Exp. Brain Res., 38: 11–17.

Ekerot, C.F. and Jorntell, H. (2001) Parallel fiber receptive fields of Purkinje cells and interneurons are climbing fiber-specific. Eur. J. Neurosci., 13: 1303–1310.

Eller, T. and Chan-Paly, V. (1976) Afferents to the cerebellar lateral nucleus. Evidence from retrograde transport of horseradish peroxidase after pressure injections through micropipettes. J. Comp. Neurol., 166: 285–302.

Fortier, P.A., Smith, A.M. and Kalaska, J.F (1993) Comparison of cerebellar and motor cortex activity during reaching: directional tuning and response variability. J. Neurophysiol., 69(4): 1136–1149.

Fushiki, H. and Barmack, N.H. (1997) Topography and reciprocal activity of cerebellar Purkinje cells in the uvula-nodulus modulated by vestibular stimulation. J. Neurophysiol., 78: 3083–3094.

Garwicz, M. and Andersson, G. (1992) Spread of synaptic activity along parallel fibers in cat cerebellar anterior lobe. Exp. Brain Res., 88: 615–622.

Gerrits, N.M. and Voogd, J. (1986) The nucleus reticularis tegmenti pontis and the adjacent rostral paramedian formation: differential projections to the cerebellum and the caudal brain stem. Exp. Brain Res., 62: 29–45.

Gerrits, N.M. and Voogd, J. (1987) The projection of the nucleus reticularis tegmenti pontis and adjacent regions of the pontine nuclei to the central cerebellar nuclei in the cat. J. Comp. Neurol., 258: 52–69.

Groenewegen, H.J. and Voogd, J. (1976) The longitudinal zonal arrangement of the olivocerebellar climbing fiber projection in the cat: An autoradiographic and degeneration study. Brain Res., Suppl. 1: 65–71.

Groenewegen, H.J., Voogd, J. and Freedman, S.L. (1979) The parasagittal zonation within the olivocerebellar projection. II. Climbing fiber distribution in the intermediate and hemispheric parts of the cat cerebellum. J. Comp. Neurol., 183: 551–602.

Gundappa-Suliur, G., De Schutter, E. and Bower, J.M. (1999) Ascending granule cell axons: an important component of cerebellar cortical circuitry. J. Comp. Neurol., 408: 580–596.

Hoddevik, G.H. (1978) The projection from nucleus reticularis tegmenti pontis onto the cerebellum in the cat. Anat. Embryol., 153: 227–242.

Isope, P. and Barbour, B. (2002) Properties of unitary granule cell-Purkinje cell synapses in adult rat cerebellar slices. J. Neurosci., 22: 9668–9678.

Ito, M. (1984) The cerebellum and the neural control. Raven, New York, pp. 235–255.

Kawamura, K. and Hashikawa, T. (1981) Projections from the pontine nuclei proper and reticular tegmental nucleus onto the cerebellar cortex in the cat. An autoradiographic study. J. Comp. Neurol., 201: 395–413.

Kunzle, H. (1975) Autoradiographic tracing of the cerebellar projections from the lateral reticular nucleus in the cat. Exp. Brain Res., 22: 255–266.

Legg, C.R., Mercier, B and Glickstein, M. (1989) Corticopontine projection in the rat: the distribution of labelled cortical cells after large injections of horseradish peroxidase in the pontine nuclei. J. Comp. Neurol., 286(4): 427–441.

Llinas, R. (1981) Electrophysiology of the cerebellar networks. In: Brooks, V.B. (Ed.), Handbook of Physiology, Section 1, The nervous system, Motor control, part 2, Vol. II. Am. Physiol. Soc., Bethesda, Maryland, pp. 831–876.

Llinas, R. (1982) General discussion: radial connectivity in the cerebellar cortex. A novel view regarding the functional organization of the molecular layer. In: Palay, S.L. and Chan-Palay, V. (Eds.), The Cerebellum: New Vistas. Springer, Berlin, pp. 189–194.

McCrea, R., Bishop, G.A. and Kitai, S.T. (1977) Electrophysiological and horseradish peroxidase studies of precerebellar afferents to the nucleus interpositus anterior. II. Mossy fiber system. Brain Res., 122: 215–228.

Mihailoff, G.A., Burne, R.A. and Woodward, D.J (1978) Projections of the sensorimotor cortex to the basilar pontine nuclei in the rat: an autoradiographic study. Brain Res., 145(2): 347–354.

Mihailoff, G.A., Lee, H., Watt, C.B. and Yates, R. (1985) Projections to the basilar pontine nuclei from face sensory and motor regions of the cerebral cortex in the rat. J. Comp. Neurol., 237(2): 251–263.

Mihailoff, G.A., McArdle, C.B. and Adams, C.E (1981) The cytoarchitecture, cytology, and synaptic organization of the basilar pontine nuclei in the rat. I. Nissl and Golgi studies. J. Comp. Neurol., 195(2): 181–201.

Mihailoff, G.A. (1993) Cerebellar nuclear projections from the basilar pontine nuclei and nucleus reticularis tegmenti pontis as demonstrated with PHA-L tracing in the rat. J. Comp. Neurol., 330: 130–146.

Miller, L.E., Holdefer, R.N. and Houk, J.C (2002) The role of the cerebellum in modulating voluntary limb movement commands. Arch. Ital. Biol., 140(3): 175–183.

Mugnaini, E. (1972) The histology and cotology of the cerebellar cortex. In: Larsell, O. and Jansen, J. (Eds.), The Comparative Anatomy and Histology of the Cerebellum, The Human Cerebellum, Cerebellar Connections, and Cerebellar cortex. University of Minnesota Press, Minneapolis, pp. 201–262.

Newman, D.B. and Ginsberg, C.Y (1992) Brainstem reticular nuclei that project to the cerebellum in rats: a retrograde tracer study. Brain Behav. Evol., 39(1): 24–68.

Parenti, R., Zappalà, A., Serapide, M.F., Pantò, M.R. and Cicirata, F. (2002) Projections of the basilar pontine nuclei and nucleus reticularis tegmenti pontis to the cerebellar nuclei of the rat. J. Comp. Neurol., 452(2): 115–127.

Perciavalle, V., Santangelo, F., Sapienza, S., Savoca, F. and Urbano, A. (1977) Motor effects produced by microstimulation of brachium pontis in the cat. Brain Res., 126: 557–562.

Potter, R.F., Ruegg, D.G. and Wiesendanger, M. (1978) Responses of neurones of the pontine nuclei to stimulation of the sensorimotor, visual and auditory cortex of rats. Brain Res. Bull., 3(1): 15–19.

Ramòn y Cajal, S. (1909–1911) Histologie du système nerveux de l'homme et des vertebres, e II, Vol. I. Maloine, Paris.

Rispal-Padel, L., Pons, C. and Cicirata, F. (1980) Simple and synergistic motor effects, induced by the stimulation of the dentate nucleus, in the awake baboon. Determination of the area of hand control. C R Seances Acad. Sci. D., 291(12): 1001–1004.

Rispal-Padel, L., Cicirata, F. and Pons, C. (1981) Contribution of the dentato-thalamo-cortical system to control of motor synergy. Neurosci. Lett., 22(2): 137–144.

Rispal-Padel, L., Cicirata, F. and Pons, C. (1982) Cerebellar nuclear topography of simple and synergistic movements in the alert baboon (Papio papio). Exp. Brain Res., 47(3): 365–380.

Rispal-Padel, L., Cicirata, F. and Pons, C. (1983) Neocerebellar synergies. In: Massion, J., Paillard, J., Schultz, W., and Wiesendanger, M. (Eds.), Neural Coding of Motor Performance. Springer, Berlin, Heidelberg, New York, pp. 213–223.

Rispal-Padel, L., Cicirata, F. and Pons, C. (1985) Topographical aspects of the cerebral cortex and dentate nucleus in the control of hand movements. Exp. Brain Res., Suppl. 10: 259–274.

Serapide, M.F., Cicirata, F., Sotelo, C., Pantò, M.R. and Parenti, R. (1994) The pontocerebellar projection: longitudinal zonal distribution of fibers from discrete regions of the pontine nuclei to vermal and parafloccular cortices in the rat. Brain Res., 644(1): 175–180.

Serapide, M.F., Pantò, M.R., Parenti, R., Zappalà, A. and Cicirata, F. (2001) Multiple zonal projections of the basilar pontine nuclei to the cerebellar cortex of the rat. J. Comp. Neurol., 430(4): 471–484.

Serapide, M.F., Parenti, R., Pantò, M.R., Zappalà, A. and Cicirata, F. (2002a) Multiple zonal projections of the nucleus reticularis tegmenti pontis to the cerebellar cortex of the rat. Eur. J. Neurosci., 15(11): 1854–1858.

Serapide, M.F., Zappalà, A., Parenti, R., Pantò, M.R. and Cicirata, F. (2002b) Laterality of the pontocerebellar projections in the rat. Eur. J. Neurosci., 15(9): 1551–1556.

Shambes, G.M., Gibson, J.M. and Welker, W. (1978) Fractured somatotopy in granule cell tactile areas of rat cerebellar hemispheres revealed by micromapping. Brain Behav. Evol., 15: 94–140.

Thach, W.T (1975) Timing of activity in cerebellar dentate nucleus and cerebral motor cortex during prompt volitional movement. Brain Res., 88(2): 233–241.

Thach, W.T., Goodkin, H.P. and Keating, J.G. (1992) The cerebellum and the adaptive coordination of movement. Annu Rev. Neurosci., 15: 403–442.

Tolbert, D.L., Alisky, J.M. and Clark, B.R (1993) Lower thoracic upper lumbar spinocerebellar projections in rats: a complex topography revealed in computer reconstructions of the unfolded anterior lobe. Neuroscience, 55(3): 755–774.

Torigoe, Y., Blanks, R.H. and Precht, W. (1986) Anatomical studies on the nucleus reticularis tegmenti pontis in the pigmented rat. I. Cytoarchitecture, topography, and cerebral cortical afferents. J. Comp. Neurol., 243(1): 71–87.

Vranesic, I., Iijima, T., Ichikawa, M., Matsumoto, G. and Knopfel, T. (1994) Signal transmission in the parallel fiber–Purkinje cell system visualized by high-resolution imaging. Proc. Natl. Acad. Sci. USA, 91: 13014–13017.

Walberg, F. and Brodal, A. (1979) The longitudinal zonal pattern in the paramedian lobule of the cat's cerebellum: an analysis based on a correlation of recent HRP data with results of studies with other methods. J. Comp. Neurol., 187: 581–588.

Wiesendanger, R. and Wiesendanger, M. (1982a) The corticopontine system in the rat. I. Mapping of corticopontine neurons. J. Comp. Neurol., 208(3): 215–226.

Wiesendanger, R. and Wiesendanger, M. (1982b) The corticopontine system in the rat. II. The projection pattern. J. Comp. Neurol., 208(3): 227–238.

CHAPTER 22

Conservation of the architecture of the anterior lobe vermis of the cerebellum across mammalian species

Roy V. Sillitoe[1], Hassan Marzban[1], Matt Larouche[1], Sepehr Zahedi[1], Jorge Affanni[2] and Richard Hawkes[1],*

[1]Department of Cell Biology & Anatomy, and Genes and Development Research Group, Faculty of Medicine, The University of Calgary, 3330 Hospital Drive N.W., Calgary, AB T2N 4N1, Canada
[2]Instituto de Neurociencia, Facultad de Medicina, Universidad de Morón, Buenos Aires, CP 1708, Argentina

Keywords: immunohistochemistry; zebrin; Purkinje cell; pattern formation

Introduction

Within the Mammalia, the morphology of the cerebellum varies dramatically both in form and size. For example, the breadth of the cerebellum of the insectivore *Sorex cinereus* (one of the smallest mammals) is approximately 5 mm whereas in the fin whale it is over 200 mm (Jansen, 1953; for further details see Larsell, 1970). However, despite numerous anatomical differences, the mammalian cerebellum, as described by Larsell (1970), is classified into ten lobules (indicated by the Roman numerals I–X). Four principal fissures, the primary fissure, secondary fissure, the prepyramidal fissure, and the posterolateral fissure, can be consistently identified across the Mammalia. However, there is a more fundamental compartmentation of the cerebellum into transverse zones and parasagittal stripes (reviewed in Voogd et al., 1996; Hawkes, 1997; Herrup and Kuemerle, 1997). Both the mediolateral and rostrocaudal patterns are highly reproducible between individuals (Ozol et al., 1999; Sillitoe and Hawkes, 2002). Furthermore, there are clear similarities between species (e.g., rat — Hawkes and Leclerc, 1987; mouse — Sillitoe and Hawkes, 2002; rabbit — Sanchez et al., 2002; cat — Sillitoe et al., 2003b; tenrec — Sillitoe et al., 2003c; hamster — Marzban et al., 2003; guinea pig — Larouche et al., 2003).

The highly conserved antigen zebrin II has proven useful in comparative studies of cerebellar compartmentation. Cloning studies have shown the zebrin II epitope is associated with the brain-specific aldolase isoenzyme, aldolase C (Ahn et al., 1994; Hawkes and Herrup, 1996). In mammals, Purkinje cells that express zebrin II form an elaborate array of parasagittal stripes. However, zebrin II is expressed by Purkinje cells in taxa ranging from fish (Lannoo

*Corresponding author. Tel.: +1 403-220-5712; Fax: +1 403-210-8109; E-mail: rhawkes@ucalgary.ca

et al., 1991a,b) to primates (Sillitoe et al., 2004), which raises the question: how conserved is cerebellar compartmentation? A whole-mount staining procedure has been developed to examine this issue in a range of mammalian and nonmammalian species. A specific focus has been placed on the anterior vermis, which has a very characteristic appearance in rats (e.g., Brochu et al., 1990) and mice (Eisenman and Hawkes, 1993; Ozol et al., 1999; Sillitoe and Hawkes, 2002). The results suggest that regardless of cerebellar size, a characteristic set of stripes is present in the anterior lobe vermis of the mammalian cerebellum and the topography of zebrin II expression is well conserved.

Materials and methods

All animal procedures conformed to institutional regulations and the *Guide to the Care and Use of Experimental Animals* from the Canadian Council for Animal Care. All cerebella were immersion or perfusion fixed in phosphate buffered 4% paraformaldehyde.

(1) Zebrafish (*Danio rerio*; a gift from Dr. Sarah Childs (University of Calgary, Canada); $N=3$)
(2) Weakly electric elephant nose fish (*Apteronotus leptorhynchus*; a gift from Dr. Ray Turner (University of Calgary, Canada); $N=3$)
(3) Frog (*Xenopus laevis*; a gift from Dr. Sarah McFarlane (University of Calgary, Canada); $N=2$)
(4) Turtle (*Pseudemys scripta elegans*; a gift from Dr. Joyce Keiffer (University of South Dakota, USA); $N=3$)
(5) Chicken (*Gallus domesticus*; a gift from Dr. W. Stell (University of Calgary, Canada); $N=10$)
(6) Mouse (*Mus musculus* (CD1); as described in Sillitoe and Hawkes 2002)
(7) Hamster (*Mesocricetus auratus*; as described in Marzban et al., 2003; $N=3$)
(8) Rat (*Rattus norvegicus*; obtained from Charles River Laboratories, St. Constant, PQ, Canada; $N=5$)
(9) Guinea pig (*Cavia porcellus*; as described in Larouche et al., 2003; $N=5$)
(10) Brazilian grey opossum (*Monodelphis domestica*; a gift from Dr. Thérèsa Cabana (Université de Montréal, Canada); $N=1$)
(11) Tenrec (*Echinops telfairi*; as described in Sillitoe et al., 2003c; $N=8$)
(12) Rabbit (*Oryctolagus cuniculus*; as described in Sanchez et al., 2002; $N=6$)
(13) Sheep (*Ovis aries*; a gift from Dr. Francine Smith (University of Calgary, Canada); $N=3$)
(14) Bush pig (*Potamochoerus porcus*; a gift from Dr. Roger Reep (University of Florida, USA); $N=1$)
(15) Deer (*Odocoileus virginianus*; a gift from Mr. Tim Baumbach; $N=1$)
(16) South American armadillo (*Chaetophractus villosus*; $N=2$)
(17) Cat (*Felis domesticus*; as described in Sillitoe et al., 2003b; $N=10$)
(18) Ferret (*Mustela putoris furo*; a gift from Dr. Keith Sharkey (University of Calgary, Canada); $N=3$)
(19) Tree shrew (*Tupaia belangerie*; as described in Sillitoe et al., 2004; $N=4$)
(20) Macaque (*Macaca mulatta*; as described in Sillitoe et al., 2004; $N=5$)

Anti-zebrin II is a mouse monoclonal antibody produced by immunization with a crude cerebellar homogenate from the weakly electric fish *Apteronotus* (Brochu et al., 1990) — it was used directly from spent hybridoma culture medium diluted 1:200. No staining was obtained when the primary antibody was omitted or replaced by myeloma-conditioned culture medium. Western blotting was performed as described in Sillitoe et al. (2003a). Whole mount immunocytochemistry was performed as described in Sillitoe and Hawkes (2002). In some cases, the incubation times in primary and secondary antibodies were extended (up to 4 days in primary antibody for the macaque: Sillitoe et al., 2004). Section immunocytochemistry was performed essentially as described in Sillitoe et al. (2003a). Whole mount photomicrographs of cerebella immersed in buffer were captured with a SPOT digital camera (Diagnostics Instruments, Sterling Heights, MI, USA) mounted on a Zeiss Stemi SV6 microscope, with incident illumination. Montages were assembled in Adobe Photoshop (Tucson, AZ, USA). Cartoons were drawn in Adobe Illustrator.

Results

The molecular weight of the zebrin II/aldolase C antigen — 36 kDa — is highly conserved throughout the vertebrates (Fig. 1A: see also Merritt and Quattro, 2002). Likewise, in the cerebellum, neuronal expression is restricted to Purkinje cells (zebrafish — Fig. 1B; mouse — Fig. 1C) and, in some species, is also detected in astrocytes (e.g., Walther et al., 1998). The only exception is the absence of zebrin II immunoreactivity in any amphibians studied to date (see below).

Fish, Amphibia, Reptiles, and Birds

Previous studies have investigated zebrin II expression in a variety of fish (e.g., *Apteronotus* — Brochu et al., 1990; *Gnathonemus* — Meek et al., 1992; sting rays — Puzdrowski, 1997; reviewed in Lannoo et al., 1991a,b). There is no zebrin II immunoreactivity in the adult cerebellum of the sea lamprey *Petromyzon marinus* (Lannoo and Hawkes, 1997: which may also lack aldolase C — see Merritt and Quattro, 2002), whereas all Purkinje cells in the cerebellar corpus of sting rays are immunopositive for zebrin II (Puzdrowski, 1997). Figure 2 shows the cerebella of two species of fish — *Brachydanio reiro* (zebrafish) and *Apteronotus leptorhynchus* (weakly electric elephant nose fish) — immunoperoxidase stained for zebrin II in whole mount. All Purkinje cells in the zebrafish cerebellum express zebrin II (Fig. 2A: see also Lannoo et al., 1991a,b). In contrast, in the weakly electric fish *Apteronotus leptorhynchus* the cerebellar cortex is compartmentalized — most Purkinje cells of the corpus cerebelli express zebrin II while those of the eminentia granularis medialis and eminentia granularis posterior are predominantly zebrin II immunonegative (Fig. 2B: see also Brochu et al., 1990; Meek et al., 1992).

Zebrin II has never been detected in the cerebellum of frogs (although aldolase C is present: Yatsuki

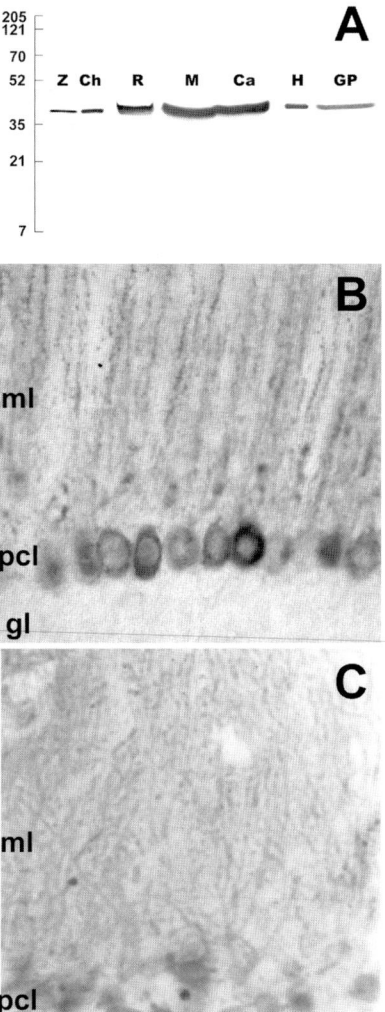

Fig. 1. Zebrin II/aldolase C is conserved across the vertebrates. A: Western blots of cerebellar homogenates from multiple species (each sample was run separately, alongside a sample from mouse, then compiled). B: Transverse section through the mouse cerebellum immunoperoxidase stained for zebrin II. Reaction product is deposited in the Purkinje cell somata and dendrites in the molecular layer (ml) but is not seen in the nuclei and is typically weak or absent from the Purkinje cell axons in the granular layer (gl). No other neurons are stained. C: Transverse section through the zebrafish cerebellum immunoperoxidase stained for zebrin II. Reaction product is deposited in the Purkinje cell somata in the Purkinje cell layer (pcl) and dendrites in the molecular layer (ml) but is not seen in the nuclei and is typically weak or absent from the Purkinje cell axons in the granular layer (gl). No other neurons are stained. Scale bar in C = 50 μm (B,C).

et al., 1998). Whole-mount immunohistochemistry (and tissue section immunohistochemistry, our unpublished data) on cerebella from the African claw-toed frog *Xenopus laevis* failed to reveal any zebrin II expression (Fig. 2C). Similarly, the zebrin II antigen was never detected in the cerebellum of the bullfrog *Rana catesbeiana* (adult or tadpoles, unpublished observations). All Purkinje cells in the cerebellum of the pond turtle *Pseudemys scripta elegans* express zebrin II (Fig. 2D).

Cerebellar compartmentation of chicken is reminiscent of that in mammals both in the parasagittal pattern and in the transverse zonal architecture (Marzban et al., in preparation). In the anterior lobe vermis a symmetrical array of zebrin II immunoreactive stripes extends across the surface of lobules II–V as seen on whole-mount immunoperoxidase-stained cerebella (Fig. 2E). At least three immunoreactive stripes are seen throughout the anterior lobe vermis. More caudally, in lobules IV–V, at least five stripes occupy the caudal anterior lobe vermis with one at the midline and two more laterally on either side. The only exception to this striped architecture of the anterior vermis is in lobule I, in which all Purkinje cells are zebrin II +.

Mammalia

The compartmentation of the anterior lobe vermis has been studied in 15 species of mammals. The stripe nomenclature is based on studies in rats (Hawkes and Leclerc, 1987) and mice (e.g., Eisenman and Hawkes, 1993; Ozol et al., 1999; Sillitoe and Hawkes, 2002). Zebrin-immunoreactive stripes are numbered from P1 + medially to P7 + laterally: zebrin-negative (P−) stripes are numbered according to the immediately medial P + stripe. The rodent nomenclature has been used below for the description of all mammals but it is speculative that stripes in different species are homologous. The pattern of zebrin II expression in the rodent cerebellum is highly reproducible between individuals and is symmetrical about the midline. In most rodents it is straightforward to distinguish two classes of Purkinje cell. However, zebrin II expression in Purkinje cells is not 'all-or-nothing' in all species. In some cases all Purkinje cell somata

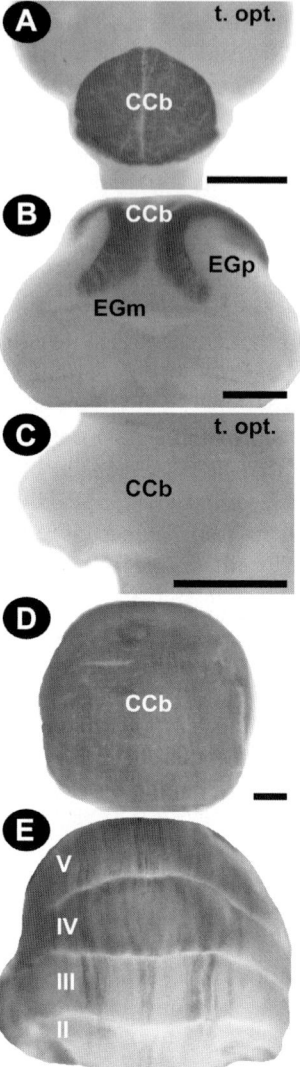

Fig. 2. Zebrin II expression in the cerebellum of a selection of non-mammalian vertebrates, each immunoperoxidase stained in whole mount. Dorsal views are shown in panels A,B,D,E; panel C is a lateral view. A: Zebrafish (*Danio rerio*). All Purkinje cells express zebrin II. No other brain tissue is stained (e.g., optic tectum — t. opt). B: Weakly-electric elephant nose fish (*Apteronotus leptorhynchus*). Purkinje cells of the corpus cerebelli (CCb) express zebrin II while those of the eminentia granularis medialis (EGm) and eminentia granularis posterior (EGp) are predominantly zebrin II immunonegative. C: *Xenopus laevis*. All Purkinje cells are zebrin II-immunonegative. D: Turtle (*Pseudemys scripta elegans*). All Purkinje cells express zebrin II immunoreactivity. E: Chicken (*Gallus domesticus*: 22 days postnatal). Parasagittal stripes of zebrin II expression are present in the anterior lobe vermis (lobules II–V are labeled). Scale bars = 1 mm.

express equal levels of the zebrin II antigen (e.g., cat — Sillitoe et al., 2003b; guinea pig — Larouche et al., 2003; macaque — Sillitoe et al., 2004). In these cerebella a parasagittal pattern is nonetheless clear in the molecular layer where Purkinje cell dendrites differentially express zebrin II. A similar complexity prevails during development, where Purkinje cells destined to be zebrin II-immunonegative in the adult may express zebrin II immunoreactivity during postnatal development (e.g., Leclerc et al., 1999: see also Rivkin and Herrup, 2003).

This chapter focuses on the anterior lobe vermis. When stripes were first identified, the midline zebrin II+ stripe was called P1+ and the first lateral stripe P2+ (Hawkes and Leclerc, 1987). However, later studies made it clear that a short stripe, typically confined to the vicinity of the primary fissure, was interpolated between P1+ and P2+. This led to a re-numbering of stripes in the anterior lobe vermis of the mouse, with the interdigitated stripe becoming P2+ and the stripe previously numbered P2+ becoming P3+ (e.g., Eisenman and Hawkes, 1993; Gallagher et al., 1998). In what follows, where a short P2+ stripe was present, the first lateral stripe in the more anterior lobules is labeled P3+. Where it was not found, we have preferred to avoid the presumption and have labeled the first lateral stripe as P2+.

Mouse

The most studied cerebellum from the perspective of Purkinje cell patterning is that of the mouse (for zebrin II expression, see Eisenman and Hawkes, 1993; Hawkes, 1997; Ozol et al., 1999; Sillitoe and Hawkes, 2002). The vermis of the mouse cerebellar cortex comprises four transverse zones — the anterior zone (AZ: ~ lobules I–V), central zone (CZ: ~ lobules VI–VII), posterior zone (PZ: ~ lobules VII–VIII) and nodular zone (NZ: ~ lobules IX–X), each of which is further subdivided into parasagittal stripes (e.g., Hawkes, 1997; Ozol et al., 1999; Armstrong and Hawkes, 2000; Armstrong et al., 2000). Zebrin II expression reveals stripes in the AZ and PZ: the CZ and NZ are uniformly immunopositive (but stripes can be revealed in the CZ and NZ in other ways, for example through HSP25 expression (Armstrong et al., 2000)).

The mouse AZ extends from lobule I to the primary fissure between lobules V and VI, where AZ and CZ Purkinje cells interdigitate. This has complicated the numbering system for stripes in the AZ (Figs. 3A,B): P1+ straddles the midline (it likely comprises two fused, zebrin II+ stripes — one derived from each hemicerebellum and projecting ipsilaterally to the cerebellar nuclei — e.g., Hawkes and Leclerc, 1987). Within the anterior lobe vermis the P2+ stripe is restricted to the region of AZ:CZ interdigitation and is probably continuous with the CZ (see e.g., Gallagher et al., 1998; Ozol et al., 1999). The next lateral AZ stripe, P3+, extends the full length of the AZ vermis, centered ~ 500 μm from the midline. The P1+/P3+ triplet is the characteristic signature of the AZ throughout the Mammalia (see below). More lateral still, a weakly immunoreactive P4+ is seen. Purkinje cells located within P4+ are loosely packed in lobules I–III with much tighter packing density in lobules IV and V.

Hamster — Mesocricetus auratus

In general, the organization of the hamster AZ is no different from that of other rodents (Marzban et al., 2003). P1+ at the midline is heavily reactive for zebrin II and is resolved in all lobules of the anterior lobe vermis as a thin stripe (Fig. 3C). In lobules III–V, P1+ is flanked by P2+ located ~ 750 μm laterally on either side. P2+ extends far beyond the primary fissure and together with P1+ they represent the most prominent stripes in the hamster AZ. Laterally, P3+ extends rostrally from lobule V and is clearly represented in lobule II as a thin stripe. The 'typical' AZ pattern, with P1+ at the midline and two P3+ stripes laterally, is only observed in lobule II.

Rat — Rattus norvegicus

The topographical organization of the rat AZ is similar to that in mouse (e.g., Brochu et al., 1990). At the midline, P1+ is heavily reactive and flanked by at least four lateral stripes (Fig. 3D). As in mouse, P2+ is only observed in lobules IV and V. The P3+ stripes are located ~ 850 μm laterally of the midline in

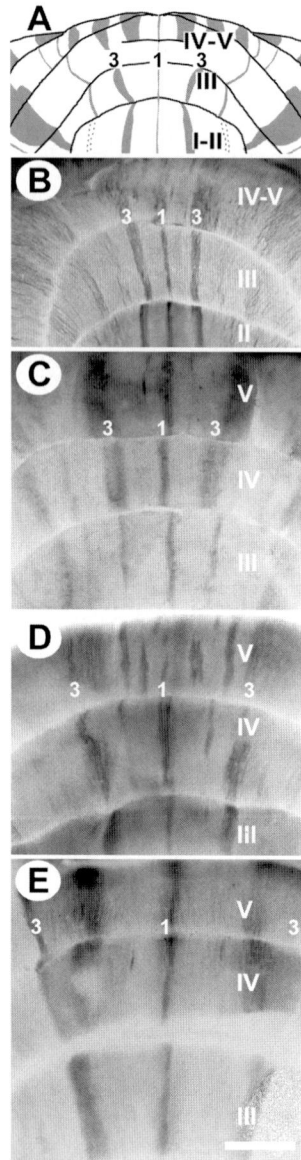

Fig. 3. Whole mount peroxidase immunocytochemistry of the anterior lobe vermis of the adult rodent cerebellum. Despite the differences in cerebellar width, the stripe pattern is conserved. A: A cartoon showing the zebrin II+/− compartmentation of the anterior lobe vermis of the adult mouse cerebellum. In lobule V, a P1+ stripe straddles the midline (labeled as '1' for clarity) and strongly immunoreactive P2+ and P3+ stripes are seen to either side. A weak P4+ lies at the interface of vermis and hemisphere. The P2+ stripe does not extend into more anterior lobules but the P1+ and P3+ remain prominent. B: Mouse anterior lobe vermis. C: Hamster anterior lobe vermis. D: Rat anterior lobe vermis. E: Guinea pig anterior lobe vermis. Scale bar in E = 1 mm (B–E).

lobules I–II and ∼1 mm laterally in lobules IV and V. P4+ is a wide stripe of loosely packed immunoreactive Purkinje cells interspersed among nonreactive cells.

Guinea Pig — Cavia porcellus

The expression pattern of zebrin II in the guinea pig AZ resembles that in mice and rats: however, rather than the P− stripes being completely zebrin II unreactive, low levels of immunoreactivity are seen (Larouche et al., 2003). Nevertheless, the differences between P+ and P− expression levels make stripes easily identifiable. The P1+ at the midline is flanked by at least four immunoreactive stripes laterally (Fig. 3E). P2+ is visible in the caudal aspects of lobule IV and in lobule V. Immunohistochemical staining in whole-mount preparations suggests the partial fusion of P2+ and P3+ laterally. Perhaps because Purkinje cells in P2− express some zebrin II immunoreactivity the P2+ and P3+ stripes appear to merge to form a single stripe with heavily immunoreactive edges (i.e., P2+ represents the medial edge and P3+ the lateral edge). In lobules I–V, P1+ is flanked by P3+, located ∼500 μm laterally in lobules I–II, ∼1.0 mm in lobule III, and ∼1.5 mm laterally in lobules IV and V. P3+ is heavily reactive for zebrin II and may be traced through all lobules of the AZ. P4+, usually visible only in lobules IV and V, contains weakly reactive Purkinje cells and distinct medial and lateral borders with the P− stripes are not defined.

The Brazilian Grey Opossum — Monodelphis domestica

The fundamental pattern of zebrin II expression is similar between marsupials and rodents (e.g., Doré et al., 1990). Whole-mount immunohistochemical staining of the *Monodelphis* cerebellum reveals the characteristic stripe organization of the AZ (Fig. 4A): P1+ straddles the cerebellar midline and at least three stripes are located laterally on either side (P2+–P4+).

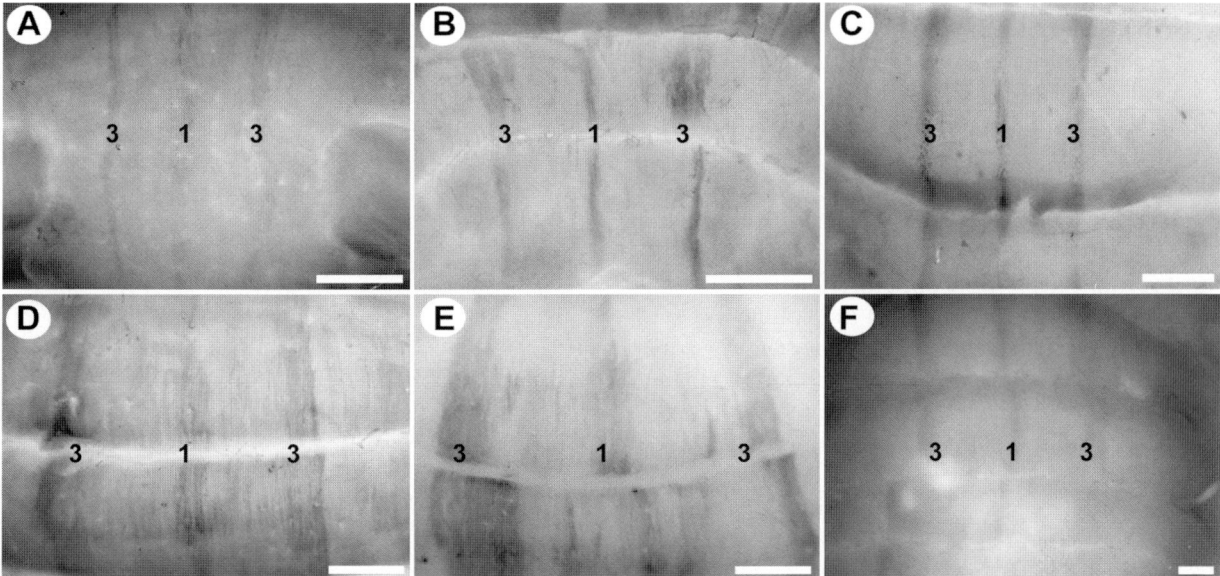

Fig. 4. Examples of the compartmentation of the mammalian anterior lobe vermis seen in whole mount peroxidase immunocytochemistry for zebrin II. The P1+ and P3+ stripes are labeled (as 1 and 3). A: Lobule II–III of the adult opossum, *Monodelphis domestica*. B: Lobule I–III of the adult tenrec, *Echinops telfairi*. C: Lobule III–IV of the adult rabbit, *Oryctolagus cuniculus*. D: Lobules II–III of the postnatal day 37 sheep, *Ovis aries*. E: Lobules III–IV of the adult wild pig, *Potamochoerus porcus*. F: Lobules II–IV of the adult deer, *Odocoileus virginianus*. Scale bars = 1 mm.

The Madagascan Hedgehog Tenrec — *Echinops telfairi*

The Madagascan lesser hedgehog tenrec (*Echinops telfairi*) is a nocturnal mammal, found exclusively in the western and southwestern regions of Madagascar (Eisenberg and Gould, 1970). Classically the tenrecs have been considered insectivores but more recent studies have placed them within the superorder Afrotheria (Stanhope et al., 1998; Mouchaty et al., 2000; Murphy et al., 2001; van Dijk et al., 2001; Douady et al., 2002; Malia et al., 2002). The tenrec cerebellum is only slightly larger than the cerebellum of mouse (~8 mm in width). In the anterior lobe, lobules I–III are fused and separated from lobule IV–V (also undifferentiated) by the preculminate fissure. Lobules IV–V and VI–VIII are separated by the deep primary fissure. A detailed description of zebrin II expression in the tenrec cerebellum is published in Sillitoe et al. (2003c). Despite the reduced lobulation, the AZ of tenrec resembles that of mouse, with a strongly zebrin II-immunoreactive midline stripe (P1+) and strong flanking P3+ stripes laterally to either side (Fig. 4B: ~625 μm from the midline in lobule I–III; ~750 μm in lobule IV–V). As in several rodents, a short P2+ stripe is seen in lobule IV–V anterior to the primary fissure, which cannot be traced anteriorly in whole-mount preparations. However, in horizontal sections through the AZ a weak stripe of zebrin II expression is seen between P1+ and P3+ that we have tentatively interpreted as the rostral extension of P2+ (Sillitoe et al., 2003c).

The Rabbit — *Oryctolagus cuniculus*

The phylogenetic placement of lagomorphs is controversial — some recent evidence suggests that they may be related to rodents (e.g., Huchon et al., 2002) whereas other investigations suggest a closer relationship to primates and scandentids (e.g., Li et al., 1990; Graur et al., 1996). Early descriptions of the gross anatomy of the rabbit cerebellum were given by Smith (1902), Bradley (1903) and Brodal (1940) amongst others, culminating in the systematic account of Larsell (1952). A detailed description of zebrin II expression in the rabbit cerebellum has been published (Sanchez et al., 2002). In the AZ a narrow midline stripe (P1+) is flanked by up to three lateral stripes (P2+ to P4+: Fig. 4C). Lobules II–IV have a medial P1+ stripe and two pairs laterally. At the most rostral extent of the anterior lobe (lobules I–II) the lateral stripes disappear. In lobule V, an additional, strongly zebrin II-immunoreactive stripe is seen, presumably derived from the CZ and homologous with the P2+ in rodents. To remain consistent with the mouse nomenclature, the stripes in lobule V are labeled P1+–P4+. There are consistent differences in the intensity of the anti-zebrin II staining: the P1+, P2+, and P4+ bands are crisply delineated and more heavily stained, P3+ has ill-defined borders and significantly lower levels of immunoreactivity

The Sheep — Ovis aries

In lobules I–III, three heavily immunoreactive stripes are present which, by location and size, are the equivalent of the P1+ and P3+ stripes in rodents (Fig. 4D). The intervening P− stripes are broader and weakly immunoreactive. Within the zebrin II-immunonegative stripe separating P1+ and P3+, a possible P2+ stripe is seen, weakly immunoreactive and difficult to resolve. P1+ and the two lateral P3+ stripes (~2 mm laterally) are prominent in lobules IV and V.

The Bush Pig — Potamochoerus porcus

At least five parasagittal stripes are visible in the AZ of the bush pig (Fig. 4E). As in rodents, P1+ at the midline is heavily reactive and easily traced from the anterior lobules of the AZ through to the CZ boundary. Laterally, P2+ on either side (the center to center distance between P1+ and P2+ is ~1 mm) is weaker in immunoreactivity but also extends through most lobules of the AZ. In certain areas the continuity of this stripe appears interrupted and often appears to be absent. This could reflect a fixation artifact, as these data were based on a single cerebellum that was immersion fixed rather than perfused. P3+ is heavily reactive (its center is located ~2 mm from the center of P1+) and is the thickest and most distinct stripe in the AZ.

The Deer — Odocoileus virginianus

Zebrin II expression in the AZ follows the typical mammalian pattern with P1+ at the midline flanked by P2+ on either side (located ~1.5 mm laterally: Fig. 4F). The level of zebrin II in Purkinje cells within P2+ is weak compared to P1+ and P3+. P3+ is heavily reactive in all lobules of the AZ, located ~3.5 mm laterally in the caudal AZ and ~2.5 mm laterally in the rostral AZ. The lateral edges of P3+ are sharp compared to the medial edges, which gradually fade into the neighboring P2− stripes. More lateral stripes proved difficult to resolve.

The Armadillo — Chaetophractus villosus

Cerebella processed for whole-mount anti-zebrin II immunocytochemistry reveal at least five stripes of high expression in the anterior lobe vermis of the South American armadillo (Fig. 5A). A thin stripe is located at the midline and two pairs of stripes lie laterally.

The Domestic Cat — Felis domesticus

Cerebella processed for whole-mount anti-zebrin II immunocytochemistry reveal at least five stripes of high expression in the anterior lobe vermis of the cat (see Sillitoe et al., 2003b: Fig. 5B). A thin midline P1+ stripe (~100 μm wide) is clearly seen in all lobules of the AZ and maintains a more or less constant width throughout. On either side of P1+ are two laterally located P2+ stripes of high intensity, which are considerably wider than P1+ and are the most prominent stripes in the AZ. More lateral stripes are seen most clearly in lobules I and II.

The Ferret — Mustela putoris furo

The topography of zebrin II in the anterior lobe vermis of the ferret cerebellum is similar to that in cat

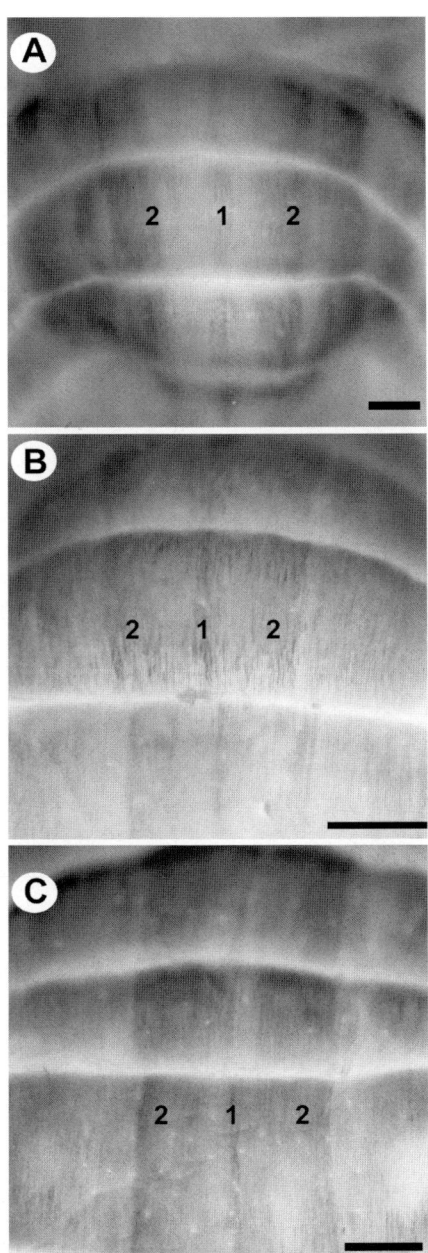

Fig. 5. The compartmentation of the anterior lobe vermis of the cerebellum in three mammals, armadillo, ferret and cat, as revealed by whole mount peroxidase immunocytochemistry for zebrin II. The P1+ and P3+ stripes are labeled (as 1 and 3). It is characteristic that in each case, the 'negative' stripe separating P1+ and P3+ exhibits moderate levels of zebrin II immunoreactivity. A: Lobules IV–V of the adult South American armadillo, *Chaetophractus villosus*. B: Lobule III–IV of the adult cat, *Felis domesticus*. C: Lobules III–IV of the adult ferret, *Mustela putoris furo*. Scale bars = 1 mm.

(Fig. 5C). A P1+ stripe at the midline is flanked by P2+ laterally (~500 μm from the midline). Unlike in rodents P2+ is continuous through all anterior vermis lobules and we do not identify an interdigitated stripe derived from the CZ (it is also possible that the P2+ here is homologous with the P3+ in rodents). More lateral stripes of the AZ are difficult to resolve.

The Tree Shrew — Tupaia belangerie

Tupaia is a generalized eutherian mammal. One phylogeny places *Tupaia* in the order Scandentia (Martin, 1990), but more recently it has been argued that the closest relatives of *Tupaia* are the lagomorphs rather than the primates (Kupfermann et al., 1999; Schmitz et al., 2000; Arnason et al., 2002). A detailed description of the *Tupaia* cerebellum is given in Sillitoe et al. (2004). At least three zebrin II immunoreactive stripes are found in the anterior lobe vermis of *Tupaia*, and are reproducible between individuals (Fig. 6A). In lobules I–V, P1+ is located at the midline with the P3+ stripe positioned approximately 500 μm from the center. The medial edges of P3+ are not sharply defined while the lateral edges are heavily reactive and form an abrupt boundary with the P3− stripe. As in mouse, the P2+ does not extend far rostrally beyond the primary fissure and is distinct only in lobule V.

The Macaque — Macaca mulatta

In contrast to the AZ in other mammals studied to date, in which the P+ stripes are narrow and the P− stripes are broad, the P+/P− stripes in macaque have approximately equal widths (Fig. 6B: P1+/P1− ~ 125 μm, P2+/P2− ~ 250 μm). The most prominent zebrin II stripes are the P1+ at the midline of the cerebellum and a second stripe on either side laterally, centered approximately 400 μm from the midline in lobules I–III and approximately 600 μm in lobules IV and V. In rodents, the first lateral stripe in the more anterior parts of the AZ has been numbered P3+ and the designation P2+ has been restricted to a short stripe in lobule V. In macaque, no interdigitated AZ:CZ boundary zone is apparent (we have argued elsewhere that this may be misleading,

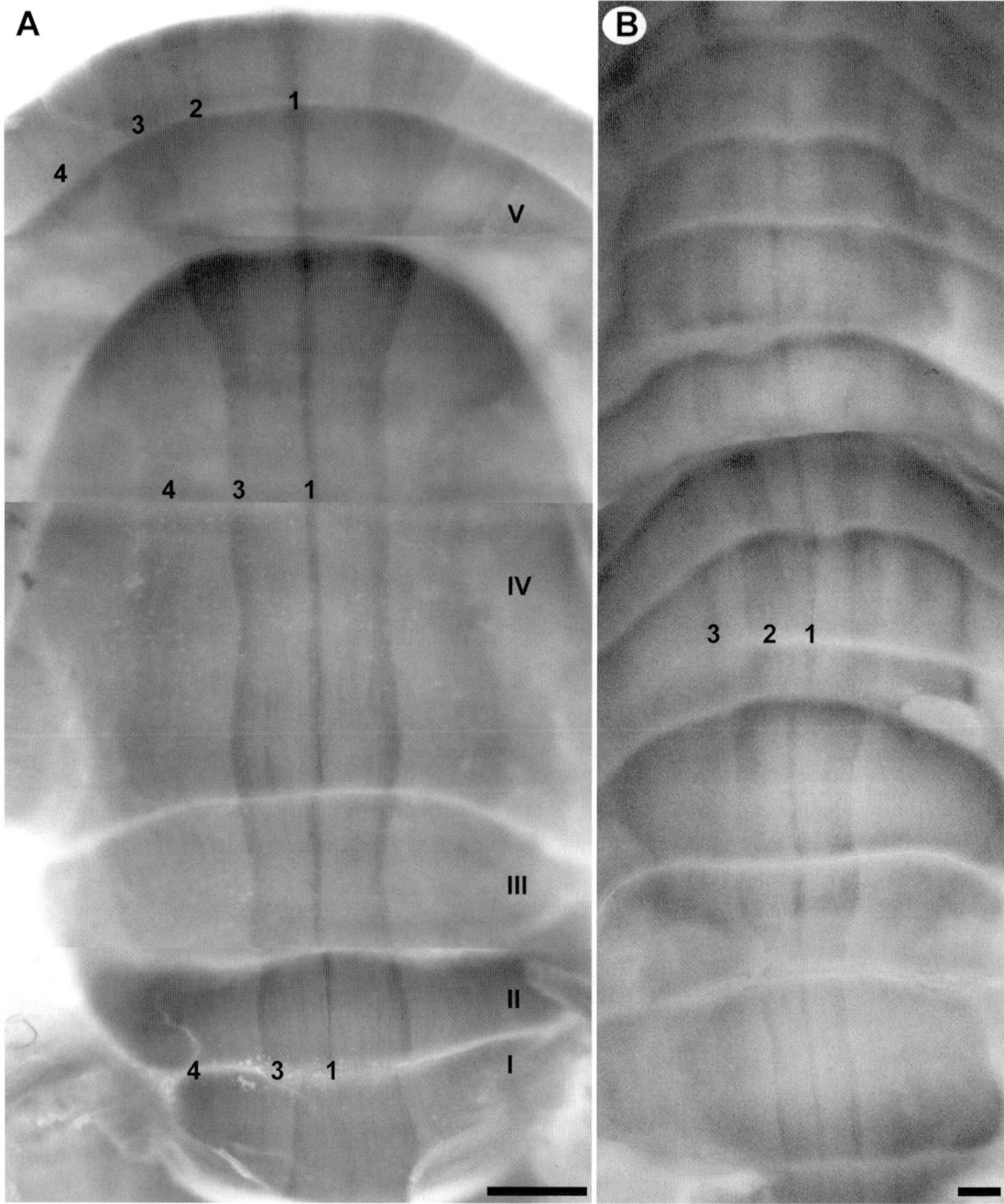

Fig. 6. The compartmentation of the anterior lobe vermis of the cerebellum in the tree shrew and the macaque, as revealed by whole mount peroxidase immunocytochemistry for zebrin II. The P1+–P4+ stripes are labeled (as 1–4). A: Whole-mount anti-zebrin II peroxidase immunohistochemistry of lobules I–V of the adult tree shrew, *Tupaia belangerie*. B: Whole-mount anti-zebrin II peroxidase immunohistochemistry of anterior lobe vermis of the adult macaque, *Macaca mulatta*. Scale bars = 1 mm.

and a CZ may be present, but also stripy: Sillitoe et al., 2004), and so the name P2+ has been used for the first lateral stripe, although this does not imply homology with P2+ in other species. These three stripes, P1+ at the midline and P2+ laterally to either side, have sharply defined edges and extend rostrocaudally through all lobules of the anterior lobe vermis. Between lobules III and IV, P3+ resolves as a distinct stripe, with its medial edge approximately 1 mm from the midline. Moving caudally, P2+ gradually increases in width in lobule IV, and in lobule V becomes the thickest stripe in the AZ. In lobule V, P1+ to P3+ are easily detected and laterally in the paravermis a P4+ can be reliably identified.

Discussion

The restriction of the zebrin II epitope to Purkinje cells and the molecular weight of the zebrin II/ aldolase C antigen are highly conserved from fish to primates. The only exception appears to be the Amphibia, in which no immunoreactivity is detected despite the presence of aldolase C (e.g., Yatsuki et al., 1998). Presumably, a mutation in the zebrin II epitope in the amphibian lineage has rendered its aldolase C unrecognized by anti-zebrin II. Zebrin II Purkinje cell stripes seem to be restricted to birds and mammals (although some fish have both zebrin II-immunoreactive and unreactive Purkinje cells, these are in clusters rather than stripes — e.g., *Apteronotus* — Fig. 2B). In turtles, all Purkinje cells are immunoreactive and no antigenic stripes are seen (Fig. 2D). However, in this case the lack of stripes is likely due to uniform expression of the antigen (i.e., similar to the CZ and NZ in rodents) rather than to the absence of compartments, as parasagittal stripes are apparent in the afferent terminal fields (e.g., mossy fibers — Künzle, 1983; climbing fibers — Künzle, 1985).

There is no way to know whether individual stripes are homologous between species (common labeling has largely been a matter of convenience). Detailed, cross-species developmental studies would be required to assess the homology between individual stripes, and to date, the lineage relations of stripes are not well understood even in mice or chicks, in which the most intensive studies have been undertaken (e.g., mouse — Hawkes et al., 1998; chicken — Lin and Cepko, 1999). Nevertheless, the main conclusion from these preliminary comparative studies is that an array of zebrin II antigenic stripes, reproducible and symmetrical about the midline, is present in all mammalian species examined. This is consistent with data from comparative studies of the corticonuclear and olivocerebellar projections in a range of different species (reviewed in Haines et al., 1982; Voogd et al., 1996). Differences in the molecular architecture of the cerebellum between phylogenetic groups of mammals are minor but consistent. For example, the pattern of zebrin II expression in the anterior vermis of ferrets (Fig. 5C) resembles that of cats (another carnivore: Fig. 5B) more than of rodents (Fig. 3). Similarly, an elaborate array of zebrin II+/− stripes can be visualized throughout all the vermis of both the macaque and the tree shrew: this distinguishes both species from other mammals, where a uniformly expressing CZ interrupts striped zebrin II expression domains. That being said, there is much similarity in the patterning of the anterior lobe vermis (AZ) between different mammals: typically an array of narrow zebrin II+ stripes, P1+ at the midline, P2+ extending a variable distance anterior of the primary fissure, P3+ laterally, and an ill-defined P4+ most laterally. The patterning is conserved irrespective of cerebellar size.

The apparent conservation of the zebrin II expression pattern in the AZ presumably derives from a common developmental program, with stripe expansion being regulated at the stem cell/germinal zone level. Several investigations have shown evidence that stripes are determined early in development and are not a response to usage. First, zebrin II+ and zebrin II− Purkinje cells are both present after cerebellar anlagen were dissected from embryos at embryonic days (E)12–15 (prior to any contact with afferents) and transplanted either into the anterior eye chamber or into cavities prepared in the neocortex (Wassef et al., 1990). Similarly, the parasagittal expression of HSP25 in the CZ and in the NZ of adult mice is unchanged after physical and genetic visual deprivation during development (Armstrong et al., 2001). Secondly, differential zebrin expression develops normally in vitro in cerebellar slice cultures in the presence of agents that disrupt normal neuronal activity (Seil et al., 1995). Likewise, embryonic cerebellar cultures lacking extracellular signals from mice transgenic for the L7/pcp2 promoter driving β-galactosidase showed a striped pattern of transgene expression (Smeyne et al., 1991). The apparent conservation of cerebellar zones and stripes appears to be consistent with a general set of rules that other CNS structures have adhered to during evolution. For example, the number of patch-matrix compartments in the striatum is conserved from rodents to human (Johnston et al., 1990). To accommodate the total increase in striatal area, the constant ratio of patch to matrix areas is maintained by an increase in the size of the individual patches.

The cerebellar vermis in adult mice consists of alternating transverse expression domains: for zebrin II, striped (AZ), homogeneous (CZ), striped (PZ) and homogeneous (NZ); for HSP25, negative (AZ), striped (CZ), negative (PZ) and striped (NZ). This suggests that there may be an homology between AZ and PZ, and CZ and NZ. The stripes revealed by using anti-zebrin II in rodents appear rather different between AZ and PZ, but this is a secondary consequence of the degree of lobulation: in mouse mutants in which lobule formation is disrupted, the AZ stripes become much wider (e.g., *disabled* — Gallagher et al., 1998; *cerebellar deficient folia* — Beierbach et al., 2001), and in cat, where lobulation of the posterior vermis is more extensive than in rodents, the PZ stripes are longer and narrower, more closely resembling those of the AZ (Sillitoe et al., 2003b). One possibility is that the cerebellar vermis evolved in three stages: first, a single transverse zone, all zebrin II + (equivalent to the cerebellum in some fish, the turtle cerebellum and the mammalian NZ); a second stage — purely hypothetical — with paired zones, one stripy (=PZ), one not (=NZ: both phenotypes are seen in fish, but there is no known example of such a cerebellum); finally, a duplication of the PZ/NZ to yield four zones (AZ/CZ/PZ/NZ). Such duplication could also account for the similarities between afferent inputs to different lobules (e.g., spinocerebellar projections to both the AZ and the PZ: reviewed in Voogd et al., 1996). A similar argument can be made for the hemispheres being lateral duplications of the vermis. For example, zebrin II expression in mouse reveals a similar alternation in vermis and hemispheres of striped (AZ = lobulus simplex), uniform (CZ = crus I), striped (PZ = crus II and paramedian lobule) and uniform (NZ = paraflocculus and flocculus).

The function of stripes in the adult cerebellum is a matter of speculation (e.g., Hawkes and Gravel, 1991; Hawkes, 1997). Several hundred distinct cerebellar modules have been identified, based upon expression patterns and afferent terminal fields (and this may be a significant underestimate — for example, see Hawkes, 1997). This modularity has at least three advantages. First, multiple representations of afferent information in different combinations can serve the parallel processing of motor information, where the timing constraints are so severe that to evaluate and integrate afferent input serially would be unrealistic. Secondly, the modular segregation of afferents and targets ensures that minor inputs to the cerebellum are properly taken into account: by segregating a particular input to a specific cerebellar subregion, it becomes possible to weigh its importance appropriately. Thirdly, segregating afferent inputs and associated interneurons opens the possibility of local structural and functional specialization (such as differential neuronal nitric oxide synthase expression to mediate long-term depression — e.g., Hawkes and Turner, 1994).

This might explain why the AZ is compartmentalized but not why it is in stripes. The embryonic origins of Purkinje cell stripes lie in an array of Purkinje cell clusters. Clusters transform into stripes in the neonate as the granular layer develops (e.g., Herrup and Kuemerle, 1997; Armstrong and Hawkes, 2000). It may be that stripes are an epiphenomenon of Purkinje cell dispersion and arise out of the expansion and lobulation of the AZ but have no significance per se. For most purposes, whether Purkinje cells are clustered or dispersed as stripes would seem to make little difference (for example, they would receive the same climbing fiber inputs and, because parallel fibers typically extend the full width of the vermis, also receive the same innervation via mossy fiber pathways). It may simply be that because lobules in the AZ run transversely, the embryonic Purkinje cell clusters are drawn out longitudinally rather than mediolaterally as the granular layer matures.

Abbreviations

AZ — anterior zone of the vermis
CCb — corpus cerebelli
CZ — central zone of the vermis
EGm — eminentia granularis medialis
EGp — eminentia granularis posterior
NZ — nodular zone of the vermis
PZ — posterior zone of the vermis
P1+–P4+ — zebrin II-immunoreactive Purkinje cell stripes
t. opt — optic tectum

Acknowledgments

These studies were supported by grants from the CIHR (RH) and the CIHR Training Program in Genetics, Child Development and Health (RVS). We are very grateful to the many colleagues who generously supplied us with cerebella (as noted in the Materials and Methods), and thus made this study possible, and for the advice of Dr. Jan Voogd.

References

Armstrong, C.L. and Hawkes, R. (2000) Pattern formation in the cerebellar cortex. Biochem. Cell Biol., 78: 551–562.

Armstrong, C., Krueger-Naug, A.M., Currie, W.C. and Hawkes, R. (2000) Constitutive expression of the 25 kDa heat shock protein Hsp25 reveals novel parasagittal bands of Purkinje cells in the adult mouse cerebellar cortex. J. Comp. Neurol., 416: 383–397.

Armstrong, C.L., Krueger-Naug, A.M.R., Currie, R.W. and Hawkes, R. (2001) Expression of heat-shock protein Hsp25 in mouse Purkinje cells during development reveals novel features of cerebellar compartmentation. J. Comp. Neurol., 429: 7–21.

Arnason, U., Adegoke, J.A., Bodin, K., Born, E.W., Esa, Y.B., Gullberg, A., Nilsson, M., Short, R.V., Xu, X. and Janke, A. (2002) Mammalian mitogenomic relationships and the root of the eutherian tree. Proc. Natl. Acad. Sci. USA, 99: 8151–8156.

Baader, S.L., Vogel, M.W., Sanlioglu, S., Zhang, X. and Oberdick, J. (1999) Selective disruption of 'late onset' sagittal banding patterns by ectopic expression of engrailed-2 in cerebellar Purkinje cells. J. Neurosci., 19: 5370–5379.

Bauchot, R. and Stephan, H. (1970) Morphologie comparée de l'encephale des insectivores tenrecidae. Mammalia, 34: 514–541.

Beierbach, E., Park, C., Ackerman, S.L., Goldowitz, D. and Hawkes, R. (2001) Abnormal dispersion of a Purkinje cell subset in the mouse mutant *cerebellar deficient folia* (cdf). J. Comp. Neurol., 436: 42–51.

Bradley, O.C. (1903) On the development and homology of the mammalian cerebellar fissures. J. Anat. Physiol., 37: 112–130.

Brochu, G., Maler, L. and Hawkes, R. (1990) Zebrin II: a polypeptide antigen expressed selectively by Purkinje cells reveals compartments in rat and fish cerebellum. J. Comp. Neurol., 291: 538–552.

Brodal, A. (1940) The cerebellum of the rabbit. A topographical atlas of the folia as revealed in transverse sections. J. Comp. Neurol., 72: 63–81.

Doré, L., Jacobson, C. and Hawkes, R. (1990) Organization and postnatal development of zebrin II antigenic compartmentation in the cerebellar vermis of the grey opossum, *Monodelphis domestica*. J. Comp. Neurol., 291: 431–449.

Douady, C.J., Catzeflies, F., Kao, D.J., Springer, M.S. and Stanhope, M.J. (2002) Molecular evidence for the monophyly of tenrecidae (Mammalia) and the timing of the colonization of Madagascar by Malagasy tenrecs. Mol. Phylogen. Evol., 22: 357–363.

Eisenberg, J.F. and Gould, E. (1970) The tenrecs: A study in mammalian behavior and evolution. Smithsonian Contrib. Zool., 27: 1–137.

Eisenman, L.M. and Hawkes, R. (1993) Antigenic compartmentation in the mouse cerebellar cortex: zebrin and HNK-1 reveal a complex, overlapping molecular topography. J. Comp. Neurol., 335: 586–605.

Gallagher, E., Howell, B.W., Soriano, P., Cooper, J.A. and Hawkes, R. (1998) Cerebellar abnormalities in the *disabled* (mdab1-1) mouse. J. Comp. Neurol., 402: 238–251.

Graur, D., Duret, L. and Gouy, M. (1996) Phylogenetic position of the order Lagomorpha (rabbits, hares and allies). Nature, 379: 333–335.

Haines, D.E. (1975) Cerebellar cortical efferents of the posterior lobe vermis in a prosimian primate (*Galago*) and the tree shrew (*Tupaia*). J. Comp. Neurol., 163: 21–39.

Haines, D.E. and Patrick, G.W. (1981) Cerebellar corticonuclear fibers of the paramedian lobule of tree shrew (*Tupaia glis*) with comments on zones. J. Comp. Neurol., 201: 99–119.

Haines, D.E. and Pearson, J.C. (1979) Cerebellar cortico-nuclear-nucleocortical topography: a study of the tree shrew (*Tupaia*) paraflocculus. J. Comp. Neurol., 187: 745–758.

Haines, D.E., Patrick, G.W. and Satrulee, P. (1982) Organization of cerebellar corticonuclear fiber systems. In: Palay, S.L. and Chan-Palay, V. (Eds.), The Cerebellum-New Vistas, Exp. Brain Res., Vol. Suppl.6. Springer-Verlag, New York, pp. 320–367.

Haines, D.E. and Whitworth, R.H. (1978) Cerebellar cortical efferent fibers of the paraflocculus of tree shrew (*Tupaia glis*). J. Comp. Neurol., 182: 137–150.

Hawkes, R., Brochu, G., Doré, L., Gravel, C. and Leclerc, N. (1992) Zebrins: Molecular markers of compartmentation in the cerebellum. In: Llinás, R. and Sotelo, C. (Eds.), The Cerebellum Revisited. Springer-Verlag, New York, pp. 22–55.

Hawkes, R. (1997) An anatomical model of cerebellar modules. Prog. Brain. Res., 114: 39–52.

Hawkes, R., Faulkner-Jones, B., Tam, P. and Tan, S.S. (1998) Pattern formation in the cerebellum of embryonic stem cell chimeras. Eur. J. Neurosci., 10: 790–793.

Hawkes, R. and Gravel, C. (1991) The modular cerebellum. Prog. Neurobiol., 36: 309–327.

Hawkes, R. and Herrup, K. (1996) Aldolase C/zebrin II and the regionalization of the cerebellum. J. Mol. Neurobiol., 6: 147–158.

Hawkes, R. and Leclerc, N. (1987) Antigenic map of the rat cerebellar cortex: the distribution of parasagittal bands as revealed by monoclonal anti-Purkinje cell antibody mabQ113. J. Comp. Neurol., 256: 29–41.

Hawkes, R. and Turner, R.W (1994) Compartmentation of NADPH-diaphorase activity in the mouse cerebellar cortex. J. Comp. Neurol., 346: 499–516.

Herrup, K. and Kuemerle, B. (1997) The compartmentalization of the cerebellum. Ann. Rev. Neurosci., 20: 61–90.

Hess, D.T. and Voogd, J. (1986) Chemoarchitectonic zonation of the monkey cerebellum. Brain Res., 369: 383–387.

Huchon, D., Madsen, O., Sibbald, M.J., Ament, K., Stanhope, M.J., Catzeflis, F., de Jong, W.W. and Douzery, E.J. (2002) Rodent phylogeny and a timescale for the evolution of Glires: evidence from an extensive taxon sampling using three nuclear genes. Mol. Biol. Evol., 19: 1053–1065.

Jansen, J. (1953) Studies on the cetacean brain. The gross anatomy of the rhombencephalon of the fin whale (*Balaenoptera physalus*, L.). Hvalradets Skrifter, 37: 1–35.

Johnston, J.G., Gerfen, C.R., Haber, S.N. and van der Kooy, D. (1990) Mechanisms of striatal pattern formation: conservation of mammalian compartmentalization. Dev. Brain Res., 57: 93–102.

Künzle, H. (1983) Spinocerebellar projections in the turtle. Observations on their origin and terminal organization. Exp. Brain Res., 53: 129–141.

Künzle, H. (1985) Climbing fiber projection to the turtle cerebellum: longitudinally oriented terminal zones within the basal third of the molecular layer. Neuroscience, 14: 159–168.

Kupfermann, H., Satta, Y., Takahata, N., Tichy, H. and Klein, J. (1999) Evolution of Mhc-DRB introns: implications for the origin of primates. J. Mol. Evol., 48: 663–674.

Lannoo, M.J., Brochu, G., Maler, L. and Hawkes, R. (1991a) Zebrin II immunoreactivity in the rat and in the weakly electric teleost *Eigenmannia* (Gymnotiformes) reveals three modes of Purkinje cell development. J. Comp. Neurol., 310: 215–233.

Lannoo, M.J. and Hawkes, R. (1997) A search for primitive Purkinje cells: zebrin II expression in sea lampreys (*Petromyzon marinus*). Neurosci. Lett., 237: 53–55.

Lannoo, M.J., Ross, L., Maler, L. and Hawkes, R. (1991b) Development of the cerebellum and its extracerebellar Purkinje cell projection in teleost fishes as determined by zebrin II immunocytochemistry. Prog. Neurobiol., 37: 329–363.

Larsell, O. (1952) The morphogenesis and adult pattern of the lobules and fissures of the cerebellum of the white rat. J. Comp. Neurol., 97: 281–356.

Larsell, O. (1970) The Comparative Anatomy and Histology of the Cerebellum from Monotremes through Apes. University of Minnesota Press, Minneapolis, 269 p.

Larouche, M., Diep, C., Sillitoe, R.V. and Hawkes, R. (2003) The topographical anatomy of the cerebellum in the guinea pig, *Cavia porcellus*. Brain Res., 965: 159–169.

Leclerc, N., Gravel, C. and Hawkes, R. (1988) Development of parasagittal zonation in the rat cerebellar cortex. MabQ113 antigenic bands are created postnatally by the suppression of antigen expression in a subset of Purkinje cells. J. Comp. Neurol., 273: 399–420.

Li, W.H., Gouy, M., Sharp, P.M., O'hUigin, C. and Yang, Y.W. (1990) Molecular phylogeny of Rodentia, Lagomorpha, Primates, Artiodactyla, and Carnivora and molecular clocks. Proc. Natl. Acad. Sci. USA, 87: 6703–6707.

Lin, J.C. and Cepko, C.L. (1999) Biphasic dispersion of clones containing Purkinje cells and glia in the developing chick cerebellum. Dev. Biol., 211: 177–197.

Malia, M.J., Adkins, R.M. and Allard, M.W. (2002) Molecular support for Afrotheria and the polyphyly of Lipotyphla based on analyses of the growth hormone receptor gene. Mol. Phylogenet. Evol., 24: 91–101.

Martin, R.D (1990) Primate Origins and Evolution: A Phylogenetic Reconstruction. Princeton University Press, Princeton, NJ, 804 p.

Marzban, H., Zahedi, S., Sanchez, M. and Hawkes, R. (2003) Antigenic compartmentation of the cerebellar cortex of the Syrian hamster, *Mesocricetus auratus*. Brain Res., 974: 176–183.

Meek, J., Hafmans, T.G.M., Maler, L. and Hawkes, R. (1992) The distribution of zebrin II in the gigantocerebellum of the mormyrid fish *Gnathonemus petersii* compared with other teleosts. J. Comp. Neurol., 316: 17–31.

Merritt, T.J.S. and Quattro, J.M. (2002) Negative charge correlates with neural expression in vertebrate aldolase isozymes. J. Mol. Evol., 55: 674–683.

Mouchaty, S.K., Gullberg, A., Janke, A. and Arnason, U. (2000) Phylogenetic position of the tenrecs (Mammalia: Tenrecidae) of Madagascar based on analysis of the complete mitochondrial genome sequence of *Echinops telfairi*. Zool. Scripta, 29: 307–317.

Ozol, K., Hayden, J.M., Oberdick, J. and Hawkes, R. (1999) Transverse zones in the vermis of the mouse cerebellum. J. Comp. Neurol., 412: 95–111.

Puzdrowski, R.L. (1997) Anti-Zebrin II immunopositivity in the cerebellum and octavolateral nuclei in two species of stingrays. Brain Behav. Evol., 50: 358–368.

Rivkin, A. and Herrup, K. (2003) Development of cerebellar modules: extrinsic control of late-phase zebrin II pattern and the exploration of rat/mouse species differences. Mol. Cell. Neurosci., 24: 887–901.

Sanchez, M., Sillitoe, R.V., Attwell, P.J., Ivarsson, M., Rahman, S., Yeo, C.H. and Hawkes, R. (2002) Compartmentation of the rabbit cerebellar cortex. J. Comp. Neurol., 444: 159–173.

Schmitz, J., Ohme, M. and Zischler, H. (2000) The complete mitochondrial genome of *Tupaia belangeri* and the phylogenetic affiliation of Scandentia to other eutherian orders. Mol. Biol. Evol., 17: 1334–1343.

Seil, F.J., Johnson, M.L. and Hawkes, R. (1995) Molecular compartmentation expressed in cerebellar cultures in the absence of neuronal activity and neuron-glia interactions. J. Comp. Neurol., 356: 398–407.

Sillitoe, R.V., Benson, M.A., Blake, D.J. and Hawkes, R. (2003a) Abnormal dysbindin expression in cerebellar mossy fiber synapses in the mdx mouse model of Duchenne muscular dystrophy. J. Neurosci., 23: 6576–6585.

Sillitoe, R.V. and Hawkes, R. (2002) Whole-mount immunohistochemistry: a high throughput screen for patterning defects in the mouse cerebellum. J. Histochem. Cytochem., 50: 235–244.

Sillitoe, R.V., Hulliger, M., Dyck, R. and Hawkes, R. (2003b) Antigenic compartmentation of the cat cerebellar cortex. Brain Res., 977: 1–15.

Sillitoe, R.V., Künzle, H. and Hawkes, R. (2003c) Zebrin II compartmentation of the cerebellum in a basal insectivore, the Madagascan hedgehog tenrec *Echinops telfairi*. J. Anat., 203: 283–296.

Sillitoe, R.V., Malz, C., Rockland, K. and Hawkes, R. (2004) Antigenic compartmentation of the primate and tree shrew cerebellum: a common topography of zebrin II in *Macaca mulatta* and *Tupaia belangerie*. J. Anatomy., in press.

Smeyne, R.T., Oberdick, J., Schilling, K., Berrebi, A.S., Mugnaini, E. and Morgan, J.I. (1991) Dynamic organization of developing Purkinje cells revealed by transgene expression. Science, 254: 710–721.

Smith, G.E. (1902) The primary subdivision of the mammalian cerebellum. J. Anat., 37: 381–385.

Stanhope, M.J., Waddell, V.G., Madsen, O., de Jong, W., Hedges, S.B., Cleven, G.C., Kao, D. and Springer, M.S. (1998) Molecular evidence for multiple origins of Insectivora and for a new order of endemic African insectivore mammals. Proc. Natl. Acad. Sci. USA, 95: 9967–9972.

Stephan, H., Baron, G. and Frahm, H.D. (1991) Insectivora with a stereotaxic atlas of the hedgehog brain. Comparative Brain Research in Mammals, Vol. 1. Springer-Verlag, New York, pp. 29–47.

van Dijk, M.A., Madsen, O., Catzeflis, F., Stanhope, M.J., de Jong, W.W. and Pagel, M. (2001) Protein sequence signatures support the African clade of mammals. Proc. Natl. Acad. Sci. USA, 98: 188–193.

Voogd, J., Jaarsma, D. and Marani, E. (1996) The cerebellum: chemoarchitecture and anatomy. In: Swanson, L.W., Björklund, A. and Hökfelt, T. (Eds.), Integrated Systems of the CNS, Part III: Cerebellum, Basal Ganglia, Olfactory System. Handbook of Chemical Neuroanatomy, Vol. 12. Elsevier Science, Amsterdam, pp. 1–369.

Walther, E., Dichgans, M., Maricich, S.M., Romito, R.R., Yang, F., Dziennis, S., Zackson, S., Hawkes, R. and Herrup, K. (1998) Genomic sequences of aldolase C (zebrin II) direct lacZ expression exclusively in non-neuronal cells of transgenic mice. Proc. Nat. Acad. Sci. USA, 95: 2615–2620.

Wassef, M., Sotelo, C., Thomasset, M., Granholm, A.C., Leclerc, N., Rafrafi, J. and Hawkes, R. (1990) Expression of compartmentation antigen zebrin I in cerebellar transplants. J. Comp. Neurol., 294: 223–234.

Yatsuki, H., Outida, M., Atsuchi, Y., Mukai, T., Shiokawa, K. and Hori, K. (1998) Cloning of the *Xenopus laevis* aldolase C gene and analysis of its promoter function in developing *Xenopus* embryos and A6 cells. Biochim. Biophys. Acta, 144: 199–217.

CHAPTER 23

Pharmacology of the metabotropic glutamate receptor mediated current at the climbing fiber to Purkinje cell synapse

Lan Zhu[1], Piergiorgio Strata[1,2] and Pavle R. Andjus[1,3,*]

[1]*Department of Neuroscience, University of Turin, Corso Raffaello 30, 10125 Turin, Italy*
[2]*Rehabilitation Hospital and Research Institute, Santa Lucia Foundation, Via Ardeatina 306, I-00179, Rome, Italy*
[3]*Department of Physiology and Biochemistry, School of Biology, University of Belgrade, Studentski trg 12-16, POB 52, 11001 Belgrade, Serbia and Montenegro*

Keywords: mGluR currents; cerebellum; Purkinje cell; climbing fiber; parallel fibers; glutamate transporters; paired pulse depression

Abstract: Different forms of synaptic plasticity in the cerebellum are mediated by metabotropic glutamate receptors (mGluRs). At parallel fiber (PF) to Purkinje cell (PC) synapses activation of mGluR gives rise to a well known slow synaptic current inhibited by antagonists of mGluR1. The distribution of mGluR types in the climbing fiber (CF) to PC synapses is not well known. However, a mGluR1α-mediated all-or-none postsynaptic current was also demonstrated at the CF-PC synapse (Dzubay and Otis, *Neuron* 36, 1159, 2002).

Using whole cell patch-clamp recording from PCs in rat cerebellar slices with AMPA receptors blocked and glutamate uptake impaired we demonstrate a more complex pharmacology of a current obtained by single or train CF stimulation. The mGluR1 specific antagonist CPCCOEt in a group of cells suppressed this response while in a similar number of other cells it induced a potentiating effect. The antagonists of mGluR groups II and III (LY341495 and MSOP, respectively) predominantly suppressed the current. The ambiguous effect of CPCCOEt was checked by measuring the paired-pulse depression of the CF EPSC, which was not changed with the antagonist in normal as well as in low (0.5 mM) external Ca^{2+} (used to prevent saturation of AMPARs), thus excluding a presynaptic effect. However, CPCCOEt induced a rise in the amplitude (by $\sim 50\%$) as well as a prolongation ($p < 0.05$) of the decay time of CF EPSCs at normal 2 mM Ca^{2+}, i.e. under conditions of AMPAR saturation, thus indicating an effect of postsynaptic origin. In 0.5 mM Ca^{2+} the decay of CF EPSCs was longer but it was also significantly prolonged ($p \ll 0.01$) by CPCCOEt. However, the CF EPSC amplitude was not significantly affected indicating an underlying Ca^{2+}-dependent mechanism.

Thus, the pharmacology of the PC mGluR-mediated response points to a dual postsynaptic role of mGluR1 giving rise to a slow postsynaptic current but also regulating other presumably mGluR-dependent currents via second messenger molecules and Ca^{2+}. The additional electrophysiological role of mGluR II & III types was also indicated. Such a complex regulatory mechanism may have an important role in the mGluR-dependent forms of homosynaptic plasticity and motor learning at the CF-PC synapse.

*Corresponding author. Tel.: +381-11-3032356; Fax: +381-11-638-500; E-mail: pandjus@bf.bio.bg.ac.yu

Introduction

In the cerebellar cortex two different excitatory inputs project to separate domains of the Purkinje cell (PC) dendrites. The distal one, rich in spines is innervated by hundreds of thousands of parallel fibers (PFs), while the proximal one has only a few spines innervated by a single climbing fiber (CF). However, this contact is extremely strong consisting of hundreds of release sites (Strata and Rossi, 1998). In these synapses glutamatergic transmission involves ionotropic (iGluRs) and metabotropic glutamate receptors (mGluRs). Different forms of synaptic plasticity expressed at both CFs and PFs require the presence of mGluRs. Such are the PF long term depression (LTD; Linden et al., 1991; Ichise et al., 2000), the CF non-associative LTD (Hansel and Linden, 2000), the short term presynaptic depression via retrograde signaling to cannabinoid receptors at CF terminals (CF cSTD; Maejima et al., 2001), and the activity-dependent pruning of the multiple CF inputs to each PC during development (Kano et al., 1997; Ichise et al., 2000).

In PCs in addition to the postsynaptic mGluR1 (Lujan et al., 1997) there is also some evidence for the expression of mGluR3, 4, and 7 (Phillips et al., 1998; Berthele et al., 1999). At PF to PC synapses, in addition to the ionotropic AMPA receptors, glutamate also activates mGluRs giving rise to a well known slow postsynaptic current (PSC) inhibited by antagonists of mGluR1 (Batchelor et al., 1994; Tempia et al., 1998; Reichelt and Knopfel, 2002). It has recently been shown that this G-protein-dependent signaling mechanism in addition to the stimulation of phospholipase C transduction pathway activates the TRPC1 cation channel that underlies the slow EPSC in PCs (Kim et al., 2003). Although the immunocytochemical studies show that mGluRs are also present at the CF synapses (Nusser et al., 1994; Petralia et al., 1997) the distribution of the different mGluR subtypes in the CF to PC synapses is not well known. Recently, a mGluR1α-mediated all-or-none postsynaptic current (mGluR PSC) was also demonstrated at the CF–PC synapse (Dzubay and Otis, 2002). In order to reveal this current in addition to iGluRs glutamate transporters also had to be blocked.

The aim of the present study is to give a more detailed information on the origin of CF-mGluR PSC isolated by a similar approach as in Dzubay and Otis (2002).

Materials and methods

Experimental animals were P22-76 Wistar rats. Cerebella were isolated in ice cold extracellular medium (in mM: 125 NaCl, 2.5 KCl, 1.25 NaH_2PO_4, 1 $MgCl_2$, 2 $CaCl_2$, 26 $NaHCO_3$, 20 glucose) gassed with 95% O_2/5% CO_2 mix. Parasagittal slices (200 μm thin) were cut and used in experiments with constant perfusion with O_2/CO_2-bubbled extracellular saline (22–26°C).

Whole-cell patch-clamp recording was performed with an EPC 7 patch-clamp amplifier (Heka, Lambrecht/Pfalz, Germany) with the ITC-16 (Instrutech, Greatneck, New York, USA) A/D converter, and a stimulus pulse generator (stimulus pulse width was 20–100 μs) with an isolator unit (WPI, Sarasota, Florida, USA). Recordings were filtered at 3 kHz and the sampling rate was 5–40 kHz. Pipettes were pulled from borosilicate glass (2–3 MΩ when filled) and contained in mM: 120 CsCl, 20 TEA, 10 HEPES, 4 Na_2ATP, 0.4 Na_2GTP, 0.1 Ca_2Cl, 2 $MgCl_2$, 10 EGTA, pH 7.3 adjusted with CsOH (all Sigma, Milan, Italy). Stimulation of CFs was performed with a soda lime glass pipette (tip diameter ∼ 10 μm) filled with extracellular solution and placed in the white matter or the molecular layer several cell diameters away from the patched PC. Recordings were performed in 20 μM bicuculline (Sigma) and 50 μM APV (Tocris Cookson Ltd, Bristol, UK) to block $GABA_A$ and NMDA receptors, respectively.

CFs were identified by a double pulse protocol at +30 mV showing characteristic paired pulse depression (PPD; Fig. 1). The stimulating electrode was adjusted so that the subthreshold response of the PFs (showing paired pulse facilitation) is minimal or absent (Fig. 1). After first identifying the CF–PC response iGluRs, i.e., AMPARs (AMPA receptors) were blocked with CNQX (6-Cyano-7-nitroquinoxaline-2,3-dione), 20 μM or NBQX (1,2,3,4-Tetrahydro-6-nitro-2,3-dioxo-benzo[f]quinoxaline-7-sulfonamide), 10 μM. Then TBOA (DL-thero-β-Benzyloxyaspartic acid), 100 μM was used to detain glutamate in the synaptic cleft by blocking glutamate transporters, thus augmenting the effect of released transmitter on mGluRs in CF–PC synapses (Fig. 2).

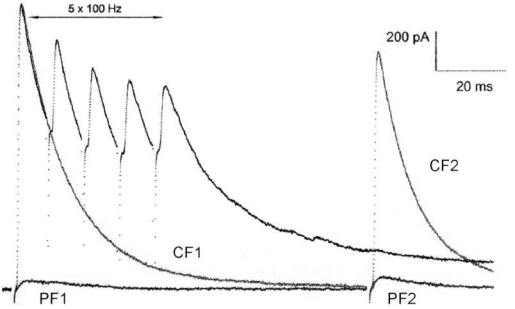

Fig. 1. Examples of CF EPSCs (superimposed recordings). The first and the last EPSCs (CF1 and CF2, respectively) are obtained by paired pulse stimulation (note depression of second response amplitude, CF2). Subthreshold stimulation induced only small PF EPSCs (PF1 and PF2). Train stimulation of 100 Hz (5 pulses; horizontal two-point arrow) could elicit 5 EPSCs, although with suppressed amplitudes, Holding voltage (V_h) was +30 mV.

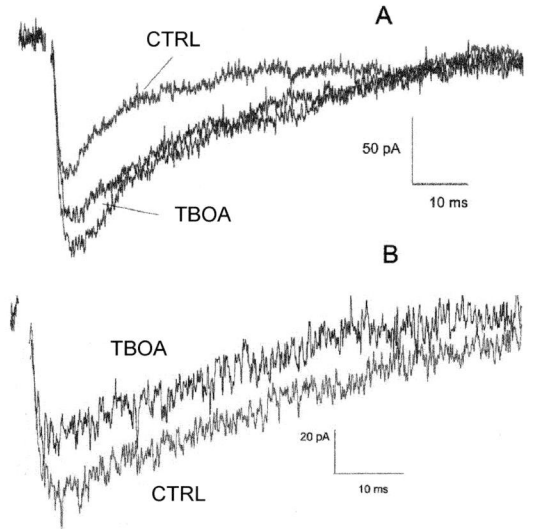

Fig. 2. Two opposite effects of TBOA on CF-mGluR PSC obtained by single stimulation (in both cases AMPA receptors were blocked). V_h was −70 and −60 mV (A and B, respectively).

In order to induce substantial glutamate release at the CF–PC synapse in some experiments a stimulus train of 10 or 100 Hz (5 or 10 pulses) was delivered. Although 100 Hz is not the natural firing frequency of CFs, such a tetanic stimulation could still elicit trains of postsynaptic currents even without TBOA and AMPARs blocked (Fig. 1).

The pharamacology of the remaining CF-induced PSC response was checked with mGluR inhibitors CPCCOEt, 7-(Hydroxyimino)cyclopropa[b]chromen-1a-carboxylate ethyl ester, (100 μM) that blocks mGluR1, the member of mGluR I group, LY341495, (2S)-2-Amino-2-[(1S,2S)-2-carboxycycloprop-1-yl]-3-(xanth-9-yl) propanoic acid, (0.2 μM) that targets group II mGluRs, i.e., mGluR 2 & 3 and MSOP, (RS)-a-Methylserine-O-phosphate, (100 μM) the antagonist of group III, i.e., mGluR 4, 6, 7, and 8. All the above antagonists were from Tocris.

Reported values are mean ± SEM.

Results and discussion

With AMPARs blocked (in CNQX or NBQX), in a number of cases (18/56 cells) TBOA induced a rise of residual CF-EPSCs (mGluR PSCs; Figs. 2A, 5, and 6A). This is expected to be due to the augmented concentration of glutamate in the synaptic cleft. However, in a similar fraction of cells (22/56 cells) it also induced a decrease of the PSC (Figs. 2B, 3B, and 5 inset), presumably due to a block of the predominant faster transporter currents (Auger and Attwel, 2000). Both effects of TBOA can be seen in the same cell depending on the type of stimulation (single vs. train), i.e., depending on the amount of released glutamate (see Fig. 5 with inset).

Upon $GABA_A$ and NMDA receptors block (with bicuculline and APV, respectively) the CF origin of the residual response in TBOA with AMPARs blocked was evidenced by its all-or-none behavior (not shown). Responses to train stimulation (10 or 100 Hz) showed a recovery tail current immediately after the last pulse (see Fig. 6C) which in some cases formed an inward current peak (see Fig. 5). With train stimulation a slow late current phase was also observed reaching its peak 40–50 ms after the train (see Fig. 6B).

The effect of the mGluR1 antagonists, CPCCOEt was ambiguous (see Table 1). In some cells it augmented the mGluR PSC amplitude more than twice (e.g., Fig. 3A) but in other cells caused its suppression by 40% (e.g., Fig. 3B). Also note in Fig. 3A the additional and strong inhibiting effect of MSOPPE, (RS)-alpha-methylserine-O-phosphate monophenyl ester, (200 μM) the antagonist of group II/III mGluRs. This ambiguous behavior was apparent

Table 1. Effect of different mGluR antagonists on the amplitude of CF-mGluR PSC

Stimulation	CPCCOEt	MSOP	LY341495
Single pulse	2.22 ± 0.34 (5)	*1.46 ± 0.10 (3)*	*1.37 (1)*
	0.62 ± 0.10 (3)	0.62 ± 0.05 (13)	0.63 ± 0.06 (4)
Train	1.82 ± 0.14 (4)	*1.55 ± 0.16 (5)*	*1.56 (1)*
	0.66 ± 0.11 (3)	0.56 ± 0.05 (10)	0.63 ± 0.07 (4)

Data are expressed as mean multiplication factors (±SEM) with number of cells in brackets. Values less than one indicate inhibiting effect. Cases of significantly smaller cell populations are indicated by italics.

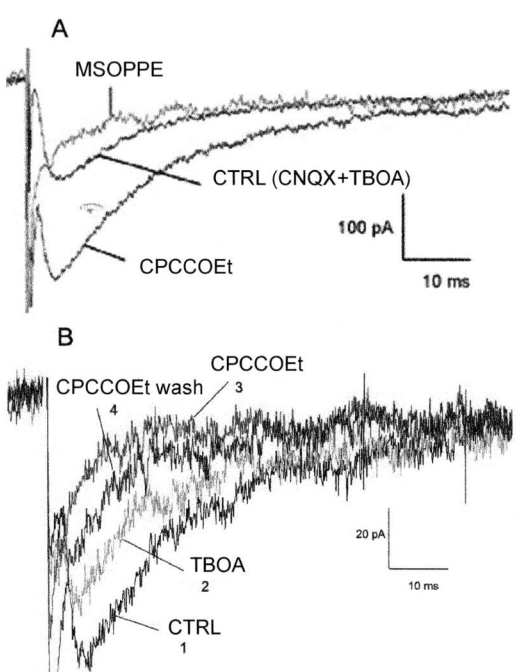

Fig. 3. Dual effect of CPCCOEt on CF mGluR PSC obtained with single pulse stimulation. (A) CPCCOEt induces a rise in amplitude. Addition of MSOPPE causes strong suppression of amplitude. $V_h = -70$ mV. (B) Suppression of PSC by CPCCOEt with subsequent partial wash. Note inhibiting effect of TBOA. Numbers indicate sequence of trace recordings. $V_h = -50$ mV.

with single-pulse stimulation (Fig. 3) as well as with trains (Table 1).

In order to reveal the mechanism underlying the dual effect of CPCCOEt we studied the effects of these antagonists on the CF EPSC and its short term plasticity without AMPAR antagonists or TBOA. Paired-pulse depression (PPD) of the CF EPSC was not changed with the CPCCOEt (Fig. 4A). In standard 2 mM external calcium, used in these experiments, AMPARs should be saturated (Foster et al., 2002) and the PPD change could be masked. Therefore, paired pulse plasticity was also checked in low (0.5 μM) external Ca^{2+} and again no change in PPD was observed (Fig. 4B). Nevertheless, even with AMPARs saturated (in 2 mM Ca^{2+}), CPCCOEt induced a rise in the CF EPSC amplitude by a factor of 1.50 ± 0.34 ($n = 6$) and a longer decay time (11.7 ± 0.7 ms vs. 15.8 ± 1.5 ms, $p < 0.05$; inset in Fig. 4A), thus pointing to a postsynaptic effect. In lower, 0.5 mM Ca^{2+} the decay of CF EPSCs was faster (7.5 ± 1.2 ms) and it was also significantly ($p \ll 0.01$) prolonged with CPCCOEt (8.8 ± 1.2 ms; inset in Fig. 4B). However, unlike the case in 2 mM Ca^{2+}, in 0.5 mM Ca^{2+} the CF EPSC amplitude was not significantly affected (mean ratio of amplitudes in CPCCOEt and in control was 1.03 ± 0.06, $n = 8$) which may indicate an underlying Ca^{2+}-dependent mechanism. Thus, in some cells the inhibition of postsynaptic mGluR1 (as in Dzubay and Otis, 2002) may cause the expected decrease of its current but in others this effect may be overcome by a rise of PSCs presumably originating from other mGluR types (since iGluRs were largely blocked). The latter mGluR currents of unknown origin would otherwise be suppressed through mGluR1 action. The experiment with the substantial MSOPPE block after CPCCOEt (Fig. 3A) indicates that non-mGluR1 currents could be originating from group II and III mGluRs. Thus, further experiments were performed in order to check the participation of the other mGluR groups.

Group II mGluRs have been shown to take part in the presynaptic inhibition via autoreceptors at the CF–PC synapse (Hashimoto and Kano, 1998; Tamaru et al., 2001; Harrison and Jahr, 2003). However, in the present experiments the effect of the group II antagonist, LY341495 on CF-mGluR PSCs induced by train and single stimuli, was clearly inhibiting (Table 1 and Fig. 5). This may point to the existence of group II mGluRs also at the CF–PC postsynaptic membrane where they mediate CF-mGluR PSCs.

Agonists of mGluR III are not known to affect significantly CF-EPSCs (Hashimoto and Kano, 1998). However, in TBOA treated cerebellar slices with AMPARs blocked the antagonists of group III mGluRs, MSOP affected CF-mGluR PSCs. The effect was predominantly inhibiting (see Table 1).

303

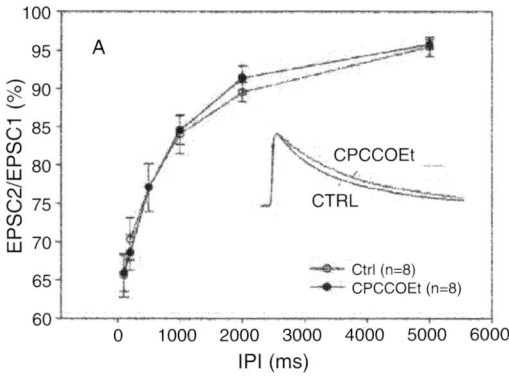

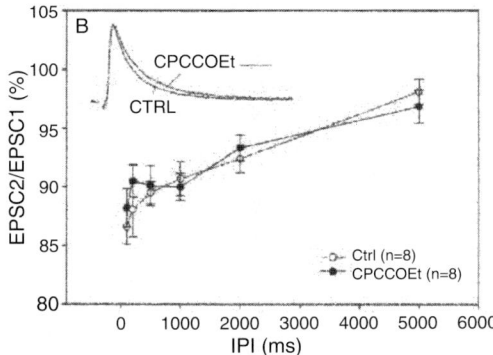

Fig. 4. Paired pulse depression recovery experiments in CPCCOEt. (A) Normal external medium (2.0 mM Ca^{2+}). (B) Low external Ca^{2+} (0.5 mM). *Ordinate*: amplitude ratio; *abscissa*: interpulse interval (IPI). Insets are examples of normalized EPSCs in CPCCOEt vs. control (each trace is average of three recordings; horizontal calibration bars indicate 5 ms). $V_h = +30$ mV.

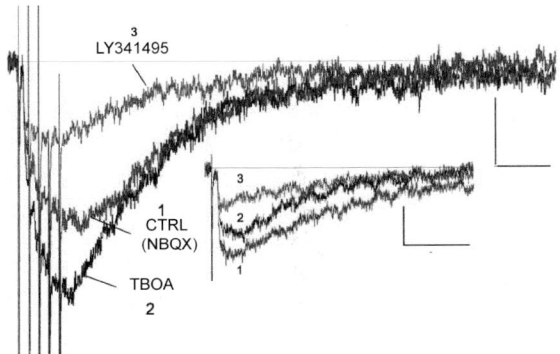

Fig. 5. The suppressing effect of LY341495. CF was stimulated with a train of 5 pulses at 100 Hz. Note the rise of PSC amplitude after TBOA. Inset: single stimulation (note suppressed amplitude in TBOA). Numbers indicate sequence of recorded traces. Scalings indicate *vertical* — 50 pA, *horizontal* —50 ms. $V_h = -50$ mV. Thin horizontal line indicates zero current level.

In addition to single pulse response (Fig. 6A) MSOP could also affect the late current and tail current phases of the mGluR response after train stimulation (Fig. 6B and C, respectively). It was observed that after a partial MSOP block, group II antagonist, LY341495 could induce further inhibition (not shown) which is congruent with the composite origin of the observed CF-mGluR PSC.

Previous work by Dzubay and Otis (2002) with single and train stimulation of CFs identified a slow mGluR1-dependent current in PCs similar to the well known mGluR current activated at PF-PC synapses (Batchelor et al., 1994; Tempia et al., 1998; Reichelt and Knopfel, 2002). We have observed such currents or their tails only with train stimulation (see Figs. 5 and 6) and we showed that their origin may not be just from group I mGluRs (see Table 1). However, in presence of TBOA and with AMPA receptors blocked, CF single as well as train stimulation gives rise to relatively faster currents. These currents may be partly ionotropic (unblocked AMPA and non-AMPA, kainate) as recently shown by Huang et al. (2004), but they are also sensitive to mGluR antagonists. Moreover, these currents are preferentially diminished with antagonists of all three mGluR groups. This behavior indicates postsynaptic origin of these mGlu receptors. On the other hand, the role of presynaptic mGluR II/III receptors may not be excluded as evidence by rare cases of current rise with the application of mGluR group-selective antagonists (see Table 1). Presynaptic group II mGluRs are well known in CF–PC synapses (Hashimoto and Kano, 1998; Tamaru et al., 2001; Huang et al., 2004), however, presynaptic inhibition via group III autoreceptors was only demonstrated in PF–PC synapses (Pekhletski et al., 1996; Miniaci et al., 2001). In the light of the recently discovered residual CF EPSC originating from non-AMPA, kainate receptors (Huang et al., 2004) and their suggested metabotropic function (Rozas et al., 2003) it would be interesting to study the possible interaction of mGluR second messenger pathways with the function of this receptor.

In conclusion, the pharamacology of the CF–PC mGluR-mediated response with glutamate transporters blocked points to a complex system of metabotropically regulated currents at the CF–PC synapses. It reveals a dual postsynaptic role of

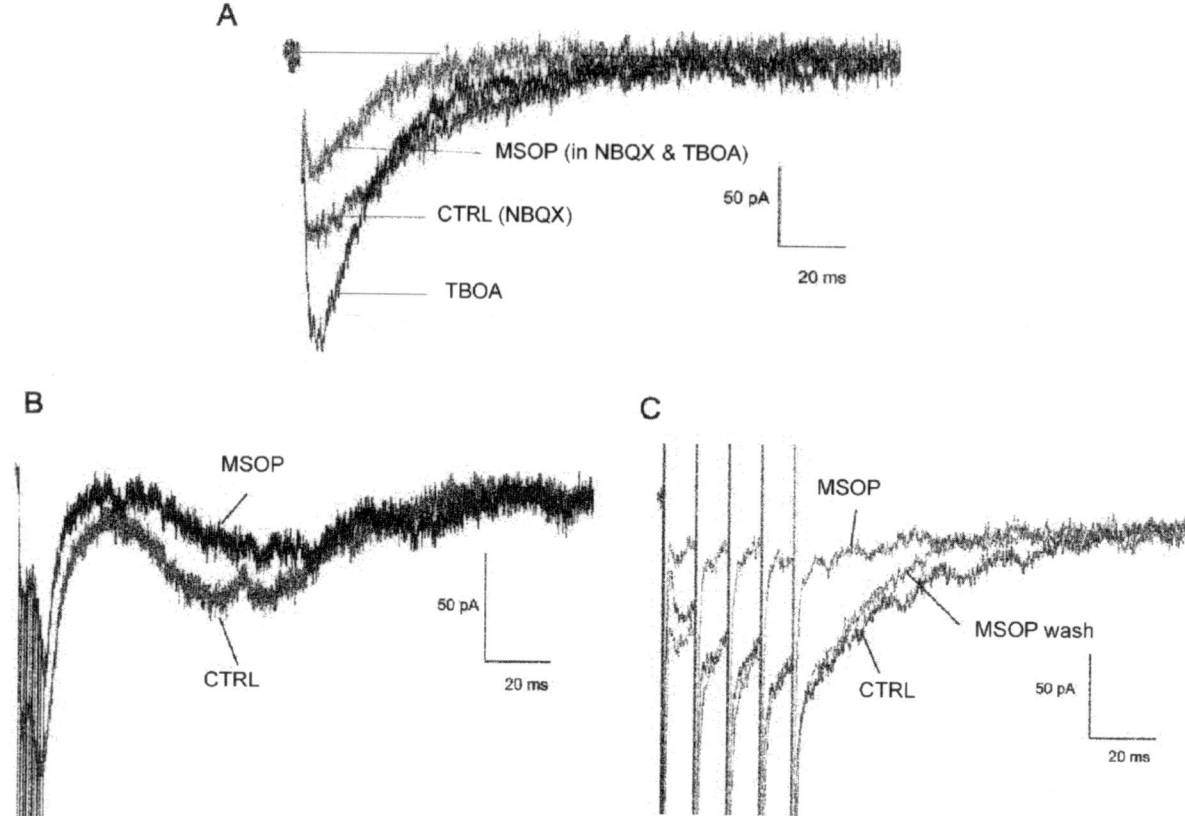

Fig. 6. MSOP effects. (A) Suppression of CF mGluR PSC obtained by single stimuli. (B) Suppressing effect of MSOP on the late current appearing after 10 pulses at 100 Hz train stimulation. (C) Inhibiting effect of MSOP and subsequent wash upon 5 pulses at 100 Hz CF stimulation. All experiments were done at $V_h = -60$ mV. Thin horizontal line indicates zero current level.

mGluR1, on one hand giving rise to a slow PSC, as already described in Dzubay and Otis (2002), but on the other hand regulating additional, presumably also metabotropic currents. These residual currents could be blocked by antagonists of mGluR II and III. In addition to postsynaptic receptor sites, presynaptic group II and III autoreceptors could not be excluded. This system of postsynaptic second messenger regulation may have an important role in the mGluR-dependent forms of CF plasticity (CF LTD, CF cSTD, and CF pruning) and motor learning. Additional modulation is supported by the action of glutamate transporters (Bergles et al., 1997; Huang et al., 2004) which can restrict the response of such a system only to trains of CF stimuli such as are known to be necessary for CF LTD.

Acknowledgments

This work was supported by grants from the MIUR and Italian Ministry of Health, and by grant 1647 of the Ministry of Science, Republic of Serbia.

Abbreviations

AMPAR	AMPA receptor
CF	climbing fiber
CNQX	6-Cyano-7-nitroquinoxaline-2,3-dione
CPCCOEt	7-(Hydroxyimino)cyclopropa[*b*]chromen-1a-carboxylate ethyl ester
cSTD	short term presynaptic depression via cannabinoid receptors

iGluR	ionotropic glutamate receptor
LTD	long term depression
LY341495	(2S)-2-Amino-2-[(1S,2S)-2-carboxycycloprop-1-yl]-3-(xanth-9-yl) propanoic acid
mGluR	metabotropic glutamate receptor
MSOP	(RS)-a-methylserine-O-phosphate
MSOPPE	(RS)-alpha-methylserine-O-phosphate monophenyl ester
NBQX	1,2,3,4-Tetrahydro-6-nitro-2,3-dioxobenzo[f]quinoxaline-7-sulfonamide
PC	Purkinje cell
PF	parallel fiber
PPD	paired pulse depression
PSC	postsynaptic current
TBOA	DL-threo-β-Benzyloxyaspartic acid

References

Auger, C. and Attwel, D. (2000) Fast removal of synaptic glutamate by postsynaptic transporters. Neuron, 28: 547–548.

Batchelor, A.M., Madge, D.J. and Garthwaite, J. (1994) Synaptic activation of metabotropic glutamate receptors in the parallel fibre–Purkinje cell pathway in rat cerebellar slices. Neuroscience, 63: 911–915.

Bergles, D.E., Dzubay, J.A. and Jahr, C.E. (1997) Glutamate transporter currents in Bergmann glial cell follow the time course of extrasynaptic glutamate. Proc. Natl. Acad. Sci. USA, 94: 14821–14825.

Berthele, A., Platzer, S., Laurie, D.J., Weis, S., Sommer, B., Zieglgansberger, W., Concard, B. and Tolle, T.R. (1999) Expression of metabotropic glutamate receptor subtype mRNA (mGluR1-8) in human cerebellum. Neuroreport, 16: 3861–3867.

Dzubay, J.A. and Otis, T.S. (2002) Climbing fiber activation of metabotropic glutamate receptors on cerebellar Purkinje neurons. Neuron, 36: 1159–1167.

Foster, K.A., Kreitzer, A.C. and Regehr, W.G. (2002) Interaction of postsynaptic receptor saturation with presynaptic mechanisms produces a reliable synapse. Neuron, 35: 1115–1126.

Hansel, C. and Linden, D.J. (2000) Long-term depression of the cerebellar climbing fiber-Purkinje neuron synapse. Neuron, 26: 473–482.

Harrison, J. and Jahr, C.E (2003) Receptor occupancy limits synaptic depression at climbing fiber synapses. J. Neurosci., 23: 377–383.

Hashimoto, K. and Kano, M. (1998) Presynaptic origin of paired-pulse depression at climbing fiber-Purkinje cell synapses in the rat cerebellum. J. Physiol. (Lond.), 506: 391–405.

Huang, Y.H., Dykes-Horberg, M., Tanaka, T., Rothstein, J.D. and Bergles, D.E. (2004) Climbing fiber activation of EAAT4 transporters and kainate receptors in cerebellar Purkinje cells. J. Neurosci., 24: 103–111.

Ichise, T., Kano, M., Hashimoto, K., Yanagihara, D., Nakao, K., Shigemoto, R., Katsuki, M. and Aiba, A. (2000) mGluR1 in cerebellar Purkinje cells essential for long-term depression, synapse elimination, and motor coordination. Science, 288: 1832–1835.

Kano, M., Hashimoto, K., Kurihara, H., Watanabe, M., Inoue, Y., Aiba, A. and Tonegawa, S. (1997) Persistent multiple climbing fiber innervation of cerebellar Purkinje cells in mice lacking mGluR1. Neuron, 18: 71–79.

Kim, S.J., Kim, Y.S., Yuan, J.P., Petralia, R.S., Worley, P.F. and Linden, D.J. (2003) Activation of the TRPC1 cation channel by metabotropic glutamate receptor mGluR1. Nature, 426: 285–291.

Linden, D.J., Dickson, M.H., Smeyne, M. and Connor, J.A. (1991) A long-term depression of AMPA currents in cultured cerebellar Purkinje neurons. Neuron, 7: 81–89.

Lujan, R., Roberts, J.D., Shigemoto, R., Ohishi, H. and Somogyi, P. (1997) Differential plasma membrane distribution of metabotropic glutamate receptors mGluR1 alpha, mGluR2 and mGluR5, relative to neurotransmitter release sites. J. Chem. Neuroanat., 13: 219–241.

Maejima, T., Hashimoto, K., Yoshida, T., Aiba, A. and Kano, M. (2001) Presynaptic inhibition caused by retrograde signal from metabotropic glutamate to cannabinoid receptors. Neuron, 31: 463–475.

Miniaci, M.C., Bonsi, P., Tempia, F., Strata, P. and Pisani, A. (2001) Presynaptic modulation by group III metabotropic glutamate receptors (mGluRs) of the excitatory postsynaptic potential mediated by mGluR1 in rat cerebellar Purkinje cells. Neurosci. Lett., 310: 61–65.

Nusser, Z., Mulvihill, E., Streit, P. and Somogyi, P. (1994) Subsynaptic segregation of metabotropic and ionotropic glutamate receptors as revealed by immunogold localization. Neuroscience, 61: 421–427.

Pekhletski, R., Gerlai, R., Overstreet, L.S., Huang, X.P., Agopyan, N., Traverse-Slater, N., Abramow-Newerly, W., Roder, J.C. and Hampson, D.R. (1996) Impaired cerebellar synaptic plasticity and motor performance in mice lacking the mGluR4 subtype of metabotropic glutamate receptor. J. Neurosci., 16: 6364–6373.

Petralia, R.S., Wang, Y.X., Singh, S., Wu, C., Shi, L., Wei, J. and Wenthold, R.J. (1997) A monoclonal antibody shows discrete cellular and subcellular localizations of mGluR1 alpha metabotropic glutamate receptors. J. Chem. Neuroanat., 13: 77–93.

Phillips, T., Makoff, A., Murrison, E., Mimmack, M., Waldvogel, H., Faull, R., Rees, S. and Emson, P. (1998) Immunohistochemical localisation of mGluR7 protein in the rodent and human cerebellar cortex using subtype specific antibodies. Brain Res. Mol. Brain Res., 57: 132–1941.

Reichelt, W. and Knopfel, T. (2002) Glutamate uptake controls expression of a slow postsynaptic current mediated by mGluRs in cerebellar Purkinje cells. J. Neurophysiol., 87: 1974–1980.

Rozas, J.L., Paternain, A.V. and Lerma, J. (2003) Noncanonical signaling by ionotropic kainate receptors. Neuron, 39: 543–553.

Strata, P. and Rossi, F. (1998) Plasticity of the olivocerebellar pathway. Trends Neurosci., 21: 407–413.

Tamaru, Y., Nomur, S., Mizuno, N. and Shigemoto, R. (2001) Distribution of metabotropic glutamate receptor mGluR3 in mouse CNS: differential location relative to pre- and postsynaptic sites. Neuroscience, 106: 481–503.

Tempia, F., Miniaci, M.C., Anchisi, D. and Strata, P. (1998) Postsynaptic current mediated by metabotropic glutamate receptors in cerebellar Purkinje cells. J. Neurophysiol., 80: 520–528.

SECTION VIII

Excitability in Cerebellar Cortex

CHAPTER 24

Nicotinic receptor modulation of neurotransmitter release in the cerebellum

Giovanna De Filippi*, Tristan Baldwinson and Emanuele Sher

Eli Lilly and Company Ltd, Lilly Research Centre, Erl Wood Manor, Sunninghill Road, Windlesham, Surrey GU20 6PH, UK

Keywords: neurotransmitter release; nicotinic acetylcholine receptors; cerebellum; granule cell; Purkinje cell; unipolar brush cell; development

Abstract: Nicotinic ACh receptors (nAChRs) are formed by pentameric combinations of α and β subunits, differentially expressed throughout the central nervous system (CNS), where they have been shown to play a role in the modulation of neurotransmitter release. nAChRs are also important during neuronal differentiation, regulating gene expression and contributing to neuronal pathfinding.

The cerebellum, which is involved in the maintenance of balance and orientation as well as refinement of motor action, in motor memory and in some aspects of cognition, undergoes a significant process of development and maturation of its neuronal networks during the first three postnatal weeks in the rat.

Autoradiographic as well as in situ hybridization and immunocytochemical studies have shown that several nicotinic receptor binding sites and subunits are expressed in the rat cerebellum from embryonic stage through to adulthood, with the highest expression levels seen during the development of the cerebellar cortex.

A diffuse cholinergic afferent projection to all lobules of the cerebellar cortex has been described, with the uvulanodulus, flocculus and lobules I and II of the anterior vermis regions receiving a particularly dense projection.

Low levels of nAChR subunit transcripts and immunoreactivity, particularly during adulthood, and the scattered distribution of immunoreactivity between neurons in the cerebellar cortex, can explain the difficulty in assessing electrophysiologically the presence of functional nAChRs in the cerebellar cortex and some contradictory results reported in the early-published papers. In recent years, several groups have shown that also in the cerebellum different nAChR subtypes modulate release of glutamate and GABA at different synapses. The possible role of these mechanisms in synaptic consolidation during development, as well as on plasticity phenomena and network activity at mature synapses, are discussed.

Introduction

Nicotinic ACh receptors (nAChRs) belong to the superfamily of neurotransmitter receptors that includes $GABA_A$, glycine, and serotonin ($5HT_3$) receptors. They are composed of pentameric combinations of α ($\alpha 2-\alpha 10$) and β ($\beta 2-\beta 4$) subunits, giving the potential of a great variety of native neuronal receptors. Nicotinic AChRs are expressed throughout the central (CNS) and peripheral (PNS) nervous system. In the PNS, they mediate synaptic transmission in autonomic ganglia and at the neuromuscular junction. In the CNS, the prevalent functional nAChRs are the $\alpha 4\beta 2$—containing ($\alpha 4\beta 2$*), which constitute over 90% of high affinity [^{3}H]nicotine and [^{3}H]cytisine binding sites in rat brain, and the $\alpha 7$—containing ($\alpha 7$*) receptors, which are generally

*Corresponding author. Tel.: +44-1276-483465; Fax. +44-1276-483525; E-mail: de_filippi_giovanna@lilly.com

DOI: 10.1016/S0079-6123(04)48024-8

thought to form homomeric nAChRs that are inhibited by α-bungarotoxin (α-BTX) and are less sensitive to nicotine (Role and Berg, 1996; Francis and Papke, 2000). The role of these central nAChRs has remained obscure for a long time. In recent years, significant information has been accumulated showing that nAChRs mainly exert a modulatory influence on the release of other neurotransmitters (reviewed in Dani, 2001; Sher et al., 2004). Activation of pre-synaptic nAChRs initiates a Ca^{2+} increase in the presynaptic terminal (either via depolarization of the terminal and subsequent opening of voltage-gated calcium channels or via direct Ca^{2+} influx through the highly Ca^{2+} permeable α7* receptor) that can enhance neurotransmitter release. Direct, fast nicotinic synaptic transmission has been detected as a small excitatory input in some areas of the brain (Dani, 2001; Sher et al., 2004). There is also evidence for significant nonsynaptic, volume transmission for ACh in the CNS (Vizi and Lendvai, 1999). Nicotinic ACh receptors also play a significant role during neuronal differentiation, regulating early gene expression and contributing to neuronal pathfinding and target selection (Role and Berg, 1996; Broide and Leslie, 1999).

The cerebellum is involved in the maintenance of the balance and orientation, the refinement of motor action, in motor memory storage and in some aspects of cognition (Fiez, 1996; Rapoport et al., 2000). The development of the cerebellar cortex takes place during the first three postnatal weeks in the rat. During this time, the granule cells migrate from the external to the internal granular layer (IGL) where they form synaptic connections with the afferent mossy fibers and other interneurons in the IGL (Altman, 1972). At this developmental stage, levels of choline acetyltransferase (ChAT; the rate limiting enzyme for the synthesis of ACh) are particularly high, both in rats (Clos et al., 1989) and humans (Court et al., 1993) also when compared to the levels of acetylcholinesterase (AChE; the ACh-degradative enzyme) (Clos et al., 1989; Court et al., 1993), suggesting that the cholinergic system may play an important role during cerebellar development.

Here we review recent evidence on nicotinic receptor localization and cholinergic innervation in the cerebellum; in light of these evidences, we interpret a body of functional findings, including our own data, for the nicotinic receptor modulation of synaptic transmission within the cerebellar cortex. Evidences for muscarinic ACh receptor presence and function in the cerebellum are not considered in this review.

Nicotinic receptor expression in the cerebellum

Autoradiographic studies

Autoradiographic studies show that in the rat, high affinity [^{3}H]nicotine binding sites become detectable in the cerebellar anlage at embryonic day (E) 15 and sparse labeling is still present at adulthood (Naeff et al., 1992). In adult rats, high affinity [^{3}H]nicotine, [^{3}H]ACh, and [^{3}H]cytisine binding sites can be detected, although the density of labeling is moderate as compared to other cortical areas (Clarke et al., 1985; Happe et al., 1994; Jaarsma et al., 1997). Binding of [^{3}H]nicotine and [^{3}H]cytisine has been reported to be identical both in distribution and relative density. Labeling is concentrated in the IGL, with no differentiation in densities between folia, and deep cerebellar nuclei, while the molecular layer and white matter display no significant binding (Happe et al., 1994; Jaarsma et al., 1997). Levels of [^{3}H]nicotine binding from the whole cerebellum reach maximal levels within the first postnatal week and then decrease to the adult ones within the third week after birth (Zhang et al., 1998). The high affinity agonist binding nicotinic receptors labeled by [^{3}H]nicotine and [^{3}H]cytisine have been proposed to be composed of the α4 and β2 subunits. [^{3}H]cytisine binding sites are immunoprecipitated by antisera against α4 and β2 subunits, but not against α2, α3, α5, β3, and β4 (Flores et al., 1992); furthermore, the largest proportion of rat brain [^{3}H]nicotine binding sites is immunoprecipitated by a monoclonal antibody specific for the α4 subunit (Whiting et al., 1987).

α-Bungarotoxin is selective antagonist of the α7 subunit containing nicotinic receptor (Seguela et al., 1993). [^{125}I]α-BTX binding is found to be very low in the cerebellum (Clarke et al., 1985; Frostholm and Rotter, 1986; Zhang et al., 1998). Maximal levels of [^{125}I]α-BTX binding in the whole cerebellum are achieved during the first postnatal week, and the adult levels are reached at postnatal day 14 (Zhang et al., 1998). Frostholm and Rotter (1986) show that small patches of very intense labeling occur in the IGL of the vestibulo-cerebellum in the vermis. These clusters have the size of glomeruli; they appear around postnatal days (P) 15–16, and reach the highest density at around P22, which is then kept constant throughout the adulthood.

In situ hybridization and immunocytochemical studies

A number of in situ hybridization and immunocytochemical studies have shown that several nicotinic receptor subunits are expressed in the rat cerebellum during development and into adulthood.

In situ hybridization studies from the whole cerebellum have demonstrated that nicotinic receptor subunit mRNA levels show a different pattern during the development of the cerebellar cortex (Zhang et al., 1998). The $\alpha 3$ mRNA signal is transient during the first week after birth; $\alpha 4$ mRNA expression is highest during the first two postnatal weeks and decreases thereafter to adult levels; $\beta 2$ mRNA level is stable during development and adulthood; $\alpha 7$ mRNA expression is 10-fold higher at P1 than at adult age, rapidly decreasing, to adult level at P14. No $\alpha 2$ mRNA is detected throughout development.

Interestingly, low to moderate expression of $\alpha 3$ and $\beta 4$ mRNA is present since E15 to birth in the cerebellar neuroepithelium and in migrating cells of the external granular layer (Zoli et al., 1995; Morley, 1997; Winzer-Serhan and Leslie, 1997; Opanashuk et al., 2001).

Neurons in the molecular layer of the adult cerebellar cortex do not express mRNA for $\alpha 2$, $\alpha 3$, $\alpha 4$, $\beta 2$ subunits (Wada et al., 1989). However, Nakayama et al. (1998) detected weak $\alpha 4$ mRNA signal in the molecular layer, where few neurons also show $\alpha 4$-like immunoreactivity (Nakayama et al., 1997). In the molecular layer, $\alpha 4$ immunoreactive axonal terminals (either basket cell axons or climbing fibers) form synapses with cell bodies and dendrites of Purkinje cells (Nakayama et al., 1998). In the same layer, virtually no immunostaining for $\alpha 7$ subunit is found other than in Purkinje cell dendrites (Dominguez del Toro et al., 1994).

Purkinje cells in adult cerebellum show strong $\alpha 4$ mRNA signals (Nakayama et al., 1998), as well as intense $\alpha 4$ immunostaining on the cell body (Nakayama et al., 1997). Also strong $\beta 2$ mRNA levels are reported (Wada et al., 1989; Hill et al., 1993) as well as intense $\beta 2$ immunoreactivity on Purkinje cell bodies and dendrites (Hill et al., 1993). Low to moderate expression of $\alpha 4$ and $\beta 2$ mRNA is reported in Purkinje cells at E17 and E19 as well as at birth. Expression of $\alpha 3$ mRNA is detected after the first postnatal week in scattered Purkinje cells (Winzer-Serhan and Leslie, 1997). Strong $\beta 4$ hybridization signals between the first and second postnatal week are also detected, which are maintained until adulthood (Dineley-Miller and Patrick, 1992; Winzer-Serhan and Leslie, 1997). Although weak $\alpha 7$ mRNA expression in Purkinje cells has been reported by Seguela et al. (1993), Dominguez del Toro et al. (1994) show intense immunostaining of Purkinje cells bodies and dendrites in the molecular layer. $\alpha 7$ immunoreactivity is not detectable for the first two days after birth. Between P3 and P5, moderate immunolabeling is observed, which increases rapidly within the second postnatal week. At this stage, a well-defined 'banding' pattern of labeling is detected in regions of the Purkinje cell layer where $\alpha 7$ subunits are being expressed (Dominguez del Toro et al., 1997).

In the IGL, $\alpha 4$ and $\beta 2$ mRNA expression can be detected (Wada et al., 1989; Hill et al., 1993; Nakayama et al., 1998). $\alpha 4$-like immunoreactivity is found in cell bodies, including granule cell bodies; moreover, immunocytochemical localization of the $\alpha 4$ subunit at the electron microscopy (EM) level shows that it is present in the plasma membrane on the soma of granule cells (Nakayama et al., 1997). A diffuse $\beta 2$ mRNA signal is detected in the IGL, with a corresponding weak $\beta 2$-like immunoreactivity (Hill et al., 1993). Low overall expression of $\beta 4$ mRNA is detectable in the granule cell layer at P5, increasing to moderate intensity during the second postnatal week. Additional scattered cells exhibiting a strong hybridization signal are visible in the IGL starting at P11 and into adulthood. In contrast, $\alpha 3$ mRNA expression is first observed at P7, but limited to scattered cells in the granule cell layer (Winzer-Serhan and Leslie, 1997). $\alpha 7$ immunoreactivity is clearly detected in the IGL, particularly in lobules IX and X (Swanson et al., 1987); however other studies show that granule cells are poorly labeled, and the very few scattered $\alpha 7$ immunoreactive neurons have been identified as Golgi cells (Dominguez del Toro et al., 1994).

Finally, in deep cerebellar nuclei, $\alpha 4$ and $\beta 2$ mRNA are detected as early as E17–E19 (Wada et al., 1989; Hill et al., 1993; Zoli et al., 1995), and moderate to relatively strong $\alpha 7$ mRNA staining is reported (Dominguez del Toro et al., 1994).

Cholinergic innervation of the cerebellum

Cholinergic innervation of the cerebellum has been a matter of controversy for a long time, despite the presence of high levels of AChE activity (Silver, 1967). The reason for this is due to the inconsistency of results from early electrophysiological studies (see next paragraph) and the reported substantial lack of stimulation of ACh release from isolated glomerulus particles and granular layer slices loaded with radiolabeled choline (Morales and Tapia, 1987). In recent years, evidence for the presence of the main components of cholinergic neurotransmission has been accumulating.

Immunohistochemistry studies with monoclonal antibodies against ChAT show that all lobules of the rat cerebellum receive a diffuse cholinergic afferent projection (Ojima et al., 1989; Barmack et al., 1992a,b; Jaarsma et al., 1996). However, three regions receive a particularly dense projection: the uvula-nodulus (lobules IX and X), the flocculus, and lobules I and II of the anterior lobe vermis. Only axon-like structures show ChAT immunoreactivity; they include: (1) a subpopulation of mossy fibers and glomerular rosettes, (2) thin beaded fibers which are morphologically similar and are closely associated with the Purkinje cell layer and found also in the molecular layer, (3) a relatively dense network of varicose fibers distributed in the cerebellar nuclei. Ojima et al. (1989) report that immunoreactive mossy fiber terminals are scattered evenly through the IGL of all lobules, except lobule VI in which they form clusters near the Purkinje cell layer. The same terminals in the hemispheres tend to be localized immediately beneath the Purkinje cell layer, suggesting that granule cell-mediated influences of ChAT positive mossy fibers on Purkinje cells are distinctive in different regions. Barmack et al. (1992a) observe that the cholinergic mossy fiber input appears to be stratified within the upper part of the IGL, particularly for cholinergic afferent projections to lobules IX and X in the rabbit and the cat. Jaarsma et al. (1996) show a parasagittal zonal enrichment of ChAT-positive mossy fibers in lobules IX and X. Such a stratification of cholinergic inputs may represent a level of functional organization to place synaptic 'weighting' on different afferent inputs. Also, immunoreactive mossy fiber terminals are reported to branch within the IGL: multiple mossy fiber rosettes appear to stem from a single axon (Barmack et al., 1992a,b; Jaarsma et al., 1996). Moreover, finely beaded axons branching off from a large mossy fiber rosette in the IGL have been observed. These fine fibers penetrate the Purkinje cell layer into the molecular layer. These observations lead to the debate whether these fibers really originate from a separate anatomical localization and/or constitute a separate class. The functional significance of different types of ChAT-positive terminals is also a matter of speculation. It is possible that different fiber morphology reflects either different anatomical origins, or, alternatively, different post-synaptic receptor specialization in the target neuron.

The ultrastructural studies of ChAT immunoreactive mossy fibers in lobules IX and X show that they form asymmetric synaptic junctions with dendritic profiles of both the granule cells and unipolar brush cells (UBC) (Jaarsma et al., 1996). Glomeruli with immunolabeled rosettes contain dendritic profiles belonging both to granule cells exclusively and to granule and UBC together. A minority of ChAT-positive rosettes form giant synapses with UBC somata. Also, UBC dendrites synapsing with immunolabeled rosettes are occasionally also found to be presynaptic to granule cell dendrites.

Ultrastructural studies of finely beaded fibers show that they form synaptic contacts with the Golgi, basket/stellate cell dendrites, as well as the Purkinje cell spines (Jaarsma et al., 1997). However, ChAT-positive boutons can also be found in close apposition to the Golgi, granule, stellate/basket cells, although no synaptic junctions are detected. Whether this could reflect a diffuse 'non-synaptic' mode of action for ACh in some areas of the cerebellar cortex, remains to be established.

ChAT-immunoreactive mossy fibers in lobules IX and X have been shown to originate from the medial vestibular nucleus and to some extent from the nucleus prepositus hypoglossi (Barmack et al., 1992c). The finely beaded ChAT-positive fibers have been shown to originate from various nuclei including the pedunculo-pontine tegmental cholinergic nucleus (PPTg), the nucleus raphe obscurus (ROb), and/or the lateral paraganto cellular nucleus (LPGi) (Jaarsma et al., 1997).

Finally, a subset of Golgi cells has been found to be ChAT positive, at least in cat (Illing, 1990) and human (De Lacalle et al., 1993) cerebella.

Nicotinic receptor activation and modulation of neurotransmitter release in cerebellum

The function of nicotinic receptor activation in the cerebellar circuitry is still largely uninvestigated, although some recent publications suggest that the modulatory role of nAChRs at various synapses could be very important in regulating cell firing and cerebellar output.

Early electrophysiological studies, mainly in vivo, report contradictory results on the action of ACh in the cerebellum.

In the pioneering studies by McCance and Phillis (1964, 1968) in anaesthetized adult cats, ACh and various cholinomimetic agents are iontophoretically applied to cells located at depths of 100 μm to 5 mm. ACh-sensitive cells are more frequently found deeper in the cortex and appear to be ubiquitously distributed in all areas of the cerebellum. Cells in the molecular layer are not sensitive to ACh, while about 37% of the Purkinje cells (identified on the basis of the depth) respond to ACh application, possibly through activation of somatic ACh receptors. In the IGL, different cells show different degrees of excitation in response to ACh application, and a higher proportion of cells located in the deeper granular layer are reported to be more responsive to ACh. The ACh-induced excitation is blocked by application of dihydro-β-erythroidine (DHβE), as well as atropine, mecamylamine, and d-tubocurarine, while acetylcholinesterase inhibitors exert an excitant action on these cells.

However, Crawford et al. (1966) report different effects of iontophoretically applied ACh in the cerebellar cortex of anaesthetized adult cats. A proportion of Purkinje cells (75%) are excited by electrophoretic application of ACh and other cholinomimetics, including nicotine. DHβE reverses this effect. No response to cholinomimetics is observed in granule cells as well as in basket cells. Moreover, the systemic administration of DHβE does not depress synaptic excitation evoked by stimulation of mossy, climbing, or parallel fibers, leading the authors to conclude that ACh is unlikely to be an excitatory neurotransmitter within the feline cerebellum.

A lack of ACh effects on lobules IX and X neurons in the IGL has also been reported in rat cerebellar slices (Crepel and Dhanjal, 1982). In these experiments, bath application of nAChR antagonists does not show any detectable effect on excitatory and inhibitory synaptic potentials evoked in Purkinje cells via mossy and climbing fiber stimulation. Although these results lead to the conclusion that the presence of a significant contingent of these fibers using ACh as the neurotransmitter is unlikely, it is not possible to exclude the presence of some cholinergic mossy fibers which may contact granule cells connected to distal parts of Purkinje cell dendrites. The contribution of these fibers to the mossy fiber-mediated excitatory postsynaptic potentials recorded at the soma would become negligible due to the electrotonic properties of Purkinje cells.

Nicotinic receptor activation in the cerebellum has been demonstrated by De La Garza et al. (1987a,b, 1988). In anaesthetized adult rats, pressure ejection of nicotine has an inhibitory effect on Purkinje cells and a strong excitatory effect on 'interneurons' in the IGL of lobules VI and VII of the vermis. These effects are shown to be due to postsynaptic mechanisms. The inhibitory effect on the Purkinje cells can be blocked by hexamethonium, or kappa-Bungarotoxin, which are known to block ganglionic-type nAChRs ($\alpha 3\beta 4$-containing) (Chiappinelli and Dryer, 1984). The excitatory effect on 'interneurons' is irreversibly blocked by α-BTX application, this being the first electrophysiological evidence suggesting that mammalian brain contains functional nicotinic receptors sensitive to α-BTX.

The difficulty to assess the presence of functional nAChRs in the cerebellar cortex electrophysiologically and the contradictory results reported in the few literature papers available can be explained by the low level of nAChR subunit transcripts and immunoreactivity in the cerebellum compared to other cortical areas. Furthermore, as discussed in the previous paragraph, the levels of nAChR transcripts appear to decrease dramatically after the first three post-natal weeks, and immunoreactivity for nAChR subunits appears to be scattered, rather than uniformly distributed, between neurons in the cerebellar cortex. An indirect evidence for the possible role of nAChRs in the cerebellar cortex comes from recently released studies. One proposed function of nAChRs in the cerebellum is the regulation of glutamate release. Nicotine-induced increase in Ca^{2+} flux in primary cultures of mouse cerebellar granule cells is decreased by N-methyl-D-aspartate (NMDA) receptor antagonist and largely inhibited by α-BTX (Didier et al., 1995), suggesting that NMDA receptor activation is mediated by the release of endogenous glutamate following α7-containing nAChR activation. Similarly, spontaneous and electrically evoked [^{3}H]D-aspartate outflow is facilitated by nicotinic receptors sensitive to α-BTX and mecamylamine in primary cultures of rat granule cells (Bianchi et al., 2000). Activation of nicotinic receptors induces [^{3}H]glutamate release in adult rat cerebellar slices, which can be prevented by α-BTX, tetrodotoxin (TTX; voltage-dependent Na^+ channel blocker) and glutamate receptor antagonists (Reno' et al., 2004). Nicotinic receptor activation has recently been reported to also modulate [^{3}H]norepinephrine release in rat cerebellar slices during development (O'Leary and Leslie, 2003). The authors emphasize the critical physiological role that the interaction of cholinergic and noradrenergic inputs may have in the regulation of maturation events during cerebellar cortex development.

The putative modulatory role of nAChRs in enhancing glutamate release in the cerebellum was shown electrophysiologically by our group (De Filippi et al., 2001). Patch-clamp experiments in the whole-cell configuration are carried out on granule cells in lobule III to VIII from P5 to P14 rats. Application of 1 mM ACh (in the presence of muscarinic and γ-amino-butyric acid (GABA) receptor antagonists) elicits a variety of effects. While a significant proportion of cells do not respond (20%), in the remaining cells ACh can either elicit a somatic current only (29%), or postsynaptic currents (PSCs) with (43%) or without (8%) the somatic response. In our study, the ACh-induced somatic current is generally small in amplitude, ranging from 3 to 70 pA, and its duration (30–40 s) outlasts the time of ACh exposure. In the majority of the cells tested, 10 mM choline (selective α7* agonist) is unable to evoke somatic currents, while 100 µM cytisine (α7* and β4* agonist, β2* partial agonist) elicits a current only in a small proportion of cells. Together with the lack of block by 1–10 nM methyllycaconitine (MLA; selective α7* antagonist) and full inhibition by 10 µM DHβE (α4β2* preferring antagonist), these results suggest that in the majority of cells tested, β2-containing receptors are more likely to be responsible for the somatic current; however the data do not completely exclude a small contribution of β4* and α7* receptors in a subpopulation of cells. PSCs are usually elicited during ACh application and they can last up to 10–30 s after ACh exposure. PSC amplitude varies from 5 to 50–60 pA (occasionally bigger events are detected), the 10–90% rise time ranges between 0.3 and 1.7 ms, while the monoexponential decay time is 6–10 ms. These kinetic parameters are similar to those of spontaneous PSCs occasionally detected. 10 mM Choline and 100 µM cytisine mimic ACh in inducing PSCs in all the cells tested. The application of a cocktail of glutamate receptor antagonists completely blocks the ACh-induced PSCs (Fig. 1A). This effect is reversible upon washout of glutamate receptor antagonists and does not affect the somatic current (De Filippi et al., 2001). This result suggests that activation of nAChRs at the synapse between mossy fiber and granule cells can modulate glutamate release. Moreover, 1–10 nM MLA (Fig. 1B) or 200 nM α-BTX block ACh-induced glutamate release, further substantiating the involvement of α7-containing receptors. Finally, ACh-induced release is largely inhibited by TTX application (Fig. 1C), suggesting that the nicotinic receptors have a 'pre-terminal' localization at these synapses. Following a prolonged bath application of a low agonist concentration, ACh modulation of glutamate release is impaired (Fig. 2A). This effect can be reversed after wash out of the α7 agonist, and it is interpreted as being due to the desensitization of the pre-terminal α7* receptors involved in the modulation of glutamate release at this synapse (Grottick et al., 2000). Recently, we have reported the potentiating effect of 5-hydroxyindole

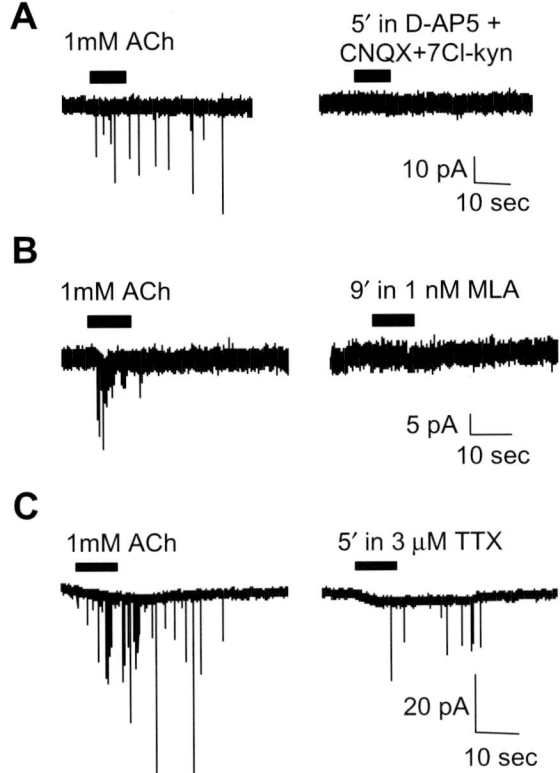

Fig. 1. ACh-induced modulation of glutamate release at the mossy fiber to granule cell synapse. The horizontal bar on top of all traces represents the duration of 1 mM ACh application. (A) ACh-induced PSCs (left trace) are completely blocked by a cocktail of glutamate receptor antagonists. (B) ACh-induced PSCs are completely blocked by α7 receptor antagonist MLA at low concentrations. (C) ACh-induced PSCs are almost completely blocked by 3 μM TTX.

(5-HI) on human and rat α7 receptors (Zwart et al., 2002). In rat cerebellar slices, we show that while 1 mM 5-HI on its own is ineffective in inducing glutamate release, co-application of 1 mM ACh with 1 mM 5-HI enhances ACh-induced release (Fig. 2B). The frequency of the excitatory PSCs during the 10 s of ACh application increases 2–3 times in the cells tested, and in few cells an increase in PSC amplitude could also be observed (Zwart et al., 2002). This modulatory effect is irreversibly blocked by 200 nM α-BTX (Fig. 2B, see also original paper).

Two recent electrophysiological papers further strengthen the evidence of modulation of synaptic transmission in the cerebellum by nAChRs. The application of ACh to Purkinje cells at early developmental stages (P5–P10) enhances both glutamate and GABA release (Kawa, 2002). ACh-induced release of glutamate and GABA is largely mediated through non-α7 nicotinic receptors and is shown to be TTX-dependent, suggesting that the nAChRs responsible for this effect are located presynaptically on the excitatory and inhibitory interneurons, presumably at preterminal axonic sites or in somato-dendritic regions of granule and basket cells. This modulatory effect of neurotransmitter release on Purkinje cells by nicotinic receptors is barely observed in older rats, suggesting that nAChRs are developmentally regulated and may have a role in the maturation of the cerebellar cortex. In the same paper, somatic nicotinic currents are detected in about 50% of the basket cells and 45% of the granule cells tested. Somatic currents in granule cells are sensitive to DHβE, substantiating our previous findings that somatic nicotinic receptors on granule cells are likely to be of the α4β2* subtype.

Finally, in the granule cells in adult rats, ACh, by acting on presynaptic nicotinic receptors possibly located on Golgi cells axon terminals, evokes a large Ca^{2+}-dependent but action potential—independent release of GABA, which in turn activates α6 subunit-containing $GABA_A$ receptors, generating a tonic inhibition on granule cells (Rossi et al., 2003). This effect can be blocked by the nAChR antagonist mecamylamine, which on its own does not alter the membrane current of granule cells, indicating that in the slice preparation, where afferents are cut, there is no spontaneous ACh release regulating GABAergic inhibition.

Discussion

The most recent electrophysiological data suggest that a major role played by nAChRs in the cerebellar cortex, as in other brain areas, is the modulation of neurotransmitter release (Dani, 2001; Sher et al., 2004).

We show that at early developmental stages, nicotinic receptor activation can induce glutamate release in a reasonable proportion of mossy fiber to granule cell synapses. The TTX sensitivity of this effect suggests that the nicotinic receptors involved (likely α7 receptors, from our characterization) are

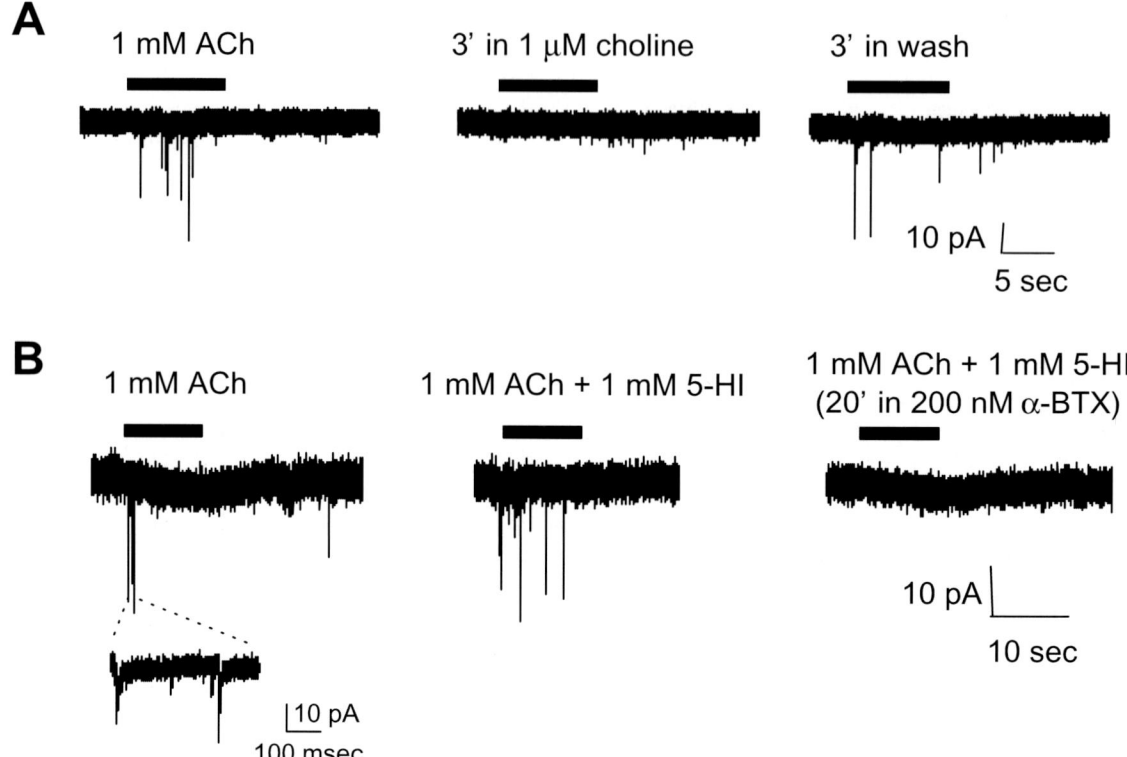

Fig. 2. Modulation of α7 nAChR-mediated glutamate release at the mossy fiber to granule cell synapse. The horizontal bar on top of all traces represents the duration of 1 mM ACh and 1 mM ACh + 1 mM 5-HI application. (A) ACh-induced glutamate release is reversibly impaired by pre-synaptic α7 receptor desensitization after prolonged bath application of a low agonist concentration. (B) Enhancement of ACh-mediated glutamate release by 5-HI. Glutamate-mediated PSCs (enlarged in the inset) are induced by application of 1 mM ACh alone (left) or by co-application of 1 mM ACh with 1 mM 5-HI (center). The potentiating effect of 5-HI is blocked by 20 min exposure to 200 nM α-BTX.

located either on somato-dendritic areas of interacting cells or preterminally in mossy fibers. One possibility is that nicotinic receptors are present on UBCs, and their activation leads to action potential generation and glutamate release on the postsynaptic granule cells. A number of evidences support this hypothesis. AChE-positive mossy fibers appear to be in close proximity of the UBCs (Harris et al., 1993). ChAT-positive mossy fiber rosettes synapse onto UBCs (Jaarsma et al., 1996). [^{125}I]α-BTX binding and ChAT reactivity distribution have been shown to correspond to the distribution of UBCs at least in the vestibulocerebellum (Jaarsma et al., 1997). UBCs in acute slices have been reported to respond to ACh application with an inward current (Rossi et al., 1995), which, however, has not been pharmacologically characterized. Alternatively, α7 receptors could be located at preterminal levels of the axon. ACh could be co-released with glutamate at the mossy fiber terminal and activate autoreceptors. ACh has been shown to colocalize with amino acid neurotransmitters (Caffe' et al., 1996); the possibility of ACh and glutamate being co-distributed in ChAT-positive mossy fibers is realistic, also in view of the solid anatomical and physiological evidence that glutamate is used for fast excitatory transmission in the majority of cerebellar mossy fibers (Crepel and Dhanjal, 1982; Somogyi et al., 1986; Barbour, 1993; D'Angelo et al., 1993; Rossi et al., 1995). Moreover, neurochemical studies in the cerebellar slices show that presynaptic nicotinic receptors (although insensitive to α-BTX and sensitive to kappa-Bungarotoxin) mediate a positive feedback control on ACh release (Lapchak et al., 1989). Finally, the

ChAT-positive fibers, distinct from mossy fibers, could also be an alternative source of ACh. Whichever the route for pre-terminal nAChR activation, the physiological relevance of this effect is unknown. At the developmental stage covered by our study, the majority of granule cells show immature electrophysiological properties (D'Angelo et al., 1994, 1997). The enhancement of glutamate release by pre-synaptic nicotinic receptors could be a mechanism for synapse consolidation. Increased excitability in granule cells via either enhanced glutamate release or cell depolarization by somatic receptor activation, could translate in increased glutamate release onto Purkinje cells. Interestingly, from a similar study carried out at the same early developmental stages in Purkinje cells, it is found that pre-terminal nicotinic receptor activation, likely of the $\alpha 4\beta 2^*$ subtype, modulates enhancement of glutamate release onto Purkinje neurons (Kawa, 2002). Although the increased granule cell excitability caused by somatic nicotinic receptor activation could account for this effect, we cannot completely rule out the possibility of nicotinic receptors being localized along the parallel fibers in pre-terminal positions. In this slice preparation, however, proximal portions of climbing fibers are removed, thus making unlikely the contribution of nAChRs eventually localized at pre-terminal positions along them. In the same study, it has been shown that at early developmental stages also the inhibitory tone of Purkinje cells via interneurons in the molecular layer can be modulated by pre-terminal nicotinic receptors, of the non-$\alpha 7$ subtype, suggesting the presence of somatic nAChRs on basket cells (which is in fact shown in the same paper). Another inhibitory synapse described to be modulated by nicotinic receptors in the cerebellar cortex, is the Golgi to granule cell synapse (Rossi et al., 2003). In the adult rat, the excitability of granule cells can be tonically inhibited by GABA released in a Ca^{2+}-dependent manner upon activation of nAChRs, likely localized on the axon terminals of Golgi cells. Therefore, depending on the timing and the duration of ACh release in the cerebellum, the fraction of granule cells excited by incoming mossy fiber inputs may be reduced, thus increasing the storage capacity of the cerebellum, as predicted by modeling work (De Schutter, 2002). Furthermore, at the mossy fiber to granule cell synapse, long-term potentiation has been described to occur reliably only when inhibition is blocked (Armano et al., 2000). Under physiological circumstances, the level of cholinergic fiber activity may profoundly affect the plasticity phenomena that occur at this synapse.

The relative abundance of nAChR transcripts at early developmental stages, compared to adult levels, suggests that these receptors may play a critical role during the ontogenesis of the cerebellum. For instance, the developmental appearance and localization of α-BTX binding sites in the vestibulocerebellum appears to be correlated in time with the arrival of the mossy fibers originating in the vestibular nuclei and the VIIIth nerve (Frostholm and Rotter, 1986). Within the same time window, the rat becomes able to walk forward and air-righting reflexes appear. Based on these observations, the authors hypothesize that α-BTX-sensitive receptors can have a role in the formation of stable vestibulocerebellar connections during development. α-BTX binding levels in the cerebellum are the highest during early postnatal development. Because of their high Ca^{2+} permeability, these receptors could regulate calcium-dependent events and ultimately modulate developmental plasticity.

In adulthood, lower but detectable levels of nicotinic receptors are present within the cerebellum, as well as a diffuse network of ChAT-positive fibers. The highest proportion of ChAT-positive fibers is detected in the vestibulocerebellum, which receives large vestibular primary and secondary afferent inputs. It has been proposed that this cholinergic pathway might be particularly important in regulating the sensitivity and processing of vestibular information (Barmack et al., 1992a,b). However, ChAT-positive fibers can be found throughout the whole cerebellar cortex. Due to their different anatomical origins, it is possible to speculate that cholinergic modulation of transmitter release could represent a fine mechanism for processing the information coming from different pathways. For instance, the cholinergic fibers arising from the PPTg may modulate the excitability of the cerebello-nuclear neurons in reaction to sleep and arousal, while fibers originating in the ROb and LPGi could be involved in the modulation of noradrenergic effects within the cerebellar cortex (O'Leary and Leslie, 2003).

Also the described stratification of cholinergic inputs (Barmack et al., 1992a; Jaarsma et al., 1996) could support the idea of ACh modulation of selected afferent inputs or functionally distinct areas.

The mechanisms of nicotinic receptor modulation at different cerebellar synapses and the consequences on the network activity are largely unknown. Hopefully, these recent findings will revamp interest in this exciting but still largely unexplored aspect of synaptic transmission modulation within the cerebellum.

Abbreviations

α-BTX	α-bungarotoxin
5-HI	5-hydroxyindole
7-Cl kyn	7-chlorokynurenic acid
ACh	acetylcholine
AChE	acetylcholinesterase
ChAT	choline acetyltransferase
CNQX	6-cyano-7-nitroquinoxaline-2,3-dione
CNS	central nervous system
D-AP5	D(−)-2-amino-5-phosphonopentanoic acid
DHβE	dihydro-β-erythroidine
E	embryonic day
EM	electron microscopy
GABA	γ-amino-butyric acid
IGL	internal granular layer
LPGi	lateral paragigantocellular nucleus
MLA	methyllycaconitine
nAChR	nicotinic ACh receptor
NMDA	N-methyl-D-aspartate
P	postnatal day
PNS	peripheral nervous system
PPTg	pedunculopontine tegmental nucleus
PSCs	postsynaptic currents
ROb	raphe obscurus
TTX	tetrodotoxin
UBC	unipolar brush cells

References

Altman, J. (1972) Postnatal development of the cerebellar cortex in rat III. Maturation of the components of the granular layer. J. Comp. Neurol., 145: 465–514.

Armano, S., Rossi, P., Taglietti, V. and D'Angelo, E. (2000) Long-term potentiation of intrinsic excitability at the mossy fiber granule cell synapse of rat cerebellum. J. Neurosci., 20: 5208–5216.

Barbour, B. (1993) Synaptic currents evoked in Purkinje cells by stimulating individual granule cells. Neuron, 11: 759–769.

Barmack, N.H., Baughman, R.W. and Eckenstein, F.P. (1992a) Cholinergic innervation of the cerebellum of rat, rabbit, cat, and monkey as revealed by choline acetyltransferase activity and immunohistochemistry. J. Comp. Neurol., 317: 233–249.

Barmack, N.H., Baughman, R.W. and Eckenstein, F.P. (1992b) Cholinergic innervation of the cerebellum of the rat, by secondary vestibular afferents. Ann. NY Acad. Sci., 656: 566–579.

Barmack, N.H., Baughman, R.W., Eckenstein, F.P. and Shojaku, H. (1992c) Secondary vestibular cholinergic projection to the cerebellum of rabbit and rat as revealed by choline acetyltransferase immunohistochemistry, retrograde and orthograde tracers. J. Comp. Neurol., 317: 250–270.

Bianchi, C., Tomasini, M.C., Antonelli, T., Marani, L. and Beani, L. (2000) Nicotinic modulation of [^{3}H]D-aspartate outflow from cultured cerebellar granule cells. Synapse, 36: 307–313.

Broide, R.S. and Leslie, F.M. (1999) The α7 nicotinic acetylcholine receptor in neuronal plasticity. Mol. Neurobiol., 20: 1–16.

Caffe', A.R., Hawkins, R.K. and De Zeeuw, C.I. (1996) Coexistence of choline acetyltransferase and GABA in axon terminals in the dorsal cap of the rat inferior olive. Brain Res., 724: 136–140.

Chiappinelli, V.A. and Dryer, S.E. (1984) Nicotinic transmission in sympathetic ganglia: blockade by the snake venom neurotoxin kappa-bungarotoxin. Neurosci. Lett., 50: 239–244.

Clarke, P.B.S., Schwartz, R.D., Paul, S.M., Pert, C.B. and Pert, A. (1985) Nicotinic binding in rat brain: autoradiographic comparison of [^{3}H]acetylcholine, [^{3}H]nicotine, and [^{125}I]α-bungarotoxin. J. Neurosci., 5: 1307–1315.

Clos, J., Ghandour, S., Eberhart, R., Vincendon, G. and Gombos, G. (1989) The cholinergic system in developing cerebellum: comparative study of normal, hypothyroid, and underfed rats. Dev. Neurosci., 11: 188–204.

Court, J.A., Perry, E.K., Johnson, M., Piggott, M.A., Kerwin, J.A., Perry, R.H. and Ince, P.G. (1993) Regional patterns of cholinergic and glutamate activity in the developing and aging human brain. Dev. Brain Res., 74: 73–82.

Crawford, J.M., Curtis, D.R., Voorhoeve, P.E. and Wilson, V.J. (1966) Acetylcholine sensitivity of cerebellar neurones in the cat. J. Physiol., 186: 139–165.

Crepel, F. and Dhanjal, S.S. (1982) Cholinergic mechanisms and neurotransmission in the cerebellum of the rat. An in vitro study. Brain Res., 244: 59–68.

D'Angelo, E., Rossi, P. and Taglietti, V. (1993) Different proportion of N-methyl-D-aspartate and non-N-methyl-D-aspartate receptor currents at the mossy fiber-granule cell synapse of developing rat cerebellum. Neurosci., 53: 121–130.

D'Angelo, E., Rossi, P., De Filippi, G., Magistretti, J. and Taglietti, V. (1994) The relationship between synaptogenesis and expression of voltage-dependent currents in cerebellar granule cells in situ. J. Physiol. (Paris), 88: 197–207.

D'Angelo, E., De Filippi, G., Rossi, P. and Taglietti, V. (1997) Synaptic activation of Ca^{2+} action potentials in immature rat cerebellar granule cells in situ. J. Neurophysiol., 78: 1631–1642.

Dani, J.A. (2001) Overview of nicotinic receptors and their roles in the central nervous system. Biol. Psych., 49: 166–174.

De Filippi, G., Baldwinson, T. and Sher, E. (2001) Evidence for nicotinic acetylcholine receptor activation in rat cerebellar slices. Pharmacol. Biochem. Behav., 70: 447–455.

De Lacalle, S., Hersh, L.B. and Caper, C.B. (1993) Cholinergic innervation of the human cerebellum. J. Comp. Neurol., 328: 364–376.

De La Garza, R., Hoffer, B.J. and Freedman, R. (1988) Heterogeneity of nicotine actions in the rat cerebellum. In: Clementi, F. et al. (Eds.), Nicotinic Acetylcholine Receptors in the nervous System, NATO ASIS Series, Vol. H25. Springer-Verlag, Berlin, Heidelberg, pp. 137–141.

De La Garza, R., McGuire, T.J., Freedman, R. and Hoffer, B.J. (1987a) Selective antagonism of nicotine actions in the rat cerebellum with α-bungarotoxin. Neurosci., 23: 887–891.

De La Garza, R., McGuire, T.J., Freedman, R. and Hoffer, B.J. (1987b) The electrophysiological effects of nicotine in the rat cerebellum: evidence for direct postsynaptic actions. Neurosci. Lett., 80: 303–308.

De Schutter, E. (2002) Cerebellar cortex: computation by extrasynaptic inhibition? Curr. Biol., 12: R363–R365.

Didier, M., Berman, S.A., Lindstrom, J. and Bursztajn, S. (1995) Characterization of nicotinic acetylcholine receptors expressed in primary cultures of cerebellar granule cells. Mol. Brian Res., 30: 17–28.

Dineley-Miller, K. and Patrick, J. (1992) Gene transcripts for the nicotinic acetylcholine receptor subunit, Beta4, are distributed in multiple areas of the central nervous system. Mol. Brain Res., 16: 339–344.

Dominguez del Toro, E., Juiz, J.M., Peng, X., Lindstrom, J. and Criado, M. (1994) Immunocytochemical localization of the α7 subunit of the nicotinic acetylcholine receptor in the rat central nervous system. J. Comp. Neurol., 349: 325–342.

Dominguez del Toro, E., Juiz, J.M., Smillie, F.I., Lindstrom, J. and Criado, M. (1997) Expression of α7 neuronal nicotinic receptors during postnatal development of the rat cerebellum. Dev. Brain Res., 98: 125–133.

Fiez, J.A. (1996) Cerebellar contribution to cognition. Neuron, 16: 13–15.

Flores, C.M., Rogers, S.W., Pabreza, L.A., Wolfe, B.B. and Kellar, K.J. (1992) A subtype of nicotinic cholinergic receptor in rat brain is composed of α4 and β2 subunits and is up regulated by chronic nicotine treatment. Molec. Pharmac., 41: 31–37.

Francis, M.M. and Papke, R.L. (2000) The functional diversity of nicotinic receptors in the nervous system: perspectives on receptor subtypes and receptor specialization. In: Clementi, F., Fornasari, D. and Gotti, C. (Eds.), Neuronal Nicotinic Receptors, Handbook of Experimental Pharmacology, Vol. 144. Springer-Verlag, Berlin, Heidelberg, pp. 301–335.

Frostholm, A. and Rotter, A. (1986) The ontogeny of α-bungarotoxin binding sites in rat cerebellar cortex: an autoradiographic study. Proc. West. Pharmacol. Soc., 29: 249–253.

Grottick, A.J., Trube, G., Corrigal, W.A., Huwyler, J., Malherbe, P., Wyler, R. and Higgins, G.A. (2000) Evidence that nicotinic α7 receptors are not involved in the hyperlocomotor and rewarding effects of nicotine. J. Pharm. Exp. Ther., 294: 1112–1119.

Happe, H.K., Peters, J.L., Bergman, D.A. and Murrin, L.C. (1994) Localization of nicotinic cholinergic receptors in rat brain: autoradiographic studies with [^{3}H]cytisine. Neurosci., 62: 929–944.

Harris, J., Moreno, S., Shaw, G. and Mugnaini, E. (1993) Unusual neurofilament composition in cerebellar unipolar brush neurons. J. Neurocytol., 22: 1039–1059.

Hill, J.A., Zoli, M., Bourgeois, J.-P. and Changeux, J.-P. (1993) Immunocytochemical localization of a neuronal nicotinic receptor: the β2-subunit. J. Neurosci., 13: 1551–1568.

Illing, R.-B. (1990) A subset of cerebellar Golgi cells may be cholinergic. Brain Res., 522: 267–274.

Jaarsma, D., Dino, M.R., Cozzari, C. and Mugnaini, E. (1996) Cerebellar choline acetyltransferase positive mossy fibers and their granule and unipolar brush cell targets: a model for central cholinergic nicotinic neurotransmission. J. Neurocytol., 25: 829–842.

Jaarsma, D., Ruigrok, T.J.H., Caffe', R., Cozzari, C., Levey, A.I., Mugnaini, E. and Voogd, J. (1997) Cholinergic innervation and receptors in the cerebellum. In: De Zeeuw, C.I., Strata, P. and Voogd, J. (Eds.), The Cerebellum: from Structure to Control, Progress in Brain Research, Vol. 114. Elsevier, Amsterdam, pp. 67–96.

Kawa, K. (2002) Acute synaptic modulation by nicotinic agonists in developing cerebellar Purkinje cells of the rat. J. Physiol., 538: 87–102.

Lapchak, P.A., Araujo, D.M., Quirion, R. and Collier, B. (1989) Presynaptic cholinergic mechanisms in the rat cerebellum: evidence for nicotinic, but not muscarinic autoreceptors. J. Neurochem., 53: 1843–1851.

McCance, I. and Phillis, J.W. (1964) The action of acetylcholine on cells in cat cerebellar cortex. Experientia, 20: 217–218.

McCance, I. and Phillis, J.W. (1968) Cholinergic mechanisms in the cerebellar cortex. Int. J. Neuropharmacol., 7: 447–462.

Morales, E. and Tapia, R. (1987) Neurotransmitters of the cerebellar glomeruli: uptake and release of labeled aminobutyric acid, serotonine and choline in a purified glomerulus fraction and in granular layer slices. Brain Res., 420: 11–21.

Morley, B.J. (1997) The embryonic and post-natal expression of the nicotinic receptor α3-subunit in rat lower brainstem. Mol. Brain Res., 48: 407–412.

Naeff, B., Schlumpf, M. and Lichtensteiger, W. (1992) Pre- and postnatal development of high affinity [^{3}H]nicotine binding sites in rat brain regions: an autoradiographic study. Dev. Brain Res., 68: 163–174.

Nakayama, H., Shioda, S., Nakajo, S., Ueno, S. and Nakai, Y. (1998) Expression of the nicotinic acetylcholine receptor $\alpha 4$ subunit mRNA in the rat cerebellar cortex. Neurosci. Lett., 256: 177–179.

Nakayama, H., Shioda, S., Nakajo, S., Ueno, S., Nakashima, T. and Nakai, Y. (1997) Immunocytochemical localization of nicotinic acetylcholine receptor in the rat cerebellar cortex. Neurosci. Res., 29: 233–239.

Ojima, H., Kawajiri, S. and Yamasaki, T. (1989) Cholinergic innervation of the rat cerebellum: qualitative and quantitative analyses of elements immunoreactive to a monoclonal antibody against choline acetyltransferase. J. Comp. Neurol., 290: 41–52.

O'Leary, K.T. and Leslie, F.M. (2003) Developmental regulation of nicotinic acetylcholine receptor-mediated [3H]norepinephrine release from rat cerebellum. J. Neurochem., 84: 952–959.

Opanashuk, L.A., Pauly, J.R. and Hauser, K.F. (2001) Effect of nicotine on cerebellar granule neuron development. Eur. J. Neurosci., 13: 48–56.

Rapoport, M., van Reekum, R. and Mayberg, H. (2000) The role of the cerebellum in cognition and behavior: a selective review. J. Neuropsych. Clin. Neurosci., 12: 193–198.

Reno', L.A.C., Zago, W. and Markus, R.P. (2004) Release of [^{3}H]glutamate by stimulation of nicotinic acetylcholine receptors in rat cerebellar slices. Neurosci., 124: 647–653.

Role, L.W. and Berg, D.K. (1996) Nicotinic receptors in the development and modulation of CNS synapses. Neuron, 16: 1077–1085.

Rossi, D.J., Alford, S., Mugnaini, E. and Slater, N.T. (1995) Properties of transmission at a giant glutamatergic synapse in cerebellum: the mossy fiber-unipolar brush cell synapse. J. Neurophysiol., 74: 24–42.

Rossi, D.J., Hamann, M. and Attwell, D. (2003) Multiple modes of GABAergic inhibition of rat cerebellar granule cells. J. Physiol., 548: 97–110.

Seguela, P., Wadiche, J., Dineley-Miller, K., Dani, J.A. and Patrick, J.W. (1993) Molecular cloning, functional properties, and distribution of rat brain $\alpha 7$: a nicotinic cation channel highly permeable to calcium. J. Neurosci., 13: 596–604.

Sher, E., Chen, Y., Sharples, T.J.W., Broad, L.M., Benedetti, G., Zwart, R., McPhie, G.I., Pearson, K.H., Baldwinson, T. and De Filippi, G. (2004) Physiological roles of neuronal nicotinic receptor subtypes: new insights on the nicotinic modulation of neurotransmitter release, synaptic transmission and plasticity. Curr. Top. Med Chem., 4: 283–297.

Silver, A. (1967) Cholinesterases of the central nervous system with special reference to the cerebellum. Int. Rev. Neurobiol., 10: 57–109.

Somogyi, P., Halashy, K., Somogyi, J., Storm-Mathisen, J. and Ottersen, O.P. (1986) Quantitation of immunogold labeling reveals enrichment of glutamate in mossy and parallel fibre terminals in cat cerebellum. Neurosci., 19: 1045–1050.

Swanson, L.W., Simmons, D.M., Whiting, P.J. and Lindstrom, J. (1987) Immunohistochemical localization of neuronal nicotinic receptors in the rodent central nervous system. J. Neurosci., 7: 3334–3342.

Vizi, E.S. and Lendvai, B. (1999) Modulatory role of presynaptic nicotinic receptors in synaptic and non-synaptic chemical communication in the central nervous system. Brain Res. Rev., 30: 219–235.

Wada, E., Wada, K., Boulter, J., Deneris, E., Heinemann, S., Patrick, J. and Swanson, L.W. (1989) Distribution of alpha2, alpha3, alpha4, and beta2 neuronal nicotinic receptor subunit mRNAs in the central nervous system: a hybridization histochemical study in the rat. J. Comp. Neurol., 284: 314–335.

Whiting, P., Esch, F., Shimasaki, S. and Lindstrom, J. (1987) Neuronal nicotinic acetylcholine receptor β-subunit is coded for by the cDNA clone $\alpha 4$. Fedn. Eur. Biochem. Socs. Lett., 219: 459–463.

Winzer-Serhan, U.H. and Leslie, F.M. (1997) Codistribution of nicotinic acetylcholine receptor subunit $\alpha 3$ and $\beta 4$ mRNAs during rat brain development. J. Comp. Neurol., 386: 540–554.

Zhang, X., Liu, C., Miao, H., Gong, Z.-H. and Nordberg, A. (1998) Postnatal changes of nicotinic acetylcholine receptor $\alpha 2$, $\alpha 3$, $\alpha 4$, $\alpha 7$ and $\beta 2$ subunits genes expression in rat brain. Int. J. Devl. Neurosci., 16: 507–518.

Zoli, M., Le Novere, N., Hill, J.A. and Changeux, J.-P. (1995) Developmental regulation of nicotinic ACh receptor subunit mRNA in the rat central and peripheral nervous system. J. Neurosci., 15: 1912–1939.

Zwart, R., De Filippi, G., Broad, L.M., McPhie, G.I., Pearson, K.H., Baldwinson, T. and Sher, E. (2002) 5-Hydroxyindole potentiates human $\alpha 7$ nicotinic receptor-mediated responses and enhances acetylcholine-induced glutamate release in cerebellar slices. Neuropharmacol., 43: 374–384.

CHAPTER 25

Role of calcium binding proteins in the control of cerebellar granule cell neuronal excitability: experimental and modeling studies

D. Gall[1], C. Roussel[1], T. Nieus[2], G. Cheron[3], L. Servais[1,3], E. D'Angelo[2] and S.N. Schiffmann[1,*]

[1]Laboratoire de Neurophysiologie (CP601), Faculté de Médecine, Université Libre de Bruxelles, Route de Lennik 808, B-1070 Brussels, Belgium
[2]Department of Cellular and Molecular Physiology and Pharmacology, University of Pavia and INFM, Via Forlanini 6, I-27100 Pavia, Italy
[3]Laboratoire d'Electrophysiologie, Université de Mons-Hainaut, B-7000 Mons, Belgium

Keywords: calretinin; calcium; binding protein; cerebellar granule cell; excitability; calcium; mathematical model

Abstract: Calcium binding proteins, such as calretinin, are abundantly expressed in distinctive patterns in the central nervous system but their physiological function remains poorly understood. Calretinin is expressed in cerebellar granule cells which provide the major excitatory input to Purkinje cells through parallel fibers. Calretinin deficient mice exhibit dramatic alterations in motor coordination and in Purkinje cell firing recorded in vivo throught unknown mechanisms. In the present paper, we review the results obtained with the patch clamp recording techniques in acute slice preparation. This data allow us to investigate the effect of a null mutation of the calretinin gene on the intrinsic electroresponsiveness of cerebellar granule cells at a mature developmental stage. Calretinin deficient granule cells exhibit faster action potentials and generate repetitive spike discharge showing an enhanced frequency increase with injected currents. These alterations disappear when 0.15 mM of the exogenous fast calcium buffer BAPTA is infused in the cytosol to restore the calcium buffering capacity. Furthermore, we propose a mathematical model demonstrating that the observed alterations of granule cell excitability can be explained by a decreased cytosolic calcium buffering capacity dur to the absence of calretinin. We suggest that calcium binding proteins modulate intrinsic neuronal excitability and may therefore play a role in the information processing in the central nervous system.

Introduction

Calcium regulates a large variety of neuronal functions, including neurotransmitter release, ionic channel permeability, enzyme activity, and gene transcription. Hence, the cytosolic calcium concentration must be tightly regulated. Cytoplasmic calcium binding proteins play a key role in this regulation leading to specific adjustments of neuronal signaling. Among these, calretinin is the only calcium binding protein known to be expressed in cerebellar granule cells (Résibois, and Rogers, 1992; Marini, et al., 1997) whereas the structurally related calbindin

*Corresponding author. Tel.: +32 2 555 6407; Fax: +32 2 555 4121; E-mail: sschiffm@ulb.ac.be

DOI: 10.1016/S0079-6123(04)48025-X

is exclusively expressed in Purkinje cells. Granule cells form the largest neuronal population in the mammalian brain. They process information entering into the cerebellar cortex through the mossy fibers (Ito, 1984) and convey major excitatory afference to Purkinje cells through the parallel fibers. The involvement of cerebellum in motor coordination has long been recognized and, interestingly, it has recently been shown that calretinin deficient mice (Cr−/−) were impaired in motor coordination tests and displayed alterations in Purkinje cell activity recorded in vivo (Schiffmann et al., 1999). As calretinin is not expressed in Purkinje cells the latter study lacked direct evidence for intrinsic cerebellar electrophysiological alterations at the cellular level due to the Cr−/− mutation. This is particularly relevant since granule cells show a calcium-dependent regulation on their intrinsic excitability (Gabbiani et al., 1994; D'Angelo, et al., 1997, 1998, 2001), so that alteration in calcium-buffering is expected to affect action potential generation. Thus, in addition to addressing a question specific to cerebellar physiology, Cr−/− mice may serve as a model for understanding the regulation of calcium dependent control of neuronal excitability. In the present chapter, we review the experimental evidence showing that the electroresponsiveness of granule cells from Cr−/− mice is altered. Furthermore, we present a mathematical model providing a link between the observed alterations in granule cell electroresponsiveness and the decreased cytosolic Ca^{2+} buffering capacity due to the absence of calretinin. Our results suggest a critical role for calretinin in regulating granule cell excitability and signal coding at the input stage of the cerebellum.

Results

Alteration of granule cell intrinsic membrane excitability

Recordings were made using the perforated patch technique in order to minimize unwanted alteration of the endogenous Ca^{2+} buffering capacity. Intrinsic granule cell electroresponsiveness was investigated in current clamp recordings. The resting potential of WT and Cr−/− granule cell was not significantly different (−64.8 ± 5.5 mV, $n=5$ vs. −65.9 ± 5.3 mV, $n=8$, NS). Neither WT nor Cr−/− granule cells generated spontaneous action potentials. Active cell membrane properties were evaluated by measuring the voltage response while injecting steps of depolarizing current of increasing intensities in the granule cell soma. Above a critical value of the injected current (4.4 ± 1.6 pA for WT, $n=5$ vs. 4.9 ± 1.0 pA for Cr−/−, $n=8$, NS) fast repetitive spiking was obtained with a threshold of spike prepotential of −58.0 ± 2.3 mV for WT granule cells ($n=5$) and of −58.7 ± 2.0 mV for Cr−/− granule cells ($n=8$, NS). Action potentials occurred in regular trains showing little or no adaptation and frequency increased with the intensity of the injected current. The average frequency was measured over the whole duration of current injection (1 s) and was used to construct current–frequency plots (Fig. 1A). At low current intensities, the current–frequency plots were interpolated with a straight line. As the threshold current and the maximal frequency varies substantially from cell to cell, the evaluation of the slope factor of the linear part of the current frequency plots was used as a normalized measure of excitability. Using such an analysis, we observed a significant increase in the slope of the current frequency plots (4.8 ± 0.2 Hz pA^{-1} for WT, $n=5$, and 6.6 ± 0.7 Hz pA^{-1} for Cr−/−, $n=8$, $p<0.05$) indicating that the excitability of Cr−/− granule cells was increased. In addition, Cr−/− granule cells showed a 23% decrease in the action potential halfwidth evaluated at the threshold potential where fast repetitive spiking is obtained (1.02 ± 0.06 ms for WT, $n=5$, and 0.78 ± 0.06 ms for Cr−/−, $n=8$, $p < 0.05$). Besides changes observed in Cr−/− mice, it should be noted that the spike shape and the frequency of spike discharge in WT mice were similar to those reported previously (D'Angelo, et al., 1998). The action potential undershoot amplitude was deeper for Cr−/− than WT granule cells, and this change did not reach statistical significance (−4.0 ± 0.7 mV for WT, $n=5$, and −6.4 ± 1.7 mV for Cr−/−, $n=7$, $p=0.2406$, NS). The fact that the latter change fails to be statistically significant probably reflects the higher sensitivity of this parameter towards cell to cell variations compared to more robust temporal parameters like the slope factor of current-frequency relationship or the action potential half-width.

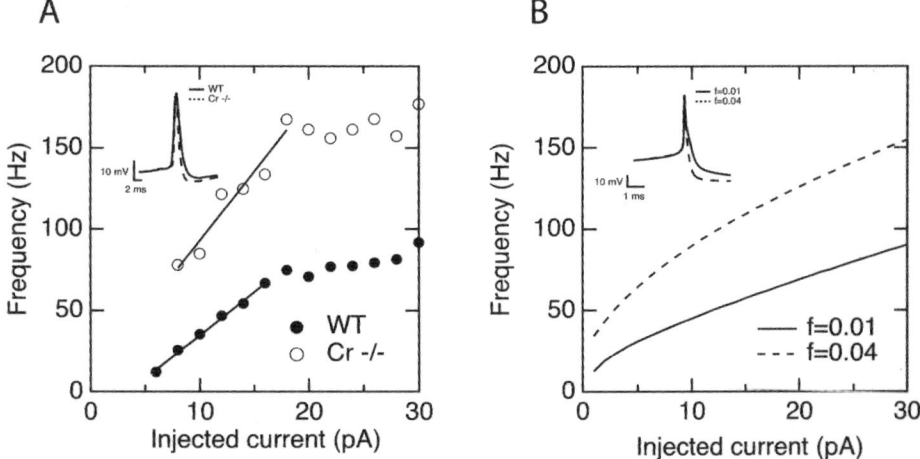

Fig. 1. Evaluation of the intrinsic excitability of a cerebellar granule cell in wild type and Cr−/− mice. Using the perforated patch configuration, the average action potential frequency was measured over the total time of current injection (1 s) at increasing intensities of the injected current. From these data, current-frequency plots could be constructed (A). At low current intensities, granule cells showed a linear encoding of stimulus intensity. The slope of the linear part of current–frequency plots is used as a measure of the intrinsic granule cell excitability. The Cr−/− granule cell showed an increased excitability and faster action potentials compared to the WT (A, inset). Panel (B) shows the effect of a reduction of the Ca^{2+} buffering capacity on the intrinsic excitability of the granule cell model. The increase in the parameter f mimicks the decrease in cytosolic Ca^{2+} buffering capacity due to the absence of calretinin (see text for details). The corresponding current–frequency plots are shown. As observed experimentally, the slope of the current–frequency plot is markedly increased and action potentials are faster (B, inset).

Relationship between calretinin deficiency and altered excitability

To investigate the role of cytosolic Ca^{2+} buffering on intrinsic granule cell excitability, we have used a mathematical model. Calcium dynamics have a profound influence on Ca^{2+}-activated K^+ current (I_{K-Ca}) activation thereby regulating spike discharge. A typical approach to investigate Ca^{2+} dynamics (Traub and Llinas, 1979) is to model Ca^{2+} and I_{K-Ca} and to adapt Ca^{2+} dilution and removal in order to match the firing pattern. This approach has previously been applied to cerebellar granule cell model (Gabbiani et al., 1994; Maex and Schutter, 1998; D'Angelo et al., 2001). Here, in order to focus our attention on Ca^{2+} dynamics, we have reduced the number of gating variables and currents involved in action potential generation. We have also modified the equation governing the Ca^{2+} dynamics to take into account variations in the concentration of endogenous Ca^{2+} buffers. As granule cells have a compact electrotonic structure (Silver et al., 1992; D'Angelo et al., 1993, 1995, 2001) a single compartment model was used. Following the classical Hodgkin and Huxley (1952) approach, the membrane can be considered as a leaky capacitor and the membrane potential dynamics are governed by the current balance equation:

$$C_m \frac{dV}{dt} = -I_{Na} - I_{K-V} - I_{Ca} - I_{K-Ca} \quad (1)$$

where C_m is the cell capacitance, I_{Na} is a voltage-dependent Na^+ current, I_{K-V} a delayed rectifier K^+ current, I_{Ca} a high-threshold voltage dependent Ca^{2+} current, and I_{K-Ca} a Ca^{2+} activated K^+ current. These ionic currents have been shown to be the core of action potential generation in cerebellar granule cells (D'Angelo et al., 1998) when the excitable response has assumed its mature pattern (D'Angelo et al., 1997). The interplay between I_{Na} and I_{K-V} is the basic mechanism giving rise to the action potentials and the presence of I_{Ca} and I_{K-Ca} allows coupling between intracellular calcium dynamics and membrane potential dynamics. The complete expression for the different ionic currents and all parameter values are the same as in Gall et al. (2003). To complete the model, the following balance equation gives the evolution of the free calcium concentration (in μM) in a submembrane shell of thickness d in a cell of surface area A.

$$\frac{dCa}{dt} = f - \left[\frac{I_{Ca}}{2FAd} - \beta_{Ca}Ca\right] \quad (2)$$

where the dimensionless parameter f represents the calcium buffering capacity of the cytosol inside the submembrane shell due to the presence of fast calcium binding proteins (binding is assumed to be instantaneous). In particular, calretinin is a fast calcium buffer, the mean free lifetime for a Ca^{2+} ion in presence of a physiological concentration of the protein being of the order of microseconds (Edmonds et al., 2000), three orders of magnitude faster than the time scale of the evolution of the variables of the system. Therefore any alteration in the calretinin level corresponds to a modification of f. Moreover, as this parameter can be seen as the ratio of *free* Ca^{2+} concentration to *bound* Ca^{2+} concentration (Chay and Keizer, 1983; Gall et al., 1999; Gall and Susa, 1999), a decrease in cytosolic Ca^{2+} buffering capacity due to the absence of calretinin can be mimicked by an increase in the value of f. In the absence of data for cerebellar granule cells, the numerical values used here for the cytosolic Ca^{2+} buffering capacity (in the order of 0.01) have been set in agreement with values reported for other neuronal types (Tatsumi and Katayama, 1993; Stuenkel, 1994; Helmchen et al., 1996), with the notable exception of cerebellar Purkinje cells which show a calcium binding ratio an order of magnitude higher (Fierro and Llano, 1996). The parameter β_{Ca} describes Ca^{2+} removal corresponding to diffusion, action of ionic pumps, and slow buffers. Calcium dynamics were adapted to yield Ca^{2+} transients in the μM range, similar to those reported from Gabbiani et al. (1994). This model should be seen as a minimal model allowing us to qualitatively understand the impact of variations in cytosolic Ca^{2+} buffering capacity on the electroresponsiveness of an excitable cell. Nevertheless, parameter values have been chosen in order to reproduce the relevant aspects of the cerebellar granule cell electroresponsiveness. In our model, calcium buffering capacity, due to the action of fast calcium buffers, is represented by the dimensionless parameter f giving the ratio of *free* Ca^{2+} concentration to *bound* Ca^{2+} concentration (Chay and Keizer, 1983; Gall et al., 1999; Gall and Susa, 1999). A decrease in cytosolic Ca^{2+} buffering capacity due to the absence of calretinin can be simulated by an increase in the value of f. The observed alterations in Cr−/− mice can be mimicked by a four-fold decrease in the cytosolic Ca^{2+} buffering capacity, raising the action potential frequency from 55.1 Hz ($f = 0.01$) to 105.7 Hz ($f = 0.04$) in good agreement with the experimental data (Fig. 1B). The linear slope increase in current frequency plots from 3.0 Hz pA^{-1} for $f = 0.01$ to 5.2 Hz pA^{-1} for $f = 0.04$ is also close to the experimental values. The action potential half-width undergoes a 41% decrease when f is increased from 0.01 to 0.04. This action potential shortening reflects a more pronounced activation of I_{K-Ca} due to faster Ca^{2+} dynamics taking place when the cytosolic Ca^{2+} buffering capacity is decreased. This greater I_{K-Ca} activation also leads to a 28% increase in the action potential undershoot. Therefore the model suggests that faster Ca^{2+} dynamics can fully explain the increased excitability observed in Cr−/− mice through increased activation of Ca^{2+}-activated K^+ current. Similar results (Fig. 2) were obtained with more complex models by Maex and Schutter (1998) and by D'Angelo et al. (2001) modified to include Ca^{2+} dynamics as described by Eq. (2). This demonstrates that the modulation of intrinsic excitability by calcium buffering properties is a robust effect and does not depend strongly on detailed assumptions underlying the different theoretical models. It is important to note that a change in the dynamics leading to the activation of I_{K-Ca} is needed to obtain hyperexcitability in the model. A simple increase in the maximal conductance due to Ca^{2+}-activated K^+ channels ($\bar{g}_{K-Ca}$) leads to the opposite effect, lowering the spike frequency (not shown).

GABA-A antagonist picrotoxin can unmask increased excitability in Cr−/− granule cells

Granule cells are excited by glutamatergic mossy fiber synapses and inhibited by GABAergic Golgi cell synapses. Without GABA-A receptor inhibitors, tonic granule cell inhibition is observed (Armano et al., 2000). Therefore, we have characterized the electroresponsiveness of cerebellar granule cells in the

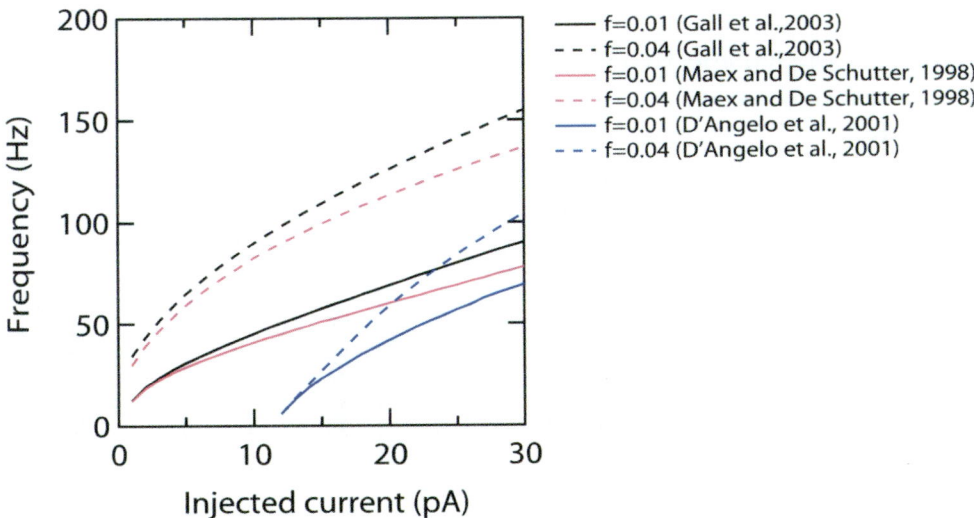

Fig. 2. Effect of a reduction of the Ca^{2+} buffering capacity on the intrinsic excitability in different granule cell models. Current–frequency plots for three granule cell models with a cytosolic Ca^{2+} buffering capacity parameter f of 0.01 (solid lines) and 0.04 (dashed line). All models shows a similar increase in excitability when the calcium beffering capacity is decreased.

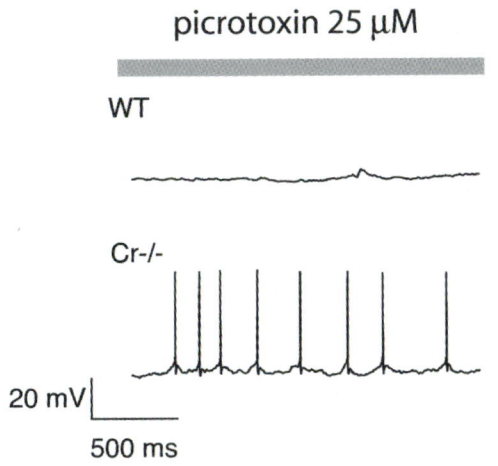

Fig. 3. GABA-A blockade by picrotoxin induces spontaneous firing in Cr−/− granule cells. In the presence of 25 μM picrotoxin, spontaneous electrical activity is seen in Cr−/− granule cells but not in wild type granule cells (WT).

presence of the GABA-A receptor blocker picrotoxin (25 μM). In these conditions, we found that six out of seven Cr−/− granule cells showed spontaneous electrical activity (Fig. 3), but no such activity was present in WT granule cells ($n = 5$, $p < 0.05$, Fischer's test). In addition, an increase in the intrinsic excitability was also observed since the slope of the current–frequency plots was significantly high in mutant mice (3.4 ± 0.6 Hz pA^{-1} for WT, $n = 6$, and 7.3 ± 1.0 Hz pA^{-1} for Cr−/−, $n = 5$, $p < 0.05$). Interestingly, the effects were not detected when the measure was repeated in the presence of 10 μM bicuculline methobromide which is also a GABA-A blocker (Fig. 4). This can be explained by the fact that bicuculline methobromide also acts as a blocker of Ca^{2+}-dependent K$^+$ channels (Seutin and Johnson, 1999). These results, therefore, reinforce the idea that this current is indeed mediating the effect of altered Ca^{2+} buffering capacity on granule cell excitability.

Discussion

In this review, we have shown that the absence of calretinin increases cerebellar granule cell excitability without altering passive electrical membrane properties. In addition, we have used a mathematical model in order to examine whether the observed alterations of Cr−/− cerebellar granule cell electroresponsiveness could be linked to the decreased Ca^{2+} buffering capacity determined by the absence of calretinin. We challenged with a decreased cytosolic Ca^{2+} buffering capacity, the model correctly predicts all the changes in Cr−/− granule cell electroresponsiveness that are

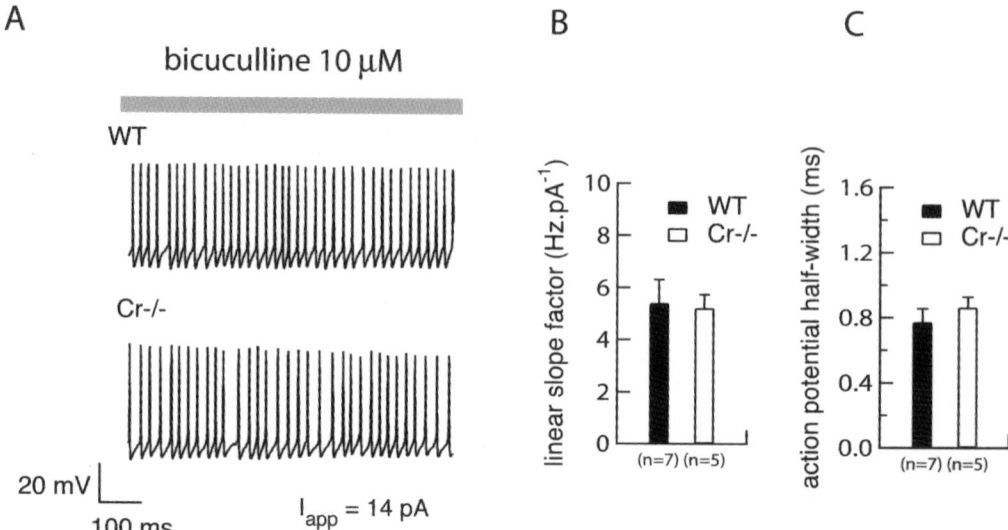

Fig. 4. No change in the intrinsic excitability of the granule cell in presence of bicuculline. In presence of 10 μM bicuculline methobromide, the Cr−/− granule cell show similar action potential frequency (A, bottom) compared to the WT (A, top) cell for the same injected current from a holding potential of −80 mV. Histograms report slope factor of the current–frequency plots for WT vs. Cr−/− (B) and corresponding durations of action potentials at half-amplitude (C) evaluated at the threshold potential where fast repetitive spiking is obtained. In these conditions, Cr−/− granule cells ($n = 7$) do not show a significantly increased excitability of faster action potentials compared to WT ($n = 5$).

observed experimentally. Our model should not be seen as a complex model of the granule cell like the one proposed by D'Angelo et al. (2001) but rather as an abstract model focusing on a specific cell property, cytosolic Ca^{2+} buffering. The purpose of this model is to demonstrate the basic mechanism linking alteration of the intrinsic excitability in Cr−/− mice to alterations in the cytosolic Ca^{2+} buffering capacity. Thus, on a broader perspective, the conclusions drawn from our simulations can be applied to other neuronal types provided that the mechanisms of excitability are essentially the same as in cerebellar granule cells and that the conductance of the Ca^{2+}-activated K^+ channels is sufficient to obtain a strong coupling between excitability and Ca^{2+} dynamics during the spike generation. Whereas it is general knowledge that increasing Ca^{2+} activated K^+ current slows down the firing rate, faster Ca^{2+} dynamics through reduced Ca^{2+} buffering has the opposite effect. The model suggests that this is due to direct control of I_{K-Ca} by the Ca^{2+} transient, speeding-up spike repolarization when the calcium buffering capacity is decreased. In this view, in addition to their obvious role in Ca^{2+} homeostasis, Ca^{2+}-binding proteins could play an active role in modulating neuronal intrinsic excitability and therefore neuronal plasticity. Although information storage is usually believed to be mediated by long-term modifications in the strength of synaptic transmission activity-dependent changes in the neuronal intrinsic excitability also take place, causing forms of nonsynaptic plasticity. Such activity-dependent changes in intrinsic excitability have been shown to occur in cerebellar granule cells (Armano et al., 2000) and neurons of the deep cerebellar nuclei (Aizenman and Linden, 2000). It would be interesting to know whether activity-dependent modifications in the localization or in the level of expression of Ca^{2+} binding proteins are involved. Whereas expression changes are still controversial (see review by Baimbridge et al. (1992)), changes in the localization of calretinin have been shown to occur in neurons (Hack et al., 2000).

In conclusion, calretinin deficiency increases the intrinsic excitability of cerebellar granule cells. The increased granule cell electroresponsiveness may explain the electrophysiological and behavioral alterations observed in vivo on alert Cr−/− mice, indicating the critical role of granule cells in the information processing in the cerebellar cortex. On a broader perspective, we suggest that modulation of neuronal excitability by Ca^{2+} binding proteins could play a functional role in the control of information coding and storage in the central nervous system.

Acknowledgments

David Gall is postdoctoral Researcher at the Belgian Fonds National de la Recherche Scientifique (FNRS). Celine Roussel is supported by the Belgian Fonds pour la Recherche dans l'Industrie et l'Agriculture (FRIA). This work was also supported by the Queen Elisabeth Medical Foundation (FMRE-Neurobiology 02–04), Fund for Medical Scientific Research (FRSM-Belgium 3.4507.02), Fondation David et Alice Van Buren, Action de Recherche Concertee (2002–2007) and by the European Commission's 5th Framework Program for research (CEREBELLUM BIO4CT98-0182 and SPIKEFORCE IST35271).

References

Aizenman, C.D. and Linden, D. (2000) Rapid, synaptically driven increase in the intrinsic excitability of cerebellar deep nuclear neurons. Nat. Neurosci., 3: 109–111.

Armano, S., Rossi, P., Taglietti, V. and D'Angelo, E. (2000) Long term potentiation of intrinsic excitability at the excitability at the mossy fiber-granule cell synapse of rat cerebellum. J. Neurosci., 20: 5208–5216.

Baimbridge, K.G., Celio, M. and Rogers, J. (1992) Calcium-binding proteins in the nervous system. Trends Neurosci., 15: 303–308.

Chay, T.R. and Keizer, J. (1983) Minimal model for membrane oscillations in the pancreatic β-cell. Biophys. J., 42: 181–190.

D'Angelo, E., Filippi, G.D., Rossi, P. and Taglietti, V. (1995) Synaptic excitation of individual rat cerebellar granule cells in situ: evidence for the role of NMDA receptors. J. Physiol. (Lond.), 484: 397–413.

D'Angelo, E., Filippi, G.D., Rossi, P. and Taglietti, V. (1997) Synaptic excitation of Ca^{2+} action potentials in immature rat cerebellar granule cells in situ. J. Neurophysiol., 78: 1631–1642.

D'Angelo, E., Filippi, G.D., Rossi, P. and Taglietti, V. (1998) Ionic mechanism of electroresponsiveness in cerebellar granule cells implicates the action of a persistent sodium current. J. Neurophysiol., 80: 493–503.

D'Angelo, E., Nieus, T., Maffei, A., Armano, S., Rossi, P., Taglietti, V., Fontana, A. and Naldi, G. (2001) Theta-frequency bursting and resonance in cerebellar granule cells: experimental evidence and modeling of a slow K^+-dependent mechanism. J. Neurosci., 21: 759–770.

D'Angelo, E., Rossi, P. and Tagiletti, V. (1993) Different proportions of N-methyl-D-aspartate and non-N-methy-D-aspartate receptor currents at the mossy fibre-granule cell synapse of developing rat cerebellum. Neuroscience, 53: 121–130.

Edmonds, B., Reyes, R., Schwaller, B. and Roberts, W. (2000) Calretinin modifies presynaptic calcium signaling in frog saccular hair cells. Nat. Neurosci., 80: 2521–2537.

Fierro, L. and Llano, I (1996) High endogenous calcium buffering in Purkinje cells from rat cerebellar slices. J. Physiol., 496: 617–625.

Gabbiani, F., Mitgaard, J. and Knoepfel, T. (1994) Synaptic integration in a model of cerebellar granule cells. J. Neurophysiol., 72: 999–1009.

Gall, D., Gromada, J., Susa, I., Herchuelz, A., Rorsman, P. and Bokvist, K. (1999) Significance of Na/Ca exchange for Ca^{2+}-buffering and electrical activity in mouse pancreatic β-cells. Biophys. J., 76: 2018–2028.

Gall, D., Roussel, C., Susa, I., D'Angelo, E., Rossi, P., Bearzatto, B., Galas, M.-C., Blum, D., Schurmans, S. and Schiffmann, S. (2003) Altered neuronal excitability in cerebellar granule cells of mice lacking calertinin. J. Neurosci., 23: 9320–9327.

Gall, D. and Susa, I. (1999) Effect of Na/Ca exchange on plateau fraction and $[Ca]_i$ in models for bursting in pancreatic β-cells. Biophys. J., 77: 45–53.

Hack, N.J., Wride, M., Charters, K., Kater, S. and Parks, T. (2000) Developmental changes in the subcellular localization of calretinin. J. Neurosci., 20: RC67.

Helmchen, F., Imoto, K. and Sakmann, B. (1996) Ca^{2+} buffering and action potential-evoked Ca^{2+} signaling in dendrites of pyramidal neurons. Biophys. J., 70: 1069–1081.

Hodgkin, A.L. and Huxley, A. (1952) A quantitative description of membrane current and its application to conduction and excitation in nerve. J. Physiol., 117: 500–544.

Ito, M. (1984) Granule cells. In: The Cerebellum and Neuronal Control. Raven, New York, pp. 74–93.

Maex, R. and Schutter, E.D. (1998) Synchronization of Golgi and granule cell firing in a detailed network of the cerebellar granule cell layer. J. Neurophysiol., 80: 2521–2537.

Marini, A.M., Strauss, K. and Jacobowitz, D. (1997) Calretinin containing neurons in rat cerebellar granule cell cultures. Brain Res. Bull., 42: 279–288.

Résibois, A. and Rogers, J. (1992) Calretinin in rat brain: an immunohistological study. Neurosci., 46: 101–134.

Schiffmann, S., Cheron, G., Lohof, A., D'Alcantara, P., Meyer, M., Parmentier, M. and Schurmans, S. (1999) Impaired motor coordination and Purkinje cell excitability in mice lacking calretinin. Proc. Natl. Sci. USA, 96: 5257–5262.

Seutin, V. and Johnson, S. (1999) Recent advances in the pharmacology of quaternary salts of bicuculline. Trends Pharmacol. Sci., 20: 268–270.

Silver, R.A., Traynelis, S. and Cull-Candy, S. (1992) Rapid time course of miniature and evoked excitatory currents at cerebellar synapses in situ. Nature, 355: 163–166.

Stuenkel, E.L. (1994) Regulation of intracellular calcium and calcium buffering properties of rat isolated neurohypophysial nerve endings. J. Physiol., 481: 251–271.

Tatsumi, H. and Katayama, Y. (1993) Regulation of the intracellular free calcium concentration in acutely dissociated neurons from rat nucleus basalis. J. Physiol., 464: 165–181.

Traub, R.D. and Llinas, R. (1979) Hippocampal pyramidal cells: significance of dendritic ionic conductances for neuronal functioning and epileptogenesis. J. Neurophysiol., 42: 476–496.

CHAPTER 26

Between in and out: linking morphology and physiology of cerebellar cortical interneurons

J.I. Simpson[1,]*, H.C. Hulscher[1], E. Sabel-Goedknegt[2] and T.J.H. Ruigrok[2]

[1]*Department of Physiology & Neuroscience, NYU Medical School, 550 First Avenue, New York, NY 10016, USA*
[2]*Department of Neuroscience, Erasmus MC, 3000 DR Rotterdam, The Netherlands*

Keywords: interneurons; juxtacellular labeling; Golgi cell; unipolar brush cell; flocculus; vestibulo-ocular reflex; nodulus; uvula

Abstract: We used the juxtacellular recording and labeling technique of Pinault (1996) in the uvula/nodulus of the ketamine anesthetized rat in an attempt to link different patterns of spontaneous activity with different types of morphologically identified cerebellar cortical interneurons. Cells displaying a somewhat irregular, syncopated cadence of spontaneous activity averaging 4–10 Hz could, upon successful entrainment and visualization, be morphologically identified as Golgi cells. Spontaneously firing cells with a highly or fairly regular firing rate of 10–35 Hz turned out to be unipolar brush cells. We also found indications that other types of cerebellar cortical neurons might also be distinguished on the basis of the characteristics of their spontaneous firing.

Comparison of the interspike interval histograms of spontaneous activity obtained in the anaesthetized rat with those obtained in the awake rabbit points to a way whereby the behaviorally related modulation of specific types of interneurons can be studied. In particular, the spontaneous activity signatures of Golgi cells and unipolar brush cells anatomically identified in the uvula/nodulus of the anaesthetized rat are remarkably similar to the spontaneous activity patterns of some units we have recorded in the flocculus of the awake rabbit. The spontaneous activity patterns of at least some types of cerebellar interneurons clearly have the potential to serve as identifying signatures in behaving animals.

Introduction

In addition to the basic cerebellar circuitry composed of mossy and climbing fibers and granule and Purkinje cells, at least five types of interneurons are found in the cerebellar cortex: i.e., Golgi-, basket-, stellate-, Lugaro-, and unipolar brush cells, most of which were described by Ramón y Cajal (1911). Apart from their distinctive morphologies, various types of interneurons can be distinguished on immunohistochemical grounds (Dino et al., 1999; Geurts et al., 2001; Nunzi et al., 2002) and all but the unipolar brush cells have been shown to be inhibitory on the basis of physiology or the presence of inhibitory neurotransmitters (Eccles et al., 1966; Aoki et al., 1986). Recently, there has been a resurgence of interest in these interneurons, acknowledging that their function is essential for cerebellar operations (Laine and Axelrad, 1998; Dieudonne, 2001; Jorntell and Ekerot, 2002; Mann-Metzer and Yarom, 2002; Jorntell and Ekerot, 2003).

Here, we have studied interneurons of the vestibulocerebellum of the rat and the rabbit. In the anesthetized rat, the juxtacellular labeling technique, developed by Pinault (1994, 1996), was used in an attempt to link different patterns of spontaneous activity with different types of morphologically identified

*Corresponding author. Tel.: +1-212-263-5428; Fax.: +1-212-689-9060; E-mail: john.simpson@med.nyu.edu

interneurons. Moreover, by extending the associations found for the rat to single cell recordings made in the flocculus of the awake, behaving rabbit, we have obtained some insights into the patterns of activity of especially the Golgi cell and the unipolar brush cell (UBC) during compensatory eye movements.

Experimental procedures

Juxtacellular technique

Male Wistar rats were anesthetized with an intraperitoneal injection of a cocktail of thiazine-hydrochloride (3 mg/kg) and ketamine (100 mg/kg) and placed in a stereotactic frame according to Paxinos and Watson (1998). The vestibulocerebellum was accessed as described previously (Ruigrok, 2003). To enhance stability of the recordings, the exposed part of the lower brain stem and cerebellum were covered with 2% agar in saline. Subsequently, fine-tipped glass pipettes (tip diameter < 1 μm impedance 8–20 MΩ) filled with 2–3% Neurobiotin (Vector Laboratories, Burlingame, CA) in 0.5 M NaCl were lowered into the cerebellar cortex, while making extracellular recordings of spontaneous activity of non-Purkinje cells using an intracellular recording amplifier (IR 283, Neuro Data Instruments Corp.) followed by an extracellular amplifier (Cyberamp 380, Axon Instruments). Relevant data was digitized and stored on videotape (Neuro-Corder, DR 484, Neuro Data). Off-line analysis was performed with the Spike2 software package (CED, Cambridge UK).

The spontaneous activity of well-isolated units was generally recorded for at least 60 s after which entrainment of their firing frequency was attempted by passing gradually increasing positive current pulses (up to 10 nA, 200 ms on − 200 ms off) through the pipette. These pulses were usually superimposed on a small negative continuous current of −0.5 nA (Fig. 1). Modulation of a unit's firing frequency was usually obtained suddenly, after which current pulses ranging from 1 to 5 nA were delivered to maintain a good entrainment for a period of at least 3 and up to 10 min. During the period of entrainment, a transfer of the Neurobiotin to the recorded cell is thought to take place (Pinault, 1996). After terminating the pulses, the pipette was very slowly withdrawn from the cell in order to reduce the chance of injury discharges. After several staining attempts (usually one per track and up to a maximum of 12/rat), the rats were deeply anesthetized and perfusion-fixed with paraformaldehyde (4% in 0.05 M phosphate buffer, pH 7.4). The brains were removed, embedded in gelatin, sectioned at 80 μm and processed for ABC histochemistry. Serially mounted sections were counterstained with thionin and examined and photographed with a Leica DMR microscope equipped with a digital camera (Leica DC-300). In some cases (e.g. see Fig. 2C), a 3-D reconstruction of the labeled neurons was made using Neurolucida software (MicroBrightField Inc., Colchester, VT).

Interneuron recordings in the flocculus of the behaving rabbit

Two Dutch belted rabbits were prepared for chronic recording from the flocculus of the cerebellum, as described by De Zeeuw and colleagues (De Zeeuw et al., 1995). The awake rabbit was placed in the prone position on a vestibular turntable and the head was centered on the vertical axis of rotation. The rabbit's head and body were moved together using sigmoidal displacement of the turntable about the vertical axis. The displacement had an amplitude of 2.5–7.5°, occurred in less than 1 s, and was presented both in the light and in the dark. The direction of the sigmoidal displacement was reversed in direction after about a 2 s stationary period. The position of the eye ipsilateral to the recorded flocculus was measured using the search coil technique. In addition to recording from Purkinje cells, identified by the presence of complex spike activity and an ensuing pause in simple spike activity, we recorded from other neurons that did not show complex spike activity. The neuronal responses were averaged as peristimulus time histograms compiled from 15 to 60 saccade-free cycles.

Results

Evaluation of entrainment success and quality of labeling

An initial series of six rats, in which 68 entrainment attempts had been made, was fully analyzed to obtain

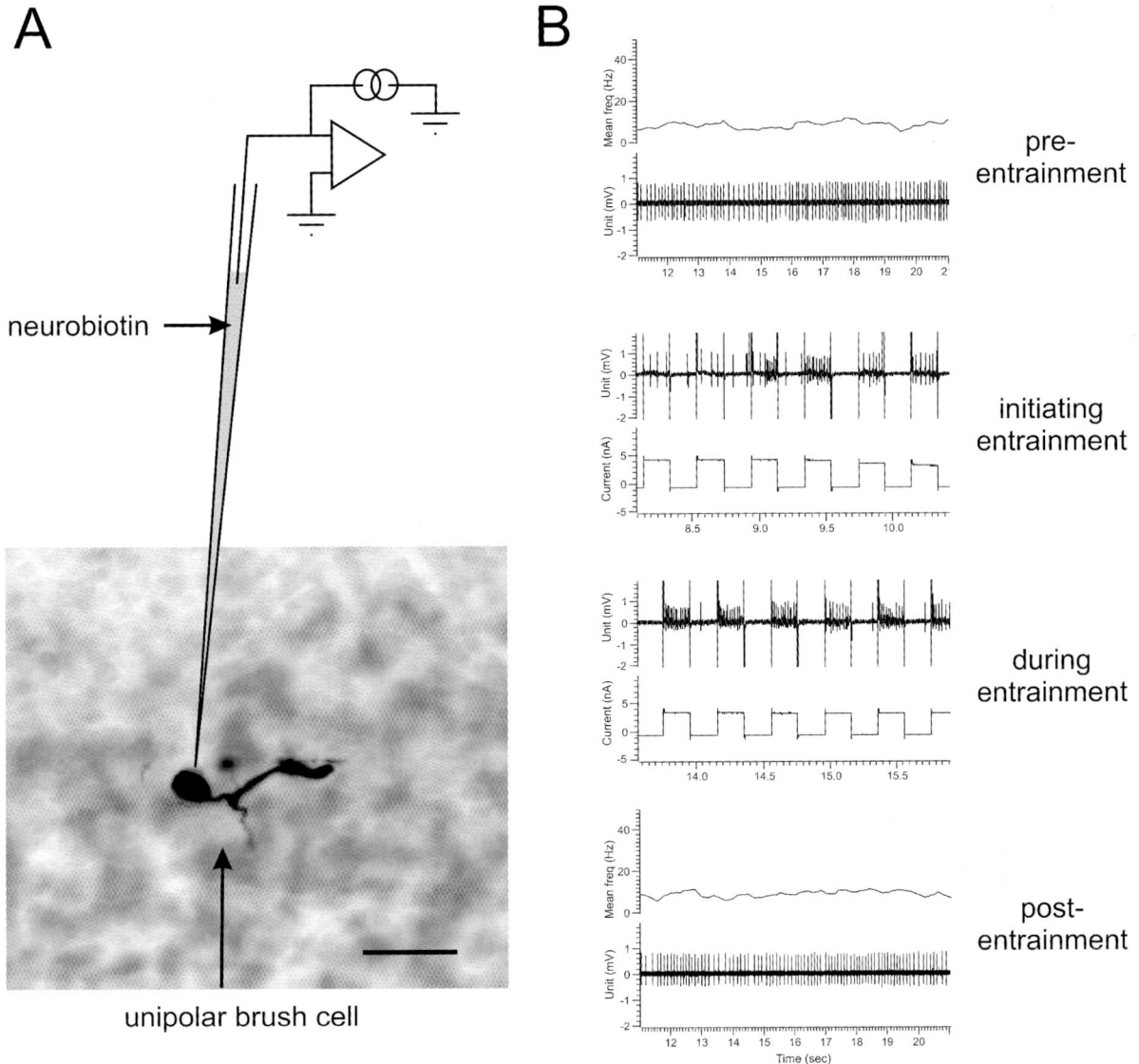

Fig. 1. Application of the juxtacellular labeling technique of Pinault (1996) to cerebellar cortical interneurons. The steps involved in using this identification technique are shown for a UBC seen labeled in A (scale bar 25 μm) with a superimposed drawing of a glass microelectrode filled with 2–3% Neurobiotin in 0.5 M NaCl. The microelectrode was used to pass both current and Neurobiotin into the juxtacellularly recorded neuron. Prior to attempting to modulate the neuronal activity, the electrode was brought close enough to the neuron so as to record the spontaneous activity as biphasic (positive–negative) spikes, as seen in the top panel of B for 10 s of the raw activity (bottom trace) together with the corresponding mean firing frequency (moving average over 0.5 s, top trace). Entrainment of the neuronal activity was initiated by applying 200 ms positive pulses (50% duty cycle) of gradually increasing amplitude (in this case up to approximately 5 nA) from a slightly negative base current (−0.5 nA) until the neuron rather suddenly started to modulate its activity in response to the current pulses (second panel in B). At the moment of initiation of entrainment the pulse amplitude was quickly reduced to prevent damage or loss of the neuron as a consequence of excessive activity. The pulse amplitude was subsequently increased to obtain a more or less stable level of modulated activity (third panel in B), which was maintained for 3–10 min. The spontaneous activity recorded after cessation of the current pulses and prior to gradually withdrawing the microelectrode is shown in the bottom panel of B.

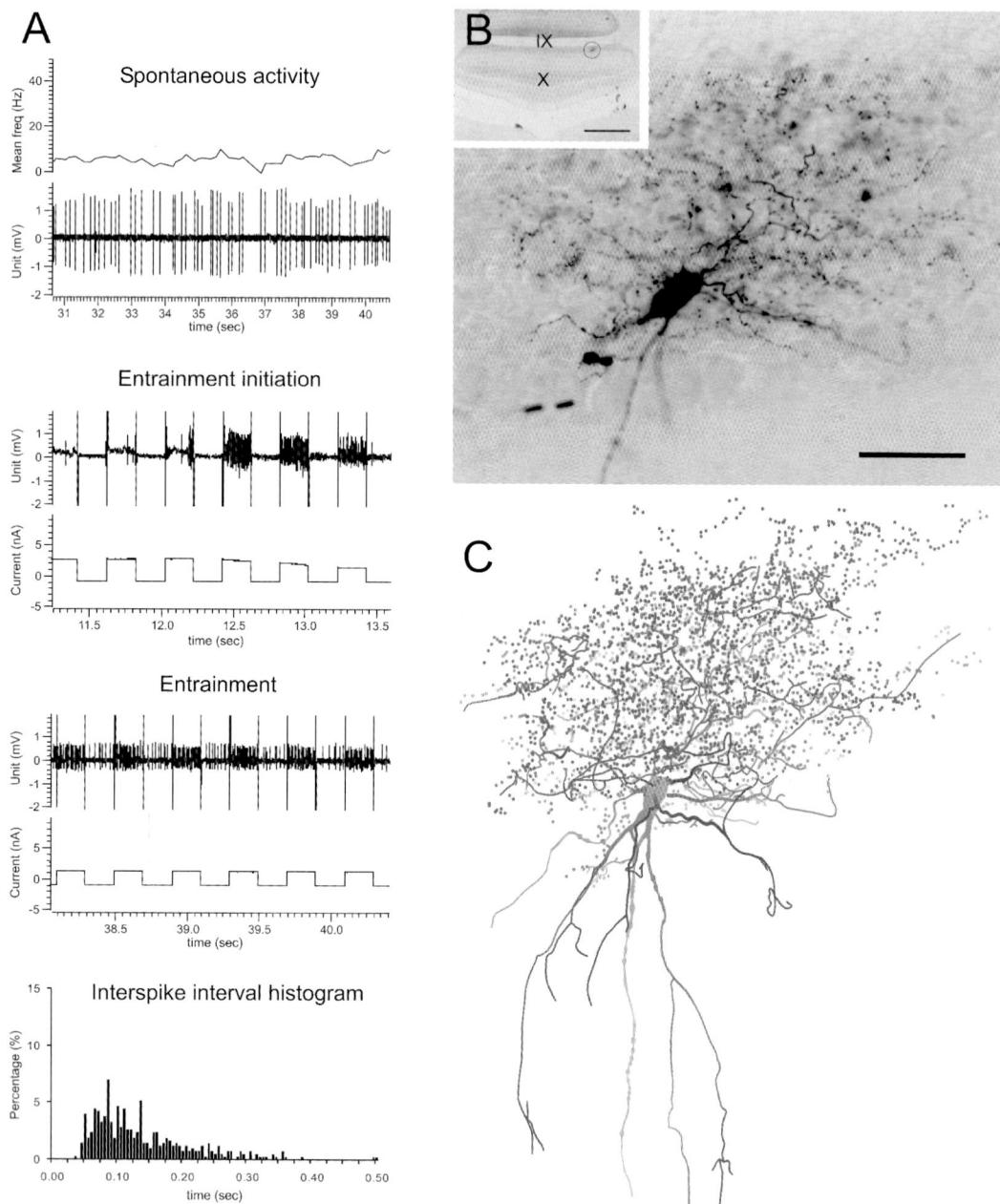

Fig. 2. Juxtacellular labeling of a representative Golgi cell. The top three panels in A illustrate the procedure, as described for the corresponding panels in Fig. 1. The bottom panel in A is the interspike interval (ISI) histogram (bin width: 5 ms) based on 60 s of spontaneous activity recorded prior to attempting entrainment. The histogram has been normalized as a percentage of the total number of intervals ($n = 429$) to facilitate comparison with the ISI histograms of other units. Note that intervals longer than 500 ms are not incorporated in the histogram. The general shape of the histogram, displaying a wide peak around 100 ms and the near absence of intervals shorter than about 50 ms, is characteristic of the large majority of the juxtacellularly labeled Golgi cells. The morphology of the labeled Golgi cell is shown in B (scale bar 50 μm) and the location of the cell in the ventral uvula (lobule IX-c) is shown at a lower magnification in the inset in B (scale bar 1 mm). The partial reconstruction of this cell is shown in C, where the dots indicate the axonal varicosities. The image of the reconstruction is rotated somewhat with respect to the view shown in B.

a measure of the reliability of the technique. Every passage of positive current pulses, however short, was considered an attempt. Eight attempts resulted in a well-labeled interneuron, meaning that the soma and fine dendritic and sometimes the axonal arborizations were visualized. In 30 additional attempts, the fine dendritic branches were not labeled, but the contour and the location of the cell body and proximal dendrites enabled identification of the recorded unit. The remaining 30 attempts were either not recovered ($n=11$) or resulted in only a more or less circumscribed spot of reaction product. In the combined categories of well-labeled and identifiable neurons, nine were classified as Golgi cells, seven as basket cells and five as stellate cells. Furthermore, five UBCs as well as five granule cells were recognized. In one case a Purkinje cell was recovered and in two cases an incomplete filling of a mossy fiber was observed. The remainder of the cells turned out to be within the brain stem ($n=4$). Our initial analysis of the spontaneous activity characteristics of the labeled neurons revealed that the interneurons in the molecular layer usually displayed a highly irregular firing pattern, with many interspike intervals (ISIs) shorter than 25 ms. The Golgi cells, on the other hand, only occasionally displayed ISIs shorter than 50 ms and their ISI histogram usually possessed a broad peak around, or somewhat above, 100 ms. In the area we recorded from, the uvula/nodulus, we found that the unipolar brush cells (UBCs) of the granule cell layer were mostly characterized by a fairly regular firing rhythm with mean frequencies varying between 10 and 35 Hz. Some indications were found that granule cells might show a considerably more intermittent type of activity than UBCs, with periods of silence of up to tens of seconds.

In the remaining pages of this chapter we focus on some of our results from the Golgi cells and UBCs obtained from the initial six rats as well as from subsequent experiments on rats and rabbits.

Spontaneous firing of Golgi cells and UBCs in the vestibulocerebellum of the ketamine anesthetized rat

Figure 2A shows the spontaneous activity pattern typical of a class of cells recorded in the granule cell layer. These cells had a rather slow, somewhat irregular, firing with an average discharge rate of 4–10 Hz. Their interspike interval histogram (ISI-histogram) possessed a rather broad peak around an interval of 100 ms and had few intervals less than 50 ms. The unit shown in Fig. 2 had an average firing frequency of 7.2 Hz (measured for a period of 60 s). It was successfully entrained and then modulated for 5 min. After cessation of modulation, the spontaneous activity was markedly increased to 35 Hz for about 15 s, but then the original activity pattern came back to an average firing frequency of 8.0 Hz at 90 s after cessation of the pulses. The subsequent detachment was successful without any sign of injury discharge.

Histological evaluation revealed, at the appropriate depth, a well-labeled cell displaying the morphological features of a classic Golgi cell as described by Ramón y Cajal (1911). The intricacies of the axonal arborizations, however, were extremely delicate and could not be reconstructed fully. Rather, the 3D-reconstruction was based on the soma, dendrites, proximal part of the axon and axonal varicosities (Fig. 2C). Similar results, that is, an ISI-histogram peaking around 100 ms with no more than a few intervals less than 50 ms and with a mean firing frequency between 4–10 Hz, always resulted in the labeling of a large cell located within the granule cell layer ($n=13$), usually with the morphological characteristics of the classical Golgi cell ($n=7$) as shown in Fig. 2B. The other cells ($n=6$) were also likely to be Golgi cells, but because of an incomplete filling or because they were variants of the classical morphology (Ramón y Cajal, 1911; Geurts et al., 2001), it was not possible to identify them further. Nevertheless, we conclude that the somewhat irregular, syncopated firing pattern, resulting in the above mentioned features of the ISI histogram is characteristic for spontaneous firing of Golgi cells in the ketamine anesthetized rat.

Another distinctive spontaneous firing pattern encountered in the granule cell layer of the vestibulocerebellum consisted of unit discharge at a continuous, fairly high, and regular rate (varying among units from 10 to 35 Hz, $n=12$). Typical units displayed an ISI histogram peaking at around 50 ms, with a sharp rising phase and a slower falling phase. Labeling revealed that all these units were UBCs. They had a rather small cell body (8–10 μm in diameter) that possessed only a single, short and

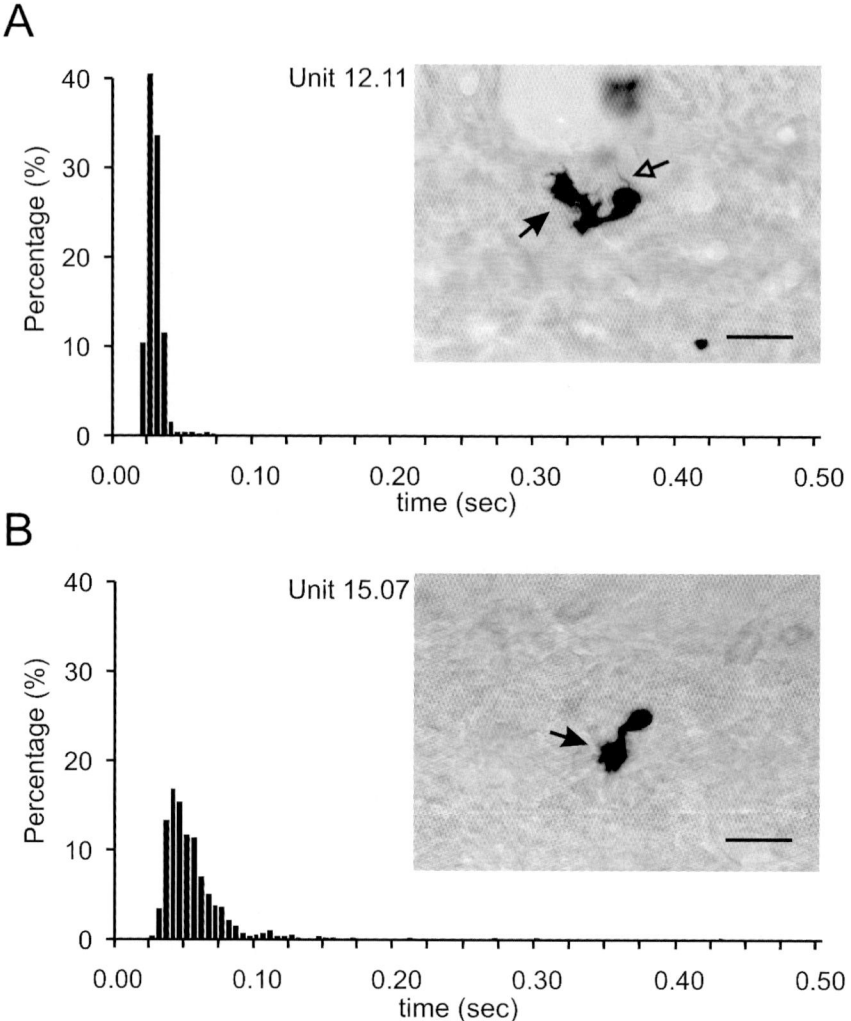

Fig. 3. Two juxtacellularly labeled UBCs and the normalized ISI histograms of their spontaneous activity. Both cells had a relatively regular pattern of firing; the two histograms depict the range found in the shapes of the UBC histograms. Unit 12.11 (A) was a very regular and fast firing unit (mean frequency 32.8 Hz). The histogram is based on 1964 consecutive intervals recorded in 60 s. The ISI histogram of unit 15.07 (B, mean frequency 17.7 Hz) was based on 707 intervals recorded in 40 s. The solid arrows in the photomicrographs indicate the brush of the UBC. The open arrow in A indicates the axon. The photomicrograph in B is a composite based on two adjacent sections. Scale bar 25 μm.

thick dendrite that ended in a wide 'brush-like' tuft from which many small protrusions or dendrioles emanated (Mugnaini and Floris, 1994). In a few cases, the axon could be seen leaving the cell soma (Fig. 3A) and was found to terminate in the same lobule in a number of rosette-like swellings as described earlier for UBCs (Dino et al., 2000). Figure 3 shows the ISI-histograms of two of these UBCs. The UBC of Fig. 3A, located in the ventral part of lobule IX-c, showed an exceptionally constant firing rate averaging 32.8 Hz. The cell of Fig. 3B was found in the dorsal part of the nodulus and averaged a rate of 17.7 Hz. However, in several other instances ($n=4$, not shown), more irregularly firing units with slower average frequencies (1–6 Hz) were also identified as UBCs. Hence, we conclude that the units displaying a fairly constant and regular rate averaging 10 to 35 Hz are likely to represent UBCs, but that not all UBCs always display such a spontaneous firing pattern.

Activity of floccular units in the awake rabbit with spontaneous activity signatures like those of Golgi cells and UBCs

The observation that in the flocculus of the awake rabbit there are several distinctive patterns of spontaneous activity of non-Purkinje cells motivated the use of the juxtacellular labeling method of Pinault to see if any of these patterns are associated with distinct types of neurons. Two patterns in particular were readily recognizable. The first consisted of a low frequency syncopated pattern that reminded some (including JIS) of castanets played in flamenco music. The second was a very regular firing with an occasional skipped spike that was reminiscent (again only to some) of the sound of a slow motorboat.

The patterns of spontaneous activity recorded in the uvula/nodulus of ketamine anesthetized rats were in several respects remarkably similar to the activity of units that were recorded in the flocculus of the awake rabbit. However, in contrast to the situation in the rat, in the awake rabbit the response of these units to natural stimuli could be investigated. Here, we have selected for illustration three floccular units from which both the spontaneous firing patterns as well as the responses to vestibular stimulation were recorded. Comparison of the spontaneous activity patterns of units in the rabbit flocculus with the associations obtained in the rat enables us to suggest which particular interneuron types were recorded in the behaving rabbit. Ultimately, the juxtacellular identification method should be used in the awake animal, but comparison of ISI histograms between anesthetized and awake animals may provide an indication of the way some types of interneurons function in the control of compensatory eye movements.

Figure 4 shows the characteristics of a unit we consider to be a candidate Golgi cell on the basis of its spontaneous firing pattern. Its firing frequency averaged 8.2 Hz and, more importantly, its ISI histogram displayed a wide peak around 100 ms and virtually no interspike intervals less than 50 ms (Fig. 4A). With vestibular stimulation, consisting of sigmoidal displacement of the animal in the horizontal plane in the dark, compensatory eye movements were produced, as can be seen in the upper panel of Fig. 4B. Under these circumstances the gain of the compensatory movements was somewhat less than 0.5. The response of the unit was predominantly a strong decrease in activity during movement of the head towards the side contralateral to the flocculus from which the cell was recorded (and coinciding with the movement of the eye to the recorded side). Since the pattern quickly returned to nearly the spontaneous level when the eye and head were not moving, this candidate Golgi cell was mainly sensitive to velocity and not to position.

Figure 5 shows the features of a unit we consider to be a candidate UBC. Its spontaneous firing was characterized by a tonic, rather high frequency level that did not show much variation. Note that the ISI histogram of spontaneous activity (averaging 19.0 Hz) resembles the histogram of the unit shown in Fig. 3B. Both histograms show a rather narrow peak at an interval of approximately 50 ms, with a rather sharp ascent and a somewhat more gradual decline. The sigmoidal head and table movement in the light resulted in a high gain compensatory eye movement. In contrast to the response of the candidate Golgi cell shown in Fig. 4, this particular unit showed a marked position-sensitive response to the vestibular stimulation. Movement of the table to the side opposite to the recorded flocculus resulted in a conspicuous depression of activity. Interestingly, the depression of activity did not begin until more than 200 ms after the eye movement began. Such a delayed modulation has not been seen in the simple spike modulation of Purkinje cells.

That a delayed response pattern may be a salient general feature of some floccular UBCs is seen in Fig. 6, which shows the response of another cell recorded in another awake rabbit. This candidate UBC was characterized by an extremely stable level of spontaneous activity (averaging 19.7 Hz) resulting in an ISI histogram very much like the one shown in Fig. 3A, although that unit had a considerably higher average firing frequency. From Fig. 6B it can be seen that this unit also had a delayed modulation of activity, but here the position-related activity was depressed when the head was moved towards the recorded side. The cell illustrated in Fig. 6 had rather less velocity sensitivity than the one shown in Fig. 5, although in both cases it was asymmetric.

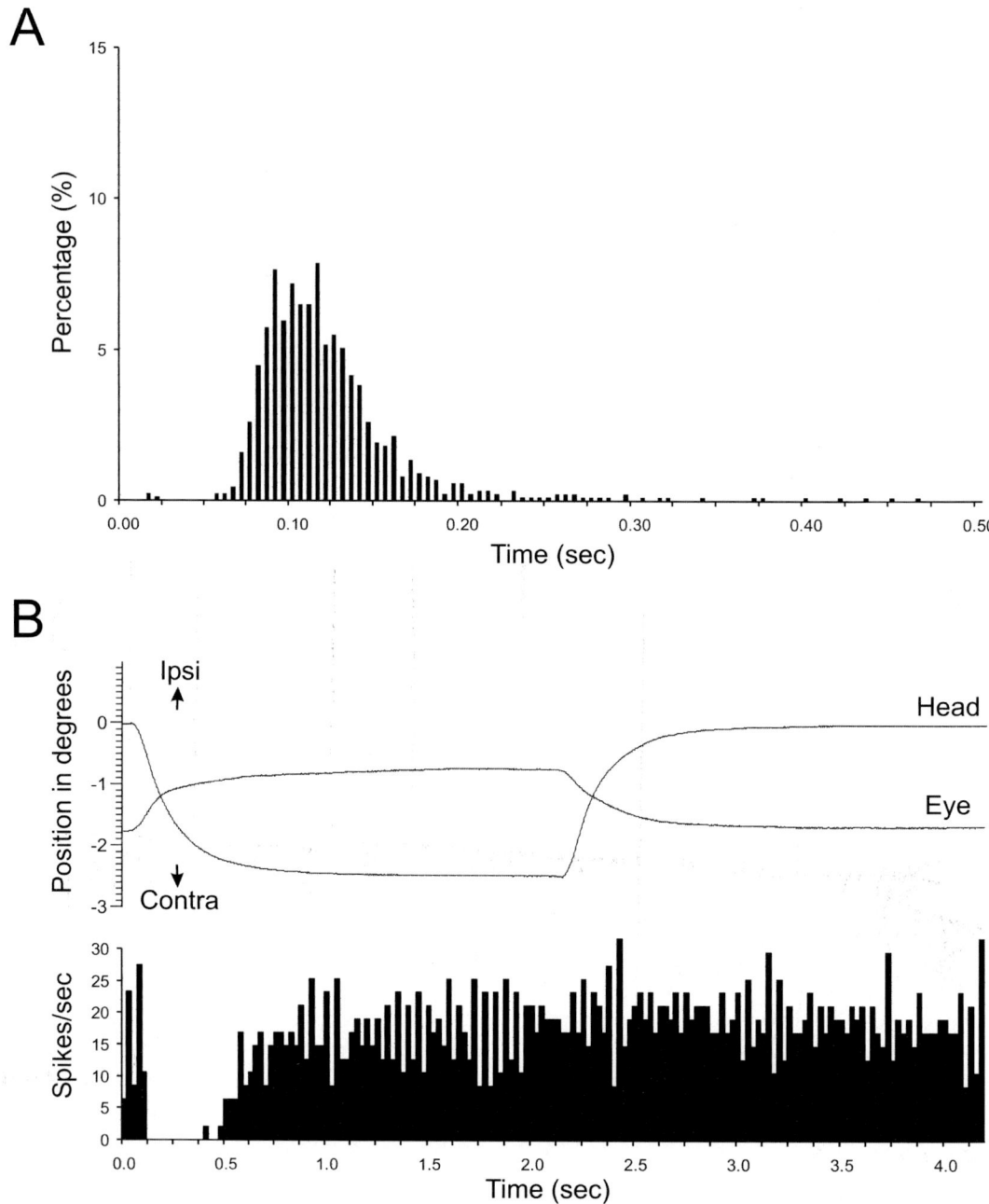

Fig. 4. Response to vestibular stimulation of a candidate Golgi cell recorded in the flocculus of an awake rabbit. This cell is considered to be a Golgi cell on the basis of the strong similarity of its normalized ISI histogram of spontaneous activity to those found for identified Golgi cells in the anesthetized rat (see Fig. 2). The histogram in A is based on 892 intervals recorded in 109 s. The vestibular stimulation shown in B consisted of a sigmoidal displacement in the dark of the head and body of the rabbit. The compensatory eye movement recorded with a scleral search coil is also shown in the top panel of B. The response of this cell is shown in the bottom panel of B as a peristimulus histogram compiled from 19 presentations of the periodic sigmoidal displacement (bin width 25 ms). Note that the response is strongly asymmetric and reflects predominantly velocity.

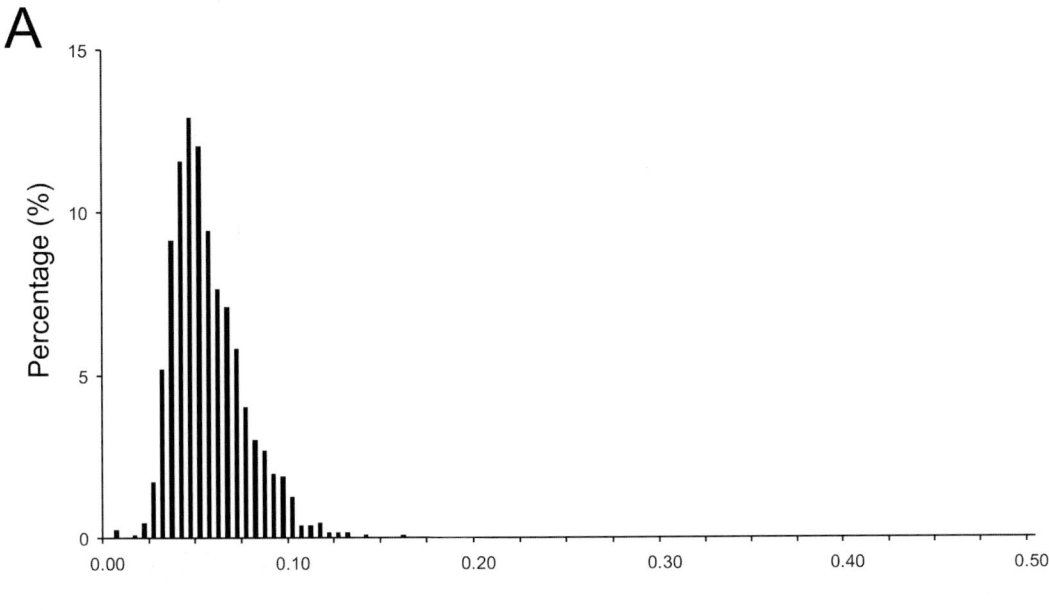
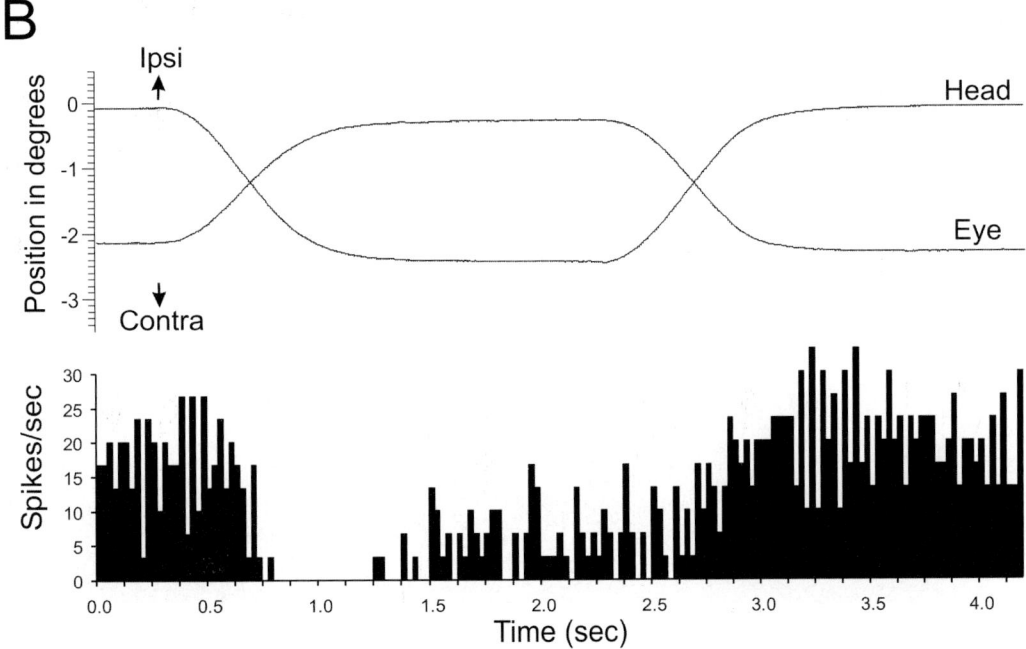

Fig. 5. Response to sigmoidal vestibular stimulation of a candidate UBC recorded in the flocculus of the same awake rabbit from which the recordings of Fig. 4 were obtained. This neuron is considered to be a UBC on the basis of the strong similarity of its normalized ISI histogram of spontaneous activity to those found for UBCs in the ventral uvula/nodulus of the anesthetized rat (see Fig. 3B). The histogram is based on 1270 intervals in 2 periods totaling 67 s. The vestibular stimulation and compensatory eye movement are shown in the top panel of B as described in Fig. 4. The response of this cell is shown in the bottom panel of B as a peristimulus time histogram compiled from 24 presentations in the light of the periodic sigmoidal displacement (bin width 25 ms). Note that this cell's response to each direction of sigmoidal vestibular stimulation did not begin until more than 200 ms after the eye movement began.

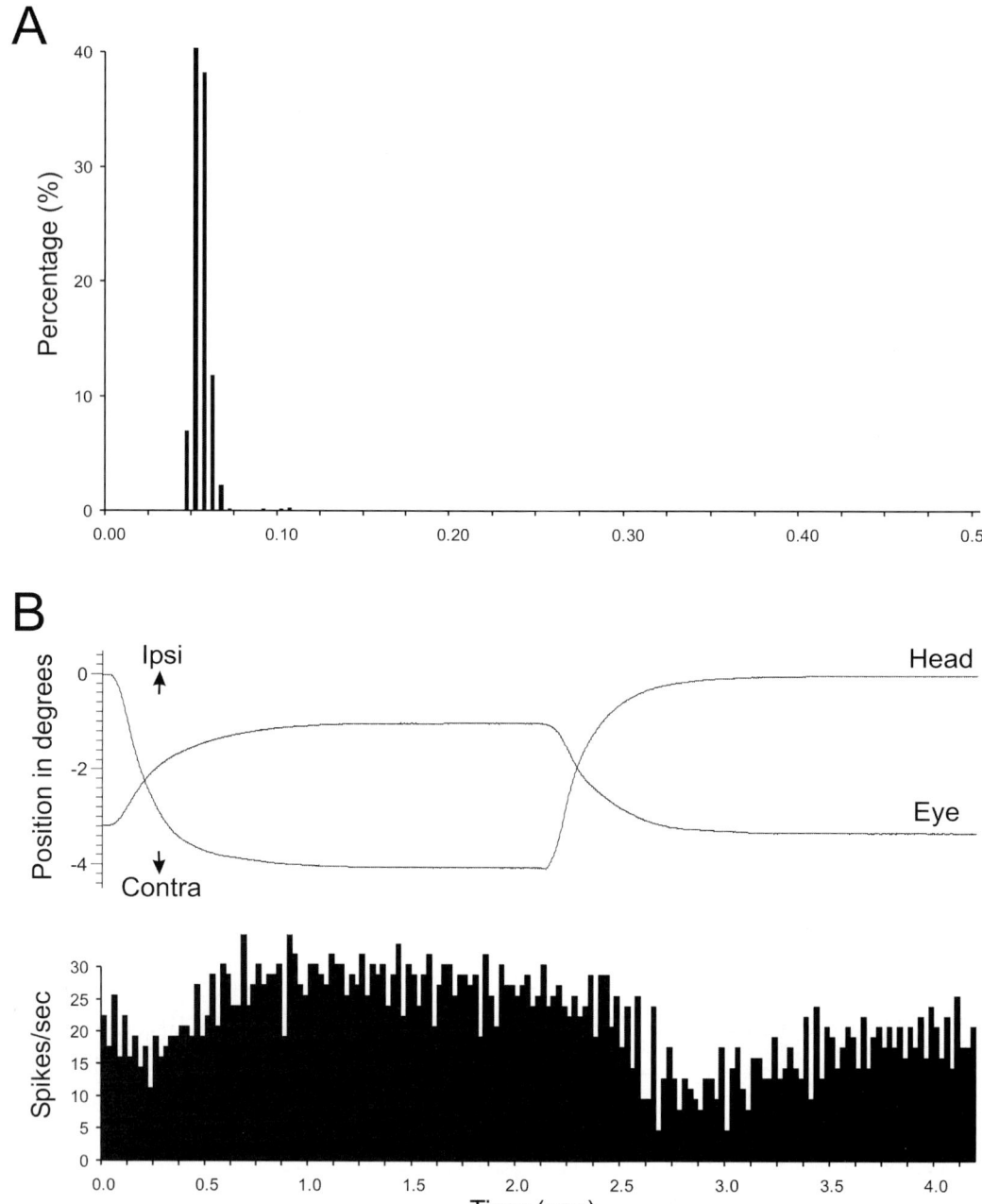

Fig. 6. Response to sigmoidal vestibular stimulation of another candidate UBC recorded in the flocculus of a second awake rabbit. This neuron is considered to be a UBC on the basis of the strong similarity of its normalized ISI histogram of spontaneous activity to those found for UBCs in the ventral uvula/nodulus of the anesthetized rat (see Fig. 3A). The histogram is based on 865 intervals in 44 s. The vestibular stimulation and the compensatory eye movements are depicted in the top panel of B in the same way as shown in Fig. 4. The response of this cell is shown in the bottom panel of B as a peristimulus time histogram compiled from 25 presentations of the periodic sigmoidal displacement in the light (binwidth 25 ms). Again, as for the cell shown in Fig. 5, the modulation in each direction did not begin until more than 200 ms after the eye movement began.

Discussion

We have shown that the juxtacellular labeling technique described by Pinault (Pinault, 1996) and previously used successfully in the hippocampus (Klausberger et al., 2003), basal ganglia (Wu et al., 2000; Mailly et al., 2003), thalamus (Pinault et al., 2001; Cetas et al., 2002) and various brain stem regions (Monconduit et al., 2002; Xi et al., 2002; Sartor and Verberne, 2003) can also be used to study interneurons of the cerebellar cortex of the rat. In this way, a correlation of cell type with some of its physiological characteristics proved possible. Labeling quality appeared to be particularly dependent on successful entrainment of the cell for at least several minutes as well as a 'clean' detachment of the pipette from the entrained cell. Small cells such as UBC's and even granule cells could be successfully recorded and stained.

We have noted that in the uvula/nodulus of the anesthetized rat the distribution of interspike intervals of spontaneously active units differed among at least Golgi cells and UBCs, which are both found in the granule cell layer. The Golgi cells discharged in a rather slow, somewhat irregular or syncopated manner with an average frequency of 4–10 Hz. Their ISI histograms typically peaked around 100 ms and only infrequently had intervals shorter than 50 ms. These characteristics confirm some of those of the recordings made from putative Golgi cells in earlier physiological studies (Edgley and Lidierth, 1987; Vos et al., 1999). On the other hand, units firing with a rather regular, fairly constant rate averaging 10–35 Hz all turned out to be UBCs. However, some UBCs may also show a different type of activity since some cells with a bursting or virtually silent type of activity also turned out to be UBCs. Interestingly, Nunzi et al. (2002) recently showed that different subsets of UBCs could be defined on the basis of their immunohistochemical characteristics. Although not shown in this chapter, we have obtained some indications that spontaneous activity-based discrimination may also be possible for other cell types such as basket-stellate cells and granule cells.

The spontaneous activity patterns of Golgi cells and UBCs identified in the uvula/nodulus of ketamine anesthetized rats are remarkably similar to those of some neurons encountered in the flocculus of the awake rabbit. Since both cell types are present in the flocculus of the rabbit (Mugnaini et al., 1997), it is attractive to propose that floccular cells displaying a somewhat irregular, syncopated activity with a broad ISI peak around 100 ms and largely without intervals shorter than 50 ms are Golgi cells, whereas cells with a regular, fairly constant firing rate with average frequencies between 10 and 35 Hz are UBCs. These proposed relations offer the possibility of opening up avenues for exploring the functional features of specific cerebellar interneurons in behaving animals, as illustrated here with examples from the awake rabbit. Verification that the spontaneous activity signatures found in the anesthetized rat are valid in the awake rabbit remains to be obtained. Nevertheless, we conclude that the juxtacellular recording and the staining technique is potentially an attractive alternative to the in vivo intracellular and patch clamp techniques and may prove valuable in the study of cerebellar interneurons.

Conclusions

We have used the juxtacellular recording and labeling technique of Pinault to study interneurons in the uvula/nodulus of the ketamine anesthetized rat. Cells displaying a somewhat irregular, syncopated cadence of spontaneous activity averaging 4–10 Hz could, upon successful entrainment and visualization, be morphologically identified as Golgi cells. Spontaneously firing cells with a fairly regular and constant firing rate turned out to be unipolar brush cells. Indications were found that other types of cerebellar cortical neurons might also be distinguished on the basis of the characteristics of their spontaneous firing. We suggest that the associations found for the rat can be extended to the flocculus of the awake rabbit to identify Golgi cells and UBCs. This method of identification has the potential to enable evaluation of the function of these specific interneurons in behavioral paradigms.

Acknowledgments

We thank B. Winkelman for the sigmoids and J. van der Burg for technical assistance. Supported by NIH grant NS-13742 (JIS and HCH) and the Dutch

Ministry of Health, Welfare and Sport (TJHR and ES-G).

References

Aoki, E., Semba, R. and Kashiwamata, S. (1986) New candidates for GABAergic neurons in the rat cerebellum: an immunocytochemical study with anti-GABA antibody. Neurosci. Lett., 68: 267–271.

Cetas, J.S., Price, R.O., Velenovsky, D.S., Crowe, J.J., Sinex, D.G. and McMullen, N.T. (2002) Cell types and response properties of neurons in the ventral division of the medial geniculate body of the rabbit. J. Comp. Neurol., 445: 78–96.

De Zeeuw, C.I., Wylie, D.R., Stahl, J.S. and Simpson, J.I. (1995) Phase relations of Purkinje cells in the rabbit flocculus during compensatory eye movements. J. Neurophysiol., 74: 2051–2064.

Dieudonne, S. (2001) Serotonergic neuromodulation in the cerebellar cortex: cellular, synaptic, and molecular basis. Neuroscientist, 7: 207–219.

Dino, M.R., Willard, F.H. and Mugnaini, E. (1999) Distribution of unipolar brush cells and other calretinin immunoreactive components in the mammalian cerebellar cortex. J. Neurocytol., 28: 99–123.

Dino, M.R., Schuerger, R.J., Liu, Y., Slater, N.T. and Mugnaini, E. (2000) Unipolar brush cell: a potential feedforward excitatory interneuron of the cerebellum. Neuroscience, 98: 625–636.

Eccles, J.C., Llinas, R. and Sasaki, K. (1966) The inhibitory interneurones within the cerebellar cortex. Exp. Brain Res., 1: 1–16.

Edgley, S.A. and Lidierth, M. (1987) The discharges of cerebellar Golgi cells during locomotion in the cat. J. Physiol., 392: 315–332.

Geurts, F.J., Timmermans, J., Shigemoto, R. and De Schutter, E. (2001) Morphological and neurochemical differentiation of large granular layer interneurons in the adult rat cerebellum. Neuroscience, 104: 499–512.

Jorntell, H. and Ekerot, C.F. (2002) Reciprocal bidirectional plasticity of parallel fiber receptive fields in cerebellar Purkinje cells and their afferent interneurons. Neuron, 34: 797–806.

Jorntell, H. and Ekerot, C.F. (2003) Receptive field plasticity profoundly alters the cutaneous parallel fiber synaptic input to cerebellar interneurons in vivo. J. Neurosci., 23: 9620–9631.

Klausberger, T., Magill, P.J., Marton, L.F., Roberts, J.D., Cobden, P.M., Buzsaki, G. and Somogyi, P. (2003) Brain-state- and cell-type-specific firing of hippocampal interneurons in vivo. Nature, 421: 844–848.

Laine, J. and Axelrad, H. (1998) Lugaro cells target basket and stellate cells in the cerebellar cortex. Neuroreport, 9: 2399–2403.

Mailly, P., Charpier, S., Menetrey, A. and Deniau, J.M. (2003) Three-dimensional organization of the recurrent axon collateral network of the substantia nigra pars reticulata neurons in the rat. J. Neurosci., 23: 5247–5257.

Mann-Metzer, P. and Yarom, Y. (2002) Jittery trains induced by synaptic-like currents in cerebellar inhibitory interneurons. J. Neurophysiol., 87: 149–156.

Monconduit, L., Desbois, C. and Villanueva, L. (2002) The integrative role of the rat medullary subnucleus reticularis dorsalis in nociception. Eur. J. Neurosci., 16: 937–944.

Mugnaini, E. and Floris, A. (1994) The unipolar brush cell: a neglected neuron of the mammalian cerebellar cortex. J. Comp. Neurol., 339: 174–180.

Mugnaini, E., Dino, M.R. and Jaarsma, D. (1997) The unipolar brush cells of the mammalian cerebellum and cochlear nucleus: cytology and microcircuitry. Prog. Brain Res., 114: 131–150.

Nunzi, M.G., Shigemoto, R. and Mugnaini, E. (2002) Differential expression of calretinin and metabotropic glutamate receptor mGluR1alpha defines subsets of unipolar brush cells in mouse cerebellum. J. Comp. Neurol., 451: 189–199.

Paxinos, G. and Watson, C. (1998) The Rat Brain in Stereotaxic Coordinates, 4th Edition. Academic Press, Sydney.

Pinault, D. (1994) Golgi-like labeling of a single neuron recorded extracellularly. Neurosci. Lett., 170: 255–260.

Pinault, D. (1996) A novel single-cell staining procedure performed in vivo under electrophysiological control: morpho-functional features of juxtacellularly labeled thalamic cells and other central neurons with biocytin or neurobiotin. J. Neurosci. Methods, 65: 113–136.

Pinault, D., Vergnes, M. and Marescaux, C. (2001) Medium-voltage 5–9-Hz oscillations give rise to spike-and-wave discharges in a genetic model of absence epilepsy: in vivo dual extracellular recording of thalamic relay and reticular neurons. Neuroscience, 105: 181–201.

Ramón y Cajal, S. (1911) Histologie du système nerveux de l'homme et des vertébrés. Maloine, Paris.

Ruigrok, T.J. (2003) Collateralization of climbing and mossy fibers projecting to the nodulus and flocculus of the rat cerebellum. J. Comp. Neurol., 466: 278–298.

Sartor, D.M. and Verberne, A.J. (2003) Phenotypic identification of rat rostroventrolateral medullary presympathetic vasomotor neurons inhibited by exogenous cholecystokinin. J. Comp. Neurol., 465: 467–479.

Vos, B.P., Volny-Luraghi, A. and De Schutter, E. (1999) Cerebellar Golgi cells in the rat: receptive fields and timing of responses to facial stimulation. Eur. J. Neurosci., 11: 2621–2634.

Wu, Y., Richard, S. and Parent, A. (2000) The organization of the striatal output system: a single-cell juxtacellular labeling study in the rat. Neurosci. Res., 38: 49–62.

Xi, M.C., Fung, S.J., Yamuy, J., Morales, F.R. and Chase, M.H. (2002) Induction of active (REM) sleep and motor inhibition by hypocretin in the nucleus pontis oralis of the cat. J. Neurophysiol., 87: 2880–2888.

SECTION IX

Cerebellar Pathology

CHAPTER 27

Sexual dimorphism in cerebellar structure, function, and response to environmental perturbations

K. Nguon[1], B. Ladd[1], M.G. Baxter[2] and E.M. Sajdel-Sulkowska[1,3],*

[1]*Department of Psychiatry, Brigham and Women's Hosp., 221 Longwood Ave., Boston, MA 02115, USA*
[2]*Department of Experimental Psychology, Oxford University, Oxford, UK*
[3]*Department of Psychiatry, Harvard Med. School, Boston, MA, USA*

Keywords: sex differences; rat; cerebellum; male; female; PCBs; hypergravity; environmental impacts

Abstract: Sexual dimorphism of CNS structure and function has been observed in humans and animals, but remains relatively unrecognized in the context of the cerebellum. Recent researh in our laboratory has examined whether these gender differences extend to cerebellar structure and function, as well as the impact of environmental factors on the developing cerebellum. Perinatal exposure to both chemical and physical perturbations in the environment (in our experiments, PCBs or hypergravity) affects growth, neurodevelopment, and motor coordination differently in males and females. These neurodevelopmental and behavioral effects are accompanied by sex-related changes in cerebellar mass and cerebellar protein expression. Exposure to chemical toxins (PCBs) resulted in more dramatic neurodevelopmental and behavioral changes in male neonates. It is possible that gender-related differences in male and female cerebellar structure and function are related to sex-specific development of the cerebellum and sex-specific distribution of specific receptors, local synthesis of trophic factors, and maturation of the pituitary hypophysial axis. These sex-related differences may underlie the sex-specific preponderance of certain neuropsychiatric disorders, and must be incorporated in the design of future basic and clinical investigations.

Introduction

Histological, neuroimaging, and behavioral evidence indicates that there is sexual dimorphism in CNS function and structure in both humans and animals. However, relatively few studies have explored gender differences in cerebellar structure and function. Considering both the established function of the cerebellum in motor coordination and its newly discovered functions in learning, motor planning, sensory processing, emotions, and cognition,

*Corresponding author. Tel.: +1 617-732-5859; Fax: +1 617-713-3078; E-mail: Esulkowska@rics.bwh.harvard.edu

acknowledgment of sex-related differences in cerebellar structure and function may have far-reaching consequences in our approach to basic research and health related issues.

Sex-related differences in cerebellar structure

Human cerebellar volume appears to be larger in men (Rhyu et al., 1999; Raz et al., 2001). While the cerebellar structure appears to be larger in men, PET evidence suggests that females have a higher rate of cerebellar metabolism (Volkov et al., 1997). Animal studies suggest more numerous dendritic segmentation in the female rat cerebellum (Juraska, 1990). Although no sex-related differences in the number of Purkinje cells were noted in normal human (Mayhew et al., 1990) and rat cerebella (Mwamengele et al., 1993), a lower number of Purkinje cells was observed in heterozygous male than in female Reeler mice (Hadj-Sahraoui et al., 1996).

The way she moves: sex-related difference in cerebellar behavior

The cerebellum's involvement in movement control is undisputed. Thus it is not surprising that the sex-related changes in cerebellar structure are associated with sex-related differences; females increase the walking speed by increasing the cadence (Yamasaki et al., 1991) and have more trouble to turn while walking (Cao et al., 1997). Animal studies suggest that motor ability, measured by rotorod performance, also differs between males and females (McFadyen et al., 2003). Young (P12) male pups also show greater improvement on repeated rotorod trial relative to female pups, an effect which is abolished by PCB treatment (personal observation).

Accumulating behavioral evidence indicates that the cerebellum is also involved in learning, sensory processing, motor planning, exploratory activity, cognitive flexibility (Pierce and Courchesne et al., 2001), and other higher cognitive (Schmahmann, 1991) and emotional (Trevarthen, 2000) processes. Thus it is of interest that sex-related differences were observed in musical processing and were associated with greater cerebellar activation in females (Gaab et al., 2003).

Cerebellar development in males and females

Cerebellar development is associated with altered protein and gene expression. Different isoforms of creatine kinase (CK) are expressed during cerebellar development with the female cerebellum expressing more CK enzymes (Ramirez and Jimenez, 2002). Cerebellar protein kinase C gamma (PKCgamma) is also regulated differently in males and females with the female cerebellar enzyme showing greater sensitivity to copper deficiency (Johnson and Prohaska, 2000). The observation of sex differences in GFAP and vimentin expression in hamster cerebella (Suarez et al., 1992) were supported by our study of the developing cerebellum, with higher GFAP expression in male cerebellum (personal observation, see below).

Male preponderance in neuropsychiatric disorders involving cerebellum

Several developmental neuropsychiatric disorders such as autism, attention deficit disorder, hyperactivity, dyslexia, as well as late onset Parkinson's disease involve the cerebellar structure. It is of interest that most neurodevelopmental disorders afflict a significantly larger proportion of males (Rutter et al., 2003), with a male-to female ratio 4:1 for autism (Baird et al., 2000; Scott et al., 2002) and a 2:1 ratio for Parkinson's disease (Van Den Eaden et al., 2003).

Autism is a heterogeneous group of developmental neuropsychiatric disorders encompassing a spectrum of neurological impairments with multiple etiologies and classified as pervasive developmental disorders (PDDs). Autism affects 0.2% of preschool children (Chakrabarti and Fombonne, 2001). Its clinical symptoms, manifested prior to the third year of life, include deficits in social and communicative skills, sensory processing, motor planning, cognitive flexibility, and in some cases cognitive ability and presence of stereotypic behavior. The disruption of sensory, motor, communicative, and cognitive functions observed in autism is associated with structural changes in several brain regions, including the brain stem, midbrain, the frontal and parietal lobes, the amygdala, and the cerebellum. The abnormalities in autistic cerebellum have been documented by many laboratories (Courchesne, 1991; Kemper and Bauman, 1993; Pierce and Courchesne et al., 2001), but the exact nature of the abnormalities is being debated. Cerebellar changes in autism most consistently include a decreased number of Purkinje cells (Townsend et al., 1999). Both imaging and autopsy data point to: (1) reduced cerebellar size (Gaffney et al., 1987; Murakami et al., 1989); (2) hypoplasia of neocerebellar vermal lobules VI and VII (Courchesne et al., 1988; Ciesielski et al., 1997) and VIII–X (Hashimoto et al., 1993; Levitt et al., 1999), but not developmentally and anatomically distinct lobules I–V (Courchesne et al., 1988); (3) a significant loss of Purkinje cells in both the vermis and hemispheres (Rivto et al., 1986; Courchesne, 1991) that can be observed in 90% of autopsy cases; and (4) a reduced number of granule cells (Heh et al., 1989). It has been suggested that increased apoptotic activity may contribute to this reduction (Fatemi et al., 2001). MRI studies indicate that the majority of autistic patients with abnormal cerebellar anatomy show vermian hypoplasia, but a minority shows vermian hyperplasia (Courchesne et al., 1994); increased cerebellar hemisphere size has also been observed. A delay in the development of the cerebellar vermis has been suggested (Hashimoto, 1994).

Cerebellar involvement in Parkinson's disease has been suggested by fMRI studies (Elsinger et al., 2003). The incidence of the disease increases rapidly after the age of 60, with the rate for men twice of that for women.

Sex-related differences in response to environmental impacts

At present, neither the etiology (etiologies) nor the underlying pathophysiological processes in autism are understood. The prevailing view is that autism (and perhaps other neuropsychiatric disorders) is caused by a pathophysiologic process arising from the interaction of early environmental insults and genetic predisposition. However, the identity of the critical environmental insults remains obscure and the mechanism(s) by which insults affect gene expression in the developing CNS are poorly understood. Studies performed in both the context of accidental, catastrophic PCB spills, and against a background of persistent PCB contamination (such as in Great Lakes Region) suggest a strong link between PCBs and neurocognitive impairments that include motor dysfunction and impairment in memory and learning.

While the developing CNS is vulnerable to environmental factors that modify genetic information and alter CNS structure and function (Sajdel-Sulkowska and Koibuchi, 2002), the cerebellum is especially vulnerable because of its protracted development, which continues after birth. In the studies described here, we compared and contrasted the response of the developing cerebellum to chemical perturbations in the form of toxic pollutants—PCBs—(Nguon et al., submitted) and to physical perturbations in the form of hypergravity (Sajdel-Sulkowska et al., 2001). By comparing the effect of such drastically different environmental factors on cerebellar development, we may begin to understand how perinatal environmental factors affect brain plasticity. We may also begin to appreciate the many ways that the environment molds the genetic message affecting cerebellar structure and function in males and females.

Furthermore, since the rat cerebellum demonstrates many of the same developmental processes observed in other brain regions and its protracted developmental stages are similar to those of the human cerebellum (Porterfield and Hendrich, 1993), the information obtained from these animal studies is potentially useful in understanding these phenomena not only in the human cerebellum, but in other brain regions as well.

Perinatal exposure to both chemical (PCBs) and physical (altered gravity) perturbations in the environment affect growth, neurodevelopment, and motor coordination differently in males and females. These neurodevelopmental and behavioral effects were accompanied by sex-related changes in cerebellar mass and cerebellar protein expression. Exposure to chemical toxins (PCBs) resulted in more dramatic neurodevelopmental and behavioral changes in male neonates. PCB-exposed male neonates had a longer righting time, a delayed onset of startle response, and dramatically decreased performance on a rotorod (Fig. 1). Exposure to altered gravity likewise increased righting time more in males, however it delayed the onset of startle response more in females and resulted in the same degree of impairment of motor coordination in males and females.

The sex-specific nature of the behavioral response to chemical insults is not unique to PCB exposure. Indeed, sex-specific responses to a number of environmental or pharmacological manipulations have been observed. These include exposure to chlorpyrifos, organophosphate pesticides contaminating the Salinas river in California (Dam et al., 2000), ethanol (Rintala et al., 2001), methylazoxymethanol acetate (MAM) (Ferguson et al., 1996), glucocorticoids (Vicedomini et al., 1986), and

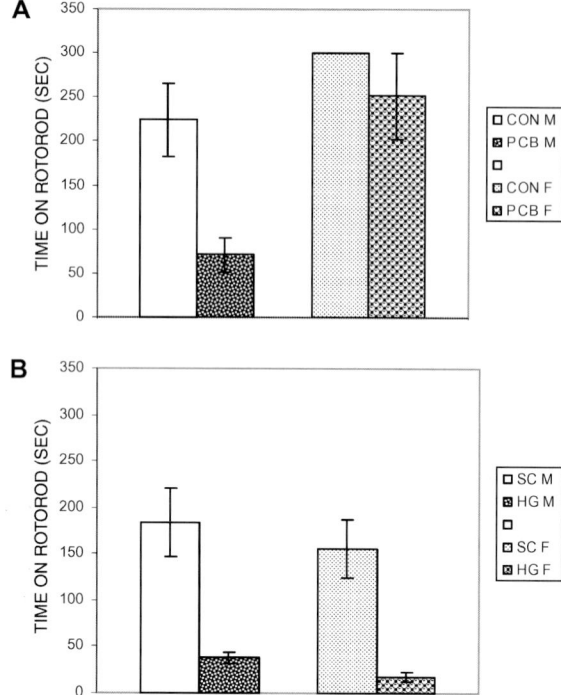

Fig. 1. Performance on rotorod test on P21. (A) Rat neonates exposed to Aroclor 1254 (10 mg/kg/day) from gestational day (G) 11 to postnatal day (P) 21 performed poorly on the rotorod test and the difference between PCB-treated rats and controls was more pronounced in the male rats (CON, dams received corn oil; PCB, dams received PCB in corn oil). (B) Exposure of rat neonates exposed to centrifuge-produced hypergravity (1.65 G) from G11 to P21 resulted in the same degree of impairment on the rotorod test in males and females (SC, stationary controls; HG, animals exposed to 1.65 G hypergravity).

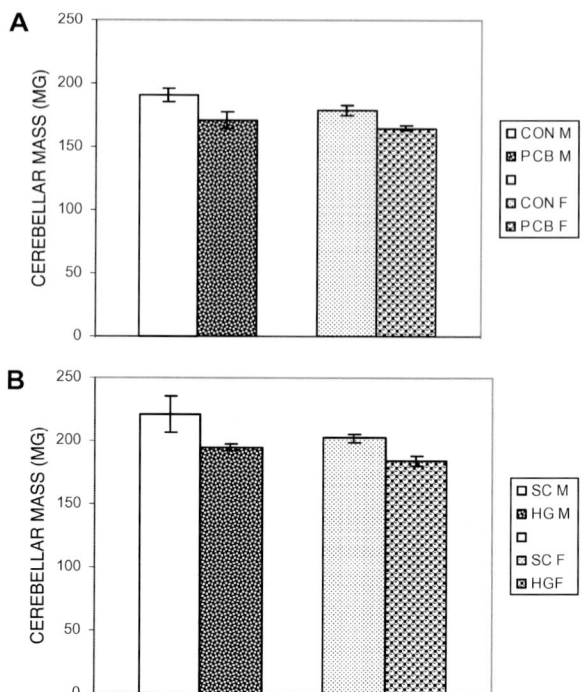

Fig. 2. Cerebellar mass on postnatal day (P) 21. (A) Rat neonates exposed to Aroclor 1254 (10 mg/kg/day) from G11 to P21 showed a decrease in cerebellar mass on P21 more pronounced in males. (B) Rat neonates exposed to hypergravity from G11 to P21 also showed a decrease in cerebellar mass on P21 more pronounced in males (B: SC, stationary controls; HG, animals exposed to 1.65 G hypergravity).

naltrexone use in ethanol detoxification (de Cabo and Paz Viveros, 1997). Rats exposed prenatally to cocaine had impaired motor coordination, with males being more affected than females (Markowski et al., 1998). On the other hand, the effect of physical insults, such as restraint during gestation, appear to be more pronounced in females (Konkle et al., 2003) as does the effect of daily neonatal handling (Rhees et al., 2001).

The greater impacts on neurodevelopment and motor coordination observed in PCB-exposed male neonates were associated with a greater reduction in cerebellar weight. In the hypergravity exposed animals, a greater reduction in cerebellar mass was observed in male neonates (Fig. 2).

Cerebellar sexual dimorphism and gene expression

We hypothesized that these changes in the cerebellum and in motor coordination (reflecting cerebellar function) may reflect altered cerebellar development and may be related to changes in cerebellar protein expression. To this end, we compared the expression of glial (GFAP) and neuronal (L1) cerebellar proteins in PCB-exposed rat neonates to that of control neonates. We drew the same comparison between hypergravity-exposed neonates and stationary controls at the time of maximal cerebellar changes. This approach was taken because data from other developmental studies suggest discordance between mRNA and protein expression (Faivre-Sarrailh et al., 1991).

The results of these analyses suggest that the exposure of the developing cerebellum to both chemical and physical perturbation is associated with changes in the expression of glial (GFAP) and neuronal (L1) proteins.

GFAP—a major cytoskeletal protein of astroglial cells (Bergmann glia) influences cell shape (Toran-Allerand et al., 1991) and facilitates neuronal migration (Hatten et al., 1984; Mason et al., 1988) during cerebellar development. GFAP expression appeared to be higher in control male than in control female cerebella, which corroborates earlier findings of higher expression of GFAP in the developing cerebella of male hamsters (Suarez et al., 1992). GFAP expression was increased in both PCB-exposed males and females, but the increase was more pronounced in males. Increased GFAP expression suggests either increased glial cell proliferation or accelerated astrocyte maturation in the PCB-exposed cerebellum. Increased GFAP expression and glial cell proliferation (gliosis) accompany diverse neurological insults, including exposure to nicotine and pesticide chlorpyrifos (Abdel-Rahman et al., 2003), chronic emotional (Sedykh et al., 1999) and physical stress (Steward et al., 1993), and has been observed following PCB exposure (Morse et al., 1996). It is possible that the increase in GFAP reflects accelerated maturation of glia rather than an increase in the number of glial cells, because of the smaller cerebellar size in PCB-treated neonates. In fact, the maximal change in GFAP coincides with the greatest deviation from control in cerebellar mass of PCB males, with both occurring on P12. On the other hand, GFAP expression was decreased in the cerebella of hypergravity-exposed male neonates on P6, with no significant changes of expression in the hypergravity-exposed females (Sulkowski et al., 2004). It is of interest that GFAP also decreases in hypothyroid rat neonates (Li et al., 2004) and there appears to be a transient hypothyroidism on P6 in hypergravity-exposed rat neonates.

The neural cell adhesion molecule L1 is a membrane-bound protein expressed predominantly by neurons (Kettenmann et al., 1983), plays a key role in neuronal–neuronal interactions in the CNS (Keilhauer et al., 1985) and is involved in several neurodevelopmental processes including granule cell migration (Asou et al., 1992; Thelen et al., 2002), axonal growth (Ignelzi et al., 1994), and synaptic plasticity which underlies learning and memory processes (Matsumoto et al., 2003). Mutations in L1 are associated with neurological abnormalities and mental retardation (Moulding et al., 2000).

L1 is a developmentally regulated neuronal protein with peak expression on P15 (Liljelund et al., 1994). In the present study we have concentrated on the expression of the 200 kDa polypeptide. In PCB-males L1-200 was not significantly altered ($t = 0.949$, $p = 0.35$) while in PCB-treated females it was decreased by 26.45% ($t = 3.185$, $p = 0.0019$). On the other hand, L1 expression was decreased in the hypergravity-exposed male cerebellum (a 16.9% decrease) but not in the HG female cerebellum.

Decreased expression of L1 has been observed following prenatal exposure to X-rays and has been associated with abnormal neuronal migration (Sun et al., 2003). Thus, a decrease in L1 would be consistent with delayed cell migration in the PCB-treated rat cerebellum. The differential effect on L1 expression suggests that cerebellar cell migration may take place on a different time schedule in female and male neonates.

Because cerebellar development coincides with the period of maturation of the hypophyseal–pituitary axis, it is tempting to suggest that the sex-specific nature of the responses to PCBs and to hypergravity may be related to sex-specific hormonal modulations. In particular, both the impact of PCBs (Seo et al., 1995) and hypergravity are associated with altered thyroid hormone levels (Sajdel-Sulkowska et al., 2001). It has been proposed that hypothyroidism induced by PCB exposure may contribute to growth deficits, motor dysfunction, and hearing disorders, because similar defects are observed in developing hypothyroid animals (Goldey et al., 1995) and can be attenuated with thyroxine (Goldey and Crofton, 1998). However, changes in thyroid hormone status following PCB exposure are limited to decreased T4 expression without the increased TSH levels typically observed in hypothyroidism, and changes in response to hypergravity are transient. Furthermore, GFAP expression decreases in hypothyroid animals (Li et al., 2004), and while it also decreases in hypergravity-exposed animals, in PCB-exposed animals we have observed an increased GFAP expression. It is thus possible that other regulatory pathways may play an important role in the response of the developing cerebellum to environmental factors.

Probing into the molecular mechanisms of cerebellar sexual dimorphism

Considerable evidence suggests that different brain regions are sexually dimorphic and undergo sex-specific neurodevelopmental processes. For example, GFAP expression is regulated by progesterone (Gotohda et al., 2002) which is actively synthesized from cholesterol and metabolized by Purkinje cells only during the neonatal period (Sakamoto et al., 2003). Furthermore, progesterone receptor isoforms are differentially expressed (Quadros et al., 2002) and regulated (Guerra-Araiza et al., 2002) by estradiol in male and female neonates. Purkinje cells also express aromatase, the key enzyme for the formation of estradiol from testosterone (Sakamoto et al., 2003). It is thus of interest that PCBs are known to alter estradiol via decreased aromatase activity (Hany et al., 1999; Kaya et al., 2002) and can thus modify the fate of testicular testosterone released in male rats at birth (Auger, 2001), and increase concentrations of estrogen in the neonatal cerebellum above both prepubertal and adult concentrations (Sakamoto et al., 2003). Thus, in the case of PCBs, the estradiol pathway may play a more important role. On the other hand, in hypergravity exposed rats the thyroid pathway may be more affected.

Conclusions

Recognition of sex-related differences in cerebellar structure and functions is long overdue. Sex-related differences in cerebellar structure may extend to establishment of sex-specific cerebellar circuitry and integration with other sex-dimorphic brain regions resulting in gender-specific behavior. Understanding the molecular mechanism which causes these gender differences may elucidate the reasons underlying the disproportionately higher incidence in males of developmental neuropsychiatric disorders involving the cerebellum.

Acknowledgments

The hypergravity experiments were performed on a 24-ft centrifuge at NASA-Ames Research Center. This work was supported by NASA-NCC2-1042 and NIES11946-01 grants awarded to EMS-S. Mark G. Baxter is an Alfred P. Sloan Research Fellow.

References

Abdel-Rahman, A., Dechkoskaia, A., Mehta-Simmons, H., Guan, X., Khan, W. and Abou-Donia, M. (2003) Increased expression of glial fibrillary acidic protein in cereberllum and hippocampus: differential effects on neonatal brain regional acetylcholinesterase following maternal exposure to combined chlorpyrifos and nicotine. J. Toxicol. Environ. Health, 66: 2047–2066.

Asou, H., Miura, M., Kobayashi, M. and Uyemura, K. (1992) The cell adhesion molecule L1 has specific role in neural cell migration. Neuroreport, 3: 481–484.

Auger, A.P. (2001) Sex differences in the developing brain: crossroads in the phosphorylation of cAMP response element binding protein. J. Neuroendocrinol., 15: 622–627.

Baird, G., Charman, T., Baron-Cohen, S., Cao, C., Ashton-Miller, J.A., Schutz, A.B. and Alexander, N.B. (1997) Abilities to turn suddenly while walking: effects of age, gender, and available response time. J. Gerontol. A. Biol. Sci. Med. Sci., 52: M88–M93.

Baird, G., Charman, T., Baron-Cohen, S., Cox, A., Swettenham, J., Wheelwright, S. and Drew, A. (2001) A screening instrument for austim at 18 months of age; a 6-year follow-up study. J. Am. Acad. Child Adolesc. Psychiatry, 40: 737–738.

Cao, C., Ashton-Miller, J.A., Schultz, A.B. and Alexander, N.B. (1997) Abilities to turn suddenly while walking: effects of age, gender, and available response time. J. Gerontol. A. Biol. Sci. Med. Sci., 52: M88–93.

Chakrabarti, S. and Fombonne, E. (2001) Pervasive developmental disorders in preschool children. JAMA, 285: 3141–3142.

Ciesielski, K.T., Harris, R.J., Hart, B.L. and Pabst, H.F. (1997) Cerebellar hypoplasia and frontal lobe cognitive deficits in disorders of early childhood. Neuropsychologia, 35: 643–655.

Courchesne, E. (1991) Neuroanatomic imaging in autism. Pediatrics, 87: 781–790.

Courchesne, E., Yeung-Courchesne, R., Press, G.A., Hesselink, J.R. and Jernigan, T.L. (1988) Hypoplasia of cerebellar vermal lobules VI and VII in autism. N. Engl. J. Med., 318: 1349–1354.

Courchesne, E., Saitoh, O., Yeung-Courchesne, R., Press, G.A., Lincoln, A.J., Haas, R.H. and Schreibman, L. (1994) Abnormality of cerebellar vermian lobules VI and VII in patients with infantile autism: identification of hypoplastic and hyperplastic subgroups with MR imaging. Am. J. Roentgenol., 162: 123–130.

Courchesne, E., Karns, C.M., Davis, H.R., Ziccardi, R., Carper, R.A., Tigue, Z.D., Chisum, H.J., Moses, P.,

Pierce, K., Lord, C., Lincoln, A.J., Pizzo, S., Schreibma, L., Haas, R.H., Akshoomoff, N.A. and Courchesne, R.Y. (2001) Unusual brain growth patterns in early life in patients with autistic disorder: an MRI study. Neurology, 57: 245–254.

Courchesne, E., Carper, R. and Akshoomoff, N. (2003) Evidence of brain overgrowth in the first year of life in autism. JAMA, 290: 337–344.

Crofton, K.M., Kodavanti, P.R., Derr-Yellin, E.C., Casey, A.C. and Kehn, L.S. (2000) PCBs, thyroid hormones, and ototoxity in rats: cross-fostering experiments demonstrate the impact of postnatal lactation exposure. Toxicol. Sci., 57: 131–140.

Dam, K., Seidler, F.J. and Slotkin, T.A. (2000) Chlorpyrifos exposure during a critical neonatal period elicits gender-selective deficits in the development of coordination skill and locomotor activity. Brain Res. Dev. Brain. Res., 121: 179–187.

de Cabo, C. and Paz Viveros, M. (1997) Effects of neonatal naltrexone on neurological and somatic development in rats of both genders. Neurotoxicol. Teratol., 19: 499–509.

Elsinger, C.L., Rao, S.M., Zimbelman, J.L., Reynolds, N.C., Blindauer, K.A. and Hoffman, R.G. (2003) Neural basis for impaired time reproduction in Parkinson's disease: fMRI studies. J. Int. Neuropsychol. Soc., 9: 1088–1098.

Faivre-Sarrailh, C., Rami, A., Fages, C. and Tardy, M. (1991) Effect of thyroid deficiency on glial fibrillary acidic protein (GFAP) and GFAP-mRNA in the cerebellum and hippocampal formation of the developing rat. Glia, 4: 276–284.

Fatemi, S.H., Halt, A.R., Stary, J.M., Realmuto, G.M. and Jalali-Mousavi, M. (2001) Reduction in anti-apoptotic protein Bcl-2 in autistic cerebellum. Neuroreport, 12: 929–933.

Ferguson, S.A., Paule, M.G. and Holson, R.R. (1996) Functional effects of methylazoxymethanol-induced cerebellar hypoplasia in rats. Neurotoxicol. Teratol., 18: 529–537.

Gaab, N., Keenan, J.P. and Schlaug, G. (2003) The effects of gender on the neural substrates of pitch memory. J. Cogn. Neurosci., 15: 810–820.

Gaffney, G.R., Tsai, L.Y., Kuperman, S. and Minchin, S. (1987) Cerebellar structure in autism. Am. J. Dis. Child, 141: 1330–1332.

Goldey, E.S., Kehn, L.S., Rehnberg, G.L. and Crofton, K.M. (1995) Effects of developmental hypothyroidism on auditory and motor function in the rat. Toxicol. Appl. Pharmacol., 135: 67–76.

Goldey, E.S. and Crofton, K.M. (1998) Thyroxine replacement attenuates hypothyroxinemia, hearing loss, and motor deficits following developmental exposure to Aroclor 1254 in rats. Toxicol. Sci., 45: 94–105.

Gotohda, T., Kuwada, A., Morita, K., Kubo, S. and Tokunaga, I. (2002) Elevation of steroid 5 alpha-reductase mRNA levels in rat cerebellum by toluene inhalation: possible relation to GFAP expression. J. Toxicol. Sci., 25: 223–231.

Guerra Araiza, C., Coyoy-Salgado, A. and Camacho-Arroyo, I. (2002) Sex differences in the regulation of progesterone receptor isoforms expression in the rat brain. Brain Res. Bull., 59: 105–109.

Hadj-Sahraoui, N., Frederic, F., Delhaye-Bouchaud, N. and Mariani, J. (1996) Gender effect on Purkinje cell loss in the cerebellum of the heterozygous reeler mouse. J. Neurogenet., 11: 45–58.

Hany, J., Lilienthal, H., Roth-Harer, A., Ostendorp, G., Heizow, B. and Winneke, G. (1999) Behavioral effects following single and combined maternal exposure to PCB 77 (3,4,3′,4′-tetrachlorobiphenyl) and PCB 47 (2,4,2′,4′-tetrachlorobiphenyl) in rats. Neurotoxicol. Teratol., 21: 147–156.

Hashimoto, T., Tayama, M., Miyazaki, M. and Kuroda, Y. (1993) Brainstem and cerebellar vermis involvement in autistic children. J. Child. Neurol., 8: 149–153.

Hashimoto, T., Tayama, M., Miyazaki, M. and Kuroda, Y. (1994) MRI measurements of the brain stem and cerebellum in high functioning autistic children. No. To. Hattatsu, 26: 3–8.

Hatten, M.E., Mason, C.A., Liem, R.K., Edmondson, J.C., Bovolenta, P. and Shelanski, M.L. (1984) Neron-astroglial interactions in vitro and their implications for repair of CNS injury. Cent. Nerv. Syst. Trauma, 1: 15–27.

Heh, C.W., Smith, R., Wu, J., Hazlett, E., Asarnow, R., Tanguay, P. and Buchsbaum, M.S. (1989) Positron emission tomography of the cerebellum in autism. Am. J. Psychiatry, 146: 242–245.

Ignelzi, M.A. Jr., Miller, D.R., Soriano, P. and Manes, P.F. (1994) Impaired neurite outgrowth of src-minus cerebellar neurons on the cell adhesion molecule L1. Neuron, 12: 873–884.

Johnson, W.T. and Prohaska, J.R. (2000) Gender influences the effect of perinatal copper deficiency on cerebellar PKC gamma content. Biofactors, 11: 163–169.

Juraska, J.M. (1990) Gender differences in the dendritic tree of granule neurons in the hippocampal dentate gyrus of weaning age rats. Brain Res. Dev. Brain Res., 53: 291–294.

Kaya, H., Hany, J., Fastbend, A., Roth-Harer, A., Winneke, G. and Lilienthal, H. (2002) Effects of maternal exposure to a reconstituted mixture of polychlorinated biphenyls on sex-dependent behaviors and steroid hormone concentrations in rats: dose-response relationship. Toxicol. Appl. Pharmacol., 178: 71–81.

Keilhauer, G., Faissner, A. and Schachner, M. (1985) Differential inhibition of nerone-neurone, neurone-astrocyte and astrocyte-astrocyte adhesion by L1, L2 and NCAM antibodies. Nature, 316: 728–730.

Kemper, T.L. and Bauman, M.L. (1993) The contribution of neuropathologic studies to the understanding of autism. Neurol. Clin., 11: 175–187.

Kettenmann, H., Wienrich, M. and Schachner, M. (1983) Antibody L1 ejected from a micropipette identifies neurons without electrical activity. Neurosci. Lett., 41: 85–90.

Konkle, A.T., Baker, S.L., Kenter, A.C., Barbagallo, L.S., Merali, Z. and Bielajev, C. (2003) Evaluation of the effects of chronic mild stressors on hedonic and physiological responses: sex and strain compared. Brain Res., 992: 227–238.

Levitt, J.G., Blanton, R., Capetillo-Cunliffe, L., Guthrie, D., Toga, S.A. and McCracken, J.T. (1999) Cerebellar vermis lobules VIII–X in autism. Prog. Neuropsychopharmacol. Biol. Psychiatry, 23: 625–633.

Li, G.-H., Post, J., Koibuchi, N. and Sajdel-Sulkowska, E.M. (2004) Effect of thyroid hormone deficiency on the expression of glial and neuronal proteins in the developing rat cerebellum. Cerebellum, 3: 100–106.

Liljelund, P., Ghosh, P. and van den Pol, A.N. (1994) Expression of the neuronal molecule L1 in the developing and adult rat brain. J. Biol. Chem., 268: 32886–32895.

Markowski, V.P., Cox, C. and Weiss, B. (1998) Prenatal cocaine exposure produces gender-specific motor effects in aged rats. Neurotoxicol. Teratol., 20: 43–53.

Mason, C.A., Edmondson, J.C. and Hatten, M.E. (1988) The extending astroglial process: development of glial shape in the growing tip, and interactions with neurons. J. Neurosci., 8: 3124–3134.

Matsumoto, Y., Noji, S. and Mizunami, M. (2003) Time course of protein synthesis-dependent phase of olfactory memory in the cricket *Gryllus bimacultus*. Zoolog. Sci., 20: 409–416.

Matsumoto-Miyai, K., Ninomiya, A., Yamasaki, H., Tamura, H., Nakamura, Y. and Shiosaka, S. (2003) NMDA-dependent proteolysis of presynaptic adhesion molecule L1 in the hippocampus by neropsin. J. Neurosci., 23: 7727–7736.

Mayhew, T.M., MacLaren, R. and Henery, C.C. (1990) Fractionator studies on Purkinje cells in human cerebellum: numbers in right and left halves of male and female brains. J. Anat., 169: 63–70.

McFadyen, M.P., Kusek, G., Bolivar, V.J. and Flaherty, L. (2003) Differences among eight inbred strains of mice in motor ability and motor learning on a rotorod. Genes. Brain Behav., 2: 214–219.

Mwamengele, G.L., Mayhew, T.M. and Dantzer, V. (1993) Purkinje cell complements in mammalian cerebella and the biases incurred by counting nucleoli. J. Anat., 183: 155–160.

Morse, D.C., Plug, A., Wesseling, W., van den Berg, K.J. and Brouwer, A. (1996) Persistent alterations in regional brain fibrillary acidic protein and synaptophysin levels following pre- and postnatal polychlorinated biphenyl exposure. Toxicol. Appl. Pharmacol., 139: 252–261.

Moulding, H.D., Martuza, R.L. and Rabkin, S.D. (2000) Clinical mutations in the L1 neural cell adhesion molecule affect cell-surface expression. J. Neurosci., 20: 5696–5702.

Murakami, J.W., Courchesne, E., Press, G.A., Yeung-Courchesne, R. and Hesseling, J.R., Arch. Neurol., 46: 689–69.

Nguon, K., Li, G.-H. and Sajdel-Sulkowska, E.M. (2004) CNS development under altered gravity: cerebellar glial and neuronal protein expression in rat neonates exposed to hypergravity. Adv. Space Res. (in press).

Pierce, K. and Courchesne, E. (2001) Evidence for a cerebellar role in reduced exploration and stereotyped behavior. Biol. Psych., 49: 655–664.

Porterfield, S.P. and Hendrich, C.E. (1993) The role of thyroid hormones in prenatal and neonatal neurological development-current perspectives. Endocr. Rev., 14: 94–106.

Quadros, P.S., Lopez, V., De Vries, G.J., Chung, W.C. and Wagner, C.K. (2002) Progesterone receptors and the sexual differentiation of the medial preoptic nucleus. J. Neurobiol., 51: 24–32.

Ramirez, O. and Jimenez, E. (2002) Sexual dimorphism in rat cerebrum and cerebellum: different patterns of catalityccally active creatine kinase isoenzymes during postnatal development and aging. Int. J. Dev. Neurosci., 20: 627–639.

Raz, N., Gunning-Dixon, F., Head, D., Williamson, A. and Acker, J.D. (2001) Age and sex differences in the cerebellum and the ventral pons: prospective MR study of healthy adults. AJNR Am. J. Neuroradiol., 22: 1161–1167.

Rhees, R.W., Lephart, E.D. and Eliason, D. (2001) Effects of maternal separation during early postnatal development on male sexual behavior and female reproductive function. Behav. Brain Res., 123: 1–10.

Rhyu, I.J., Cho, T.H., Lee, N.J., Uhm, C.S., Kim, H. and Suh, Y.S. (1999) Magnetic resonance image-based cerebellar volumetry in health Korean adults. Neurosci. Lett., 270: 149–152.

Rintala, J., Jaatinen, P., Kiianmaa, K., Iikonen, J., Kemppainen, O., Sarviharju, M. and Hervonen, A. (2001) dose-dependent decrease in glial fibrillary acidic protein-immunoreactivity in rat cerebellum after lifelong ethanol consumption. Alcohol, 23: 1–8.

Rivto, E.R., Freeman, B.J., Scheibel, A.B., Duong, T., Robinson, H., Guthrie, D. and Ritvo, A. (1986) Lower Purkinje cell counts in the cerebella of four autistic subjects: initial findings of the UCLA-NSAC Autopsy Research report. Am. J. Psychiatry, 143: 862–866.

Rutter, M., Caspi, A. and Moffitt, T. (2003) Using sex differences in psychopathology to study causal mechanisms: unifying issues and research strategies. J. Child Psychol. Psychiatry, 44: 1092–1115.

Sajdel-Sulkowska, E.M., Li, G.-H., Ronca, A.E., Baer, L.A., Sulkowski, G.M., Koibuchu, N. and Wade, C.E. (2001) Effects of hypergravity exposure on the developing central nervous system: possible involvement of thyroid hormone. Exp. Biol. Med., 226: 790–798.

Sajdel-Sulkowska, E.M. and Koibuchi, N. (2002) Impact of thyroid status-disrupting environmental factors on brain development. Recent Res. Devel. Endocr., 3: 101–117.

Sakamoto, H., Mezaki, Y., Shikimi, H., Ukena, K. and Tsutsui, K. (2003) Dendritic growth and spine formation in response to estrogen in the developing Purkinje cell. Endocrinology, 1444: 4466–4477.

Schmahmann, J.D (1991) An emerging concept. The cerebellar contribution to higher function. Arch. Neurol., 48: 1178–1187.

Scott, F.J., Baron-Cohen, S., Bolton, P. and Brayne, C. (2002) Brief report: prevalence of autism spectrum conditions in children aged 5–11 years in Cambridgeshire, UK. Autism, 6: 231–237.

Sedykh, A.I., Leshchinskaia, I.A., Duka, T.I., Nerush, P.A. and Pavlov, V.A. (1999) Effect of chronic emotional stress and low dose ethanol on the level of neuronal cell adhesion molecule and glail fibrillary acidic protein in rat brain. Ukr. Biokhim. Zh., 71: 97–103.

Seo, B.W., Li, M.H., Hansen, L.G., Moore, R.W., Peterson, R.E. and Schantz, S.L. (1995) Effects of gestational and lactational exposure to coplanar polychlorinated biphenyl (PCB) congeners or 2,3,7,8-tetrachlorodibenzo-p-dioxin (TCDD) on thyroid hormone concentrations in weanling rates. Toxicol Lett., 78: 253–262.

Steward, O., Kelley, M.S. and Torre, E.R. (1993) The process of reinnervation in the dentate gyrus of adult rats: temporal relationship between changes in the levels of glial fibrillary acidic protein (GFAP) and GFAP mRNA in reactive astrocytes. Exp. Neurol., 124: 1676–1683.

Suarez, I., Bodega, G., Rubio, M. and Fernandez, B. (1992) Sexual dimorphism in the hamster cerebellum demonstrated by glial fibrillary acidic protein (GFAP) and vimentin. Glia, 5: 10–16.

Sulkowski, G.M., Li, G.-H. and Sajdel-Sulkowska, E.M. (2004) Environmental impacts on the developing CNS: CD15, NCAM-L1, and GFAP expression in rat neonates exposed to hypergravity. Adv. Space Res., 33: 1423–1430.

Sun, X.Z., Zhang, R., Cui, C., Takahashi, S., Kubota, Y., Sawada, K. and Fukui, Y. (2003) Expression of neural cell adhesion molecule L1 in the brain of rats exposed to X-irradiation in utero. J. Med. Invest., 50: 187–191.

Thelen, K., Kedar, V., Panicker, A.K., Schmid, R.S., Midkiff, B.R. and Maness, P.F. (2002) The neural cell adhesion molecule L1 potentiates integrin-dependent cell migration to extracellular matrix protein. J. Neurosci., 22: 4918–4931.

Toran-Allerand, C.D., Bentham, W., Miranda, R.C. and Anderson, J.P. (1991) Insulin influences astroglial morphology and glial fibrillary acidic protein (GFAP) expression in organotypic cultures. Brain Res., 558: 296–304.

Townsend, J., Corchesne, E., Covington, J., Westerfield, M., Harris, N.S., Lyden, P., Lowry, T.P. and Press, G.A. (1999) Spatial attention deficits in patients with acquired developmental cerebellar abnormality. J. Neurosci., 19: 5632–5643.

Trevarthen, C. (2000) Autism as neurodevelopmental disorder affecting communication and learning in early childhood: prenatal origins, postnatal course and effective educational support. Prostglandins Leukot. Essent. Fatty Acids, 6: 41–46.

Van Den Eaden, S.K., Tanner, C.M., Bernstein, A.L., Fross, R.D., Leimpeter, A., Bloch, D.A. and Nelson, L.M. (2003) Incidence of Parkinson's disease: variation by age, gender, and race/ethnicity. Am. J. Epidemiol., 157: 1015–1022.

Vicedomini, J.P., Nonneman, A.J., DeKosky, S.T. and Scheff, S.W. (1986) Perinatal glucocorticoids disrupt learning: a sexually dimorphic response. Physiol. Behav., 36: 145–149.

Volkov, N.D., Wang, G.J., Fowler, J.S., Hitzemann, R., Pappas, N., Pascani, K. and Wong, C. (1997) Gender differences in cerebellar metabolism: test-retest reproducibility. Am. J. Psychiatry, 154: 119–121.

Yamasaki, M., Sasaki, T. and Torii, M. (1991) Sex difference in the pattern of lower limb movement during treadmill walking. Eur. J. Appl. Physiol. Occup. Physiol., 62: 99–103.

CHAPTER 28

Cerebellar dysfunction in multiple sclerosis: evidence for an acquired channelopathy

Stephen G. Waxman*

Department of Neurology and PVA/EPVA Center for Neuroscience Research, Yale University School of Medicine, New Haven, CT 06510 and Rehabilitation Research Center, VA Hospital, West Haven, CT 06516, USA

Keywords: Purkinje cells; sodium channels; action potential electrogenesis; acquired channelopathy

Abstract: Cerebellar dysfunction in multiple sclerosis (MS) is a significant contributor to disability, is relatively refractory to symptomatic therapy, and often progresses despite treatment with disease-modifying agents. Thus, there is a need for better understanding of its pathophysiology. This chapter reviews a growing body of evidence which suggests that mis-tuning of Purkinje cells, due to expression of an abnormal repertoire of sodium channels, contributes to cerebellar deficits in MS. Within the normal nervous system, sodium channel $Na_v1.8$ is expressed in a highly specific manner within spinal sensory and trigeminal neurons, and is not present within Purkinje cells. $Na_v1.8$ mRNA and protein are, however, expressed within Purkinje cells both in models of MS (experimenal autoimmume encephalomyelitis; EAE), and in postmortem tissue from humans with MS. Expression of $Na_v1.8$ within Purkinje cells in vitro alters electrogenesis in these cells in several ways: first, by increasing duration and amplitude of action potentials; second, by decreasing the proportion of action potentials that are conglomerate and the number of spikes per conglomerate action potential; and third, by supporting sustained, pacemaker-like impulse trains in response to depolarization, which are not seen in the absence of $Na_v1.8$. Similar changes are observed in recordings from Purkinje cells in vivo from mice with EAE. Taken together, these results suggest that expression of $Na_v1.8$ within Purkinje cells distorts their pattern of firing in MS.

Cerebellar dysfunction in multiple sclerosis is a significant contributor to disability. It is relatively refractory to symptomatic therapy and slowing of its progression by disease-modifying agents has not been demonstrated. Thus there is a need for a better understanding of the pathophysiology of cerebellar dysfunction in multiple sclerosis.

From a clinical perspective there are several features which distinguish cerebellar deficits from other neurological deficits in multiple sclerosis. First, clinical abnormalities due to cerebellar dysfunction tend to be persistent, even early in the course of the disease, in contrast to other types of clinical deficits which tend to be remitting (Matthews et al., 1991). Second, clinical abnormalities attributable to cerebellar dysfunction can occur in multiple sclerosis patients in the absence of cerebellar lesions that are manifested on neuroimaging. Third, occasional multiple sclerosis patients exhibit paroxysmal bouts of ataxia with a temporal profile (sudden, brief, stereotyped attacks) similar to the paroxysmal episodes seen in the episodic ataxias, a group of disorders caused by hereditary channelopathies. Interestingly, paroxysmal ataxia in multiple sclerosis has been reported to respond favorably to treatment with sodium channel blockers such as carbamazepine (Andermann et al., 1959; Espir et al., 1966). This paper will review a

*Tel.: +1 (203) 785-5947; Fax: +1 (203) 785-7826
E-mail: stephen.waxman@yale.edu

growing body of evidence which suggests that a previously unrecognized molecular and functional abnormality, mistuning of Purkinje cells due to expression of an abnormal repertoire of sodium channels (an acquired transcriptional channelopathy) contributes to cerebellar deficits in multiple sclerosis.

Plasticity and lability of sodium channel gene transcription

Neurophysiological doctrine traditionally referred to *the* sodium channel as if it were a singular entity. Analysis at the molecular level, however, has shown that there are at least nine different genes which encode distinct voltage-gated sodium channels with a common overall structural motif but with different amino acid sequences (Catterall, 1992; Plummer and Meisler, 1999; Goldin et al., 2000). The physiological signatures of most of these channels have now been studied and this analysis has demonstrated that different sodium channels can exhibit different voltage-dependences, activation and inactivation kinetics, and recovery properties. Different repertoires of sodium channels are deployed in different types of neurons and this contributes to their different functional characteristics, determining properties as fundamental for neuronal function as threshold, refractory period, and the temporal patterning of action potential generation (see, e.g., Stuart and Sakmann, 1995; Pennartz et al., 1997; Waxman, 2000; Taddese and Bean, 2002). In view of this it is not unexpected that changes in sodium channel transcription can have significant effects on neuronal function.

Even within the normal neuron system, the transcription of sodium channels within neurons is not a static process. On the contrary, it is dynamic. The levels of transcription for some sodium channels (e.g., $Na_v1.6$) increase during the course of development of some types of neurons, while transcription of others (e.g., $Na_v1.3$) is reduced (Beckh et al., 1989; Brysch et al., 1991; Felts et al., 1997). The multiple factors that regulate channel transcription, translation, and deployment are only beginning to be understood. Neurotrophic factors exert a complex set of effects on channel expression and appear to play important roles. Both NGF (Black et al., 1997; Dib-Hajj et al., 1998; Fjell et al., 1999a; Cummins et al., 2000) and GDNF (Fjell et al., 1999b; Boucher et al., 2000; Cummins et al., 2000) increase transcription of the $Na_v1.8$ and $Na_v1.9$ sodium channel genes, while decreasing transcription of the $Na_v1.3$ gene in spinal sensory neurons. Electrical activity also appears to modulate the transcription of sodium channels within myocytes (Offord and Catterall, 1989) and neurons (Sashihara et al., 1996, 1997; Klein et al., 2003). In addition, there are regulatory controls on post-transcriptional aspects of sodium channel expression. For example, an oligodendrocyte-derived soluble factor contributes to the regulation of clustering of $Na_v1.2$ channels along axons (Kaplan et al., 2001), while myelination is a prerequisite for clustering of $Na_v1.6$ channels at nodes of Ranvier (Boiko et al., 2001; Kaplan et al., 2001).

There is also plasticity in the transcription of ion channel genes within neurons in the uninjured adult nervous system. A striking example is provided by the magnocellular neurons within the hypothalamic supraoptic nucleus which, in response to increase in the osmolarity of the extracellular milieu, convert from a quiescent mode to a bursting mode in which they generate high-frequency bursts of action potentials that trigger the release of vasopressin at their terminals within the pituitary (Andrew and Dudek, 1983; Inenaga et al., 1993; Li and Hatton, 1996). Tanaka et al. (1999) used in situ hybridization, immunocytochemistry, and patch clamp methods to examine the expression of various sodium channel subtypes as these neurons undergo a transition from the quiescent to the bursting state. Their in situ hybridization results clearly show that, in association with the transition to the bursting state, transcription of mRNA for the $Na_v1.2$ and $Na_v1.6$ sodium channels (but not for $Na_v1.1$ or $Na_v1.3$) is up-regulated. Immunocytochemical studies showed that the increase in channel mRNA is accompanied by an increase in channel protein. Patch clamp recordings show that the newly produced channels are inserted into the neuronal cell membrane and are functional, producing both fast, transient currents (which underlie the depolarizing upstroke of the action potential) and persistent currents (which can be activated by small depolarizations close to resting potential, thus lowering the threshold for action potential generation) (Tanaka et al., 1999). Thus, even in the uninjured adult neuron system, plasticity

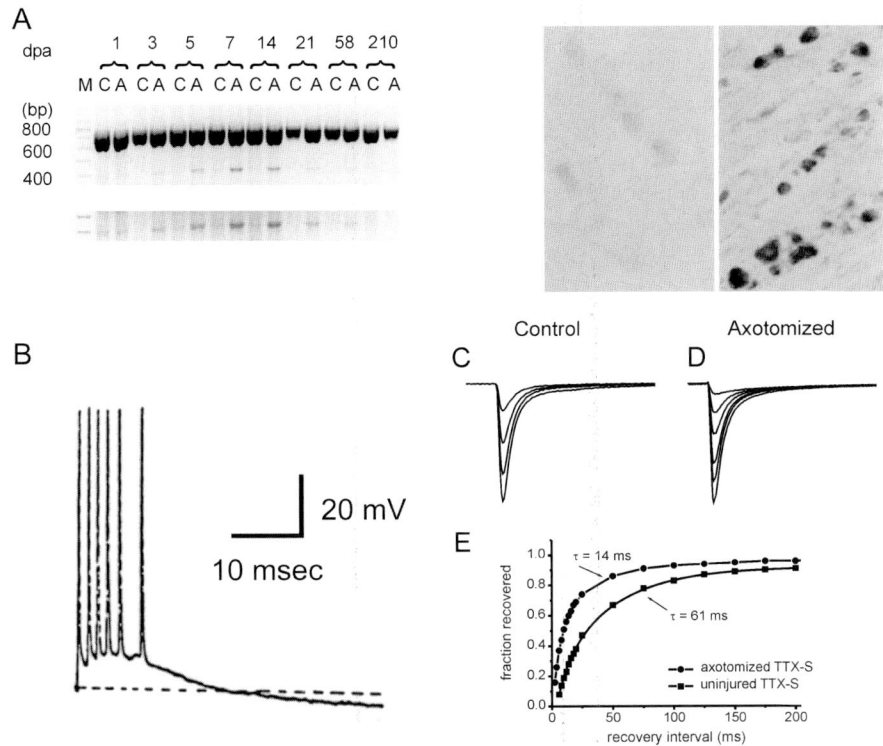

Fig. 1. A: Neuronal sodium channel expression can change following axonal injury. A, C–E: $Na_v1.3$ sodium channel expression is up-regulated in spinal sensory neurons following axonal transection. Gels (A, left) show RT-PCR products from controls (C) and after axotomy (A) following co-amplification of $Na_v1.3$ mRNA together with β-actin transcripts (days post-axotomy indicated above gels), with computer-enhanced images of amplification products shown below gels. Micrographs (A, right) show in situ hybridizations showing $Na_v1.3$ signal in control DRG and 5–7 days following axotomy. Modified from Dib-Hajj et al. (1996). B: Intra-axonal recording showing repetitive action potential activity in a previously transected, regenerating axon from rat sciatic nerve (one year post-crush), following block of potassium channels with 4-aminopyridine. The abnormal burst activity arises from a prolonged depolarization that follows the first action potential. This bursting, and the slow depolarization, are not seen in uninjured axons. Modified from Kocsis and Waxman (1983). C–E: A rapidly-repriming tetrodotoxin-sensitive current produced by $Na_v1.3$ emerges in DRG neurons following peripheral axotomy. C: Family of TTX-sensitive sodium current traces (with recovery times indicated) showing time course of recovery from inactivation at −80 mV, from a control DRG neuron. D: Similar family of traces showing accelerated repriming 7 days after peripheral axotomy. E: Single exponential fits showing accelerated recovery from inactivation following sciatic nerve transection. Modified from Black et al. (1999a).

of sodium channel expression can lead to a functional remodeling of neuronal membranes, which tunes neurons by regulating their electrogenicity.

Given the dynamic changes in channel expression that occur within the normal nervous system, it is not surprising that pathological events can trigger changes in ion channel expression within neurons.

Peripheral nerve injury and spinal cord injury

A precedent for the acquired channelopathy in multiple sclerosis is provided by studies on peripheral nerve injury and spinal cord injury. These studies indicate that altered sodium channel expression, triggered by axonal transection, can produce altered firing patterns in dorsal root ganglion neurons which contribute to clinically significant phenomena including paraesthesia and neuropathic pain. The intra-axonal recording in Fig. 1B (Kocsis and Waxman, 1983), from a previously transected, long-term regenerated axon in rat sciatic nerve, illustrates the tendency for injured axons to produce aberrant repetitive action potential activity which is not seen in uninjured axons. The bursting activity arises from a prolonged or 'slow' depolarization, lasting for more than 10 ms, with a configuration that suggests a contribution of sodium channels, not seen in normal axons, to repetitive, high-frequency firing.

Recent studies demonstrate that nerve injury does, in fact, trigger dysregulation of sodium channel gene transcription. The changes triggered by nerve injury include the up-regulated expression of $Na_v1.3$ sodium channel mRNA that is normally expressed only at very low levels in spinal sensory neurons (Waxman et al., 1994; see Fig. 1A). Newly formed $Na_v1.3$ mRNA is translated so that $Na_v1.3$ sodium channel protein, which is not detectable in uninjured spinal sensory neurons or their axons, is produced (Black et al., 1999a). Concomitant with the up-regulation of $Na_v1.3$ expression, patch-clamp studies show the emergence of a new sodium current, characterized by rapid repriming (i.e., rapid recovery from inactivation) within the axotomized neurons (Cummins and Waxman, 1997) (Fig. 1C–E). Confirmation that $Na_v1.3$ channels produce a significant part of this current has been provided by patch clamp recordings on heterologously expressed $Na_v1.3$ channels and on $Na_v1.3$ channels experimentally expressed in DRG neurons (Cummins et al., 2001). Rapid repriming of the inappropriately expressed $Na_v1.3$ channels produces a decrease in refractory period which leads to high-frequency action potential activity, thus contributing to hyperexcitability of nociceptive spiral sensory neurons following nerve injury (Cummins and Waxman, 1997) and spinal cord injury (Hains et al., 2003). Knockdown of $Na_v1.3$ via antisense oligodeoxynucleotides reduces the hyperexcitability and ameliorates the pain associated with experimental spinal cord injury (Hains et al., 2003)

In the case of nerve injury, several lines of evidence point to loss of access to peripheral pools of neurotrophic factors including NGF and GDNF as factors that contribute to altered channel expression. For example, DRG neurons respond differently to axonal transection within peripheral nerves (which project to target tissues that produce neurotrophins), compared to transection within dorsal roots (which project to the spinal cord). Up-regulated transcription of $Na_v1.3$ is triggered by transection of the peripherally directed branch of DRG neurons within the sciatic nerve, but not by central axotomy within the dorsal root (Black et al., 1999a). Moreover, experimental delivery of NGF and GDNF down-regulates $Na_v1.3$ expression in spinal sensory neurons (Black et al., 1997; Boucher et al., 2000; Cummins et al., 2000; Leffler et al., 2002). Thus, nerve injury— in part by interrupting access to peripheral pools of trophic factors—can produce changes in the pattern of sodium channel gene transcription in spinal sensory neurons that alter the excitability of these cells, and these changes in channel gene expression and the resultant changes in neuronal activity can be clinically significant, contributing to the generation of neuropathic pain.

A cerebellar channelopathy in multiple sclerosis and its models

In addition to demyelination and axonal degeneration, recent evidence suggests that a transcriptional channelopathy may also contribute to neuronal dysfunction in multiple sclerosis (Waxman, 2002). To determine whether there are changes in sodium channel transcription with surviving neurons in demyelinating diseases, Black et al. (1999b) initially studied the mutant *Taiep* rat, a model in which myelin initially ensheaths CNS axons normally but subsequently degenerates due to an abnormality of myelin-forming oligodendrocytes (Duncan et al., 1992). These studies focused on expression of the $Na_v1.8$ sodium channel (which was originally termed SNS, **S**ensory **N**euron **S**pecific), a slowly-inactivating tetrodotoxin (TTX)-resistant sodium channel which is normally detectable only in spinal sensory neurons and trigeminal neurons (Akopian et al., 1996; Sangameswaren et al., 1996). Using in situ hybridization Black et al. (1999b) showed that, following loss of myelin, there was markedly enhanced expression of $Na_v1.8$ mRNA within Purkinje cells in this model system. Immunocytochemical studies showed that the mRNA was translated, i.e., that the up-regulation of $Na_v1.8$ mRNA is accompanied by the production of $Na_v1.8$ protein in Purkinje cells.

Black et al. (2000) then examined the expression of $Na_v1.8$ in the brains of mice with chronic-relapsing experimental allergic encephalomyelitis (CR-EAE), an inflammatory model of MS, and in postmortem tissue from humans with MS. These studies used the Biozzi mouse CR-EAE model (Baker et al., 1990)

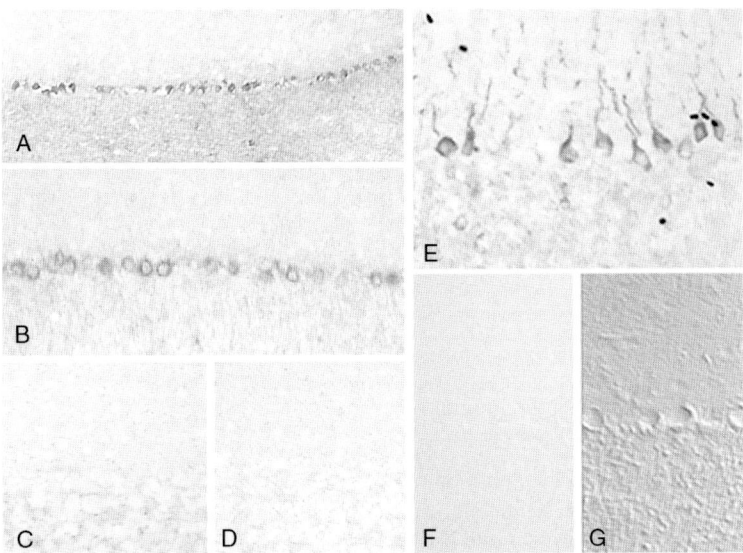

Fig. 2. Sodium channel $Na_V1.8$ (originally termed SNS, Sensory Neuron Specific) is abnormally expressed within cerebellar Purkinje neurons in mice with EAE. A–D: in situ hybridization showing $Na_V1.8$ mRNA within Purkinje cells in EAE (A, B) but not in healthy control mice (C) or after hybridization with sense riboprobes (D). E–G: Immunostaining with $Na_V1.8$-specific antibodies, showing up-regulated expression of $Na_V1.8$ protein in EAE (E) compared to controls (F, bright field; G, Nomarski; image). A, 120; B–D, × 200; E–G, × 220. From Black et al. (2000).

which reproducibly provides lesions within the cerebellum, a characteristic that is not commonly observed in monophasic EAE (Baker et al., 2000). In situ hybridization demonstrated significantly increased expression of $Na_V1.8$ mRNA within Purkinje cells in CR-EAE (Fig. 2A–D). As shown in Fig. 2E–G, expression of $Na_V1.8$ protein within Purkinje cells was also up-regulated (Black et al., 2000). The up-regulated $Na_V1.8$ transcription was not part of a global increase in expression of sodium channels, because there was no up-regulation of expression of $Na_V1.9$, another TTX-resistant sodium channel that is normally expressed in a preferential pattern, like $Na_V1.8$, in spinal sensory neurons. Study of postmortem brain tissue, from patients with disabling progressive multiple sclerosis with cerebellar deficits on neurological examination, also demonstrated up-regulation of $Na_V1.8$ mRNA and protein within human Purkinje cells (Fig. 3) (Black et al., 2000).

These observations in two animal models of multiple sclerosis, and in postmortem tissue from humans with multiple sclerosis, demonstrate that expression of $Na_V1.8$ is up-regulated in Purkinje neurons, producing $Na_V1.8$ sodium channel protein which cannot normally be detected in these cells.

Annexin II/p11 is a protein that binds to the N-terminus of $Na_V1.8$ and facilitates the insertion of functional channels in the neuronal cell membrane (Okuse et al., 2002). Immunocytochemical studies have demonstrated (Fig. 4) that annexin II/p11 is up-regulated within Purkinje cells, and co-localized with $Na_V1.8$, in EAE and MS (Craner et al., 2003a). The demonstration of coordinated up-regulation of both $Na_V1.8$ and its binding partner annexin II/p11 suggests that functional $Na_V1.8$ channels are inserted into Purkinje cell membranes in multiple sclerosis.

Functional consequences of $Na_V1.8$ up-regulation

What are the functional consequences of up-regulated expression of $Na_V1.8$ in Purkinje cells in multiple sclerosis? One possibility is that $Na_V1.8$ up-regulation might be an adaptive change, providing a substrate for the restoration of action potential conduction along demyelinated Purkinje cell axons. Renganathan et al. (2001) found that $Na_V1.8$ channels produce a substantial fraction of the inward current that flows during the upstroke of the action potential in the DRG neurons in which these

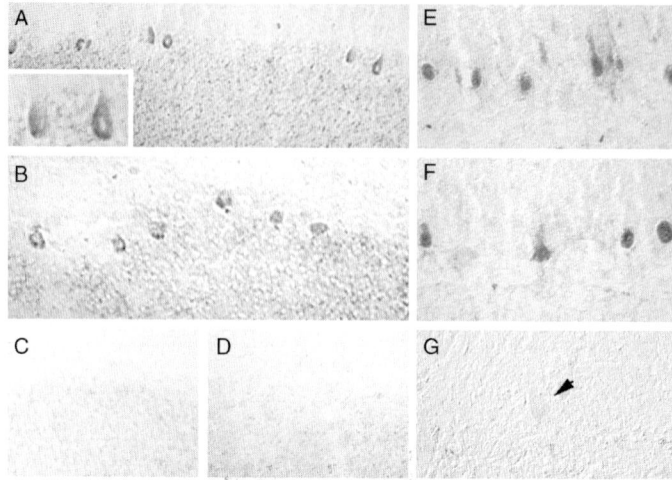

Fig. 3. Expression of the Sensory Neuron Specific (SNS) sodium channel $Na_v1.8$ is up-regulated in cerebellar Purkinje neurons from patients with multiple sclerosis. In situ hybridization with $Na_v1.8$ specific antisense riboprobes demonstrates increased $Na_v1.8$ mRNA in Purkinje cells from multiple sclerosis patients obtained at post-mortem (A, B), compared to controls without neurological disease (C). No signal is present following hybridization with sense riboprobe (D). Immunocytochemistry with $Na_v1.8$-specific antibodies demonstrates up-regulation of $Na_v1.8$ channel protein in Purkinje cells from multiple sclerosis patients (E, F) compared to controls (G; arrowhead indicates Purkinje cell). A: × 120, inset × 280; B, C, D: × 165; E, F, G: × 175. From Black et al. (2000).

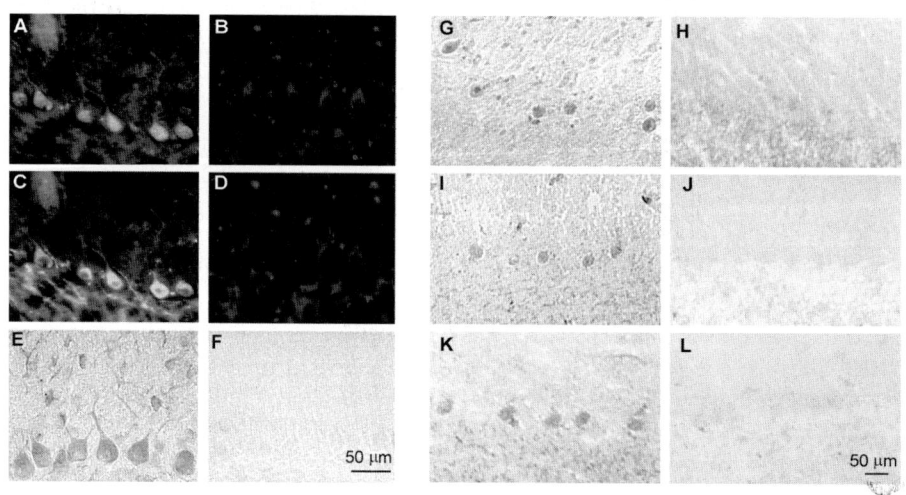

Fig. 4. AnnexinII/p11, which facilitates insertion of functional $Na_v1.8$ channels into the cell membrane, is up-regulated and co-expressed with $Na_v1.8$ in Purkinje cells in EAE and multiple sclerosis. Left panels: $Na_v1.8$ (A) and annexinII/p11 (C) are up-regulated in Purkinje neurons in EAE vs control (B and D, respectively). Note the co-localization of annexinII/p11 and $Na_v1.8$ in the same neurons (compare panels A & C). AnnexinII/p11 immunostaining extends along the proximal portion of the dendritic tree in EAE (E) but not in control tissue (F). Right: $Na_v1.8$ is up-regulated in Purkinje neurons in postmortem MS tissue (G) vs control (H). Up-regulated expression of annexinII/p11 is shown for two MS cases (I and K) in comparison to controls (J and L). Modified from Craner et al. (2003b).

channels are normally present. Thus, it might be predicted that if $Na_v1.8$ channels were inserted into the demyelinated axon membrane, they might contribute to restoration of action potential conduc-tion in demyelinated axons. However, the physiological results described below suggest that $Na_v1.8$ would alter the accommodative properties of the axons. Moreover, $Na_v1.8$ protein has not been detected along demyelinated Purkinje cell axons, in contrast to $Na_v1.2$ and $Na_v1.6$ protein which can be clearly detected along axons that have been demyelinated (Craner et al., 2003b). These results argue against an adaptive role of $Na_v1.8$ in restoration of conduction along demyelinated axons.

The available evidence suggests that up-regulated expression of $Na_v1.8$ in EAE and MS is a maladaptive change. Electrogenesis within Purkinje cells is known to depend, in part, on the activity of sodium channels (Llinas and Sugimori, 1980; Stuart and Hausser, 1994; Raman and Bean, 1997). Mutations of the sodium channels that are normally expressed in Purkinje cells can produce substantial changes in patterns of impulse generation in these neurons which can result in clinical signs of cerebellar dysfunction including ataxia (Kohrman et al., 1996; Raman et al., 1997). $Na_v1.8$ exhibits a unique physiological signature including depolarized voltage-dependence of inactivation, slow development of inactivation (Akopian et al., 1996; Sangameswaren et al., 1996), and rapid recovery from inactivation (Elliott and Elliott, 1993; Dib-Hajj et al., 1997). Because of these properties $Na_v1.8$ channels are predicted to be available over a wider range of dynamic activity and membrane potential than other sodium channels (Schild and Kunze, 1997), and cells expressing $Na_v1.8$ should be more slowly adapting than cells lacking $Na_v1.8$ (Elliott and Elliott, 1993).

A number of experimental observations support the prediction that expression of $Na_v1.8$ does, indeed, alter the firing patterns of neurons and show, in particular, that $Na_v1.8$ expression in Purkinje cells distorts their temporal pattern of activity. Renganathan et al. (2001) used voltage clamp and current clamp recording to examine the pattern of electrogenesis in DRG neurons from transgenic $Na_v1.8$ $-/-$ DRG neurons in which functional $Na_v1.8$ channels are not present (Akopian et al., 1999), and to compare it with electrogenesis in $Na_v1.8$ $+/+$ neurons. This study showed that the expression of $Na_v1.8$ channels within DRG neurons markedly influences both the configuration of the action potentials [a change that can effect activation of N-type channels and thus transmitter release (Scroggs and Fox, 1992) if $Na_v1.8$ is deployed to axon terminals; see Fig. 5D] and the temporal pattern of firing in response to depolarizing stimuli. As shown in Fig. 5A,B, current clamp recordings demonstrated that $Na_v1.8$ $+/+$ C-type DRG neurons can produce

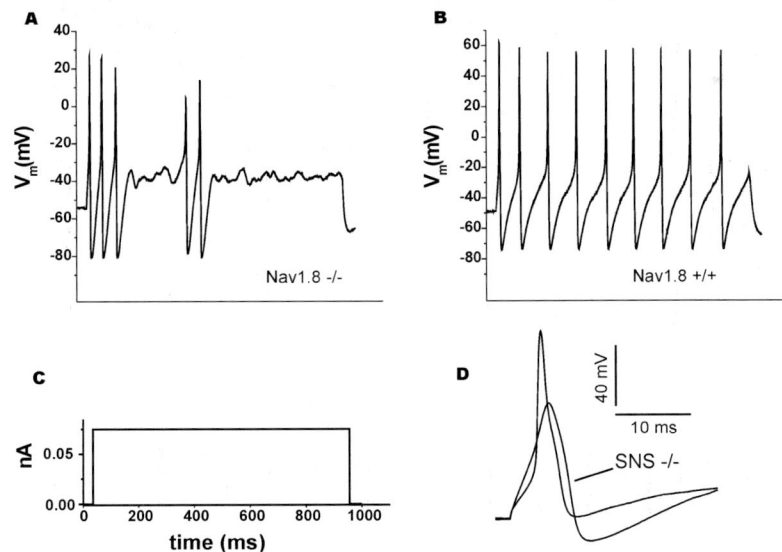

Fig. 5. Expression of $Na_v1.8$ channels influences the firing pattern and action potential configuration of DRG neurons. A–C: The firing pattern in response to an identical depolarizing stimulse (C) is different in a $Na_v1.8$ $+/+$ dorsal root ganglion (DRG) neuron (B) compared with a $Na_v1.8$ $-/-$ DRG neuron from a $Na_v1.8$ knockout mouse (A). D: Action potential configuration is also different in $Na_v1.8$ $+/+$ and $Na_v1.8$ $-/-$ neurons. Note the greater amplitude of the action potential from a $Na_v1.8$ $+/+$ neuron. From Renganathan et al. (2001).

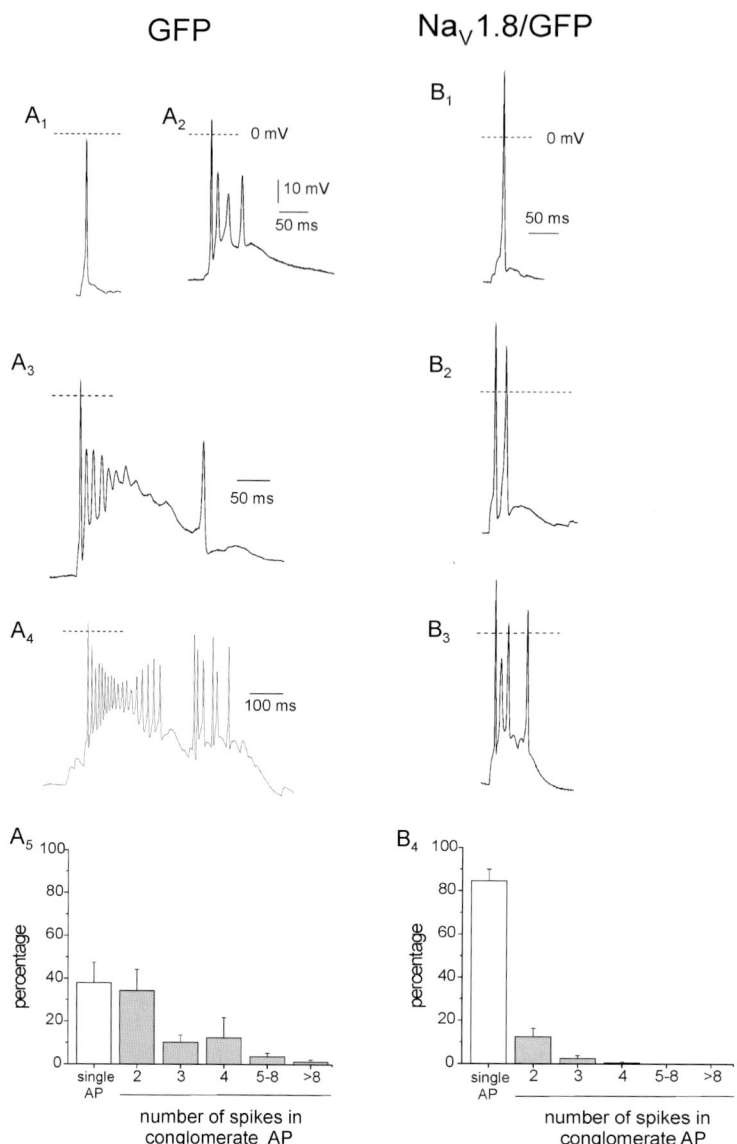

Fig. 6. Expression of $Na_v1.8$ alters action potential electrogenesis within Purkinje cells. These current clamp recordings show spontaneous action potentials recorded in Purkinje neurons two days after biolistic expression of GFP which provided a marker of transfection (without $Na_v1.8$) (A_1–A_4) or of $Na_v1.8$/GFP (B_1–B_3). A_1–A_4: Action potentials in control neurons lacking $Na_v1.8$ (A_1–A_4) show little if any overshoot (dotted lines indicate 0 mV) and tend to be conglomerate (62%; A_2–A_4). B_1–B_3: Action potentials in neurons expressing $Na_v1.8$ display larger overshoot. Conglomerate action potentials are less common after expression of $Na_v1.8$ (15%) and, when present, tend to consist of doublets (B_2), only rarely consisting of > 2 spikes (B_3). Time calibration in A_2 applies to A_1; time calibration in B_1 applies to B_1–B_3. The mV calibration in A_2 applies to all panels. A_5, B_4: Percentage of action potentials that were single, or conglomerate with 2, 3, 4, 5–8 or > 8 spikes, in Purkinje cells lacking $Na_v1.8$ and with $Na_v1.8$ respectively. There is a lower percentage of conglomerate action potentials and smaller number of spikes per conglomerate action potential in Purkinje neurons expressing $Na_v1.8$. Modified from Renganathan et al. (2003).

sustained pacemaker-like trains of action potentials in response to depolarizing stimuli, which are not present within $Na_v1.8$ $-/-$ C-type DRG neurons, consistent with the suggestion that cells expressing $Na_v1.8$ should be slowly adapting.

More recent studies have directly examined the effect of expression of $Na_v1.8$ within Purkinje neurons. One study (Renganathan et al., 2003) used patch clamp to examine Purkinje cells transfected in vitro with $Na_v1.8$. This study demonstrated that $Na_v1.8$ channels can be functionally expressed at physiological levels (4.5 nA) within Purkinje cells which subsequently display slowly inactivating, TTX-resistant current with properties identical to those of native $Na_v1.8$ channels, and with current amplitudes similar to those observed within DRG neurons (where $Na_v1.8$ is normally expressed). Current clamp recordings demonstrated that, following the expression of $Na_v1.8$ within Purkinje cells, their pattern of electrogenesis is altered in several ways: first, by increasing the duration and amplitude of action potentials (compare Fig. 6B and 6A), second, by decreasing the proportion of action potentials that are conglomerate (15% in $Na_v1.8$-transfected neurons, compared with 62% in controls) and the number of spikes per conglomerate action potential which was 2.13 in $Na_v1.8$-transfected neurons compared to 3.38 in controls (again, compare Fig. 6B and 6A); and third, by supporting sustained, pacemaker-like impulse trains in response to depolarization, which are not seen in the absence of $Na_v1.8$ (Fig. 7).

These results demonstrate that aberrant expression of $Na_v1.8$ within Purkinje cells in vitro distorts their pattern of activity. More recent studies have begun to provide evidence for similar changes in the firing patterns of Purkinje cells in vivo. Saab et al. (2004) observed changes in the firing patterns of Purkinje cells in vivo in mice with EAE, with fewer spikes per conglomerate active potential. Runs of abnormal, non-adapting high-frequency firing were also seen in EAE (Fig. 8). Taken together, these results show that expression of $Na_v1.8$ within Purkinje neurons in vitro distorts the pattern of activity of these cells, a change that probably interferes with cerebellar function, and indicate that similar changes occur in vivo in EAE. Whether the abnormal expression of $Na_v1.8$ leads to degeneration

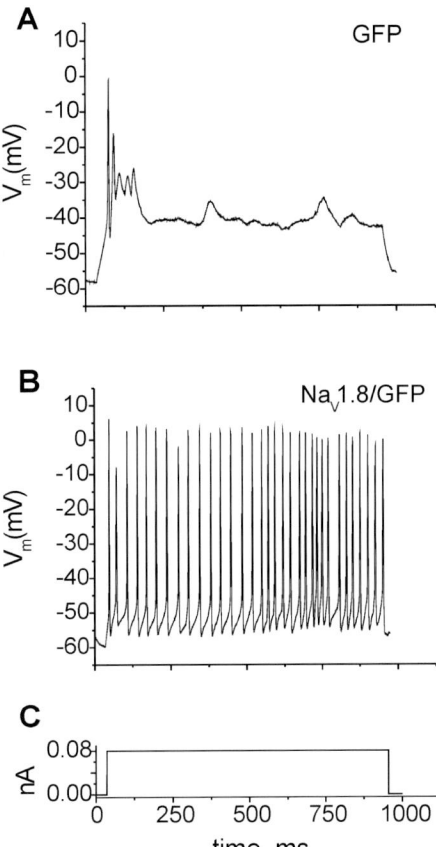

Fig. 7. Purkinje neurons transfected with $Na_v1.8$ show sustained repetitive firing, not present in the absence of $Na_v1.8$, on injection of depolarizing current. A: Control Purkinje neuron lacking $Na_v1.8$ produces a conglomerate action potential consisting of five spikes, but no sustained firing, in response to a sustained depolarizing stimulus (80 pA, 1 s). B: Purkinje neuron expressing $Na_v1.8$ produces larger-amplitude action potentials and shows sustained pacemaker-like activity in response to identical stimulus. The current pulse protocol is shown in C. Modified from Renganathan et al. (2003).

of Purkinje cells, in addition to changes in their pattern of firing, is not known at this time.

Because biopsy of cerebellar tissue is usually not performed in multiple sclerosis, it will not be easy to directly establish whether these physiological changes occur in Purkinje cells in humans with multiple sclerosis. Consistent with a role of $Na_v1.8$ in producing cerebellar dysfunction, Craner et al. (2003c) observed, in a relapsing-remitting model of

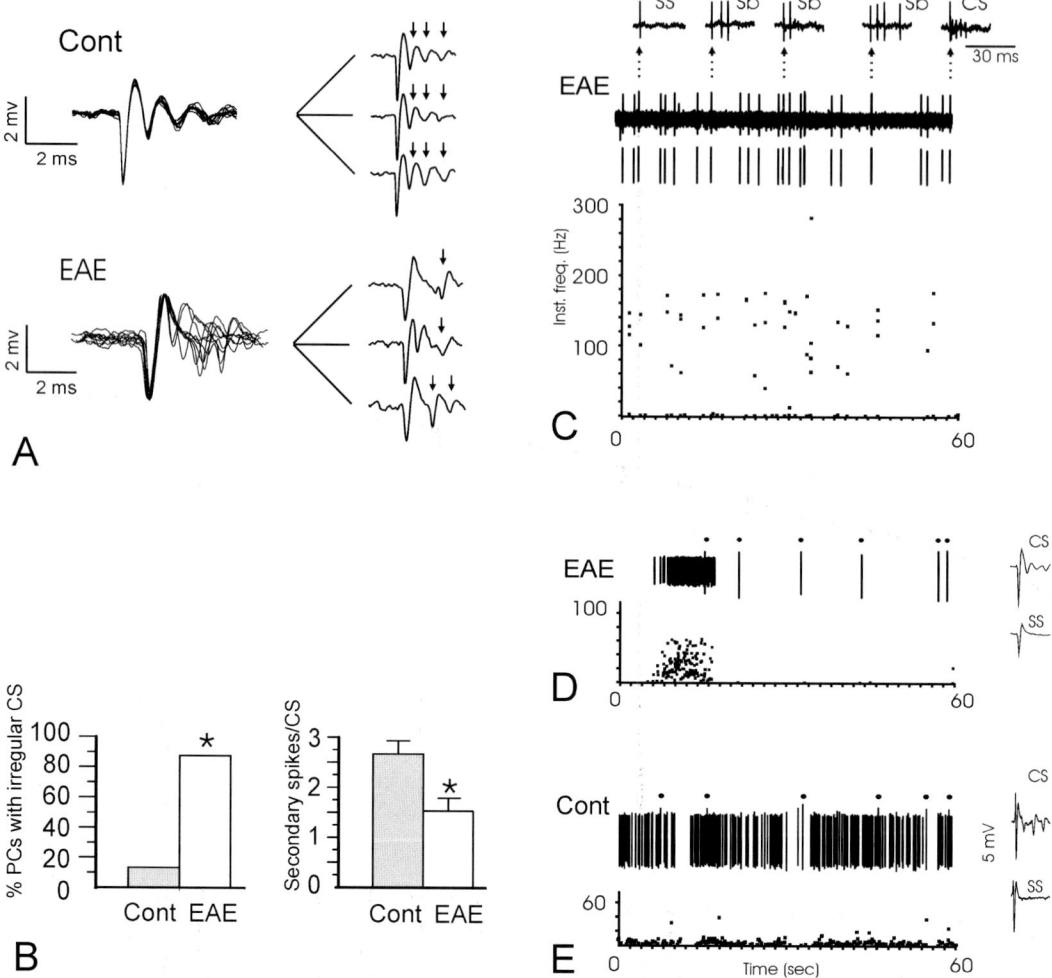

Fig. 8. The firing patterns of Purkinje cells within their native cerebellar environment in vivo are altered in EAE. A: Conglomerate action potentials in Purkinje cells from control (upper traces) and EAE (lower traces). Superimposed conglomerate action potentials from single cells show regularity of secondary spikes in controls, and irregularity in EAE. Individual conglomerate action potentials display irregularity in latencies of secondary spikes (downward arrows in the right traces in EAE). B, Left: Histogram showing the percentage of Purkinje cells with irregular temporal organization of conglomerate action potentials in controls and EAE (*$p < 0.001$). B, right: Histogram showing the average number of secondary spikes per conglomerate action potential in Purkinje cells isolated from control mice and mice with EAE (*$p < 0.005$). C–E: Abnormal high-frequency bursting in Purkinje cell in EAE. C: Action potentials recorded from Purkinje cell in EAE which produced brief bursts (Sb) of repetitive single spikes consisting of doublets, triplets or quadruplets, not seen in controls, interspersed with isolated single spikes (SS) and conglomerate action potentials. (C, top). The second and third rows show a continuous recording from one of these cells on a slow time base, and isolated action potentials plotted below using template matching techniques, which were also used to plot instantaneous frequency of action potentials (C, D, E bottom row). In another Purkinje cell (D), sustained high-frequency SS bursts (up to 60 Hz) are followed by an extended period (more than 3 min) of no recorded SS activity (conglomerate action potentials are indicated by dots; individual conglomerate action potentials and SS are shown to the right). E: These activity patterns were not observed in Purkinje cells recorded in control mice. From Saab et al. (2004).

EAE, that the level of $Na_v1.8$ expression within Purkinje cells increases progressively and is correlated with the severity of non-remitting (including cerebellar) deficits.

The hypothesis that $Na_v1.8$ channels, aberrantly expressed within Purkinje neurons, contribute to symptomatology in multiple sclerosis would be strengthened if it could be shown that $Na_v1.8$ blocking drugs ameliorate ataxia or other cerebellar abnormalities. Drugs that specifically block $Na_v1.8$ are not yet available, but this is likely to change since the deployment of $Na_v1.8$ within nociceptive neurons has made it an attractive molecular target. When $Na_v1.8$-specific blocking drugs are developed, a next step will be to examine the effect of these drugs on animal models of multiple sclerosis such as EAE and, if indicated, in humans with multiple sclerosis. Improvement in cerebellar function in response to $Na_v1.8$-specific blockade would provide strong evidence for a cerebellar channelopathy due to up-regulated expression of $Na_v1.8$, and might provide a new approach to the treatment of cerebellar deficits in multiple sclerosis.

Acknowledgments

Research in the author's laboratory has been supported, in part, by grants from the National Multiple Sclerosis Society, the Medical Research Service and Rehabilitation Research Service, Department of Veterans Affairs, the Eastern Paralyzed Veterans Association, the Paralyzed Veterans of America and the Nancy Davis Foundation.

References

Akopian, A.N., Sivilotti, L. and Wood, J.N. (1996) A tetrodotoxin-resistant voltage-gated sodium channel expressed by sensory neurons. Nature, 379: 257–262.

Akopian, A.N., Souslova, V., England, S., Okuse, K., Ogata, N., Ure, J., Smith, A., Kerr, B.J., McMahon, S.B., Boyce, S., Hill, R., Stanfa, L.C., Dickenson, A.H. and Wood, J.N. (1999) The tetrodotoxin-resistant sodium channel SNS has a specialized function in pain pathways. Nat. Neurosci., 2: 541–548.

Andermann, F., Cosgrove, J.B.R., Lloyd-Smith, D. and Walters, A.M. (1959) Paroxysmal dysarthria and ataxia in multiple sclerosis. Neurology, 9: 21–216.

Andrew, R.D. and Dudek, F.E. (1983) Burst discharge in mammalian neuroendocrine cells involves an intrinsic regenerative mechanism. Science, 221: 1050–1052.

Baker, D., O'Neill, J.K., Gschmeissner, S.E., Wilcox, C.E., Butter, C. and Turk, J.L. (1990) Induction of chronic relapsing experimental allergic encephalomyelitis in biozzi mice. J. Neuroimmunology, 28: 261–270.

Baker, D., Pryce, G., Croxford, J.L., Brown, P., Pertwee, R.G., Huffman, J.W. and Layward, L. (2000) Cannabinoids control spasticity and tremor in a multiple sclerosis model. Nature, 404: 84–87.

Beckh, S., Noda, M., Lubbert, H. and Numa, S. (1989) Differential regulation of three sodium channel messenger RNAs in the rat central nervous system during development. EMBO J., 8: 3611–3616.

Black, J.A., Langworthy, K., Hinson, A.W., Dib-Hajj, S.D. and Waxman, S.G. (1997) NGF has opposing effects on Na^+ channel III and SNS gene expression in spinal sensory neurons. NeuroReport, 8: 2331–2335.

Black, J.A., Cummins, T.R., Plumpton, C., Chen, Y.H., Hormuzdiar, W., Clare, J.J. and Waxman, S.G. (1999a) Upregulation of a silent sodium channel after peripheral, but not central, nerve injury in DRG neurons. J. Neurophysiol., 82: 2776–2785.

Black, J.A., Fjell, J., Dib-Hajj, S., Duncan, I.D., O'Connor, L.T., Fried, K., Gladwell, Z., Tate, S. and Waxman, S.G. (1999b) Abnormal expression of SNS/PN3 sodium channel in cerebellar Purkinje cells following loss of myelin in the taiep rat. NeuroReport, 10: 913–918.

Black, J.A., Dib-Hajj, S., Baker, D., Newcombe, J., Cuzner, M.L. and Waxman, S.G. (2000) Sensory neuron specific sodium channel SNS is abnormally expressed in the brains of mice with experimental allergic encephalomyelitis and humans with multiple sclerosis. Proc. Natl. Acad. Sci., 97: 11598–11602.

Boiko, T., Rasband, M.N., Levinson, S.R., Caldwell, J.H., Mandel, G., Trimmer, J.S. and Matthews, G. (2001) Compact myelin dictates the differential targeting of two sodium channel isoforms in the same axon. Neuron, 30: 91–104.

Boucher, T.J., Okuse, K., Bennett, D.L.H., Munson, J.B., Wood, J.N. and McMahon, S.B. (2000) Potent analgesic effects of GDNF in neuropathic pain states. Science, 290: 124–127.

Brysch, W., Creutzfeldt, O.W., Luno, K., Schlingensiepen, R. and Schlingensiepen, K.-H. (1991) Regional and temporal expression of sodium channel messenger RNAs in the rat brain during development. Exp. Brain Res., 86: 562–567.

Catterall, W.A. (1992) Cellular and molecular biology of voltage-gated sodium channels. Physiol. Rev., 72: 515–548.

Craner, M.J., Lo, A.C., Black, J.A. and Waxman, S.G. (2003a) Abnormal sodium channel distribution in optic nerve axons in a model of inflammatory demyelination. Brain, 126: 1552–1561.

Craner, M.J., Lo, A.C., Black, J.A., Baker, D., Newcombe, J., Cuzner, M.L. and Waxman, S.G. (2003b) Annexin II/p11 is up-regulated in Purkinje cells in EAE and MS. NeuroReport, 14: 555–558.

Craner, M.J., Kataoka, Y., Lo, A.C., Black, J.A., Baker, D. and Waxman, S.G. (2003c) Temporal course of upregulation of Nav1.8 in Purkine neurons parallels the progression of clinical deficit in EAE. J. Neuropath. Exper. Neurol., 62: 968–976.

Cummins, T.R., Black, J.A., Dib-Hajj, S.D. and Waxman, S.G. (2000) Glial-derived neurotrophic factor upregulates expression of functional SNS and NaN sodium channels and their currents in axotomized dorsal root ganglion neurons. J. Neurosci., 20: 8754–8761.

Cummins, T.R., Aglieco, F., Renganathan, M., Herzog, R.I., Dib-Hajj, S.D. and Waxman, S.G. (2001) $Na_v1.3$ sodium channels: rapid repriming and slow closed-state inactivation display quantitative differences following expression in a mammalian cell line and in spinal sensory neurons. J. Neurosci., 21: 5952–5961.

Cummins, T.R. and Waxman, S.G. (1997) Down-regulation of tetrodotoxin-resistant sodium currents and up-regulation of a rapidly repriming tetrodotoxin-sensitive sodium current in small spinal sensory neurons following nerve injury. J. Neurosci., 17: 3503–3514.

Dib-Hajj, S.D., Black, J.A., Felts, P. and Waxman, S.G. (1996) Down-regulation of transcsripts for Na channel α-SNS in spinal sensory neurons following axotomy. Proc. Natl Acad. Sci., 93: 14950–14954.

Dib-Hajj, S.D., Black, J.A., Cummins, T.R., Kenney, A.M., Kocsis, J.D. and Waxman, S.G. (1998) Rescue of alpha-SNS/PN3 sodium channel expression in small dorsal root ganglion neurons after axotomy by nerve growth factor in vivo. J. Neurophysiol., 79: 2668–2676.

Dib-Hajj, S.D., Ishikawa, I., Cummins, T.R. and Waxman, S.G. (1997) Insertion of an SNS-specific tetrapeptide in the S3–S4 linker of D4 accelerates recovery from inactivation of skeletal muscle voltage-gated Na channel μ1 in HEK293 cells. FEBS Letts., 416: 11–14.

Duncan, I.D., Lunn, K.F., Holmgren, B., Urba-Holmgren, R. and Brignolo-Holmes, L. (1992) The taiep rat: a myelin mutant with an associated oligodendrocyte microtubular defect. J. Neurocytology, 21: 870–884.

Elliott, A.A. and Elliott, J.R. (1993) Characterization of TTX-sensitive and TTX-resistant sodium currents in small cells from adult rat dorsal root ganglia. J. Physiol. (Lond.), 463: 39–56.

Espir, M.L.E., Watkins, S.M. and Smith, H.V. (1966) Paroxysmal dysarthria and other transient neurological disturbances in MS. J. Neurol. Neurosurg. Psychiat., 29: 323–330.

Felts, P.A., Yokoyama, S., Dib-Hajj, S., Black, J.A. and Waxman, S.G. (1997) Sodium channel α-subunit mRNAs I, II, III, NaG, Na6 and hNE: Different expression patterns in developing rat nervous system. Molec. Brain Res., 45: 71–83.

Fjell, J., Cummins, T.R., Fried, K., Black, J.A. and Waxman, S.G. (1999a) In vivo NGF deprivation reduces SNS/PN3 expression and TTX-R sodium currents in IB4-negative DRG neurons. J. Neurophys., 81: 803–911.

Fjell, J., Cummins, T.R., Dib-Hajj, S.D., Fried, K., Black, J.A. and Waxman, S.G. (1999b) Differential role of GDNF and NGF in the maintenance of two TTX-resistant sodium channels in adult DRG neurons. Mol. Brain Res., 67: 267–282.

Goldin, A.L., Barchi, R.L., Caldwell, J.H., Hofmann, F., Howe, J.R., Hunter, J.C., Kallen, R.G., Mandel, G., Meisler, M.H., Netter, Y.B., Noda, M., Tamkun, M.M., Waxman, S.G., Wood, J.N. and Catterall, W.A. (2000) Nomenclature of voltage-gated sodium channels. Neuron, 2: 365–368.

Hains, B.C., Klein, J.P., Saab, C.Y., Craner, M.J., Black, J.A. and Waxman, S.G. (2003) Upregulation of sodium channel $Na_v1.3$ and functional involvement in neuronal hyperexcitability associated with central neuropathic pain after spinal cord injury. J. Neurosci., 23: 8881–8892.

Inenaga, K., Nagamoto, T., Kannan, H. and Yamashita, H. (1993) Inward sodium current involvement in regenerative bursting activity of rat magnocellular supraoptic neurons in vitro. J. Physiol. Lond., 465: 289–301.

Kaplan, M.R., Cho, M.H., Ullian, E.M., Isom, L.L., Levinson, S.R. and Barres, B.A. (2001) Differential control of clustering of the sodium channels Na(v)1.2 and Na(v)1.6 at developing CNS nodes of Ranvier. Neuron, 30: 105–119.

Klein, J.P., Tendi, E.A., Dib-Hajj, S.D., Fields, R.D. and Waxman, S.G. (2003) Patterned electrical activity modulates sodium channel expression in sensory neurons. J. Neurosci. Res., 74: 192–198.

Kocsis, J.D. and Waxman, S.G. (1983) Long-term regenerated nerve fibres retain sensitivity to potassium channel blocking agents. Nature, 304: 640–642.

Kohrman, D.C., Smith, M.R., Goldin, A.L., Harris, J. and Meisler, M.H. (1996) A missense mutation in the sodium channel Scn8a is responsible for cerebellar ataxia in the mouse mutant jolting. J. Neurosci., 16: 5993–5999.

Leffler, A., Cummins, T.R., Dib-Hajj, S.D., Hormuzdiar, W.N., Black, J.A. and Waxman, S.G. (2002) Glial-derived neurotrophic factor and nerve growth factor reverse changes in repriming of TTX-sensitive Na^+ currents following axotomy of dorsal root ganglion neurons. J. Neurophysiol., 88: 650–660.

Li, Z. and Hatton, G.I. (1996) Oscillatory bursting of physically firing rat supraoptic neurones in low-Ca^{2+} medium: Na^+ influx, cytosolic Ca^{2+} and gap junctions. J. Physiol. Lond., 496: 394–397.

Llinas, R. and Sugimori, M. (1980) Electrophysiological properties of in vitro Purkinje cell somata in mammalian cerebellar slices. J. Physiol., 305: 171–195.

Matthews, W.B., Compston, A., Allen, I.V. and Martyn, C.N. (1991) McAlpine's Multiple Sclerosis. Churchill Livingstone, New York.

Offord, J. and Catterall, W.A. (1989) Electrical activity, cAMP, and cytosolic calcium regulate mRNA encoding sodium channel α subunits in rat muscle cells. Neuron, 2: 1447–1452.

Okuse, K., Malik-Hall, M., Baker, M.D., Poon, W.Y.L., Kong, H., Chao, M.V. and Wood, J.N. (2002) Annexin II light chain regulates sensory neuron-specific sodium channel expression. Nature, 47: 653–656.

Pennartz, C.M.A., Bierlaagh, M.A. and Geurtsen, A.M.S. (1997) Cellular mechanisms underlying spontaneous firing in rat suprachiasmatic nucleus: imvolvement of a slowly inactivating component of sodium current. J. Neurophysiol., 78: 1811–1825.

Plummer, W. and Meisler, M.H. (1999) Evolution and diversity of sodium channel genes. Genomics, 57: 323–331.

Raman, I.M. and Bean, B.P. (1997) Resurgent sodium current and action potential formation in dissociated cerebellar purkinje neurons. J. Neurosci., 17: 4517–4526.

Raman, I.M., Sprunger, L.K., Meisler, M.H. and Bean, B.P. (1997) Altered subthreshold sodium currents and disrupted firing patterns in Purkinje neurons of Scn81 mutant mice. Neuron, 19: 881–891.

Renganathan, M., Cummins, T.R. and Waxman, S.G. (2001) The contribution of $Na_v1.8$ sodium channels to action potential electrogenesis in DRG neurons. J. Neurophysiol., 86: 629–640.

Renganathan, M., Gelderblom, M., Black, J.A. and Waxman, S.G. (2003) Expression of $Na_v1.8$ sodium channels perturbs the firing patterns of cerebellar Purkinje cells. Brain Res., 959: 235–243.

Saab, C.Y., Craner, M.J., Kataoka, Y., Waxman, S.G. (2004) Abnormal Purkinje cell activity *in vivo* in experimental allergic encephalomyelitis, in preparation. Exper. Brain Res., in press.

Sangameswaren, L., Delgado, S.G., Fish, L.M., Koch, B.D., Jakeman, L.B., Stewart, G.R., Sze, P., Hunter, J.C., Eglen, R.M. and Herman, R.C. (1996) Structure and function of a novel voltage-gated tetrodotoxin-resistant sodium channel specific to sensory neurons. J. Biol. Chem., 271: 5953–5956.

Sashihara, S., Greer, C.A., Oh, Y. and Waxman, S.G. (1996) Cell-specific differential expression of $Na+$ channel β1 subunit mRNA in the olfactory system during postnatal development and following denervation. J. Neurosci., 16: 702–714.

Sashihara, S., Waxman, S.G. and Greer, C.A. (1997) Down-regulation of Na^+ channel mRNA following sensory deprivation of tufted cells in the neonatal rat olfactory bulb. NeuroReport, 8: 1289–1293.

Schild, J.H. and Kunze, D.L. (1997) Experimental and modeling study of Na^+ current heterogeneity in rat nodose neurons and its impact on neuronal discharge. J. Neurophysiol., 78: 3198–3209.

Scroggs, R.S. and Fox, A.P. (1992) Multiple Ca^{2+} currents elicited by action potential waveforms in acutely isolated adult rat dorsal root ganglion neurons. J. Neurosci., 12: 1789–1801.

Stuart, G. and Hausser, M. (1994) Initiation and spread of sodium action potentials in cerebellar purkinje cells. Neuron, 13: 703–712.

Stuart, G. and Sakmann, B. (1995) Amplification of epsps by axosomatic sodium channels in neocortical phramidal neurons. Neuron, 15: 1065–1076.

Taddese, A. and Bean, B. (2002) Subthreshold sodium current from rapidly inactivating sodium channels drive spontaneous firing of tuberomammillary neurons. Neuron, 33: 587–600.

Tanaka, M., Cummins, T.R., Ishikawa, K., Black, J.A., Ibata, I. and Waxman, S.G. (1999) Molecular and functional remodeling of electrogenic membrane of hypothalamic neurons in response to changes in their input. Proc. Natl. Acad. Sci. USA, 96: 1088–1093.

Waxman, S.G. (2000) The neuron as a dynamic electrogenic machine: Modulation of sodium channel expression as a basis for functional plasticity in neurons. Phil. Trans. Roy. Soc. Lond. B, 355: 199–213.

Waxman, S.G. (2001) Transcriptional channelopathies: an emerging class of disorders. Nature Rev. Neurosci., 2: 652–659.

Waxman, S.G. (2002) Ion channels and neuronal dysfunction in multiple sclerosis. Arch. Neurol., 59: 1377–1380.

Waxman, S.G., Kocsis, J.D. and Black, J.A. (1994) Type III sodium channel mRNA is expressed in embryonic but not adult spinal sensory neurons, and is re-expressed following axotomy. J. Neurophysiol., 72: 466–471.

CHAPTER 29

Don't get too excited: mechanisms of glutamate-mediated Purkinje cell death

Jennifer E. Slemmer, Chris I. De Zeeuw and John T. Weber*

Department of Neuroscience, Erasmus Medical Center, Dr. Molenwaterplein 50, P.O. Box 1738, 3000 DR Rotterdam, The Netherlands

Keywords: Bergmann glia; cytoskeleton; dendritic spines; excitotoxicity; GLAST; microtubules; S-100β; traumatic brain injury

Abstract: Purkinje cells (PCs) present a unique cellular profile in both the cerebellum and the brain. Because they represent the only output cell of the cerebellar cortex, they play a vital role in the normal function of the cerebellum. Interestingly, PCs are highly susceptible to a variety of pathological conditions that may involve glutamate-mediated 'excitotoxicity', a term coined to describe an excessive release of glutamate, and a subsequent over-activation of excitatory amino acid (NMDA, AMPA, and kainite) receptors. Mature PCs, however, lack functional NMDA receptors, the means by which Ca^{2+} enters the cell in classic hippocampal and cortical models of excitotoxicity. In PCs, glutamate predominantly mediates its effects, first via a rapid influx of Ca^{2+} through voltage-gated calcium channels, caused by the depolarization of the membrane after AMPA receptor activation (and through Ca^{2+}-permeable AMPA receptors themselves), and second, via a delayed release of Ca^{2+} from intracellular stores. Although physiological levels of intracellular free Ca^{2+} initate vital second messenger signaling pathways in PCs, excessive Ca^{2+} influx can detrimentally alter dendritic spine morphology via interactions with the neuronal cytoskeleton, and thus can perturb normal synaptic function. PCs possess various calcium-bining proteins, such as calbindin-D28K and parvalbumin, and glutamate transporters, in order to prevent glutamate from exerting deleterious effects. Bergmann glia are gaining recognition as key players in the clearence of extracellular glutamate; these cells are also high in S-100β, a protein with both neurodegenerative and neuroprotective abilities. In this review, we discuss PC-specific mechanisms of glutamate-mediated excitotoxic cell death, the relationship between Ca^{2+} and cytoskeleton, and the implications of glutamate, and S-100β for pathlogical conditions, such as traumatic brain injury.

Introduction

Purkinje cells (PCs) present a unique cellular profile in both the cerebellum and the brain. Because they represent the only output cell of the cerebellar cortex, they play a vital role in the normal function of the cerebellum. Interestingly, PCs are highly susceptible to a variety of pathological conditions that may involve glutamate-mediated 'excitotoxicity,' a term coined to describe an excessive release of glutamate, and a subsequent over-activation of excitatory amino acid (NMDA, AMPA, and kainate) receptors. Mature PCs, however, lack functional NMDA receptors, the means by which Ca^{2+} enters the cell in classic hippocampal and cortical models of excitotoxicity. In PCs, glutamate predominantly mediates its effects first via a rapid influx of Ca^{2+} through voltage-gated calcium channels, caused by the depolarization of the membrane after AMPA receptor activation (and through Ca^{2+}-permeable AMPA receptors themselves), and second, via a

*Corresponding author. Tel.: +31-10-4089252; Fax: +31-10-4089459; E-mail: j.weber@erasmusmc.nl

DOI: 10.1016/S0079-6123(04)48029-7

delayed release of Ca^{2+} from intracellular stores. Although physiological levels of intracellular free Ca^{2+} initiate vital second-messenger signaling pathways in PCs, excessive Ca^{2+} influx can detrimentally alter dendritic spine morphology via interactions with the neuronal cytoskeleton, and thus can perturb normal synaptic function. PCs possess various calcium-binding proteins, such as calbindin-D28K and parvalbumin, and glutamate transporters, in order to prevent glutamate from exerting deleterious effects. Bergmann glia are gaining recognition as key players in the clearance of extracellular glutamate; these cells are also high in S-100β, a protein with both neurodegenerative and neuroprotective abilities. In this review, we discuss PC-specific mechanisms of glutamate-mediated excitotoxic cell death, the relationship between Ca^{2+} and the cytoskeleton, and the implications of glutamate and S-100β for pathological conditions, such as traumatic brain injury.

Purkinje cells (PCs) of the cerebellar cortex receive two types of excitatory synaptic inputs. The first is from climbing fibers (CFs), arising from the inferior olive (IO); the second is from numerous parallel fibers (PFs), the axons of granule cells, also in the cerebellar cortex. The dendrites of granule cells receive their afferent projections from mossy fibers, which are also excitatory. The axons of PCs project to cerebellar and vestibular nuclei and, because they represent the only output cell of the cerebellar cortex, PCs play a vital role in normal cerebellar function, such as fine-tuning movement and posture.

Although PCs may be innervated by multiple CFs during development, they receive input from a single CF in adulthood. Despite being innervated by only a single CF, there are approximately 1500 CF–PC synaptic contacts (Strata and Rossi, 1998), which are characterized by a high probability of release (Dittman and Regehr, 1998). CFs primarily contact the more proximal portions of the PC dendritic tree, while PFs form synaptic contacts at the spines of the distal portion of the PC dendrites. The distal dendrites of PCs receive input from upwards of 100,000 PFs. Neurotransmission at both CF–PC and PF–PC synapses is mediated by the excitatory amino acid, glutamate.

Excess release of glutamate and over-activation of excitatory amino acid receptors, known as glutamate-mediated 'excitotoxicity,' is believed to play a large role in several pathological processes, such as ischemia, traumatic brain injury (TBI) and neurodegenerative disorders like Alzheimer's, Huntington's and Parkinson's disease (Farooqui and Horrocks, 1994; Weber, 2004). The fact that each CF innervates one PC with hundreds of synapses not only provides a secure synaptic connection, but the synchronous release of glutamate at these synapses can put PCs at risk for excitotoxic damage. In addition, the vast number of glutamatergic PF–PC synapses also represents a risk of excessive glutamate neurotransmission. Indeed, a very recent review (Sarna and Hawkes, 2003) highlights the susceptibility of PCs to a variety of pathological conditions that may involve excitotoxic mechanisms. Given the unique role of PCs as the only output cell from the cerebellar cortex, understanding the mechanisms underlying their susceptibility to excitotoxic mechanisms is important for developing possible treatment strategies for various disease states.

In this review, we discuss the mechanisms underlying glutamate-mediated death in PCs. To this end, we first review the types of glutamate receptors in the central nervous system (CNS), and general mechanisms of excitotoxicity in neurons and specifically in PCs. The roles of glutamate transporters and calcium-binding proteins in PCs, and of Bergmann glia (BG), are also examined. We then turn our attention to glutamate-induced cytoskeletal alterations in dendrites and dendritic spines, again first in neurons in general and in PCs in specific. Neuronal microtubules and their associated proteins are reviewed in detail before discussing data gathered on the interactions between cytoskeletal and postsynaptic proteins in PCs. Lastly, we discuss these interactions and the role of glutamate excitotoxicity on cerebellar traumatic injury, focusing on PCs in vivo and adding our own in vitro data. We also discuss novel theories on the role of glial S-100β protein in cell dysfunction. The complex interactions among PCs, presynaptic input from PFs and CFs, and BG, can be appreciated by referring to Fig. 1 throughout this review.

Types of glutamate receptors in the CNS

Glutamate activates a variety of functionally distinct receptors in the CNS (for reviews, see Ozawa et al.,

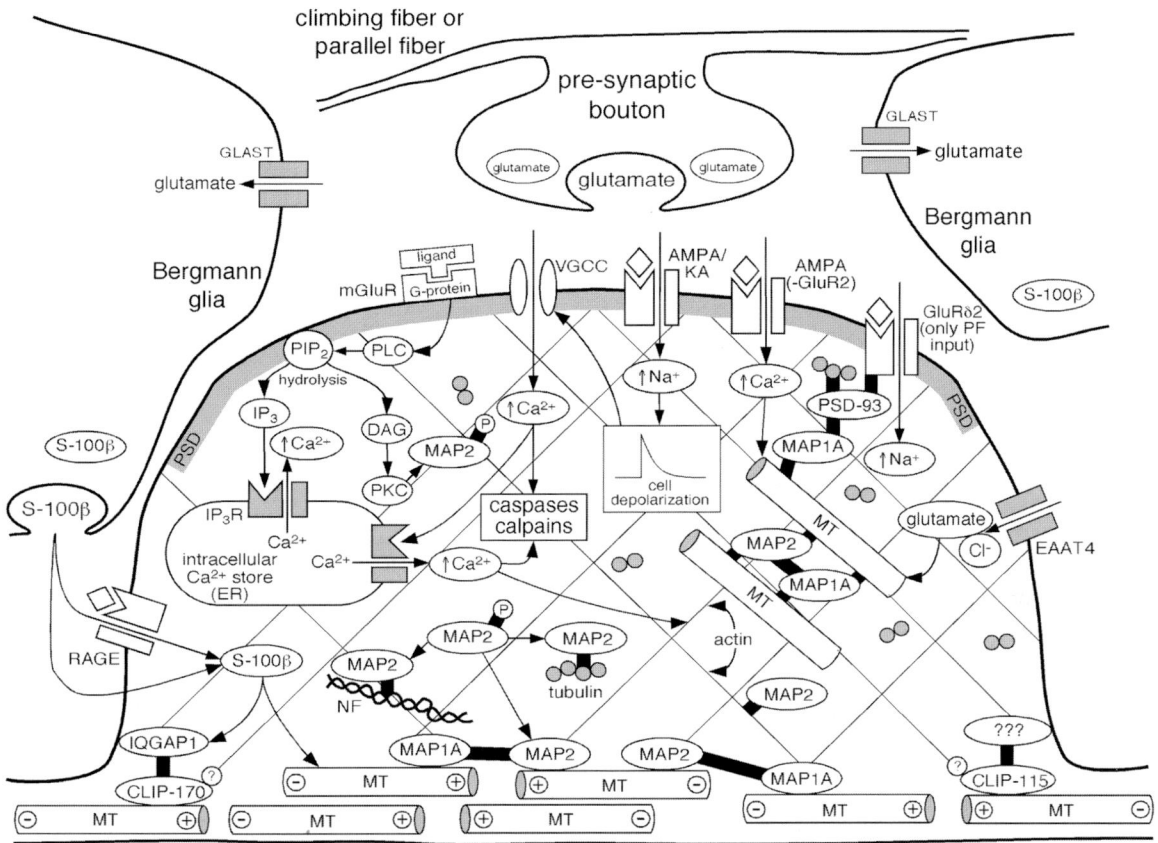

Fig. 1. **Glutamate-mediated excitotoxicity in a Purkinje cell spine.** Glutamate released from a pre-synaptic bouton from either a climbing fiber or a parallel fiber into the synaptic cleft initiates several pathways in Purkinje cells. Involvement of Bergmann glia is also shown, as these cells can uptake glutamate via GLAST, and can release S-100β. A putative scaffold protein that associates with CLIP-115 (similar to the association between IQGAP1 and CLIP-170) is indicated by question marks. Putative contact points between CLIP-115 and actin, and between CLIP-170 and actin, are also indicated by question marks. See text for further explanations. Abbreviations: AMPA/KA: Na^+-permeable AMPA/KA receptor; AMPA (-GluR2): Ca^{2+}-permeable AMPA receptor; CLIP-115: cytoplasmic linker protein of 115 kDa; CLIP-170: cytoplasmic linker protein of 170 kDa; DAG: diacylglycerol; EAAT4: excitatory amino acid (glutamate) and chloride co-transporter; GLAST: glutamate-aspartate transporter; GluRδ2: PC-specific glutamate receptor subtype which receives input from parallel fibers (PFs) only; IP_3: inositol-(1,4,5)-trisphosphate; IP_3R: IP_3 receptor; IQGAP1: scaffold protein; MAP1A: microtubule associated protein 1A; MAP2: microtubule associated protein 2; MAP2 with P: phosphorlyated MAP2; MT: microtubule; NF: neurofilament; PIP_2: phosphatidylinositol bisphosphate; mGluR: metabotropic glutamate receptor; PLC: phospholipase C; PKC: protein kinase C; PSD: post-synaptic density; PSD-93: post-synaptic density protein 93; RAGE: receptor for advanced glycation end-products; VGCC: voltage-gated calcium channel.

1998, and Coutinho and Knöpfel, 2002). These include ionotropic receptors, such as the α-amino-3-hydroxy-5-methyl-4-isoxazolepropionate (AMPA), kainate, and N-methyl-D-aspartate (NMDA) receptors, as well as metabotropic glutamate receptors (mGluRs). The ionotropic receptors are composed of various subunits that form membrane channels, allowing a flux of ions into the cell from the extracellular space. NMDA receptors (which consist of the subunits NR1, NR2A–D and NR3A–D) are well distributed in the brain, and activation of this receptor subtype causes an influx of Ca^{2+} into cells. AMPA (GluR1-4 subunits) and kainate (GluR5-7; KA1 and 2 subunits) receptor activation leads primarily to Na^+ influx, although some types of these receptors are also permeable to Ca^{2+} depending on their subunit combination. AMPA receptor activation can also indirectly lead to Ca^{2+} influx by causing membrane depolarization which activates voltage-gated Ca^{2+} channels (VGCCs), as well as by increasing the activation of NMDA receptors, which are sensitive to the membrane potential of the cell.

There are also several types of mGluRs, divided into three groups, which are activated directly by binding of glutamate (for review, see Coutinho and Knöpfel, 2002). Group I mGluRs (mGluR1 and mGluR5) are linked through G-proteins which are coupled to phospholipase C (PLC). PLC cleaves phosphatidylinositol bisphosphate (PIP_2) from the cell membrane, producing diacylglycerol (DAG), which activates the enzyme protein kinase C (PKC), and inositol 1,4,5-trisphosphate (IP_3), which binds to IP_3 receptors on intracellular Ca^{2+} stores located on the endoplasmic reticulum (ER), resulting in a release of Ca^{2+} from the stores and an elevation of intracellular free Ca^{2+} ($[Ca^{2+}]_i$). In many types of neurons, the depletion of Ca^{2+} stores also stimulates influx of extracellular Ca^{2+} through channels on the plasma membrane, a process termed 'capacitative Ca^{2+} influx' (Bouron, 2000; Weber et al., 2001; Baba et al., 2003). Group II (mGluR2 and 3) and III (mGluR4 and 6–8) mGluRs are also G-protein-linked receptors that inhibit adenylate cyclase, which leads to a decrease in the second messenger, cAMP. In general, activation of group II and III mGluRs leads to a decrease in $[Ca^{2+}]_i$ (Coutinho and Knöpfel, 2002). mGluRs can also have other types of modulatory effects on membrane ion channels. For example, some studies have shown that activation of group I mGluRs can enhance the elevation of $[Ca^{2+}]_i$ mediated by NMDA receptors (Bruno et al., 1995; Rahman and Newman, 1996; Pisani et al., 2001).

General mechanisms of glutamate-mediated toxicity

Glutamate excitotoxicity occurs primarily in two phases: first, a rapid phase in which excessive glutamate receptor activation leads to increased $[Na^+]_i$, $[Cl^-]_i$, and $[H_2O]_i$, inducing cell swelling (Rothman, 1985); and second, a more delayed phase in which increased Ca^{2+} influx and release of Ca^{2+} from intracellular stores leads to activation of Ca^{2+}-dependent enzymes, inducing degradative and apoptotic cell death pathways (Trump and Berezesky, 1995). These phases have been well characterized in a wide range of in vivo and in vitro preparations, in both cortical and hippocampal neurons. In general, excitotoxicity is linked to overactivation of NMDA receptors in most areas of the brain, for example, in the hippocampus, and in particular, in the CA1 region (Rosenmund and Westbrook, 1993; Arias et al., 1997; Blaabjerg et al., 2001). In cortical cells, NMDA activation and Ca^{2+} influx couples directly to cell death pathways involving excess nitric oxide (NO) production from neuronal nitric oxide synthase (nNOS) (Sattler et al., 1999; Aarts et al., 2002).

Although PCs express various NMDA receptor subunits, they apparently do not express functional NMDA receptors in mature animals (Llano et al., 1991; Kataoka and Ohmori, 1996). Therefore, unraveling the pathways that lead to PC death after exposure to a high concentration of glutamate, may not be as clear as for other types of neurons. AMPA receptors represent the major class of glutamatergic ionotropic receptors in PCs, which are expressed at both CF (Zhang et al., 1990) and PF synapses (Elias et al., 1993). Upon activation, there is an influx of Na^+ and a subsequent depolarization of the membrane, causing an influx of Ca^{2+} through VGCCs. The mGluR types 1, 3, 4, and 7 have been reported in PCs (for review, see Knöpfel and Grandes, 2002). PCs also express GluRδ2 receptors, which are located in the dendritic spines of PF–PC synapses (Takayama et al., 1996). The GluRδ2 subunit is uniquely expressed in PC spines from early development, where it stabilizes synapses between PCs and PFs, but not between PCs and CFs (Kurihara et al., 1997). The deletion of GluRδ2 in mice leads to the persistence of multiple CF innervations, a loss of PC–PF synapses, and impaired coordination (Kashiwabuchi et al., 1995). Kurihara et al. (1997) investigated the role of GluRδ2 in early development, and found that, whereas 98–99% of PC spines form synapses with PFs in wild-type mice, that number is reduced to almost half in GluRδ2 mutant mice. Kurihara et al. (1997) postulated that signaling downstream normally triggers the switch in CF innervation from multiple to a single fiber per PC. Although glutamate receptors in PCs are critical in maintaining normal synaptic neurotransmission, excitotoxic levels of glutamate can alter PC morphology and function, leading to altered cerebellar output.

Glutamate action on Purkinje cells

The release of glutamate into the synaptic cleft by single-shock activation of PFs or CFs causes a depolarization of the PC membrane, which is mediated by AMPA receptor activation (Eilers et al., 1995; Schmolesky et al., 2002). The repetitive stimulation of PFs appears to activate two signaling pathways in PCs: (1) a rapid influx of Ca^{2+} through VGCCs caused by the depolarization of the membrane by AMPA receptor activation; and (2) a delayed release of Ca^{2+} from intracellular stores caused by the production of IP_3 by mGluR activation (Finch and Augustine, 1998). CNQX, an AMPA receptor antagonist, is able to block the initial, but not the late, rise in Ca^{2+}. Llano et al. (1991) have also shown that glutamate causes a large but transient increase in dendritic $[Ca^{2+}]_i$ in both Ca^{2+}-containing and Ca^{2+}-free media, indicating that part of the increase in $[Ca^{2+}]_i$ is attributable to release from intracellular Ca^{2+} stores.

Recently it has been shown that mGluR antagonists have little effect on the glutamate-induced production of IP_3 in cultured PCs, but that the AMPA receptor antagonist CNQX could block most of the glutamate-mediated IP_3 production (Okubo et al., 2001). In this preparation, AMPA receptor activation, and the resultant depolarization and activation of VGCCs, is specifically required for IP_3 production. These results were corroborated in acute cerebellar slices, in which electrical stimulation of CFs induced IP_3 production. However, only PF and CF co-activation would produce sufficient IP_3 to activate IP_3 receptors, leading to an increase in $[Ca^{2+}]_i$ in the spines of PCs (Okubo et al., 2001). In another study, the depletion of ER Ca^{2+} stores instigated the docking of the ER to the plasma membrane, where IP_3 receptors could open store-operated Ca^{2+} channels (Ma et al., 2000), similar to a capacitative mechanism reported in other neuronal types (Bouron, 2000; Weber et al., 2001; Baba et al., 2003). Under certain experimental conditions, activation of PFs or CFs produces an mGluR-mediated slow membrane conductance (Tempia et al., 2001; Dzubay and Otis, 2002). Recently, it has been demonstrated that mGluR1 activation is coupled directly to activation of TRPC1 membrane cation channels at PF synapses, which allow additional influx of Ca^{2+} into PCs (Kim et al., 2003). This TRPC1 channel appears to be responsible for the mGluR-mediated slow membrane potential conductance at PF–PC synapses. Therefore, glutamate receptor activation appears to stimulate a wide variety of mechanisms, which can lead to Na^+ and Ca^{2+} elevation in PCs.

Mechanisms of excitotoxicity in Purkinje cells: basic findings

Exactly which of the aforementioned pathways can lead to toxicity in PCs is not known, but much data has been pieced together from a variety of studies. Brorson et al. (1994) found that activation of non-NMDA receptors was sufficient to cause excitotoxic death in cultured PCs independent of Na^+ (and thus independent of activation of VGCCs after depolarization) but was dependent on direct Ca^{2+} permeation, apparently through non-NMDA receptors. Kainate application in Ca^{2+}-free solution slightly attenuated cell death, but the removal of Na^+ did not block kainate-induced toxicity. CNQX, NOS inhibitors, and calpain inhibitors, however, were effective in reducing PC death without blocking kainate-induced Ca^{2+} influx. Therefore, Ca^{2+}-mediated toxicity appeared to be dependent on activation of degradative enzymes, such as phospholipases, endonucleases, and proteases, as reported for other types of cells (Weber, 2004). Further studies in cerebellar cultures found that PCs displayed different reactions to excitotoxic agonists. Whereas Ca^{2+}-dependent toxicity via AMPA or glutamate was selective for PCs as compared to other cerebellar neurons, PCs were spared from NMDA-mediated toxicity (Brorson et al., 1995). Neuritic beading caused by kainite application has also been found at regions with large, localized increases in $[Ca^{2+}]_i$, and in close proximity to synaptic contact points (Bindokas and Miller, 1995). Increased $[Ca^{2+}]_i$ hindered the restoration of normal $[Na^+]_i$, leading to both localized swelling (i.e., beads) and the activation of Ca^{2+}-dependent degradative enzymes (Bindokas and Miller, 1995). An example of a beaded axon in a cultured PC can be seen in Fig. 4B. Other previous reports have also theorized that the excitotoxic degeneration of PCs is caused by an exacerbated increase in $[Ca^{2+}]_i$, leading to activation of calcium-dependent enzymes (Choi and Rothman, 1990; Schmidt-Kastner and Freund, 1991; for review, see Weber, 2004).

Mechanisms of excitotoxicity in Purkinje cells: mutant rodent models

Several transgenic and mutant rodent models have shed further light on the mechanisms of glutamate-mediated PC death (for extensive review, see Sarna and Hawkes, 2003). For example, the spastic *Han-Wistar* mutant rat model has proven useful in studying some cause and effect relationships in PC degeneration. The original characterization of these mutants indicated that this rat might constitute a model for excitotoxicity (Cohen et al., 1991). These rats display glutamate dysfunction in both the hippocampus and the cerebellum, and by P60, there is a reduction of PCs of approximately 30% (Nisim et al., 1999). Non-NMDA receptor antagonists, such as CNQX, provided neuroprotection in a dose-dependent manner by attenuating PC loss. Several intrinsic factors could lead to the ataxia and, ultimately, to the premature death that occur in these mutants. These rats express very low levels of the GluR2 subunit (Margulies et al., 1993), and decreased GluR2 makes AMPA receptors permeable to Ca^{2+} (Carriedo et al., 1998). This subunit is developmentally down-regulated, leading to Ca^{2+} influx, and allowing levels of $[Ca^{2+}]_i$ similar to that after NMDA receptor activation. Also, experiments in *Xenopus* oocytes injected with mutant *Han-Wistar* rat cerebellar mRNA elicited significantly enhanced responses to glutamate or kainate (Cohen et al., 1991). Taken together, these findings implicate glutamate-mediated toxicity, which leads to high Ca^{2+} influx, as a causal factor for PC death in this model.

In another study, *Han-Wistar* rats were given injections of the non-competitive NMDA receptor antagonists, MK-801 and ketamine, and their neuroprotective abilities were measured in both the hippocampus and the cerebellum (Brunson et al., 2001). Both compounds, but more importantly MK-801, which has a much higher affinity for the NMDA receptor than ketamine, were able to attenuate degeneration in the hippocampus, likely by binding to the NMDA receptor and reducing the influx of Ca^{2+} into the cell. Interestingly, these compounds also reduced PC death. The ability of these compounds to provide neuroprotection to PCs, however, requires a more complex explanation and cannot be exerted through the same mechanism as in the hippocampus, considering that PCs do not express functional NMDA receptors. Therefore, a more indirect mechanism of NMDA receptor antagonism must be involved, such as by decreasing presynaptic glutamate release. Similar findings have been demonstrated in vitro, where application of MK-801 increased PC survival (Mount et al., 1993).

In an extensive study, Tolbert and Clark (2000) studied the effects of ablation of the IO in normal and *Shaker* mutant rats, which exhibit a hereditary form of ataxia. *Shaker* rats display a predictable pattern (both spatially and temporally) of PC degeneration, starting at around the seventh week, and lasting for a period of six to eight weeks. The authors hypothesized that IO ablation in mutant rats would rescue PCs that were destined to die. Although ablation had no effect on PCs in normal rats, ablation surprisingly accelerated PC loss in mutant rats, indicating that the *Shaker* phenotype may be due to olivocerebellar deafferentation, which caused the loss of a trophic signal necessary for PC survival. All of these studies in mutant rats indicate that PCs are vulnerable to glutamate-mediated excitotoxicity even downstream of the initial action of glutamate.

Two mutant mice complement the PC degeneration seen in the *Shaker* rat. First, the ataxic *Lurcher* mouse, is characterized anatomically by atrophic PC dendritic trees, with branches ending in 'stubs,' and overall reductions in size of more than 60% as compared to controls (Caddy and Herrup, 1990). The *Lurcher* phenotype is caused by a mutation in the GluRδ2, leading to specific PC apoptosis or delayed cell death. As mentioned above, GluRδ2 was shown to stabilize spines in order to form mature synapses between PCs and PFs (Kurihara et al., 1997). The altered GluRδ2 in these mice made PCs more susceptible to excitotoxic mechanisms by creating a persistent state of membrane depolarization, leading to increased $[Ca^{2+}]_i$ (Zuo et al., 1997). Second, the Purkinje cell degeneration (pcd) mouse, also presents a clear pattern of PC degeneration, but on a much earlier time scale than the *Shaker* rat. These mice lose almost all PCs between P15 and P45 (Landis and Mullen, 1978), apparently via PC-specific apoptotic pathways (Gillardon et al., 1995). These mutant rodent models have provided important further evidence of the unique susceptibility of PCs to excitotoxic damage.

Mechanisms of excitotoxicity in Purkinje cells: glutamate transporters and glia

Once glutamate is released from presynaptic terminals and exerts its actions on postsynaptic receptors, it needs to be removed from the synaptic cleft by glutamate transporters, which are located on PCs themselves, as well as on surrounding glia. Disturbances to transporters could lead to elevated levels of glutamate and excitotoxic mechanisms. The PC-specific EAAT4 transporter is located largely on postsynaptic spines found in parasagittal zebrin II(+) stripes in the cerebellum, which may make EAAT4-containing PCs more capable of handling excitotoxic levels of glutamate (Yamada et al., 1996; Nagao et al., 1997; Welsh et al., 2002). EAAT4 has been shown to take up glutamate together with Cl^- (Fairman et al., 1995). The findings of Dehnes et al. (1998) and others (Barbour et al., 1994; Tong and Jahr, 1994; Fairman et al., 1995; Takahashi et al., 1996; Otis et al., 1997) corroborate the proposal that glutamate transporters, including EAAT4, may do more than simply remove glutamate from the extracellular space—they may also serve as glutamate-gated chloride channels and assist in the regulation of synaptic transmission. Therefore, alterations in EAAT4 could have adverse effects on normal PC physiology as well as make PCs more susceptible to excitotoxic damage (Takahashi et al., 1996).

Ruiz and Ortega (1995) identified a Na^+-dependent glutamate–aspartate transporter (GLAST) specific to Bergmann glia (BG) in the cerebellum, which is responsible for the majority of glutamate uptake (Lopez-Bayghen et al., 2003). Many studies indicate an important interaction between PCs and GLAST-expressing BG. Immunogold localization has identified that the vast majority of EAAT4 is found on PC spine membranes that are contacted by BG (Dehnes et al., 1998). BG fibers almost completely seal PC synapses (Grosche et al., 1999) and GLAST has been localized to the portions of BG membranes facing excitatory PC synapses (Chaudhry et al., 1995). In normal PCs, postnatal developmental processes subsequently increase the need for GLAST expression in BG; the rapid creation of elaborate dendritic arborizations and the formation of glutamatergic PF–PC synapses coincides with the up-regulation of GLAST in BG (Takacs and Hamori, 1994; Shibata et al., 1996; Kurihara et al., 1997). There are higher levels of GLAST expression in BG which are associated with PCs versus BG not associated with PCs in the cerebella of wild-type and mutant *reeler* and *weaver* mice; when comparing these populations of BG, however, the level of GLAST in BG from mutant cerebella was significantly reduced as compared to controls (Fukaya et al., 1999).

Because of the important interaction between BG and PCs, aberrant BG could have detrimental implications for PC synaptic functions (for review, see Watanabe, 2002). For example, alterations in the GluR2 subunit in BG could have adverse effects on PCs. BG have Ca^{2+}-permeable AMPA receptors. Inserting a GluR2 subunit into BG, making them Ca^{2+}-impermeable, has a distinct effect on BG morphology, resulting in deficient ensheathing of PC synapses (Iino et al., 2001). This retraction of BG fibers after the insertion of GluR2 subunits increases the distance between GLAST and PC synapses, thus reducing the ability of the transporter to take up glutamate from the synaptic cleft. Interestingly, Sjöbeck and Englund (2001) noted gliosis in the molecular layer, and a significantly reduced PC density in the cerebellar vermis of Alzheimer's patients, and gliosis in the molecular layer. In fact, decreased glutamate transporter activity and increased excitotoxicity has been linked to the synaptic damage seen in Alzheimer's disease (Masliah et al., 1996).

Glia can also contribute to glutamate toxicity in PCs in other indirect ways. For example, mGluR and AMPA receptor activation lead to $[Ca^{2+}]_i$ oscillations in cultured astrocytes, causing them to release glutamate (Pasti et al., 2001). Glutamate release from astrocytes can have similar kinetic and exocytotic characteristics as neuronal transmitter release, and can elevate $[Ca^{2+}]_i$ repeatedly in neurons (Pasti et al., 2001). The electrical stimulation of PFs raised $[Ca^{2+}]_i$ in BG, which Grosche et al. (1999) identified as Ca^{2+} microdomains in BG that ensheathed PC spines, indicating that cerebellar glia may not form an integrated network similar to that seen in the hippocampus. Alterations in BG, however, could affect their ability to maintain local, autonomous pools of $[Ca^{2+}]_i$. Indeed, conditional ablation of astrocytes in the cerebellum can have indirect effects on PCs (Delaney et al., 1996). Although there was no reduction in PC number, ablation of astrocytes caused PC dendrites to display an aberrant morphology (the so-called 'weeping willow' appearance). There was also a significant loss of granule cells in this study, so the abnormal development of PCs could also be due to this finding, as the maturation of granule cells is also required for normal growth and elaboration of PC dendrites (Caddy and Herrup, 1990). In either case, BG are key players in the clearance of extracellular glutamate, and thus in the maintenance of normal cerebellar function.

Mechanisms of excitotoxicity in Purkinje cells: role of calcium-binding proteins

As already described, previous reports have theorized that the excitotoxic degeneration of PCs is caused by an exacerbated increase in $[Ca^{2+}]_i$ to such an extent that Ca^{2+}-dependent enzymes are activated (Choi and Rothman, 1990; Schmidt-Kastner and Freund, 1991; Trump and Berezesky, 1995). PCs have a unique presence of Ca^{2+}-binding proteins (CBPs) in the cerebellum, which include high expression levels of the two Ca^{2+} buffers, calbindin-D28K and parvalbumin. The fact that calbindin comprises 15% of total protein in PCs (Baimbridge et al., 1982) has led to the hypothesis that it plays a crucial role in PC function, presumably by buffering large Ca^{2+}-influxes induced by PF and CF activation, and thereby maintaining normal Ca^{2+} homeostasis. Further evidence for this suggestion stems from the findings that a decrease of CBPs in vitro is correlated with PC death and dysfunction, and there are also decreased amounts of CBPs in spinocerebellar ataxia (Vig et al., 2001). Although PCs are rich in calbindin, the ability of this protein to buffer large amounts of Ca^{2+} does not always appear to reduce death in these cells. Similar findings have demonstrated the insufficient buffering capabilities of calretinin and parvalbumin in other cerebellar cell types (Schwaller et al., 2002). Single CF stimulation does not appear to saturate parvalbumin and calbindin in PCs (Schmidt et al., 2003), however, repetitive stimulation may saturate these proteins. This could also occur when there are excess levels of glutamate in the synaptic cleft under certain pathological circumstances (i.e., ischemia or TBI).

Several studies have been conducted using knockout mice lacking various CBPs (for review, see Schwaller et al., 2002). Mice deficient in calbindin present an interesting profile, in that they display normal development without concurrent up-regulation of other CBPs, such as parvalbumin and calretinin, and yet they suffer from severe ataxia when presented with complex coordination tests (Airaksinen et al., 1997). Further investigations using confocal imaging revealed that the postsynaptic dendritic Ca^{2+} transients were altered in these mutant mice. Airaksinen et al. (1997) concluded that calbindin rapidly buffers $[Ca^{2+}]_i$ with a high affinity during the initial phase of Ca^{2+} influx, and that this protein is involved in cerebellar movement control. Further studies by Barski et al. (2003) in PC-specific calbindin knock-out mice indicated that Ca^{2+} transients had increased amplitudes and faster decay rates. These mice also display impaired motor function, again suggesting that calbindin may play an important role in motor coordination. Sayer et al. (2000) have also postulated that calbindin may not act solely as a Ca^{2+} buffer, but may have additional roles, such as signal transduction in the nucleus. In addition, calbindin and parvalbumin may also serve to regulate Ca^{2+} dynamics that are used as temporal and spatial signals in neurons, similar to calmodulin (Braun and Schulman, 1995).

The ability of a single dendritic spine to make morphological modifications in response to synaptic input allows each spine to isolate the synaptic signal, and any local increases in $[Ca^{2+}]_i$, from other spines and from the parent dendrite. Harris and Stevens (1988) described the intimate relationship between spine morphology and the size of the postsynaptic density (PSD) in PCs. The PSD structure is located below the postsynaptic membrane in dendritic spines, and contains specialized proteins that assist in the membrane targeting of receptors (for review, see Kennedy, 1997). They found that PC spines with large heads and large PSDs had shorter necks than spines with smaller heads and PSDs. Spine volume overall was proportional to the volume of the ER, which occupied a large portion of both spine and neck cytoplasm (Harris and Stevens, 1988). A larger ER could increase the uptake capacity of intracellular Ca^{2+} stores. Spine size was also proportional to PSD size and to the number of associated presynaptic vesicles (Harris and Stevens, 1988). Indeed, anatomical differences have also been noted in studies of knock-out mice. Whereas dendritic spines in PCs from parvalbumin knock-out mice were indistinguishable from controls, spines from calbindin knock-out mice had slightly longer necks, and spines from double knock-out animals had longer, thinner spine necks and heads that were doubled in size (Vecellio et al., 2000). These changes in spine morphology may have been compensatory mechanisms to account for the lack of sufficient Ca^{2+}-buffering proteins (Vecellio et al., 2000).

Glutamate-induced alterations in dendritic spines

Glutamate exerts its effects primarily at dendritic spines, the contact points between PCs and CFs or PFs. Spines are a dynamic component of the neuronal cytoskeleton that form, grow, mature, and shrink in response to synaptic and intracellular signals. In PCs, dendritic spines are dramatically affected by the release of presynaptic glutamate, and their ability to gather and disseminate reliable information to the soma, and then to the axon, is crucial for normal cerebellar function. Thus, it is not surprising that PC spines contain a broad variety of synaptic proteins and cytoskeletal elements, some of which are either PC-specific, such as calbindin, or are very highly expressed in PCs, such as microtubule-associated protein 1A (MAP1A). A large portion of the studies regarding spine dynamics have been conducted in cell types residing outside of the cerebellum; therefore, we must rely on much of this data when making inferences about these processes in PCs. In the following sections, we describe and compare normal synaptic and spine physiology in hippocampal/cortical neurons and in PCs, and discuss the relationship between glutamate excitotoxicity and its effects on cytoskeletal elements such as microtubules (MTs), MAPs and actin.

Microtubules and MAPs

MTs are highly polarized structures, with static minus-ends and growing plus-ends. Neuronal MTs in axons and dendrites lend structural stability as well as provide tracks for intracellular transport. MT organization differs between axons and dendrites: in axons, MTs are oriented with their minus-ends towards the soma, and their growing plus-ends towards the growth cones; in dendrites, MT plus-ends are oriented in both directions (Baas et al., 1988). These data, originally gathered in hippocampal neurons, have been reconfirmed in PCs (Stepanova et al., 2003). In order to maintain cytoskeletal structure and thus synaptic efficacy, PC spines contain a wide range of MAPs that serve to stabilize polymerized tubulin. These proteins can only attach themselves to tubulin polymers in their unphosphorylated form. One MAP that has received considerable attention is MAP2, a dendrite-specific MAP. Several studies have highlighted the importance of MAP2 in neuronal, and in PC, survival and function. Tubulin does not readily self-assemble without MAP2; the addition of MAP2, however, causes polymerization (Murphy and Borisy, 1975). Decreased MAP2 and abnormalities of MAP2 in PCs leads to beading, or blebbing, and destabilization of dendritic processes, resulting in abnormal functioning. In particular, PCs display a loss of viable synapses due to resorption of postsynaptic spines (Abdel-Rahman et al., 2001).

Studies by Matus and colleagues have demonstrated that, although MAP2 accumulates distally within the growing dendritic tree, there is no analogous accumulation of tubulin, suggesting that both the distribution and the function of MAP2 may be related to factors other than its binding to MTs (Matus et al., 1990). As both MAP2 and a related protein, MAP1, are dendrite-specific in PCs, these proteins may play a role in establishing neuronal polarity by specifying dendritic and axonal routes of transport from the cell body (Matus et al., 1981). Bernhardt and Matus (1982) suggested that the synthesis of MAP2 in PCs could be a prerequisite for the formation of dendritic MTs and thus for the stabilization of growing dendrites. Indeed, PC degradation without concurrent cell death was correlated with decreased MAP2 expression in the *nervous mouse* model, though it was not possible to draw a causal relationship between these two events (Brion et al., 1988).

Glutamate-induced $[Ca^{2+}]_i$ elevations cause the disassembly of established MTs and inhibit the assembly of new MTs. In hippocampal neurons, short applications of glutamate (4 h) cause dendritic outgrowth, but longer applications (12 h) instigate dendritic retraction, as well as fewer MTs in growth cones and more MTs in shafts (Wilson and Keith, 1998). In addition, high Ca^{2+} (mM level) causes MT depolymerization in vitro (Schliwa et al., 1981). High Ca^{2+} after glutamate application may initiate a variety of downstream signaling pathways, resulting in MT growth or retraction. Ca^{2+}-induced activation of calpain has been shown to inhibit MT growth (Johnson et al., 1991), whereas activation of calmodulin/CaMKII leads to phosphorylation of MAPs and neurite outgrowth (Diéz-Guerra and Avila, 1995). Bigot et al. (1991) found that MAP2, which exhibited diffuse staining in unstimulated cultures, became bound to MTs following the administration of a variety of excitatory amino acids (e.g., NMDA and AMPA); MAP2-bound MTs were also more resistant to depolymerization by nocodazole. Ca^{2+} imaging, utilizing the indicator dye fura-2, indicated that hippocampal neurons pretreated with taxol (a MT stabilizing agent) had a significantly attenuated $[Ca^{2+}]_i$ response. This finding indicated that MT depolymerization played a role in the glutamate-induced elevation of $[Ca^{2+}]_i$ in untreated neurons (Furukawa and Mattson, 1995). Taxol did not reduce Ca^{2+} influx through NMDA receptors, whereas it did suppress influx through AMPA receptors directly, as $[Ca^{2+}]_i$ levels did not differ between taxol-treated and untreated neurons after a depolarizing dose of KCl. Colchicine, a selective MT disruptor, caused a significant decrease in cell survival, a situation that was significantly improved by pretreatment with taxol (Furukawa and Mattson, 1995). Taxol, already used extensively in the treatment of cancer, exerts its effects through the stabilization of MTs, thereby halting the natural process of cell division. The fact that MTs may contribute to the mechanisms behind excitotoxicity merits further investigation into the usefulness of taxol and related compounds in treating pathological conditions such as ischemia and TBI.

It was previously assumed that cytoskeletal alterations, caused by agents that raised $[Ca^{2+}]_i$, such as glutamate, were detrimental to cell survival. It appears, however, that various members of the neuronal cytoskeleton react differently to elevated levels of $[Ca^{2+}]_i$. Recent data indicate that actin depolymerizing agents, such as cytochalasin D, can protect hippocampal neurons against excitotoxic damage by reducing Ca^{2+} influx through both NMDA receptors and VGCCs (Johnson and Byerly, 1993; Rosenmund and Westbrook, 1993; Furukawa and Mattson, 1995; Furukawa et al., 1995). Spines have a heterogenous population of actin. One form, called filamentous actin, or F-actin, is highly localized in spines as compared with dendritic shafts. Halpain et al. (1998) studied the early effects of NMDA and glutamate application on dendritic spines in cultured hippocampal neurons. Brief applications of either agonist caused a loss of spines and a reorganization of spine actin, but not a loss of synaptic markers. Blocking NMDA receptors with D-APV or MK-801 prohibited the loss of F-actin punctae. The influx of Ca^{2+} through many types of glutamate receptors led to spine actin destabilization; Ca^{2+} influx through mGluRs, through L-type Ca^{2+} channels or after depolarization, however, had no effect on actin. Thus, specific pathways of Ca^{2+} influx are required to destabilize actin. Jasplakinolide, an actin-stabilizing compound, prevented the loss of F-actin punctae when co-applied with NMDA, and Halpain et al. (1998) postulated that the specific activation of NMDA receptors stimulated calcineurin, a Ca^{2+}-dependent phosphatase abundant in neurons, which may in turn regulate F-actin stability. Indeed, calcineurin was shown to label both axonal and dendritic growth cones in immature cells, but its expression shifted to neuritic processes in mature cells (Ferreira et al., 1993). Its ability to regulate the phosphorylation of the axon-specific tau was crucial for the determination of neuronal polarity. Dendritic MAP2 can also be dephosphorylated by calcineurin (Goto et al., 1985). In this state, MAP2 would display a decreased affinity for MTs.

In a similar study, Allison et al. (1998) studied the contribution of AMPA receptors to actin depolymerization, a phenomenon with more relevance to PC spine dynamics. They found that F-actin in hippocampal neurons was resistant to depolymerization by cytochalasin D, but not to latrunculin A; application of the latter compound resulted in spine collapse but not total synapse failure, as only 40% of synapses lost AMPA receptors and NMDA receptor clusters. Thus, one major function of dendritic spines may be to sequester AMPA receptors, but they are less tightly attached to the cytoskeleton and the PSD than NMDA receptors (Allison et al., 1998). AMPA receptor activation by glutamate was shown to inhibit actin activity, leading to firmer, rounder, more stable spines, without the involvement of NMDA receptors (Fischer et al., 2000). The blockade of spine motility by AMPA receptor activation is dependent on depolarization and the resultant influx of Na^+ into the cell. Blocking voltage-gated Na^+ channels, however, did not prevent spine rounding and stability, indicating that glutamate-induced AMPA activation specifically provides the required Na^+ influx (Fischer et al., 2000).

MT or MAP2 alterations have profound effects on neuronal growth and survival. For example, Paula-Barbosa and Tavares (1985) found that the number of MTs in rats fed alcohol for one month was significantly reduced compared to controls, a trend which continued for rats fed alcohol for much longer periods. Previous data had indicated that rat PC dendritic trees are morphologically altered after alcohol consumption (Tavares et al., 1983), which the authors correlated with a loss of PCs (Paula-Barbosa and Tavares, 1985). In the hippocampus, injections of both dihydrokainate (a glutamate transport inhibitor) and kainate induced alterations in MAP2, likely due to excitotoxic activation of NMDA and non-NMDA (AMPA and kainate) receptors (Arias et al., 1997). Initially, there was a loss of MAP2 immunoreactivity at 3 h post-injection, then a redistribution of MAP2 from dendrites to soma at 12 h, and finally a decrease or total loss of MAP2 at 24 and 48 h. This late-phase protein loss was likely due to the activation of calpains or other calcium-dependent proteases (Siman et al., 1989; Johnson et al., 1991). Similarly, almost all of the MAP2 staining in hippocampal dendrites disappeared in slice cultures exposed to the mGluR agonist, trans-ACPD, as compared to control slices; many of those dendrites that did remain demonstrated clear signs of degeneration (Blaabjerg et al., 2001). The neurotoxic effects of millimolar doses of trans-ACPD appeared to result from activation of NMDA receptors, as co-application of trans-ACPD with MK-801 abolished this effect. The co-application of NBQX, an AMPA receptor antagonist, however, did not attenuate the effects of trans-ACPD.

Although reports of MTs, and associated proteins such as MAP2, in spines have been contradictory (Bernhardt and Matus, 1984; Kaech et al., 1997), marked amounts of free tubulin have been found in PSD fractions (van Rossum et al., 1999) and van Rossum and Hanisch (1999) have provided a model of glutamatergic spine dynamics involving both MTs and MAPs. In the hippocampus, NMDA receptor activation, and subsequent Ca^{2+} influx, may cause both the dissolution of the actin framework, and the polymerization of MTs, stabilized by MAPs. Although Ca^{2+} has been shown to cause MT disassembly in vitro (for review, see Donato, 2001), it is possible that the rich abundance of CBPs in spines could minimize the destabilizing effect of Ca^{2+}. (Indeed, NMDA receptor activation leads to increased MT stability by modulating the phosphorylation state of MAP2.) These MTs could serve as temporary tracks for the transport of newly synthesized proteins into nascent spines. The fact that MTs have been difficult to locate in spines (for example, see Westrum et al., 1980; Fig. 2) may be due to their transient nature in these cellular structures (van Rossum and Hanisch, 1999).

Interactions between different cytoskeletal elements

Mounting evidence indicates that various cytoskeletal networks are tightly connected (for a recent review, see Dehmelt and Halpain, 2004). For example, MAP2 has been shown to bind to actin (Selden and Pollard, 2002) and to neurofilaments (NFs) (Aamondt and Williams, 1984). The relationship between MAP1A and MAP2 may result in the elaborate network of cross-bridges that exist among dendritic MTs (Shiomura and Hirokawa, 1987). The MT and actin cytoskeletons may also work in concert to transport receptors and other key components to the synapse. Washbourne et al. (2002) demonstrated that packets of NMDA receptors were transported along dendritic MTs in cortical neurons. Application

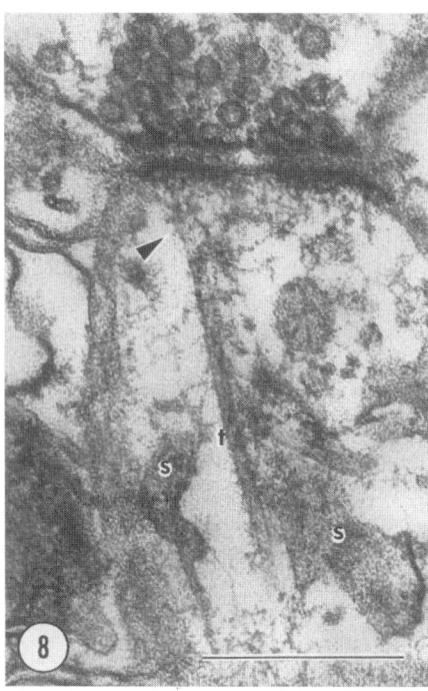

Fig. 2. Microtubule in a dendritic spine. Westrum et al. (1980) located a microtubule (*t*), surrounded by sacs (*s*) in a dendritic spine derived from a 33-day-old rat. The postsynaptic density is indicated by an arrowhead. Scale bar = 0.25 µm. (Copied with permission from Westrum et al., 1980.)

receptors and VGCCs. Actin depolymerization protected neurons against excitotoxic damage, suggesting that the protective mechanism involves suppression of Ca^{2+} influx through plasma membrane channels. MT depolymerization, however, does not suppress increased $[Ca^{2+}]_i$ in neurons. Although Ca^{2+} can cause a selective loss of F-actin (Halpain et al., 1998), local increases in Ca^{2+} in PC dendritic spines are critical for cell signaling. Collapsing spines after NMDA or glutamate application may make the neuron more vulnerable to a second excitotoxic event (Halpain et al., 1998). Proteins like espin, whose ability to bundle actin is not inhibited by Ca^{2+}, can help PC spines retain their shape, and thus their function (Sekerkova et al., 2003).

Postsynaptic proteins in Purkinje cells

As mentioned above, spine morphology and PSD size are tightly coupled in PCs (Harris and Stevens, 1988). PCs were shown to contain PSD-93 (Brenman et al., 1996; Kim et al., 1996), a relative of the well-characterized PSD-95 protein, at both the PSD, where PCs receive excitatory input from PFs, and along dendritic MTs (Brenman et al., 1998). PSD proteins may assist in synaptic efficacy by maintaining the proximity between receptors and signaling enzymes (van Rossum and Hanisch, 1999). PSD-93 highly co-localized with MAP1A in both soma and proximal dendrites in PCs (Brenman et al., 1998). Pedrotti et al. (1994) demonstrated that MAP1A was also able to bind to actin and neurofilaments. PSD-93 was shown to link nNOS with membrane-bound receptors (Brenman et al., 1996). PSD-93 was not required for either the development or the function of PC–PF synapses, but may serve as a link between postsynaptic receptors in PC spines and signaling pathways (McGee et al., 2001). Indeed, PSD-93 was shown to cluster GluRδ2, as PSD-95 clustered NMDA receptors (Roche et al., 1999). GluRδ2 was found at both PC–PF and PC–CF synapses in early development, but only at PC–PF synapses in adult animals (Roche et al., 1999). Hirai (2000) demonstrated that these receptors are bound to the actin cytoskeleton. Actin-disrupting agents decreased the number of immunoreactive receptor clusters, indicating that cytoskeletal alterations could have profound effects on GluRδ2 localization and function.

of vincristine, an MT depolymerizing agent, abolished NMDA receptor transport, whereas latrunculin A, an actin depolymerizing drug, had no effect. Imaging experiments showed that these NMDA receptor packets were transported via MTs to contact sites, either between growth cone filipodia and a dendritic shaft, or between a dendritic shaft and an axon, where they remained (Washbourne et al., 2002). Sergé et al. (2003) demonstrated that mGluR5 receptors can bind to MTs (though evidence was lacking for either direct or MAP-associated binding), and that the MTs are transported by actin. Similarly, Kim and Lisman (2001) demonstrated that AMPA receptors were transported first via MTs, and then directed from dendritic shafts to spines by actin.

Furukawa et al. (1995) found that glutamate did not reduce phalloidin staining of actin in hippocampal cultures incubated in Ca^{2+}-free media, indicating that Ca^{2+} influx is crucial for the actin-depolymerizing action of glutamate. The application of cytochalasins decreased Ca^{2+} influx through NMDA

The fact that PSD-93 localizes to PC dendritic MTs, as well as associates with MAP1A, may give evidence of a putative tubulin-based cytoskeleton in spines. Indeed, other MAPs have been implicated in linking cytoskeletal elements with synaptic proteins. For example, CRIPT (cysteine-rich interactor of PDZ three) was shown to interact with both NMDA receptors and MTs, and the co-expression of CRIPT and PSD-95 in COS7 cells caused PSD-95 to be redistributed to MTs. CRIPT also recruited PSD-93 in a similar fashion (Niethammer et al., 1998). CRIPT may bind to MTs and PSD-95 independently, and mediate the interaction of MTs and PSD-95, as an abbreviated form of CRIPT, which could not bind with PSD-95, could still reorganize MTs. Like PSD-93, CRIPT was also expressed in the somadendritic region of PCs (Niethammer et al., 1998).

Brain specific cytoplasmic linker protein of 115 kDa (CLIP-115), like its more ubiquitous relative, CLIP-170 (Pierre et al., 1992), was also shown to reorganize MTs in a similar fashion (De Zeeuw et al., 1997). Specifically, CLIP-115 was responsible for both the transport and the localization of dendritic lamellar bodies (DLBs), structures that the authors hypothesized may be a special form of intracellular Ca^{2+} store. Nocodazole-induced MT depolymerization disrupted the translocation of DLBs. CLIP-115 was expressed in MAP2-positive dendrites in cultured hippocampal cells, in BG fibers (De Zeeuw et al., 1997), and in cultured PCs (Stepanova et al., 2003). Mice lacking the *CYLN2* gene, which codes for CLIP-115, demonstrated hippocampal and cerebellar dysfunction, possibly due to altered MT-dependent intracellular transport (Hoogenraad et al., 2002). CLIPs and other related proteins, which interact specifically with growing MT plus-ends, could play a vital role in actin-MT interaction, as the haploinsufficiency for the *CYLN2* gene in humans could partially account for the marked neurodevelopmental deficits seen in William's syndrome (Hoogenraad et al., 1998; Francke, 1999).

Long-term depression in Purkinje cells: a neuroprotective mechanism?

Normal glutamate neurotransmission appears to stabilize dendritic spines, however what happens to spines during glutamate excitotoxicity is not completely known. It is possible that these highly stable spines could be continually activated by glutamate, leading to continued ion influx. Spines contain a form of actin that is resistant to depolymerizing drugs (Allison et al., 1998), which could be resistant to the effects of Ca^{2+}. Indeed, the fact that spines serve to isolate a high level of Ca^{2+} from the remainder of the neuron indicates that they may serve a neuroprotective role (Segal, 1995).

Dendritic spines in PCs may also play a neuroprotective role in other ways. For example, reductions in the synaptic strength of the PF and CF inputs, as measured by postsynaptic currents in PCs, are well described. In Marr–Albus–Ito models of cerebellar function, long-term depression (LTD) at PF–PC synapses provides a cellular substrate of some forms of cerebellar motor learning (Marr, 1969; Albus, 1971; Ito, 1984). PF–LTD can be induced by simultaneous activation of PFs and CFs at low frequencies (Ito, 1984). LTD can also be obtained at the CF–PC synapse following tetanization of the CF alone (Hansel and Linden, 2000). CF–LTD is accompanied by a long-term reduction in the amplitude of complex spike-associated Ca^{2+} transients in PCs (Weber et al., 2003). It is possible that a reduction in CF-evoked Ca^{2+} signaling in PCs may provide a neuroprotective mechanism. This idea has previously been suggested for PF–LTD (De Schutter, 1995; Llinás et al., 1997). This hypothesis is even more attractive at the CF input as it has been shown that prolonged periods of CF firing at elevated frequencies can lead to PC death (O'Hearn and Molliver, 1997), despite the high Ca^{2+} buffering capacity of these cells (Fierro and Llano, 1996). Although speculative, a long-term reduction in Ca^{2+} signals may provide a mechanism by which PCs adapt in order to handle a potentially high and toxic Ca^{2+} load. Other potential forms of naturally occurring neuroprotection may occur in the cerebellum. For example, the substitution of choline chloride for NaCl attenuates AMPA-induced dark cell degeneration (DCD), an apoptotic form of cell death, in PCs (Strahlendorf et al., 2001). Substitution of Na^+ and a blockade of Na^+ channels were ineffective, however, indicating that choline itself was responsible for the decreased DCD. Choline potently agonizes the $\alpha 7$ nicotinic receptor subtype, and may protect cells from excitotoxic damage. Therefore, endogenous choline in the cerebellum may provide neuroprotection via the α7 receptor (Strahlendorf et al., 2001).

Traumatic injury of Purkinje cells

One of our interests related to glutamate-mediated excitotoxicity is the study of TBI, which also has profound effects on the cytoskeletal elements of neurons (Kanayama et al., 1996; Folkerts et al., 1998; Saatman et al., 1998). PCs have been shown to be particularly vulnerable to TBI, after either a cerebral or a direct cerebellar insult in vivo (see Fig. 3). For example, Fukuda et al. (1996) subjected rats to cerebral injury, and found profound PC loss and microglial activation in the cerebellum by 3–7 days

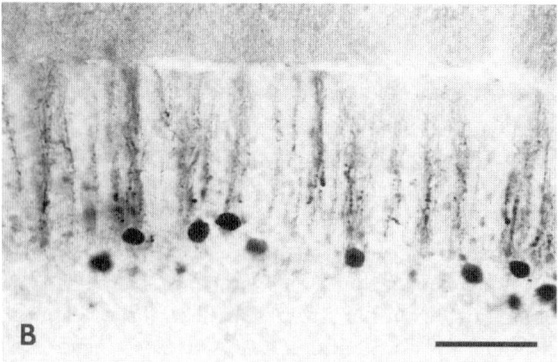

Fig. 3. Degenerating Purkinje cells after traumatic injury in the rat. Immunolocalization of PEP-19, a PC-specific protein, indicates the massive loss of PC soma (arrowheads) and degeneration of processes at 7 days postinjury, in both sagittal (A) and coronal (B) sections. Scale bar = 80 μm. (Copied with permission from Fukuda et al., 1996.)

postinjury. The reactive microglia corresponded to the zebrin stripes created by the dendritic trees of PCs. The authors theorized that the microglia were activated by apoptotic signals derived from dying PCs, and that the zebrin-like stripes were caused by excitotoxic mechanisms derived from excess activation of CFs and PFs. Similarly, Ai and Baker (2002) found presynaptic hyperexcitability in mossy fibers and PFs in rats after fluid percussion injury, leading to extensive PC death at 7 days postinjury. Like Fukuda et al. (1996), they postulated that excitotoxic mechanisms, derived from the extremely high number of synaptic contacts between PFs and PCs, caused the massive PC death. Similar findings have been reported in rats treated with the psychoactive compound ibogaine (O'Hearn and Molliver, 1997). Ibogaine treatment caused PC degeneration in parasagittal bands, which corresponded with regions of activated microglia. Harmaline, which also causes neuronal hyperactivity in the IO, caused a similar pattern of degeneration, indicating that these two compounds likely act on IO neurons with similar mechanisms (O'Hearn and Molliver, 1997). The chemical ablation of the inferior olive, however, spared PCs and did not cause microglial activation. Interestingly, both ibogaine and harmaline caused a loss of MAP2 and calbindin concomitantly with a loss of PCs that degenerated in parasagittal stripes (O'Hearn and Molliver, 1993).

We have recently characterized the response of cerebellar cells to mechanical stretch injury in vitro (Slemmer et al., 2004). We found a reduction in both MAP2-positive neurons and in calbindin-positive PCs following mechanical injury. Figure 4 depicts prototypical images of both control and injured PCs in which axonal beading and dendritic degeneration are readily apparent. We also noted reduced levels of GLAST antibody staining after in vitro injury in cerebellar cultures (Slemmer et al., 2004). Although the reasons for the reduced GLAST expression were unclear (e.g., the transporter itself has been down-regulated after injury or is physically altered in some manner), reduced GLAST activity would have marked effects on the function and survival of PCs, as described above. In addition, stretch injury in vitro has been previously shown to cause the release of glutamate from astrocytes (Ahmed et al., 2002). Therefore, glutamate levels are elevated after injury,

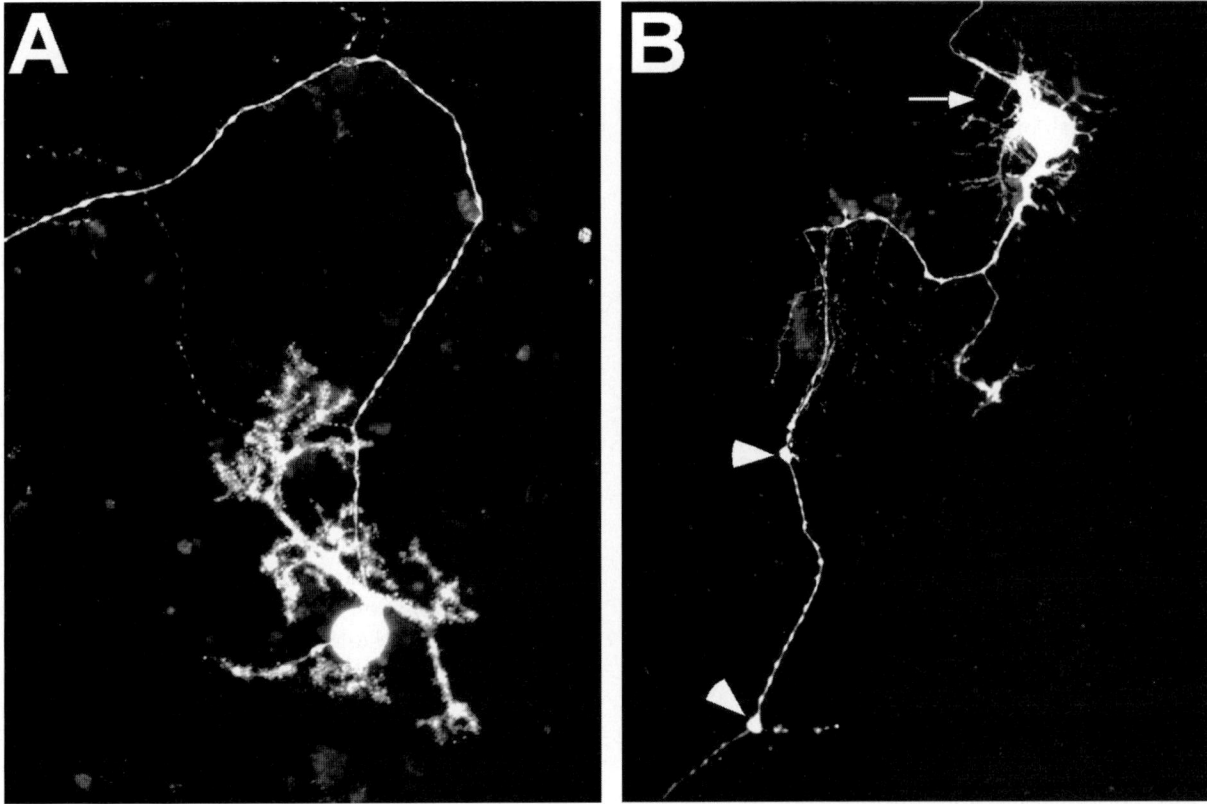

Fig. 4. Axonal beading and dendritic retraction after in vitro stretch injury in cultured Purkinje cells stained against calbindin-D28K. A: Control PC showing a full dendritic arbor and long axon. B: Injured PC with a beaded axon (beads indicated by arrowheads) and a retracted dendritic tree (arrow).

in addition to a lower capacity to clear it from the extracellular space. This correlated well with in vivo findings. For example, Watase et al. (1998) subjected GLAST mutant mice to cold-induced cerebellar injury. They postulated that the significantly larger edema in these mice was caused by the inability of mutant BG to reuptake glutamate from the extracellular space. Indeed, cortical protein levels of GLAST were significantly reduced for up to 3 days after controlled cortical impact brain injury in vivo before returning to control levels at 7 days postinjury (Rao et al., 1998). If GLAST is critical in preventing excitotoxic damage after injury, as Rao et al. (1998) claimed, then the reduction we observed in GLAST fluorescence after stretch injury in cultured cerebellar cells could be indicative of faulty glutamate transport, causing excitotoxic levels of glutamate to accumulate in the extracellular space.

Role of S-100β

Another possible contributor to secondary damage after trauma and glutamate-induced excitotoxicity could be S-100β protein. S-100β is a calcium-binding protein found in astrocytes and Schwann cells, and clinical studies (for example, see Herrmann et al., 2001) have reported elevated levels of S-100β following TBI. Although the mechanisms for elevated S-100β following TBI are unknown, increased S-100β levels in peripheral blood after trauma may be indicative of damage to the blood-brain barrier or could indicate the activation of secondary damage pathways (Raabe and Seifert, 1999; Herrmann et al., 2001). We found elevated S-100β levels after hippocampal injury in vitro (Slemmer et al., 2002). We also saw a significant release of S-100β by cerebellar glia after in vitro injury (Slemmer et al.,

2004), the relative amount of which was several-fold higher than what we had previously reported for hippocampal glia (Slemmer et al., 2002). S-100β stimulates glial proliferation (Reeves et al., 1994), which could possibly lead to brain swelling after trauma. However, it is also possible that S-100β is playing a somewhat protective role. For example, S-100β may be an important mediator of glia-neuronal interactions, and it has been shown to stimulate neurite extension (Reeves et al., 1994). Therefore, glial cells may release S-100β in an attempt to save, or to repair, dying or damaged neurons. The cerebellum has been shown to contain high amounts of S-100β, specifically in BG (Lossi et al., 1995). Because BG play such a critical role in cerebellar development, they may require high levels of S-100β for their intercellular functions.

S-100β has been increasingly linked to a variety of neuronal processes as well (for review, see Donato, 2001); for example, S-100β disrupts MTs, and has been shown to instigate neuritic outgrowth, a necessary element in developing neurons (Azmitia et al., 1990). Excessive glial S-100β release after TBI, however, may prove detrimental to mature neurons. Thus, we applied exogenous S-100β to control (uninjured) cerebellar cultures, and found that the number of PCs (but not the number of MAP2-positive neurons) was significantly reduced at 24 h postapplication (Slemmer et al., 2004). Decades ago, Cajal (1913–14, 1933) described a 'neurotrophic factor' found in Schwann cells, and 'neurobiones,' living entities which formed a 'conductive pathway' in the protoplasm, which Azmitia (2002) proposed matched the properties of S-100β and MTs, respectively. Indeed, S-100β has been shown to cause Ca^{2+}-dependent depolymerization of MTs in a dose-dependent manner (Sorci et al., 2000). Barger and Van Eldik (1992) demonstrated that exogenous S-100β stimulated large increases in $[Ca^{2+}]_i$ in cultured neurons; the authors suggested that abnormal amounts of S-100β could result in levels of $[Ca^{2+}]_i$ commonly associated with excitotoxicity. In our hands, PCs were also susceptible to stretch injury and therefore, it was possible that there was some direct form of S-100β-mediated toxicity that caused significant PC death (Slemmer et al., 2004). S-100β may have interacted with MTs in PCs, causing these cells to retract their neuritic processes and lift from the substrate. Consistent with this latter hypothesis, early studies conducted by Cajal on traumatically injured cats showed that PCs developed highly retracted dendritic trees as early as two days postinjury (Cajal, 1933; see also Azmitia, 2002; Fig. 5).

Mouse models have demonstrated the importance of S-100β for glial-neuronal communication. For

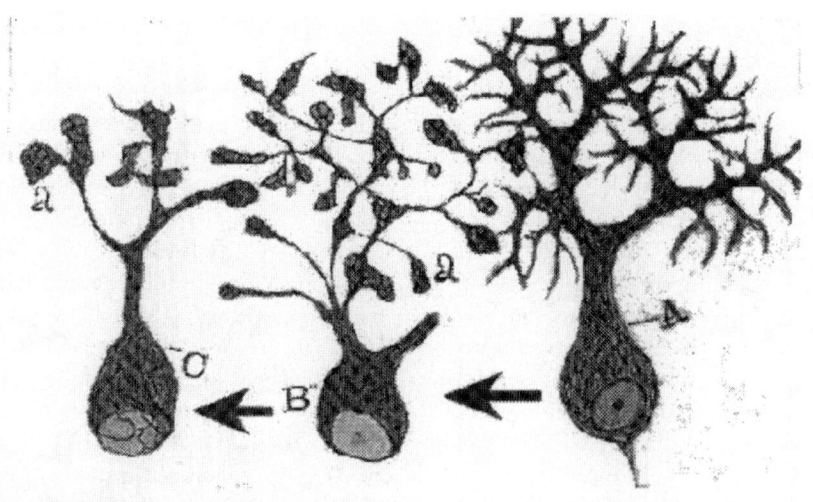

Fig. 5. Purkinje cell of cat of twenty-five, sacrificed two days after traumatic lesion. (A) Normal, (B, C) cells which retracted dendrites terminate in reticulated clubs. (Copied with permission from Cajal, 1933.)

example, the deletion of S-100β enhanced long-term potentiation (LTP) in hippocampal neurons (Nishiyama et al., 2002), whereas the over-expression of S-100β impaired hippocampal LTP (Gerlai et al., 1995). Also, the addition of exogenous S-100β reduced the level of LTP in S-100β knock-out mice to the level of LTP in wild-type mice, indicating that S-100β, normally released by astrocytes, has a direct effect on the surrounding neurons (Nishiyama et al., 2002). Cerebellar astrocytes from S-100β knock-out mice had larger Ca^{2+} transients in response to KCl or caffeine as compared to controls, whereas transients were normal in granule cells. The reduced ability of mutant astrocytes to buffer Ca^{2+} could affect intercellular communication between BG and neurons through aberrant Ca^{2+} waves (Xiong et al., 2000). Although S-100β is found in control BG, and not in PCs (Xiong et al., 2000), PCs have been shown to selectively take up antibodies to S-100β (Karpiak and Mahadik, 1987). S-100β has been shown to bind to RAGE (receptor for advanced glycation end-products) with nanomolar affinity (Hofmann et al., 1999), though it may exert its effects on neurons via other means (i.e., direct entry into cells; Azmitia and Whitaker-Azmitia, 1997).

Interestingly, S-100β has been shown to regulate IQGAP1, a scaffold protein that localizes complexes involved in actin- and MT-based functions to the plasma membrane (Mbele et al., 2002). Also, Rac1 and Cdc42, members of the Rho family of GTPases, were shown to label areas for the targeting of IQGAP1 and CLIP-170, leading to MT polarization (Fukata et al., 2002). The interaction of S-100β and IQGAP1 is regulated by both Ca^{2+} and Zn^{2+}; the presence of Zn^{2+} caused S-100β to have an affinity for Ca^{2+} within a physiological range (Mbele et al., 2002). Glutamate-mediated Ca^{2+} influx during excitotoxicity could activate S-100β, which in turn could alter the ability of CLIP-170 or IQGAP1 to bind to MTs, and thus perturb actin–MT interactions at the spine level. Although S-100β is not an excitatory amino acid, these reports suggest that it can have similar actions to glutamate in PCs, such as a reorganization of the cytoskeleton in dendrites and spines. Such findings could have implications not only for TBI, but also for other types of pathologies that involve excitotoxic mechanisms in PCs.

Conclusions

In this review, we have demonstrated that understanding the mechanisms underlying glutamate-mediated excitotoxicity in PCs cannot be accomplished simply by extrapolating data gathered in other brain regions, partially because PCs lack functional NMDA receptors. PCs are highly susceptible to a variety of pathologies with an excitotoxic component, such as TBI, and many of these conditions display cytoskeletal alterations. The ability of Ca^{2+} to instigate a variety of acute and downstream signaling pathways has been well documented, and many treatment strategies have focused upon reducing Ca^{2+} influx. The influence of BG and S-100β, however, has yet to be fully comprehended or appreciated. Although S-100β has been shown to play a neuroprotective role at times, the high release of S-100β after TBI could be initiating a series of cell death pathways similar to glutamate. The interaction between PCs and BG could provide a new model with which to test novel treatment strategies, especially considering the number of excitotoxic animal models that show selective PC degeneration.

Abbreviations

AMPA	α-amino-3-hydroxy-5-methyl-4-isoxazolepropionate
BG	Bergmann glia
$[Ca^{2+}]_i$	intracellular free calcium
CBP	Ca^{2+}-binding protein
CF	climbing fiber
CLIP-115	cytoplasmic linker protein of 115 kDa
CLIP-170	cytoplasmic linker protein of 170 kDa
CNQX	6-cyano-7-nitroquinoxaline-2,3-dione
CNS	central nervous system
DAG	diacylglycerol
DCD	dark cell degeneration
DLB	dendritic lamellar body
EAAT	excitatory amino acid transporter
ER	endoplasmic reticulum
GLAST	glutamate-aspartate transporter
IO	inferior olive
IP_3	inositol-(1,4,5)-trisphosphate
LTD	long-term depression
LTP	long-term potentiation
MAP	microtubule-associated protein

mGluR	metabotropic glutamate receptor
MT	microtubule
NBQX	2,3-dihydroxy-6-nitro-7-sulfamoyl-benzo-(F)-quinoxaline
NF	neurofilament
NMDA	N-methyl-D-aspartate
nNOS	neuronal nitric oxide synthase
NO	nitric oxide
PC	Purkinje cell
PIP$_2$	phosphatidylinositol bisphosphate
PKC	protein kinase C
PLC	phospholipase C
PF	parallel fiber
PSD	postsynaptic density
RAGE	receptor for advanced glycation end-products
TBI	traumatic brain injury
trans-ACPD	trans-($\pm$)-1-amino1,3-cyclopentanedicarboxylic acid
VGCC	voltage-gated calcium channel

References

Aamondt, E.J. and Williams, R.C. (1984) Microtubule-associated proteins connect microtubules and neurofilaments in vitro. Biochemistry, 23: 6023–6031.

Aarts, M., Liu, Y., Liu, L., Besshoh, S., Arundine, M., Gurd, J.W., Wang, Y.T., Salter, M.W. and Tymianski, M. (2002) Treatment of ischemic brain damage by perturbing NMDA receptor-PSD-95 protein interactions. Science, 298: 846–850.

Abdel-Rahman, A., Shetty, A.K. and Abou-Donia, M.B. (2001) Subchronic dermal application of N,N-diethyl m-toluamide (DEET) and permethrin to adult rats, alone or in combination, causes diffuse neuronal cell death and cytoskeletal abnormalities in the cerebral cortex and the hippocampus, and Purkinje neuron loss in the cerebellum. Exp. Neurol., 172: 153–171.

Ahmed, S.M., Weber, J.T., Liang, S., Willoughby, K.A., Sitterding, H.A., Rzigalinski, B.A. and Ellis, E.F. (2002) NMDA receptor activation contributes to a portion of the decreased mitochondrial membrane potential and elevated intracellular free calcium in strain-injured neurons. J. Neurotrauma, 19: 1619–1629.

Ai, J. and Baker, A. (2002) Presynaptic hyperexcitability at cerebellar synapses in traumatic injury rat. Neurosci. Lett., 332: 155–158.

Airaksinen, M.S., Eilers, J., Garaschuk, O., Thoenen, H., Konnerth, A. and Meyer, M. (1997) Ataxia and altered dendritic calcium signaling in mice carrying a targeted null mutation of the calbindin D28k gene. Proc. Natl. Acad. Sci. USA, 94: 1488–1493.

Albus, J.S. (1971) A theory of cerebellar function. Math. Biosci., 10: 25–61.

Allison, D.W., Gelfand, V.I., Spector, I. and Craig, A.M. (1998) Role of actin in anchoring postsynaptic receptors in cultured hippocampal neurons: differential attachment of NMDA versus AMPA receptors. J. Neurosci., 18: 2423–2436.

Arias, C., Arrieta, I., Massieu, L. and Tapia, R. (1997) Neuronal damage and MAP2 changes induced by the glutamate transport inhibitor dihydrokainate and by kainate in rat hippocampus in vivo. Exp. Brain Res., 116: 467–476.

Azmitia, E.C., Dolan, K. and Whitaker-Azmitia, P.M. (1990) S-100β but not NGF, EGF, insulin or calmodulin is a CNS serotonergic growth factor. Brain Res., 516: 354–356.

Azmitia, E.C (2002) Cajal's hypotheses on neurobiones and neurotropic factor match properties of microtubules and S-100β. In: Azmitia, E.C., DeFelipe, J., Jones, E.G., Rakic, P., Ribak, C.E. and Ribak, C.E. (Eds.), Progress in Brain Research, Vol. 136. Elsevier, Amsterdam, pp. 87–100.

Azmitia, E.C. and Whitaker-Azmitia, P.M (1997) Development and neuroplasticity of central serotenergic neurons. In: Baumgarten, H.G. and Gothert, M. (Eds.), Handbook of Experimental Pharmacology, Serotenergic Neurons and 5-HT Receptors in the CNS. Springer-Verlag, Berlin, pp. 1–39.

Baas, P.W., Deitch, J.S., Black, M.M. and Banker, G.A. (1988) Polarity orientation of microtubules in hippocampal neurons: uniformity in the axon and nonuniformity in the dendrite. Proc. Natl. Acad. Sci. USA, 85: 8335–8339.

Baba, A., Yasui, T., Fujisawa, S., Yamada, R.X., Yamada, M.K., Nishiyama, N., Matsuki, N. and Ikegaya, Y. (2003) Activity-evoked capacitative Ca^{2+} entry: implications in synaptic plasticity. J. Neurosci., 23: 7737–7741.

Baimbridge, K.G., Miller, J.J. and Parkes, C.O. (1982) Calcium-binding protein distribution in the rat brain. Brain Res., 239: 519–525.

Barbour, B., Keller, B.U., Llano, I. and Marty, A. (1994) Prolonged presence of glutamate during excitatory synaptic transmission to cerebellar Purkinje cells. Neuron, 12: 1331–1343.

Barger, S.W. and Van Eldik, L.J. (1992) S100β stimulates calcium fluxes in glial and neuronal cells. J. Biol. Chem., 267: 9689–9694.

Barski, J.J., Hartmann, J., Rose, C.R., Hoebeek, F., Morl, K., Noll-Hussong, M., De Zeeuw, C.I., Konnerth, A. and Meyer, M. (2003) Calbindin in cerebellar Purkinje cells is a critical determinant of the precision of motor coordination. J. Neurosci., 23: 3469–3477.

Bernhardt, R. and Matus, A. (1982) Initial phase of dendrite growth: evidence for the involvement of high molecular weight microtubule-associated proteins (HMWP) before the appearance of tubulin. J. Cell Biol., 92: 589–593.

Bernhardt, R. and Matus, A. (1984) Light and electron microscopic studies of the distribution of microtubule-associated protein 2 in rat brain: a difference between dendritic and axonal cytoskeletons. J. Comp. Neurol., 226: 203–221.

Bigot, D., Matus, A. and Hunt, S.P. (1991) Reorganization of the cytoskeleton in rat neurons following stimulation with excitatory amino acids in vitro. Eur. J. Neurosci., 3: 551–558.

Bindokas, V.P. and Miller, R.J. (1995) Excitotoxic degeneration is initiated at non-random sites in cultured rat cerebellar neurons. J. Neurosci., 15: 6999–7011.

Blaabjerg, M., Kristensen, B.W., Bonde, C. and Zimmer, J. (2001) The metabotropic glutamate receptor agonist 1S,3R-ACPD stimulates and modulates NMDA receptor mediated excitotoxicity in organotypic hippocampal slice cultures. Brain Res., 898: 91–104.

Bouron, A. (2000) Activation of a capacitative Ca^{2+} entry pathway by store depletion in cultured hippocampal neurones. FEBS Lett., 470: 269–272.

Braun, A.P. and Schulman, H. (1995) The multifunctional calcium/calmodulin-dependent protein kinase: from form to function. Annu. Rev. Physiol., 57: 417–445.

Brenman, J.E., Christopherson, K.S., Craven, S.E., McGee, A.W. and Bredt, D.S. (1996) Cloning and characterization of postsynaptic density 93, a nitric oxide synthase interacting protein. J. Neurosci., 16: 7407–7415.

Brenman, J.E., Topinka, J.R., Cooper, E.C., McGee, A.W., Rosen, J., Milroy, T., Ralston, H.J. and Bredt, D.S. (1998) Localization of postsynaptic density-93 to dendritic microtubules and interaction with microtubule-associated protein 1A. J. Neurosci., 18: 8805–8813.

Brion, J.P., Guilleminot, J. and Nunez, J. (1988) Dendritic and axonal distribution of the microtubule-associated proteins MAP2 and tau in the cerebellum of the nervous mutant mouse. Brain Res. Dev. Brain Res., 44: 221–232.

Brorson, J.R., Manzolillo, P.A., Gibbons, S.J. and Miller, R.J. (1995) AMPA receptor desensitization predicts the selective vulnerability of cerebellar Purkinje cells to excitotoxicity. J. Neurosci., 15: 4515–4524.

Brorson, J.R., Manzolillo, P.A. and Miller, R.J. (1994) Ca^{2+} entry via AMPA/KA receptors and excitotoxicity in cultured cerebellar Purkinje cells. J. Neurosci., 14: 187–197.

Bruno, V., Copani, A., Knöpfel, T., Kuhn, R., Casabona, G., Dell'Albani, P., Condorelli, D.F. and Nicoletti, F. (1995) Activation of metabotropic glutamate receptors coupled to inositol phospholipid hydrolysis amplifies NMDA-induced neuronal degeneration in cultured cortical cells. Neuropharmacology, 34: 1089–1098.

Brunson, K.L., Khanna, A., Cromwell, H.C. and Cohen, R.W. (2001) Effect of the Noncompetitive NMDA antagonists MK-801 and ketamine on the spastic Han-Wistar mutant: a rat model of excitotoxicity. Dev. Neurosci., 23: 31–40.

Caddy, K.W. and Herrup, K. (1990) Studies of the dendritic tree of wild-type cerebellar Purkinje cells in lurcher chimeric mice. J. Comp. Neurol., 297: 121–131.

Cajal, S.R. (1913–14) Degeneration and Regeneration of the Nervous System. Translation by R.M. May (1928). Oxford University Press, London.

Cajal, S.R. (1933) Neuron theory or reticular theory? Objective evidence of the anatomical unity of nerve cells. Translated by M. Ubeda-Purkiss and C.A. Fox. Madrid, CSIC. In: DeFelipe, J. and Jones, E.G. (Eds.), Cajal's Degeneration and Regeneration of the Nervous System. Oxford University Press, New York, pp. 91–100.

Carriedo, S.G., Yin, H.Z., Sensi, S.L. and Weiss, J.H. (1998) Rapid Ca^{2+} entry through Ca^{2+}-permeable AMPA/kainate channels triggers marked intracellular Ca^{2+} rises and consequent oxygen radical production. J. Neurosci., 18: 7727–7738.

Chaudhry, F.A., Lehre, K.P., van Lookeren Campagne, M., Ottersen, O.P., Danbolt, N.C. and Storm-Mathisen, J. (1995) Glutamate transporters in glial plasma membranes: highly differentiated localizations revealed by quantitative ultrastructural immunocytochemistry. Neuron, 15: 711–720.

Choi, D.W. and Rothman, S.M. (1990) The role of glutamate neurotoxicity in hypoxic-ischemic neuronal death. Annu. Rev. Neurosci., 13: 171–182.

Cohen, R.W., Fisher, R.S., Duong, T., Handley, V.W., Campagnoni, A.T., Hull, C.D., Buchwald, N.A. and Levine, M.S. (1991) Altered excitatory amino acid function and morphology of the cerebellum of the spastic Han-Wistar rat. Brain Res. Mol. Brain Res., 11: 27–36.

Coutinho, V. and Knöpfel, T. (2002) Metabotropic glutamate receptors: electrical and chemical signaling properties. The Neuroscientist, 8: 551–561.

Dehmelt, L. and Halpain, S. (2004) Actin and microtubules in neurite initiation: are MAPs the missing link. J. Neurobiol., 58: 18–33.

Dehnes, Y., Chaudhry, F.A., Ullensvang, K., Lehre, K.P., Storm-Mathisen, J. and Danbolt, N.C. (1998) The glutamate transporter EAAT4 in rat cerebellar Purkinje cells: a glutamate-gated chloride channel concentrated near the synapse in parts of the dendritic membrane facing astroglia. J. Neurosci., 18: 3606–3619.

Delaney, C.L., Brenner, M. and Messing, A. (1996) Conditional ablation of cerebellar astrocytes in postnatal transgenic mice. J. Neurosci., 16: 6908–6918.

De Schutter, E. (1995) Cerebellar long-term depression might normalize excitation of Purkinje cells: a hypothesis. Trends Neurosci., 18: 291–295.

De Zeeuw, C.I., Hoogenraad, C.C., Goedknegt, E., Hertzberg, E., Neubauer, A., Grosveld, F. and Galjart, N. (1997) CLIP-115, a novel brain-specific cytoplasmic linker protein, mediates the localization of dendritic lamellar bodies. Neuron, 19: 1187–1199.

Diéz-Guerra, F.J. and Avila, J. (1995) An increase in phosphorylation of microtubule-associated protein 2 accompanies dendrite extension during the differentiation of cultured hippocampal neurones. Eur. J. Biochem., 227: 68–77.

Dittman, J.S. and Regehr, W.G. (1998) Calcium dependence and recovery kinetics of presynaptic depression at the climbing fiber to Purkinje cell synapse. J. Neurosci., 18: 6147–6162.

Donato, R. (2001) S100: a multigenic family of calcium-modulated proteins of the EF-hand type with intracellular and extracellular functional roles. Int. J.. Biochem. Cell Biol., 33: 637–668.

Dzubay, J.A. and Otis, T.S. (2002) Climbing fiber activation of metabotropic glutamate receptors on cerebellar Purkinje neurons. Neuron, 36: 1159–1167.

Eilers, J., Augustine, G.J. and Konnerth, A. (1995) Subthreshold synaptic Ca^{2+} signalling in fine dendrites and spines of cerebellar Purkinje neurons. Nature, 373: 155–158.

Elias, S.A., Yae, H. and Ebner, T.J. (1993) Optical imaging of parallel fiber activation in the rat cerebellar cortex: spatial effects of excitatory amino acids. Neuroscience, 52: 771–786.

Fairman, W.A., Vandenberg, R.J., Arriza, J.L., Kavanaugh, M.P. and Amara, S.G. (1995) An excitatory amino-acid transporter with properties of a ligand-gated chloride channel. Nature, 375: 599–603.

Farooqui, A.A. and Horrocks, L.A. (1994) Involvement of glutamate receptors, lipases, and phospholipases in long-term potentiation and neurodegeneration. J. Neurosci. Res., 38: 6–11.

Ferreira, A., Kincaid, R. and Kosik, K.S. (1993) Calcineurin is associated with the cytoskeleton of cultured neurons and has a role in the acquisition of polarity. Mol. Biol. Cell., 4: 1225–1238.

Fierro, L. and Llano, I. (1996) High endogenous calcium buffering in Purkinje cells from rat cerebellar slices. J. Physiol. (Lond.), 496: 617–625.

Finch, E.A. and Augustine, G.J. (1998) Local calcium signalling by inositol-1,4,5 trisphosphate in Purkinje cell dendrites. Nature, 396: 753–756.

Fischer, M., Kaech, S., Wagner, U., Brinkhaus, H. and Matus, A. (2000) Glutamate receptors regulate actin-based plasticity in dendritic spines. Nat. Neurosci., 3: 887–894.

Folkerts, M.M., Berman, R.F., Muizelaar, J.P. and Rafols, J.A. (1998) Disruption of MAP-2 immunostaining in rat hippocampus after traumatic brain injury. J. Neurotrauma, 15: 349–363.

Francke, U. (1999) Williams-Beuren syndrome: genes and mechanisms. Hum. Mol. Genet., 8: 1947–1954.

Fukuda, K., Aihara, N., Sagar, S.M., Sharp, F.R., Pitts, L.H., Honkaniemi, J. and Noble, L.J. (1996) Purkinje cell vulnerability to mild traumatic brain injury. J. Neurotrauma, 13: 255–266.

Fukaya, M., Yamada, K., Nagashima, M., Tanaka, K. and Watanabe, M. (1999) Down-regulated expression of glutamate transporter GLAST in Purkinje cell-associated astrocytes of reeler and weaver mutant cerebella. Neurosci. Res., 34: 165–175.

Fukata, M., Watanabe, T., Noritake, J., Nakagawa, M., Yamaga, M., Kuroda, S., Matsuura, Y., Iwamatsu, A., Perez, F. and Kaibuchi, K. (2002) Rac1 and Cdc42 capture microtubules through IQGAP1 and CLIP-170. Cell, 109: 873–885.

Furukawa, K., Smith-Swintosky, V.L. and Mattson, M.P. (1995) Evidence that actin depolymerization protects hippocampal neurons against excitotoxicity by stabilizing $[Ca^{2+}]_i$. Exp. Neurol., 133: 153–163.

Furukawa, K. and Mattson, M.P. (1995) Taxol stabilizes $[Ca^{2+}]_i$ and protects hippocampal neurons against excitotoxicity. Brain Res., 689: 141–146.

Gerlai, R., Wojtowicz, J.M., Marks, A. and Roder, J. (1995) Overexpression of a calcium-binding protein, S100 beta, in astrocytes alters synaptic plasticity and impairs spatial learning in transgenic mice. Learn. Mem., 2: 26–39.

Gillardon, F., Baurle, J., Wickert, H., Grusser-Cornehls, U. and Zimmermann, M. (1995) Differential regulation of bcl-2, bax, c-fos, junB, and krox-24 expression in the cerebellum of Purkinje cell degeneration mutant mice. J. Neurosci. Res., 41: 708–715.

Goto, S., Yamamoto, H., Fukunaga, K., Iwasa, T., Matsukado, Y. and Miyamoto, E. (1985) Dephosphorylation of microtubule associated protein 2, tau factor and tubulin by calcineurin. J. Neurochem., 45: 276–283.

Grosche, J., Matyash, V., Möller, T., Verkhratsky, A., Reichenbach, A. and Kettenmann, H. (1999) Microdomains for neuron-glia interaction: parallel fiber signaling to Bergmann glial cells. Nat. Neurosci., 2: 139–143.

Halpain, S., Hipolito, A. and Saffer, L. (1998) Regulation of F-actin stability in dendritic spines by glutamate receptors and calcineurin. J. Neurosci., 18: 9835–9844.

Hansel, C. and Linden, D.J. (2000) Long-term depression of the cerebellar climbing fiber—Purkinje neuron synapse. Neuron, 26: 473–482.

Harris, K.M. and Stevens, J.K. (1988) Dendritic spines of rat cerebellar Purkinje cells: serial electron microscopy with reference to their biophysical characteristics. J. Neurosci., 8: 4455–4469.

Herrmann, M., Curio, N., Jost, S., Grubich, C., Ebert, A.D., Fork, M.L. and Synowitz, H. (2001) Release of biochemical markers of damage to neuronal and glial brain tissue is associated with short and long term neuropsychological outcome after traumatic brain injury. J. Neurol. Neurosurg. Psychiatry, 70: 95–100.

Hirai, H. (2000) Clustering of delta glutamate receptors is regulated by the actin cytoskeleton in the dendritic spines of cultured rat Purkinje cells. Eur. J. Neurosci., 12: 563–570.

Hofmann, M.A., Drury, S., Fu, C., Qu, W., Taguchi, A., Lu, Y., Avila, C., Kambham, N., Bierhaus, A., Nawroth, P., Neurath, M.F., Slattery, T., Beach, D., McClary, J., Nagashima, M., Morser, J., Stern, D. and Schmidt, A.M. (1999) RAGE mediates a novel proinflammatory axis: a central cell surface receptor for S100/calgranulin polypeptides. Cell, 97: 889–901.

Hoogenraad, C.C., Eussen, B.H., Langeveld, A., van Haperen, R., Winterberg, S., Wouters, C.H., Grosveld, F., De Zeeuw, C.I. and Galjart, N. (1998) The murine CYLN2 gene: genomic organization, chromosome localization, and comparison to the human gene that is located within the 7q11.23 Williams syndrome critical region. Genomics, 53: 348–358.

Hoogenraad, C.C., Koekkoek, B., Akhmanova, A., Krugers, H., Dortland, B., Miedema, M., van Alphen, A., Kistler, W.M., Jaegle, M., Koutsourakis, M., Van Camp, N., Verhoye, M., van der Linden, A., Kaverina, I., Grosveld, F., De Zeeuw, C.I. and Galjart, N. (2002) Targeted mutation of *Cyln2* in the Williams syndrome critical region links CLIP-115 haploinsufficiency to neurodevelopmental abnormalities in mice. Nat. Genet., 32: 116–127.

Iino, M., Goto, K., Kakegawa, W., Okado, H., Sudo, M., Ishiuchi, S., Miwa, A., Takayasu, Y., Saito, I., Tsuzuki, K. and Ozawa, S. (2001) Glia–synapse interaction through Ca^{2+}-permeable AMPA receptors in Bergmann glia. Science, 292: 926–929.

Ito, M. (1984) The cerebellum and neural control. Raven Press, New York.

Johnson, G.V.W., Litersky, J.M. and Jope, R.S. (1991) Degradation of microtubule-associated protein 2 and brain spectrin by calpain: A comparative study. J. Neurochem., 56: 1630–1638.

Johnson, B.D. and Byerly, L. (1993) A cytoskeletal mechanism for Ca^{2+} channel metabolic dependence and inactivation by intracellular Ca^{2+}. Neuron, 10: 797–804.

Kaech, S., Fischer, M., Doll, T. and Matus, A. (1997) Isoform specificity in the relationship of actin to dendritic spines. J. Neurosci., 17: 9565–9572.

Kanayama, G., Takeda, M., Niigawa, H., Ikura, Y., Tamii, H., Taniguchi, N., Kudo, T., Miyamae, Y., Morihara, T. and Nishimura, T. (1996) The effects of repetitive mild brain injury on cytoskeletal protein and behavior. Methods Find. Exp. Clin. Pharmacol., 18: 105–115.

Karpiak, S.E. and Mahadik, S.P. (1987) Selective uptake by Purkinje neurons of antibodies to S-100 protein. Exp. Neurol., 98: 453–457.

Kashiwabuchi, N., Ikeda, K., Araki, K., Hirano, T., Shibuki, K., Takayama, C., Inoue, Y., Kutsuwada, T., Yagi, T., Kang, Y., Aizawa, S. and Mishina, M. (1995) Impairment of motor coordination, Purkinje cell synapse formation, and cerebellar long-term depression in GluRδ2 mutant mice. Cell, 81: 245–252.

Kataoka, Y. and Ohmori, H. (1998) Of known neurotransmitters, glutamate is the most likely to be released from chick cochlear hair cells. J. Neurophysiol., 76:1870–1879.

Kennedy, M.B. (1997) The postsynaptic density at glutamatergic synapses. Trends Neurosci., 20: 264–268.

Kim, E., Cho, K-O., Rothschild, A. and Sheng, M. (1996) Heteromultimerization and NMDA receptor clustering activity of Chapsyn-110, a novel member of the PSD-95 family of synaptic proteins. Neuron, 17: 103–113.

Kim, C.H. and Lisman, J.E. (2001) A labile component of AMPA receptor-mediated synaptic transmission is dependent on microtubule motors, actin, and N-ethylmaleimide-sensitive factor. J. Neurosci., 21: 4188–4194.

Kim, S.J., Kim, Y.S., Yuan, J.P., Petralia, R.S., Worley, P.F. and Linden, D.J. (2003) Activation of the TRPC1 cation channel by metabotropic glutamate receptor mGluR1. Nature, 426: 285–291.

Knöpfel, T. and Grandes, P. (2002) Metabotropic glutamate receptors in the cerebellum with a focus on their function in Purkinje cells. Cerebellum, 1: 19–26.

Kurihara, H., Hashimoto, K., Kano, M., Takayama, C., Sakimura, K., Mishina, M., Inoue, Y. and Watanabe, M. (1997) Impaired parallel fiber—Purkinje cell synapse stabilization during cerebellar development of mutant mice lacking the glutamate receptor δ2 subunit. J. Neurosci., 17: 9613–9623.

Landis, S.C. and Mullen, R.J. (1978) The development and degeneration of Purkinje cells in pcd mutant mice. J. Comp. Neurol., 177: 125–143.

Llano, I., Dreessen, J., Kano, M. and Konnerth, A. (1991) Intradendritic release of calcium induced by glutamate in cerebellar Purkinje cells. Neuron, 7: 577–583.

Llinás, R., Lang, E.J. and Welsh, J.P. (1997) The cerebellum, LTD, and memory: alternative views. Learn. Mem., 3: 445–455.

Lopez-Bayghen, E., Espinoza-Rojo, M. and Ortega, A. (2003) Glutamate down-regulates GLAST expression through AMPA receptors in Bergmann glial cells. Mol. Brain Res., 115: 1–9.

Lossi, L., Ghidella, S., Marroni, P. and Merighi, A. (1995) The neurochemical maturation of the rabbit cerebellum. J. Anat., 187: 709–722.

Ma, H.T., Patterson, R.L., van Rossum, D.B., Birnbaumer, L., Mikoshiba, K. and Gill, D.L. (2000) Requirement of the inositol trisphosphate receptor for activation of store-operated Ca^{2+} channels. Science, 287: 1647–1651.

Margulies, J.E., Cohen, R.W., Levine, M.S. and Watson, J.B. (1993) Decreased GluR2$_B$ receptor subunit mRNA expression in cerebellar neurons at risk for degeneration. Dev. Neurosci., 15: 110–120.

Marr, D. (1969) A theory of cerebellar cortex. J. Physiol. (Lond.), 202: 437–470.

Masliah, E., Alford, M., DeTeresa, R., Mallory, M. and Hansen, L. (1996) Deficient glutamate transport is associated with neurodegeneration in Alzheimer's disease. Ann. Neurol., 40: 749–766.

Matus, A., Bernhardt, R. and Hugh-Jones, T. (1981) High molecular weight microtubule-associated proteins are preferentially associated with dendritic microtubules in brain. Proc. Natl. Acad. Sci. USA, 78: 3010–3014.

Matus, A., Delhaye-Bouchaud, N. and Mariani, J. (1990) Microtubule-associated protein 2 (MAP2) in Purkinje cell dendrites: evidence that factors other than binding to microtubules are involved in determining its cytoplasmic distribution. J. Comp. Neurol., 297: 435–440.

Mbele, G.O., Deloulme, J.C., Gentil, B.J., Delphin, C., Ferro, M., Garin, J., Takahashi, M. and Baudier, J. (2002) The zinc- and calcium-binding S100B interacts and co-localizes with IQGAP1 during dynamic rearrangement of cell membranes. J. Biol. Chem., 277: 49998–50007.

McDonald, J.W. and Johnston, M.V. (1990) Physiological and pathophysiological roles of excitatory amino acids during central nervous system development. Brain Res. Brain Res. Rev., 15: 41–70.

McGee, A.W., Topinka, J.R., Hashimoto, K., Petralia, R.S., Kakizawa, S., Kauer, F., Aguilera-Moreno, A., Wenthold, R.J., Kano, M. and Bredt, D.S. (2001) PSD-93 knock-out mice reveal that neuronal MAGUKs are not required for development or function of parallel fiber synapses in cerebellum. J. Neurosci., 21: 3085–3091.

Mount, H.T., Dreyfus, C.F. and Black, I.B. (1993) Purkinje cell survival is differentially regulated by metabotropic and ionotropic excitatory amino acid receptors. J. Neurosci., 13: 3173–3179.

Murphy, D.B. and Borisy, G.G. (1975) Association of high-molecular-weight proteins with microtubules and their role in microtubule assembly in vitro. Proc. Natl. Acad. Sci. USA, 72: 2696–2700.

Nagao, S., Kwak, S. and Kanazawa, I. (1997) EAAT4, a glutamate transporter with properties of a chloride channel, is predominantly localized in Purkinje cell dendrites, and forms parasagittal compartments in rat cerebellum. Neuroscience, 78: 929–933.

Niethammer, M., Valtschanoff, J.G., Kapoor, T.M., Allison, D.W., Weinberg, T.M., Craig, A.M. and Sheng, M. (1998) CRIPT, a novel postsynaptic protein that binds to the third PDZ domain of PSD-95/SAP90. Neuron, 20: 693–707.

Nishiyama, H., Knöpfel, T., Endo, S. and Itohara, S. (2002) Glial protein S100B modulates long-term neuronal synaptic plasticity. Proc. Natl. Acad. Sci. USA, 99: 4037–4042.

Nisim, A.A., Hernandez, C.M. and Cohen, R.W. (1999) The neuroprotective effects of non-NMDA antagonists in the cerebellum of the spastic Han Wistar mutant. Dev. Neurosci., 21: 76–86.

O'Hearn, E. and Molliver, M.E. (1993) Degeneration of Purkinje cells in parasagittal zones of the cerebellar vermis after treatment with ibogaine or harmaline. Neuroscience, 55: 303–310.

O'Hearn, E. and Molliver, M.E. (1997) The olivocerebellar projection mediates ibogaine-induced degeneration of Purkinje cells: a model of indirect, trans-synaptic excitotoxicity. J. Neurosci., 17: 8828–8841.

Okubo, Y., Kakizawa, S., Hirose, K. and Iino, M. (2001) Visualization of IP_3 dynamics reveals a novel AMPA receptor-triggered IP_3 production pathway mediated by voltage-dependent Ca^{2+} influx in Purkinje cells. Neuron, 32: 113–122.

Otis, T.S., Kavanaugh, M.P. and Jahr, C.E. (1997) Postsynaptic glutamate transport at the climbing fiber-Purkinje cell synapse. Science, 277: 1515–1518.

Ozawa, S., Kamiya, H. and Tsuzuki, K. (1998) Glutamate receptors in the mammalian central nervous system. Prog. Neurobiol., 54: 581–618.

Pasti, L., Zonta, M., Pozzan, T., Vicini, S. and Carmignoto, G. (2001) Cytosolic calcium oscillations in astrocytes may regulate exocytotic release of glutamate. J. Neurosci., 21: 477–484.

Paula-Barbosa, M.M. and Tavares, M.A. (1985) Long term alcohol consumption induces microtubular changes in the adult rat cerebellar cortex. Brain Res., 339: 195–199.

Pedrotti, B., Colombo, R. and Islam, K. (1994) Microtubule associated protein MAP1A is an actin-binding and cross-linking protein. Cell Motil. Cytoskeleton, 29: 110–116.

Pierre, P., Scheel, J., Rickard, J.E. and Kreis, T.E. (1992) CLIP-170 links endocytic vesicles to microtubules. Cell, 70: 887–900.

Pisani, A., Gubellini, P., Bonsi, P., Conquet, F., Picconi, B., Centonze, D., Bernardi, G. and Calabresi, P. (2001) Metabotropic glutamate receptor 5 mediates the potentiation of N-methyl-D-aspartate responses in medium spiny striatal neurons. Neuroscience, 106: 579–587.

Raabe, A. and Seifert, V. (1999) Fatal secondary increases in serum S-100B protein after severe head injury. Report of three cases. J. Neurosurg., 91: 875–877.

Rahman, S. and Newman, R.S. (1996) Characterization of metabotropic glutamate receptor-mediated facilitation of N-methyl-D-aspartate depolarization of neocortical neurones. Br. J. Pharmacol., 117: 675–683.

Rao, V.L.R., Baskaya, M.K., Dogan, A., Rothstein, J.D. and Dempsey, R.J. (1998) Traumatic brain injury down-regulates glial glutamate transporter (GLT-1 and GLAST) proteins in rat brain. J. Neurochem., 70: 2020–2027.

Reeves, R.H., Yao, J., Crowley, M.R., Buck, S., Zhang, X., Yarowsky, P., Gearhart, J.D. and Hilt, D.C. (1994) Astrocytosis and axonal proliferation in the hippocampus of *S100b* transgenic mice. Proc. Natl. Acad. Sci. USA, 91: 5359–5363.

Roche, K.W., Ly, C.D., Petralia, R.S., Wang, Y.X., McGee, A.W., Bredt, D.S. and Wenthold, R.J. (1999) Postsynaptic density-93 interacts with the δ2 glutamate receptor subunit at parallel fiber synapses. J. Neurosci., 10: 3926–3934.

Rosenmund, C. and Westbrook, G.L. (1993) Calcium-induced actin depolymerization reduces NMDA channel activity. Neuron, 10: 805–814.

Rothman, S.M. (1985) The neurotoxicity of excitatory amino acids is produced by passive chloride influx. J. Neurosci., 5: 1483–1489.

Ruiz, M. and Ortega, A. (1995) Characterization of an Na$^+$-dependent glutamate/aspartate transporter from cultured Bergmann glia. NeuroReport, 6: 2041–2044.

Saatman, K.E., Graham, D.I. and McIntosh, T.K. (1998) The neuronal cytoskeleton is at risk after mild and moderate brain injury. J. Neurotrauma, 15: 1047–1058.

Sarna, J.R. and Hawkes, R. (2003) Patterned Purkinje cell death in the cerebellum. Prog. Neurobiol., 70: 473–507.

Sattler, R., Xiong, Z., Lu, W.Y., Hafner, M., MacDonald, J.F. and Tymianski, M. (1999) Specific coupling of NMDA receptor activation to nitric oxide neurotoxicity by PSD-95 protein. Science, 284: 1845–1848.

Sayer, R.J., Turnbull, C.I. and Hubbard, M.J. (2000) Calbindin 28 kDa is specifically associated with extranuclear constituents of the dense particulate fraction. Cell Tissue Res., 302: 171–180.

Schliwa, M., Euteneuer, U., Bulinski, J.C. and Izant, J.G. (1981) Calcium lability of cytoplasmic microtubules and its modulation by microtubule-associated proteins. Proc. Natl. Acad. Sci. USA, 78: 1037–1041.

Schmidt, H., Stiefel, K.M., Racay, P., Schwaller, B. and Eilers, J. (2003) Mutational analysis of dendritic Ca^{2+} kinetics in rodent Purkinje cells: role of parvalbumin and calbindin D28k. J. Physiol., 551: 13–32.

Schmidt-Kastner, R. and Freund, T.F. (1991) Selective vulnerability of the hippocampus in brain ischemia. Neuroscience, 40: 599–636.

Schmolesky, M.T., Weber, J.T., De Zeeuw, C.I. and Hansel, C. (2002) The making of a complex spike: ionic composition and plasticity. Ann. N.Y. Acad. Sci., 978: 359–390.

Schwaller, B., Meyer, M. and Schiffmann, S. (2002) 'New' functions for 'old' proteins: the role of the calcium-binding proteins calbindin D-28k, calretinin and parvalbumin, in cerebellar physiology. Studies with knockout mice. Cerebellum, 1: 241–258.

Segal, M. (1995) Dendritic spines for neuroprotection: a hypothesis. Trends Neurosci., 18: 468–471.

Sekerkova, G., Loomis, P.A., Changyaleket, B., Zheng, L., Eytan, R., Chen, B., Mugnaini, E. and Bartles, J.R. (2003) Novel espin actin-bundling proteins are localized to Purkinje cell dendritic spines and bind the Src homology 3 adapter protein insulin receptor substrate p53. J. Neurosci., 23: 1310–1319.

Selden, S.C. and Pollard, T.D. (2002) Phosphorylation of microtubule-associated proteins regulates their interaction with actin filaments. J. Biol. Chem., 258: 7064–7071.

Sergé, A., Fourgeaud, L., Hémar, A. and Choquet, D. (2003) Active surface transport of metabotropic glutamate receptors through binding to microtubules and actin flow. J. Cell Sci., 116: 5015–5022.

Shibata, T., Watanabe, M., Tanaka, K., Wada, K. and Inoue, Y. (1996) Dynamic changes in expression of glutamate transporter mRNAs in developing brain. NeuroReport, 7: 705–709.

Shiomura, Y. and Hirokawa, N. (1987) Colocalization of microtubule-associated protein 1A and microtubule-associated protein 2 on neuronal microtubules in situ revealed with double-label immunoelectron microscopy. J. Cell Biol., 104: 1575–1578.

Siman, R., Noszek, J.C. and Kegerise, C. (1989) Dynamic interaction between soluble tubulin and C-terminal domains of N-methyl-D-aspartate receptor subunits. J. Neurosci., 9: 1579–1590.

Sjöbeck, M. and Englund, E. (2001) Alzheimer's disease and the cerebellum: a morphologic study on neuronal and glial changes. Dement. Geriatr. Cogn. Disord., 12: 211–218.

Slemmer, J.E., Matser, E.J.T., De Zeeuw, C.I. and Weber, J.T. (2002) Repeated mild injury causes cumulative damage to hippocampal cells. Brain, 125: 2699–2709.

Slemmer, J.E., Weber, J.T., De Zeeuw, C.I. (2004) Cell death, glial protein alterations and elevated S-100β release in cerebellar cell cultures following mechanically induced trauma. Neurobiol. Dis.,15:563–572.

Sorci, G., Agneletti, A.L. and Donato, R. (2000) Effects of S100A1 and S100B on microtubule stability. An *in vitro* study using triton-cytoskeletons from astrocyte and myoblast cell lines. Neuroscience, 99: 773–783.

Stepanova, T., Slemmer, J., Hoogenraad, C.C., Lansbergen, G., Dortland, B., De Zeeuw, C.I., Grosveld, F., van Cappellen, G., Akhmanova, A. and Galjart, N. (2003) Visualization of microtubule growth in cultured neurons via the use of EB3-GFP (end-binding protein 3-green fluorescent protein). J. Neurosci., 23: 2655–2664.

Strahlendorf, J.C., Acosta, S., Miles, R. and Strahlendorf, H.K. (2001) Choline blocks AMPA-induced dark cell degeneration of Purkinje neurons: potential role of the α7 nicotinic receptor. Brain Res., 901: 71–78.

Strata, P. and Rossi, F. (1998) Plasticity of the olivocerebellar pathway. Trends Neurosci., 21: 407–413.

Takacs, J. and Hamori, J. (1994) Developmental dynamics of Purkinje cells and dendritic spines in rat cerebellar cortex. J. Neurosci. Res., 38: 515–530.

Takahashi, M., Sarantis, M. and Attwell, D. (1996) Postsynaptic glutamate uptake in rat cerebellar Purkinje cells. J. Physiol., 497: 523–530.

Takayama, C., Nakagawa, S., Watanabe, M., Mishina, M. and Inoue, Y. (1996) Developmental changes in expression and distribution of the glutamate receptor channel δ2 subunit according to the Purkinje cell maturation. Brain Res. Dev. Brain Res., 92: 147–155.

Tavares, M.A., Paula-Barbosa, M.M. and Gray, E.G. (1983) A morphometric Golgi analysis of the Purkinje cell dendritic tree after long-term alcohol consumption in the adult rat. J. Neurocytol., 12: 939–948.

Tempia, F., Alojado, M.E., Strata, P. and Knöpfel, T. (2001) Characterization of the mGluR$_1$-mediated electrical and calcium signaling in Purkinje cells of mouse cerebellar slices. J. Neurophysiol., 86: 1389–1397.

Tolbert, D.L. and Clark, B.R. (2000) Olivocerebellar projections modify hereditary Purkinje cell degeneration. Neuroscience, 101: 417–433.

Tong, G. and Jahr, C.E. (1994) Block of glutamate transporters potentiates postsynaptic excitation. Neuron, 13: 1195–1203.

Trump, B.F. and Berezesky, I.K. (1995) Calcium-mediated cell injury and cell death. FASEB J, 9: 219–228.

van Rossum, D. and Hanisch, U.-K. (1999) Cytoskeletal dynamics in dendritic spines: direct modulation by glutamate receptors? Trends Neurosci., 22: 290–295.

van Rossum, D., Kuhse, J. and Betz, H. (1999) Dynamic interaction between soluble tubulin and C-terminal domains of N-methyl-D-aspartate receptor subunits. J. Neurochem., 72: 962–973.

Vecellio, M., Schwaller, B., Meyer, M., Hunziker, W. and Celio, M.R. (2000) Alterations in Purkinje cell spines of calbindin D-28K and parvalbumin knock-out mice. Eur. J. Neurosci., 12: 945–954.

Vig, P.J.S., Subramony, S.H. and McDaniel, D.O. (2001) Calcium homeostasis and spinocerebellar ataxia-1 (SCA-1). Brain Res. Bull., 56: 221–225.

Washbourne, P., Bennett, J.E. and McAllister, A.K. (2002) Rapid recruitment of NMDA receptor transport packets to nascent synapses. Nat. Neurosci., 5: 751–759.

Watanabe, M. (2002) Glial processes are glued to synapses via Ca^{2+}-permeable glutamate receptors. Trends Neurosci., 25: 5–6.

Watase, K., Hashimoto, K., Kano, M., Yamada, K., Watanabe, M., Inoue, Y., Okuyama, S., Sakagawa, T., Ogawa, S., Kawashima, N., Hori, S., Takimoto, M., Wada, K. and Tanaka, K. (1998) Motor discoordination and increased susceptibility to cerebellar injury in GLAST mutant mice. Eur. J. Neurosci., 10: 976–988.

Weber, J.T., Rzigalinski, B.A. and Ellis, E.F. (2001) Traumatic injury of cortical neurons causes changes in intracellular calcium stores and capacitative calcium influx. J. Biol. Chem., 276: 1800–1807.

Weber, J.T., De Zeeuw, C.I., Linden, D.J. and Hansel, C. (2003) Long-term depression of climbing fiber-evoked calcium transients in Purkinje cell dendrites. Proc. Natl. Acad. Sci. USA, 100: 2878–2883.

Weber, J.T (2004) Calcium homeostasis following traumatic neuronal injury. Curr. Neurovascular Res., 1: 151–171.

Welsh, J.P., Yuen, G., Placantonakis, D.G., Vu, T.Q., Haiss, F., O'Hearn, E., Molliver, M.E. and Aicher, S.A. (2002) Why do Purkinje cells die so easily after global brain ischemia? Aldolase C, EAAT4, and the cerebellar contribution to posthypoxic myoclonus. Adv. Neurol., 89: 331–359.

Westrum, L.E., Jones, D.H., Gray, E.G. and Barron, J. (1980) Microtubules, dendritic spines and spine apparatuses. Cell Tissue Res., 208: 171–181.

Wilson, M.T. and Keith, C.H. (1998) Glutamate-modulation of dendrite outgrowth: alterations in the distribution of dendritic microtubules. J. Neurosci. Res., 52: 599–611.

Xiong, Z., O'Hanlon, D., Becker, L.E., Roder, J., MacDonald, J.F. and Marks, A. (2000) Enhanced calcium transients in glial cells in neonatal cerebellar cultures derived from S100B null mice. Exp. Cell Res., 257: 281–289.

Yamada, K., Watanabe, M., Shibata, T., Tanaka, K., Wada, K. and Inoue, Y. (1996) EAAT4 is a postsynaptic glutamate transporter at Purkinje cell synapses. NeuroReport, 7: 2013–2017.

Zhang, N., Walberg, F., Laake, J.H., Meldrum, B.S. and Ottersen, O.P. (1990) Aspartate-like and glutamate-like immunoreactivities in the inferior olive and climbing fibre system: a light microscopic and semiquantitative electron microscopic study in rat and baboon (*Papio anubis*). Neuroscience, 38: 61–80.

Zuo, J., De Jager, P.L., Takahashi, K.A., Jiang, W., Linden, D.J. and Heintz, N. (1997) Neurodegeneration in Lurcher mice caused by mutation in $\delta2$ glutamate receptor gene. Nature, 388: 769–773.

SECTION X

Epilogue

CHAPTER 30

Epilogue

by Rodolfo Llinás

*NYU School of Medicine, Department of Physiology and Neuroscience, 550 First Avenue, New York, NY 10016, USA;
E-mail: llinar01@med.nyu.edu*

Participants of the Symposium 'Creating Coordination in the Cerebellum' held October 2–4, 2003 in Catania, Italy. The symposium was organized in honor of the lifetime work of Constantino Sotelo (who is standing in the center).

And so, the feast of science, discussion, and friendship commemorating Costantino Sotelo's pseudo-retirement party in beautiful Sicily having come to a close, this is now to be followed by the writing of papers. Rather than the summary paper that usually adorn the books published on such occasions I have decided to write a few closing remarks about Costantino, and our cerebellar field.

Costantino Sotelo, friend, colleague, and survivor of many cerebellar meetings, was a young post-doctoral student with Sandy Palay when we first met in 1969 at the AMA Institute for Biomedical Research in Chicago. For the next few years of our lives we were not sure if our relationship was going to be one of friendly collaboration, simple co-existence or a no holds barred all out dragged out personality clash. Our mutual interest and curiosity for cerebellar structure and function quickly put an end to personal bravados, and culminated in the gradual development of a wonderful, if prickly, friendship, and many co-authored publications that, on the whole, turned out to be quite positive. At the time of our first meeting I was heading the Department of Neurosciences at the now defunct AMA Institute and had decided to organize an international symposium with all cerebellar luminaries of the day. Our charge was to generate a volume to encompass all that was known at that time concerning the cerebellum from a phylogenetic and ontogenetic perspective with emphasis on morphology and function. Papers presented in the symposium ranged from gross comparative morphology through ultrastructure, electrophysiology, and behavior. It was clear to all participants that the cerebellar system was about to be understood in the next few years. It was also suspected that such a victory was going to be the first in a rapid succession of strategically designed commando-type incursions into the rest of the brain that would mark the final blow of reason over the ultimate scientific question, the workings of the brain.

At this meeting 33 years later, we are still 'on the verge' of understanding the cerebellar anatomy and function. The main difference from our naïve early intellectual posture is that we now know that the cerebellar functions can only be understood in the context of the function of the rest of the CNS. Thus, that the cerebellum is to be considered as the center for motor or sensory control or a learning center seems today, in all seriousness, a rather tired approach. We all now know, or should know, that sensory and motor functions can only be clearly defined at the sensory entry and at the motoneuronal output, the final common path. As we move away from such sites, designation of sensory or motor is dangerous if not foolish. As for the cerebellum to be the seat for the motor memory engram 'the holy grail' of 20th century neuroscience, much needs to be discussed still, since the evidence for this is even thinner on the ground that the less contentious motor versus sensory perspective mentioned above. Much water will pass before this is finally understood.

In conclusion, we, the participants want to thank our dear colleagues Federico Cicirata and Chris de Zeeuw for organizing and bringing to final fruition a memorable meeting in wonderful Catania.

Subject Index

$\alpha 4$ and $\beta 2$, 311
α-amino-3-hydroxy-5-methyl-4-isoxazolepropionate (AMPA), 368
α-bungarotoxin (α-BTX), 309, 313
$\beta 2$ mRNA, 311
β-actin promoter, 60
β-galactosidase, 33
γ-aminobutyric acid, 82, 313
122-channel MEG probe, 143
129/S1 genetic background, 23
129/S1 wild-type animals, 25
129/S1, 24
129/Sv, 23
20 μM bicuculline methiodide, 82
3-acetylpyridine (3-AP), 49, 208
3-acetylpyridine, 54
3-AP, 50
300 pS membrane conductance, 184
6-well plates, 38

A-type K^+ currents, 90
Abnormal repolarization, 136
Acetylcholinergic afferents, 95
ACSF, 83
Actin-disrupting agents, 378
Action potential generation, 136
Active caspase-3, 41
Activity-dependent
  competition, 47
  regulation, 50
Address selection, 47
Administration of mifepristone (RU486), 38
ADP, 83, 86
Adult rats, 115
Afrotheria, 289
Afterdepolarization, 81, 85
Afterhyperpolarization (AHP), 81–83, 86
Agonist, 238
Agonist–antagonist recruitment, 228

AHP amplitude, 83, 86
AHP, 90
Aldolase C, 283, 293
alanine-rich, 49
Alert mutant mice, 168
Alert transgenic mice, 185
Alzheimer's disease (AD), 105
Amelio-rates, 356
Aminoacid similarity, 51
AMPA
  mediated depolarization, 88
  receptor, 72, 119, 121, 367, 370
AMPA, 75, 190, 367
AMPARs, 301
Amphibia, 285
Amphibians, 285
Amphibian lineage, 293
Anatomy, 140
Ankle passive cyclic movements, 239
Annexin II/p11, 357
ANOVA, 83
Antagonist, 85, 238
Anterior
  cingulate gyrus (ACG), 102
  cingulate sulcus, 102
  lobe vermis, 292
Anteroposterior compartments, 13
Anti-apoptotic Bcl-2 protein, 41
Anti-CaBP antibodies, 39
Anti-GAD antibodies, 4
Anti-rabbit FITC, 38
Anti-zebrin II, 127, 294
Antisense oligodeoxynucleotides, 356
Apoptosis, 41
Apoptotic cell, 370
APV, 301
Architecture, 29, 71
Artificial cerebrospinal fluid (ACSF), 82
Ascending axons, 268

Asperger syndrome, 104
Astrocytes, 59, 285
Asynchronous state, 182
Ataxic patients, 227, 238
Atrophy, 141
Attention deficit disorder, 344
Auditory
    hallucination characteristic, 104
    region, 206
Autism, 104, 344
Autistic patients, 104
Autofluorescence imaging, 127
Autofluorescence, 125
Automatism, 95
Awake rabbit, 334
Axio-proximal body, 276
Axoaxonal gap junctions, 186
Axonal
    degeneration, 356
    plexus, 33, 175, 176
    regeneration, 37, 38
Axonogenesis, 6
Axosomatic membrane, 90
Axotomized olivary neurons, 37
Axotomized Purkinje cells, 37
Axotomy, 37, 40

Baboon's cerebellar nuclei, 276
BAPTA, 82, 88
Basal
    ganglia, 105, 177, 210
    pontine nuclei (BPN), 216
Basilar
    pontine nuclei (BPN), 261
Basket cells, 95
Basket/stellate neuron, 61
Bcl-2
    expression, 39
    overexpression, 38
    protection, 37
    protects, 41
Bcl-2, 37, 40–42
BCM, 77
Behavior, 106
Behavioral development, 121
BEN, 16
Bergman glia, 31, 367
BG glia, 382

Biceps femoris muscle, 247
Bicuculline
    methiodide, 85
    methobromide, 325
Bicuculline, 174
Binding protein, 321
Binomial analysis, 75
Biochemical and anatomical morphogenesis of the cerebellum, 23
Biotinylated dextrane amine (BDA), 264
Birds, 285
BK channel, 90
Block preparation, 114
Blocking
    electrical activity, 53
    Kv1.1 channels, 136
    mitochondrial respiration, 128
Box-car reference function, 154
BPN, 262
Brain-specific aldolase isoenzyme, 283
Brainstem evoked potentials, 229
Brainstem-cerebellum, 61
Brainstem, 97
Bursting state, 182

C-fos, 59
C-terminal truncation, 76
C57BL/6 background, 23–25
C57BL/6-En1$^{+/-}$ mice, 25
C57BL/6J ova, 38
Ca^{2+}
    activated K^{2+} channels, 325
    binding proteins (CBPs), 374
    chelation, 77
    dependent mechanism, 302
    homeostasis, 167
    influx, 88
    mediated depolarization, 88
    sensitive K, 86, 90
    spikes, 115
    transient, 326
Ca^{2+}/calmoduline, 176
CaBP positive cells, 15
CaBP-immunostained cultures, 38
CaBP-positive Purkinje cells, 41
CaBP, 38
$CaCl_2$, 82
Cadherins, 16

Calbindin
    D28k (CB), 167, 367, 374
    immunostaining, 15
Calbindin, 10, 24
Calcitonin gene-related peptide (CGRP), 11
Calcium
    activated potassium (KCa^{2+}), 82
    activated potassium current, 81
    binding proteins, 167, 321, 374
    channels, 190
    chelation, 90
    chelator, 86
Calcium, 71, 321
Calretinin (CR), 167, 321, 323, 325
    deficiency, 323
    immunoreactivity, 176
Cannabinoid receptors, 299
Carbamazepine, 353
Carbenoxolone, 174
Cardiovascular control, 97
Caspase-3 activation, 41
Cavernous angioma, 229
Cavia porcellus, 288
Cb or Cr genes, 168
Cb, 168, 176
Cell
    adhesion molecules, 16
    death, 37, 41
    intrinsic programs, 57
    lineage, 57
    Physiological Cerebellar Plasticity, 69
    specification, 57
Central
    fixation (Fix), 153
    pattern generator, 243
Cerebellar
    Activity, 111
    architecture, 61
    ataxia, 227
    channelopathy, 356
    circuitry, 71
    cortex, 29, 47, 96, 113, 117, 125, 135, 165, 167, 176,
        181, 185
    cortical afferents, 22
    cortical interneurons, 329
    cortical neurons, 213
    corticonuclear microcomplex, 96
    damage, 202
    development, 57
    dysfunction, 353
    embryogenesis, 64
    functional architectures, 125
    glomeruli cells, 3
    granule cells, 323
    hemisphere, 227
    hemispheres, 161, 213
    lamella, 14
    maps, 3
    metabolism, 343
    Motor Control, 199
    neurons, 58
    nuclear neurons, 4, 96
    nuclei (CN), 269
    nuclei, 113, 182, 214, 269
    nucleofugal axons, 101
    ontogenesis, 65
    oscillations, 185
    parenchyma, 60
    prefrontal loop, 140
    primordia, 29
    primordium, 30
    progenitors, 57, 60, 62
    Purkinje cells, 42
    ventricular zone, 58
    vermis, 294
Cerebellopetal fibers, 262
Cerebellum, 37, 47, 58, 71, 81, 89, 95, 100, 104, 105,
        113, 139, 151, 201, 210, 243
    activations in schizophrenia, 145
    function, 140
    regionalization, 31
Cerebellum–pons–medulla block preparation,
        115, 121
Cerebral association cortex, 102
Cerebral
    cortical input, 161
    glucose metabolism, 105
    visuomotor functions, 103
Cerebrocerebellar
    communication, 101
    network, 151
    pathways, 261
Cereboponto-cerebellar, 261
CF, 207, 302
    evoked Ca^{2+} transient, 90
    input, 83

CF (*Continued*)
  input–output, 83
  LTD, 90
  mGluR PSC, 300
  non-associative LTD, 299
  stimulation, 121
  tetanization, 85
CF–LTD, 83
CGRP+ olivocerebellar axons, 13
Chaetophractus villosus, 290
Channel-mediated transient hyperpolarization, 90
Channelopathies, 353
Chemical synapses, 174
Chemoaffnity theory, 16
Chemotactic molecules, 8
Chemotactic, 8
Chick embryonic cerebellum, 15
Chick embryos BEN/SCI/DM-GRASP, 16
Chlorpyrifos, 346
Cholinergic innervation, 77
Chronic-relapsing experimental allergic encephalomyelitis (CR-EAE), 356
CIO, 127
Climbing fiber
  activation, 89
  afferents, 96
  innervated spine, 52
  regression, 48
  synaptic plasticity, 81
  tetanization protocol, 83
Climbing fibers (CF), 12, 47, 81, 83, 113, 208, 299
CLIP-170, 383
CMC, 229
CNQX, 119, 129, 370
CNS, 21, 58, 62, 225, 368
CNS axons, 356
CNS embryogenesis, 62
Coarse-grained to a fine-grained olivocerebellar map, 17
Cocontraction, 239
Cognitive deficits, 202
Colchicine, 376
Collective oscillation, 176
Colorectal Cancer-DCC-Unc5H2, 8
Commitment, 57
Communication system, 261

Competition, 47
Complementary (push–pull) inputs, 186
Complex goal-directed movements, 239
Complex spike afterhyperpolarization, 83
Complex spike, 81
Conjunctive stimulation, 115
Connexin36 knockout mice, 195
Conscious episodic memory retrieval, 141
Contralateral
  forelimbs, 215
  hemisphere, 160
  inferior olive, 127
Control, 95
Control condition, 83
Controller, 228
Cortex, 89, 104
Cord-based central program, 243
Cortical
  folia, 60
  gamma, 177
  interneurons, 61, 120
  molecular layer, 4
  stripes, 265
  synapses, 77
Cortico-cerebellar pathway, 237
Cortico-nuclear circuitry, 61
Cortico-ponto-cerebellar pathway, 278
Corticonuclear
  microcomplex, 207
  projection, 33, 293
Corticonuclear, 120
Corticopontine system, 263
Corticospinal
  nucleofugal axons, 101
  tract, 244
Costantino Sotelo, 394
Coupled and uncoupled projections from BPN and NRTP, 272
CPCCOEt, 301, 302
Cr, 176
Creatine kinase (CK), 344
CRIPT (cysteine-rich interactor of PDZ three), 379
Cross-species developmental studies, 293
Crus I, 214
Crus I projections, 22
Crus II, 267
Crus II/pml projections, 22

Crystalline arrangement, 29
cSTD, 299
Cuneate nuclei, 6
Current clamp recordings, 322
Cutaneous reflexes, 243
Cycles, 26
Cyclic flexion–extensions, 228, 229
Cysteine-rich, 379
Cytochalasin D, 376
Cytoplasm, 73
Cytoplasmic organelles, 6
Cytoskeletal elements, 378
Cytoskeleton, 367
Cytosolic resistance, 193

D-APV, 376
D2-cyclin, 134
DAG, 73
Damping, 234
Damping factor, 236
DCN, 31, 34
DCN neurons, 33
Death of mouse Purkinje cells, 41
Declarative learning, 201
Dedicated progenitors, 31
Deep nuclei, 60
Deep nuclei neurons, 61
Deficient parallel fibers, 48
Degradative, 370
Deiter's nucleus, 273
Delay-to-phase conversion, 238
Delay, 181
Delayed-rectifier K^+, 134
Delayed-rectifying potassium current, 190
Demyelination, 356
Demyelinated axons, 359
Demyelinating diseases, 356
Dendritic appendages, 4
Dendritic compartments, 50
Dendritic depolarizations, 89
Dendritic domain, 51
Dendritic membrane, 90
Dendritic spines, 50, 52, 367
Dendrodendritric gap junctions, 176
Dentate nucleus, 275
Depolarization plateau, 84
Developing CNS, 57
Development, 32

Developmental potential, 57
Dextran amines, 214
Di-2 ANEPEQ, 115
DiA caudolateral territory, 15
DiA-labeled neurons, 15
Diacylglycerol (DAG), 82
Differentiation, 57
Digital manipulation (skilled movements), 278
Dihydro-β-erythroidine (DHβE), d-tubocurarine, 313
Diphenyleneiodonium (DPI), 128
Distal dendrites, 51
DIV, 40
DMR microscope, 38
DNA double helix, 21
Donor cells, 61
Dopaminergic afferents, 95
Dorsal cochlear nucleus, 63
Dorsal peripeduncular nucleus (dPd), 270
DRG neurons, 356
Drosophila, 57
Drosophila homologue, 22
DTX-K (toxin K), 135
Dynamics, 181
Dynamic decoupling, 4
Dysfunctional prefrontal–thalamic–cerebellar circuitry, 141
Dyslexia, 103
Dysmetria, 107

E3, 14
E5, 14
E5–E7, 14
E7.5–E8 chick embryos, 14
E8.5–E9 in mouse, 29
E9, 14
E10, 14
E10.5 mid-hindbrain progenitors, 64
E10.5, 31
E10.5–E11.5, 33
E11.5, 34
E11.5–12.5, 8
E11.5 mice embryos, 31
E12 primordium, 60
E12.5, 31
E13.5, 9, 31
E16, 13, 32
E16 rats, 34

E17.5, 13
E18 to P7 slices, 40
E18.5, 34
E18, 32, 37
EAAT4 transporter, 373
Earle's salts, 38
Early cerebellar development, 21
Early primordium, 29
EBNA-293 cells, 9
ECD analysis, 143
Echinops telfairi, 289
Eckhorn system, 171
ECR and TA EMG activation, 230
Ectopic grafting, 8
Ectopic locations, 58
ECW, 145
EEG, 229
EEG data analysis, 142
EGFP-positive cells, 61
EGL, 30, 60
   cells, 59
   proliferation and expansion, 31
Electrical
   activity block, 50
   activity, 50, 115
   block, 52
   synapses, 174
Electro-oculographic recordings, 154
Electromyogram (EMG), 229
Electronmicrograph, 52
Electrooculogram (EOG), 145
Electrophysiological
   analysis, 51
   mapping techniques, 214
   recording, 49, 77, 140
   response, 128
   studies, 125
Electrophysiology, 82, 115
Electrophysiology recordings, 207
Electroresponsiveness, 325
Electroresponsiveness of granule cells, 322
Embryonic
   cerebellar precursors, 65
   cerebellar tissue, 60
   cerebellum, 60
   CNS, 62, 64
   development, 65
   donors, 63
   hippocampus, 59
   life, 58
   progenitors, 57, 62
   Purkinje cells, 49
EMG, 227, 229, 243
EMG burst, 174
EMG response, 174
EMG signals, 229
EMG→MC phase-difference, 231
Emotion, 106
$En1^{-/-}$, 23
$En1^{-/-}$ phenotype, 24
$En2^{-/-}$, 23
En-1, 13
En-1 genes, 13
En-2, 13
En-2 genes, 13
En-2 alleles, 25
Endogenous rhythms, 77
Endoplasmic reticulum (ER), 370
Engrailed-1, 21, 23
Engrailed genes, 21, 25
Engrailed-1 phenotype, 24
Engrailed-2, 21, 23
   gene, 22
   mutant, 22
Enhanced Green Fluorescent Protein (EGFP), 60
Environmental perturbations, 343
Enzyme, 73
Ependymal
   domain, 30
   layer, 29
   surface, 29, 30
Eph, 16
Eph receptor tyrosine kinases, 16
EphA3, 16
EphA4, 9, 16, 17
Ephrins, 16
Ephrin-A2, 16, 17
Ephrin-A5, 16
Eph–ephrin system, 17
Epifluorescence microscope, 214
Epifluorescence signals, 126
Epilepsy phenotype, 135
Episodic ataxia, 125
Episodic ataxia type 1 (EA1), 134, 135
EPSC, 72, 75, 302
EPSC-amplitude, 51

EPSPs, 121
Estradiol, 348
Ether-anesthesia, 114
Etiologies, 345
Excitatory postsynaptic current (EPSC), 82, 184
Excitatory postsynaptic potentials (EPSPs), 115
Excitotoxicity, 367
Extension, 49
Extensor Carpi Radialis (ECR), 227, 229
External cuneate nuclei (ECN), 7
External cuneatus neurons, 6
External granular layer (EGL), 58
External stimuli, 105
Extra-cellular blockage, 73
Extracellular signals, 57
Extrinsic signals, 57
Eye hand coordination, 154, 157
Eye movements, 97, 139

F-values, 83
F1 heterozygotes, 25
Factor, 234
Fast-Blue, 15
Fastigial nucleus, 207
Fate restriction, 57
Feedback, 181
Felis domesticus, 290
Ferret, 290
Fertilized cell, 57
FFA, 147
Fgf8, 21
Fiber-innervated spines, 51
Fiber, 51
Fiber–Purkinje cell synapses, 52, 129
Fish, 285
Flavoprotein autofluorescence, 128, 129
Flavoproteins, 127
Floccular units, 334
Flocculus, 114, 121, 329
Floor plate, 8
Fluoro-Emerald, 214, 218
Fluoro-Ruby, 214, 218
FMRI study, 151
FMRI/MEG experiment, 143
FMRI, 142, 151
Freely walking rat, 243
Frontal association cortices, 105

Functional magnetic resonance imaging (fMRI), 139, 151

G-protein-dependent, 299
G-proteins, 369
GABA immunocytochemistry, 4
GABA-A antagonist picrotoxin, 325
GABA-ergic neurons, 262
GABA, 82, 190
$GABA_A$
  blocker bicuculline, 129
  receptor blocker, 85
  receptor synapses, 176
  receptors, 301
$GABA_A$ 174, 309
GABAergic, 4, 194
GABAergic synapse, 168
Gait ataxia, 229
Gamma-frequency oscillations, 186
Ganglionic-type nAChRs, 313
Gap junctions, 4, 189
GAP-43, 37, 38, 49
Gastrocnemius muscle, 247
GBSS-Glu, 38
Gbx2, 21
GDNF, 356
Gender-specific behavior, 348
Genetic manipulations, 97
Germinative neuroepithelia, 6, 59
GFAP, 344, 346
GLAST
  activity, 380
  antibody, 380
  expression, 380
GLAST, 367
Gli genes, 13
Glia, 58, 63
Glial, 61
Glial type, 59
Gliosis, 347
Glomeruli, 5
Glomerulus, 192
GluRδ2, 48, 51, 52
Glutamate
  action, 370
  receptors, 81
  receptor δ2 subunit (GluRδ2), 47, 51, 54
  transporters, 299, 367

Glutamate-gated ion channel, 51
Glutamate-induced [Ca^{2+}]i, 375
Glutamate-mediated Purkinje cell, 367
Glutamate-mediated toxicity, 370
Glutamate, 51, 82
Glutamate–aspartate transporter (GLAST), 373
Glutamic acid decarboxylase-positive, 262
Glycine, 309
GMP-cyclic dependent protein kinase, 10
GMP-protein, 130
Golgi
  cells (GoC), 71
  cells, 30, 91, 95, 113, 176, 329, 334, 335
  cell spikes, 170
  neurons, 58
Golgi–granule cell IPSC, 186
Goniometric, 229
Granular cells, 267
Granular layer interneurons, 58
Granule
  cells (GrC), 71
  cells, 49, 59–61, 64, 71, 95, 181, 184
  cell layer, 83, 119
  cell neuronal excitability, 321
Grapheme-to-phoneme conversion, 104
Grasping, 238
GrC E-S potentiation, 76
Group II antagonist, 303
Group III autoreceptors, 304
Guinea Pig, 288

H-current, 190
Hamster, 287
Han-Wistar mutant rat model, 371
Hand, 236
Hand and foot voluntary oscillations, 227
Hand curve, 236
Handleg co-ordination, 236
HCN1 channel, 90
Hemi-cerebellectomized rats, 204, 208
Hemicerebellectomy, 204
Hemicerebellum, 6
Hemispheric, 227
Hemodynamic flow, 139
Hemodynamic intrinsic optical signal, 130
HEPES, 82
Heterochronic transplantation, 58, 60

Heterogeneity of Purkinje cells, 10
Heterologous afferent fiber competition, 52
Heterophilic
  binding, 16
  transplants, 64
  transplantation, 58
  transplantation experiments, 63
HH10–12 in chick, 29
High-Mg^{2+} solution, 117
Hippocampal
  CA1 pyramidal neurons, 90
  LTP, 74
  oscillations, 167
  pyramidal cells, 194
Hippocampus, 63, 64, 75, 89, 177, 186, 205
Histochemical markers, 125
HNK-1 genes, 13
Homeobox DNA binding region, 22
Homo- and heterosynaptic LTD, 209
Homologous competition, 48
Homophilic binding, 16
Hotfoot, 51
HRP, 34
Hu-bcl-2, 41–42
  cerebellar slices, 40
  transgenic cerebellar slices, 39
  transgenic mice, 38, 40
Human cerebellar generators, 140
Hybridization, 89, 354
Hydroxy-5-methyl-4-isoxazolepropionate, 368
Hyperactivity dyslexia, 344
Hyperexcitability, 356
Hypergravity, 343, 345
Hypergravity-exposed, 347
Hyperpolarizing, 115
Hyperspiny transformation, 50
Hypnosis, 105
Hypoplasia, 344
Hypothalamic supraoptic nucleus, 354
Hypothyroid, 347
Hypothyroidism, 347
Hypoxia, 114

I_{K-Ca} activation, 324
I_{AHP}, 90
IGluRs, 51, 300
Imaging cerebellum activity, 139
Imaging recording, 77

Imaging, 111
Imagining auditory-cued hand movements, 105
Immunochemistry, 89
Immunocytochemical
    markers, 125
    studies, 354
Immunocytochemistry, 38, 354
Immunoglobulin-like cell adhesion molecules, 16
Immunoreactive receptor, 378
In utero transplantation, 57
In vitro preparation, 113
In vitro transplantation experiments, 61
Inertia of the hand, 234
Inferior olivary
    neurons, 49, 50, 127
    nucleus, 189
    nucleus, 4
    perikarya, 6
Inferior olive, 52, 113, 165, 182
    neurons, 96
Inhibition, 181
Inhibition of PKC, 38
Inhibitory postsynaptic
    current, 181
    potentials (IPSPs), 183
Inositol-trisphosphate (IP3), 82
Intercalating synapses, 225
Interlimb curves, 232
Interlimb phase-relation, 230
Intermixed DiI neurons, 15
Internal feedback, 104
Internal model, 95, 104
    mechanism, 95
    of a motor apparatus, 104
Interneurons, 184, 329
Interpeduncular nucleus, 262
Intra-cellular blockage, 73
Intracellular Ca^{2+}, 73
Intracerebellar injections, 49
Intradendritic PC, 90
Intralimb control, 228
Intraperitoneal injection, 49
Intrinsic excitability of a cerebellar granule cell, 322
Intrinsic excitability, 75, 81, 90, 326
Intrinsic nature of cerebellar compartmentation, 13
Intrinsic plastic potentiality, 50
Inverse model, 101
IO stimulation, 115

IO-stimulation-evoked excitation, 115
Ionotropic
    AMPA receptors, 299
    glutamate receptor, 51, 73, 81, 129
    iGluRs, 299
    receptors, 368
IPSC, 184
IPSC kinetics, 184
Ipsilateral, 215
Ipsilateral periorbital region, 215
IQGAP1, 383
ISI-histograms, 333
Isthmic, 29

Juxtacellular labeling, 329

K gluconate, 82
K^+ channels, 325
K^+ channel blocker TEA, 76
Kainate receptors, 304
Kainate, 367
Kainic acid, 49
Kainic acid lesioned, 49
$K_{Ca^{2+}}$ channels, 88, 90
KCl, 82
Ketamine/xylazine anesthesia, 127
Kinaesthetic a.erent signals, 239
Knock-out mice, 51, 8
KOH, 82
Krox-24 transcription factor, 49
Kv1 potassium channels, 125
Kv1.1 blocker, 135
Kv1.1 potassium channels, 135

L7-PKCI, 208
L7/Pcp-2/lacZ hybrid gene, 13
L7/Pcp2, 13
LacZ, 31, 60
LacZ-adenovirus, 31
Late onset Parkinson's disease, 344
Latencies, 174
Lateral
    cerebellum, 213, 220
    malleolus induced cutaneous reflexes, 255
    malleolus, 244
    reticular (LRN), 7
Leaky capacitor, 323
Learning, 95

LEDs, 220
Leg, 236
LFP oscillations (LFPOs), 168
LFPO inhibition, 174
Lidocaine injection, 127
Lifting, 238
Ligands named ephrins, 16
Ligands, 16
Limbic, 58
Lobules I–X, 95
Lobuli IV–VI, 154
Lobuli VIIB–VIII, 154
Lobuli VII–VIII, 152
Locomotion, 254
Long term
  depression (LTD), 81
  depression, 81, 130
  recordings, 84
  depression, 379
Low-Ca^{2+}, 117
LTD, 95, 113, 115, 121, 201, 207, 299
LTP, 71, 73, 75
  expression, 74–76
  impairment, 76
  mechanism, 76
Lugaro cells (LC), 71, 95, 113, 176
Lurcher, 51
LY341495, 301, 303
Lysine, 38

Macaca mulatta, 291
Magnetic field tomography (MFT), 142
Magnetoencephalographic data, 139
Magnetoencephalography (MEG), 139
Mammalian cerebellum, 214
MAP1A, 375
Marr's model, 73
Matching hypothesis, 10
Mature
  cerebellar cortex, 53
  cerebellum, 29, 49, 54
  neurons, 63
McIlwain tissue chopper, 38
Mechanisms of physical and mental exercises, 106
Medial accessory olive, 115
Medial gastrocnemius muscle, 249
Mediolateral alternating parasagittal binding pattern, 16

Mediolateral compartments, 13
Medulla, 114
Medulloblastoma, 229
MEG, 139, 141, 142, 145, 237
MEG methodology, 140
Membrane, 115
  depolarization, 73
  potential, 83
Memory guided, 151
Mental model, 95
Mesencephalic descendents, 23
Mesocricetus auratus, 287
Metabotropic glutamate receptor mediated current, 299
Metabotropic glutamate receptors (mGluRs), 73, 81, 368
Metastasis, 229
Metencephalic descendents, 23
Metencephalon, 141
Methylsulfate, 85
Metronome, 230
MF, 207
MFT, 139
MFT analysis, 143, 145
$MgCl_2$, 82
MGlu receptor, 72, 73, 76
MGluRs, 299
MGluR II/III, 304
MGluR1, 130
MGluR3, 4, and 7, 299
MGluR1α-mediated, 300
MHB, 21
  junction, 30
  landscape, 25
MiCAM01 system, 115
Microcomplexes, 107
Microelectrodes, 171
Microglia, 41
Microglia-derived superoxide ions, 41
Microtubules, 367, 375
Microtubule-associated proteins (MAPs), 375
Mid-stance, 248
Midbrain, 344
Millipore culture, 38
Mitochondrial metabolism, 127
Mixed synapses, 174
MK-801, 376

Modern control theories, 106
Molecular
  candidates, 16
  interneurons, 174
  layer, 184
  layer interneurons, 59, 60
  mechanisms, 47
Monkeys, 161
Monodelphis domestica, 288
Monosynaptic mossy-fiber excitation, 184
Morphogenetic
  movements, 29
  proteins, 57
Morphological
  analysis, 51
  growth, 121
  studies, 49
Morris water maze (MWM), 204
Mossy
  fiber, 71
    inputs, 213
    intrinsic excitability, 76
  fiber–granule cell LTP, 71, 73, 76
Motoneuron, 54
Motor, 230
  command, 230
  coordination, 227
  cortex, 100, 206
  representations, 278
  sequences, 153, 154
Mouse, 287
Mouse embryo, 6
Movement frequency, 232
Movement system, 230
Movement-related mental model, 104
Mowiol, 38
MRI scan, 229
MRI studies, 159
MRNA, 354, 356
MSOP, 301, 302
MSOP block, 303
MSOPPE, 301, 302
MT disruptor, 376
Multijoint arm movements, 239
Multiple cerebellar phenotypes, 65
Multiple ion channel, 127
Multiple sclerosis, 353
Multipotent progenitors, 65

Multisensorial, 206
Murine cerebellar organotypic cultures, 40
Muscle fiber, 54
Mustela putoris furo, 290
Mutant mice, 76
Myelination, 354
Myeloma-conditioned culture medium, 284

N-methyl-D-aspartate (NMDA), 368
N-terminus, 357
Na_2-ATP, 82
Na_3-GTP, 82
Na^+ and K^+ channels, 182
Na^+ influx, 82
NAChRs, 309
NaCl, 38, 82
Nascent EGL, 62
National Instruments AT-MIO16-L, 229
$Na_v1.8$ +/+ C-type DRG, 361
$Na_v1.8$ mRNA, 357
$Na_v1.8$ up-regulation, 357
Navigation, 201
NBQX, 300
Neocerebellar
  vermal lobules VI, 344
  lobules VII, 344
Neocerebellum, 275
Neocortex, 63, 64, 177, 182, 186
Neocortical
  association areas, 105
  cells, 64
  layers, 64
  progenitors, 64
Neonatal
  animals, 114
  hippocampus, 59
  rat, 113, 121
Neoplastic lesion, 229
Nervous system, 47
Netrin-1, 8
Netrin-1 knock-out mice, 8
Neural
  circuitry, 47
  controller, 228, 239
  graft, 57
  phenotypes, 60
  progenitor, 57

Neuroanatomical studies, 214
Neurobiotin, 330
Neurochemistry, 140
Neuroepithelia, 58
Neuroepithelial stripes, 29
Neuroepithelium, 6, 29, 31
Neurofilaments, 378
Neuroimaging, 343
Neuroimaging techniques, 139
Neurological, 203
Neuromuscular
  force controller, 239
  junction, 48, 54
  synapse differentiation, 52
Neuron system, 354
Neuronal
  activity, 127
  activity patterns, 221
  architectures, 125
  circuits, 106
  death, 37, 41
  degeneration, 37
  migration, 4
  precursors, 6
  rhythmogenesis, 174
  systems, 37
  type, 59
Neuropathic pain, 354
Neurophysiological doctrine, 353
  monitoring, 140
  studies, 220
Neuropsychiatric disorders, 344, 348
Neuropsychological
  evaluation, 203
  functioning, 141
Neuroscience, 106
Neurotoxin, 49
Neurotransmission, 77
Neurotransmitter, 75
  activity, 141
Neutral red, 125
  fluorescence, 127
  imaging, 126
NGF, 356
Nicotinamide adenine dinucleotide (NADH), 127
Nicotinic receptor, 309
Nicotinic receptor abnormalities, 104
Nitric oxide synthase (NOS), 73

NLm, 270
NMDA, 71, 75, 190, 367
  receptor blockage, 77
  receptor stimulation, 75
  receptor, 72, 73, 76
  receptors, 71–73, 76, 119, 174, 301
NMDA-receptor-mediated mossy fiber, 176
NO
  donor, 76
  production, 76
  scavengers, 76
Nociceptive neurons, 363
Nodulus, 97, 121, 329
Non-AMPA, 304
Noradrenergic afferents, 95
Noradrenergic innervation, 77
NOS inhibitors, 76
Noun-to-verb conversion test, 103
Novel phenomena, 126
NR2A C-terminal chain, 76
NR2A-$C^{\Delta C/\Delta C}$ mice, 76
NR2C C-terminal chain, 76
NRTP nucleus reticularis tegmenti pontis, 279
NRTP, 270
NSE-bcl-2 DNA, 38
Nuclei neurons, 61
Nucleocortical mossy fiber projections, 120
Nucleokinesis, 6
Nucleotide sequence, 21
Nucleus interpositalis anterioris (NIA), 270
Nucleus interpositalis posterioris (NIP), 270
Nucleus interpositus posterior (NIP), 218
Nucleus reticularis tegmenti pontis (NRTP), 216, 261
Nucleus reticularis tegmenti pontis, 261
Null alleles, 25
Nurtured patterns, 105

Occulus, 97
OCS, 145
Odocoileus virginianus, 290
Off-beam decrease in fluorescence, 129
Olfactory bulb, 63, 64
Oligodendrocyte-derived soluble factor, 354
Oligo-dendrocytes, 59
Olivary
  gap junctions, 190
  glomeruli, 4, 193
  nuerons, 4, 189

Olivo-cerebellar
  axons, 49
  fibers, 120
  pathways, 214
  projection, 115, 125, 293
  system, 3, 4
Olivo-cerebellar-cerebello-olivary feedback loop, 4
Olivo-cortico-nuclear connections, 215, 218
Olivocortico-nuclear zones, 214
Omosynaptic activity, 77
On-beam increase in fluorescence, 129
Oncogene-immortalized cells, 59
Onset latencies, 261
Ontogenetic phases, 60
Optic chiasm, 215
Optical
  images, 115
  imaging, 113, 125
  recording, 115
  responses, 117
  signals, 127
Organism's genome, 21
Organophosphate pesticides, 346
Organotypic
  cerebellar cultures, 38
  culture, 37
  explants, 61
Orphan receptor, 51
Oryctolagus cuniculus, 289
Oscillations, 177, 181, 231
Oscillation frequency, 184, 232
Otx2, 21
Ovis aries, 290
Oxidation/reduction, 127
Oxide–cyclic GMPprotein, 130

P compartmentation, 34
P0, 38
P10, 38
P1, 38
P21, 13
P4 cerebellar precursors, 62
P5 block preparation, 122
P5 rats, 121
P7, 38
Paired pulse depression, 299
Paired pulse ratio (PPR), 75
Paradoxical calretinin-immunoreactivity, 176

Paraesthesia, 354
Paraflocculus, 114, 121, 266
Paraformaldehyde, 38
Parallel
  anatomical loops, 210
  fibers (PF), 47, 52, 71, 113, 208
    beams, 125, 170
    synapses, 51
  fiber–Purkinje cell synaptic activity, 129
Parallel-fiber conduction, 184
Parallel-fiber synapses, 182
Parameter β_{Ca}, 324
Parasagittal
  bands, 125, 127
  section, 275
  stripes, 264, 283
Parietal cortex, 105
Parkinson's disease, 177
Parvalbumin (PV), 167
Parvalbumin immunostaining, 34
Parvalbumin, 11, 367, 374
Pathophysiology of schizophrenia, 141
Pathway selection, 47
Pathways, 26
Pattern analysis, 143
Pax-2, 13, 21, 25
Pax2 expression, 25
PBSGTA, 38
PC
  bodies, 174
  collaterals, 175
  cytoplasm, 177
  firing rate modulation, 174
  membrane potential, 82
  morphology, 370
  response, 174
PCBs, 345
PCB treatment, 344
PC–molecular interneurons, 177
PC–PF synapses, 370
Pentameric combinations, 309
PEP-19, 10
Perfusion chamber, 115
Perinatal lethal condition, 23
Peripheral nerve injury, 354
Perturbations, 243
Pervasive developmental disorders (PDDs), 344
Pesticide chlorpyrifos, 347

PET, 142, 343
PET studies, 159
Pf-PC LTD, 76
Pf-PC synapses, 76
PF-Purkinje cells synapses, 207
PF–LTD induction, 81
Phantom limb sensation, 105
Pharmacological
    actions, 130
    manipulations, 97
    tools, 51
Phase-relation between the EMG onsets, 232
Phenotypic
    distributions, 64
    diversity, 58
    profile, 54
    repertoires, 57
Phosphate buffer, 38
Phosphatidylinositol bisphosphate (PIP2), 369
Phospholipase C (PLC), 369
Phospholipase C transduction pathway, 299
Phosphorylation, 176
Phylogenetic groups of mammals, 293
Physiological
    alterations, 176
    experiments, 214
    functioning, 174
    properties of area 1, 218
Picrotoxin, 82, 85
Pilocytic astrocytoma, 229
Pinceau ensheathing, 176
Pituitary, 354
PKC, 42, 73, 76
PKC inhibition, 38
PKC inhibitor transgenic mouse, 131
Plasticity, 47, 181
PMA, 102
PMC, 229
Polysynaptically interconnecting, 185
Pontine nuclei, 113, 206
Pontine projections, 270, 273
Pontobulbar, 6
Pontocerebellar pathway, 278
Pontocerebellar projections, 265
Pontomedullary stream, 6
Positron emission tomography (PET), 139
Post-glomerular neurons, 194
Post-junctional cells, 192

Posterior, 152
    cingulate, 105
    parietal cortex (PPC), 143
    parietal cortex, 105
    vermis, 115, 143
Postmitotic P, 30, 34, 41
Postnatal
    development, 287
    progenitors, 62
Postsynaptic
    current, 299
    domain, 52
    domains, 48
    modulation, 75
    molecule, 51
    proteins, 378
    selection, 48
    $[Ca^{2+}]i$ levels, 82
Potamochoerus porcus, 290
Potassium channels, 134, 190
PPD, 302
Pre-SMA, 102
Precerebellar
    neurons, 6
    nuclei, 113
    nucleus, 4
Prefrontal cortex, 102, 105, 141, 210
Prepyramidal fissure, 283
Presumptive DCN, 31
Primary fissure, 283
Primary visual, 105
Primordium, 29
Principal fissures, 283
Pro-apoptotic *bax* gene, 41
Procedural learning, 201
Progenitor cells, 58
Propidium iodide, 49
Protein kinase C (PKC), 82
Protein kinase C gamma (PKCgamma), 344
Protein kinase C-dependent long-term depression, 210
Protein of 43 kD, 37
Protein phosphatase 2A, 122
Protracted development, 345
protein kinase C myristoylated alanine-rich C kinase substrate (MARCKS), 49
Pseudosubstrate of PKC, 131

Psychology, 106
Pulmonary adenocarcinoma, 229
Purkinje
   axon collaterals, 186
   cell (P) subtypes, 29
   cells (PCs), 71
   cells PEP 19 negative, 13
   cells, 3, 22, 37, 50, 81, 91, 95, 104, 125, 168, 213, 267
      axons, 32
      degeneration, 204
      dendrites, 48
      excitability, 81
      regeneration, 42
   dendritic arbor, 51
Putative role, 206
Pyloric circuit, 174
Pyramidal neurons, 64, 186

Rabbit polyclonal antibody, 38
Radial glia, 31
Rapid eye movement (REM) sleep, 145
Rat embryos, 32
Rat, 287
Rattus norvegicus, 287
Reactive synaptogenesis, 50
Receptor blockers, 127
Receptor tyrosine kinases, 16
Reciprocal inhibition, 184
Reciprocal trophic interactions, 49
Recording temperature, 184
Reduced ratio of peak velocity to maximum amplitude, 238
Reflex modulation, 254
Relative, 234
Relative damping factor, 234
Renshaw cells, 239
Repeated trials of learning, 103
Repressing action, 50
Reptiles, 285
Rescued $En1^{-/-}$, 24
Resonance, 181
Resonance frequency, 234
Respiration, 121
Reticulotegmental nucleus, 261
Retinal ganglion cells (RGCs), 41
Rhombencephalic neurons, 30

Rhombic lip, 6, 58
   precursors, 64
   progenitor cells, 6
Rhombomere 1, 29
Rhythm generator, 181, 228
Rhythmic movements, 243
Rhythmic muscle activations, 255
Rodents, 207
RORα, 31
Rostral half of the medial accessory olive (rMAO), 215
Rostral part of the dorsal accessory olive (rDAO), 215
Rotorod performance, 343
Rotorod test, 345

S-100β, 367, 381
S/M phase, 31
S1/M1, 229
Saccades, 145, 151, 153, 154
Saccade-specific activation, 152
Saccadic eye movements, 152
Sagittal stripes of Purkinje cells, 269
Sagittal zones, 4
Saline, 174
Schizophrenia, 139, 141
Schizophrenic patients, 104
Schizophrenic patients response, 141
Selection, 47
Sensorimotor
   arrest rhythm, 167
   cortices, 105
   processing, 171
Sensory cancellation, 101
Sensory Neuron Specific, 356
Sequential movements, 151
Sequential reaching, 273
Serotonergic afferents, 95
Serotonin (5HT3), 309
Serotoninergic innervation, 77
Sex differences, 343
Sexual dimorphism, 343
SGC inhibitors, 76
Shaker, 372
Shaker mutant rats, 372
Short term presynaptic depression, 299
Signal plus noise, 143
Single C57BL/6 recessive gene, 26

Single trial analysis, 139
SK channel, 90
Slice cultures, 38
Slice preparation and electrophysiology, 82
Slit-1/2/3, 8
SMA, 102, 229
SNS, 356
Sodium channel blockers, 353
Sodium cyanide (NACN), 128
Somato-motor cortex, 58
Somato-sensory, 229
Somatodendritic region, 49
Somatosensory region, 206
Somatosensory stimulation, 140
Sonic hedgehog, 57, 65
Spatial navigation, 201
Spatial network, 207
Spatiotemporal mapping, 170
Spatiotemporal schedule, 58
Spinal cord, 97
Spinal cord injury, 354
Spinal cord-based central program generator, 243
Spines, 50, 52
Spinocerebellar ataxia, 374
Spinocerebellar projections, 294
Spinocerebellum, 237
Spinogenesis, 54
SPM, 145
SPM foci, 144
Spreading acidification, 125, 126, 131
SPS, 145
SSLP markers, 25
Staining procedures, 38
Stance, 249
Startle response, 345
Statistical analysis, 143
STDP, 77
Stellate cells, 95
Step cycles, 246, 249
Stimulus, 83
Stimulus-induced volley, 254
Structural Cerebellar Plasticity, 45
Subventricular zones, 6
Sucrose, 82
Surface stimulation electrode, 132
Swing phase, 249, 254
Synaptic conductance changes, 74

Synaptic
  excitability, 90
  inhibition, 77
  organization, 54
  plasticity, 49, 51, 82, 89, 90, 95, 106
  remodeling, 47
  transmission, 71, 113
Synaptogenesis, 48
Synchronous movements, 237
Synchronous state, 182
Synchrony, 181

T-maze paradigms, 204
TA muscles, 230
Tactile stimulus, 101
Talairach space, 146, 157
Tangential direction, 6
Tangential migration, 4
Target deletion, 49
Target selection, 47
TBOA, 300, 301
TC, 229
Telencephalic phenotypes, 64
Telencephalic regions, 63, 64
Temporoparietal, 105
Temporoparietal cortex, 102, 104
Teratoma, 229
Territory infarct, 229
Testicular testosterone, 348
Testosterone, 348
Tetanization, 86
Tetanus, 83
Tetrodotoxin (TTX), 50, 54
Thalamus, 182
The Armadillo, 290
The Brazilian Grey Opossum, 288
The Bush Pig, 290
The cerebello-olivary projection, 5
The Deer, 290
The Domestic Cat, 290
The Ferret, 290
The frontal and parietal lobes, 344
The Macaque, 291
The Madagascan Hedgehog Tenrec, 289
The Rabbit, 289
The Sheep, 290
The Tree Shrew, 291
Theory of Mind, 104

Threshold potential, 323
Thumb opposition movements, 153
Tibialis Anterior (TA), 227, 229
Time-to-phase delay, 237
Tomographic analysis, 140
Toxicity, 130
Transcerebellar pathways, 239
Transient cerebellar dysfunction, 135
Transplantation experiments, 64
Transplantation studies, 57
Transverse zones, 283
Traumatic brain injury, 367
Traumatic injury, 380
Trilayered cortical folia, 60
Triple-step saccades (TR), 153
Triton-X, 38
TRPC1 cation channel, 299
TTX, 50
   completely blocks SAD, 134
   dependent, 316
   experiments, 53
   resistant, 356, 361
TUJ-1 labeled neurons, 31
TUNEL, 41
Tupaia belangerie, 291
Two-degrees-of-freedom control, 97
Type-1 rotation of 180°, 14
Type-2 rotation, 14
Type-2 rotation experiments, 14

UBC, 333
Unc5H3, 8
Unc5H3 mutants, 31
Unidentified neoplasia, 229
Unipolar brush cells (UBC), 71, 58, 95, 329, 339
Upper granular layer, 50
Uvula, 97, 121, 329
Uvula-nodulus, 329

Vascular, 229
Vasopressin, 354
Velum medullaris, 30
Ventral bank, 102
Ventral lamella of the principal olive (vlPO), 215
Ventricular layer progenitors, 30
Vermian hypoplasia, 344
Vermian patients, 227

Vermis, 114
Vermis nucleus, 207
Vermis-paravermis, 234
   region, 227
Vestibular nuclear neurons, 96
Vestibular region, 206
Vestibulo-cerebellum, 333, 310
Vestibulo-ocular reflex (VOR), 207, 244, 255, 329
   function, 121
VGCCs, 370
Visual, 229
Visual corticopontine input, 263
Visual region, 206
Visually guided saccades (VG), 153
Visuomotor control, 213
Visuospatial attention, 161
Visuospatial memorization, 161
Voltage gated calcium channels (VGCCs), 81, 369
Voltage-gated sodium channels, 353
Voltage-sensitive dye (Di-4 ANEPPS), 115
Voluntary movement, 101

Walking speed, 249
Walking, 246
Weakly-electric elephant nose fish, 286
Whole-mount staining, 284
Wistar mutant, 371
Wnt1, 6, 21
Wnt-7B, 13
Wnt-7B genes, 13
Working memory task (WM), 153
WTA, 174
WT mice, 168

Xenopus retinotectal topography, 14
Xenopus spinal axons, 8

Zebrin, 10, 34
Zebrin I, 10, 22
Zebrin II, 22
Zebrin I and II antibodies, 17, 33
Zebrin label, 33
Zebrin-II antibodies, 17
Zebrin-immunoreactive stripes, 286
Zebrin-positive Purkinje cells, 13
Zonal labeling, 266